HUMAN
BIOLOGY,
ANATOMY &
PHYSIOLOGY
FOR THE HEALTH SCIENCES

HUMAN
BIOLOGY,
ANATOMY &
PHYSIOLOGY
FOR THE HEALTH SCIENCES

WENDI ROSCOE

Fanshawe College

NELSON
EDUCATION

Human Biology, Anatomy, and Physiology for the Health Sciences
by Wendi Roscoe

Vice President, Editorial Higher Education:
Anne Williams

Senior Publisher:
Paul Fam

Marketing Manager:
Leanne Newell

Developmental Editor:
Candace Morrison

Photo Researcher/Permissions Coordinator:
Jessie Coffey

Senior Production Project Manager:
Imoinda Romain

Production Service:
Integra Software Services Pvt. Ltd.

Copy Editor:
Carolyn Jongeward

Art Editor:
Jackie Dulson

Proofreader:
Integra Software Services Pvt. Ltd.

Indexer:
Integra Software Services Pvt. Ltd.

Design Director:
Ken Phipps

Managing Designer:
Franca Amore

Art Coordinator:
Suzanne Peden

Interior Design:
Peter Papayanakis Design

Cover Design:
Cathy Mayer

Cover Image:
© Ronald Martinez/Getty Images

Compositor:
Integra Software Services Pvt. Ltd.

Library and Archives Canada Cataloguing in Publication

Roscoe, Wendi A. (Wendi Annette), author

Human biology, anatomy, and physiology for the health sciences / Wendi A. Roscoe, Fanshawe College.

Includes index.

ISBN 978-0-17-650717-6 (pbk.)

1. Human biology—Textbooks.
2. Human anatomy—Textbooks.
3. Human physiology—Textbooks.
I. Title.

QP34.5.R68 2015
612   C2014-906363-6

ISBN-13: 978-0-17-650717-6
ISBN-10: 0-17-650717-5

To my sons, Adam and Josh.

Brief Contents

Table of Contents

Preface

NOTE TO THE STUDENT

This text includes the foundational knowledge of human biology, anatomy, and physiology that is relevant for students entering health care programs. I wrote this text so students could learn about and understand how their own body functions: from the macromolecules that we eat to the structures and function of our cells and to how groups of cells work together to form our organ systems. As well, I've included many examples of what can happen when our molecules, cells, or systems do not function properly. It is well known that people learn best when the information matters to them, when it is interesting, personal, and they can form connections to information they already know. When students understand their own health, they will live a healthier lifestyle and also become better health care professionals. My goal was to write this text for students to learn and understand that what we eat, think, and do affects us at every level, from the receptors on our cell membranes to the development of disease. I tried to continually interrelate the topics from each chapter so that previous information is reinforced, and so that by the end of the text students can see how everything fits together into a big picture.

NOTE TO THE INSTRUCTOR

In health care programs, it is most common for students to learn anatomy and physiology. In general or introductory programs, it is common for students to learn biology, which usually includes topics such as biodiversity, cells, genetics, plant structure and function, evolution of organisms, organ systems, and ecology. In this textbook I have chosen biology topics that specifically relate to human health, human evolution and inheritance, metabolism, genetics, biotechnology in medicine, infections and the importance of normal flora, as well as the anatomy and physiology of the body systems. In each chapter, I have included "Did You Know?" boxes that provide interesting facts that can increase students' interest and understanding of the content. Throughout the text I have included the most recent research related to nutrition that affects cell chemistry and physiology so students can see more clearly how environmental factors play a role in the regulation of our body systems and the imbalances that lead to disease. For example, because many students have difficulty with the cellular metabolism unit, I have included a discussion about how the vitamins we eat in foods act as the co-enzymes that allow our cells to produce energy. So the content is not just a list of chemical reactions that they have to memorize; they can start to connect what they do know about food with their own energy level and body weight, making the topic much more interesting and relevant. I have written this text in a way that's clear, concise, engaging, and easy to understand. I have included review questions in each chapter after each section so that students can frequently test their knowledge and level of understanding. I have also included at the end of the book many diagrams for students to label so this text has the feel of a workbook (see Appendix G). It is my hope that the combination of topics covered in this text as well as the student-focused approach will give students a good starting point for understanding human health in their health care program.

ABOUT THE NELSON EDUCATION TEACHING ADVANTAGE (NETA)

The **Nelson Education Teaching Advantage (NETA)** program delivers research-based instructor resources that promote student engagement and higher-order thinking to enable the success of Canadian students and educators. Be sure to visit Nelson Education's **Inspired Instruction** website at **http://www.nelson.com/inspired** to find out more about NETA.

INSTRUCTOR RESOURCES

Downloadable Instructor Supplements

All NETA and other key instructor ancillaries can be accessed through **http://www.nelson.com/login** and **http://login.cengage.com**, giving instructors the ultimate tools for customizing lectures and presentations.

NETA Test Bank This resource was written by Trisha Morrow and Steve Morrow of St. Lawrence College. It includes over 800 multiple-choice and true/false questions written according to NETA guidelines for effective construction and development of higher-order questions. The test bank was copyedited by a NETA-trained editor.

The NETA Test Bank is available in a new, cloud-based platform. **Testing Powered by Cognero®** is a secure online testing system that allows you to author, edit, and manage test bank content from any place you have Internet access. No special installations or downloads are needed, and the desktop-inspired interface, with its drop-down menus and familiar, intuitive tools, allows you to create and manage tests with ease. You can create multiple test versions in an instant, and import or export content into other systems. Tests can be delivered from your learning management system, your classroom, or wherever you want. Testing Powered by Cognero for *Human Biology, Anatomy, and Physiology for the Health Sciences* can also be accessed through **http://www.nelson.com/login** and **http://login.cengage.com**.

NETA PowerPoint Microsoft® PowerPoint® lecture slides for every chapter have been created by Wendi Roscoe of Fanshawe College. There is an average of 50 slides per chapter, many featuring key figures, tables, and photographs from *Human Biology, Anatomy, and Physiology for the Health Sciences*. NETA principles of clear design and engaging content have been incorporated throughout, making it simple for instructors to customize the deck for their courses.

Image Library This resource consists of digital copies of figures, short tables, and photographs used in the book. Instructors may use these jpegs to customize the NETA PowerPoint or create their own PowerPoint presentations.

NETA Instructor's Manual This resource was written by Kari Draker-Fortis from Fleming College. It is organized according to the textbook chapters and addresses key educational concerns, such as typical stumbling blocks students face and how to address them. Other features include Learning Objectives and Additional Resources.

DayOne Day One—Prof InClass is a PowerPoint presentation that instructors can customize to orient students to the class and their text at the beginning of the course.

TurningPoint® Another valuable resource for instructors is **TurningPoint® classroom response software** customized for *Human Biology, Anatomy, and Physiology for the Health Sciences*, created by Wendi Roscoe of Fanshawe College. Now, you can author, deliver, show, access, and grade, all in PowerPoint, with no toggling back and forth between screens. With JoinIn, you are no longer tied to your computer. You can walk about your classroom as you lecture, showing slides and collecting and displaying responses with ease. If you can use PowerPoint, you can use JoinIn on TurningPoint.

CengageNOW

This resource was written by Trisha Morrow and Steve Morrow of St. Lawrence College. CengageNOW™ connects students to assignable content matched to their text. CengageNOW™ is an interactive learning solution that helps students focus on what they need to learn. It improves academic performance by increasing students' time on task and giving them prompt feedback. This online tutorial and diagnostic tool identifies each student's unique needs with a pre-test that generates a Personalized Study Plan for each chapter. The CengageNOW study plan helps students focus on concepts they're having the most difficulty mastering. It refers to the accompanying ebook and provides a variety of learning activities designed to appeal to diverse ways of learning. After completing the study plan, students take a post-test to measure their understanding of the material. Instructors can track and monitor student progress by using the instructor Gradebook.

With a focus on active learning, concept mastery, and automatic grading, CengageNOW is an easy-to-use digital resource designed to get students involved in their learning progress and better prepared for class participation and assessment. CengageNOW for *Human Biology, Anatomy, and Physiology for the Health Sciences* can be used for self-study or assigned as homework. CengageNOW is enhanced with a **MindTap Reader**. Visualization and interactivity are elevated to a new level by integrating text features that connect the print content to digital assets.

STUDENT ANCILLARIES

CengageNOW

CengageNOW™ for *Human Biology, Anatomy, and Physiology for the Health Sciences* is enhanced with a **MindTap Reader**. Visualization and interactivity are elevated to a new level by integrating text features that connect the print content to digital assets, such as the MindTap Reader.

Go to CengageNOW for *Human Biology, Anatomy, and Physiology for the Health Sciences*. Using this resource, you can:

- evaluate your knowledge of the material covered in your textbook
- prepare for exams with a prep quiz
- generate a **Personalized Learning Plan** to identify the areas you should study and target resources for your review
- gain access to rich, interactive learning modules tied to each chapter of your textbook

Visit **www.nelson.com/student** to start using **CengageNOW**. Enter the access code included with your text. If a code is not provided, you can buy instant access at NELSONbrain.com.

ACKNOWLEDGMENTS

I would like to thank everyone that has helped in the development of this textbook. Thank you to Candace Morrison, developmental editor; Imoinda Romain, senior production project manager; Jackie Dulson, art editor; Sue Peden, art coordinator; Carolyn Jongeward, copyeditor; Alex Antidius Arpoudam, project manager; and Jessie Coffey, permissions researcher. Thanks to Karen Nancekivell, senior sales and editorial representative, and Paul Fam, senior publisher, who encouraged me to write this text and supported me during every step of the way. Thanks to Kristen Phalen and Dave Leonard for all of their help with the review questions and writing the glossary, and Dave for all of the great conversations about the order of the chapters and interesting "Did You Know?" ideas to include. Thanks to Tara Lawrence of Fanshawe College for the feedback about the topics and order of topics included in the text and for helping me write the medical terminology appendix. Thanks to Robin Smith for doing the histology staining on the tissue sections that were used in Chapter 11. I would like to thank my father, Garry Roscoe, for the great photos that are included throughout the text. Finally, I would like to thank the reviewers:

Joyce Myers, Durham College
Barry Weese, Georgian College
Kari Draker-Fortis, Fleming College
Naman Sharma, University of Ottawa
Rachel Steels, Cambrian College
Barbara Czaban, York University
Lynn Connaty, Durham College
Denise Nelson-Mogaji, George Brown College
Ruthanna Dyer, York University
Jayson Parker, University of Toronto
Coral L. Murrant, University of Guelph

Their comments were honest, valuable, and helpful. They challenged me to do my best work. They encouraged me to find the best information I could. Their willingness to share their ideas and resources has made this book a first-class resource.

Wendi Roscoe, 2015
London

About the Author

Wendi Roscoe (H.B.Sc., M.Sc., B.Ed., Ph.D.) is a faculty member in the Department of Health Sciences at Fanshawe College, in London, Ontario. Wendi has an H.B.Sc. (genetics), a M.Sc. (physiology), a B.Ed., and a Ph.D. (physiology and pharmacology) from the University of Western Ontario, London, Ontario. Her Ph.D. research was focused on neuroinflammation in a mouse model of multiple sclerosis. She has published in several peer-reviewed journals, such as *Journal of Neuroimmunology* and *Journal of Neuroscience Research*. Wendi has been teaching physiology, biology, anatomy, and nutrition courses at the university and college level since 2004, and is currently teaching biology and physiology at Fanshawe College. She is also taking a holistic nutrition program at the Canadian School of Natural Nutrition, with the goal of becoming a registered holistic nutritionist. Wendi has an excellent understanding of student challenges and needs within this course.

Courtesy of Dream Catcher Photography

1 Introduction to Biology, Anatomy, and Physiology

Steve Debenport/E+/Getty Images

Education is the most powerful weapon which you can use to change the world.

Nelson Mandela

1.1 OVERVIEW

Biology, anatomy, and physiology are areas of science that are often studied together because of the overlap of the content in each subject. **Biology** is the study of all living things, including their structure, function, growth, and interrelationships. The focus of this textbook is on humans and microorganisms. **Anatomy** is the study of physical structures within an organism, and **physiology** is the study of how those anatomical parts function individually as well as in conjunction with other structures of the body. For example, the respiratory system consists of the anatomical structures of the bronchi, bronchioles, and alveoli, which contain specific types of epithelial and connective tissues. Because those cells and tissues are structured in a particular way, they allow oxygen to diffuse into the bloodstream. The respiratory system functions together with the cardiovascular system and the digestive system to bring oxygen and nutrients to the cells. Throughout this text, we look at different aspects of molecular and cellular biology, anatomy, and physiology to learn how the human body works—and learn how to maintain optimal health. Although this introductory textbook is intended for students entering a health care program, most of the information will provide a useful guide for anyone who wants to understand their own health.

CONCEPT REVIEW 1.1

1. Define biology, anatomy, and physiology.

1.2 ORGANIZATION OF LIVING THINGS

Organisms within each species, whether human, worm, potato, or bacterium, must have **genetic differences** in order for the species as a whole to survive the constantly changing environmental conditions. As soon as organisms within a species become too genetically similar, the probability of their extinction increases. This is because genetic differences among individuals enable a species to survive environmental change. **Diversity** of living organisms is a very important concept that will be discussed throughout this text.

Approximately 8.7 million species are currently known to be living on earth, and an estimated 86% of the total has not yet been identified. (Some estimate the total number of species to be close to 100 million.) Even considering the vast amount of genetic variation that exists, all organisms can be categorized based on certain characteristics into one of the following six **kingdoms**: Archaebacteria, Eubacteria, Protista, Fungi, Plants, or Animals.

The kingdoms

Each kingdom includes organisms with specific traits (Figure 1.1). **Archaebacteria** and **Eubacteria** are the two kingdoms that consist of **prokaryotic** organisms, which are simple, unicellular, and do not contain membrane-bound organelles, such as a nucleus. All other kingdoms consist of **eukaryotic** organisms that have a complex cellular structure (Chapter 3). Prokaryotes are also generally termed *bacteria*. Bacteria are essential to life on earth; they live cooperatively together on and within every other species, including humans. Earth is approximately 4.5 billion years old, and prokaryotes colonized Earth approximately 3.5 billion years ago. Bacteria (cyanobacteria) were the first organisms to release oxygen into the atmosphere, and they play many vital roles in maintaining life on earth.

Archaebacteria

Archaebacteria are prokaryotic unicellular organisms that were most likely the first living cells on earth (Figure 1.2). They still exist today, some surviving in extreme environments such as hot springs, volcanoes, the highly salty Dead Sea, soil, and the human digestive tract. Archaea have an independent evolutionary history and, compared to prokaryotic organisms, have very different DNA sequences, lipid structures, cell wall structures, and metabolism.

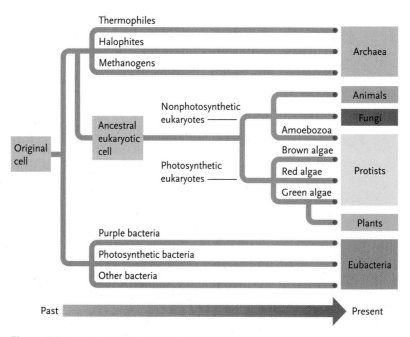

Figure 1.1 The Six Kingdoms

All living organisms can be categorized into one of six kingdoms.

Source: From DIGIUSEPPE/FRASER. Biology 11U. © 2011 Nelson Education Ltd. Reproduced by permission. www.cengage.com/permissions

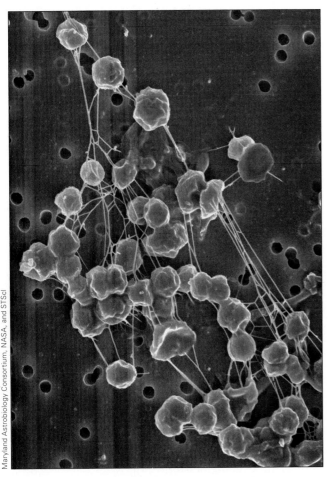

Figure 1.2 Archaebacteria

Archaebacteria are prokaryotes that live in many extreme environments.

Eubacteria

Eubacteria are also prokaryotic, single-celled organisms, although sometimes they cluster together, such as the common *staphylococcus aureus* that inhabits human skin and nasal passages (Figure 1.3). Eubacteria survive in many different environments, including the human body; they constitute much of our body's normal flora that is extremely beneficial and important for our health. A small percentage of bacteria can cause infection in humans, such as tuberculosis, salmonella, *E. coli*, and Chlamydia (Chapter 23).

DID YOU KNOW?

The human body contains 10 times more bacterial cells than human cells.

Protists

Organisms in the protista kingdom are the most numerous and most diverse of any kingdom and can be found in any environment containing water. **Protists** are

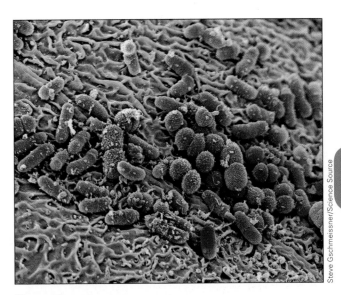

Figure 1.3 Eubacteria

Eubacteria are prokaryotes that live in many environments, including inside our digestive tract, and some species can cause disease.

mostly unicellular and are eukaryotic (Figure 1.4). Some species of protist are multicellular and very large, such as the giant kelp, which grows to 65 feet or 20 metres. Eukaryotic cells are complex and contain many membrane-bound organelles—such as mitochondria, a nucleus, and endoplasmic reticulum (Chapter 3)—that are not found in prokaryotic cells. Species in the protista kingdom can have features of the other three eukaryotic kingdoms, and can be divided into fungus-like, animal-like, and plant-like. It is thought that the multicellular kingdoms arose through the evolution of protists. As do bacteria, protists survive almost everywhere on earth, in the oceans, fresh water, and on land. Some examples of protists include algae, slime moulds, and plankton.

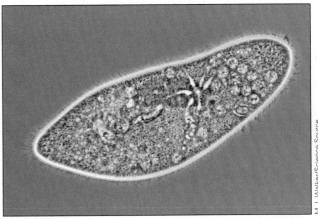

Figure 1.4 Protists

Protists are the eukaryotic, single-celled organisms that gave rise to the eukaryotic, multicellular organisms.

Fungi

Organisms in the fungi kingdom are eukaryotic and multicellular, with the exception of yeast, which is unicellular (Figure 1.5). They are not photosynthetic and most often absorb their nutrients from other organisms, such as trees or dead organisms. Some examples of fungi include mould, mildew, mushrooms, and yeast. Fungi play an important role in **decomposition** (as do some archaea and eubacteria) and nutrient recycling. Yeast is beneficial to humans in the production of bread, alcohol, and soy sauce (*Saccharomyces cerevisiae*), and other fungal species are used to produce antibiotics, such as penicillin. Some fungal species can cause infection in humans, for example, athlete's foot or yeast infections (*Candida albicans*).

DID YOU KNOW?

Crimini mushrooms are small, nutritious, portabella mushrooms that contain vitamins C, B6, B12, and D.

Plants

Species in the plant kingdom are eukaryotic and multicellular (Figure 1.6). The plant kingdom includes a wide range of organisms, many of which are important food sources for humans. **Plants** have **chloroplasts** in their cells, which are membrane-bound organelles that

Figure 1.5 Fungi

Fungi are eukaryotic, multicellular organisms that include yeasts, moulds, and mushrooms.

Courtesy Garry Roscoe

Figure 1.6 Plants

Plants are eukaryotic, multicellular organisms that use photosynthesis to acquire energy.

contain the pigment **chlorophyll** that is used in **photosynthesis**. During photosynthesis, plants combine carbon dioxide from the air with hydrogen from water, and energy from sunlight to produce glucose. Many plants provide a healthy source of food, including broccoli, spinach, carrots, apples, and blueberries. Some plants or parts of plants can be poisonous to humans; for example, the leaves of cherry trees and peach trees contain toxic molecules that can be converted into cyanide poison.

DID YOU KNOW?

Many bean species contain toxic chemicals and can cause severe and possibly fatal effects in humans. Raw beans, such as lima beans and kidney beans, must be soaked in water for at least five hours and then boiled for a minimum of 10 minutes to ensure the toxin is degraded. Canned beans have already undergone this process and can be eaten without cooking.

Animals

The **animal** kingdom includes organisms that are eukaryotic, multicellular, and are capable of **locomotion**; they can move from one location to another (Figure 1.7). Examples of organisms in the animal kingdom include humans, dogs, fish, sponges, roundworms, and birds. Many animal species are food

Courtesy Garry Roscoe

Figure 1.7 Animals
Animals are eukaryotic multicellular organisms that have locomotion.

sources for humans—and this varies among cultures around the world—such as cows, chickens, turkey, fish, pigs, lamb, shark, snake, buffalo, moose, duck, and deer.

By looking at the organization of living things in terms of six kingdoms, we can gain perspective on where humans fit within this classification. The rest of this chapter and the textbook as a whole focus on human cell biology, anatomy, and physiology; Chapter 23 describes some organisms from other kingdoms that can potentially cause human disease.

DID YOU KNOW?

Dogs have been called "man's best friend" and have been domesticated for thousands of years. The companionship and loyalty between dogs and humans may be due in part to some specific genetic similarities. For example, humans and dogs live cooperatively in groups, and both are exceptionally good at reading body language and bonding; this is due to their high levels of secretion of oxytocin (Chapter 16). Humans have used selective breeding extensively with dogs (Chapter 7) to perpetuate certain desired traits in dogs.

CONCEPT REVIEW 1.2

1. Make a chart to organize the characteristics of organisms in each kingdom.

2. Give examples of species from each kingdom.

1.3 ORGANIZATION OF THE HUMAN BODY

The human body can be organized into different levels from the microscopic to the entire organism: chemicals, molecules, cells, tissues, organs, organ systems, and organism (Figure 1.8). **Chemicals** are the individual elements, such as nitrogen, hydrogen, carbon, and oxygen, that combine to form small molecules, such as glucose, water, amino acids, or nucleotides, that also combine to form **macromolecules**, such as complex carbohydrates, DNA, proteins, and fats (Chapter 2). The small molecules combine in various ways to produce the large macromolecules that form part of every **cell** of every living organism (Chapter 3). C ells combine to form the various **tissues** in the body: epithelial, connective, muscle, and nerve (Chapter 11). Different combinations of those tissues form **organs**. For example, the innermost layer of the stomach is composed of epithelial tissue, which is surrounded by strong collagen-containing connective tissue, which binds to three smooth muscle layers that contract to digest food; neurons that send signals to the smooth muscle cells cause them to contract when food is in the stomach. All organs are composed of different amounts of some or all tissue types. Certain organs function with other organs and tissues to form **organ systems**, such as the digestive system, respiratory system, or cardiovascular system (Chapters 12–22). All organ systems combined make up the entire human **organism**.

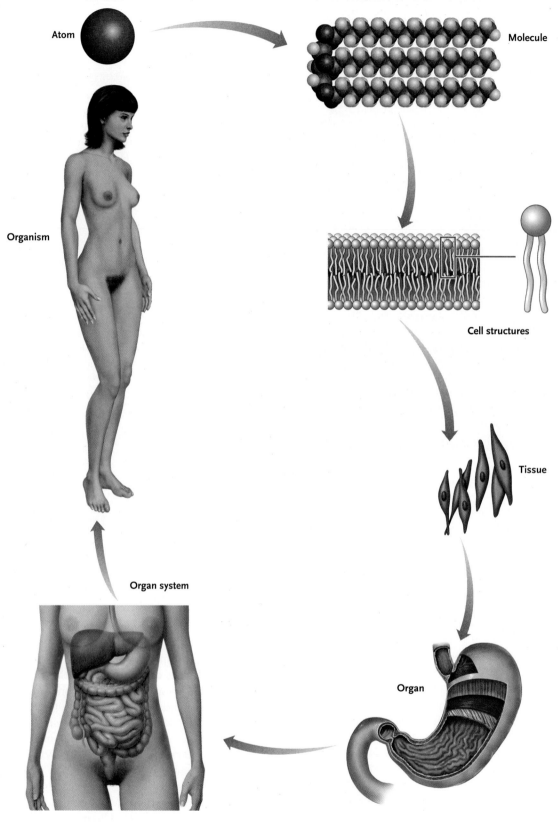

Figure 1.8 Organization of the Body

The body is organized from the smallest atoms and molecules to form tissues, organs, and organ systems.

1. Describe how the human body is organized into levels.

1.4 ORGAN SYSTEMS

There are 11 organ systems in the body, and they all interact with each other. Note that you will need to be able to relate information from one chapter to the next, so by the end of the book you will have a working knowledge of how the various parts of the body work *together*. For example, when we breathe air, the oxygen diffuses across the alveoli (respiratory system) into the blood stream (cardiovascular system), which carries that oxygen to all the cells in the body so that each cell can produce energy through metabolism in the form of ATP. The cardiovascular system also carries nutrients (macromolecules) absorbed by the digestive system, which is regulated by the nervous system and the endocrine system. Some organs are considered part of more than one organ system: for example, the pancreas is part of the digestive system because it produces digestive enzymes that are secreted into the small intestine, and it is also part of the endocrine system because it produces insulin that regulates blood sugar levels. This section gives a brief overview of the structures and functions of each organ system, which will be discussed in detail in later chapters.

The integumentary system

The integumentary system (Chapter 12) is made up of the skin, hair, nails, sebaceous glands, and sweat glands (Figure 1.9). The major functions of this system are to protect the body from infection and physical damage; help regulate body temperature through production of sweat and dilation or constriction of blood vessels in the dermis layer of the skin; excrete some waste products; produce vitamin D from sunlight; and detect sensations, such as touch, pain, pressure, vibration, and temperature.

The skeletal system

The skeletal system (Chapter 13) consists of bones, cartilage, and the joints (Figure 1.10). The function of this system is to provide support for the body and to protect internal organs: for example, the skull protects the brain. It works in combination with the **muscular system** to allow body movement. The skeletal system also stores minerals, such as calcium, and plays an important role in blood calcium homeostasis. It stores fats in the yellow bone marrow and produces red and white blood cells in the red bone marrow.

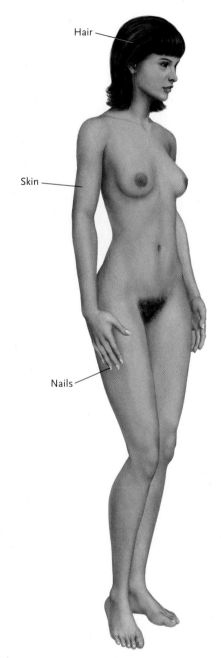

Hair

Skin

Nails

Figure 1.9 **The Integumentary System**

The muscular system

The muscular system (Chapter 14) consists of skeletal muscles, smooth muscle (in organs), and cardiac muscle (the heart) (Figure 1.11). The function of the skeletal muscles is to enable the body to move and maintain

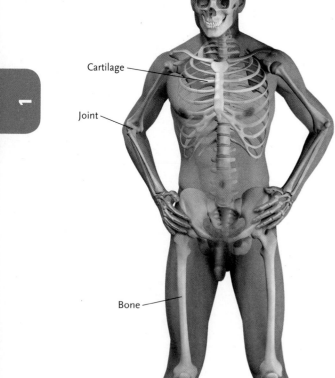

Figure 1.10 The Skeletal System

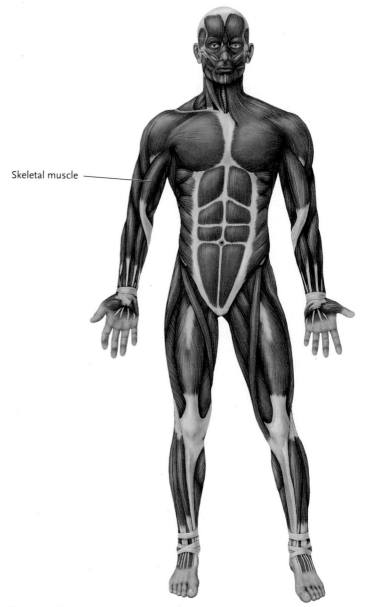

Figure 1.11 The Muscular System

posture. These muscles also function in heat production because heat is produced as a byproduct of cellular respiration, and, compared to all other cell types in the body, muscle cells produce the most ATP.

The nervous system

The nervous system (Chapter 15) consists of the brain, spinal cord, nerves, and receptors, which include those of the special senses: visual, auditory, equilibrium, gustatory (taste), and olfactory (smell) (Figure 1.12). The function of the nervous system is to *detect* changes within the body and the environment, *interpret* those

stimuli, and then *respond* to those changes. The nervous system plays an important role in regulating certain biological factors, such as body temperature, blood pressure, and hormone levels; this regulation is called **homeostasis** (Section 1.6).

The endocrine system

The endocrine system (Chapter 16) consists of the glands that produce hormones that travel through the bloodstream to highly specific target cells that have very specific functions (Figure 1.13). Like the nervous system, the endocrine system plays an important role in

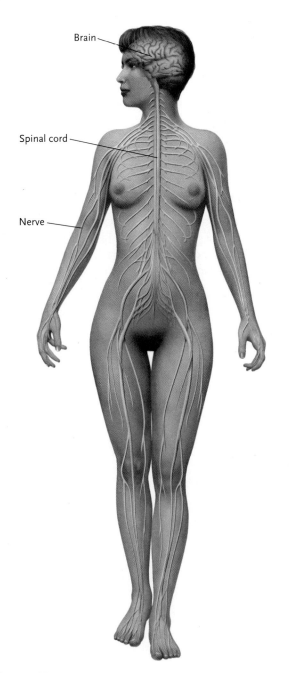

Figure 1.12 The Nervous System

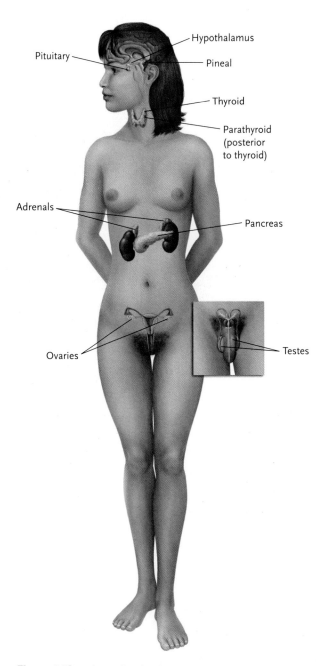

Figure 1.13 The Endocrine System

homeostasis. The glands of the endocrine system include the hypothalamus, pituitary gland, pineal gland, thyroid gland, parathyroid gland, adrenal glands, pancreas, ovaries, and testes. There are also non-endocrine tissues that produce and secrete hormones, such as those of the digestive system. The numerous functions of the endocrine system depend on which gland is secreting which hormone. Very briefly, some functions include the regulation of the autonomic nervous system (hypothalamus), regulation of growth and reproduction (hypothalamus and pituitary gland), regulation of metabolism (thyroid gland), regulation of blood calcium levels (parathyroid gland), regulation of stress hormones (adrenal gland), regulation of blood sugar (pancreas), and production of sperm or eggs (testes or ovaries).

The cardiovascular system

The cardiovascular system (Chapter 17) consists of the heart, blood vessels, and blood (Figure 1.14). The function of the cardiovascular system is to transport oxygen, nutrients, hormones, and waste products in the bloodstream to the various tissues of the body, or to the kidneys for excretion. The blood is composed

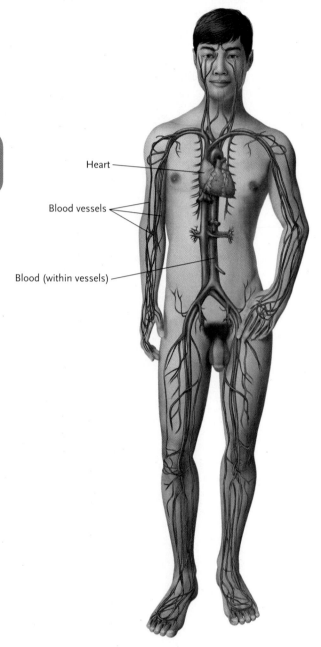

Figure 1.14 **The Cardiovascular System**

The respiratory system

The respiratory system (Chapter 18) consists of mouth, nose, pharynx, larynx, trachea, bronchi, bronchioles, and alveoli (air sacs) (Figure 1.15). The function of the respiratory system is to take up oxygen that diffuses

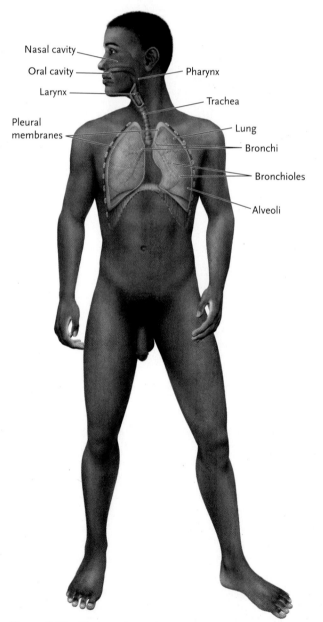

Figure 1.15 **The Respiratory System**

of red blood cells that carry oxygen; white blood cells that perform immune functions (Chapter 22); platelets, which function in blood clotting; and plasma, the water portion of the blood that contains many dissolved ions, nutrients, and molecules that act as chemical buffers: for example, bicarbonate regulates blood pH. The cardiovascular system—in conjunction with the endocrine system and the nervous system—plays an important role in the regulation of body temperature by dilating or constricting certain blood vessels, and regulating the body's water balance and blood pressure.

across the alveoli into blood vessels; give off carbon dioxide; regulate blood acidity by expelling carbon dioxide; and produce vocal sounds through the larynx, commonly called the voice box.

The digestive system

The digestive system (Chapter 19) consists of many organs that are directly involved with transporting and digesting food, including the mouth, esophagus, stomach, small intestine, large intestine, rectum, and anus (Figure 1.16). The digestive system also includes accessory organs that assist food digestion, such as the salivary glands, liver, pancreas, and gallbladder. The functions of the digestive system are to break down food macromolecules (Chapter 2) such as proteins, complex carbohydrates, and fats into their smaller components—amino acids, simple carbohydrates, and fatty acids—that can then be absorbed into the blood stream or the lymphatic vessels for circulation to all cells in the body; it also functions in the elimination of undigested waste.

The urinary system

The urinary system (Chapter 20) consists of the kidneys, ureters, bladder, and urethra (Figure 1.17). The functions

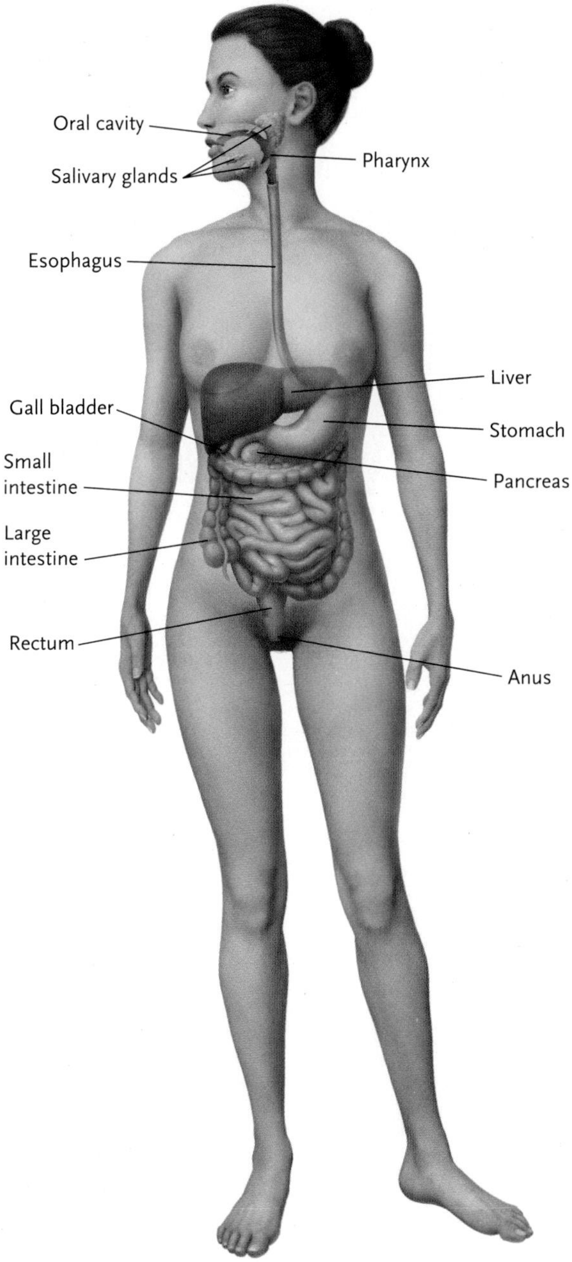

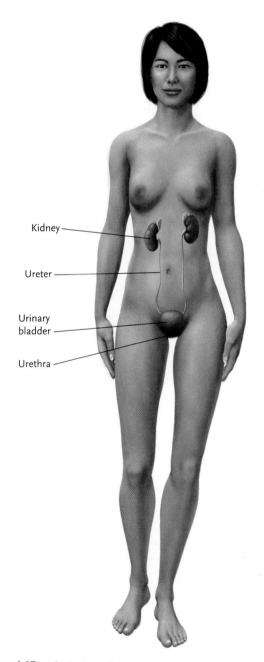

Figure 1.16 **The Digestive System**

Figure 1.17 **The Urinary System**

of the urinary system are to filter the blood that circulates through the kidneys, reabsorb essential nutrients and water, and excrete waste such as excess ions, water, and urea (produced from the breakdown of proteins). In conjunction with hormonal signals from the endocrine system, the kidneys regulate blood pressure by excreting or reabsorbing water and ions. The kidneys also help regulate blood pH by the excretion of excess hydrogen ions. The urinary system is involved in regulating the production of red blood cells—by producing the hormone erythropoietin—and the production of vitamin D3 (the active form of vitamin D), which regulates blood calcium levels (Chapters 16 and 20).

The reproductive system

The female reproductive system (Chapter 21) consists of mammary glands, ovaries, fallopian tubes, uterus, and vagina. The male reproductive system includes the testes, vas deferens, epididymis, prostate gland, and penis (Figure 1.18). The functions of the reproductive system are to produce gametes (oocytes or sperm) in the

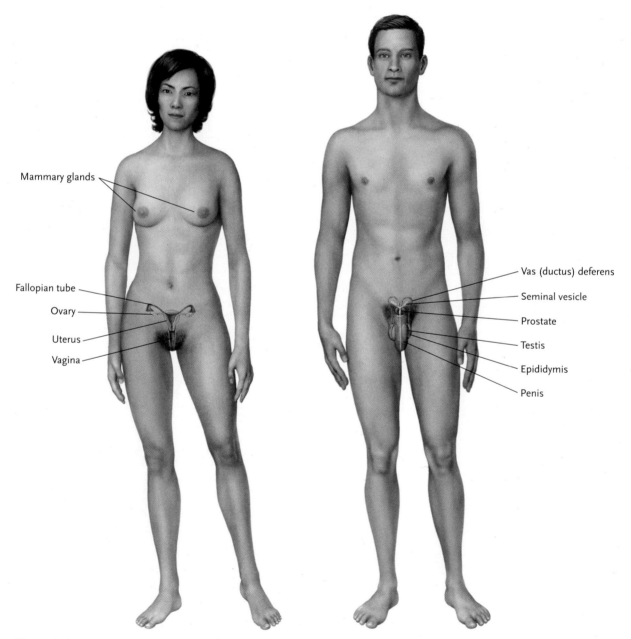

Mammary glands

Fallopian tube

Ovary

Uterus

Vagina

Vas (ductus) deferens

Seminal vesicle

Prostate

Testis

Epididymis

Penis

Figure 1.18 The Reproductive System

gonads (ovaries or testes) and to produce milk in the mammary glands. The reproductive system is part of the endocrine system because the ovaries produce the hormones estrogen and progesterone, and the testes produce testosterone. Both males and females produce each of these hormones, but males make more testosterone, and females make more estrogen and progesterone.

DID YOU KNOW?

Prostate cancer is the most common type of cancer in men in Canada; it affects one in seven men and kills over 300,000 men in Canada and the U.S. every year. Men at highest risk are over the age of 65, overweight, have a family history of prostate cancer, and/or are of African descent. Research shows that diet is a major factor in this and many other cancers (Chapters 2 and 5).

The immune system and the lymphatic system

The immune system (Chapter 22) consists of the white blood cells, bone marrow, lymph nodes, spleen, thymus, and lymphatic system, which is composed of lymph nodes, lymph vessels, lymphoid tissue (such as tonsils), and lymph fluid (Figure 1.19). The main functions of the lymphatic system are to take up excess fluids that have leaked from the cardiovascular system (plasma) into the interstitial or intercellular space (interstitial fluid) and to carry that fluid through the lymphatic vessels (lymph) to be deposited back into the circulatory system. The lymphatic system is also involved in absorbing fats from the digestive tract and then circulating them to the blood stream. The immune system also functions to destroy infectious microorganisms such as bacteria, viruses, fungi, or parasites (Chapter 23).

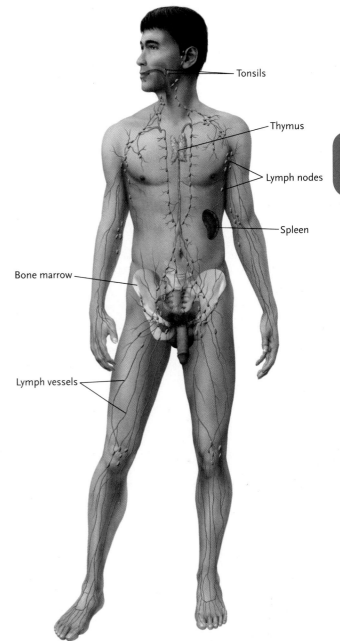

Figure 1.19   The Immune and Lymphatic System

Tonsils

Thymus

Lymph nodes

Spleen

Bone marrow

Lymph vessels

CONCEPT REVIEW 1.4

1. Compile a list of the structures that make up each organ system.

2. Are any organs associated with more than one organ system? If so, which ones?

3. What are the main functions of each organ system?

4. Which organ system is involved in temperature regulation, absorption of nutrients, regulation of blood sugar, elimination of waste, body movement, storage of minerals and fat, transportation of oxygen, production of vitamin D?

1.5 ANATOMICAL TERMS

The parts of the human body are named in combination with directional terms (Figure 1.20). This makes it possible for specific regions of the body to be consistently identified among health care professionals and in patient records. For example, if someone has a burn on their right forearm near the wrist, the location of the burn would be identified as the right, anterior, antebrachial region, proximal to the carpal region. An understanding of medical anatomical terminology is very important in the health field. The terms applied to any region of the body refer to the body in what's known as the **anatomical position**: standing up, facing forward, toes forward, and palms facing forward.

For the purpose of anatomical terminology the body is divided into general regions: head; neck; trunk, including thoracic, abdominal, and pelvic regions; upper limbs; and lower limbs. In each region of the body, the underlying bones, muscles, nerves or blood vessels often have similar names. For example, the thigh is the femoral region, the thigh bone is the femur, and the thigh contains a femoral artery and a femoral nerve. The temporal region of the head contains a temporal bone, and the temporal lobe of the brain. An understanding of these anatomical terms will help in your study of the organ systems.

Directional terms

Directional terms are used to describe the regions of the body in relation to each other, and these terms refer to the body in the anatomical position (Figure 1.21). For example, the skin is superficial to the muscles, and the patellar region is distal to the inguinal region.

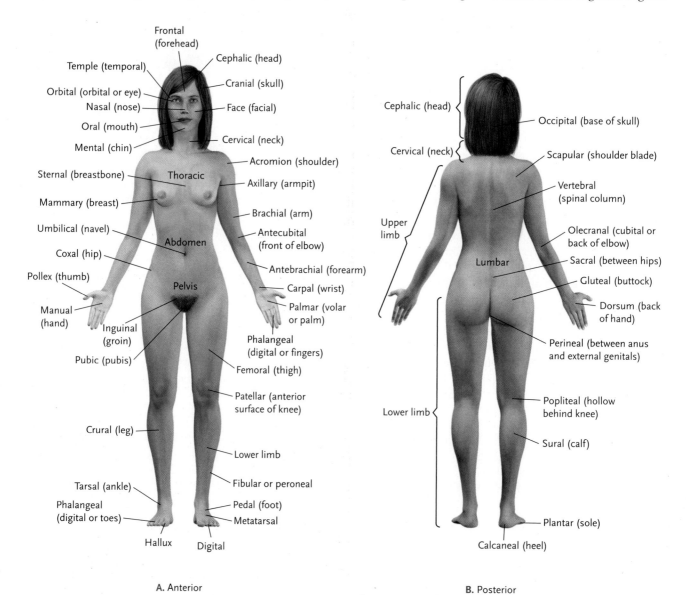

A. Anterior

B. Posterior

Figure 1.20 Anatomical Terms
Anatomical terms are applied to various regions of the body as shown in the anatomical position.

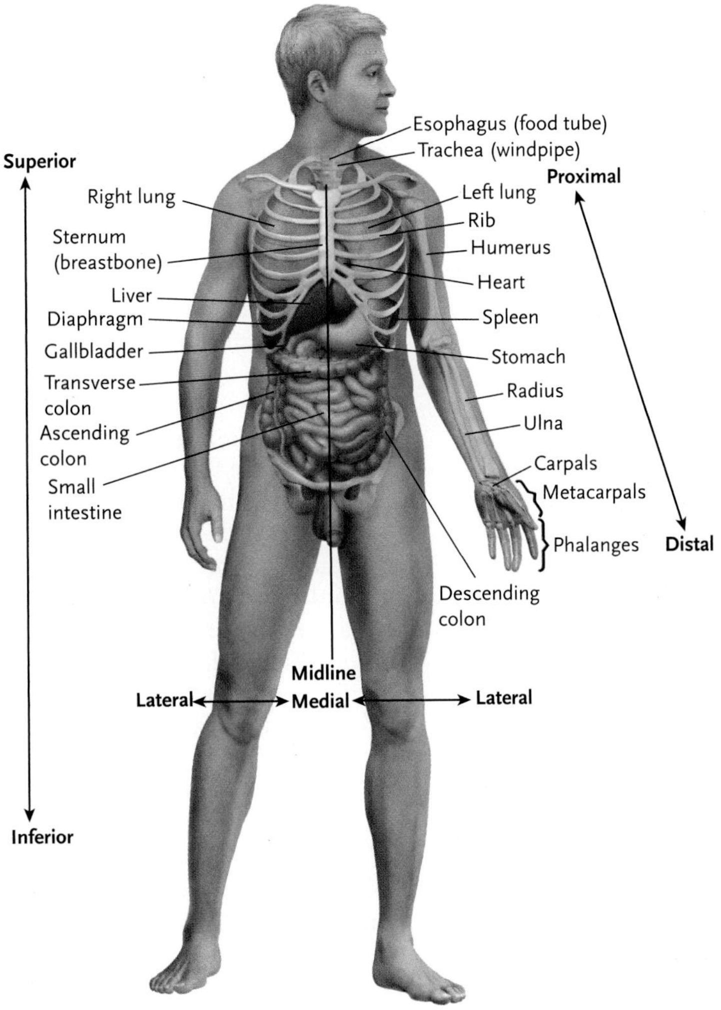

Figure 1.21 Directional Terms
The location of each body part can be described relative to other body parts by using directional terms.

See Table 1.1 for a description of each directional term and to develop an understanding of where the major body organs are located in relation with each other. For example, the esophagus is posterior to the trachea: you could also say the trachea is anterior to the esophagus. Most of the liver is located in the right abdominal region, the heart is intermediate to the lungs, and the pancreas is inferior (below) and posterior (behind) to the stomach. This information is medically significant, especially if you or your patient has pain in a certain location; for example, pain in the upper right or lower right abdominal region could indicate either appendicitis or gall stones.

Body planes

The body, or parts of the body, can also be divided into sections: sagittal, frontal, transverse, or oblique. A **sagittal plane** divides the body into left and right sections; equal sections are **midsagittal**, and unequal sections are **parasagittal**. A **frontal plane** divides the body into anterior and posterior sections. A **transverse plane**

TABLE 1.1

Directional Terms

Directional Term	Definition
Superior	Above, over, toward the head
Inferior	Below, under, further from the head
Anterior	In front, toward the front of the body
Posterior	Behind, toward the back of the body
Proximal	Closer to the attachment point of a limb to the trunk
Distal	Further from the attachment point of a limb to the trunk
Medial	Toward the midline
Lateral	Further from the midline
Intermediate	In between two structures
Superficial	Toward the surface of the body
Deep	Toward the interior of the body

divides the body into superior and inferior sections. Transverse sections are also called a **cross-sectional** or **horizontal plane,** and these do not necessarily divide the body or organ into equal halves. An **oblique plane** divides the body on an angle (Figure 1.22). This textbook describes body cavities, organs, or tissues in a variety of sections. It is important for you to learn what type of section you are looking at so you can understand the relationships among the body parts.

Body cavities

The body can be divided into specific **body cavities,** which each contain specific organs or tissues. Figure 1.23 shows a sagittal and a frontal view of the body cavities: **cranial; vertebral; thoracic,** including the **pleural cavity, pericardial cavity,** and the **mediastinum;** and **abdominopelvic.** The cranial cavity contains the brain, and the vertebral cavity contains the spinal cord. The thoracic cavity contains the lungs, heart, trachea, upper esophagus, and two other cavities: the pleural cavity (space between pleural membranes that surround the lungs); the pericardial cavity (space between the pericardial membranes that surround the heart); and the mediastinum (anatomical central region of the thoracic cavity). The diaphragm separates the thoracic cavity from the abdominopelvic cavity. The abdominopelvic cavity contains the organs of the digestive system, the spleen, and the pelvic region, which contains the bladder, parts of the large intestine, and the internal reproductive organs.

Abdominopelvic regions and quadrants

The abdominopelvic cavity can be further divided into either nine regions or four quadrants. The nine regions, shown in Figure 1.24, include the right hypochondriac, epigastric, left hypochondriac, right lumbar, umbilical, left lumbar, right inguinal, hypogastric, and left inguinal. The upper horizontal division line is at the distal end of the rib cage; the lower horizontal line is just superior to the pelvis (iliac crest). The two vertical division points are approximately one inch medial to the nipples. Figure 1.24 shows some of the organ structures in each region. A patient with epigastric pain could have a problem with several different organs, including the liver, stomach, distal esophagus, or transverse colon.

The abdominopelvic region is also commonly divided into quadrants. Figure 1.25 shows the quadrants; the right and left sides are divided into equal halves, and the horizontal plane is divided at the umbilicus.

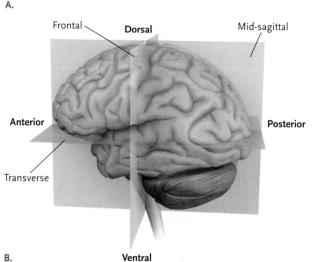

Midsagittal plane (through midline)

Frontal plane

Parasagittal plane

Transverse plane

Oblique plane

A.

Frontal — **Dorsal** — Mid-sagittal

Anterior — **Posterior**

Transverse

B. **Ventral**

Figure 1.22 **Planes of the Body and Planes of an Organ**
The body (A) and the organs (B) can be described in terms of frontal, sagittal, transverse, or oblique planes.

DID YOU KNOW?

Sometimes pain in a certain organ can be felt in very different areas of the body; this is called *referred pain*. Pain associated with a heart attack can be felt in the neck, upper arm, or back, even though the heart is located in the intermediate thoracic region. A "brain freeze" from eating or drinking cold foods quickly can be felt in the frontal sinuses because the trigeminal nerve carries sensory information from the hard palate, which is the back of the roof of the mouth.

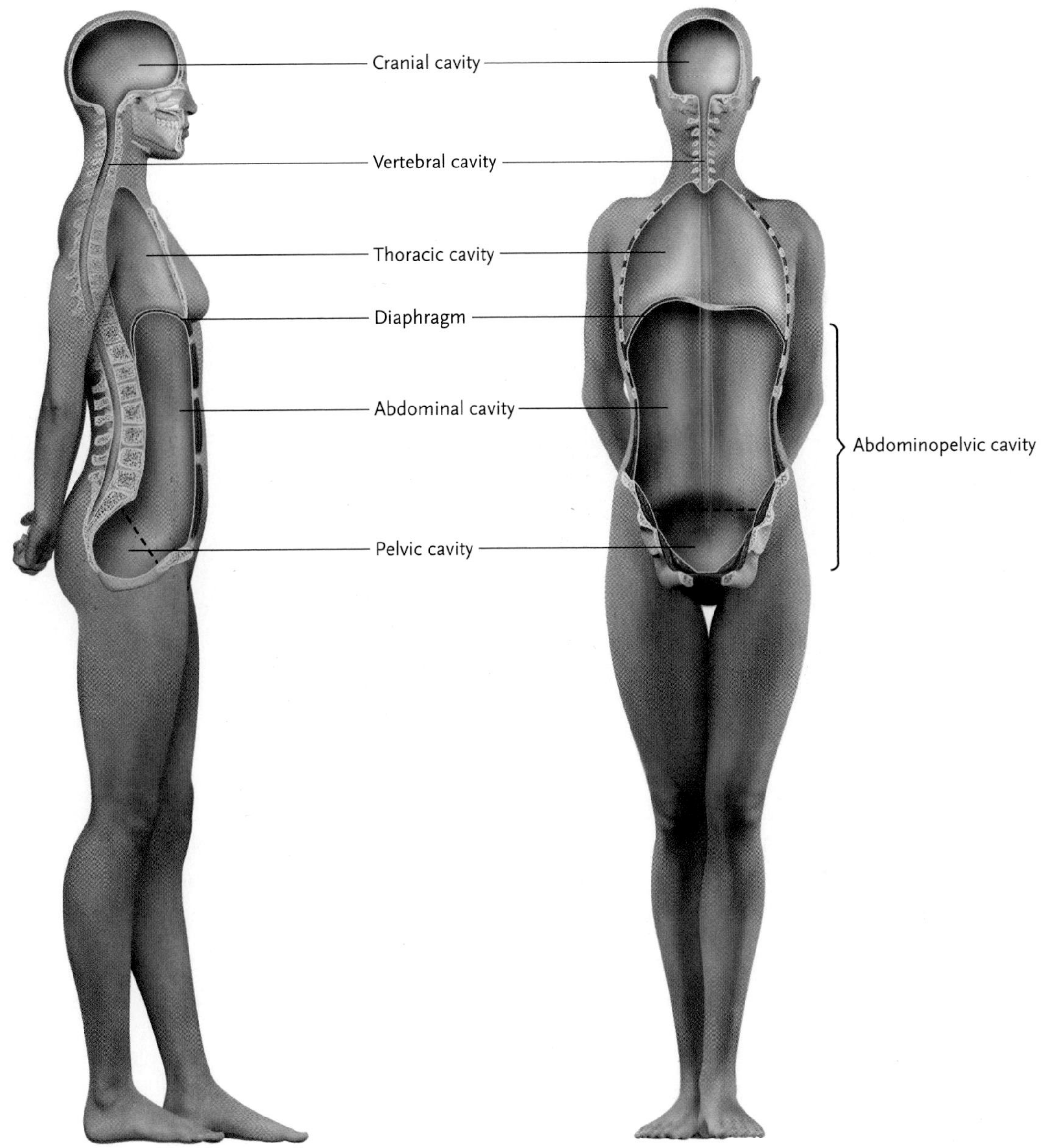

Figure 1.23   Body Cavities and Thoracic Cavity

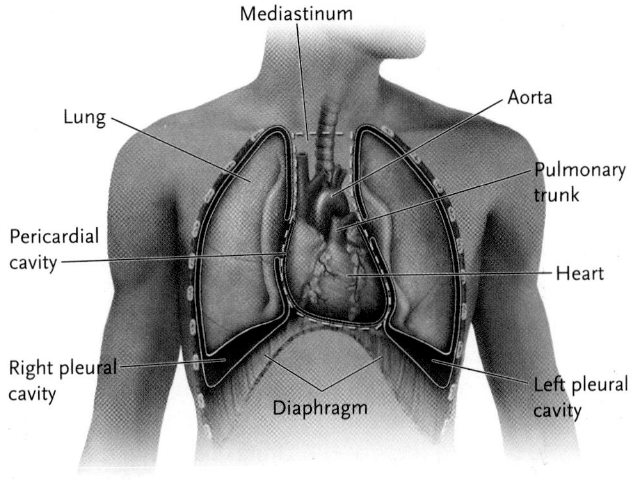

The body cavities (A) include the cranial, vertebral, thoracic, and abdominopelvic cavities; the thoracic cavity (B) includes the pleural and pericardial cavities.

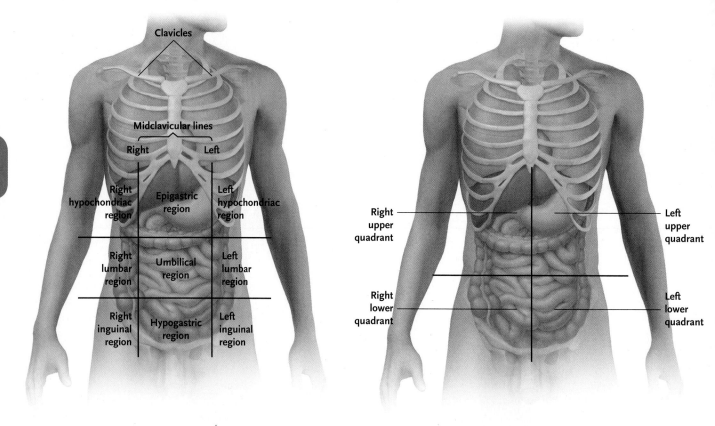

Figure 1.24 Abdominopelvic Cavity Regions
The abdominopelvic cavity can be divided into nine regions.

Figure 1.25 Abdominopelvic Cavity Quadrants
The abdominopelvic cavity can also be divided into four quadrants.

CONCEPT REVIEW 1.5

1. Why is the anatomical position important?

2. Make a chart comparing the common name and anatomical name for each body region.

3. Describe the different planes that are used to divide the body or a body part.

4. What organs are found in each of the following cavities: cranial, vertebral, thoracic, and abdominopelvic?

5. Where would you find pleural membranes, pericardial membranes, and the peritoneum?

6. What is the mediastinum?

7. List the major organs that are located within each of the nine abdominopelvic regions, or the four quadrants.

1.6 PROPERTIES OF LIVING THINGS

Living organisms have certain characteristics that differentiate them from non-living things. These characteristics are introduced below and covered in detail in this textbook as follows: all living things are made up of the same macromolecules (Chapter 2); they have a cellular structure (Chapter 3); they grow and metabolize nutrients (Chapter 4); they have mechanisms of homeostasis (below); they reproduce at a cellular and organism level (Chapters 5 and 21); they have DNA that they pass on through inheritance (Chapter 6);

and all species change over time through a process called **evolution** (Chapter 7).

Macromolecules

Every living thing is made up of the basic elements, carbon, oxygen, hydrogen, and nitrogen. All living things combine these basic chemical elements into small molecules, such as sugars, amino acids, fatty acids, and nucleotides. The small molecules combine into various macromolecules: polysaccharides (complex carbohydrates), proteins, fats, and nucleic acids.

Cellular structure

All living organisms consist of one or more cells, which are composed of phospholipids, proteins, carbohydrate groups, and cholesterol. Single-cell organisms can be prokaryotes or eukaryotes (Chapter 1, Section 2) and complex multicellular organisms are eukaryotic. Viruses have a protein capsule instead of a cell membrane and, therefore, are not technically considered living things, but viruses are significant to all living things and will be discussed in more detail in Chapter 23.

Growth and metabolism

All living organisms grow and undergo metabolism, producing energy in the form of ATP. Depending on the particular species and conditions, ATP can be produced either with oxygen—aerobically—or without oxygen—anaerobically. Humans produce ATP both with and without oxygen.

Reproduction

All living organisms reproduce at a cellular level through mitosis and at an organism level by producing sperm or eggs, spores, or seeds.

Hereditary material

Every living organism passes on some of their genetic material to their offspring. Humans have 46 DNA molecules (chromosomes): 23 from one parent and 23 from the other parent. Offspring are similar and yet different from their parents, and certain traits are passed on through generations.

Evolution

All populations of organisms continuously change genetically from one generation to the next. Humans have 46 chromosomes, but the particular genes on those chromosomes vary: such as the genes that encode skin pigmentation, or body height. Although humans today differ from humans 100,000 years ago, the distinctly human DNA remains intact.

Homeostasis

Homeostasis is the regulation of multiple factors within the **internal environment** of the body. The internal environment includes whatever is in close contact with the body's cells, such as the blood and the interstitial fluid that surrounds all cells and contains certain amounts of ions and nutrients. The **external environment** includes everything exterior to the body that the body cannot regulate, such as temperature, food acidity, humidity, air pollutants, and microorganisms that may become inhaled or swallowed. Any part of the body that is in contact with the outside world is considered part of the external environment, including the digestive tract, external ear canals, nose and sinuses, inside the trachea, bronchi, and air sacs. No matter what kind of changes occur in that external environment—such as changes in oxygen levels, nutrients ingested, and inhaled or ingested toxins—the internal environment is highly regulated so that the body's cells function optimally.

Homeostasis is a continual process that's necessary for the normal functioning of every living thing. Some factors are strictly regulated within a narrow range for the human body, including oxygen, carbon dioxide, pH, H^+, blood glucose, core body temperature, blood pressure, water, and electrolytes (sodium, potassium, chloride, calcium, etc.). When these factors are in balance, cells function properly, and we feel healthy. When homeostasis can no longer maintain a balance, disease occurs. Every organ system is involved in homeostasis in some way, and this will be discussed in each of the textbook's chapters on organ systems.

Components of feedback systems

Homeostasis regulates most of the factors discussed above through a process of **negative feedback** (Figure 1.26), which brings about a reverse in the level of a controlled factor. As well, a process of **positive feedback** sometimes occurs, in which a response caused by an initial stimulation continues to increase.

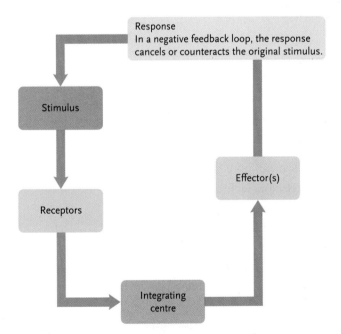

Figure 1.26 **Negative Feedback Mechanism**
A negative feedback mechanism consists of receptors that detect a change and signal the integrating centre, which then sends a signal to effectors to correct the change.

All of the body's feedback systems consist of the following components: **receptors** that detect the level of some factor (such as temperature); an **integrating centre** that compares information from the receptors to a standard set point (such as 37°C for normal body temperature); and **effectors**, which respond to signals from the integrating centre: for example, sweat glands. The receptors constantly send information about the body's internal environment to the integrating centre, which constantly signals the effectors to increase or decrease a specific response.

Receptors

The receptors are either extensions of a sensory neuron that sends information to the integrating centre about a specific stimulus, or individual receptor cells that communicate directly with a sensory neuron. The following list gives the names and functions of the common types of sensory receptors: **chemoreceptors** detect chemical concentrations such as neurotransmitters, drugs, or hormones; **osmoreceptors** detect changes in osmolarity, that is, water and ion concentrations; **tactile receptors** detect touch, pressure, and vibration; **baroreceptors** detect blood pressure; **photoreceptors** detect light and are found only in the retina; **mechanoreceptors** detect stretching (e.g., spindle fibres in muscles); **proprioceptors** detect body position; **nociceptors** detect pain; and **thermoreceptors** detect temperature.

Integrating centre

The integrating centre interprets the signals coming from the receptors and determines when any deviation occurs from the standard set point. Certain parts of the brain act as an integrating centre, and certain glands also act as an integrating centre: for example, the **hypothalamus** assesses temperature, thirst, hunger, and sleep; the **medulla oblongata** regulates breathing, blood pressure, and heart rate; and **pancreas** determines blood glucose levels.

Effectors

The effectors include any part of the body that responds to stimulation from the integrating centre and brings an altered factor back to normal. They include **skeletal muscles** (e.g., shivering when cold); **smooth muscles** (e.g., vasodilation, bronchodilation, increased gastric motility); **cardiac muscle** (e.g., change in heart rate or its force of contraction); and the **endocrine glands**.

Negative feedback mechanism

Negative feedback means there is a reverse in the level of a controlled factor. For example, exercise causes the body temperature to increase, which is then regulated by the production of sweat that cools the body. The higher the increase in body temperature, the more sweat is produced in an effort to maintain the normal body temperature. When normal body temperature cannot be maintained in extreme conditions, illness and possibly death will occur.

The following example describes what's involved in a negative feedback mechanism to regulate body temperature: stimulus (cold external environment)—receptors (thermoreceptors)—integrating centre (hypothalamus determines and signals that body temperature has dropped below the 37°C set point)—effectors (skeletal muscles shiver and produce heat)—new stimulus (increased body temperature)—receptors (thermoreceptors)—hypothalamus (stops signalling that body temperature is below normal)—effectors (skeletal muscles stop shivering).

Positive feedback mechanism

A positive feedback mechanism continues to increase a response caused by an initial stimulation (Figure 1.27). There are very few examples of positive feedback systems in the human body. One example is childbirth, where the stimulus is pressure of the head of the fetus, the receptors are mechanoreceptors in the uterus, and the integrating centre is hypothalamus, which produces oxytocin that signals the effectors—the uterine muscles—to contract; this response leads to further stimulation of the mechanoreceptors, which signals the hypothalamus to produce more oxytocin.

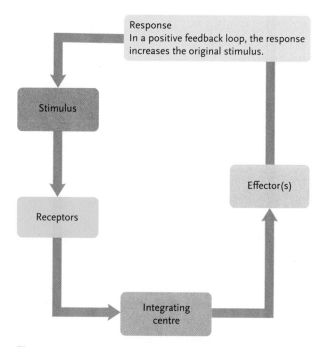

Figure 1.27   Positive Feedback Mechanism

Childbirth is an example of a positive feedback mechanism, where the response is continually increased by the stimulus until the baby is born.

CONCEPT REVIEW 1.6

1. Make a list of the different features that are found in all living organisms.

2. What is homeostasis? Describe the difference between negative feedback and positive feedback mechanisms, and give examples.

3. Explain the function of a receptor, an integrating centre, and an effector.

4. What are the different types of receptors, and what kind of stimulus do they detect?

5. Name some common effectors in the body.

2 Biological Molecules

It is the mark of an educated mind to be able to entertain a thought without accepting it.

Aristotle

2.1 WATER

Universal solvent

All living things from the smallest single-celled bacterium to multicellular plants and animals are composed of anywhere from 45 to 95% water. In humans, infants are approximately 75% water, and this amount decreases with age: mostly in the first 10 years of life, but continuing into old age. Healthy adult males are on average 60% water, and females are 55% water. Healthy females compared to healthy males of appropriate body weight have slightly less water because they tend to carry more body fat. Anyone who is obese can have as little as 45% body water content.

Many important nutrients, ions, and molecules are dissolved in the water inside and outside of cells. Water is the fluid of the body, the **solvent**, and it contains dissolved molecules, the **solutes**. Due to its unique properties, water is the universal solvent in all living things. Also, the specific concentration of different molecules within the cell or surrounding the cell is important to the cell's ability to function properly.

2

DID YOU KNOW?

Losing just 1% of your normal water content will stimulate thirst. During extensive exercise you can lose about 5% or 2 L of body water.

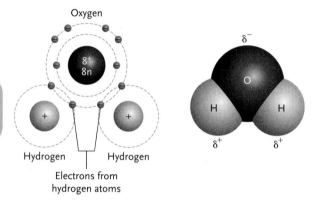

Figure 2.1   Water Molecule

Oxygen and hydrogen atoms share electrons in a water molecule and form covalent bonds.

Polarity

Oxygen has six positively charged protons compared to hydrogen's single proton. Oxygen and hydrogen form a **covalent bond** in which electrons are shared between the two atoms. With the formation of the covalent bond, the negatively charged electrons are pulled slightly closer to the protons in the oxygen atom, giving the oxygen atoms a slightly negative charge and the

hydrogen atoms a slightly positive charge (Figure 2.1). This charge difference causes water molecules to be **polar**. Because of this polarity, water molecules form **hydrogen bonds**, where the slight positive and slight negative charges attract (Figure 2.2). Hydrogen bonds cause water molecules to form droplets and stay close together rather than spread apart; this is called **surface tension**.

Hydrophilic and hydrophobic molecules

Hydrophilic means "water loving." Any molecule or atom that has a slight charge or a full charge, such as a sodium ion or a glucose molecule, which is polar, can dissolve in water (Figure 2.3a). Substances such as oils that do not have a charge are **nonpolar**, and they cannot dissolve easily in water; these molecules are **hydrophobic**, which means "water fearing." Both hydrophilic and

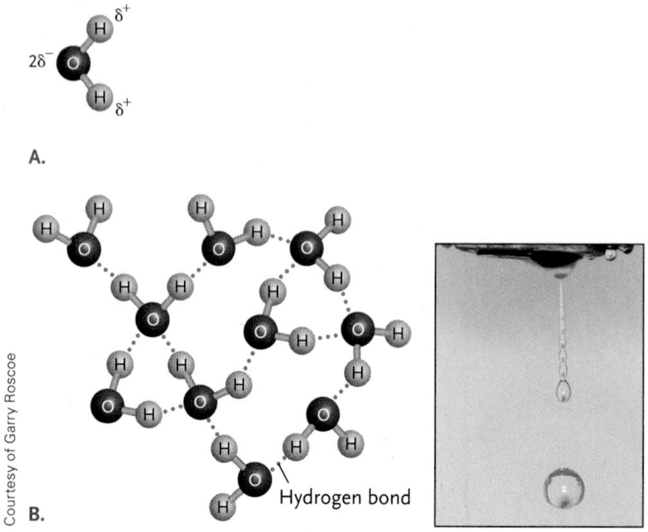

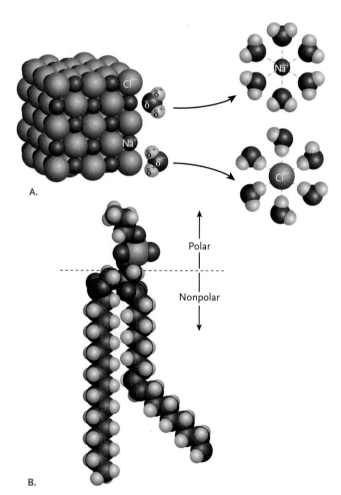

Figure 2.3   (A) Charged and Polar Molecules, (B) Amphipathic Molecules

(A) Charged molecules dissolve easily in water because their charges are attracted to slight charges in the water molecules. (B) Molecules that contain both polar and nonpolar regions are amphipathic.

Art sources: (top) From SANADER, MILAN. Chemistry 11U. © 2011 Nelson Education Ltd. Reproduced by permission. www.cengage.com/permissions; (bottom) From DIGIUSEPPE/FRASER. Biology 12U. © 2012 Nelson Education Ltd. Reproduced by permission. www.cengage.com/permissions

Figure 2.2   (A) Hydrogen Bonds, (B) Water Droplet

Because of the slight charge differences in the oxygen and hydrogen of water molecules, the atoms stay closely connected by hydrogen bonds.

Art sources: (A and B): From DIGIUSEPPE/SANADER. Chemistry 12U. © 2012 Nelson Education Ltd. Reproduced by permission. www.cengage.com/permissions

Courtesy of Garry Roscoe

hydrophobic molecules are essential to the specific functioning of cells, the formation of cell membranes, and the ways that nutrients and wastes move across those membranes. Sometimes molecules can have regions that are polar and regions that are nonpolar—such as soap. These molecules are called **amphipathic** because they have *both* hydrophilic and hydrophobic components (Figure 2.3b).

CONCEPT REVIEW 2.1

1. Approximately how much of the human body is composed of water?

2. What is the difference between a solute and a solvent?

3. Explain the difference between hydrophobic and hydrophilic, polar and nonpolar.

4. What is surface tension?

5. Describe the properties of an amphipathic molecule.

6. Why is water so biologically important?

2.2 MACROMOLECULES

Macromolecules are very large molecules made up of combinations of many smaller molecules; for example, protein macromolecules are composed of many amino acids. All living things are based on four primary atoms—carbon, oxygen, hydrogen, and nitrogen—that combine to form small molecules. Molecules combine to form macromolecules by a process called **dehydration synthesis**. The four macromolecules that make up all living cells are proteins, carbohydrates, nucleic acids, and fats. These are all formed from the combination of the small individual molecules of amino acids, monosaccharides, nucleotides, and fatty acids. There are many different kinds of proteins, carbohydrates, fats, and nucleic acids in different organisms. The proteins in a fish or an almond are different from the proteins formed in our cells after digestion and absorption, but the individual amino acids that a protein is made up of are the same in every living thing. The differences are in how those small molecules combine together. We will look at each macromolecule in more detail in the following sections.

When we eat food, we eat the polymers that plants or animals produced. These are called **organic molecules** because they come from living organisms. Our digestive system breaks these down into **monomers** that are absorbed into the bloodstream and circulated to the cells. Our cells use these molecules either to produce energy in the form of ATP (adenosine triphosphate) (Chapter 4) or to build human proteins, carbohydrates, nucleic acids, and fats, which all have a multitude of cellular functions in various body systems.

Molecules combine by the removal of a water molecule, hence the term, *dehydration synthesis*. For example, during protein synthesis in our cells, the protein is the macromolecule, and it is made up of **amino acids**. A hydroxyl group is removed from one amino acid, and hydrogen is removed from the other amino acid, and then a covalent bond forms (Figure 2.4).

When **polymers** are broken down in our cells or in our digestive system, the reverse reaction of dehydration synthesis occurs. A water molecule is added with the aid of a hydrolytic enzyme, and the covalent bond is broken. Hydrogen and a hydroxyl group are added to form the individual monomers (Figure 2.5).

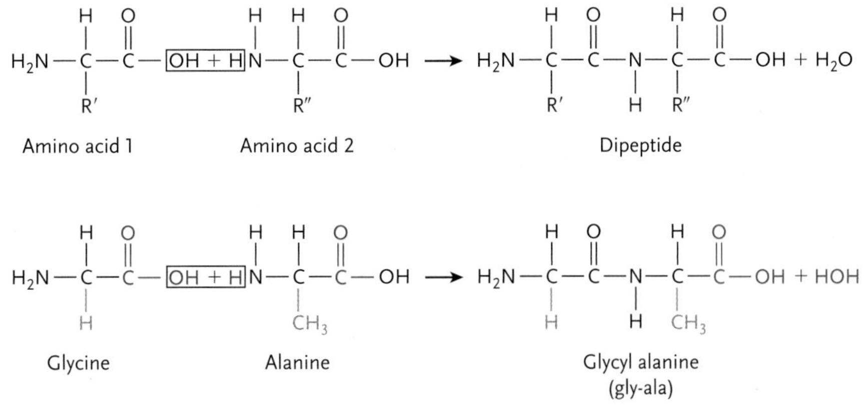

Figure 2.4 **Dehydration Synthesis**

Macromolecules are formed by a process called dehydration synthesis, where a water molecule is produced and a covalent bond is formed.

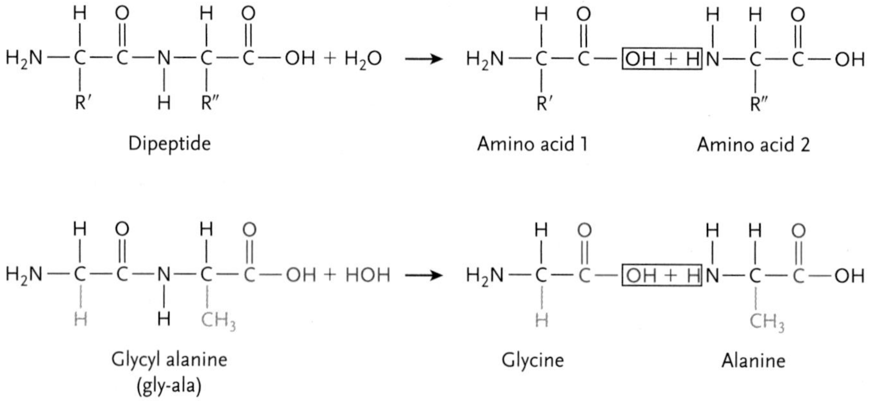

$$H_2N-\overset{\overset{\displaystyle H}{|}}{\underset{\underset{\displaystyle R'}{|}}{C}}-\overset{\overset{\displaystyle O}{\|}}{C}-\overset{\overset{\displaystyle H}{|}}{\underset{\underset{\displaystyle H}{|}}{N}}-\overset{\overset{\displaystyle H}{|}}{\underset{\underset{\displaystyle R''}{|}}{C}}-\overset{\overset{\displaystyle O}{\|}}{C}-OH + H_2O \longrightarrow H_2N-\overset{\overset{\displaystyle H}{|}}{\underset{\underset{\displaystyle R'}{|}}{C}}-\overset{\overset{\displaystyle O}{\|}}{C}-\boxed{OH + H}N-\overset{\overset{\displaystyle H}{|}}{\underset{\underset{\displaystyle R''}{|}}{C}}-\overset{\overset{\displaystyle O}{\|}}{C}-OH$$

Dipeptide   Amino acid 1   Amino acid 2

$$H_2N-\overset{\overset{\displaystyle H}{|}}{\underset{\underset{\displaystyle H}{|}}{C}}-\overset{\overset{\displaystyle O}{\|}}{C}-\overset{\overset{\displaystyle H}{|}}{\underset{\underset{\displaystyle H}{|}}{N}}-\overset{\overset{\displaystyle H}{|}}{\underset{\underset{\displaystyle CH_3}{|}}{C}}-\overset{\overset{\displaystyle O}{\|}}{C}-OH + HOH \longrightarrow H_2N-\overset{\overset{\displaystyle H}{|}}{\underset{\underset{\displaystyle H}{|}}{C}}-\overset{\overset{\displaystyle O}{\|}}{C}-\boxed{OH + H}N-\overset{\overset{\displaystyle H}{|}}{\underset{\underset{\displaystyle CH_3}{|}}{C}}-\overset{\overset{\displaystyle O}{\|}}{C}-OH$$

Glycyl alanine   Glycine   Alanine
(gly-ala)

Figure 2.5   Hydrolysis

Macromolecules are broken down by a process called *hydrolysis*, where a water molecule is added and a covalent bond is broken.

Art source: From JENKINS ET AL. Nelson Chemistry 12, 1E. © 2003 Nelson Education Ltd. Reproduced by permission. www. cengage.com/permissions

CONCEPT REVIEW 2.2

1. Explain the difference between dehydration synthesis and hydrolysis.

2.3 PROTEINS

Amino acids

The human body produces over 100,000 different proteins. Amino acids are the monomers that are the building blocks of the thousands of different types of proteins. There are 20 different amino acids. These amino acids differ only by their functional group, and they can be arranged in multiple ways to produce proteins (Figure 2.6). Every cell in the body can build the correct proteins in the correct amount in the correct cell type throughout our lives. For example, after we eat, the pancreas beta islet cells are the only cells that synthesize the hormone protein insulin (Chapters 16 and 19), and our blood sugar level rises. Every protein is formed by very specific amino acid sequences. A typical protein contains anywhere from 100 to 1000 amino acids. The *order* of amino acids in a protein is critical for the proper folding of the protein, which is based on interactions between the functional groups. The protein can function only if it contains the correct amino acids and is folded into the correct conformational structure. For example, a single amino acid change in a chloride-channel protein causes improper folding, with the result that the membrane protein cannot function, and this causes the disease cystic fibrosis. However, not all changes in the amino acid sequence will cause disease (Chapters 6 and 8).

An individual amino acid has a **carboxyl group**, an **amino group**, and a functional group. The functional group is what differs among all 20 amino acids and gives each amino acid its distinctive characteristics (Figure 2.7). The covalent bond that links the amino acids is called a **peptide bond**. When many amino acids are covalently bonded by peptide bonds, a **polypeptide**—a protein—is formed.

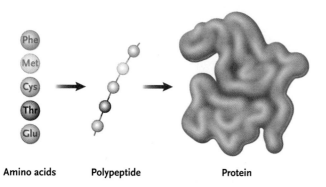

Amino acids   Polypeptide   Protein

Figure 2.6   Protein Synthesis

Proteins are macromolecules made up of amino acids, and the order of amino acids in a protein determines the structure and function of that protein.

Art source: From DIGIUSEPPE/FRASER. Biology 12U. © 2012 Nelson Education Ltd. Reproduced by permission. www.cengage.com/permissions

DID YOU KNOW?

People with curly hair have more cysteine amino acids in their hair proteins. Cysteine amino acids have a sulfur atom that can form disulfide bonds with other cysteine molecules; this causes kinks in the proteins and, therefore, curly hair.

Essential amino acids

The human body can synthesize 11 of the 20 amino acids from other molecules. The remaining nine, which are called **essential amino acids**, cannot be synthesized by the human body. We have to obtain the following nine

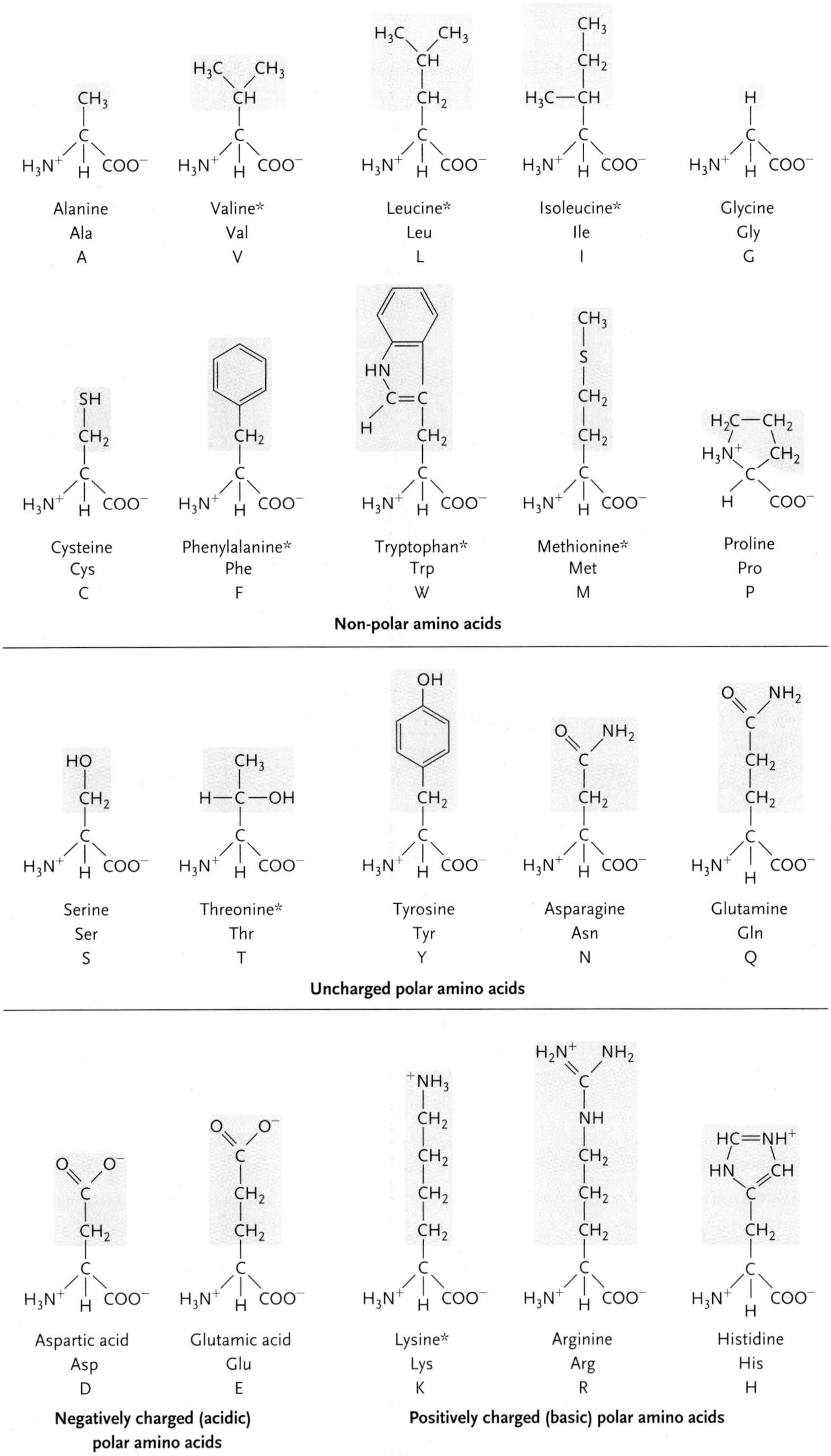

Figure 2.7 Amino Acids
The 20 amino acids differ only in their functional groups.

essential amino acids from our diet: phenylalanine, valine, threonine, tryptophan, isoleucine, methionine, leucine, histidine, and lysine. The most common food sources that contain *all* amino acids are meat, fish, eggs, chicken, soy products, and dairy products. Other foods contain some, but not all, of the essential amino acids. See Table 2.1.

DID YOU KNOW?

Vegetarians can become deficient in essential amino acids if they do not eat adequate combinations of plant protein sources. Legumes, beans, nuts, seeds, and grains each contain some of the essential amino acids, which is why it is important for vegetarians to eat these food groups in combinations. Amino acid deficiencies can cause numerous symptoms, including chronic fatigue, hair loss, irritability, and reproductive problems.

Protein structure

The sequence of amino acids in a protein is called the *primary structure* (Figure 2.8). For example, the formation of the protein insulin requires the correct amino acids in the correct order, and the protein has to fold into the correct shape so that it will bind specifically to insulin receptors. The body's cells know the correct order of amino acids because of the DNA sequences called genes, which is discussed in the next section.

As amino acids become arranged in the correct order on the ribosomes in each cell, they take on a *secondary structure*; this is based on the properties of the functional groups of the individual amino acids. Depending on which amino acids are polar, nonpolar, or acidic, for example, the protein starts to fold in a way that corresponds to the interactions between the amino acids. Hydrogen bonds form between polar amino acids, just like hydrogen bonds form between water molecules. Hydrophobic amino acids are attracted to other hydrophobic amino acids, and the secondary structure develops. Most of the secondary structures formed are called **alpha helices** or **beta sheets**.

The protein takes on a three-dimensional shape based on the primary amino acid sequence and the secondary structures that are formed. Eventually, the protein forms a three-dimensional *tertiary structure*, which constitutes the functional structure of the protein. To function

TABLE 2.1

Major Categories of Proteins in the Human Body		
Type of Protein	**Function**	**Examples**
Enzyme	Increases the rate of chemical reactions in all cells of the body	Lipase breaks down lipids (Chapter 19). Polymerase combines nucleotides during DNA replication (Chapter 8).
Protection	Involved in protecting the body from pathogens or cellular damage	IgG is an antibody involved in acute infections (Chapter 22). Fibrinogen is a protein involved in blood clotting (Chapter 17).
Membrane proteins	Allows various molecules to cross cell membranes, act as signalling molecules, or adhere to surrounding tissues	Na+/K+ pump transports sodium out of the cell and potassium into the cell (Chapter 3). Calcium channels transport calcium into neurons to release neurotransmitters (Chapter 15).
Cell recognition	Acts as markers that allow immune cells to recognize self-cells and pathogens	Red blood cells contain markers known as blood type, A, B, AB, and O (Chapter 6). Major histocompatibility proteins are involved in presenting foreign antigens to immune cells (Chapter 22).
Contractile	Involved in muscle contraction and movement	Actin and myosin are the contractile proteins in muscle tissue (Chapter 14).
Transporter	Involved in moving substances throughout the body	Lipoproteins transport fats through the bloodstream (Chapters 17 and 19). Hemoglobin transports oxygen through the bloodstream (Chapters 6, 17, and 18).
Structural	Maintains the structure of the cytoskeleton or extracellular matrix	Collagen is an extracellular matrix protein that holds cells together (Chapters 3 and 11). Actin is a protein that forms intracellular microfilaments that allow cells to move (Chapter 3).
Regulatory	Involved in signalling molecules that regulate cell functions	Insulin is a hormone that regulates blood sugar (Chapter 16). Growth hormone is involved in maintaining bone and muscle density (Chapter 16). Troponin and tropomyosin are regulatory proteins involved in muscle contraction (Chapter 14).

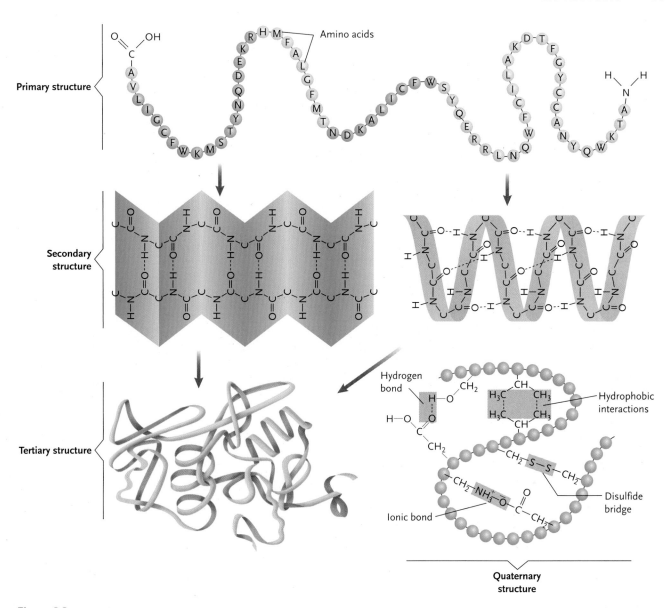

Figure 2.8 **Protein Structure**

Art source: From DIGIUSEPPE/FRASER. Biology 12U. © 2012 Nelson Education Ltd. Reproduced by permission. www.cengage.com/permissions

properly, every distinctive protein has to have not only the correct amino acid sequence but also the correct tertiary structure, which is determined by the amino acid functional groups. Nonpolar amino acids tend to be attracted to one another, and polar or charged amino acids are attracted to each other. Sometimes multiple tertiary structures combine to form a final functional protein product, called a *quaternary structure*. An example of a quaternary protein in our body is the hemoglobin protein where four protein subunits combine to form the final functional macromolecule. Hemoglobin is found in red blood cells and carries oxygen from the lungs to the tissues. Some proteins, such as microtubules or microfilaments, are multiprotein complexes composed of hundreds of individual proteins (Chapter 3).

Denaturation

When you cook an egg, the increased temperature causes the egg proteins to unfold; this explains why the egg changes from a semi-liquid to a semi-solid state. The increased temperature adds enough energy to break the weak hydrogen bonds that form the secondary and tertiary protein structures, but the covalent peptide bonds remain intact. A change in the pH can also affect the bonds that hold proteins in their conformational shape: for example, the acidic environment of the stomach. This process of breaking the hydrogen bonds in a protein is called **denaturation**. Denaturation is an irreversible reaction—for example, cooling a cooked egg does not change it back to its original state—and denatured proteins no longer function (Figure 2.9). Their folded

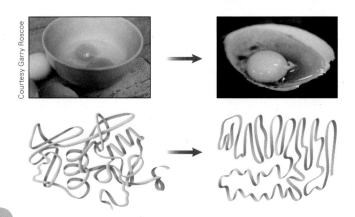

Figure 2.9 Denaturation

Proteins unfold and become denatured when exposed to heat during cooking or to hydrochloric acid in the stomach.

Art source (bottom): From DIGIUSEPPE/FRASER. Biology 12U. © 2012 Nelson Education Ltd. Reproduced by permission. www.cengage.com/permissions

structure is critical for protein function. However, denaturation does not change the structure of individual amino acids. When we eat proteins, digestive enzymes break the proteins into the amino acids that we absorb, so cooking food does not affect the quality of these amino acids.

Enzymes

The amino acid sequence and the way it folds determine the function of a protein. **Enzymes** are very important globular proteins that facilitate thousands of chemical reactions in our body's cells. In humans, more than 75,000 different enzymes are produced in various types of cell and perform a myriad of essential functions. Enzymes have grooves or specific tertiary structures that form **active sites** whereby an enzyme can **catalyze**, or speed, the rate of chemical reactions. Enzymes increase the contact between specific regions of molecules so that a particular chemical reaction can occur: for example, bringing amino acids in close proximity so that dehydration synthesis can occur (Figure 2.10).

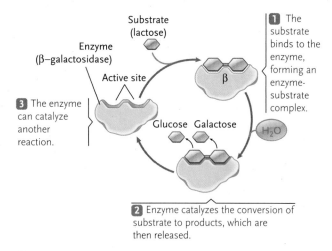

Figure 2.10 Enzyme Function

Art source: From DIGIUSEPPE/FRASER. Biology 12U. © 2012 Nelson Education Ltd. Reproduced by permission. www.cengage.com/permissions

CONCEPT REVIEW 2.3

1. How many amino acids are there? How many of them are essential?

2. Name three components of an amino acid. Which of these components are the same in every amino acid?

3. What types of bonds are formed in proteins? Which bonds hold amino acids together, and which bonds are important for the formation of secondary and tertiary structures?

4. Describe the different structures of proteins.

5. What are alpha helices and beta sheets?

6. Give three examples of quaternary proteins in the human body.

7. What is denaturation?

8. Why are enzymes important?

DID YOU KNOW?

Having an extremely high fever is fatal because cellular proteins can start to denature. Without the thousands of proteins doing their normal cellular functions, death could occur quite rapidly.

2.4 NUCLEIC ACIDS

Nucleic acids provide cellular "instructions" that determine the correct amino acid sequence for every protein produced by every cell in the body. This is true for every living organism.

DNA

Every cell in the human body (except mature red blood cells that lack a nucleus) contains 46 **chromosomes**.

We have two copies of 23 chromosomes: one set received from each parent. Each chromosome consists of one nucleic acid macromolecule called **DNA**: deoxyribonucleic acid. Each DNA macromolecule contains hundreds to thousands of **genes**, which give the blueprint for a specific **protein**. Large chromosomes can have over 2000 genes, such as chromosome #1. Smaller chromosomes, such as chromosome 21, have less than 400 genes (Figure 2.11).

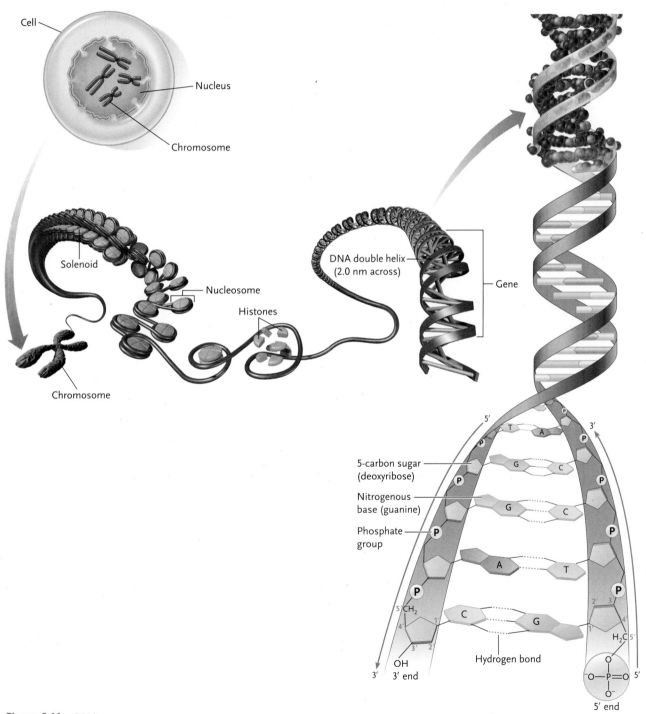

Figure 2.11   DNA

Each chromosome in every cell is a single macromolecule of DNA, which contains hereditary material.

The X chromosome has over 2000 genes, and the male Y chromosome has only 78 genes. In total, the human genome is made up of 25,000 genes.

DNA structure

As discussed in Section 2.3, the *order* of amino acids in the protein determines the structure and function of each of the thousands of proteins in the human body. Every cell produces the correct proteins with the correct order of amino acids because of the sequence of **nucleotides** in each gene. When red blood cells produce the hemoglobin protein, the order of amino acids is determined by the nucleotides in the hemoglobin gene, which all humans have on chromosome #11. Exactly how nucleotide sequences lead to amino acid sequences is discussed in detail in Chapter 8. Each DNA nucleotide is an organic

monomer made up of three components: a 5-carbon sugar (deoxyribose), a phosphate group, and a nitrogen-containing base (Figure 2.12a). Adenine and guanine are nitrogenous bases with a double-ring structure; they are called **purines**. Cytosine and thymine are single-ring bases called **pyrimidines** (Figure 2.12b).

It is often said that DNA contains the "stored information" for protein synthesis. This is because it carries the specific order of the nucleotides (A, T, C, and G) in every gene, and this translates into the amino acid sequence of every protein that's synthesized. Imagine that our nucleotides are like letters of the alphabet; from letters, we make words. The sequence of nucleotide "letters" has to be correct in order to spell amino acid "words" accurately and to make the proper protein "sentences." A misspelled word can dramatically change the meaning of a sentence. Such a misspelling of an amino acid sequence corresponds to a DNA mutation (discussed in Chapters 6 and 8).

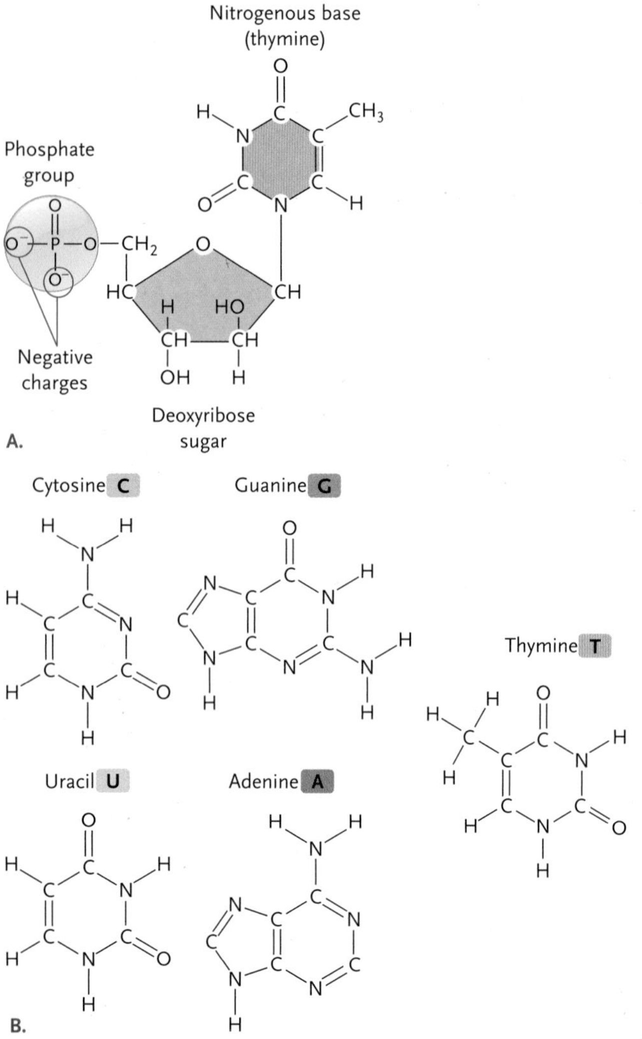

Figure 2.12 (A) Nucleotide Structure, (B) Nitrogenous Bases

(A) Each nucleotide contains a five-carbon sugar, a phosphate, and a nitrogen base. (B) The five nitrogenous bases of the nucleotides are adenine, cytosine, guanine, thymine, and uracil (found only in RNA).

Art source: From DIGIUSEPPE/FRASER. Biology 12U. © 2012 Nelson Education Ltd. Reproduced by permission. www.cengage.com/permissions

DID YOU KNOW?

Some genes are expressed only at certain times in our life, such as the gene for fetal hemoglobin, which is useful for transporting oxygen in a fetus. After birth, the red blood cells make a different hemoglobin protein from a different gene.

DNA is a **double helix** whereby two strands of DNA bind together by the nitrogenous bases. A gene can be as small as 100 nucleotides, or as long as 2000 nucleotides on one strand of DNA. On each strand of DNA the sugar and the phosphates are bound together by very strong **phosphodiester bonds**; these make up the "backbone" of the DNA strand and the nitrogenous bases extend from each sugar molecule. The sugar is bound to the base by a covalent **glycosidic bond**, also very strong. Glycosidic bonds are found between any sugar and some other group that could be another sugar or something else, such as a nitrogenous base. Each base binds with only one specific base on the opposite strand. Adenine always binds with thymine, and cytosine always binds with guanine (Figure 2.13). The nucleotide bases bind to the opposite nucleotide by weak hydrogen bonds. Two hydrogen bonds form between adenine and thymine, and three hydrogen bonds hold guanine and cytosine together. The two strands of DNA are antiparallel because they are oriented in opposite directions. Note that each carbon on the deoxyribose sugar shown in Figure 2.13 is numbered. The first carbon is bound to the base, and the third and fifth carbons are bound to phosphate groups. The DNA strand on the left has a free phosphate group attached to carbon number 5: this is called the 5′ (five prime) end. The opposite strand has a free phosphate on carbon number 3: this is called the 3′ (three prime) end. How DNA polymerase enzymes replicate DNA molecules is covered in Chapter 8.

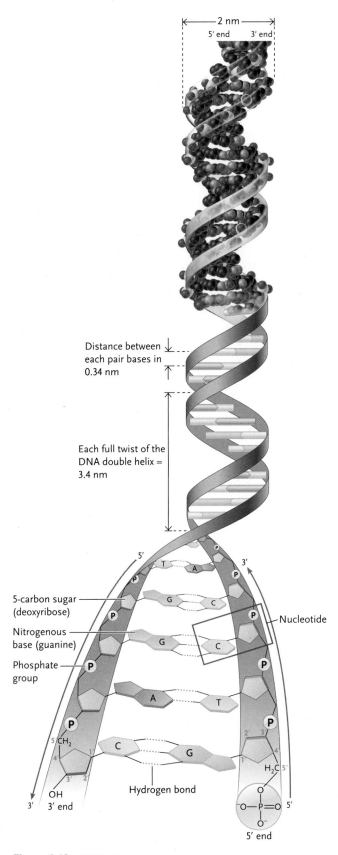

Figure 2.13   **DNA Structure**

DNA is a double helix; adenine from one strand always binds with thymine on the opposite strand, and guanine always binds with cytosine on the opposite strand.

Art source: From RUSSELL/HERTZ/STARR/FENTON. Biology, 2E. © 2013 Nelson Education Ltd. Reproduced by permission. www.cengage.com/permissions

RNA

The other type of nucleic acid macromolecule found in cells is called **RNA**: ribonucleic acid. The function of DNA and RNA is quite different. RNA is the intermediate step between the nucleotide sequence of each gene and the eventual completed protein. Suppose that a pancreas cell is stimulated by increased blood sugar levels to produce the protein hormone insulin. Inside each pancreatic beta islet cell nucleus the insulin gene sequence becomes *transcribed* into an RNA sequence (Figure 2.14). That RNA molecule then leaves the nucleus and moves into the cytoplasm of the cell to a ribosome, where it acts as the template for the assimilation of the amino acid sequence that produces insulin. The insulin gene consists of 153 nucleotides that lead to the production of the small, 51-amino acid, insulin protein. The process of RNA transcription is discussed in detail in Chapter 9. See Table 2.2.

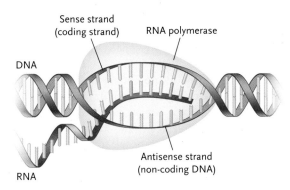

Figure 2.14   **RNA**

RNA is a single-stranded nucleic acid molecule transcribed from a DNA gene sequence that codes for the synthesis of a protein.

TABLE 2.2

Comparison of DNA and RNA		
	DNA	**RNA**
Structure	double-stranded helix	single-stranded and can have various folding patterns
Nucleotides	A, T, C, and G	A, U, C, and G
Sugar	deoxyribose	ribose
Location	nucleus only	transcribed in the nucleus and moves into cytoplasm
Function	stores genetic information	used as a template for protein synthesis

CONCEPT REVIEW 2.4

1. What is a chromosome? How many do humans have?

2. What is the function of DNA? Where is it located?

3. Describe the components of a nucleotide.

4. What three types of chemical bonds would you find in nucleic acids?

5. What is the complementary sequence of the following strand of DNA?

   A T G G C T A G T C

6. What is the function of RNA?

7. What are the differences between DNA and RNA?

2.5 CARBOHYDRATES

Simple and complex carbohydrates

Carbohydrates can be generally classified as simple carbohydrates (sugars)—the **monosaccharides** and **disaccharides**—and complex carbohydrates, the **polysaccharides**. (*Saccharide* is derived from the Greek, *saccharin*, which means sweet.) All carbohydrates have a ratio of 1:2:1, referring to the ratio of carbon:hydrogen:oxygen. The carbon-hydrogen bonds are what make carbohydrates particularly useful for energy production (Chapter 4). The molecular formula for each monosaccharide is $C_6H_{12}O_6$. However, the structures are slightly different for each of the three monosaccharides: **glucose**, **fructose**, and **galactose** (Figure 2.15). Fructose gives fruit its sweet taste. Honey consists of both glucose and fructose monosaccharides, and it is sweeter than table sugar, which is composed of glucose-fructose disaccharides.

DID YOU KNOW?

Fructose is the sweetest-tasting sugar. As high-fructose corn syrup, it is added to many foods, particularly soft drinks, canned fruits, and concentrated juices. Fructose is processed by the liver and converted directly into fat. It does not decrease the hunger response and is a significant contributor to weight gain, cardiovascular disease, and type II diabetes.

Disaccharides

Each of the monosaccharides can be paired by glycosidic bonds to form disaccharides. Table sugar, or **sucrose**, is formed from the dehydration synthesis reaction between glucose and fructose. **Maltose**, found in grains, is formed from two glucose molecules, and **lactose**, milk sugar, is formed from glucose and galactose. These are the most common disaccharides (Figure 2.16). They have different physical properties; some disaccharides are crystals (sucrose), and some are sticky (fructose).

Lactose intolerance

Why do some people become bloated and have stomach pains and diarrhea after drinking milk? Lactose is a disaccharide that requires a specific enzyme, **lactase**, to break it down into glucose and galactose. Some people lose the ability to produce this enzyme as they get older. The lactose stays in the digestive tract instead of being broken down and absorbed. The lactose molecules are then taken up by the normal bacteria that reside in the digestive tract; these bacteria break down the lactose to make ATP and produce methane gas as a waste product. The lactose remaining in the intestine also increases osmotic concentration in the

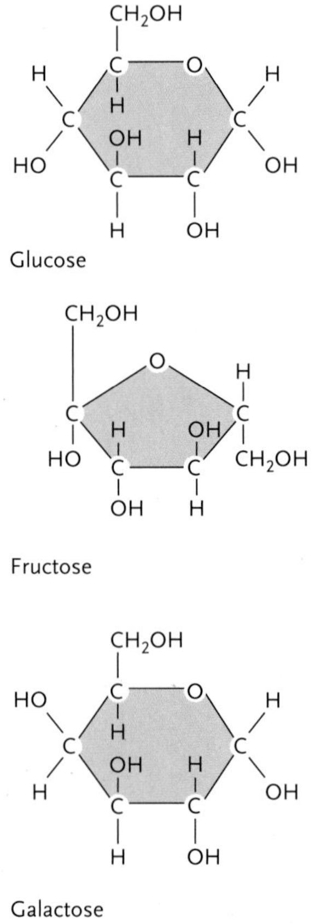

Figure 2.15 Monosaccharides

The three monosaccharides—glucose, fructose, and galactose—have the same chemical formula $C_6H_{12}O_6$ and slightly different chemical structures.

Art source: From DIGIUSEPPE/FRASER. Biology 12U. © 2012 Nelson Education Ltd. Reproduced by permission. www.cengage.com/permissions

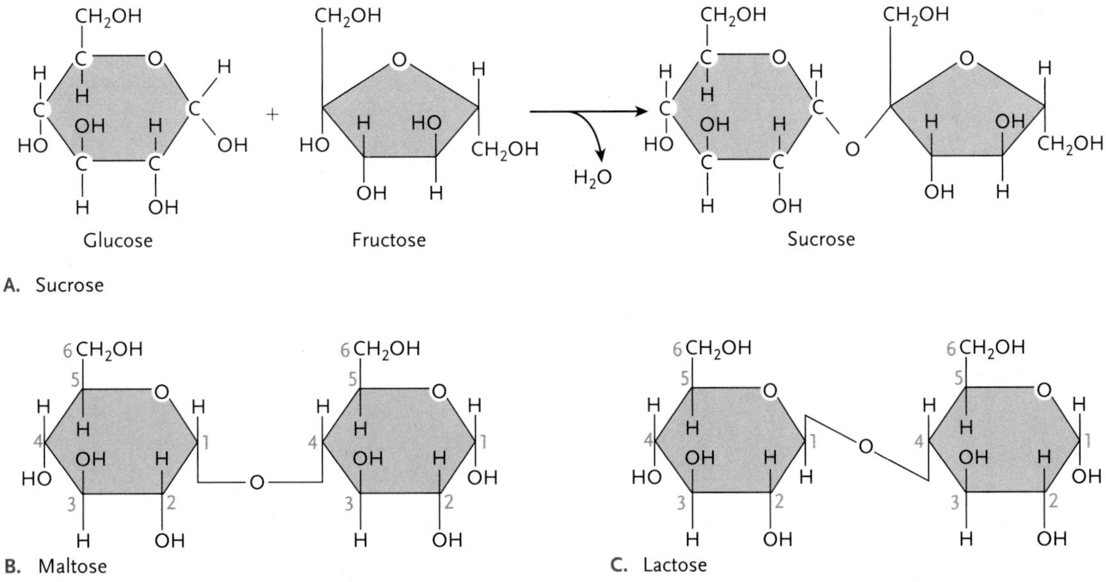

Figure 2.16 Disaccharides
Three important disaccharides are (A) sucrose, (B) maltose, and (C) lactose.

intestine, which causes water to move into the intestine, and this causes diarrhea.

Polysaccharides

Polysaccharides, the complex carbohydrates, are long chains of glucose. Three major complex carbohydrates are relevant to humans: starch, glycogen, and cellulose (Figure 2.17). Starch, a polysaccharide that we can eat, is the *stored energy* in plant cells. Starchy plant foods include rice, oatmeal, potatoes, yams, and grains such as corn, wheat, barley, rye, millet, and buckwheat. The carbohydrates in our diet are an important source of energy. The digestive system breaks down the long chains of glucose molecules through **hydrolysis**, and individual glucose molecules are absorbed into the bloodstream. The body's cells then take up these glucose molecules and use them to make ATP (Chapter 4). When we eat more carbohydrates than needed for ATP production, the excess is stored as either glycogen or fat. The polysaccharide **glycogen** is the stored energy in animal cells. The liver and muscle cells store some excess carbohydrates as glycogen, which can be used later when blood sugar levels start to decrease with exercise or between meals.

Another important plant polysaccharide is **cellulose**. Cellulose is a structural plant carbohydrate and is not used to produce energy. When we eat plants such as fruit, vegetables, whole grains, and legumes, we cannot digest the cellulose. However, cellulose is important in our diet as **fibre**. The consumption of fibre contributes significantly to our health: it regulates bowel movements and prevents constipation; it plays a role in the prevention of bowel cancer, hemorrhoids, and diverticulitis; it lowers blood cholesterol; and it helps regulate blood sugar.

DID YOU KNOW?

Chitin is a structural carbohydrate found in crustaceans and insects; it gives them a tough exoskeleton.

All polysaccharides are formed by dehydration synthesis between glucose molecules. However, each polysaccharide has very different properties due to the way the bonds are formed. For example, in cellulose the glucose molecules form bonds between the first carbon of one glucose and the fourth carbon of the next glucose, which leads to a straight chain structure; in glycogen, the glucose molecules form bonds between the first and fourth or the first and sixth carbons of adjacent glucose molecules, which leads to a branched structure (Figure 2.17).

DID YOU KNOW?

Carrageenan is found in many processed products such as chocolate bars and granola bars. It is a polysaccharide formed with galactose sugars instead of glucose. This gives foods a smooth texture and adds chewiness.

Dietary carbohydrates

Since starches and glycogen break down to become glucose, why isn't it just as healthy to eat chocolate bars and candy? One reason why candy is less healthy than other foods composed of complex carbohydrates

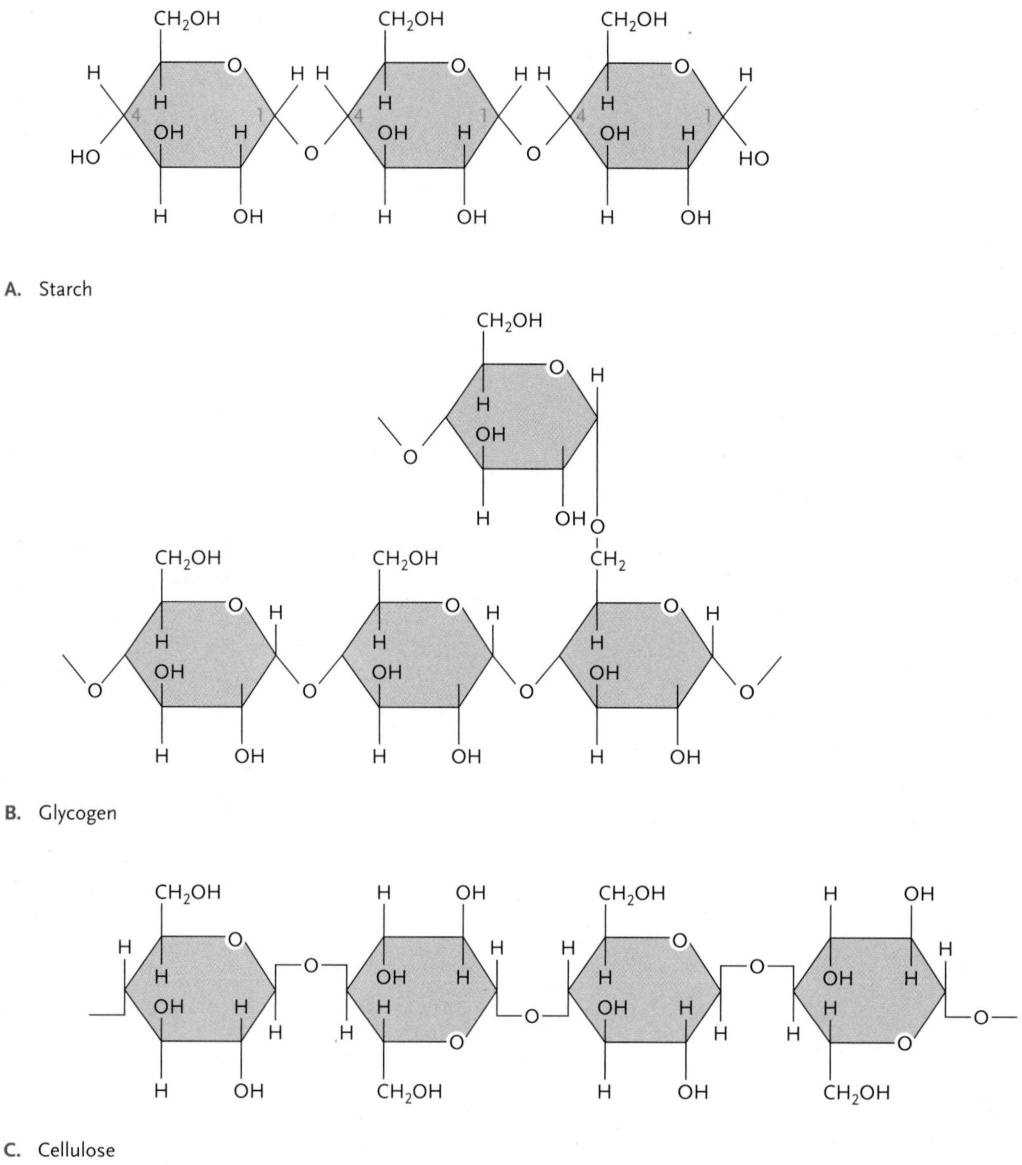

A. Starch

B. Glycogen

C. Cellulose

Figure 2.17 **(A) Starch, (B) Glycogen, (C) Cellulose**
Three polysaccharides are important to humans: (A) starch, (B) glycogen, and (C) cellulose. All three are composed of glucose, but each one is a very different molecule because of slight differences in their bond formation.

relates to the rate of digestion and absorption from the intestines into the bloodstream. Candy is a simple sugar that can be readily absorbed into the blood, and this causes a rapid increase in blood sugar level. When this occurs, the pancreas secretes insulin, causing a lot of that sugar to be converted into fat, rather than used for energy production. Starch takes much longer to digest because every bond must be broken by enzymes in the digestive tract before each glucose molecule can be absorbed. A second reason why sugar is not as healthy as starch relates to the composition of the sugar and its usefulness in energy production. Most simple carbohydrates are a combination of glucose and fructose, whereas starch is made up of only

glucose. The body's cells prefer to use glucose for ATP production, and the liver converts most fructose into fat. Therefore, eating a lot of simple sugars results in weight gain.

DID YOU KNOW?

People who have high amounts of fructose in their diet can develop fatty liver disease, which has symptoms that are similar to cirrhosis of the liver caused by alcoholism.

CONCEPT REVIEW 2.5

1. What are the primary atoms that make up carbohydrates?

2. What is the difference between a monosaccharide, a disaccharide, and a polysaccharide?

3. What are the three primary monosaccharides, and how can they combine to form disaccharides?

4. What three polysaccharides are important for human health? Describe how they differ from each other and how they are important.

5. Give examples of foods that contain maltose, sucrose, lactose, starch, and cellulose.

2.6 LIPIDS

Function and solubility

Lipids, also called **fats**, are an important component of our body and therefore important to include in our daily diet. The body uses fats for many vital functions in every cell. Fats consist of many carbon-hydrogen bonds, which make them very useful molecules for storing energy. See Table 2.3.

Recall from Section 1.1 that water molecules are polar because of a slightly unequal sharing of electrons in the covalent bonds. In fatty acids, the carbon and hydrogen covalent bonds have electrons equally distributed between the atoms. These bonds are nonpolar, or hydrophobic. When fats are placed in water they cluster together as droplets in the water. This property is extremely important in the formation of cell membranes from fats.

TABLE 2.3

Important Functions of Fats	
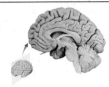	**Brain.** Fats compose 60% of the brain and are essential to its functioning, including learning, memory, and mood.
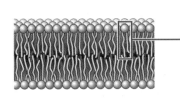	**Cells.** Fatty acids and cholesterol are responsible for building cell membranes and help cells stay flexible.
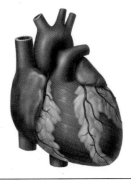	**Heart.** Sixty percent of the energy for the heart to contract comes from burning fats.

Courtesy of Mark Nielsen, AnatBooks, LTD

(Continued)

TABLE 2.3

Important Functions of Fats *(continued)*

Art source: From RUSSELL/HERTZ/STARR/ FENTON. Biology, 2E. © 2013 Nelson Education Ltd. Reproduced by permission. www.cengage.com/ permissions

Nerves. Fats compose the material that insulates and protects the nerves, thereby isolating electrical impulses and speeding their transmission.

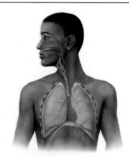

Lungs. Lung **surfactant** enables the lungs to work properly.

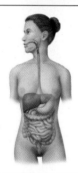

Digestion. Eating fats slows down the digestion process so the body has more time to absorb nutrients and help provide a constant level of energy. Fats also keep the body satiated for longer. Fat-soluble vitamins (A, D, E, and K) can be absorbed only if fat is present in the digestive tract when the vitamins are ingested.

Organs. Fats cushion and protect the internal organs.

Immune system. Fats decrease inflammation and help the metabolism and immune system stay healthy.

Types of fats

There are three main categories of fats: triglycerides, cholesterol, and phospholipids. Triglycerides are composed of one **glycerol** molecule bonded to the carboxylic acid end of three **fatty acids** (Figure 2.18). Triglycerides, the body's main source of stored energy, are found in fat cells (**adipocytes**). Groups of these cells are called **adipose tissue** (Chapter 11)—also called **subcutaneous fat** because it occurs mainly in the subcutaneous layer under the dermis of the skin (Chapter 12). Many people have an excess of this type of fat, and it can only be decreased when the cells use this fat as an energy source. The cells that use the most stored energy to make ATP are the skeletal muscle cells; this explains why exercise and a healthy diet are crucial for maintaining a healthy body weight.

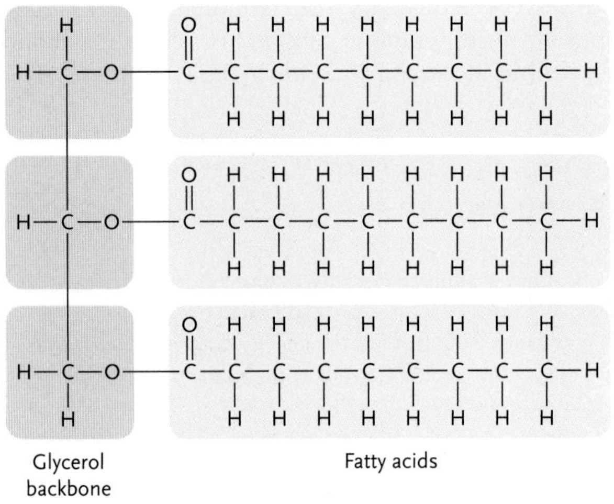

Figure 2.18 Triglyceride

Glycerol backbone

Fatty acids

Saturated and unsaturated fats

Fats can be solid or liquid at room temperature. Generally, fats that are liquid at room temperature are **unsaturated** and fats that are solid at room temperature are **saturated**. The reason for this is the chemical structure of the fatty acids. If there are only single bonds between the carbon atoms of the fatty acid chain (Figure 2.19), the structure of the fat molecules will be straight and can pack together easily; this makes the fatty acid a solid. If there are one or more double bonds between the carbons, the fat molecules will be bent and unable to pack closely together; this makes the fatty acid a liquid.

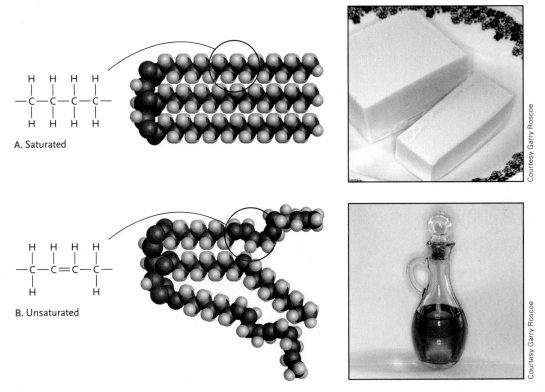

A. Saturated

B. Unsaturated

Figure 2.19 (A) Saturated Fat, (B) Unsaturated Fat
Comparison of (A) saturated and (B) unsaturated fats and their chemical structures.

Foods that contain healthy saturated fats are meats, fish, avocados, coconut oil, and dairy products. Foods that contain healthy unsaturated fats—those in the cis formation (Figure 2.20a)—include plant oils, such as olive oil and flaxseeds, nuts, seeds, and vegetables, and also fish oil. Unsaturated fats that have one double bonded carbon are called **monounsaturated**. Those that have two or more double bonds in the fatty acid chains are called **polyunsaturated**.

Margarine

Margarine is a vegetable fat that has been converted into a solid by a process called **hydrogenation**. During this process hydrogen is added to the fatty acids to essentially convert the double bonded carbon atoms into single bonded atoms. This changes the fat into a saturated fat that becomes solid at room temperature. The problem is that not every double bond is fully hydrogenated, so margarine is still partially unsaturated. Furthermore, the way the hydrogen atoms are added is not in a chemical structure that can be easily used by the body. These incompletely hydrogenated fats are called **trans fats** (Figure 2.20b); they are the very worst fats for human health. Some labels claim the margarine is non-hydrogenated, and so it is healthier. However, the margarine remains semi-solid at room temperature because it contains palm oils, which are saturated fats. This type of margarine is better than hydrogenated margarines, but not different from butter. Trans fats are found in fast foods and deep fried foods because the high cooking temperature converts the unsaturated vegetable oils into trans fats.

Trans fats

The enzyme in our digestive system that digests fats, lipase, functions specifically for unsaturated fats in the *cis* conformation, not those in the *trans* conformation. Therefore, trans fats cannot be broken down and they remain circulating in the blood stream much longer than any other fat. Trans fats can also bind to cell membrane receptors on endothelial cells that line the blood vessels; this results in these fats being taken into the lining of the blood vessels. Trans fats then stimulate the immune response, causing immune cells to also enter the blood vessel wall in an attempt to rid the body of these toxic

DID YOU KNOW?

Very healthy unsaturated fats such as olive oil can be transformed into trans fats if they are exposed to high temperatures during cooking. Saturated fats are best for cooking because they cannot be converted into trans fats: for example, coconut oil (which is saturated even though it is called an oil). Some oils, such as peanut oil, have a higher "smoking" point and can be cooked at higher temperatures. If the oil produces smoke when you cook, your food will contain trans fats.

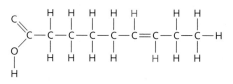

A. Cis fatty acid

B. Trans fatty acid

Figure 2.20 (A) Cis Fatty Acid, (B) Trans Fatty Acid

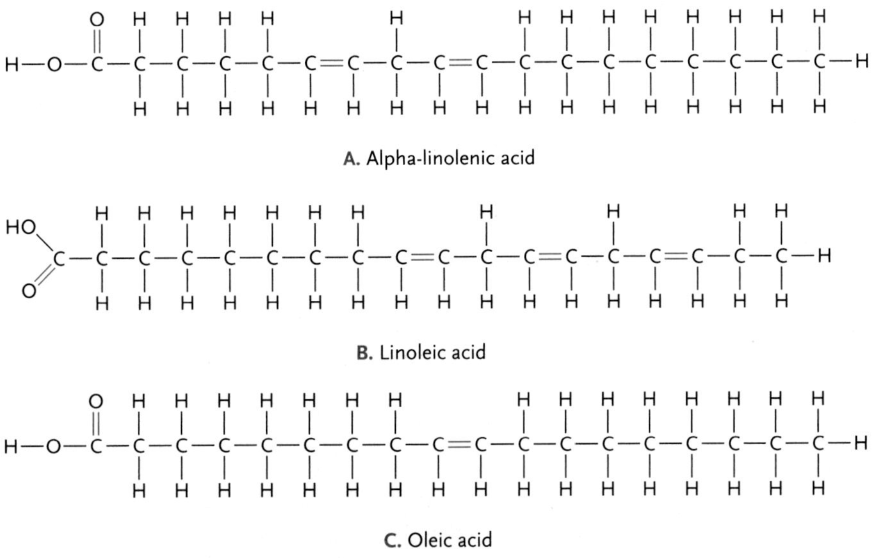

A. Alpha-linolenic acid

B. Linoleic acid

C. Oleic acid

Figure 2.21 **(A) Alpha-linolenic Acid, (B) Linoleic Acid, (C) Oleic Acid**
The three essential fatty acids are (A) alpha-linolenic acid (omega 3), (B) linoleic acid (omega 6), and (C) oleic acid (omega 9).

molecules. The combination of fat and immune cells in vessel walls is the plaque that causes heart disease (Chapter 17).

Essential fatty acids

We need **essential fatty acids** in our diet because the body's cells cannot produce them from other sources. There are three essential fatty acids: alpha-linolenic acid, which is **omega 3**; linoleic acid, which is **omega 6**; and oleic acid, which is **omega 9** (Figure 2.21). These fatty acids are essential for immune-cell regulation and for the production of mood hormones; deficiency in essential fatty acids can cause immune disorders and depression. Alpha-linolenic acid is an 18-carbon fatty acid chain with three double bonds. This makes it a polyunsaturated fatty acid: having its first double bond at the third carbon gives the term *omega 3*. Linoleic acid is also an 18-carbon fatty acid chain, but it has two double bonds, is polyunsaturated, and the first double bond is at the sixth carbon, making it omega 6. Dietary sources of essential fatty acids include fish, shellfish, olive oil, peanuts, flaxseed, leafy green vegetables, walnuts, and eggs.

Cholesterol

Most cholesterol in our body is produced by the liver (Figure 2.22). Cholesterol plays very important roles in human physiology: (1) it is a major constituent of cell membranes; (2) it is used by the liver to make bile, which helps in the digestion of fats; (3) it is used to produce low-density lipoproteins (LDLs) that transport fats through the bloodstream; and (4) it is used as the starting material for the production of steroid hormones, such as estrogen, progesterone, testosterone, and aldosterone (Chapter 16).

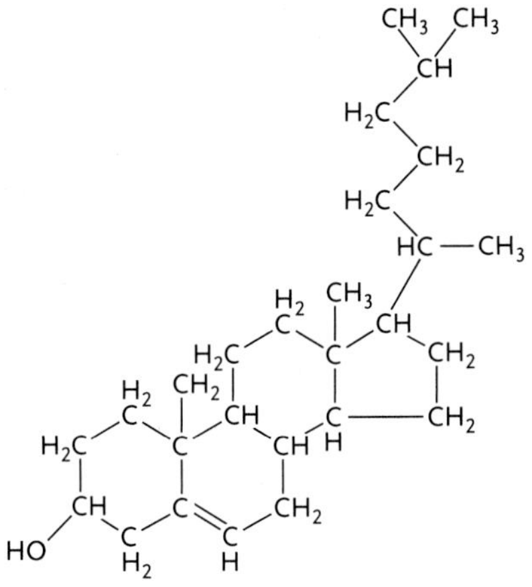

Figure 2.22 **Cholesterol**

Art source: From RUSSELL/HERTZ/STARR/FENTON. Biology, 2E. © 2013 Nelson Education Ltd. Reproduced by permission. www.cengage.com/permissions

DID YOU KNOW?

Blood cholesterol levels increase more by eating trans fats than by eating cholesterol. This occurs because the liver produces more cholesterol when it makes LDLs to transport the trans fats that are not as easily taken up by cells. Therefore, cholesterol stays in the circulatory system longer and plays a significant role in the development of cardiovascular disease (Chapter 17).

Phospholipids

The last category of fats is **phospholipids,** the primary fats found in **cell membranes.** Phospholipids are unique because they have two nonpolar fatty acid chains, called *tails,* and a polar *head* that contains a phosphate group. Because they contain both polar and nonpolar regions, phospholipids are amphipathic. This important feature of these fats explains why they can produce cell membranes. All cell membranes consist of a bilayer of phospholipids: the polar heads are directed toward the water-filled cell **cytoplasm** and the exterior of the cell, the **interstitial fluid**; and the tails are directed toward the interior of the membrane (Figures 2.23 and 2.24). Since phospholipids are mostly unsaturated, the cells are liquid. In addition to the phospholipids, most animal cell membranes contain cholesterol, which helps the membrane remain flexible. The cell membrane is discussed in more detail in Chapter 3.

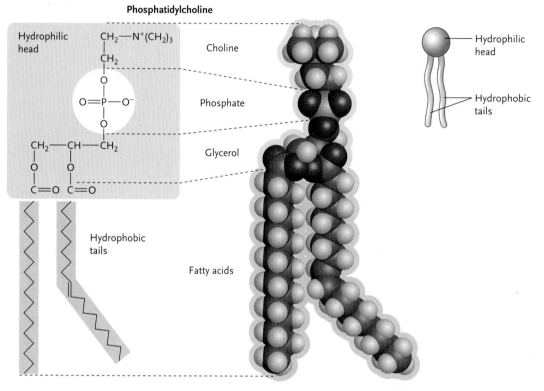

Figure 2.23   Structure of a Phospholipid

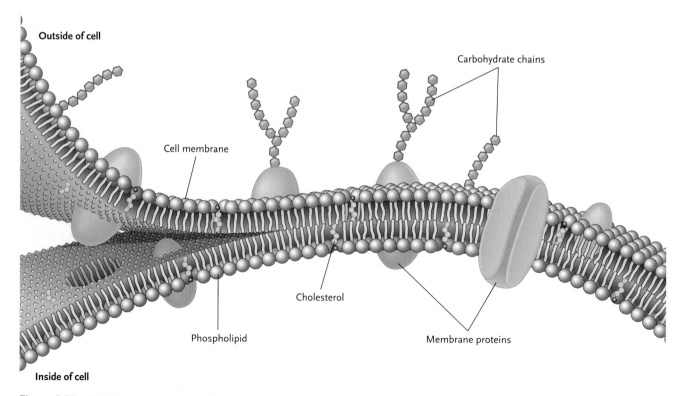

Figure 2.24   Cell Membrane

CONCEPT REVIEW 2.6

1. Explain how lipids are important for the functioning of three parts of the body (refer to Table 2.3).

2. What is adipose tissue?

3. What is the difference between saturated and unsaturated triglycerides?

4. List food sources of healthy fats, and name the fats you should minimize in your diet. Why?

5. What is hydrogenation?

6. Name four functions of cholesterol.

7. What is the primary type of fat that makes up cell membranes?

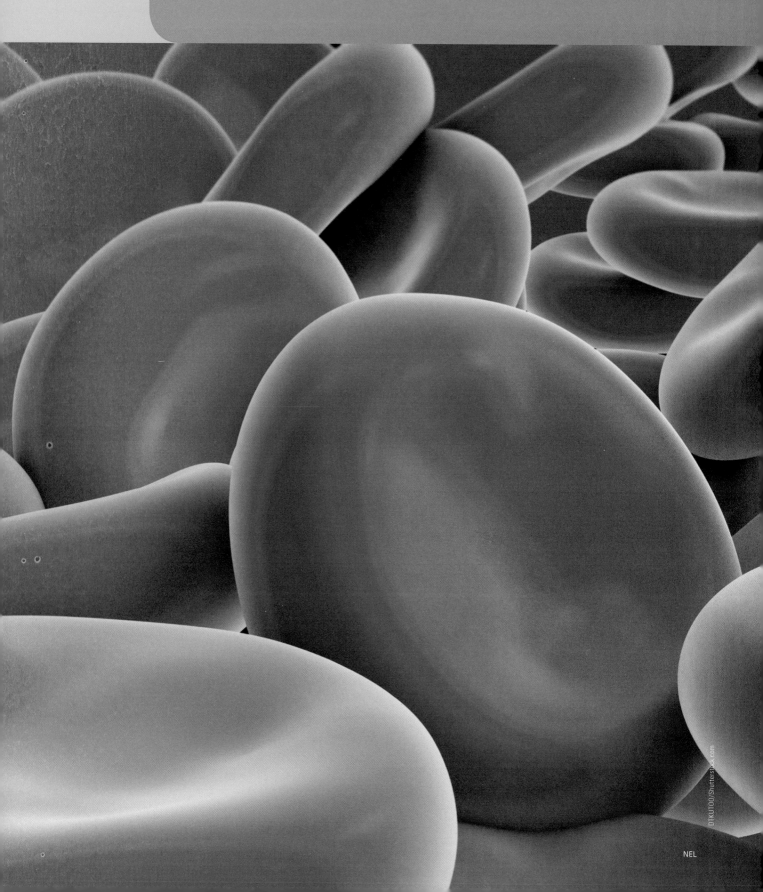

Education is not the filling of a pail, but the lighting of a fire.

William Butler Yeats

3.1 CELL THEORY

Many different shapes and sizes of cells exist in humans and every living organism. Most human cells are between 5 and 20 μm in diameter (a micrometre (μm) is 1/1000th of a millimetre) (Figure 3.1). The head of a pin is approximately 2 mm wide, which is 2000 μm.

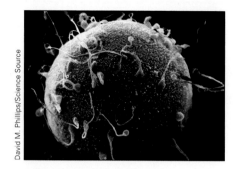

David M. Phillips/Science Source

Figure 3.1 Human Sperm and Egg Cell

In humans, the female egg cell is the largest cell type, and sperm cells are the smallest.

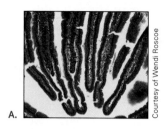

A.

Courtesy of Wendi Roscoe

B.

Figure 3.2 (A) Microvilli, (B) Neuron

Cells such as microvilli (A) and neurons (B) have specialized structures that give them a greatly increased surface area.

Art source (bottom): From RUSSELL/HERTZ/STARR/FENTON. Biology, 2E. © 2013 Nelson Education Ltd. Reproduced by permission. www.cengage.com/permissions

Cell size

Cells need oxygen and nutrients, and they need to get rid of waste. As a cell gets bigger its volume increases, and oxygen and CO_2 have to diffuse further to enter and leave the cell. Therefore, the *number* of cells in larger organisms increases, not the *size* of their cells. The largest cell in the human body is approximately 100 μm, which allows diffusion of molecules. However, some cells have special features that increase surface area; for example, extensions such as microvilli or dendrites increase the surface area of the cell. Cells called neurons transmit electrical signals within the nervous system and can be as long as one metre in length; however, because they are very narrow, the diffusion distance to the cell surface remains very small (Figure 3.2). Epithelial cells in the small intestine have microvilli, which dramatically increase the surface area of those cells, and so nutrients can be absorbed into the bloodstream efficiently.

Cell theory

Cells were seen for the first time under the microscope invented by Robert Hooke in 1665. Through years of research by many scientists, cell theory has come to mean several things:

1. All living organisms are composed of cells (including bacteria, protists, fungi, plants, and animals, but not including viruses, which are surrounded by protein capsules and are not considered "living").

2. The cell is the basic unit of all living organisms and is the smallest single unit that can survive independently.

3. All cells come from pre-existing cells.

4. All cells contain genetic material, DNA, which is passed on to new cells that arise from cell division and offspring.

5. All cells are either prokaryotic or eukaryotic and are composed of basically the same types of phospholipids, proteins, carbohydrates, and nucleic acid molecules.

CONCEPT REVIEW 3.1

1. Why do some cells have structures such as microvilli or dendrites?

2. List the five main components of the cell theory.

3.2 PLASMA MEMBRANE

All cell membranes, also called **plasma membranes**, are formed from amphipathic phospholipids arranged in a **bilayer**. The bilayer is formed from two rows of phospholipids arranged with the hydrophobic tails directed inward. The membranes that surround organelles in eukaryotic cells also consist of the same phospholipid bilayer. There may be some subtle differences between plant, fungus, and animal membranes, but they are all made of essentially the same types of molecules. Recall the structure of a phospholipid; it has a hydrophilic (polar) head, including phosphate, choline, and glycerol, and two hydrophobic (nonpolar), unsaturated fatty acid tails (Figure 3.3). Polar molecules are attracted to other polar molecules, so the polar end of the phospholipid is attracted to the water of the cytoplasm inside the cell and the water in the extracellular space—the interstitial fluid. The nonpolar fatty acid tails are attracted to other non-polar molecules, and so the phospholipid bilayer forms spontaneously mainly through these hydrophobic interactions where the tails make up the inner portion of the membrane (Figure 3.4).

Fluid mosaic model

Since cell membranes are fluid and contain many other molecules in addition to the phospholipids, we use the term **fluid mosaic model** to describe a typical

cell membrane. Membranes have cholesterol molecules embedded within the membrane between bent, unsaturated fatty acid tails. Cholesterol, which is mostly hydrophobic, increases the strength of the membrane and decreases its fluidity. Cell membranes also contain proteins, glycolipids (lipids with carbohydrate groups), and **glycoproteins** (proteins with carbohydrate groups). Different molecules are found within membranes of different cell types and serve many functions depending on the tissue type and regulatory factors going on in the body. The following describes some of the most common membrane molecules:

- **Receptors**—proteins that bind regulatory molecules such as neurotransmitters, hormones, cell-signalling molecules, or nutrients
- **Antigens**—usually glycoproteins that act as cell markers
- **Enzymes**—proteins that act as catalysts for assorted chemical reactions that may occur on cell membranes

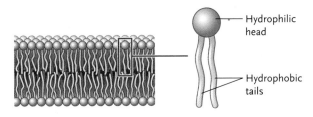

Figure 3.3   Structure of a Phospholipid

Phospholipids are made up of a five-carbon sugar, a phosphate group, and two fatty acid tails.

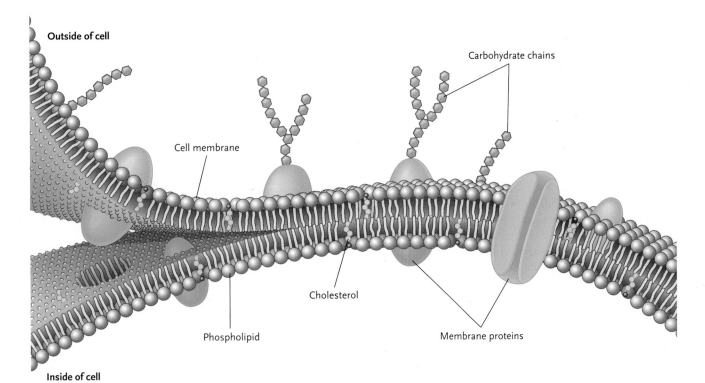

Figure 3.4   The Cell Membrane

The cell membrane is formed by a bilayer of phospholipids and contains cholesterol and many different proteins.

- **Ion channels**—proteins that span the membrane and allow the transport of specific ions into or out of the cell
- **Carrier proteins**—proteins that move substances like nutrients across cell membranes by changing their conformational structure
- **Adhesion molecules**—membrane proteins that play an important role in intercellular interactions
- **Gap junction channels**—protein membrane channels that play an important role in intercellular communication

CONCEPT REVIEW 3.2

1. Explain the difference between nonpolar and polar (hydrophobic vs. hydrophilic).
2. Draw a simple phospholipid, and label fatty acids, choline, glycerol, and phosphate.
3. Describe the general structure of a cell membrane.
4. List all of the components of a typical cell membrane and their main function.

3.3 PROKARYOTIC AND EUKARYOTIC CELLS

There are two major types of cells, prokaryotic and eukaryotic. The prokaryotic kingdoms include the archaebacteria and eubacteria, and the eukaryotic kingdoms include the protists, fungi, plants, and animals.

Prokaryotic cells

Prokaryotic cells—the simplest cell type—lack a nucleus and any other membrane bound organelles (Figure 3.5). A nucleoid region in the centre of the cell contains the genetic material, which is usually a single circular DNA molecule. Prokaryotic cell membranes consist of a phospholipid bilayer and many different membrane proteins, glycoproteins, and glycolipids as described in Section 3.2.

Most bacteria have extra protective layers. The cell wall of most eubacteria consists of sugars and amino acids, called **peptidoglycan**, which is important for maintaining cell strength and regulating osmotic pressure inside the cell. Eubacteria can be classified into two main groups based on the structure of the cell wall: (1) **gram positive** bacteria have many layers of peptidoglycan and stain purple when exposed to the stain crystal violet; (2) **gram negative** bacteria have a smaller layer of peptidoglycan in addition to an external **lipopolysaccharide** layer and stain pink (Chapter 23). It is important to note that bacterial cell walls are not the same as the cellulose cell wall in plants. Some bacteria may also have an external protein or carbohydrate **capsule** that further protects them from environmental changes such as ion concentrations, temperature, dehydration, and pH. This capsule makes their destruction by our immune cells more difficult and therefore increases the ability of certain bacteria to cause human diseases.

DID YOU KNOW?

Because some antibacterial drugs such as penicillin block the ability of bacteria to produce a cell wall, the bacteria cannot reproduce. Animal cells do not have a cell wall, so penicillin does not damage our cells.

Prokaryotic cells contain cytoplasm with dissolved ions and nutrients. Prokaryotes also have **ribosomes**, which are essential for protein synthesis. Ribosomes consist of RNA and protein and do not have a membrane. Prokaryotes may also have **pili** (singular is pilus) and/or **flagella** (singular is flagellum). A pilus is a hairlike

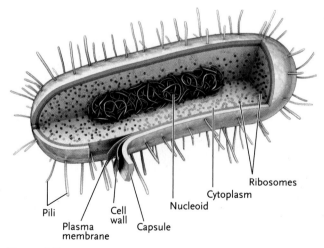

Figure 3.5   Prokaryotic Cell Structure

Prokaryotes have the simplest cell structure.

Art source: From RUSSELL/HERTZ/STARR/FENTON. Biology, 2E. © 2013 Nelson Education Ltd. Reproduced by permission. www.cengage.com/permissions

structure that helps bacteria adhere to structures. Also, through a process called **conjugation,** the pili function in the transfer of small pieces of DNA between bacteria. A flagellum is an external protein structure used for locomotion.

Eukaryotic cells

Eukaryotic cells are much larger, approximately 10 times larger, than most prokaryotic cells. They have a more complex cell structure, including a phospholipid bilayer membrane (Section 3.2), cytoplasm, ribosomes, and membrane-bound organelles (Section 3.4). Not all eukaryotic cells are exactly the same; there are some difference between protist, fungal, plant, and animal cells. Protists, fungi, and plants have a cellulose cell wall that provides extra protection against dehydration and other environmental factors. The cells of plants and some protists contain organelles called chloroplasts; these contain the pigment chlorophyll, which allows plant cells to undergo photosynthesis.

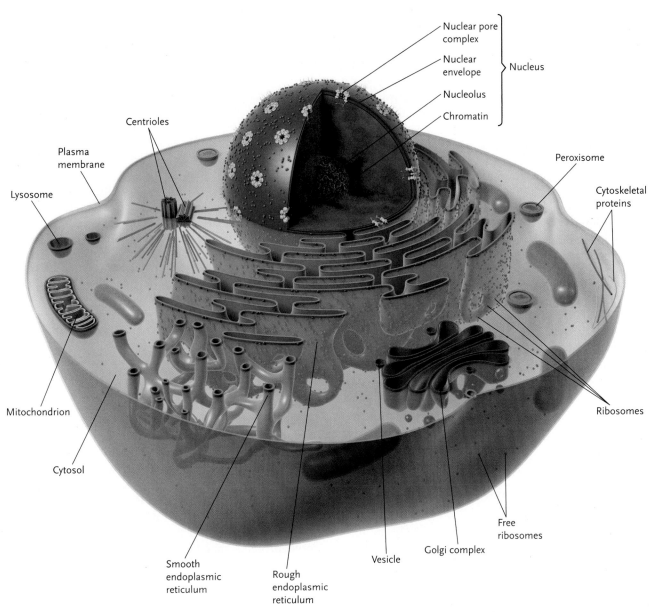

Figure 3.6 **Eukaryotic Cell**
Eukaryotic animal cells are complex and contain many membrane-bound organelles.

CONCEPT REVIEW 3.3

1. Name the structures in prokaryotic cells.

2. Describe the functions of each structure in prokaryotic cells.

3. Why is the cell wall and capsule important for prokaryotes?

4. What is the difference between gram positive and gram negative bacteria?

5. How do eukaryotes differ from prokaryotes?

3.4 ORGANELLES OF THE CELL

This section describes the typical structures and organelles in animal cells (see Figure 3.6).

Nucleus

The **nucleus** is a membrane-bound organelle that houses the genetic material. In humans, there are 46 chromosomes of DNA within every nucleus of every cell in the body: except for mature red blood cells, which have no nucleus. The nucleus is considered the control centre of the cell because it contains the DNA and is therefore the location of transcription of the genes. During cell division, the DNA becomes condensed into tightly packed chromosomes (Chapter 5). During the regular growth phase when transcription occurs, the DNA exists as loose strands called **chromatin**. The fluid within the nucleus—called **nucleoplasm**—is very similar to the composition of the cytoplasm (Figure 3.7).

The membrane surrounding the nucleus is similar to the plasma membrane: a phospholipid bilayer containing cholesterol and proteins. In addition, the **nuclear membrane**, also called the **nuclear envelope**, is double-layered and contains pores that allow large molecules such as RNA to move out of the nucleus. After transcription, messenger RNA molecules move into the cytoplasm to connect with ribosomes so that protein synthesis can occur.

The nucleus contains a small non-membrane-bound organelle called the **nucleolus**. The nucleolus contains DNA that consist of genes that code for ribosomal RNA transcripts and proteins that are important in the production of a new ribosome subunit.

Endoplasmic reticulum

The **endoplasmic reticulum** (ER) is a membrane network found within the cytoplasm of every cell type (Figure 3.8). There are two types of endoplasmic reticulum: smooth and rough. The function of both types varies depending on the cell type. **Smooth ER** has a smooth surface and does not have any ribosomes. The function of smooth ER is to synthesize lipids, such as

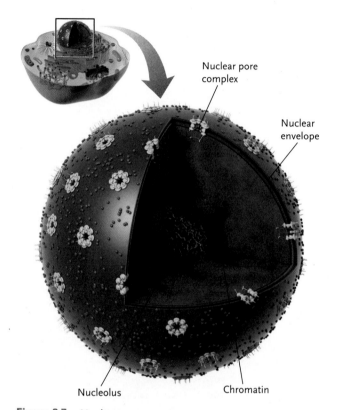

Figure 3.7 Nucleus

The nucleus consists of a double membrane that contains pores and houses the nucleolus and DNA.

phospholipids and cholesterol, for the maintenance of the cell membrane. The smooth ER also produces steroid molecules and metabolizes carbohydrates. Some drugs can be broken down in the smooth ER, mainly in liver cells. In muscle cells the sarcoplasmic reticulum plays a role in the regulation of calcium levels during muscle contraction. **Rough ER** is mainly involved in protein synthesis on the ribosomes; it is also involved in protein folding and the packaging of newly synthesized proteins that become transported to the Golgi bodies for further processing and modification. Messenger RNA (mRNA) molecules—transcribed in the nucleus—are transported through the nuclear pores to a ribosome that's either free in the cytoplasm or on rough ER; then protein synthesis (translation)

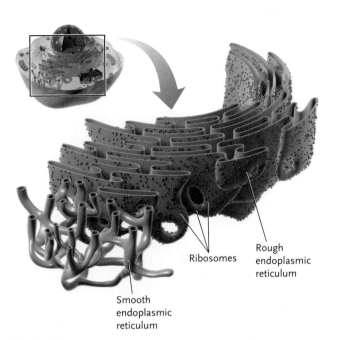

Figure 3.8 **Endoplasmic Reticulum**

The rough endoplasmic reticulum contains ribosomes and is the site of protein synthesis; the smooth endoplasmic reticulum is the site of carbohydrate and fat synthesis.

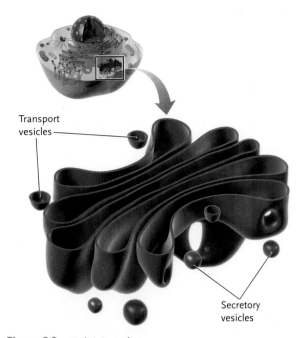

Figure 3.9 **Golgi Complex**

In the Golgi complex, proteins are modified and packaged into vesicles.

can occur. In this sense, the ER is continuous with the nuclear envelope.

Golgi body

The Golgi body, also called the Golgi apparatus or Golgi complex, is a separate membrane network found in the cell near the ER. It looks like a stack of semi-circular membranes (Figure 3.9). Proteins that have been newly synthesized on the rough ER are transported to the Golgi for further processing. The proteins are transported within **vesicles** that are membrane-bound sacs that can merge with the membranes of the Golgi. The side of the Golgi closest to the ER is called the **cis Golgi**, and the side farthest from the ER is the **trans Golgi**. When new proteins arrive within the Golgi several things happen, depending on the cell type. After translation occurs, most proteins become modified to include carbohydrate groups or phosphate molecules, so the finished functional proteins are often combinations of amino acids plus carbohydrate or lipid groups. **Glycosylation** enzymes in the Golgi cause the addition of **oligosaccharides** (small strands of monosaccharides), and **phosphorylation enzymes** add phosphate groups to some proteins. The Golgi body also functions in transporting lipids and creating **lysosomes**. It is very important in the production of **proteoglycan**, a glycosylated protein that constitutes a significant portion of the extracellular matrix of most tissues.

Mitochondria

Mitochondria are organelles that produce cell energy, **ATP**, in the presence of oxygen, which makes them the powerhouse of the cell. Mitochondria consist of two membrane layers; **cristae** is the name of the folded inner membrane (Figure 3.10). It is widely believed that mitochondria originated through a process of **endosymbiosis**—where primitive bacteria engulfed by another cell gave rise to eukaryotic organisms—approximately 1.5 billion years ago. The inner membrane probably derives from the prokaryotic cell membrane, and the outer mitochondrial membrane likely derived from the plasma membrane of the cell that engulfed the prokaryote. Mitochondria share many similarities with prokaryotic organisms:

1. They are approximately the same size, 1–5 μm.
2. They have a small circular DNA molecule that can self-replicate.
3. Mitochondria replicate by binary fission as do bacteria.
4. Mitochondria have ribosomes and are similar to bacterial ribosomes.

The mitochondrial DNA molecule has 37 genes, and 13 of these are responsible for producing molecules related to **cellular metabolism** and ATP production (Chapter 4). Processes that involve the 24 other mitochondrial genes include regulation of calcium, programmed cell death (apoptosis), cellular aging, the cell cycle, and cell growth.

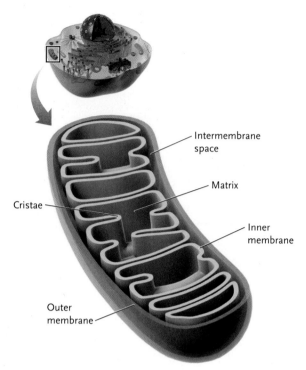

Figure 3.10   Mitochondrion

Mitochondria are organelles that produce ATP.

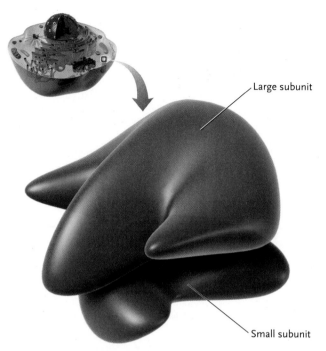

Figure 3.11   Ribosomes

Ribosomes are the site of protein synthesis.

DID YOU KNOW?

All mitochondria in your body, and therefore the mitochondrial DNA, came from your mother's egg cell during fertilization. The sperm only rarely contributes mitochondria. Mitochondrial DNA can be studied to determine relationships and migration of humans.

Ribosomes

Ribosomes contain subunits of protein and the ribosomal RNA (rRNA) molecules produced in the nucleolus region of the nucleus (Figure 3.11). The small subunit and the large subunit join together in the presence of a messenger RNA (mRNA) transcript. Ribosomes are the site of **translation**, which is protein synthesis (Chapter 9). Ribosomes are found in all prokaryotic and eukaryotic cells, although the structure in each cell type is slightly different. The ribosomes in mitochondria more closely resemble the ribosomes in prokaryotes. In eukaryotic cells the small and large subunits consist of different proportions of rRNA and proteins; these enable the mRNA transcript to fit between the subunits during translation, and they contain enzymatic sites that catalyze the formation of the peptide bonds between amino acids during translation.

Lysosomes

Lysosomes can be called the digestive system of the cell. Lysosomes are small membrane-bound organelles that contain digestive enzymes and play an important role in breaking down molecules, macromolecules, and old organelles inside the cell. Lysosomes are produced by the Golgi bodies. Membrane proteins that pump hydrogen ions into the lysosome make the pH inside the lysosome more acidic than the cytoplasm. Many molecules that come from this breaking down process are recycled and used by other parts of the cell. Any resulting waste products that are not reused are excreted from the cell either by diffusion or exocytosis; they then enter the bloodstream and are eliminated from the body through the kidneys. Also, in our white blood cells, a large number of lysosomes are involved in breaking down infectious organisms such as bacteria and viruses.

Peroxisomes

A major function of **peroxisomes** is to break down long chain fatty acids, which are then shuttled to the mitochondria for oxidation and the production of ATP. Peroxisomes in cells of the nervous system, specifically, oligodendrocytes and Schwann cells, break down fatty acid molecules to produce myelin—the protective covering on many neurons (Chapter 15). Peroxisomes also remove hydrogen ions from certain substrates and, in the presence of oxygen, produce **hydrogen peroxide** (H_2O_2), which then helps to break down other substances, such as alcohol and formaldehyde, and can be effective in destroying infectious organisms. We can buy hydrogen peroxide to use on cuts and scrapes to help decrease infection. In plants, peroxisomes play an important role in seed germination and photosynthesis.

Centrioles

Centrioles, found in most eukaryotic cells except for plants and fungi, are involved in the *organization* of spindle fibres (microtubules) that transport chromosomes during cell division (Chapter 5). It was previously believed that the centrioles produced spindle fibres; however, cells that have their centrioles removed can still produce spindle fibres. The position of the centrioles within a cell determines the position of the nucleus and other organelles in the cell (Figure 3.12).

Microtubule triplet

Figure 3.12 **Centrioles**
Centrioles are composed of microtubules and are involved in spindle formation during mitosis.

CONCEPT REVIEW 3.4

1. Make a chart and summarize the main function of each eukaryotic organelle.

2. What characteristics of mitochondria support the endosymbiosis theory?

3.5 INTRACELLULAR AND EXTRACELLULAR PROTEINS

All cells have an internal structure composed mainly of intracellular proteins and molecules that hold organelles in place, give the cell shape, and help transport other molecules within the cell; these proteins make up the **cytoskeleton.** The three main categories of cytoskeletal proteins are **microtubules, intermediate filaments,** and **microfilaments.**

Microtubules

The largest intracellular protein structures, microtubules, are large hollow tubes formed from individual proteins called **tubulin.** Approximately 25 nm thick, microtubules are attached at each end onto the cell membrane. Their role is to help maintain cell structure and to provide a means for transporting molecules throughout the cell. The spindle fibres that help transport chromosomes during cell division are composed of microtubules. Many different motor proteins can bind to microtubules, which help move molecules or

vesicles around the cell. Figure 3.13 shows the three main cytoskeletal proteins, and Figure 3.14 shows the motor protein, dynein, transporting a vesicle along a microtubule.

Intermediate filaments

Intermediate filaments, slightly smaller than microtubules and approximately 10 nm in diameter, are composed of intertwined "ropes" of protein. Different types of intermediate filaments consisting of many different proteins occur in different kinds of cells. The main function of each type is to support cell structure and help organelles remain in place. Some examples of intermediate filaments include **keratin,** found in hair and nails; **vimentin,** found in most cells; and **neurofilaments,** found in the axons of neurons.

Microfilaments

The smallest cytoskeletal proteins, the microfilaments, are approximately 7 nm in diameter and composed mostly of **actin** proteins. The primary role of microfilaments is cell movement, such as crawling. Our white

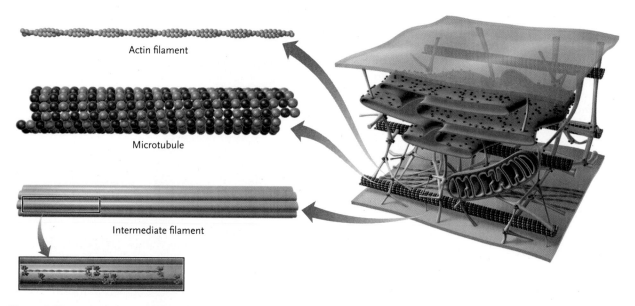

Figure 3.13 Cytoskeletal Proteins

Cytoskeletal proteins are important for giving cells their shape, supporting the organelles, and helping molecules move around within the cell.

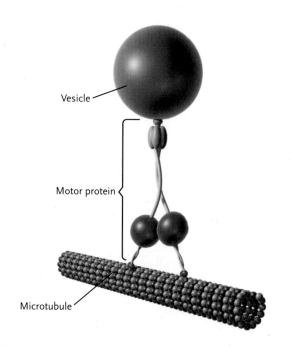

Figure 3.14 Microtubules and Motor Proteins

Microtubules and motor proteins are used for transporting vesicles within the cell.

blood cells have a large number of microfilaments; they continually move into and out of the bloodstream to locations of inflammation and cellular damage. Movement, also called migration, is a property of many different cell types. For example, astrocytes in the central nervous system move toward damaged areas, stem cells migrate to new locations in the body during growth and repair of tissues, and cancer cells can migrate to different locations in the body—a process called **metastasis**.

Extracellular proteins

Every cell produces **extracellular proteins** that are either secreted out into the extracellular matrix or integrated into the cell membrane; they are involved with intercellular interactions such as adhesion, signalling, or cellular communication (Figure 3.15). **Collagen** surrounds many cell types, particularly cells in connective tissues (Chapter 11), and mostly in cartilage, skin, tendons, and ligaments. Collagen provides strength to many tissues and resists stretching. **Elastin** is an extracellular matrix protein that allows tissues to be flexible and stretchy and also to reform to their original shape after being stretched. Tissues with high elastin content include the skin, bladder, lungs, and blood vessels. **Integrins** are membrane-spanning proteins that are involved in attaching the cell membrane to proteins in the extracellular matrix. They also function in transferring signals from outside to inside the cell, and vice versa, which allows for very rapid responses to environmental changes. When a blood clot forms, integrins on platelets bind to the extracellular fibrin protein during clot formation. **Fibronectin** is a protein that connects integrins to other extracellular matrix proteins. Fibronectin is very important during human growth and development and also in the healing of wounds; it aids cell adhesion, migration, and cell differentiation.

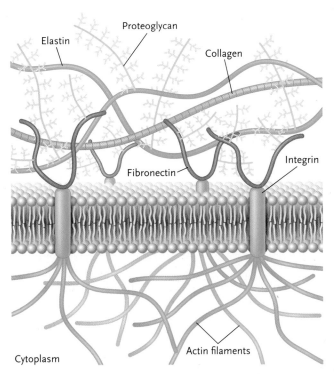

Figure 3.15 **Extracellular Matrix Proteins**

Extracellular matrix proteins are very important for connecting cells together, giving tissues strength and flexibility. Different tissues have different amounts and types of extracellular matrix proteins, and this image shows some of the most common ones.

Cellular junctions

Desmosomes consist of cadherin proteins that make cell membranes adhere together by connecting to the intermediate filaments of the cytoskeleton (Figure 3.16). Desmosomes, found in stratified epithelial tissue and in cardiac muscle, prevent shearing damage to tissues. **Tight junctions,** composed of proteins called **occludens,** also connect epithelial cells. They bring cells very close together so there is minimal extracellular fluid in that area. In this way they create a barrier that prevents molecules moving between cells—this means intracellular movement can occur only via membrane transport proteins. Tight junctions are important in the formation of the blood-brain barrier and in controlled absorption in the small intestine (Chapters 15 and 19). **Gap junctions** are channels composed of **connexin** proteins that allow the direct transfer of small molecules between cells. Gap junctions, important in cardiac muscle cells, allow the direct and rapid transfer of ions so that heart muscle cells can contract together (Chapter 17).

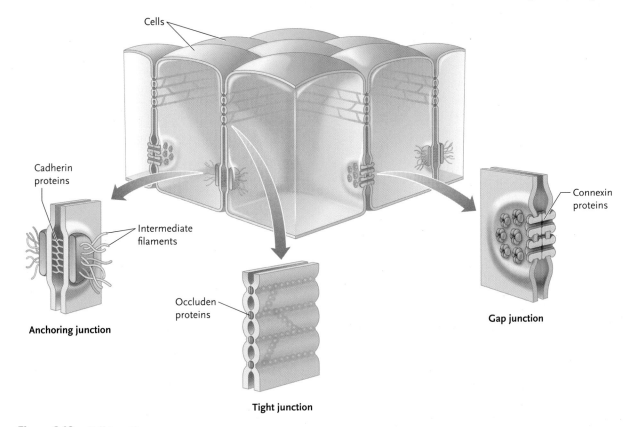

Figure 3.16 **Cell Junctions**

The main types of cell junctions are desmosomes (anchoring junctions), tight junctions, and gap junctions.

CONCEPT REVIEW 3.5

1. Describe the three main cytoskeletal proteins.
2. What are the main functions of integrins, fibronectin, collagen, and elastin?
3. What are the main functions of desmosomes, tight junctions, and gap junctions?
4. Give an example of a tissue that contains each type of cellular junction.

3.6 MOVEMENT ACROSS MEMBRANES

The function of cell membranes is to regulate what goes into and out of the cell. Recall that membranes are composed of phospholipids that have a polar head (facing toward the cytoplasm or extracellular space) and nonpolar tails that make up the inner portion of the membrane. Phospholipids are formed in the cytosol adjacent to the endoplasmic reticulum (ER). The ER adds choline and phosphate and then a vesicle buds off the ER and transports them to the cell membrane. Membranes are **semi-permeable** because only certain molecules can easily cross the hydrophobic tails of the inner membrane; everything else is regulated by proteins in the cell membrane. Molecules that can cross easily include small, nonpolar molecules such as O_2, CO_2, N_2, glycerol, and H_2O. Water can cross the membrane even though it is polar because the water molecule is very small, and osmotic pressure is a strong force that regulates the water balance inside cells. Water can also move across cell membranes through protein pores called aquaporins (Chapter 20).

Diffusion

Diffusion refers to the movement of any molecule from an area of high concentration to an area of low concentration until equilibrium is reached without any input of energy (Figure 3.17). Once equilibrium is reached, the molecules continue to move, but there is no further *net* movement in any direction (all molecules constantly move due to thermal energy). Diffusion in our body includes the movement of small, nonpolar molecules across a cell membrane and down a concentration gradient (Figure 3.18). For example, oxygen can easily move from a point of higher concentration in the blood stream toward that of lower concentration inside cells.

Factors that affect the rate of diffusion across a membrane

- Concentration gradient. A large difference in solute concentration on either side of the membrane increases the rate of diffusion
- Temperature. An increase in temperature increases the rate of diffusion.

- Surface area. An increased surface area on a cell membrane increases the rate of diffusion; for example, microvilli in the small intestine increase the surface area for nutrients to diffuse into the bloodstream.
- Diffusion distance. A shorter diffusion distance increases the rate of diffusion; tissue may have many or few capillaries for oxygen and nutrients to diffuse from.
- Size of molecule. Small molecules can diffuse faster than large molecules through the lipid bilayer.

Courtesy Garry Roscoe

Figure 3.17 Diffusion
Diffusion is the movement of molecules from an area of higher concentration to an area of lower concentration.

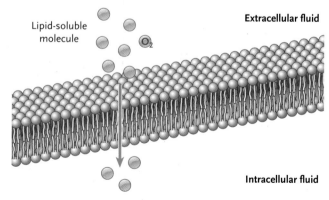

Figure 3.18 Diffusion across a Membrane
Diffusion across a membrane involves the movement of molecules from a region of higher concentration to that of lower concentration without requiring any membrane proteins.

Osmosis

Osmosis is the movement of water across a semi-permeable membrane to an area of low water concentration without the input of energy: the movement is toward the higher solute concentration. The movement of water into or out of a cell depends on the concentration of other molecules (solutes) inside and outside the cell. Charged and polar molecules attract water molecules. The higher the concentration of solutes on one side of a membrane, the higher the osmotic concentration, and water moves toward the solute. If we compare the solute concentration in the interstitial fluid surrounding a cell (extracellular space) to the solute concentration inside the cell, the **osmotic concentration** is either (1) the same in both, **isotonic;** (2) higher in the interstitial fluid, **hypertonic;** or (3) less in the interstitial fluid, **hypotonic** (see Figure 3.19). The movement of water into a cell due to the osmotic concentration is called the osmotic pressure. If the osmotic pressure is very great the cell may burst. Likewise, an increase in osmotic concentration outside of cells can cause cells to lose so much water that they shrivel up. Because the ion concentration surrounding single-celled organisms, such as protists or bacteria, can be extremely variable, those organisms have cell walls that help control the osmotic pressure inside the cell. In our body, many different organ systems play a role in homeostasis to ensure that osmotic pressure is highly regulated.

Facilitated diffusion

Facilitated diffusion is the movement of molecules across a cell membrane, down a concentration gradient, and without the input of energy, but with the help of a membrane protein (Figure 3.20). The membrane protein can be a channel or a carrier. Channels are specific protein openings in the membrane that allow very specific ions to move across. A sodium channel will not allow calcium to move across the membrane. Many ions are very important for certain cell functions, primarily, Na+, K+, Ca++, and Cl-. For various reasons, many cell functions require higher concentrations of certain ions either inside or outside of the cell, and certain mechanisms ensure

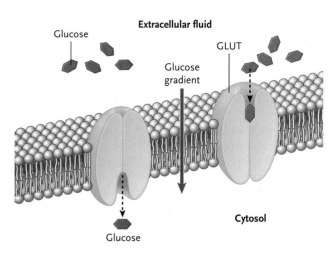

Figure 3.20 Facilitated Diffusion

Facilitated diffusion involves the movement of polar or charged molecules across a cell membrane by means of a membrane transporter protein that is either a channel or a carrier.

those gradients remain intact (Figure 3.21). Cell types that require an influx of sodium ions—such as neurons in an action potential—increase the expression of sodium-channel membrane proteins. Later chapters discuss this concept in more detail. The molecules and their concentrations inside and surrounding the cells in every tissue type are vital to the functioning of the whole body. For example, eating salty food (NaCl) affects the Na+ ion concentration gradient in our bloodstream and the interstitial fluid of all tissues. Some of that sodium enters the cells; some contributes to the solute concentration surrounding the cells (due to osmosis, water tends to move out of the cells); and some is excreted by the kidneys.

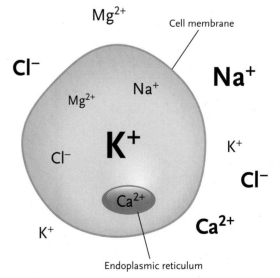

Figure 3.21 Ion Concentrations Within and Surrounding a Typical Cell

Every cell must have a specific balance of ions inside and outside the cell. The concentration of sodium and chloride ions is higher outside the cell, that of potassium is higher inside the cell, and that of calcium is higher inside the smooth ER within the cell.

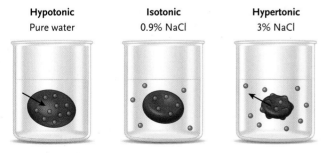

Figure 3.19 Hypertonic, Hypotonic, and Isotonic Solutions

In a hypertonic solution, water moves out of the cell. In a hypotonic solution, water moves into the cell. In an isotonic solution there is no net movement of water into or out of the cell.

Carrier proteins

Similar to ion channels, carrier proteins are membrane proteins that regulate the movement of certain molecules into or out of the cell, down a concentration gradient, without ATP. The difference with carrier proteins is that they are not channels that can allow a continuous stream of molecules to cross the membrane. Instead, they are proteins that bind a specific molecule and then change conformational shape; this causes the specific molecule to move across the membrane. Most membrane proteins that regulate the movement of substances such as nutrients, amino acids, glucose, and fatty acids are membrane carrier proteins. For example, after eating a meal, our pancreas secretes the hormone **insulin** (Chapters 16 and 19), which stimulates cells to increase the expression of a membrane carrier protein called GLUT (glucose transporter); there are several types of GLUT proteins. These proteins then bind to and carry glucose molecules into cells; glucose can then be used to produce ATP (Chapter 4). All the factors that affect the rate of diffusion are true for facilitated diffusion, except for carrier molecules, which can reach **saturation.** At that point, further increases in the concentration gradient do not further increase the rate of diffusion.

Active transport

Active transport refers to the movement of molecules across a membrane *against* the concentration gradient, by means of a membrane protein, and *with* the input of energy (ATP). For molecules to move against the concentration gradient, energy is required. Although a membrane protein similar to a carrier protein serves this function, in this case, ATP causes a conformational change in the protein and brings the molecule or ion across the membrane. The most common and important example is the **sodium-potassium pump.** The sodium-potassium pump ensures that a gradient is

maintained—*not* equilibrium—across the membrane. Na⁺ ions need to be in higher concentration outside every cell type for many reasons. For example, neurons and muscles cells use sodium to cause action potentials, and other cells use the sodium gradient to help bring nutrients into the cell: a process called **coupled transport.** All cells in the entire body have sodium-potassium pumps to keep sodium concentrations higher in the interstitial fluid than inside the cell.

Sodium-potassium pump

The sodium-potassium pump is a membrane protein that uses one molecule of ATP to move exactly three sodium ions out of the cell and two potassium ions into the cell (Figure 3.22). Sodium moves into the cell gradually through a number of mechanisms, and the Na-K pump ensures that the concentration gradients remain intact. Almost one-third of all the ATP used in our body fuels the Na-K pumps! That is equivalent to at least 500 calories per day; this varies based on body weight because more cells mean more Na-K pumps to fuel. Along with maintaining an ion concentration gradient across the cell, an increase in positively charged sodium ions outside the cell causes an **electrical gradient** to occur. Therefore, the inside of every cell is slightly more negative compared to the outside; this is called the **resting membrane potential** and is approximately –70 millivolts in most cell types. *All* cells have a resting membrane potential, even though only neurons and muscle cells have action potentials or graded potentials (Chapter 15).

Proton pump

Membrane proteins called **proton pumps** occur within the inner membrane of every mitochondrion (Figure 3.23). The proton pump is another example of an active transport membrane protein that moves protons (H⁺) *against* its concentration gradient. As protons

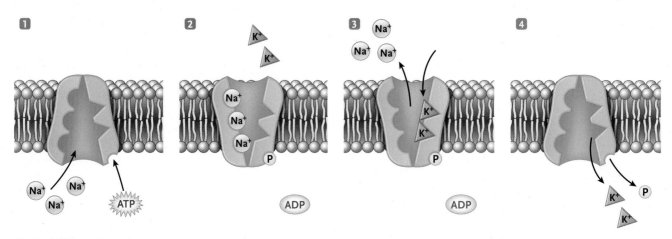

Figure 3.22 Sodium-Potassium Pump
The sodium-potassium pump is an important example of active transport. Every cell in the body must maintain a higher concentration of sodium outside the cell and potassium inside the cell.

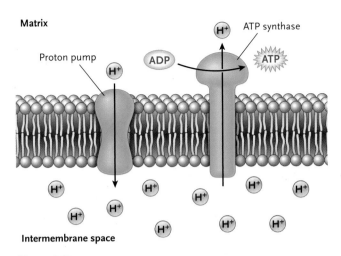

Figure 3.23   Proton Pump

The proton pump is an example of active transport. Proton pumps are membrane proteins located in the inner mitochondrial membrane and are important for the production of ATP.

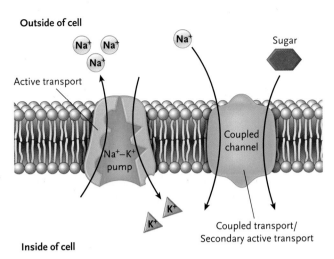

Figure 3.24   Coupled Transport

Coupled transport involves the simultaneous movement of two molecules through the same membrane protein. Two molecules moving in the same direction is called symport, and two molecules moving in opposite directions is called antiport.

move into the inter-membrane space, an electrochemical gradient forms due to the protons' positive charge. Whenever a charge or concentration difference occurs across a membrane, the molecules "want" to reach equilibrium, so the H⁺ ions tend to move back across the membrane into the matrix of the mitochondria. However, this can only occur by means of very specific membrane proteins called **ATP synthase**. When protons move down their concentration gradient and back into the matrix through ATP synthase enzymatic activity phosphate groups are added onto the ADP molecules, and new ATP molecules are formed; this process is called **chemiosmosis** (Chapter 4). Note that the purpose of the concentration gradient in this situation is to have the electrochemical gradient perform work, which in this example produces ATP molecules.

Coupled transport

Another example of an electrochemical gradient involves the movement of nutrients into the cells by means of the sodium ion gradient across the plasma membrane (Figure 3.24). Since sodium is always in higher concentration outside the cell than inside due to the Na-K pump, sodium naturally tends to move back into the cell. The cell uses this energy to carry nutrients such as amino acids or glucose into the cell from the blood stream, and this involves another membrane protein, so the process is called *co-transport* or coupled transport. This example again shows that cell membranes are semi-permeable and allow only certain substances to cross the membrane because they regulate which types and how many membrane proteins will be integrated. Cell membrane proteins are very dynamic and change rapidly depending on the environmental conditions surrounding the cells. If blood sugar levels

are high after a meal, more coupled-transport membrane proteins will be transcribed and translated and incorporated into the cell membrane during that time.

Transport of large molecules

Some molecules must enter or exit the cell, but they are too big to move through membrane proteins (Figure 3.25). For example, the insulin produced by our pancreatic cells and made up of 51 amino acids is transported from the ER to the Golgi and then from the Golgi to the plasma membrane in membrane-bound vesicles. These insulin-carrying vesicles merge with the plasma membrane because the membrane composition of the vesicle and the cell is the same, and the insulin proteins are then released from the cell and transported by the bloodstream to various tissues. The release of large molecules from vesicles across the cell membrane is called **exocytosis**. Movement of large molecules into the cell is called **endocytosis**, and this can involve **pinocytosis**—the movement of fluids—or **phagocytosis**—the movement of particles. When a large particle enters a cell, it is engulfed by a portion of the plasma membrane. This process forms a vesicle that usually merges with lysosomes, which break down the contents of the vesicle. For example, phagocytosis occurs when white blood cells such as **macrophages** engulf bacteria and viruses and any other dead cells in order to break down and destroy such invading microorganisms (Chapter 22). Pinocytosis involves the non-specific engulfing of large amounts of **extracellular fluid** brought into the cell by the vesicle. Most of the content of these vesicles is already broken down or consists of small nutrients that can be immediately used by the cell. Many cell types bring large amounts of nutrients into the cell in this manner.

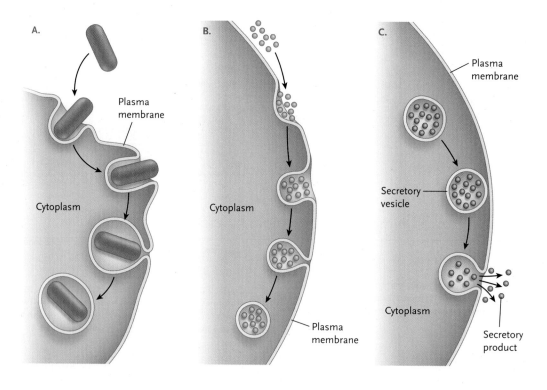

Figure 3.25 Movement of Large Molecules across a Cell Membrane

Phagocytosis (A), pinocytosis (B), and exocytosis (C) are processes that allow the movement of large molecules or large amounts of fluid to enter or leave the cell.

CONCEPT REVIEW 3.6

1. What does semi-permeable mean?
2. Explain how diffusion works. What kinds of molecules can cross the membrane easily?
3. What kinds of molecules require a membrane protein to cross into or out of a cell?
4. What is osmosis?
5. What is solute and solvent?
6. Explain why water would move into or out of a cell.
7. What are hypertonic, hypotonic, and isotonic solutions? What happens to cells if they are placed in each of those solutions?
8. What is facilitated diffusion? Explain the difference between channels and carrier proteins.
9. What does it mean when carrier proteins become saturated?

10. What is active transport? Explain how it differs from facilitated diffusion.
11. Give two examples of active transport in our cells.
12. Explain the steps involved in moving sodium out of cells and potassium into cells. Why is ATP required?
13. What are the two main reasons why our cells need Na⁺/K⁺ pumps? Which cells have Na⁺/K⁺ pumps?
14. Where are proton pumps located in cells? What is their purpose? What is chemiosmosis?
15. Explain how sugars and amino acids move into the cell using coupled transport.
16. Explain the difference between endocytosis (two types) and exocytosis.
17. What is receptor-mediated endocytosis?

Education is not the preparation for life; education is life itself.

John Dewey

4.1 OVERVIEW OF CELLULAR RESPIRATION

Cellular energy

Organisms from every kingdom use **adenosine triphosphate (ATP)** as the energy supply for thousands of chemical reactions that occur in every cell. ATP consists of an adenine (one of the four nitrogenous bases), a ribose sugar, and three phosphates (Figure 4.1). Energy is transferred from the ATP when the last phosphate bond is broken and then the phosphate (inorganic phosphate, Pi) binds to another molecule; this converts the ATP into ADP (adenosine diphosphate): $ATP \longrightarrow ADP + Pi$.

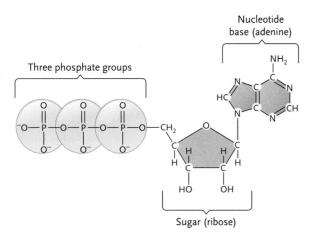

Figure 4.1 Adenosine Triphosphate (ATP)

Adenosine triphosphate (ATP) is the energy molecule used by cells.

Art soruce: From RUSSELL/HERTZ/STARR/FENTON. Biology, 2E. © 2013 Nelson Education Ltd. Reproduced by permission. www.cengage.com/permissions

When the phosphate from ATP binds to a protein, such as the sodium-potassium pump membrane protein (Chapter 3), the protein changes its conformational shape, which then aids a cellular process, such as moving sodium ions out of the cell.

Cellular production of ATP

Our cells make ATP from the foods we eat. Recall from Chapter 2 that we eat macromolecules: proteins, carbohydrates (polysaccharides), and fats (lipids). When we digest these foods they are broken down into their building blocks—amino acids, monosaccharides, and glycerol and fatty acids—and then taken up by our cells. **Food energy** (such as glucose) is converted into **chemical energy** (ATP) through the process of cellular respiration. Cells use ATP to perform thousands of functions every second. Excess food energy that is not used for energy production becomes stored as triglycerides (fat) until needed. Therefore, to maintain our body weight, we need to eat the same amount of food energy that our cells use in chemical energy.

DID YOU KNOW?

There are nine calories in one gram of fat, compared to only four calories in one gram of carbohydrate or protein. This is because fatty acids have long chains of carbon-hydrogen bonds, which produce many ATP molecules after being oxidized. By contrast, monosaccharides and amino acids have fewer carbon-hydrogen bonds to produce ATP molecules.

Food energy

We measure the amount of energy in the food we eat in terms of <u>C</u>alories (kilocalorie). A <u>c</u>alorie is the amount of energy required to raise the temperature of 1 ml of

water 1°C. This is a very small amount so we use kilocalorie (<u>C</u>alories) as the unit of measure. Depending on their age, adults require anywhere from 1500 to 3000 calories per day to maintain their body weight (fewer calories are required as we get older), state of health, genetic factors, and activity level.

DID YOU KNOW?

Eating 3500 calories *more* than required to maintain body weight leads to a weight gain of approximately one pound (454 g) (a block of butter is one pound). For example, if you drink one extra can of regular cola per day (130 calories per can), without changing your food intake or amount of exercise, you will gain one pound of body weight after approximately 23 days.

Conversion of food energy into ATP molecules

Eukaryotic cells primarily use the monosaccharide glucose to produce ATP; however, our cells can also use fatty acids, amino acids, and nucleotides (Section 4.5). The **carbon-hydrogen bonds** are important because through a series of **oxidation** and **reduction** reactions these bonds are broken, the hydrogen moves to another molecule, and the energy in these bonds leads to the production of ATP. The carbon-hydrogen bonds in methane are the source of energy in natural gas, just as the carbon-hydrogen bonds in glucose are used in our cells (Figure 4.2). In oxidation reactions, electrons become transferred from the nutrient being broken down, such as glucose. When glucose

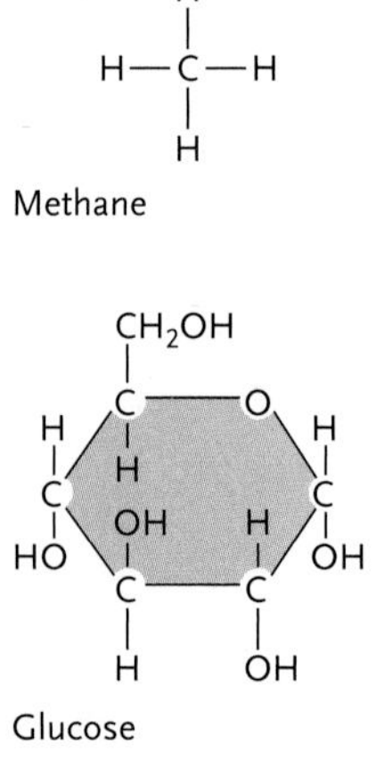

Methane

Glucose

Figure 4.2 Methane and Glucose

Both methane and glucose can be used to make energy because they both have carbon-hydrogen bonds.

Art soruce: From DIGIUSEPPE/SANADER. Chemistry 12U. © 2012 Nelson Education Ltd. Reproduced by permission. www.cengage.com/permissions

is **oxidized**, the electrons move to electron carriers (NAD$^+$ or FAD$^+$); these are then **reduced** to produce NADH or FADH$_2$. Oxidation and reduction reactions always go together, and are referred to as **redox reactions** (Figure 4.3).

The process of oxidizing food molecules (glucose, amino acids, fatty acids, glycerol, and nucleotides) and reducing NAD$^+$ ions leads to the production of ATP; this process is called **cellular respiration** and also **oxidative phosphorylation**. The steps in this process are discussed in this chapter. The oxidation of glucose in the presence of oxygen is called **aerobic respiration**. This explains why we need to breathe oxygen. When the level of oxygen in a cell is temporarily low, ATP is produced through **anaerobic respiration**: for example, when muscle cells have low oxygen levels during strenuous exercise (Section 4.6).

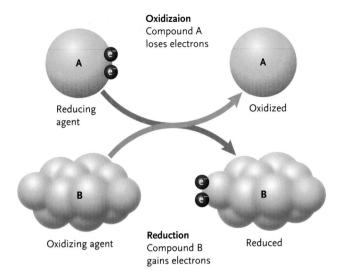

Oxidizaion
Compound A loses electrons

Reducing agent

Oxidized

Oxidizing agent

Reduction
Compound B gains electrons

Reduced

Figure 4.3 Oxidation-Reduction (Redox) Reaction

An oxidation-reduction (redox) reaction involves the transfer of electrons from one molecule to another. Losing electrons is oxidation, and gaining electrons is reduction.

CONCEPT REVIEW 4.1

1. What molecule is used as energy in our cells?
2. How does that molecule give us energy?
3. What nutrients are used in our cells to produce energy?
4. What does oxidation and reduction mean?
5. Why are redox reactions important for cellular respiration?

6. What is a Calorie?
7. Why are the carbon-hydrogen bonds in food molecules important?
8. What is the difference between aerobic and anaerobic respiration?

4.2 GLYCOLYSIS

Location of ATP production

Aerobic respiration occurs in four main stages, and each stage occurs in different areas of the cell (Figure 4.4). The first stage, glycolysis, occurs in the cytoplasm; the rest occur in specific regions of the mitochondria. During the four stages, one glucose molecule becomes entirely oxidized (broken down) and yields many ATP molecules. The reaction equation is as follows:

$$C_6H_{12}O_6 + 6O_2 \longrightarrow 6CO_2 + 6H_2O + \text{Energy (ATP or heat)}$$

Stages of aerobic cellular respiration

1. Glycolysis occurs in the cytoplasm.
2. Pyruvate oxidation occurs in the matrix.
3. Krebs cycle occurs in the matrix.
4. Electron transport proteins are found in the inner membrane.

Oxidization of glucose in the cytoplasm

When glucose is the nutrient that produces ATP in the cell, it undergoes a process called glycolysis (Figure 4.5). Glycolysis occurs in the cytoplasm of all cells. It is the oldest evolutionary process used by every organism to make ATP: including prokaryotic organisms (bacteria) that do not have mitochondria. Glycolysis—a 10-step process of chemical reactions— involves the oxidation of glucose (six carbon atoms) and the reduction of NAD$^+$, which results in the production of two pyruvate molecules (three carbon atoms each). Once glucose enters the cell (see coupled transport from Chapter 2), it is phosphorylated with two phosphates from two ATP molecules; therefore, two ATP molecules are required to start glycolysis.

Once glucose is phosphorylated, oxidation-reduction (redox) reactions result in the production of two three-carbon pyruvate molecules. During this process, two NAD$^+$ molecules are reduced to NADH, and four ATP molecules are formed. Since two ATP molecules are required to begin the process, a *net* amount of two ATP molecules are produced during glycolysis. The ATP molecules formed

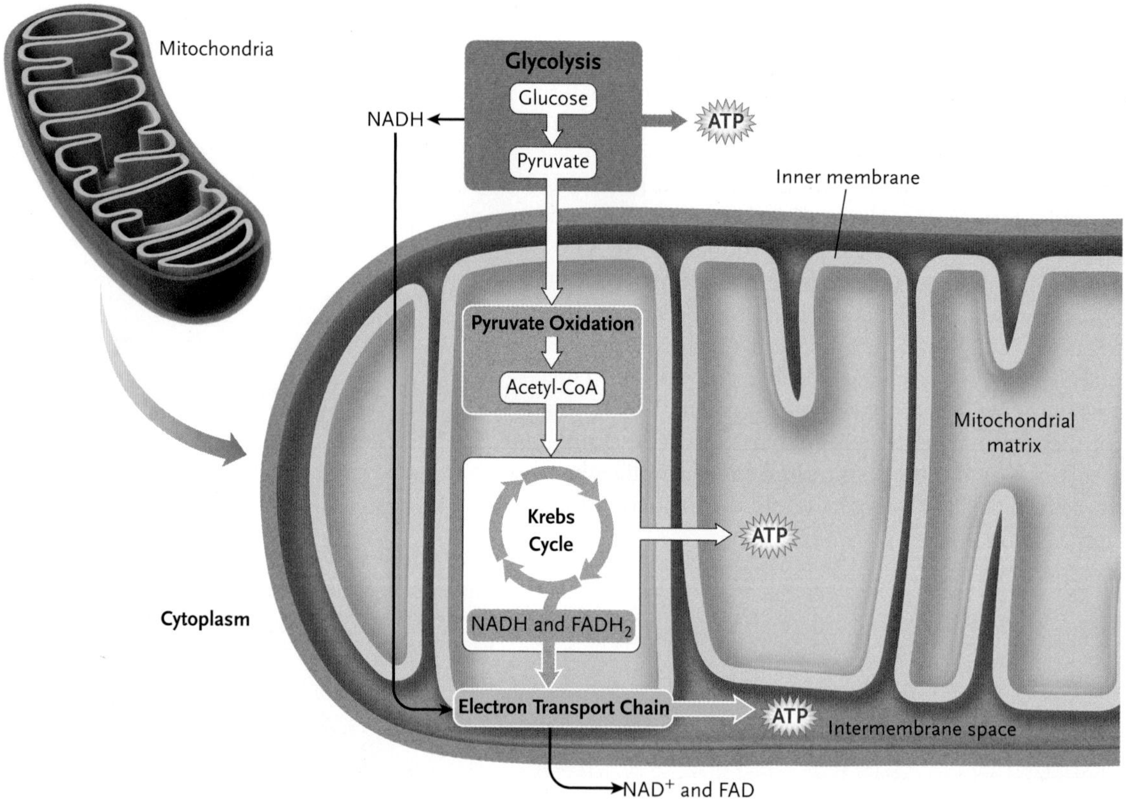

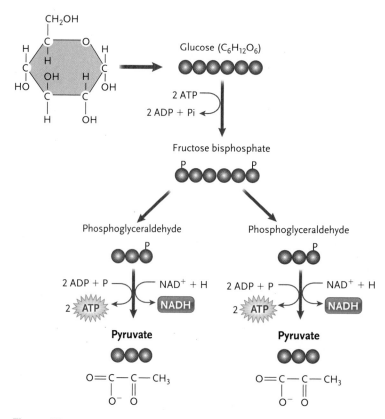

Figure 4.4 Overview of Aerobic Cellular Respiration

Cellular respiration involves the following stages: glycolysis, pyruvate oxidation, Krebs cycle, and the electron transport chain.

Glucose ($C_6H_{12}O_6$)

2 ATP
2 ADP + Pi

Fructose bisphosphate

Phosphoglyceraldehyde

Phosphoglyceraldehyde

2 ADP + P — NAD$^+$ + H
2 ATP — NADH

2 ADP + P — NAD$^+$ + H
2 ATP — NADH

Pyruvate

Pyruvate

Figure 4.5 Glycolysis

Glycolysis involves the oxidation of one molecule of glucose to form two molecules of pyruvate.

during glycolysis are formed with a specific enzyme found in the cytoplasm; this enzyme can add a phosphate to **adenosine diphosphate** (ADP) through a process called **substrate-level phosphorylation**. The NADH molecules are used in the final step, the electron transport chain, to produce many more ATP molecules.

CONCEPT REVIEW 4.2

1. What molecules are required to initialize glycolysis?
2. Where does glycolysis occur?
3. Why is glycolysis important?
4. What are the final products of glycolysis?
5. What is substrate-level phosphorylation?
6. What kinds of organisms undergo glycolysis?

4.3 PYRUVATE OXIDATION AND KREBS CYCLE

Pyruvate oxidation

The end result of glycolysis is two pyruvate molecules, each containing three carbons. These pyruvate molecules can diffuse through the mitochondrial membrane into the matrix, where further oxidation can occur. Pyruvate oxidation results in electrons and hydrogen ions moving from pyruvate to the electron acceptor NAD^+ to produce more NADH molecules (Figure 4.6). Also during this process, a carbon atom is removed from each pyruvate, which leaves a two-carbon molecule (an acetyl group) that combines with coenzyme A. The final product is a molecule called acetyl-Coenzyme A (acetyl-CoA). The enzyme that removes the carbon atom from pyruvate is called pyruvate dehydrogenase. The carbon atom is removed as a carbon dioxide (CO_2) molecule. The process of removing this carbon as CO_2 is called decarboxylation. This explains why we exhale carbon dioxide.

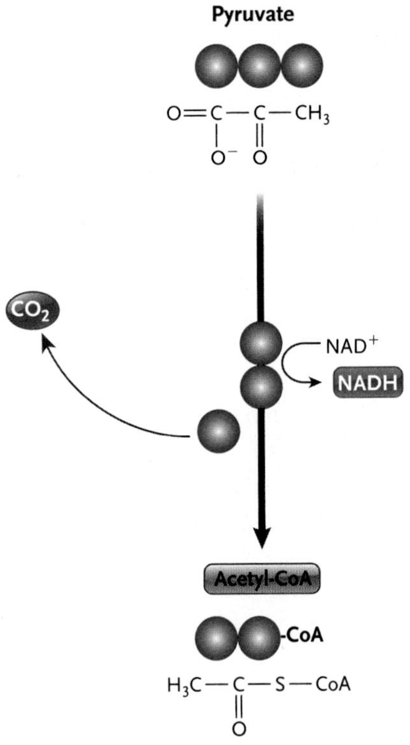

Cellular ATP capacity

Cells require more or less ATP at different times, depending on what cell functions are occurring. During exercise, for example, muscle cells require very high levels of ATP for muscle contraction; however, during periods of rest, those cells do not need to make as much ATP. In this situation, the cell stores the extra nutrients for use later when more ATP is needed. The body stores extra nutrients (glucose, amino acids, and fatty acids) as either glycogen or fat molecules (triglycerides) (Chapter 2). When fats are used as energy, they do not go through glycolysis. Because they are long carbon-hydrogen chains, fats can be converted into two-carbon acetyl groups. Acetyl-CoA, formed in combination with coenzyme A, enters the Krebs cycle (the next stage of the cellular respiration process).

Pyruvate

$$O=C-C-CH_3$$

Acetyl-CoA

$$H_3C-C-S-CoA$$

Figure 4.6 Pyruvate Oxidation
Pyruvate oxidation involves the oxidation of pyruvate to form acetyl-CoA.

DID YOU KNOW?

B vitamins are essential for our cells to make energy. NAD^+ stands for nicotinamide adenine dinucleotide, formed from the vitamin B3, **niacin**. FAD^+ is flavin adenine dinucleotide, formed from the vitamin B2, **riboflavin**. And coenzyme A is produced from vitamin B5, **pantothenic acid**. A deficiency in B vitamins leads to extreme fatigue because insufficient ATP will be produced.

Krebs cycle

The Krebs cycle involves a series of nine reactions that can be broken down into three steps. In the first step, acetyl-CoA enters the cycle and binds to a four-carbon molecule, **oxaloacetate**, forming a six-carbon molecule, **citrate**, also called **citric acid**. Because of the formation

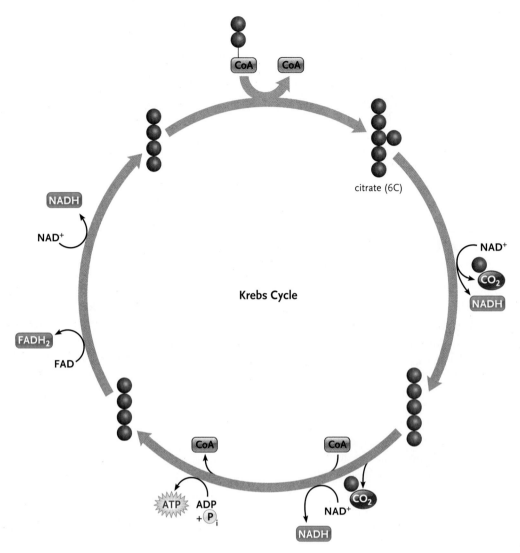

Figure 4.7 Krebs Cycle

The Krebs Cycle involves the production of one FADH$_2$ and three NADH molecules for each acetyl-CoA that enters the Krebs cycle. These molecules carry the hydrogens that enter the electron transport chain.

of this molecule, the Krebs cycle is also called the **citric acid cycle** (Figure 4.7).

In the second step, two carbon molecules are removed as carbon dioxide. Recall from the initial equation that six CO$_2$ molecules are formed for each glucose molecule that is oxidized: two are formed when pyruvate is oxidized, and the other four are formed at this point during the Krebs cycle. During the process of decarboxylation, electrons and hydrogen ions are transferred to NAD$^+$,

forming NADH and FADH$_2$. Since two carbons have been removed as CO$_2$, in the third step one four-carbon molecule remains: oxaloacetate. This is the initial starting material, which is regenerated and ready for the next cycle.

The most important result of the Krebs cycle is the production of the NADH and FADH$_2$ molecules. These reduced coenzymes are important because they carry the electrons and hydrogen ions that are required for the final step in ATP production, the electron transport chain.

CONCEPT REVIEW 4.3

1. Where do pyruvate oxidation and the Krebs cycle occur?

2. What is the end product of pyruvate oxidation?

3. What waste product is formed during pyruvate oxidation and the Krebs cycle?

4. What is the important result of the Krebs cycle?

5. How much ATP is formed during the Krebs cycle?

6. Why are NADH and FADH$_2$ molecules so important?

4.4 ELECTRON TRANSPORT CHAIN AND CHEMIOSMOSIS

Importance of NADH and FADH₂

The coenzymes NAD^+ and FAD^+ that were reduced to form NADH and $FADH_2$ during the previous processes provide the necessary electrons and hydrogen ions (protons) for the electron transport chain. Recall from Chapter 3 that proton pumps are membrane proteins that move hydrogen ions across a membrane. The proton pump membrane proteins are located in the cristae—the inner layer of the mitochondrial membrane. These hydrogen ions move across the inner mitochondrial membrane to form a H^+ gradient in the **intermembrane space** (Figure 4.8). It is important to have a high concentration of hydrogen ions in the intermembrane space because the concentration difference between outside and inside the inner mitochondrial membrane is used as energy for the production of ATP (secondary active transport). Because molecules always go toward equilibrium (Chapter 3), the high concentration of H^+ in the intermembrane space tends to move the protons down the concentration gradient and back into the matrix of the mitochondria. The only way these H^+ can move back into the matrix is through a specific membrane protein called ATP synthase. Chemiosmosis is the term used to describe the process of using a chemical (H^+) gradient for moving across the membrane, with the result of producing many ATP molecules. The enzyme activity of ATP synthase phosphorylates ADP to form ATP.

Electron movement through membrane proteins

When NADH and $FADH_2$ split so that the H^+ can be pumped into the intermembrane space, the electrons are transferred through the membrane proteins (hence the term, **electron transport chain**). Recall from Chapter 3 that active transport requires energy to move molecules toward an area of higher concentration. The movement of H^+ ions into the intermembrane space is **active transport**, and the energy to move the molecules comes from the transfer of electrons through the electron transport proteins (Figure 4.8), instead of through ATP as in the sodium-potassium pump.

Electrons at the end of the electron transport chain

After being transported through the membrane proteins, the electrons need to combine with a molecule; otherwise they would build up in the mitochondria and react with other molecules. The final electron acceptor in our cells is oxygen. This is why we need to breathe. The electrons, the protons (H^+) that have moved through the ATP synthase from the intermembrane space into the matrix through ATP synthase, and the oxygen molecules combine to form water molecules (see Figure 4.9).

DID YOU KNOW?

People die in fires primarily due to inhaling toxic chemicals produced from burning vinyl, fabrics, and plastics. One such chemical is cyanide, which is fatal within minutes if inhaled, because it inhibits the enzymes involved in the production of ATP.

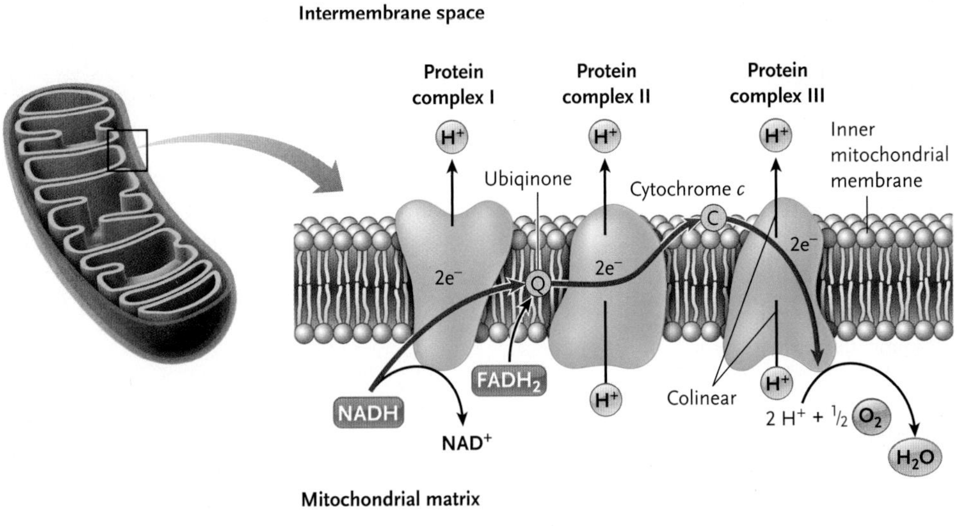

Figure 4.8 **The Electron Transport Chain**
The electron transport chain contains the proton pumps that create the proton gradient in the intermembrane space. This hydrogen ion gradient provides the energy for the production of ATP through ATP synthase.

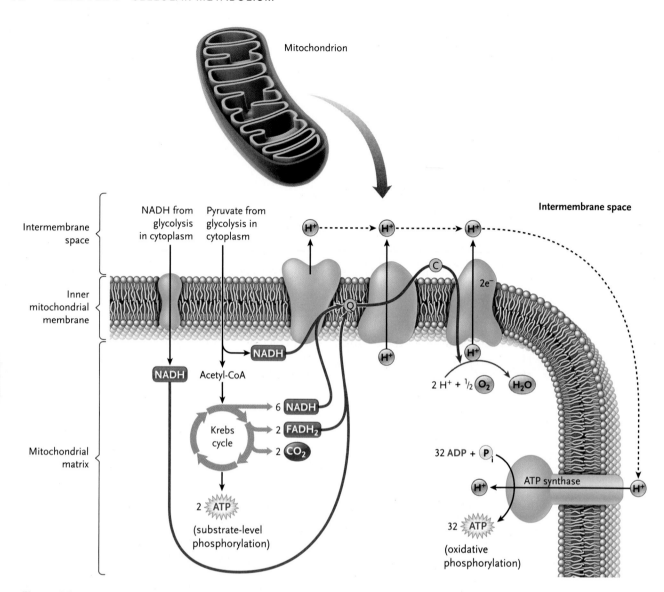

Figure 4.9 **Overview of Electron Transport Chain**

Art soruce: From DIGIUSEPPE ET AL. Nelson Biology 12, 1E. © 2003 Nelson Education Ltd. Reproduced by permission. www.cengage.com/permissions

CONCEPT REVIEW 4.4

1. What is the purpose of the NADH and $FADH_2$ molecules that are created during glycolysis, pyruvate oxidation, and the Krebs cycle?

2. Where are the H^+ ions transported to? Why?

3. What do the proton pumps use as energy for actively transporting the H^+ ions?

4. What happens to the electrons after they move through the electron transport chain?

5. What is the function of ATP synthase?

6. Define chemiosmosis.

4.5 OTHER NUTRIENTS THAT PRODUCE ATP

Our cells primarily use glucose as the starting material for ATP production, but most cell types can produce ATP from fatty acids, glycerol, amino acids, and nucleotides (Figure 4.10). Glucose is the only nutrient that undergoes glycolysis to produce pyruvate. Other nutrients enter the process at various stages.

When eaten, proteins are digested into their building blocks, amino acids. Amino acids in our cells function to produce proteins—on the rough endoplasmic reticulum (Chapter 3)—or to produce NADH and $FADH_2$ molecules

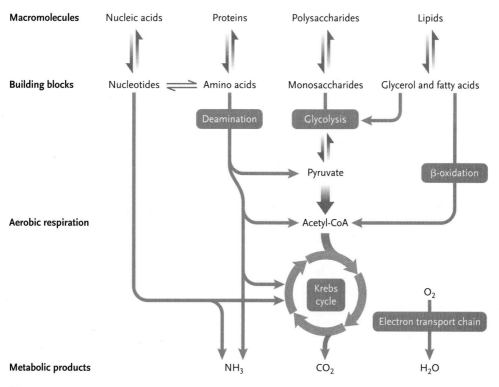

Figure 4.10 Overview of the Breakdown of Molecules to Produce ATP

through oxidation, which leads to ATP production. Amino acids contain **nitrogen groups** (amino groups, see Chapter 2), that have to be removed through a process called **deamination**. The nitrogen groups are removed as **ammonia (NH₃)**, which is then converted to **urea** and excreted by the kidneys.

The deaminated amino acids can then be converted into pyruvate or acetyl-CoA, which will be oxidized as previously described.

Nucleic acids, found in all foods made of cells—vegetables and meats—are digested into their building blocks, nucleotides. Nucleotides also contain nitrogen and must go through deamination before entering the Krebs cycle as acetyl-CoA.

Fats in the diet are broken down in the digestive system into fatty acids and glycerol. Since fatty acids are long chains of carbon-hydrogen bonds, they can be oxidized into two-carbon acetyl-CoA molecules and then enter the Krebs cycle. Recall that this is also the stage where fatty acid chains can be formed if ATP levels are sufficient in the cell and excess nutrients are present.

CONCEPT REVIEW 4.5

1. What molecules can be used by cells to produce ATP?

2. Explain how each macromolecule is broken down into building blocks and describe how they enter the process differently from glucose.

3. What is deamination? Which nutrients undergo this process?

4. What waste products are produced during cellular respiration?

4.6 ANAEROBIC RESPIRATION

ATP production without oxygen

In some situations, certain body cells may temporarily lack oxygen: for example, during sprinting when muscle contraction requires bursts of energy. The muscle cells quickly deplete oxygen during strenuous activity, and the cell relies only on glycolysis for ATP production. Recall that a small amount of ATP is produced through substrate-level phosphorylation in the cytoplasm during glycolysis. When oxygen is absent, ATP is produced through anaerobic respiration.

Since oxygen is required as the final electron acceptor in the electron transport chain, when oxygen is absent the electron transport process cannot occur. Therefore, the pyruvate produced during glycolysis is not be oxidized to produce acetyl-CoA; instead, the 3-carbon pyruvate is converted into a three-carbon **lactate**, also called **lactic acid**. The NADH molecules formed in this process must be converted back into NAD+ because the H+ is not used to build a concentration gradient. The process of moving the H+ from NADH to pyruvate to form lactate is called **fermentation**. Fermentation in humans can be summarized by the following equation:

$$Pyruvate + NADH \longrightarrow lactate + NAD^+$$

Once oxygen is again available, the lactate can be converted back to pyruvate, and aerobic respiration can resume.

Anaerobic respiration in yeast

In yeast, the three-carbon pyruvate goes through a decarboxylation step, where a carbon is removed as carbon dioxide, producing **acetaldehyde**. Then the H+ from NADH is transferred to acetaldehyde, forming a two-carbon **ethanol** molecule. Ethanol is the alcohol molecule found in wine and beer (Figure 4.11). The production of CO_2 by yeast makes it a useful organism for baking bread; the CO_2 makes the bread fluffy, and the ethanol evaporates during baking (Figure 4.12).

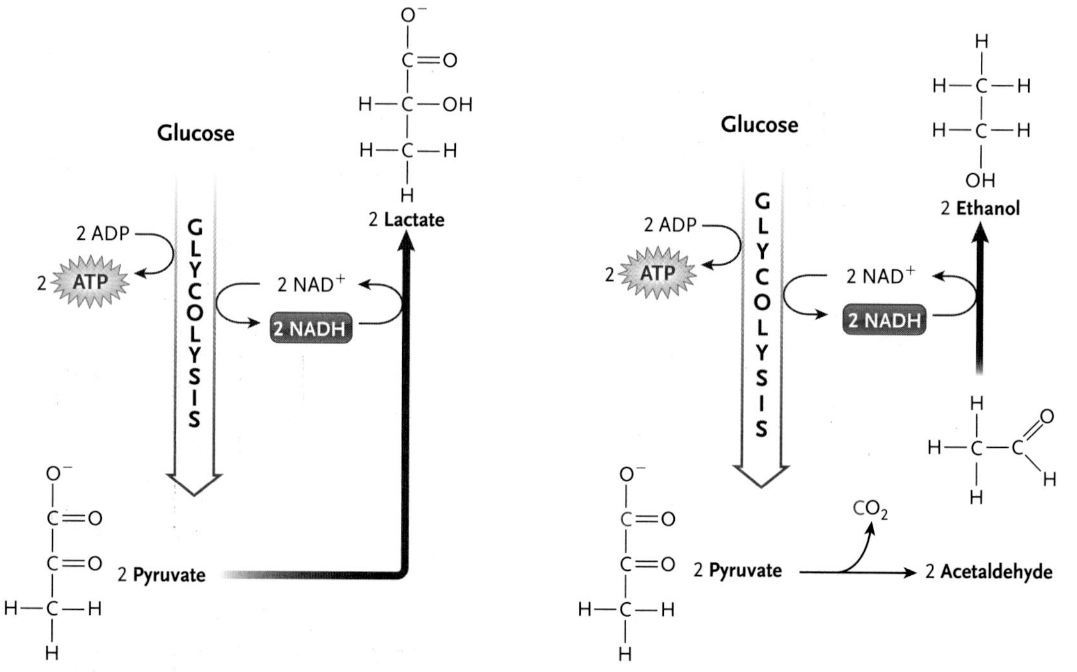

Lactic acid fermentation in muscle cells

Alcohol fermentation in yeast

Figure 4.11 Anaerobic Respiration

In humans, anaerobic respiration leads to the formation of lactic acid; in yeast, ethanol is produced.

Courtesy Garry Roscoe

Figure 4.12 **Examples of Fermentation**

CONCEPT REVIEW 4.6

1. Describe how anaerobic respiration is similar or different in humans and yeast.

2. When does anaerobic respiration occur in human cells?

3. What is fermentation?

4. What happens to lactate formed in muscle cells?

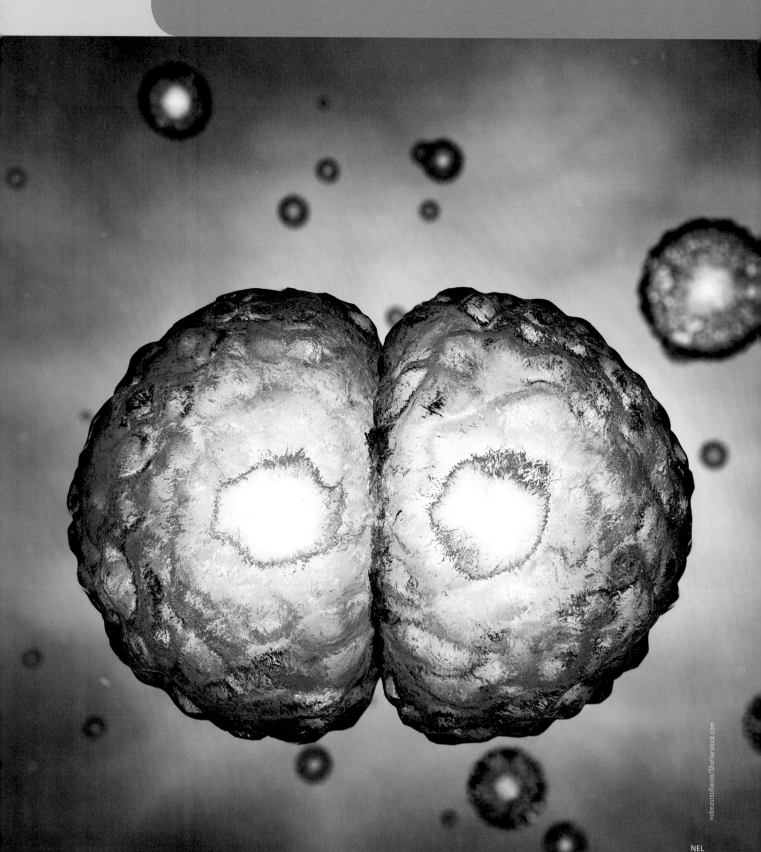

The only person who is educated is the one who has learned how to learn and change.

Carl Rogers

5.1 PROKARYOTIC REPRODUCTION

Bacteria (prokaryotes) replicate asexually, producing identical copies of themselves. Bacteria have one circular DNA molecule and sometimes a small extra piece of DNA called a plasmid (Chapters 10 and 23) (Figure 5.1). The bacterial chromosome may have millions of nucleotides, for example, *E. coli* bacteria have 4.6 million nucleotides, and a plasmid is much smaller and may contain approximately 2000 to 3000 nucleotides. Often the plasmid contains antibiotic resistance genes (Chapter 23).

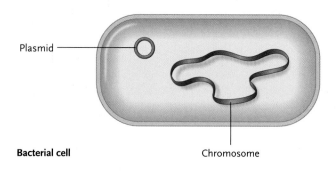

Figure 5.1 Structures in a Typical Bacterial Cell

Prokaryotic cell division involves two steps: **DNA replication** and **binary fission.** This is the simplest type of cell division because prokaryotes have only one main DNA molecule to replicate, and there are no membrane-bound organelles to copy and separate into two new cells. Eukaryotic cells have a much more complex cell division process due to the presence of multiple chromosomes and organelles.

Prokaryotic DNA replication begins at the point of origin where enzymes separate the strands of DNA—hydrogen bonds are broken between bases (Chapter 2). Then **DNA polymerase** proceeds in both directions in a process that adds new complementary nucleotides until two new strands of DNA are formed (see also Chapter 8). After the DNA is copied, the cytoplasm separates through binary fission, forming two new identical cells (Figure 5.2).

Genetic variation is extremely important and necessary for the survival of every population of organisms (Chapters 1 and 7), so every living thing must have a method for acquiring new DNA. Humans acquire half of their DNA from each parent during fertilization (meiosis, Section 5.5). Bacteria acquire genetic variation through the following mechanisms of **DNA transfer:**

- When bacteria directly exchange pieces of DNA with another bacterium—through a pilus that is usually a plasmid, but may also be a piece of their chromosomal DNA—the process is called conjugation.
- When bacteria acquire new DNA fragments from viruses, the process is called transduction.
- When bacteria acquire fragments of DNA or plasmids that happen to be near them, the process is called transformation.

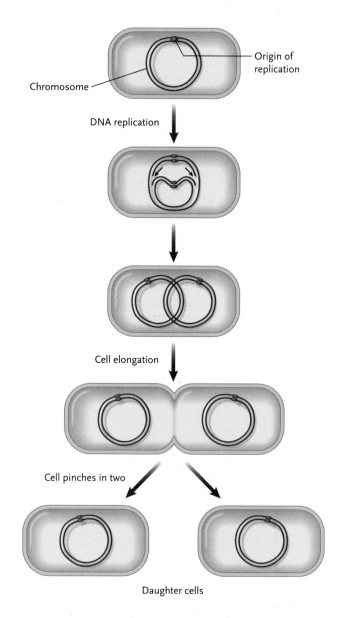

Figure 5.2 Binary Fission

Binary fission, the process of bacterial cell division, involves replication of the DNA and the splitting of the cytoplasm to form two new identical cells.

For example, many different species of bacteria live in our digestive tract, and they can take up DNA from each other: from DNA that might be in our food, from other bacteria that have died and left DNA behind, or from viruses that infect bacteria. So although bacterial reproduction (binary fission) is **asexual,** bacteria have extensive resources for constantly increasing their genetic variation (Figure 5.3). Bacterial genetic variation is discussed further in Chapters 7, 10, and 23.

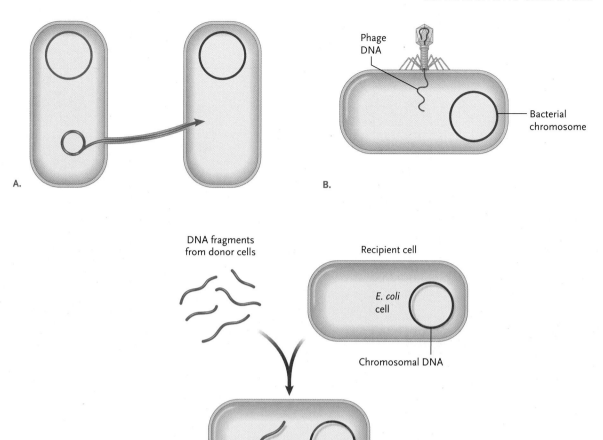

Figure 5.3 Conjugation, Transduction, and Transformation

Conjugation (A), transduction (B), and transformation (C) are the main mechanisms that allow bacteria to acquire new DNA.

Art source: From RUSSELL/HERTZ/STARR/FENTON. Biology, 2E. © 2013 Nelson Education Ltd. Reproduced by permission. www.cengage.com/permissions.

CONCEPT REVIEW 5.1

1. Describe the two main components of bacterial replication.

2. What is a plasmid?

3. Explain the steps in bacterial DNA replication.

4. How do bacteria acquire genetic variation if they replicate asexually?

5. Explain the difference between conjugation, transduction, and transformation.

5.2 EUKARYOTIC CELL CYCLE

All cells, whether prokaryotic or eukaryotic, have a cell cycle that involves a growth phase, followed by DNA replication, followed by cell division. Eukaryotic cells contain more DNA than prokaryotic cells, and their DNA must be condensed into transportable chromosomes for the more complex cell division process called **mitosis**. In eukaryotic cells, the cell cycle consists of **interphase** (G_1, S, G_2) and mitosis (M). Cells spend 98% of the time in interphase, either growing and performing normal cell functions or preparing for cell division; only 2% of the time is spent in cell division (Figure 5.4). The following outlines the steps of the cell cycle in interphase and mitosis:

- **G_1 phase.** This is the main **gap phase**, also called the **growth phase** of the cell. During G1 normal cell functions occur and also preparation for DNA replication. Most cells spend most of their lifespan in this phase of the cell cycle. During this phase, the cell replicates organelles such as mitochondria and centrioles; however, some organelles don't necessarily replicate before each cell division, instead they are simply redistributed into the two new cells.

- **S phase.** During this phase, DNA replication occurs, also called DNA synthesis (Chapter 8). In humans, all 46 chromosomes are duplicated before the cell divides, so each new cell has a copy of all 46 chromosomes.

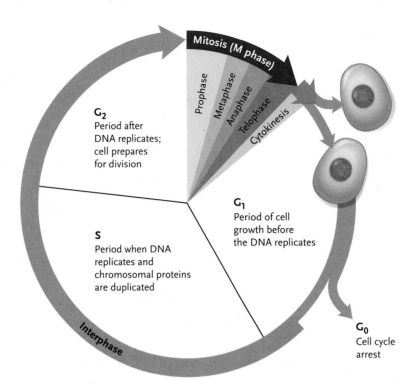

Figure 5.4   The Cell Cycle

The steps of the cell cycle include G₁, S, G₂, mitosis, and cytokinesis.

Art source: From RUSSELL/HERTZ/STARR/FENTON. Biology, 2E. © 2013 Nelson Education Ltd. Reproduced by permission. www.cengage.com/permissions.

- **G₂ phase.** During the second gap phase or growth phase, the cell continues to prepare for mitosis. In some cells this phase can be very short, and in some cases G₂ does not occur at all. The main event that takes place is the **G₂ checkpoint**, where DNA is checked for mutations after the replication process. Many different proteins help with this process, specifically the one called **p53** (Section 5.4). Once the DNA has been replicated during the S phase, it quickly starts to condense; this continues through G₂, and the DNA is fully condensed by the time mitosis begins.

- **M phase.** During mitosis protein synthesis stops, and no more cell growth occurs; all of the cell's energy is directed toward the separation of the chromosomes and organelles into two new daughter cells. At the end of mitosis, the process of splitting the cytoplasm occurs: **cytokinesis.** Mitosis results in two new daughter cells that are exactly identical to the original cell.

Chromosomes

To understand mitosis, it is important to take a closer look at the DNA. Humans have 46 chromosomes; we acquire 23 chromosomes from each parent. Those chromosomes are the largest macromolecules in the body, measuring up to two inches or 50 mm in length if they were to be removed from a cell and stretched to full length. After DNA replicates during S phase, each human cell nucleus (~10 μm) has a total of 92 chromosomes. For this to be possible, the DNA must be highly condensed (Chapter 2, Figure 2.14). All DNA must not only be condensed but also be organized so that each new daughter cell receives an exact duplicate of the original 46 chromosomes. Each set of 23 chromosomes are called **homologous chromosomes**, or **homologues.** After DNA replication, each chromosome has an identical copy, which is called a **sister chromatid** (Figure 5.5). It is very important to recognize the difference between homologous chromosomes and sister chromatids in order to understand how chromosomes are separated during mitosis and how this process differs from meiosis (Section 5.5).

When all the human chromosomes and their homologous pairs from one cell are organized according to size from chromosome 1 to 23, the result is a **karyotype** (Figure 5.6). The longest chromosome is designated number 1, and the rest are approximately ordered in decreasing size to chromosome 22; the Y chromosome is the smallest, and the X chromosome is almost as large as chromosome number 1. Every chromosome can be identified based on its size, location of the centromere, and banding pattern.

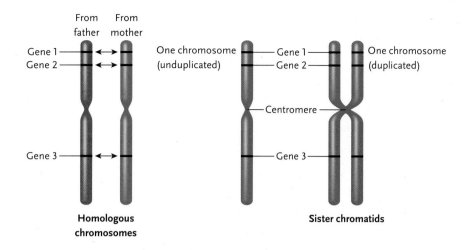

Figure 5.5 Homologous Chromosomes and Sister Chromatids

Homologous chromosomes contain the same genes, but they may have different variations of those genes, which are called alleles. One homologous chromosome is obtained from each parent. Sister chromatids are exactly the same and are a result of DNA replication.

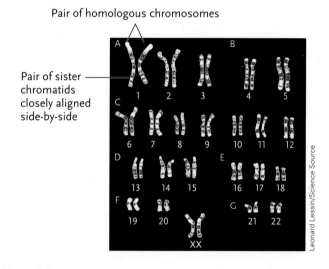

Figure 5.6 Karyotype

A karyotype is the organization of all of a cell's chromosomes.

CONCEPT REVIEW 5.2

1. What is the cell cycle?
2. In what phase do most cells spend the most time?
3. Explain generally what is occurring in G_1, S, G_2, and mitosis.
4. What is cytokinesis?
5. What is the main purpose of mitosis?

6. How many chromosomes do humans have in each cell nucleus, and how many do we receive from each parent?
7. What is the difference between homologous chromosomes and sister chromatids?
8. What is a karyotype?

5.3 MITOSIS

Somatic cells and germ cells

Mitosis occurs in eukaryotic **somatic cells**, which are all the body's cells that are not destined to become sperm or eggs: for example, liver cells, bone cells, and skin cells. Meiosis refers to the process of generating either sperm or eggs (also called **gametes**) from **germ cells** (Section 5.5). Note that somatic cells have all 46 chromosomes—both copies of the 23 homologous pairs—and therefore are **diploid (2n)**. Germ cells in the testes and ovaries are diploid; they have all 46 chromosomes, and during meiosis the diploid germ cells divide to produce **haploid (n)** gametes that have 23 chromosomes (Table 5.1).

Stages of Mitosis

During mitosis, five distinct phases can be recognized: prophase, metaphase, anaphase, telophase, and cytokinesis (Figure 5.7). The first four phases involve division of the contents of the nucleus, primarily the chromosomes. In the final stage, called cytokinesis, the cytoplasm divides and two new *identical* cells are produced; these are known as *daughter cells*.

During mitosis, the stages from prophase to telophase involve the separation of the chromosomes, and during the stage of cytokinesis the division of the cytoplasm occurs. The following sections describe the specific events that occur in each stage.

Prophase

The DNA replicated during S phase and the sister chromatids remain attached at the centromere, and there are 92 chromosomes in the cell. Condensation of the DNA began at the end of G_2, and by the end of prophase, DNA is visible under a microscope and individual compact chromosomes can be seen. The nuclear membrane and the nucleolus break down. The centrioles that replicated in G_1 begin to migrate to opposite poles of the cell, and **spindle fibres** (microtubules) form.

Metaphase

Spindle fibres connect to a protein complex called a **kinetochore**, which is located at the centromere region of each set of sister chromatids. The chromosomes line up in the centre of the cell: the **equatorial plane**. Spindle fibres consist of microtubules (Chapter 3), and the centrioles (also composed of microtubules) are involved in organizing the spindle fibres.

Anaphase

Spindle fibres begin to break down at the centriole end, which causes the shortening spindles to pull the chromosomes toward the poles of the cell. During this phase, an enzyme called *separase* is produced, which breaks down the *cohesin* proteins that hold the sister chromatids together, causing the centromeres to split, and then each sister chromatid can move to opposite ends of the cell.

Telophase

Sister chromatids are at each end of the cell, and the spindle fibres break down. A new nuclear membrane forms around each set of 46 chromosomes at each end of the cell, the chromosomes begin to uncondense, and the nucleolus reforms.

Cytokinesis

The cytoplasm divides equally so that each new cell receives a portion of the organelles and the fluid. In animal cells the membrane indents because of contracting **actin** microfilaments (Chapter 3) that form a **cleavage furrow** (Figure 5.8). ATP is required for the contraction of these proteins that constrict the cell membrane. Two new identical, somatic, diploid daughter cells are produced.

> ### DID YOU KNOW?
>
> Red blood cells live for approximately three to four months, and mature bone cells, osteocytes, can live for over 25 years. Cells that never go through mitosis include mature red blood cells, skeletal muscle cells, neurons, osteocytes, gametes, and the outermost layer of skin cells (stratum corneum); those cells are replaced by adult stem cells.

TABLE 5.1

Cell Division				
Cell Type	**Cell Division**	**Purpose**	**Diploid or Haploid**	**Number of Chromosomes**
somatic cells	mitosis	producing two new identical cells	diploid (2n)	46
germ cells	meiosis	producing gametes	diploid (2n)	46
gametes (sperm or egg)	none	fertilization	haploid (n)	23

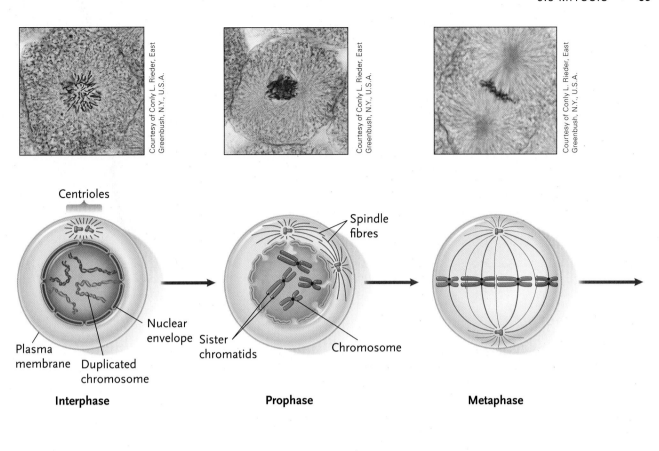

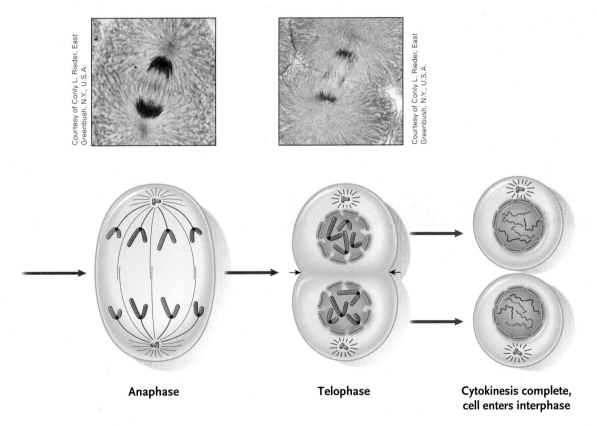

Figure 5.7 Stages of Mitosis

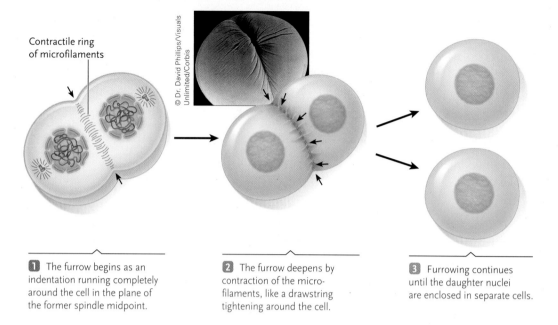

Contractile ring of microfilaments

© Dr. David Phillips/Visuals Unlimited/Corbis

1 The furrow begins as an indentation running completely around the cell in the plane of the former spindle midpoint.

2 The furrow deepens by contraction of the micro-filaments, like a drawstring tightening around the cell.

3 Furrowing continues until the daughter nuclei are enclosed in separate cells.

Figure 5.8   Cleavage Furrow

A cleavage furrow is formed when actin microfilaments in the cell contract to form two new cells.

Art source: From RUSSELL/HERTZ/STARR/FENTON. Biology, 2E. © 2013 Nelson Education Ltd. Reproduced by permission. www.cengage.com/permissions

CONCEPT REVIEW 5.3

1. Describe the main events that occur in each of the five stages of mitosis.

2. What is the role of the spindle fibres?

3. How do chromosomes become equally distributed?

4. Does it matter which sister chromatid moves into which new cell?

5. How many chromosomes are in each new daughter cell?

6. Give examples of some cell types that undergo mitosis.

5.4 CELL-CYCLE REGULATION AND CANCER

The cell cycle is controlled by **checkpoints** that ensure the full completion of one phase before advancing to the next phase. During the checkpoint stage of cell division, specific proteins become involved in searching the DNA for mutations. Some mutations may be obvious, for example, if two purines have combined certain proteins recognize a bulge in the DNA helix. If these proteins do not recognize any DNA mutations the cell cycle proceeds to the next stage. If they do find a mutation, the cell cycle becomes arrested at that stage. Then the mutation is either fixed, or the cell dies in a programmed process called **apoptosis**. Apoptosis is important for preventing damaged cells from dividing and perpetuating more mutations. Three principal checkpoints control the cycle in eukaryotes during G_1, G_2, and M (Figure 5.9). During the G_1 checkpoint the cell determines whether DNA rep-

lication can occur; if there are no DNA mutations the cell progresses to S phase. During the G_2 checkpoint, the cell determines whether any DNA mutations occurred during DNA replication. Replication mistakes can be quite high, and the body has many DNA repair enzymes that help fix mistakes (Chapter 8). The final checkpoint occurs during metaphase of mitosis to ensure the cell cycle should be completed. (Figure 5.9).

DNA mutations that go unrecognized

Cancer occurs when mutated cells continue to divide and are not regulated by the cellular checkpoints (Figure 5.10). This process results in a growth of cells called a **tumour**. Sometimes an overgrowth of cells is not cancerous and is called a **benign tumour.** Since these cells are usually surrounded by a healthy layer of cells, they are known as **encapsulated,** and they do not spread to other areas of the body. Cancer cells generally have two or more of the following characteristics:

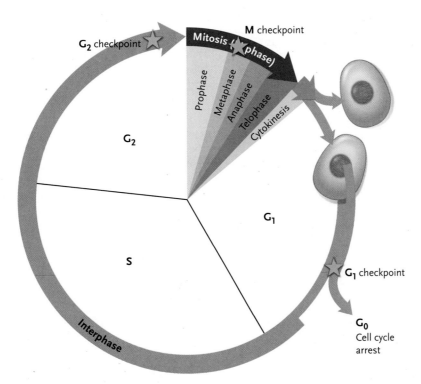

Figure 5.9 **Cell-Cycle Checkpoints**
The cell-cycle checkpoints are important for stopping the cell cycle when DNA mutations occur.
Art source: From RUSSELL/HERTZ/STARR/FENTON. Biology, 2E. © 2013 Nelson Education Ltd. Reproduced by permission.
www.cengage.com/permissions

- They form a malignant tumour that invades nearby tissue.
- Up-regulated genes, oncogenes, increase the rate of the cell cycle.
- They do not undergo cell death but become immortalized.
- They spread to other regions of the body—metastasis.
- They cause the growth of new blood vessels so they can access more nutrients—angiogenesis.

The genes that normally regulate the rate of the cell cycle are called **proto-oncogenes** and tumour-suppressor genes. Proto-oncogenes code for proteins that increase the rate of the cell cycle during periods of growth or wound healing or pregnancy. Tumour-suppressor genes code for proteins that act at the three checkpoints and stop the cell cycle if DNA damage is detected. If proto-oncogenes or tumour-suppressor genes are themselves mutated, the rate of cell division cannot be regulated and cancer can occur. A mutated proto-oncogene that up-regulates cell division more than normal is an oncogene. The study of cancer is called **oncology**.

Causes of DNA mutations

DNA mutations can be inherited or caused by environmental factors, such as radiation, viruses, chemicals. Errors during DNA replication occur quite frequently;

however, most are fixed either by DNA polymerase or DNA repair enzymes (Chapter 8). Of all cancers, 5% are inherited. Anything that causes cancer is called a **carcinogen**. Note that the following factors may be carcinogenic in different doses in different people; the overall health of an individual affects the capacity of cells to remove toxins and repair DNA.

Most common carcinogens
- radiation—x-rays, nuclear power plant radiation
- asbestos—not so common now that it is no longer used as insulation
- dioxins—herbicides and pesticides, and many cleaning products
- tobacco smoke
- benzene—in gasoline, paint remover, and rubber, and also released by volcanoes
- aflatoxin—fungus that grows on raw grain and nuts
- formaldehyde—plastics, and also in aspartame
- industrial smoke and pollution
- viruses—three most common: HIV, HPV, and hepatitis
- charred meat—contains benzopyrene
- acrylamide—can be found in starchy foods cooked at high temperatures, such as French fries or potato chips

- pressure-treated wood
- excessive UV light (but some sunlight is important for good health)
- chronic inflammation—causes overgrowth of cells and angiogenesis (growth of blood vessels)

Cancer cells may develop if any of the following categories of genes are mutated: (1) genes that regulate the cell cycle, such as proto-oncogenes and tumour-suppressor genes; (2) DNA repair enzymes; (3) genes that increase angiogenesis; (4) genes that prevent apoptosis; and (5) genes that regulate the immune response.

One very important tumour-suppressor gene is p53. It prevents cell division at the G_1 checkpoint if it detects any abnormal DNA before the DNA replication phase. Mutations of the p53 gene have been shown to play a significant role in the development of many cancers. The important functions of p53 include preventing the cell from moving into S phase, recruiting DNA repair enzymes to fix DNA mutations, and initiating apoptosis if the mutation can't be fixed (Figure 5.11).

What happens to cancer cells that get passed the checkpoints? Our cells have multiple mechanisms for managing DNA mutations. We have many genes that

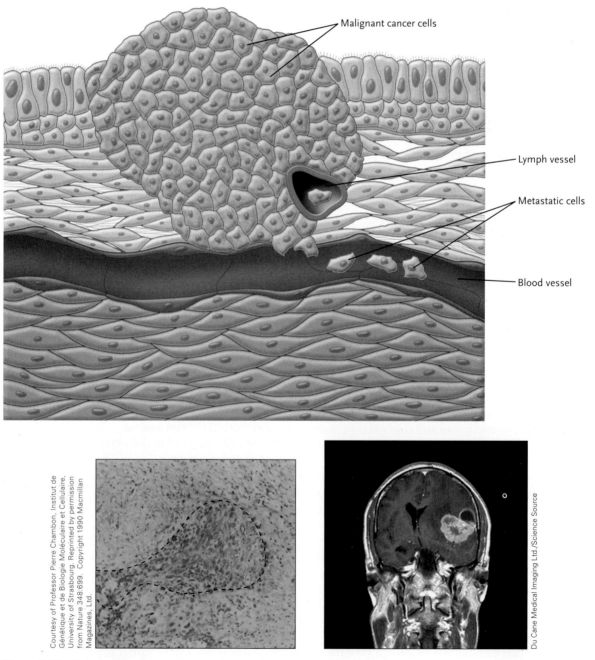

Malignant cancer cells

Lymph vessel

Metastatic cells

Blood vessel

Courtesy of Professor Pierre Chambon, Institut de Génétique et de Biologie Moléculaire et Cellulaire, University of Strasbourg. Reprinted by permission from Nature 348:699. Copyright 1990 Macmillan Magazines, Ltd.

Du Cane Medical Imaging Ltd./Science Source

Figure 5.10   Cancer Cells
Cancer cells are groups of cells that grow more rapidly than normal and form a tumour.

code for proteins that either fix mutations—such as DNA repair enzymes—or initiate apoptosis if mutations cannot be fixed. But if cells with mutations continue to divide unregulated our immune system can kill them. We have **natural killer cells**: white blood cells that specifically recognize cancer cells and can destroy them. Everyone develops some cancer cells all the time. However, if they develop faster than they can be destroyed then malignant tumours can form and spread throughout the body.

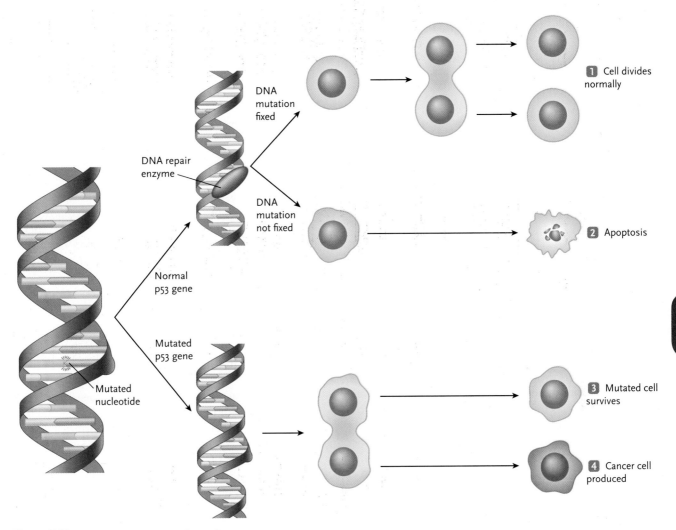

Figure 5.11   **Tumour-Suppressor Gene: P53**

P53 is a tumour-suppressor gene that stops the cell cycle at the G_1 checkpoint when DNA mutations are detected.

CONCEPT REVIEW 5.4

1. What is apoptosis?

2. What is angiogenesis?

3. Describe the three checkpoints and how they are involved in cell-cycle regulation.

4. What is a carcinogen?

5. What types of factors can cause cancer? Give a few examples.

6. Explain the difference between a proto-oncogene, an oncogene, and a tumour-suppressor gene. Give one example of a tumour-suppressor gene.

7. What kinds of DNA mutations may be involved in cancer?

8. Describe how p53 prevents cancer cell development.

9. How does our immune system play a role in preventing cancer?

DID YOU KNOW?

Anti-cancer foods, teas, and spices contain important chemicals that regulate our immune response, decreases angiogenesis, or are antioxidants that prevent free radical molecules from damaging our DNA. Some great ideas to add to your diet include vitamin E foods (fish and nuts), green tea, lavender tea, blueberries, red grapes (also red wine), kale, turmeric, garlic, onion, pumpkin and pumpkin seeds, tomato, pineapple, coconut oil, olive oil, and dark chocolate.

5.5 MEIOSIS

During mitosis, somatic cells divide in order to replace the body's old and damaged cells. During meiosis, haploid sperm or eggs (also called gametes) are generated from diploid germ cells so the organism can reproduce. The stages of meiosis are similar to mitosis because the DNA replicates and then the chromosomes become distributed into the new cells. The main difference between meiosis and mitosis is that meiosis involves two cell divisions and mitosis only one. During meiosis haploid cells are produced with 23, instead of 46 chromosomes. During replication, a sperm and egg that each has 23 chromosomes form an organism with the normal diploid number of 46 (Figure 5.12).

A very important feature of meiosis is that the separation of the homologous chromosomes into the gametes is random, and therefore different combinations of homologues can be found in every gamete, which means that any combination of gametes that leads to fertilization will lead to a genetically unique individual. The first cell is called a **zygote** and it undergoes mitosis many times to eventually produce an embryo. Genetic variation is an important characteristic of every population, and this is discussed further in Chapters 6 and 7.

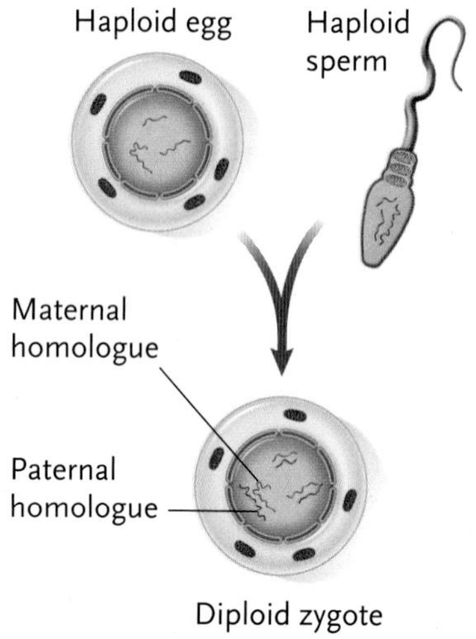

Figure 5.12 Haploid Gametes

Haploid gametes have 23 chromosomes, one of each of the homologues.

Art source: From DIGIUSEPPE/FRASER. Biology 11U. © 2011 Nelson Education Ltd. Reproduced by permission. www.cengage.com/permissions

Unlike mitosis, meiosis involves two cell divisions, meiosis I and meiosis II. The difference between these divisions relates to which chromosomes separate into the new cells. During the meiosis I, the homologous pairs separate and during meiosis II, the sister chromatids separate. DNA replication occurs only once: during S phase before the first meiotic division.

Stages of meiosis

Prophase I

The first prophase involves all the same events that occur in prophase of mitosis. DNA replicated during S phase and sister chromatids remain attached at the centromere, and the cell has 92 chromosomes. Condensation of DNA occurs, the nuclear membrane and the nucleolus break down, the centrioles migrate to opposite poles of the cell, and spindle fibres (microtubules) form.

Metaphase I

Spindle fibres connect to the kinetochore at the centromere region of each set of sister chromatids. The chromosomes line up at the equatorial plane. However, instead of all 46 chromosomes lining up with their attached sister chromatid, in meiosis I the homologous pairs with their attached sister chromatids line up (Figure 5.13). This difference is important because during this first cell division the daughter cells are haploid even though they have 46 chromosomes because the sister chromatids are still connected, they only contain *one* of each homologous pair.

Anaphase I

Spindle fibres begin to break down at the centriole end, which causes the shortening spindles to pull the homologous chromosomes toward the poles of the cell, with sister chromatids still connected together.

Telophase I

Homologous chromosomes are situated at each end of the cell and the spindle fibres break down. A new nuclear membrane forms around each set of chromosomes at each end of the cell, the chromosomes begin to uncondense, and the nucleolus reforms.

Cytokinesis

The cytoplasm becomes divided equally so that each new cell receives a portion of the organelles and fluid. Two haploid cells are produced, but because each

MEIOSIS I

MEIOSIS II

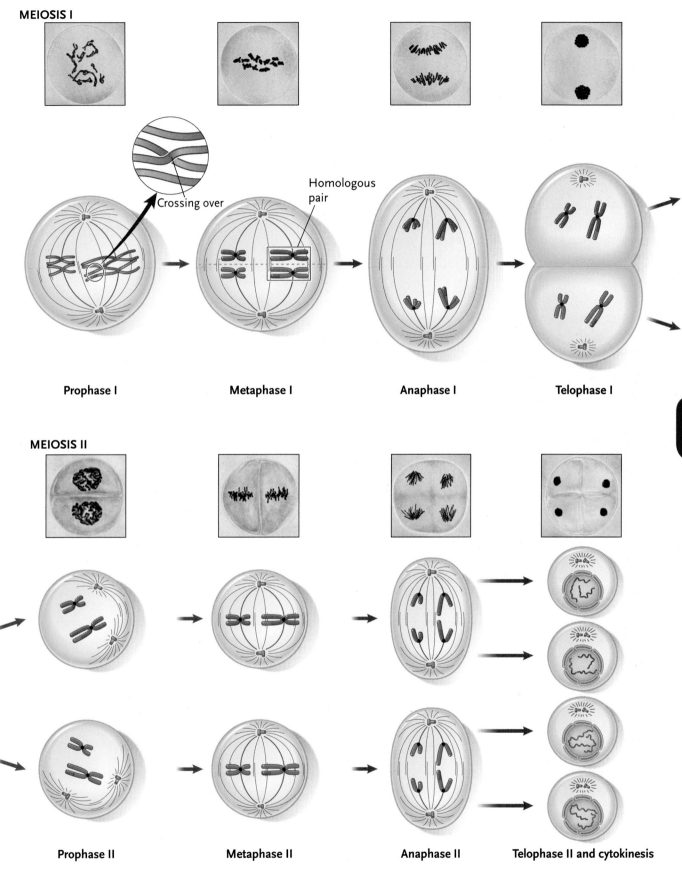

Figure 5.13 Stages of Meiosis

Art source: From DIGIUSEPPE/FRASER. Biology 11U. © 2011 Nelson Education Ltd. Reproduced by permission. www.cengage.com/permissions

homologous chromosomes remains connected to its sister chromatid, there are 46 chromosomes in each cell. Therefore, one more cell division occurs so that each gamete contains only 23 chromosomes.

Prophase II

DNA does *not* replicate before the next prophase. The cells produced from meiosis I move directly into meiosis II, so there is no second S phase. Condensation of the DNA occurs, the nuclear membrane and the nucleolus break down, centrioles migrate and spindle fibres form.

Metaphase II

Spindle fibres connect to the kinetochore, and the chromosomes line up at the equatorial plane, but now in meiosis II, the chromosomes with their sister chromatid all line up (Figure 5.13).

Anaphase II

Spindle fibres begin to break down at the centriole end, which causes the sister chromatids to separate toward each end of the cell.

Telophase

Sister chromatids are situated at each end of the cell, and the spindle fibres break down. A new nuclear membrane forms around each set of 23 chromosomes at each end of the cell, the chromosomes begin to uncondense, and the nucleolus reforms.

Cytokinesis

Cytokinesis in meiosis I and meiosis II occurs differently in males and females. In males, by the end of meiosis II, there are four sperm cells, each with 23 chromosomes. In females, each cell division occurs so that more of the cytoplasm goes into one cell and the other becomes a polar body that dissolves, leaving only one egg cell. To further complicate this process, females don't complete the last meiotic division unless fertilization occurs. Eggs stay arrested in metaphase II until ovulation, and do not complete the last stage of cell division if unfertilized.

Human genetic variation

Considering all the possible combinations of homologues that could separate during the first meiotic cell division into each new cell, there are two possible homologues for each cell, and since we have 23 pairs of chromosomes, there are 2^{23} possible combinations of chromosomes that can occur in each gamete from one individual. Therefore, each person can produce over 8 million possible different gametes. This scope of possible variation increases further because every individual is formed from two different gametes from two different people. The probability of two parents having two children with the same genetic material (just looking at random assortment of chromosomes) is one in 2.5 billion.

Crossing over

Another element ensures that genetic variation predominates. There is not only random separation of homologues—also called **independent assortment** (Chapter 6)—but also a process called **crossing over**, which occurs during prophase I. When homologous chromosomes start to pair up at the beginning of meiosis, the homologous chromosomes overlap. Fragments of non-sister chromosomes break in the same place, and a section of each chromosome is swapped, resulting in a hybrid chromosome (Figure 5.14).

Meiosis has unique features that are not seen in mitosis. Meiosis has two cell divisions so that haploid gametes are produced, and meiosis has crossing over, which ensures that every individual is unique (except in the case of identical twins when the zygote splits to form two embryos that have identical DNA).

DID YOU KNOW?

Females produce all of their eggs before they are born. By week 20 of gestation, females have around 7 million primary oocytes arrested in mid-meiosis. The second meiotic division occurs only if fertilization occurs after the female reaches reproductive age.

DID YOU KNOW?

During pregnancy, a sample can be taken of the amniotic fluid surrounding the embryo, which contains cells from the embryo. These cells can be tested for any chromosome abnormalities, such as Down syndrome when there is an extra chromosome 21. This test is called an **amniocentesis**.

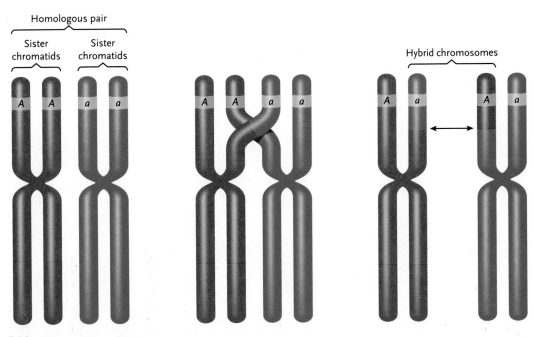

Homologous pair

Sister chromatids   Sister chromatids

A A   a a

Hybrid chromosomes

A A a a

A a   A a

Figure 5.14   Crossing Over

Crossing over involves the transfer of sections of homologous chromosomes during prophase I and increases genetic variation.

Art source: From RUSSELL/HERTZ/STARR/FENTON. Biology, 2E. © 2013 Nelson Education Ltd. Reproduced by permission. www.cengage.com/permissions.

CONCEPT REVIEW 5.5

1. What is the purpose of meiosis?

2. Which cells are haploid? Diploid? How many chromosomes are in each?

3. When do homologous chromosomes and sister chromatids separate during meiosis?

4. What is a gamete? Where are they formed in males and females?

5. Why is genetic variation important?

6. How does meiosis ensure that every individual has a unique combination of genes?

7. How is meiosis different in males and females?

8. What is crossing over and why is it important?

5

6 Genetic Inheritance

Universal Image Group/Getty Images

6.1 MENDEL'S THEORY OF INHERITANCE

Heredity refers to the transfer of traits from parent to offspring. Humans acquire 23 chromosomes from each parent, resulting in a total of 46 chromosomes that contain all 25,000 genes in every nucleated cell. Recall from Chapter 5 that each pair of chromosomes are homologous pairs, so they are *almost* exactly the same; whereas sister chromatids that result from DNA replication are *exactly* identical, assuming no mutations occurred during S phase (Figure 6.1).

Each gene codes for an RNA molecule that acts as the template for protein synthesis. The resulting proteins constitute a particular trait. Recall from Chapter 2 that proteins can be membrane transporters such as a chloride channel, hemoglobin that carries oxygen in red blood cells, hormones that signal cells to perform various functions, enzymes that catalyze reactions, and many more. The visible appearance or function related to a specific protein—such as brown hair pigmentation, or freckles, or a disease such as cystic fibrosis—is called a phenotype. Note that a phenotype can also be due to the combination of genes expressed; for example, eye colour is actually due to the combined expression of many genes.

The genes that code for a phenotype are the genotype. Each of the two copies of every gene, one from each parent, is called an allele. When the alleles are the same, they are homozygous; when they are different, they are heterozygous. Sometimes one allele is dominant and another one is recessive. For example, in humans the disease allele that causes sickle cell anemia is recessive to the healthy allele. In this case, a person has the sickle cell anemia phenotype if they have both disease alleles making their genotype homozygous recessive. The convention is to use a capital letter (e.g., *A*) to denote a dominant allele and a lower case letter (e.g., *a*) to denote a recessive allele. Table 6.1 shows that four genotypes are possible because we have two alleles for each trait.

In the 1800s, a monk by the name of Gregor Mendel studied the phenotypes of generations of peas. However, he did not know that organisms have two copies of each gene, or that DNA carries traits from parents to offspring. He called homozygous traits "true-breeding" and discovered the inheritance patterns of dominant-recessive traits. He was fortunate to study peas that have many single-gene traits. In humans, most phenotypes are determined

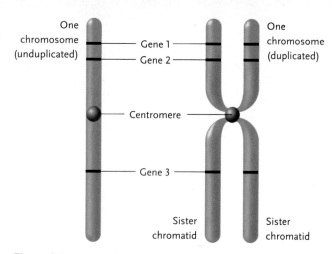

Figure 6.1 Sister Chromatids

Sister chromatids are exact copies of each other formed during DNA replication.

Art source: From DIGIUSEPPE/FRASER. Biology 11U. © 2011 Nelson Education Ltd. Reproduced by permission. www.cengage.com/permissions

by multiple genes (Section 6.4). Some phenotypes of pea plant are as follows (Figure 6.2):

- purple flowers or white flowers
- yellow seeds or green seeds
- wrinkled seeds or round seeds
- tall or short plants

TABLE 6.1

Genotype and Phenotype

Genotype	Phenotype
homozygous dominant (AA)	healthy
heterozygous (Aa)	healthy
homozygous recessive (aa)	sickle cell anemia

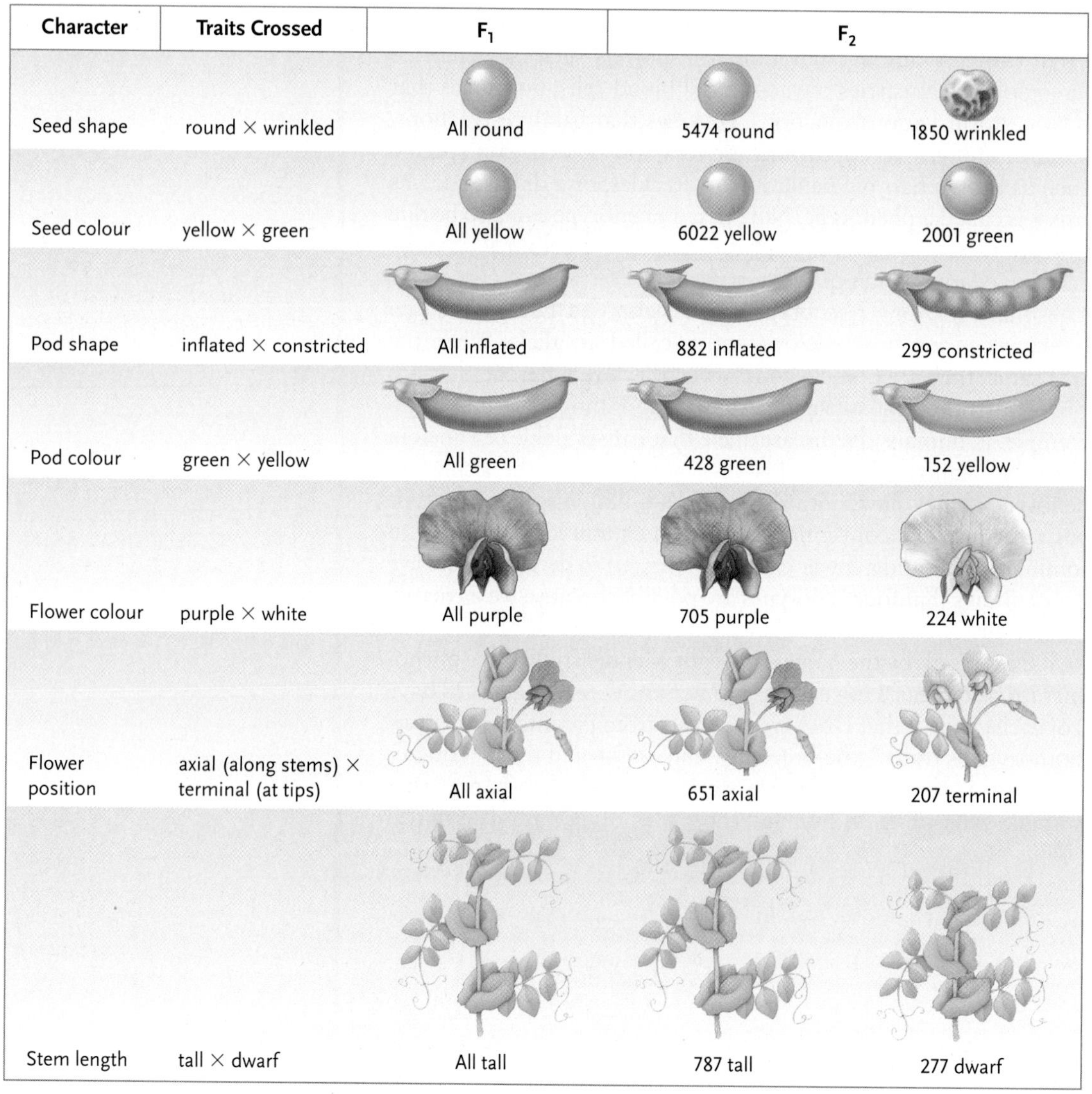

Character	Traits Crossed	F₁	F₂	
Seed shape	round × wrinkled	All round	5474 round	1850 wrinkled
Seed colour	yellow × green	All yellow	6022 yellow	2001 green
Pod shape	inflated × constricted	All inflated	882 inflated	299 constricted
Pod colour	green × yellow	All green	428 green	152 yellow
Flower colour	purple × white	All purple	705 purple	224 white
Flower position	axial (along stems) × terminal (at tips)	All axial	651 axial	207 terminal
Stem length	tall × dwarf	All tall	787 tall	277 dwarf

Figure 6.2 Examples of Dominant and Recessive Pea Plant Phenotypes

Art source: From RUSSELL/HERTZ/STARR/FENTON. Biology, 2E. © 2013 Nelson Education Ltd. Reproduced by permission. www.cengage.com/permissions

Mendel cross-fertilized different pea plants with different traits and studied the frequency of the phenotypes in the offspring. The true-breeding plants always gave rise to offspring with the same traits, such as all purple flowers or all white flowers. When he crossed true-breeding purple-flowered pea plants with true-breeding white-flowered pea plants, the first generation, called F1, always had purple flowers. When he cross-fertilized that F1 generation, the next generation, F2, contained plants with white flowers again (Figure 6.3). He called the phenotype of the F1 generation the **dominant** trait, and the phenotype that reappeared in the F2 generation, the **recessive** trait. Crossing purple-flowered plants with white-flowered ones resulted in 100% with purple flowers—that is, dominant—in the F1 generation; crossing two purple-flowered plants from the F1 generation resulted in 75% with purple flowers and 25% with white flowers. Breeding many F1 generation pea

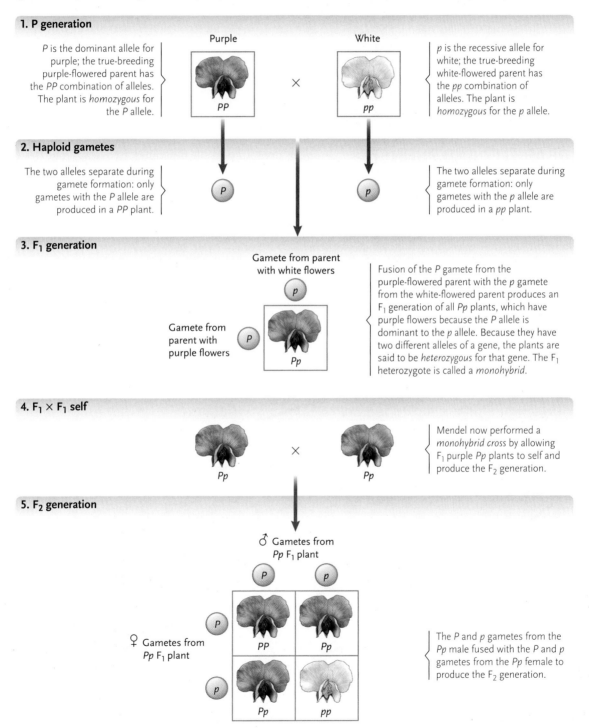

1. P generation

P is the dominant allele for purple; the true-breeding purple-flowered parent has the *PP* combination of alleles. The plant is *homozygous* for the *P* allele.

Purple   White

PP   ×   pp

p is the recessive allele for white; the true-breeding white-flowered parent has the *pp* combination of alleles. The plant is *homozygous* for the *p* allele.

2. Haploid gametes

The two alleles separate during gamete formation: only gametes with the *P* allele are produced in a *PP* plant.

P   p

The two alleles separate during gamete formation: only gametes with the *p* allele are produced in a *pp* plant.

3. F₁ generation

Gamete from parent with white flowers

p

Gamete from parent with purple flowers

P

Pp

Fusion of the *P* gamete from the purple-flowered parent with the *p* gamete from the white-flowered parent produces an F₁ generation of all *Pp* plants, which have purple flowers because the *P* allele is dominant to the *p* allele. Because they have two different alleles of a gene, the plants are said to be *heterozygous* for that gene. The F₁ heterozygote is called a *monohybrid*.

4. F₁ × F₁ self

Pp   ×   Pp

Mendel now performed a *monohybrid cross* by allowing F₁ purple *Pp* plants to self and produce the F₂ generation.

5. F₂ generation

♂ Gametes from *Pp* F₁ plant

P   p

♀ Gametes from *Pp* F₁ plant

P

p

PP   Pp

Pp   pp

The *P* and *p* gametes from the *Pp* male fused with the *P* and *p* gametes from the *Pp* female to produce the F₂ generation.

Figure 6.3   Inheritance Pattern

The inheritance pattern for breeding dominant-recessive, single gene traits in pea plants can be determined using a Punnett square.

Art source: From RUSSELL/HERTZ/STARR/FENTON. Biology, 2E. © 2013 Nelson Education Ltd. Reproduced by permission. www.cengage.com/permissions

plants always resulted in the same proportion of 3:1 (dominant:recessive) in the F2 generation. Now that we know the "true-breeding" plants are homozygous and the "not-true-breeding plants" are heterozygous, we can predict the phenotype ratios for any combination of genotypes by using a **Punnett square** (Section 6.2).

The possible genotypes and phenotypes for pea plant flower colour are as follows:

- PP (homozygous dominant)—purple
- Pp (heterozygous)—purple
- Pp (homozygous recessive)—white

CONCEPT REVIEW 6.1

1. Define the following: homozygous, heterozygous, dominant, recessive, allele, phenotype, and genotype.

2. Consider two pea plants that are heterozygous for flower colour, with purple and white being the possible alleles.

   a. What colour would the flowers be?

   b. What are the possible gametes they could form?

   c. What are the possible offspring they could produce?

   d. What is the probability that those pea plants would have offspring with white flowers?

6.2 PROBABILITY AND PUNNETT SQUARES

Probability refers to the likelihood that some event will occur. If we flip one coin, there are two possible outcomes; it could land on heads or tails. If we flip two coins there are four possible outcomes; two heads, one head and one tail, one tail and one head, or two tails (Figure 6.4). The probability of each possible outcome is as follows:

- 2 heads—¼ or 25%
- 1 head and 1 tail—½ or 50%
- 2 tails—¼ or 25%

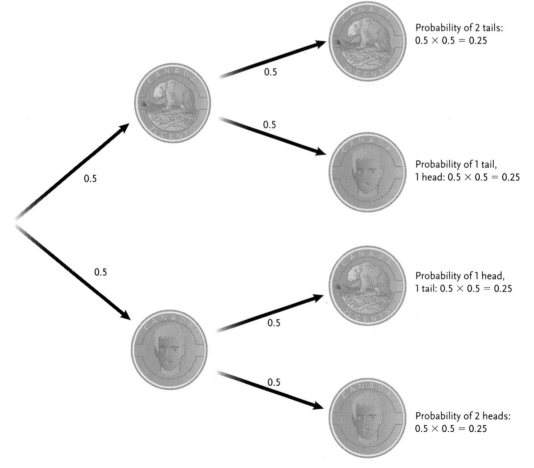

Probability of 2 tails:
0.5 × 0.5 = 0.25

Probability of 1 tail,
1 head: 0.5 × 0.5 = 0.25

Probability of 1 head,
1 tail: 0.5 × 0.5 = 0.25

Probability of 2 heads:
0.5 × 0.5 = 0.25

Figure 6.4 **Probability Tree**
A probability tree shows all of the possible outcomes from flipping a coin twice.

Considering genetic inheritance and each individual having two alleles, the possible outcomes are the same:

- 2 dominant—¼ or 25%
- 1 dominant and 1 recessive—½ or 50%
- 2 recessive—¼ or 25%

A Punnett square provides a method for determining the possible offspring genotypes and phenotypes from the alleles of each parent. If we look at only a single trait,

this is considered a **monohybrid cross**. Since each person has two copies of each gene—alleles—on homologous chromosomes, and since homologous chromosomes separate during meiosis into different gametes (Chapter 5), we can determine the probability of each possible genotype and phenotype in the offspring (Figure 6.4): the possible genes in the gametes are shown outside the boxes, and the possible combinations of genes of the offspring are shown inside the boxes. For example (Figure 6.5), if a parent pea

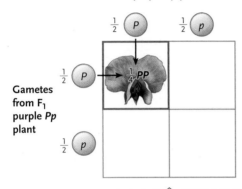

A. To produce an F_2 plant with the PP genotype, two P gametes must combine. The probability of selecting a P gamete from one F_1 parent is $\frac{1}{2}$, and the probability of selecting a P gamete from the other F_1 parent is also $\frac{1}{2}$. Using the product rule, the probability of producing purple-flowered PP plant from a $Pp \times Pp$ cross is $\frac{1}{2} \times \frac{1}{2} = \frac{1}{4}$.

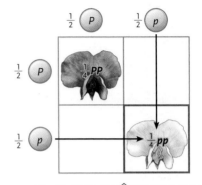

B. To produce an F_2 plant with the pp genotype, two p gametes must combine. The probability of selecting a p gamete from one F_1 parent is $\frac{1}{2}$, and the probability of selecting a p gamete from the other F_1 parent is also $\frac{1}{2}$. Using the product rule, the probability of producing white-flowered pp plant from a $Pp \times Pp$ cross is $\frac{1}{2} \times \frac{1}{2} = \frac{1}{4}$.

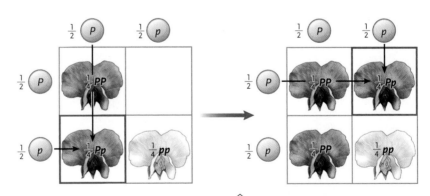

C. To produce an F_2 plant with the Pp genotype, a P gamete must combine with a p gamete. The cross $Pp \times Pp$ can produce Pp offspring in two different ways: (1) a P gamete from the first parent can combine with a p gamete from the second parent; or (2) a p gamete from the first parent can combine with a P gamete from the second parent. We apply the sum rule to obtain the combined probability: each of the ways to get Pp has an individual probability of $\frac{1}{4}$, so the probability of Pp, purple-flowered offspring is $\frac{1}{4} + \frac{1}{4} = \frac{1}{2}$.

Figure 6.5   **Monohybrid Cross**

This example shows a monohybrid cross between heterozygous pea plants and the possible genotypes and phenotypes of the offspring.

plant is heterozygous for flower colour, one allele is dominant and the other allele is recessive. When the homologous chromosomes separate during meiosis, it is possible for a gamete to have a dominant allele or a recessive allele. If both parents are heterozygous, the following possible offspring can arise:

- genotype ratio—1:2:1—homozygous dominant: heterozygous:homozygous recessive
- phenotype ratio—3:1—purple:white

DID YOU KNOW?

Some human traits are passed on by dominant-recessive inheritance (Table 6.2). However, in these examples, other genes can act on these features. So these traits can also be considered polygenic, and many variations will be seen in the population. For example, face shape can vary from round, oval, oblong, to square, but oval shapes tend to be dominant over square shapes.

Testcross

Mendel devised the **testcross** to determine the genotype of individuals expressing the dominant phenotype. In pea plants, yellow peas are dominant over green peas, so a yellow pea could be homozygous dominant (YY)

TABLE 6.2

Dominant-Recessive Human Traits

Trait	Dominant	Recessive
Shape of face	Oval	Square
Cleft in chin	No cleft	Cleft
Hairline	Widow peak	Straight
Eyebrow size	Broad	Slender
Eyebrow shape	Separated	Joined
Eyelash length	Long	Short
Earlobes	Free	Attached
Freckles	Freckles	No freckles
Tongue rolling	Rolling	Non-rolling
Hitch-hiker's thumb	Straight	Hitch-hiker
Little finger	Bent	Straight
Interlaced fingers	Left thumb over right	Right thumb over left

or it could be heterozygous (Yy). If a yellow pea plant is crossed with a homozygous recessive green pea plant, different outcomes will occur depending on whether the yellow plant is YY or Yy (Figure 6.6). If the yellow plant is homozygous dominant, the phenotype ratios of the offspring will be 100% yellow. If the yellow plant is heterozygous, the phenotypes ratios of the offspring will be 50% yellow and 50% green.

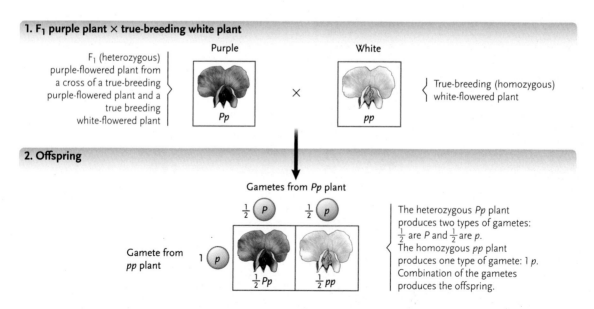

1. F₁ purple plant × true-breeding white plant

F₁ (heterozygous) purple-flowered plant from a cross of a true-breeding purple-flowered plant and a true breeding white-flowered plant

Purple
Pp

×

White
pp

True-breeding (homozygous) white-flowered plant

2. Offspring

Gametes from *Pp* plant

½ P   ½ p

Gamete from *pp* plant   1 p

½ Pp   ½ pp

The heterozygous *Pp* plant produces two types of gametes: ½ are *P* and ½ are *p*. The homozygous *pp* plant produces one type of gamete: 1 *p*. Combination of the gametes produces the offspring.

Figure 6.6 Testcross (*continued*)

A testcross is used to determine the genotype of an organism expressing the dominant phenotype.

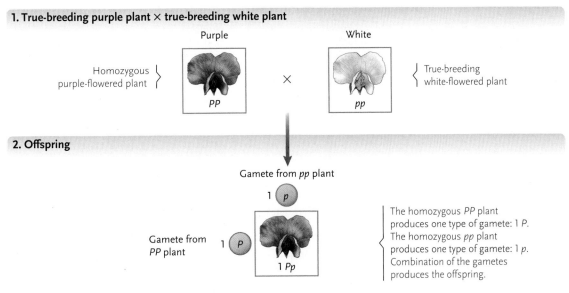

1. True-breeding purple plant × true-breeding white plant

Purple   White

Homozygous
purple-flowered plant }   *PP*   ×   *pp*   { True-breeding
white-flowered plant

2. Offspring

Gamete from *pp* plant

1 (*p*)

Gamete from
PP plant   1 (*P*)

1 *Pp*

The homozygous *PP* plant
produces one type of gamete: 1 *P*.
The homozygous *pp* plant
produces one type of gamete: 1 *p*.
Combination of the gametes
produces the offspring.

Figure 6.6   *(Continued)*

CONCEPT REVIEW 6.2

1. Yellow peas are dominant over green peas. Given the case that a heterozygous yellow pea plant was crossed with a green pea plant, use a Punnett square to determine the probability of the offspring pea plant being the following:

   a. heterozygous
   b. yellow
   c. green
   d. homozygous dominant

2. A pea plant has purple flowers, but the genotype is not known. Explain how a testcross can be used to determine whether the purple plant is PP or Pp.

3. The ability to curl your tongue up on the sides is dominant over the inability to curl your tongue on the sides. A woman who can curl her tongue marries a man who cannot curl his tongue. Their first child cannot curl their tongue like their dad. What is the genotype of the mom? Dad? Child?

4. In rats, long tails are dominant over short tails. Can two long-tailed rats have short-tailed offspring?

5. Cystic fibrosis is a recessive disease. If a mother is a carrier of the CF allele but does not have the disease, and the father is homozygous healthy, is it possible for these parents to have children with cystic fibrosis?

6.3 DIHYBRID CROSS

Mendel's **law of segregation** states that the two alleles of a trait separate from each other during the formation of gametes, so half the gametes will carry one copy and half will carry the other copy. This is because diploid germ-line cells form haploid gametes; each gamete has one of each homologous chromosome. Mendel's **law of independent assortment** states that how the homologous pairs separate into gametes is completely random, so any possible combination can occur.

If the genes for *two* different traits are found on different chromosomes, we can use a **dihybrid cross** to look at the possible genotype and phenotype outcomes. For example, in peas, the genes that code for round and yellow are dominant over the genes that code for wrinkled and green. If a homozygous round and yellow pea plant is crossed with a homozygous wrinkled and green pea plant, all the offspring will be heterozygous round and yellow peas (Figure 6.7). If those heterozygous offspring are crossed, all the possible genotypes and phenotypes can be determined using a Punnett square (Figure 6.7). As for a monohybrid

1. P generation

The genotype of the true-breeding round, yellow parent is *RR YY*, where *R* is the dominant allele for round, and *Y* is the dominant allele for yellow. The plant is homozygous for both the *R* and *Y* alleles.

The genotype of the true-breeding wrinkled and green parent is *rr yy*, where *r* is the recessive allele for wrinkled, and *y* is the recessive allele for green. The plant is homozygous for both the *r* and *y* alleles.

2. Haploid gametes

Only gametes with the *R* and *Y* alleles are produced in an *RR YY* plant.

Only gametes with the *r* and *y* alleles are produced in an *rr yy* plant.

3. F₁ generation

Fusion of an *R Y* gamete from the round, yellow parent with an *r y* gamete from the wrinkled, green parent produces an F₁ generation all of which have the genotype *Rr Yy*, phenotype round, yellow seeds. The doubly heterozygous individual is called a *dihybrid*. The seeds are round because the *R* allele is dominant to the *r* allele, and yellow because the *Y* allele is dominant to the *y* allele.

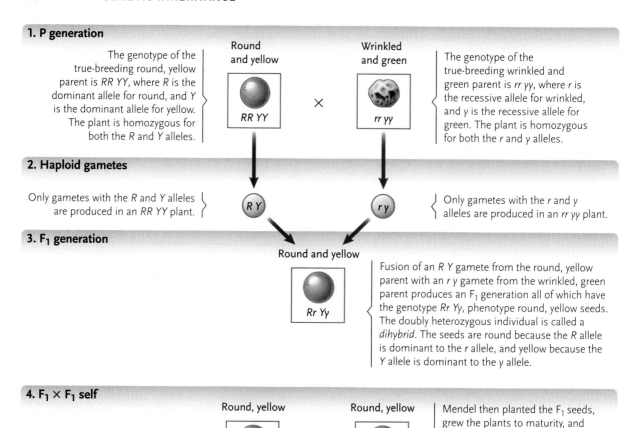

4. F₁ × F₁ self

Mendel then planted the F₁ seeds, grew the plants to maturity, and selfed them; that is, he crossed the F₁ to themselves. A cross such as this of two double heterozygotes is called a *dihybrid cross*.

5. F₂ generation

If the alleles that control seed shape and seed colour assort independently, each F₁ plant grown from the seeds would produce four types of gametes: the *R* allele for seed shape would go to a gamete with either the *Y* or *y* allele for seed colour, and similarly, the *r* allele would go to a gamete with either the *Y* or *y* allele. Thus, independent assortment of genes from the *Rr Yy* parents is expected to produce four types of gametes with equal probability: $\frac{1}{4} R Y$, $\frac{1}{4} R y$, $\frac{1}{4} r Y$, and $\frac{1}{4} r y$. Random fusion of the four different male gametes with the four different female gametes produces the F₂ generation.

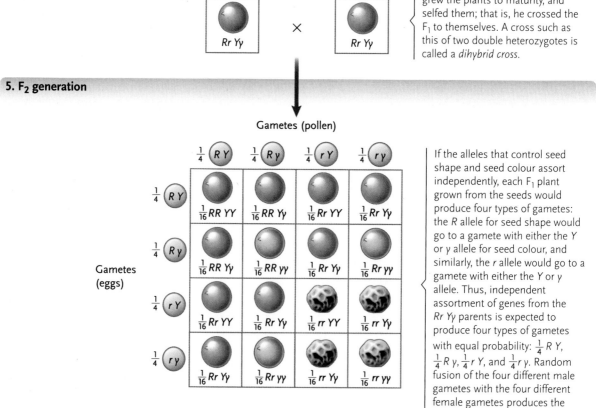

Figure 6.7 Dihybrid Cross

A dihybrid cross is used to determine the probability of offspring acquiring two traits.

Art source: From RUSSELL/HERTZ/STARR/FENTON. Biology, 2E. © 2013 Nelson Education Ltd. Reproduced by permission. www.cengage.com/permissions.

cross, the possible gametes are written outside the squares, and the possible offspring are written inside the squares.

To determine the possible gametes when looking at two different traits, we can use the acronym **FOIL** (first, outer, inner, last). If a pea plant is heterozygous for round and yellow traits, the genotype is RrYy. The R and r are homologous pairs that separate randomly into different gametes during the first meiotic division; similarly, the Y and y are homologous chromosomes that separate randomly into different gametes during meiosis. Each gamete

must include either the R or r AND a Y or y. Therefore, a diploid parent pea plant can form the following possible haploid gametes:

RrYy

First—RY

Outer—Ry

Inner—rY

Last—ry

CONCEPT REVIEW 6.3

1. Round and yellow are dominant over wrinkled and green peas. Given the case that one parent plant is homozygous for the round pea trait and heterozygous for the yellow trait, and the other parent plant is heterozygous for the round trait and is green, answer the following questions:

   a. What are the genotypes of the parent plants?
   b. What are the possible gametes those parent plants can form?
   c. What are the possible genotypes of the offspring?

   d. What are the possible phenotypes of the offspring?
   e. What are the phenotypic and genotypic *ratios* for the above cross?

2. Freckles and brown hair are dominant over no freckles and blond hair. Suppose mom has no freckles and blond hair, and dad is heterozygous for both traits. What is the probability that their children will have brown hair and freckles?

6.4 NON-MENDELIAN INHERITANCE

Polygenic inheritance

Often a phenotype does not express a simple dominant or recessive inheritance pattern. Multiple genes determine most traits and the phenotype varies from one extreme to another: for example, human height ranges from very short to very tall, and most people's height is somewhere in the middle (Figure 6.8). When a phenotype shows **continuous variation** in a population, it is known as **polygenic** because multiple genes play a role in that phenotype. Other examples of polygenic traits in humans include body weight, skin colour, eye colour, intelligence, and diseases such as cancer or schizophrenia.

Pleiotropic

When one allele causes multiple phenotypes in a population, it is considered **pleiotropic**. For example, cystic fibrosis (CF), an inherited gene mutation in the chloride channel gene, can cause a wide variety of symptoms in different people with that allele. Cystic fibrosis is a recessive disease, so a person must have both mutated alleles to have the disease. A person heterozygous for

CF does not have cystic fibrosis. The symptoms of CF range from mild to severe and can affect the lungs, liver, pancreas, and sweat glands.

Incomplete dominance

The alleles for each trait are neither fully dominant nor fully recessive in heterozygotes. Sometimes a heterozygote has an intermediate trait in comparison to the homozygous dominant and the homozygous recessive (Figure 6.9). For example, if red flowers are incompletely dominant to white flowers, a heterozygote would be pink. In incomplete dominance inheritance the following three genotypes and three phenotypes are possible:

- homozygous dominant (RR)—red
- heterozygous (Rr)—pink
- homozygous recessive (rr)—white

Codominance

A gene may have more than two alleles in a population, and sometimes in heterozygotes *both* alleles are expressed. For example, in humans, there are three possible alleles for blood type (A, B, or O), and each

A. Students at Brigham Young University, arranged according to height

B. Actual distribution of individuals in the photo according to height

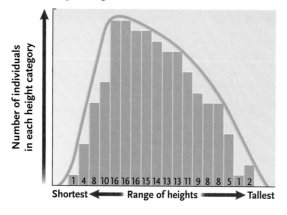

Shortest ◄──── Range of heights ───► Tallest

C. Idealized bell-shaped curve for a population that displays continuous variation in a trait

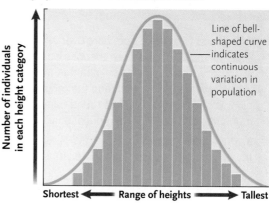

Line of bell-shaped curve indicates continuous variation in population

Shortest ◄──── Range of heights ───► Tallest

If the sample in the photo included more individuals, the distribution would more closely approach this ideal.

Figure 6.8 Polygenic Inheritance

Polygenic inheritance occurs when a single trait, such as body height, is determined by multiple genes, usually showing a continuous variation distribution in a population.

person can have any two alleles. If a heterozygote has an A and a B allele, both the A and B proteins will be expressed, giving a phenotype of blood type AB; this is an example of **codominance** (Figure 6.10). The ABO blood types (not including the Rh factor) represent genes that code for an enzyme that adds sugar groups to the cell membranes of red blood cells. These sugar groups act as cell recognition markers called antigens. The gene that encodes the enzyme, designated I (for immunoglobulin), has three alleles: I^A, I^B, which code for either antigen A or B, and i, which does not code for an antigen at all. Since antigens can stimulate an immune response (Chapter 22), it is important for people to receive the correct blood type during transfusions.

Different combinations of the three blood type alleles produce four different possible phenotypes, or blood types (A, B, AB, and O) (Table 6.3). I^A and I^B are completely dominant over i, and I^A is codominant with I^B.

More than 40 other antigens are expressed on red blood cells, the most significant being Rh factor (antigen D). If this antigen is present the blood type is *positive*, if this antigen is not present the blood type is *negative*. Positive is dominant over negative because if one of the alleles expresses the antigen, that antigen will be present on the red blood cells. When an individual is exposed to an antigen that they themselves do not express, an immune reaction can occur and can be fatal.

DID YOU KNOW?

Chimpanzees have blood type A and are thought to have a common ancestor with the Cro-Magnon. Gorillas have blood type B and are thought to have a common ancestor with the Neanderthal.

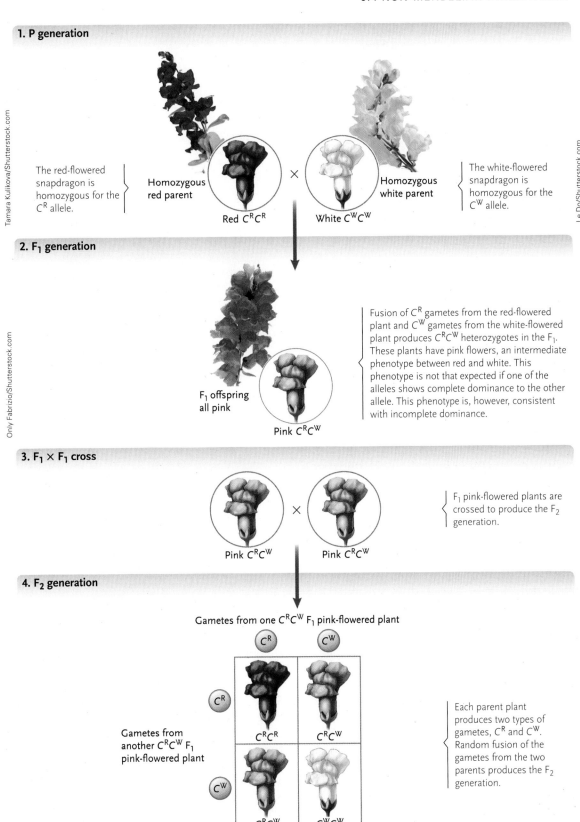

Figure 6.9   Incomplete Dominance

Incomplete dominance occurs when the heterozygote phenotype is a blend of the dominant and recessive traits, such as pink flowers.

Art source: From RUSSELL/HERTZ/STARR/FENTON. Biology, 2E. © 2013 Nelson Education Ltd. Reproduced by permission. www.cengage.com/permissions

Possible alleles in
gametes from father

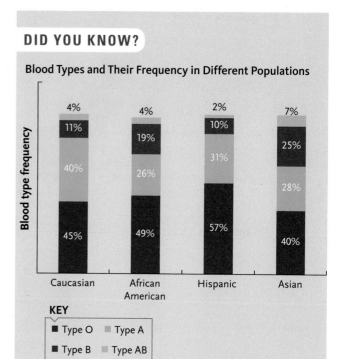

Figure 6.10 Codominance—Blood Types

Codominance occurs when the heterozygote expresses *both* traits; for example, both A and B alleles are present in type AB blood.

Art source: From RUSSELL/HERTZ/STARR/FENTON. Biology, 2E. © 2013 Nelson Education Ltd. Reproduced by permission. www.cengage.com/permissions

Blood Types and Their Frequency in Different Populations

Source: Adapted from http://www.redcrossblood.org/learn-about-blood/blood-types

TABLE 6.3

Blood Type Genotypes and Phenotypes

Genotype	Phenotype
$I^A I^A$	type A
$I^A i$	type A
$I^B I^B$	type B
$I^B i$	type B
$I^A I^B$	type AB
$i i$	type O

Example: If mom has blood type A and dad has blood type B, is it possible for them to have a child that is type O? The answer is yes, because the parents could be heterozygous $I^A i$ and $I^B i$. If the gametes containing i alleles combined, then the baby can be ii, which is type O.

Epistasis

Epistasis refers to the phenomenon where the effects of one gene are modified by one or more other genes; these are sometimes called **modifier genes**. For example,

Labrador retrievers can have the phenotype black, brown, or yellow. For these dogs to be black or brown, they must have one or both dominant alleles for dark fur pigmentation (E); another allele will determine if that colour is black (BB or Bb) or brown (bb). A dog with EE or Ee will be black or brown, and a dog with ee will be yellow, regardless of the other colour alleles (Figure 6.11). A dog with the BBee or Bbee or bbee genotype will be yellow, because the epistasis (E) allele can mask the expression of the B or b allele.

Environmental influences

Sometimes environmental factors determine how genes are expressed: for example, temperature, or the season, or the length of daylight. Hydrangea flowers have blue flowers if the soil is acidic and pink flowers if the soil is more alkaline (Figure 6.12). Another example is the arctic fox, which has white fur in the winter when it is cold and brown fur in the summer when it is warm. The sex of many reptiles depends on the temperature of the eggs; eggs incubated at temperatures higher than 30 degrees will be female, lower temperatures will be male. In humans, environmental factors such as nutrition or exposure to infectious organisms, affects phenotypes such as height, weight, and immune function.

A. Black Labrador

B. Chocolate brown Labrador

C. Yellow Labrador

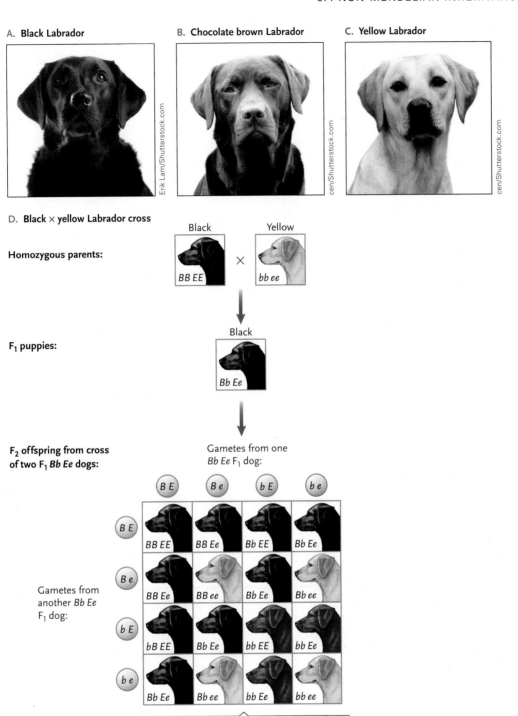

D. Black × yellow Labrador cross

Homozygous parents:

Black   Yellow

BB EE × *bb ee*

F₁ puppies:

Black

Bb Ee

F₂ offspring from cross of two F₁ *Bb Ee* dogs:

Gametes from one *Bb Ee* F₁ dog:

Gametes from another *Bb Ee* F₁ dog:

F₂ phenotypic ratio is 9 black:3 chocolate:4 yellow

Figure 6.11   Epistasis

Epistasis occurs when the phenotype of one gene masks the phenotype of another gene.

Art source: From RUSSELL/HERTZ/STARR/FENTON. Biology, 2E. © 2013 Nelson Education Ltd. Reproduced by permission. www.cengage.com/permissions.

X-linked influences

Some inheritance patterns depend on which chromosomes the gene is located on. We have 23 pairs of homologous chromosomes: numbered 1–22 and also an X or Y. Even though X and Y are considered homologous, they do not have similar genes. Females have XX and males have XY. The Y chromosome contains the sex-determining factor that causes testosterone production and male development. The X chromosome contains many

Figure 6.12 Environmental Factors

Environmental factors can influence the expression of certain genes; for example, the pH level in the soil alters the colour of hydrangeas, and the colour of arctic fox fur changes with the seasons.

Female $X^R X^w$

	X^R	X^w Gametes
X^R	Female, red eyes $X^R X^R$	Female, red eyes $X^R X^w$
X^w	Male, red eyes $X^R Y$	Male, white eyes $X^w Y$

Male $X^R Y$

Gametes

Figure 6.14 X-linked Inheritance

X-linked inheritance is related to genes on the X chromosome. Males are more likely than females to express an X-linked mutation because males have only one X chromosome and females would need to have the mutation on both of their X chromosomes.

Art source: From RUSSELL/HERTZ/STARR/FENTON. Biology, 2E. © 2013 Nelson Education Ltd. Reproduced by permission. www.cengage.com/permissions

genes that are not related to sex characteristics, and those genes are called **X-linked**. For example, fruit flies have a mutant eye colour gene that causes white eyes instead of the normal red eyes (Figure 6.13), mostly in males. A female fruit fly with red eyes could have the genotype $X^R X^R$ or $X^R X^r$, with red eyes (R) being dominant over white eyes (r). A male fruit fly with red eyes has the genotype $X^R Y$, and with white eyes has the genotype $X^r Y$. It is possible for a female to have white eyes, but she must have both X chromosomes with the eye colour mutation ($X^r X^r$).

Human X-linked, recessive traits include hemophilia, which is a blood clotting disorder; red-green colour blindness; and Duchene muscular dystrophy. Each of these traits are much more common in males than females because males have only one X chromosome and females need to inherit *two* mutated alleles to have the disease (Figure 6.14).

Linkage

Some genes tend to be frequently inherited together. When genes are located close together on the same chromosome, they are considered **linked** (Figure 6.15). The further two genes are away from each other on a chromosome, the liklihood of linkage decreases; this is due to the posssibility that those genes may be separated by crossing-over during meiosis.

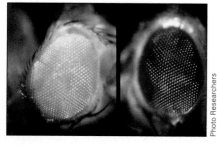

Figure 6.13 Drosophila Melanogaster

Drosophila melanogaster normally have red eyes, but sometimes males can have white eyes.

DID YOU KNOW?

Black and orange calico cats can only be female. The fur colour gene is located on X chromosomes, so male calicos can only be black or orange. However, females have two X chromosomes, and the black and orange gene is co-dominant, so females can have a black and an orange allele. This is an example of epistasis because a separate gene determines the pattern of fur colour distribution, but only in females. This is also an example of a polygenic trait because another gene codes for white fur. The overall phenotype of fur colour and distribution is determined by multiple genes.

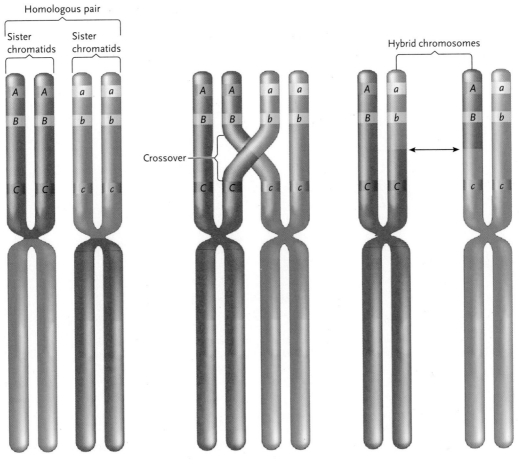

Figure 6.15   Linkage

Linkage occurs when two genes are located close together on a chromosome, and so they are much less likely to be separated during crossing over; the traits tend to be inherited together.

Art source: From RUSSELL/HERTZ/STARR/FENTON. Biology, 2E. © 2013 Nelson Education Ltd. Reproduced by permission. www.cengage.com/permissions.

CONCEPT REVIEW 6.4

1. Define polygenic inheritance and give three examples.

2. Define pleiotropic and give one example.

3. If bird feather colour is incompletely dominant, and one allele causes red feathers and another allele gives blue feathers, what will a heterozygote look like? If two heterozygous birds mate, how many of their offspring will be blue?

4. If a mom has blood type A, and her baby has blood type O, which of the following is the dad: Larry with type O, Bill with type B, or Fred with type AB. What is mom's genotype? Explain how you know that.

5. Mom has type O blood, and dad is AB. What are the possible genotypes and phenotypes of their children?

6. What is epistasis? Explain how a dog can have the dominant black fur gene but is a yellow lab?

7. Give two examples of environmental factors that affect gene expression.

8. Colour blindness is X-linked. If dad is colour blind and mom is a carrier, what is the probability that they will have a child with colour blindness?

9. What is linkage?

6.5 GENETIC DISORDERS

Errors can occur during meiosis. **Nondisjunction** refers to the failure of chromosomes to separate correctly during either meiosis I or meiosis II. Recall from Chapter 5 that during the first meiotic division, the homologous chromosomes separate, and during the second division, the sister chromatids separate. When one gamete has more or fewer chromosomes than it is supposed to, this is called **aneuploidy**: an abnormal chromosome number. If gametes with abnormal chromosome numbers are involved in fertilization, normal development is usually

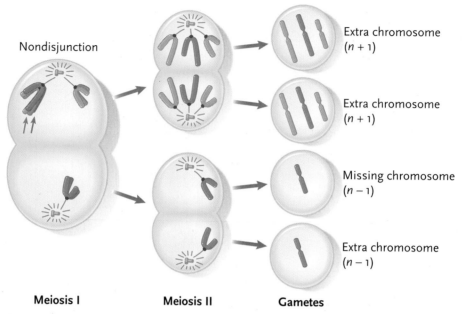

Figure 6.16 Nondisjunction

Nondisjunction occurs when the chromosomes do not separate properly during meiosis and gametes have more or fewer chromosomes than normal.

Art source: From DIGIUSEPPE/FRASER. Biology 11U. © 2011 Nelson Education Ltd. Reproduced by permission. www.cengage.com/permissions

not possible and early death occurs. However, if there is an extra chromosome 21 or, more rarely, an extra chromosome 22 development can continue (Figure 6.16). A child with an extra chromosome 21—**trisomy 21**—has **Down syndrome** and will have delayed development and mental impairment.

Nondisjunction of the **sex chromosomes** can also lead to three viable genotypes (Table 6.4).

Single-gene hereditary diseases

Hereditary mutations that do not cause death almost always result in recessive alleles. Recessive alleles are not eliminated from the population because usually they don't cause disease in heterozygous individuals, who

TABLE 6.4

Nondisjunction of the Sex Chromosomes

Genotype	Phenotype
XXX—Trisomy X	female, tall stature, learning disabilities, limited fertility
XO—Turner syndrome	female, short stature, sex organs don't mature at adolescence, may be infertile
XXY—Klinefelter syndrome	male, severity varies, small gonads, breast development, may be infertile
XYY	normal male

have one recessive and one dominant allele. When a gene mutation occurs on chromosomes 1–22, it is **autosomal**; when it occurs on the X chromosome, it is X-linked. If individuals only have the disease if they have both mutated alleles, the disease is recessive. If individuals have the disease with only one mutated allele, the disease is dominant.

The following list describes seven examples of single-gene hereditary diseases.

• Sickle cell anemia is an autosomal recessive disease, where the **hemoglobin** gene is mutated, causing abnormal red blood cell formation and reduced capacity to carry oxygen (Figure 6.17). Sickle cells are stiff and sticky and tend to block blood vessels,

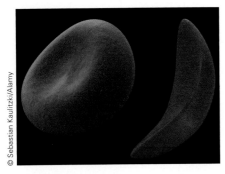

Figure 6.17 Sickle Cell Anemia

Sickle cell anemia is an autosomal recessive disorder caused by a mutation in the hemoglobin gene.

and they can damage many organs. Symptoms begin in children around four months old; with advancing medical treatment, life expectancy is now into the 40s or 50s.

- Tay Sachs is an autosomal recessive disease, where the HEXA gene that codes for a lysosomal enzyme is mutated, causing brain deterioration and leading to death most often around age four. Different HEXA gene mutations cause variations in the disease and populations that are affected. There are infantile, juvenile, and adult-onset variations. For example, a 4-base insertion in exon 11 causes the infantile form, which is found in Ashkenazi Jews and a population of Cajuns in Louisiana.

- Cystic fibrosis is an autosomal recessive disease caused by a mutation in the chloride channel membrane protein, called the *cystic fibrosis transmembrane conductance regulator (CFTR) gene*. The mutated protein cannot properly regulate the movement of chloride and sodium ions across epithelial cells membranes, and this leads to an altered water balance that primarily affects the lungs, liver, pancreas, and sweat glands. Early symptoms include salty sweat, poor growth, and frequent respiratory infections, but many body systems can be affected (Figure 6.18).

- Huntington's disease is an autosomal dominant disease, where the CAG nucleotide repeats are inserted into the Huntington gene, which increases the number of glutamine amino acids incorporated into the protein. Huntington's is a neurodegenerative disease that affects mood, cognitive ability, muscle coordination, and involuntary muscle movements called chorea. Huntington's primarily affects people of European descent.

- Hemophilia is a recessive X-linked disease that affects a clotting factor gene (either factor VIII or factor IX) and prevents the formation of fibrin, which is required to form a clot whenever a blood vessel has been damaged. People with hemophilia can have

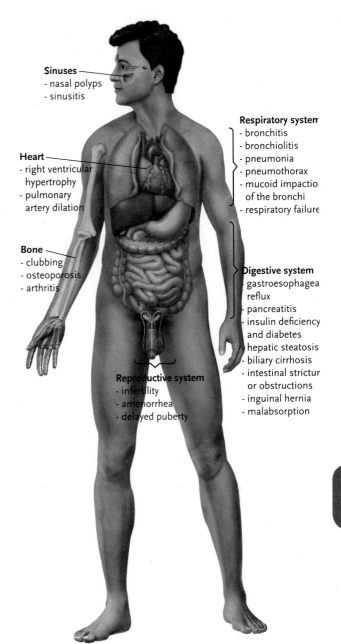

Figure 6.18 Cystic Fibrosis

Cystic fibrosis is an autosomal recessive disorder caused by a mutation in the chloride channel gene, called the CFTR gene, which stands for cystic fibrosis transmembrane conductance regulator.

internal or external bleeding episodes that can be treated with infusions of the missing clotting factor.

- Duchene muscular dystrophy is a recessive X-linked disease caused by a mutation in the **dystrophin** gene that holds myofibrils of muscle tissue together (Chapter 14). Without proper functioning of this protein, muscles degenerate and the disease progresses from muscle weakness to paralysis by age 10 to 12. Life expectancy is usually not longer than 25 years.

- Familial hypercholesterolemia is an autosomal dominant disease, where a mutation occurs in the **LDLR (low-density lipoprotein receptor)** gene. This gene is required for removing cholesterol from the blood stream so cells can use it to make energy or the steroid hormones derived from cholesterol, such as testosterone or estrogen (Chapter 16). Those affected can have cardiovascular disease by age 30.

DID YOU KNOW?

Before 1985, donated blood was not tested for diseases such as HIV or hepatitis. It is estimated that 50% of hemophiliacs that received blood transfusions contracted HIV: approximately 10,000 people in North America.

CONCEPT REVIEW 6.5

1. Which chromosome is affected in Down syndrome?
2. Which sex chromosome nondisjunctions are viable?

3. List the seven most common single-gene hereditary diseases, the inheritance pattern, the gene affected, and the main phenotype for each.

Education's purpose is to change an empty mind to an open one.

Malcom Forbes

7.1 DARWIN'S THEORY OF EVOLUTION

Charles Darwin (Figure 7.1) published his book *On the Origin of Species by Means of Natural Selection* in 1859. The theory of evolution was a revolutionary idea in the 1800s, and it was not accepted by many people. Darwin's theory can be divided into two major components: microevolution and macroevolution. **Microevolution**—small changes within a single species—has been confirmed with many different experiments. **Macroevolution**—the appearance of new species over thousands to millions of years—cannot be proven and therefore remains a "theory." This portion of Darwin's work remains highly debated today. The focus of this chapter is on microevolution and its significant implications to humans and health; the main ideas of macroevolution are also briefly described (see Section 7.3).

Figure 7.1 Charles Darwin

Figure 7.2 Genetic Variation

Genetic variation is important for the survival of populations of organisms.

Evolution is defined as the change in the inherited traits in a **population** of organisms over time. A single organism does not *evolve* because individuals are born with the DNA they acquired through the process of reproduction. Individuals can *adapt* to some environmental changes, such as increasing red blood cell production at higher altitudes when oxygen concentrations are lower. Adaptation involves changing the *regulation* of the expression of certain genes. If a person living at high altitudes moved back to sea level, they would decrease the expression of erythropoietin (Chapters 10, 16, and 20), which causes the increased production of red blood cells. Adaptations can continually change in an individual. Evolution involves passing on different genes to offspring, such as blood type A versus blood type O. Every living bacterium, protist, fungus, plant, and animal has their own unique combination of genetic material that will prove to be beneficial or deleterious to their survival and subsequent ability to reproduce. Organisms that survive will pass on their DNA to their offspring. Recall from Chapters 5 and 6 that during **meiosis**, every gamete and, therefore, every zygote has a different combination of genes compared to their parents. Through sexual reproduction and **genetic variation** over many generations, new combinations of traits continually emerge within all populations (Figure 10.2).

Populations evolve by means of **natural selection**. Darwin believed that all species over-reproduce, and only a small percentage of those born will survive. Herbert Spencer, a philosopher and biologist,

coined the phrase "**survival of the fittest**" in 1864, after reading Darwin's *On the Origin of Species*. This term has persisted and is used frequently to describe the process of evolution: individuals born with the most beneficial traits for a *particular environment* will survive, reproduce, and pass on their DNA. The most obvious examples involve predator–prey relationships. The predators that are able to catch prey acquire the nutrients and good health for mating and reproducing and fighting off disease and competitors. Predators born with the DNA that gives them, for example, the ability to run fast, see in the dark, hear movement, or have a heightened olfactory sense for detecting prey will have an advantage over other individuals in that species and will more likely pass on their DNA to offspring (Figure 7.3). Likewise, prey that can evade predators, for example, by running quickly, or having a keen sense of hearing, or camouflage colouring will also be more likely to survive and pass on their DNA. Organisms born with genetic defects that prevent them

Figure 7.3 Predator–Prey Relationships

Predator–prey relationships act as selection pressures and cause natural selection to occur in both the predators and the prey.

from finding food, evading predators, competing for a mate, or being able to produce offspring will not pass on their DNA. In this way, the "fittest" organisms perpetuate certain genes in future generations.

Survival of the fittest does not necessarily mean that the biggest, fastest organisms will be the most "fit." In many circumstances and depending on the environmental conditions, it may be more beneficial to be small, or nocturnal, or have camouflage-coloured fur. For us, and depending on the environment we live in, having access to healthcare, or clean water, or an education may be more important factors than being able to run fast or fight off predators,.

Why don't organisms reach a point when they have the best combination of traits in relation to their environment and then just stay the same? Why does evolution continue to occur in all living things? Organisms live in continually changing circumstances, and **ecology**—the relationship between organisms and their environment—determines evolution. Since Earth first formed 4.5 billion years ago, conditions in every part of the planet have continued to change: such as changes in seasons, weather patterns, ice ages, forest fires, food availability, predators, and infectious organisms (Figure 7.4a). Each organism has different combinations of traits that are more or less beneficial in their particular environment. Therefore, depending on the type of environment change and the DNA of the organism, certain individuals are more likely to survive, reproduce, and pass on their DNA. This explains why genetic variation is important (Chapters 5 and 6).

Environmental changes do not necessarily impact large ecosystems. An environmental change can be as simple as putting an electric bug zapper in your backyard (Figure 7.4b). Most moths (and many other insects) have a gene that causes them to be drawn to light, but some have a mutation in this gene so that they are not drawn toward light (your bug zapper will kill only the moths that have this gene without this mutation). In this example, the environmental change is the addition of the bug zapper, and the fittest moths in your backyard environment have the mutated version of the light-drawing gene; they are not attracted to light and therefore will survive.

Figure 7.4 **Environmental Change**
Environment changes can be large (A) or small (B).

Darwin's theory of natural selection

Darwin's idea of how organisms change is based on the following steps:

1. All individuals have a unique combination of the genetic traits they acquired from their parents. All populations have genetic variation.

2. The organisms that survive pass on their DNA to their offspring, so closely related individuals have more similar DNA than do less closely related individuals.

3. All living organisms over-reproduce, and only some of those offspring will survive, depending on their DNA and the current environmental conditions (Figure 7.5).

4. The organisms that do survive and pass on their DNA provide the basis for future generations to also be able to survive.

5. Over time, populations of organisms change because of new combinations of traits in conjunction with changes in the environment.

Figure 7.5 **Over-Reproduction**
All organisms over-reproduce, which ensures that some will survive any given environment.

DID YOU KNOW?

Insects have a gene that draws them toward light because it is beneficial for them to be able to navigate at night when the moon is out. The light from the moon helps insects travel in straight directions. Since we use artificial lights, insects get trapped when they fly around light bulbs at night, instead of following the moonlight.

CONCEPT REVIEW 7.1

1. Define evolution.

2. What does "survival of the fittest" mean? Give an example.

3. What is the relationship between evolution and ecology?

4. Explain the major factors involved in natural selection.

5. Give some examples of environmental factors that affect evolution.

6. Give examples of some human inherited traits that you think would be beneficial for survival today.

7.2 MICROEVOLUTION

Microevolution is defined as the change in inherited traits in a population of organisms within a single species over time. For example, putting a bug zapper in your backyard will change the insect populations so that more individuals will carry the mutated gene and will not be drawn toward light. At the same time, other environmental changes, for example, in the weather, the seasons, maybe a drought, or infection will affect the evolution of other traits in this insect population. Over very long periods of time many small changes can lead to large-scale changes that may give rise to a new species; this is macroevolution. A **species** is a group of organisms that can breed only with each other and produce viable offspring. Therefore, if many small changes in moth DNA occur over time so that they are no longer able to breed with the previous group of moths, they would be considered a new species of moth. There are currently 160,000 moth species.

had to continually stretch to reach food, which caused those individual giraffes to acquire slightly longer necks during their lifetime. Then, the offspring of those giraffes would be born with longer necks than the offspring of giraffes that didn't stretch to reach food: leading to the long-necked giraffes of today.

DID YOU KNOW?

Giraffes are the tallest living terrestrial mammal and can be 6 m tall (20 ft) and weigh 1600 kg (3500 lb.)

DID YOU KNOW?

There are over 400,000 species of beetle, 5000 species of frog, and 400 species of primate. Primates first existed approximately 65 million years ago.

Darwin is not the only person who questioned where organisms come from and why so much diversity exists among organisms. The French scientist **Jean Baptiste de Lamarck** (Figure 7.6) presented the first ideas of evolution in 1809. The fossil record shows that today's giraffes are much taller than their ancestors, indicating that microevolution has occurred in the giraffe population over time. Darwin and Lamarck had different ideas about why these changes occurred.

Lamarck believed that organisms inherited characteristics that were *acquired* by their parents during their lifetime. Figure 7.7 illustrates Lamarck's view that giraffes

Figure 7.6 Jean Baptiste de Lamarck

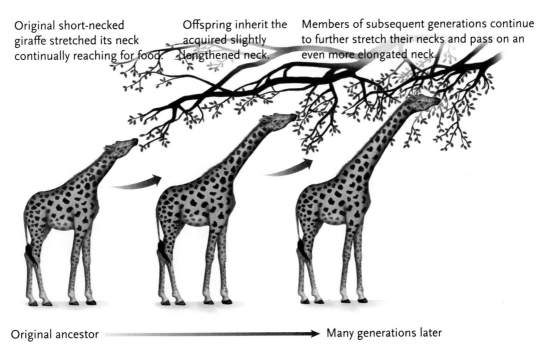

Original short-necked giraffe stretched its neck continually reaching for food.

Offspring inherit the acquired slightly lengthened neck.

Members of subsequent generations continue to further stretch their necks and pass on an even more elongated neck.

Original ancestor ————————————————————→ Many generations later

Figure 7.7   Lamarck's Giraffes
Lamarck believed that giraffes evolved to have longer necks than their ancestors because they had to continually stretch to reach higher sources of food.

Art source: From DIGIUSEPPE/FRASER. Biology 11U. © 2011 Nelson Education Ltd. Reproduced by permission. www.cengage.com/permissions

Darwin believed that giraffes are taller now because of natural selection. Several genes are involved in determining the height of a giraffe: similar to the multiple genes that determine height in humans (polygenic trait, Chapter 6). Genetic variation among giraffes allows for a range of heights within their population at any time. In ancestral giraffes, the tall giraffes were not as tall as they are now, but some of them were slightly taller than others at that time. These giraffes could reach more food higher in the trees, compared to the shorter giraffes that perhaps had to compete with other animals that ate the same food. If food supply was a continual

DID YOU KNOW?

Lamarck wasn't as far off as it may seem. Current evidence shows that behavioural changes in many organisms, including humans, can affect the regulation of genes in their children; this is called *epigenetics*. For example, rats that are not licked by their mother at birth will have an impaired stress response when they are older, not because of a change in their DNA sequences, but because of a change in how those stress response genes are expressed.

selection pressure, over many years the tallest giraffes were the most likely to survive and pass on their "tall" genes to their offspring, leading to the very tall giraffes that exist today.

Importance of microevolution to humans

Microevolution occurs all the time in our ecosystems, and we interact with and influence many species of bacteria, fungi, plants, and animals. Since bacteria have lived everywhere on earth for over 3.5 billion years, humans co-evolved with bacteria, and 95% of bacteria are beneficial to humans. There are many reasons why we could not survive without them (Chapters 19 and 23).

An example of microevolution in bacteria involves **antibiotic resistance**: an extremely important health care concern. Bacteria can reproduce every 30 minutes in perfect conditions and, therefore, have a very rapid generation time. Bacteria undergo genetic changes based on changes in the environment, such as lack of a nutrient, or change in temperature or level of moisture. Since penicillin was discovered in 1928, bacterial microevolution has occurred at a rapid rate. And since that time, many beneficial **antibiotics** have been created to fight a myriad of bacterial infections in people and animals. However, antibiotics have been overused.

Antibiotics kill most, but not all, bacteria in any given population, making antibiotics a **selection pressure** that affects the microevolution of bacteria. Some bacteria have antibiotic **resistance** genes that allow them to survive in the presence of antibiotics; these bacteria are the fittest in an environment where antibiotics occur (Figure 7.8). Through natural selection, the fittest bacteria survive the antibiotic environment and pass on the resistant gene(s) to their bacterial offspring. Over the

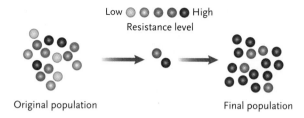

Low ⬤ ⬤ ⬤ ⬤ ⬤ High
Resistance level

Original population   Final population

Figure 7.8   Example of Microevolution

Bacteria that have antibiotic resistance genes will survive in an environment where antibiotics are present.

Figure 7.9   Artificial Selection—Dogs

Humans have bred many varieties of dogs based on traits that were appealing. When humans determine which organisms should reproduce, this is called artificial selection.

last few decades, more and more bacterial species that cause infection in humans and animals have become resistant to one or more antibiotic drugs. These bacterial populations have undergone microevolution in a relatively short period of time. Resistance genes have existed in bacterial populations for millions of years as a defence mechanism against fungus and other species that produce natural antibiotic molecules. Now that scientists have figured out how to isolate these molecules for medical use, the large amounts of antibiotics consumed by people and farm animals has brought about an increase in resistant bacterial populations. This happens because the non-resistant bacteria die off,

DID YOU KNOW?

In 1928 Alexander Fleming isolated the first antibiotic ever used, penicillin, from the common indoor fungus found on salted food products, *Penicillium chrysogenum*.

but the resistant bacteria survive. This may become an extremely critical health care issue in the coming years. There are already several multi-resistant bacterial species that can cause fatal infections in humans: "super bugs" (Chapter 23).

Artificial selection

People have an impact on the microevolution of many species, specifically domesticated animals and food plants. Human influence is a selection pressure that generates a variety of new species from the same ancestral species. As a result of this selection pressure, microevolution leads to genetic variation that in the long term leads to macroevolution.

For example, all dogs from Chihuahuas to Great Danes to the gray wolf are the same species, *Canis lupus*. Domesticated pet dogs of all breeds are considered a subspecies of the gray wolf, *Canis lupus familiaris*. Being part of the same species means that organisms can interbreed and produce viable, fertile offspring. The current enormous variation in breeds

of dog is much more extreme than would ever naturally occur. This variation is due to **artificial selection**, whereby breeders choose which traits are most desirable and then breed those traits: either in animals or in plants (Figure 7.9).

Humans began growing and artificially selecting food crops and domesticating animals approximately 8500 BC, the beginning of the agricultural revolution. Plants or animals with desirable characteristics were selected for reproduction, which caused rapid microevolutionary changes. Some examples of selected traits include cows that produce more milk; sheep with more wool; chickens that lay more eggs; bigger, stronger work animals; pigs that produce the most meat; seedless grapes or bananas, which can only be produced with human intervention. Also, broccoli, cauliflower, brussels sprouts, cabbage, kale, and collard greens are all variations of the ancestral species *Brassica oleracea* (Figure 7.10). These vegetables exist because of the artificial selection of traits such as taste, appearance, colour, texture, size, and also type of leaf and, flower. Although all these vegetables came from the same ancestral species, this variety is due to human influence on genetic variation, and microevolution leads to macroevolution (see Section 7.3).

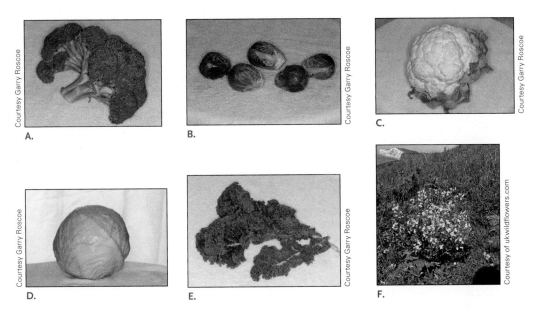

Figure 7.10 Artificial Selection—Vegetables

All our food plants have been artificially selected over thousands of years to give us the variety that we eat today. (A) broccoli, (B) Brussels sprouts, (C) cauliflower, (D) cabbage, (E) kale, and (F) ancestral wild mustard species.

CONCEPT REVIEW 7.2

1. Using bacteria and antibiotic resistance as an example, explain how microevolution occurs.

2. How do antibiotic-resistant bacteria become a selection pressure that affects humans?

3. What is the difference between microevolution and macroevolution?

4. How would Darwin and Lamarck each explain why modern-day giraffes have longer necks than ancestral giraffes?

5. How is natural selection different from artificial selection?

7.3 MACROEVOLUTION

Macroevolution refers to the production of new species that occurs through thousands or millions of years of microevolution.

Darwin's finches

When Darwin travelled between South America and the Galapagos Islands, he observed the similarities and differences of several species of finch. An important difference between the species is the shape and size of the beak, which is highly adapted to different types of food in South America and the Galapagos. The finches in the two geographical locations are different enough to make them different species; however, they also have a number of similarities even though they live hundreds of kilometres apart. Darwin hypothesized that microevolution led to macroevolution, and this created all the finch species that exist today.

Two genes play a major role in the development of the shape and size of the beak: bone morphogen-etic protein 4 (BMP4) and calmodulin. Changes in the timing and amount of expression of these genes can account for variations in beak shape and size in the population (Figure 7.11). Depending on the environment and food supply, certain finches will be more fit than others to eat certain foods: such as large, hard seeds; small seeds in small cracks; berries; worms; or insects. The theory of macroevolution is based on the idea that small changes in genes or gene expression over time—that is, microevolution— cause great changes over long periods of time, and this leads to the formation of new species.

Rate of evolution

The general view is that evolution occurs over very long periods of time, perhaps millions of years. In different species, and depending on the environmental factors involved, macroevolution occurs at different rates. Two key theories explain the rate of evolution: gradualism, and punctuated equilibrium. **Gradualism** is a slow and gradual process whereby very small changes occur over

Geospiza conirostris

Birds with long bills open cactus fruits to feed on the fleshy pulp.

Birds with intermediate bills may be favoured during nondrought years when many types of food are available.

Birds with deep bills strip bark from trees to locate insects.

© Krystyna Szulecka/Alamy

Figure 7.11 Macroevolution—Darwin's Finches

Darwin studied the geographical distribution of finches that helped lead him to his ideas about macroevolution.

Art source: From RUSSELL/HERTZ/STARR/FENTON. Biology, 2E. © 2013 Nelson Education Ltd. Reproduced by permission. www.cengage.com/permissions.

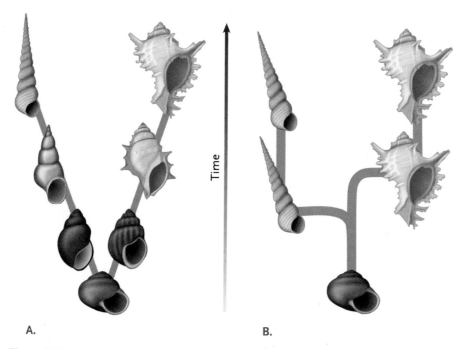

Time

A.

B.

Figure 7.12 Gradualism and Punctuated Equilibrium

Two theories that relate to the rate of evolution are *gradualism* and *punctuated equilibrium*.

Art source: From DIGIUSEPPE/FRASER. Biology 11U. © 2011 Nelson Education Ltd. Reproduced by permission. www.cengage.com/permissions

millions of years. **Punctuated equilibrium** is a process whereby periods of no change are followed by periods of rapid change, as can be seen in the fossil record (Figure 7.12).

Some organisms hardly change over millions of years; for example, alligators have few noticeable changes throughout their existence of at least 200 million years (Figure 7.13).

Evidence for macroevolution

Three main lines of evidence suggest that existing species came from a common ancestor. Fossils, homologous anatomy, and homologous molecules each show evidence of relationship among all living things.

© Bmoore81/Dreamstime.com

Figure 7.13 Alligator

Alligators have not changed significantly over 200 million years.

Fossils

Fossils are considered the most direct evidence of macroevolution. Any part of an organism that contains minerals cannot be degraded by microorganisms and can form rock; this includes bones, feathers, hooves, scales, horns, teeth, and shells. Also, an organism that is encased in tree sap (amber) can also be preserved: this idea was used in the film *Jurassic Park*. Fossilization is rare and occurs only under certain circumstances, which is why the fossil record is incomplete. Fossils show similarities between organisms over time.

Using the example of primate skulls, Figure 7.14 shows a range of skulls from modern chimpanzees to primitive human primates that existed millions of years ago and to current modern human skulls. Viewing the skull shape you can see a trend in skull change that could possibly indicate that modern humans came from an ancestor in common with the modern chimpanzee. Figure 7.15 shows some other famous fossil examples.

Figure 7.16 shows the approximate order of the existence of primates. Figure 7.17 shows the general time periods when some of the main hominid species existed on earth. Note that *Homo sapiens* has been on earth for only 200,000 years and that includes the time period of the earlier Cro-Magnons. "Modern humans" with our current anatomical structures, have been on earth for approximately 50,000 years. Also note that Neanderthals and *Homo sapiens* co-existed for a few hundred thousand years; Neanderthals became extinct around the end of the last ice age, 20,000 years ago, and now we represent the remaining species of the *Homo* genus.

Homologous anatomy

Recall from Chapter 5 that there are two copies of each chromosome, one from each parent. These chromosomes are very similar but not exactly the same because different alleles come from each parent;

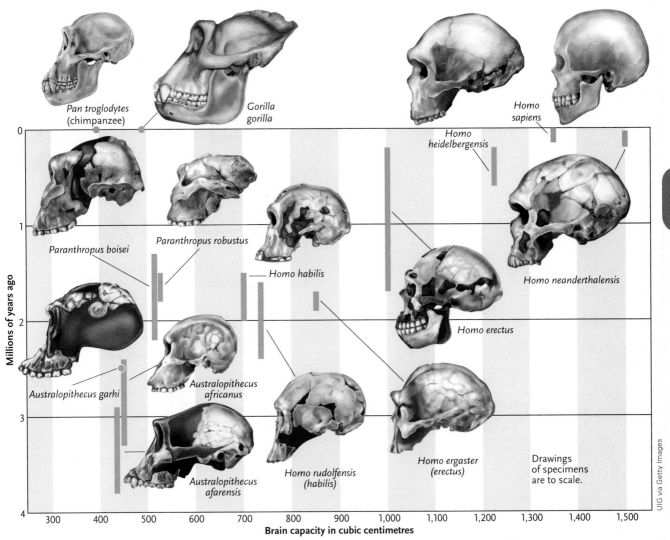

Figure 7.14   Fossil Hominid Skulls

Figure 7.15   Examples of Fossils

to show that similarities indicate a common ancestor. Numerous similarities between different organisms appear at various times during embryonic development. For example, all vertebrate embryos have **pharyngeal pouches**. These differentiate into gills in fish but disappear in mammals at a later stage in their embryonic development. All vertebrate embryos have a bony tail, which disappears in many mammals. Humans even have fur—called **lanugo**—at certain stages of embryonic development. Comparing the anatomical structures of a variety of vertebrates, we can see that most have the same basic bone structure, but with some differences in size and shape. A very few genes have modified basic structures, as in Darwin's example of finch beaks. Figure 7.18 shows examples of homologous structures in numerous vertebrate animals. Humans, cats, whales, bats, horses, and many other species all have a humerus, radius, ulna, carpals, and phalanges, they all have the same bones but with modifications.

therefore, these chromosomes are called *homologous*. Evolutionists use the term **homologous structures** to compare anatomy between different organisms and

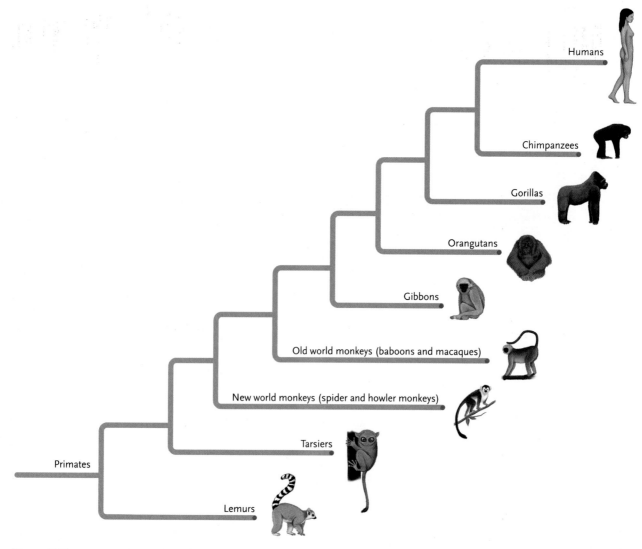

Figure 7.16   Approximate Order of the Establishment of Various Primate Species

Art source: From DIGIUSEPPE/FRASER. Biology 11U. © 2011 Nelson Education Ltd. Reproduced by permission. www.cengage.com/permissions

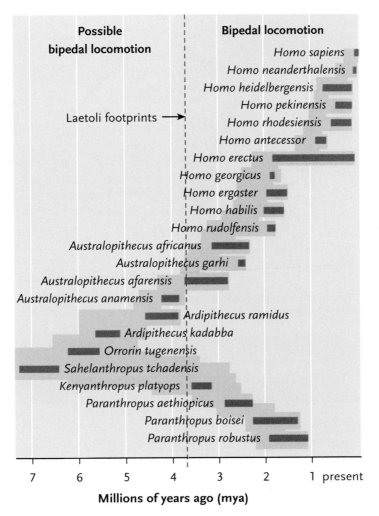

Figure 7.17 Human Evolution Timeline

Art source: From DIGIUSEPPE/FRASER. Biology 11U. © 2011 Nelson Education Ltd. Reproduced by permission. www.cengage.com/permissions

Homologous molecules

To prove that macroevolution occurred, caused the formation of all species on earth today, and continues to occur, researchers use the evidence that the primary molecules of all living things are homologous. Every organism contains DNA, which codes for RNA and that codes for amino acid sequences to produce functional proteins (Chapter 2). Every living thing has the same four nucleotides arranged into genes. The difference in nucleotide sequences causes different proteins to be produced. In the human body, different arrangements of nucleotides code for insulin, or collagen, or hemoglobin, or any of the 100,000 different proteins that humans produce. In finches, birds with bigger beaks have increased expression of the gene responsible for bone morphogenetic protein 4 (BMP4).

Many common genes exist in many organisms, and the closer relationship between species the more similarities exist. Scientists use mice to research many

human diseases because mice possess many of the same genes and proteins that humans do; the genes and proteins might be slightly different but have the same function. For example, mice and humans both have a gene that codes for a cytochrome C protein that is required for the production of ATP. Figure 7.19 shows the homologous DNA sequence for the beginning of the cytochrome C gene in mice and humans. The sequences are almost identical, but there are some small differences. However, a comparison between the chimpanzee cytochrome C gene and that of humans shows more similarity than the comparison between the human and the mouse cytochrome C gene sequences. The more similar the species are to each other, the more similar are their anatomy and their DNA and protein molecules. Today, with DNA sequencing technology, it is possible to compare the genes of many organisms, and this may help to more accurately determine the relationships and commonalities between organisms.

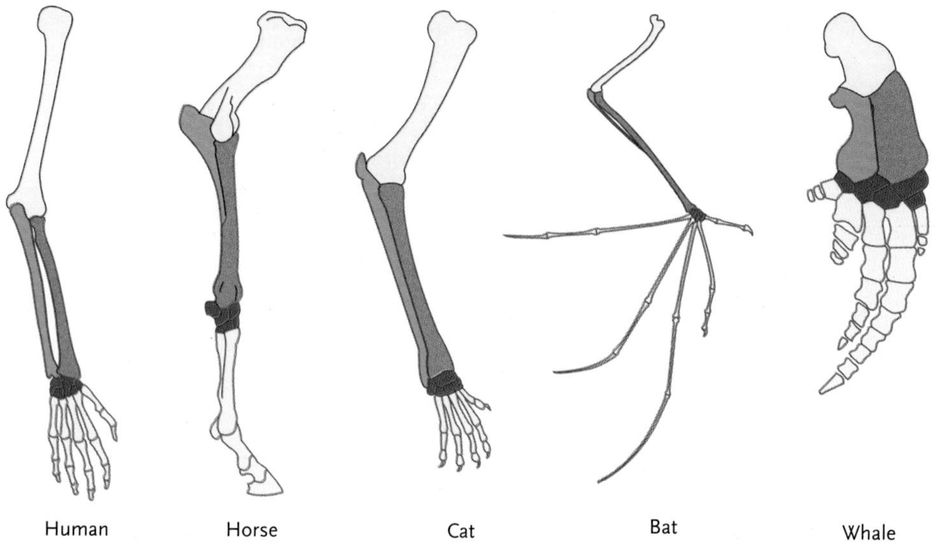

| Human | Horse | Cat | Bat | Whale |

Figure 7.18 Examples of Homologous Structures

Art source: From DIGIUSEPPE/FRASER. Biology 11U. © 2011 Nelson Education Ltd. Reproduced by permission, www.cengage.com/permissions.

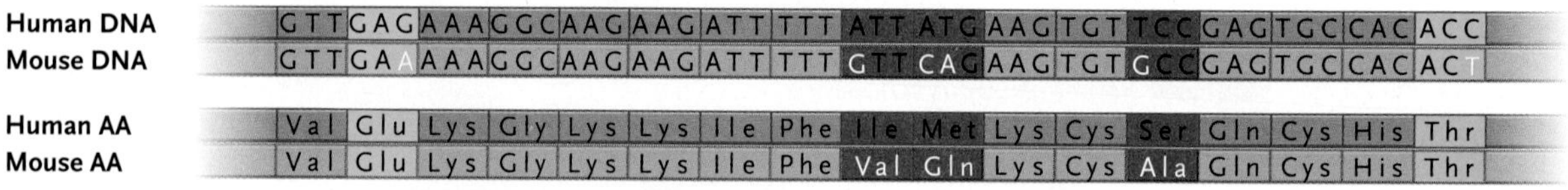

Human DNA	GTT	GAG	AAA	GGC	AAG	AAG	ATT	TTT	ATT	ATG	AAG	TGT	TCC	GAG	TGC	CAC	ACC
Mouse DNA	GTT	GAA	AAA	GGC	AAG	AAG	ATT	TTT	GTT	CAG	AAG	TGT	GCC	GAG	TGC	CAC	ACT

Human AA	Val	Glu	Lys	Gly	Lys	Lys	Ile	Phe	Ile	Met	Lys	Cys	Ser	Gln	Cys	His	Thr
Mouse AA	Val	Glu	Lys	Gly	Lys	Lys	Ile	Phe	Val	Gln	Lys	Cys	Ala	Gln	Cys	His	Thr

Figure 7.19 Homologous Molecules

A comparison of a human and mouse gene sequence shows homologous molecules.

CONCEPT REVIEW 7.3

1. Using Darwin's finches as an example, explain how macroevolution occurs.

2. What is the difference between gradualism and punctuated equilibrium?

3. Name the three main methods used as evidence of macroevolution.

4. Humans are more closely related to bats than to birds. How can anatomical structures or molecular sequences be used to determine if this is true or not?

5. What is a selection pressure? Give an example of a selection pressure that affects bacteria.

7.4 POPULATION GENETICS

Population genetics is the study of the frequency of **alleles** in a population. Every organism has at least two copies of every gene, one from each parent—although plants can sometimes have more than two copies (Chapter 6). Each copy of the gene is not necessarily the same; for example, a pea plant can have a purple flower allele and a white flower allele, or it can have two alleles for the same colour of flower. When the two alleles are the same, the genotype is **homozygous,** and when they are different, the genotype is **heterozygous** (Chapter 6). Remember that homozygous or heterozygous refers to the pea plant's **genotype**, and having the trait of purple or white flowers describes the **phenotype**.

Within a population as a whole, there are a certain number of dominant and recessive alleles, and this is known as the **gene pool**. It is sometimes assumed that dominant traits will eventually predominate in a population and recessive traits will disappear, but this does not happen in a stable, non-evolving population. Two scientists, Hardy and Weinberg studied allele frequencies in stable populations and determined that the amount of dominant and recessive alleles stays the same over several generations. For a population to be considered stable or in equilibrium, the following assumptions apply:

1. The population has to be large. Small populations can have random changes in **allele frequency**.

2. Mating has to be random. The population will not be stable if certain phenotypes determine mating behaviour and evolution occurs.

3. There are no DNA mutations. Mutations can create new alleles that alter the gene pool.

4. There is no migration into or out of the population. If alleles are entering or leaving the population the allele frequencies change and evolution occurs.

5. There is no natural selection, that is, no selection pressure on a specific trait.

6. There are no environmental changes that cause random changes in allele frequencies, which is called genetic drift.

The **Hardy-Weinberg equilibrium** refers to populations where the above assumptions are true—the population is stable, not evolving, and allele frequencies stay the same over successive generations. The following example illustrates what can be determined about a given population based on the Hardy-Weinberg equilibrium.

Example 1

Suppose 100 mice live in a field, and they have black or white fur. In a stable population, fur colour does not influence mating, no mice are entering or leaving the field, there are no DNA mutations, no selection pressures such as predators, and no environmental changes. If black fur is dominant, the mice with two black fur alleles (homozygous dominant) and the heterozygous mice will have black fur; only homozygous recessive mice will have white fur. How many total fur colour alleles exist in that field of mice? If there are 100 mice and each mouse has two alleles, then 200 fur colour alleles exist in that population. If we suppose that exactly half of those alleles code for white fur and half for black fur, how many mice in that population would have white fur? Did you guess 50%? Did you guess 25%? If you guessed 25% you would be correct. This is because heterozygous mice have a white fur allele, but they have a black fur phenotype. This is explained with the following two equations.

The letters p and q represent the dominant and recessive alleles:

p = dominant allele, black fur gene

q = recessive allele, white fur gene

Every mouse has two copies of every gene. Table 7.1 shows the possible genotypes and phenotypes in this population of mice

In the total mouse population, the sum of the dominant and recessive alleles is 100% of the population

TABLE 7.1

Genotypes and Phenotypes in Population of Mice (Example 1)

Genotype	Phenotype
pp or p^2 (homozygous dominant)	black fur
pq (heterozygous)	black fur
qq or q^2 (homozygous recessive)	white fur

(assuming that the trait is dominant-recessive inheritance only).

p + q = 1 (all the dominant alleles + all the recessive alleles = 100%)

In the total mouse population, the sum of all the genotypes is 100% of the population.

$p^2 + 2pq + q^2 = 1$ (homozygous dominant + heterozygous + homozygous recessive = 1)

Continuing with the fur colour example, if half the alleles in the mouse population are black, p = 0.5 and, therefore, q = 0.5.

p + q = 1

0.5 + 0.5 = 1

How many mice are white?

$p^2 + 2pq + q^2 = 1$

$(0.5)^2 + 2(0.5)(0.5) + (0.5)^2 = 1$

0.25 + 0.5 + 0.25 = 1

Since q^2 represents the white mice, and q^2 equals 0.25, then 25% of the mouse population is white. According to the Hardy-Weinberg equilibrium, the frequency of alleles would not be changing over time; therefore, the mouse population is not evolving.

Example 2

What is the expected number of mice with black fur in a population where the black fur allele frequency is 0.6 (60% of all alleles in the population are for black fur) and black fur is dominant over white fur?

We know that p + q = 1, and since black fur is dominant, p must equal 0.6 and, therefore, q must equal 0.4. We also know that the possible genotypes for mice with black fur are homozygous dominant (p^2) *and* heterozygous (2pq). Therefore, the number of black mice in this population would be equal to $p^2 + 2pq$, which is 0.36 + 0.48 = 0.84. Therefore, 84% of this population of mice will have black fur, and 16% will have white fur.

Example 3

Determine the white fur allele frequency in a population of mice if 9% of the population has white fur?

We know that $p^2 + 2pq + q^2 = 1$, and q^2 represents homozygous recessive white mice, which is 9% of the population, so $q^2 = 0.09$. In order to determine the white fur allele frequency, we need to solve for q:

$q^2 = 0.09$

q = square root of 0.09

q = 0.3, or 30%

Therefore, 30% of the alleles in this population are the white-fur allele.

DID YOU KNOW?

Cystic fibrosis is a genetically inherited mutation in the gene that codes for cell membrane chloride ion channels. When this gene is mutated the protein channel does not form properly and causes disease in individuals that are homozygous recessive for the mutation. In North America, the frequency of the mutant allele is 0.025, which means that 2.5% of the alleles for cell membrane chloride channels are mutated, and 97.5% of those alleles are normal.

Factors that affect evolution of populations

Hardy-Weinberg equilibrium is not the usual situation in any given population at any time. Several factors are involved in altering allele frequencies over time: natural and artificial selection, DNA mutations, migration, non-random mating, and genetic drift.

Types of selection

Artificial selection occurs when people choose the traits in an animal or plant that they deem desirable: such as good-natured dogs, fast race horses, large tasty tomatoes. Natural selection refers to the process where the organisms that are most fit for an environment survive and reproduce and pass on their DNA to their offspring. There are three types natural selection: stabilizing, disruptive, and directional (Figure 7.20).

Stabilizing selection occurs when the extremes of a phenotype are eliminated from a population. It is impossible to say for certain that a specific individual will have a better or worse chance of survival because many factors play a role; however, some traits that on

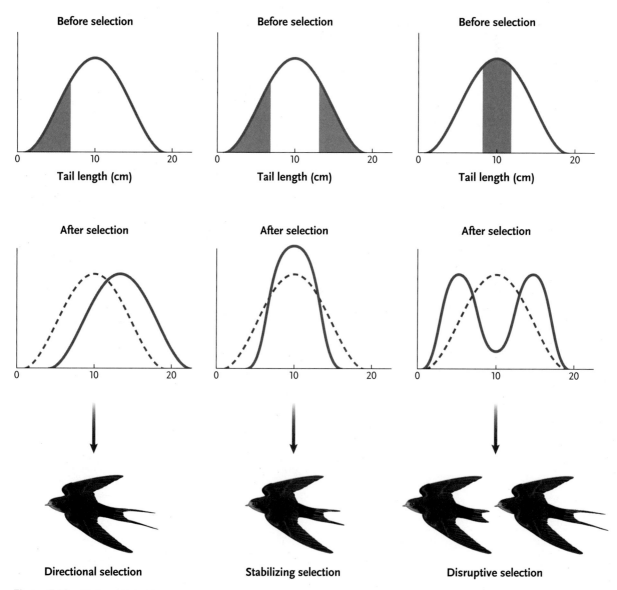

Figure 7.20 Natural Selection

Natural selection can be divided into three categories: stabilizing, disruptive, and directional. The dotted red line indicates the original population, and the blue solid line shows the change in the population after the particular selection pressure has occurred.

average benefit survival in a whole population can usually be predicted. In stabilizing selection, genetic variation decreases as one specific trait is selected for. Human birth weight provides an example: a normal distribution of birth weights varies from very small to very large, with most babies being somewhere in the middle. Very small babies are less likely to survive because they may be underdeveloped, less likely to feed properly, and lose body heat easily. Very large babies are less likely to survive the birthing process. Babies born of intermediate size are the most likely to survive and pass on their intermediate-size genes.

Disruptive selection, the opposite of stabilizing selection, favours the extremes of a trait, and the intermediate trait is eliminated. This type of selection forms two distinct groups within the population; it may be responsible for the early changes involved in macroevolution where, eventually, the two groups become different species. For example, the Galapagos finches that Darwin studied had a range of beak sizes and shapes in different environments, and these were adapted for different types of food. Darwin believed the finches came from a common ancestor population. Perhaps in one environment the food supply was such that birds could survive if they had very large beaks to eat large hard seeds or if they had very small beaks that could reach food in small cracks. The birds with intermediate beak sizes would have been less adapted to find food in that environment, leading to a population of finches with two distinct beak sizes.

Directional selection favours one extreme trait in a population so that the opposite extreme is eliminated. Recall the example from Section 7.2. In an environment with many backyard electric bug zappers, insects with a mutation in the gene that draws them toward light are more fit than insects without this mutation. Eventually, the population of insects in that area will be altered so that the allele frequency of the mutant gene will be more prevalent. Another example is antibiotic resistance in bacterial populations. In a population of bacteria that has a range of resistance to antibiotics, only the most resistant bacteria will survive during exposure to those drugs, leading to future generations with greatly increased expression of antibiotic resistance traits (Figure 7.8).

Sickle cell anemia is an example of an inherited disease. It is caused by a mutation in the hemoglobin gene so that the hemoglobin protein doesn't fold properly and is much less efficient at carrying oxygen (Chapter 6) (Figure 7.21). Individuals need both mutated alleles to have the disease. If people that have sickle cell anemia have a shortened life span, why is the sickle cell mutation still prevalent in certain populations? The possible genotypes in a population are as follows:

- homozygous dominant—two healthy hemoglobin alleles, so no sickle cell anemia occurs

Figure 7.21 Blue-Footed Booby

The blue-footed booby has blue feet and a special dance that it uses to attract females.

- heterozygous—one healthy allele and one mutated allele, so no sickle cell anemia occurs
- homozygous recessive—two mutated alleles, so sickle cell anemia occurs

Recall that individuals considered more or less fit are the ones that survive to reproduce and pass on their DNA to their offspring, and this depends on the environment they live in. In Africa, the sickle cell anemia allele is much more prevalent than in North America because certain environmental factors play a role in each geographical location. In North America, the sickle cell allele has no benefit, so homozygous recessive individuals have a shorter life span, and directional selection occurs so that people without the disease are more fit, and the allele continues to decrease in the population. However, in Africa, the sickle cell allele continues to be prevalent in the population because heterozygous individuals are less likely to die from malaria. Malaria is a *Plasmodium* parasite that infects red blood cells and it is a leading cause of death in Africa. In Africa, homozygous dominant individuals without a sickle cell allele are more likely to die from malaria, and homozygous recessive individuals are more likely to die from sickle cell anemia. Therefore, the heterozygous individuals with one healthy allele and one mutated allele are the most "fit" in that environment because they do not have the sickle cell symptoms and they are less likely to die from malaria. This stabilizing selection in Africa acts on the sickle cell allele; this is also referred to as the **heterozygote advantage**. In central Africa the frequency of the sickle cell allele is 0.12 (12%) compared to 0.1 (1%) in North America.

DNA mutations

Most DNA mutations are not beneficial, but they are considered the ultimate source of genetic variation in every population. All living organisms have genetic material made up of specific sequences of nucleotides that form genes that code for proteins (Chapter 2). Some DNA mutations have no effect on the structure or function of the protein (Chapter 8), but some mutations cause disease (Chapter 6), such as the mutation in the hemoglobin gene that causes sickle cell anemia, or the mutation in the chloride channel gene that causes cystic fibrosis. It is thought that some DNA mutations can cause the formation of a new allele that is a new protein that does not cause disease. An example is the evolution of the human blood cell types. Blood type A and B have existed throughout the existence of primate species, and continue to exist in humans. When parents of each blood type have children, sometimes that child has an A and a B allele, giving them blood type AB (Chapter 6). It is thought that type O came about later than type A and B as a mutation in the type A allele. So a DNA mutation led to the formation of a new allele and therefore a new phenotype that is not deleterious. It is also important to note that any DNA mutations that occur in somatic cells during an individual's life—for example, due to chemical exposure or any kind of mutagen—will not be passed onto their offspring unless it is also found in the gametes.

DID YOU KNOW?

Chimpanzees have only blood type A, and gorillas have only blood type B. Some researchers believe that chimpanzees and Cro-Magnons have a common ancestor that gave rise to Caucasians, and that gorillas and Neanderthals have a common ancestor that gave rise to Asians and Africans. Interestingly, more Caucasians have blood type A, and more Asians and Africans have blood type B. Type O blood is more common among First Nations.

Migration

The movement of individuals to new locations is known as migration. This movement affects the allele frequencies of many populations. **Gene flow** is the transfer of alleles from one population to another, which occurs when groups of people or animals migrate and join other populations. Migration can also increase genetic variation in a population. For example, now with increased ease of transportation around the globe, people migrate to different countries to live and have families. Consider a simple phenotype, skin colour: increased migration of people from different countries has dramatically increased the genetic variation and frequency of various skin colour alleles in North America.

Non-random mating

Random mating occurs when there are no preferred genes for reproduction: for example, pollination. Flowering plants pollenate any nearby flowers that are the same species. This is almost never the case in the animal kingdom, where genotypes are usually highly linked to reproduction, such as dancing rituals (blue-footed booby, Figure 7.21); strength (many animal males fight for a female); bird songs; feather colour; pheromone scent. Think of some genes that may be involved in human mating.

Genetic drift

Genetic drift is the process of random changes in allele frequencies that are not due to an organism being fit: for example, as the impact of major environmental changes such as a hurricane that kills many organisms in a population. This is also called the **bottleneck effect**. The individuals that survive and then continue to pass on their DNA may not necessarily be the ones with the most beneficial traits for the next generation. Also, the small population that is left may not be representative of the original population (Figure 7.22).

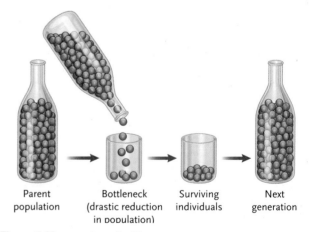

Parent population | Bottleneck (drastic reduction in population) | Surviving individuals | Next generation

Figure 7.22 **Bottleneck Effect**

Environmental changes can cause a bottleneck effect, which alters the frequency of alleles in a population when many organisms in that population die.

Art source: From DIGIUSEPPE/FRASER. Biology 11U. (c) 2011 Nelson Education Ltd. Reproduced by permission. www.cengage.com/permissions

Example of microevolution possibly leading to macroevolution

Guppies are small brightly coloured and spotted fish often kept in aquariums because of their attractive colouring. Depending on their environment in the wild, populations of guppies show altered phenotypes over short periods of time. Microevolution involves the change in allele frequencies within a population over time, where organisms with the most beneficial traits for that environment survive and reproduce, passing along their alleles to their offspring. In one study, a guppy population was divided so that one group remained in a predator-free environment and the other group existed with a predator, such as pike. As a result, the allele frequencies and phenotypes of those populations changed within six months to a year. The types of changes observed for both populations are as follows:

1. Guppies in the predator-free environment
   - Brightly coloured and spotted males attracted females and were more likely to reproduce, so the number of guppies with spots and bright colours increased.
   - Reproduction age was slightly later than reproductive age of guppies living with predators.
   - Swimming speed was lower.

2. Guppies in the high-predation environment
   - Males with fewer spots and camouflage colouring were less likely to get eaten by the predator, so they survived to reproduce.
   - Due to predation stress, age of reproduction decreased, and body size was smaller.
   - Swimming speed increased.

The guppy population changed over time in the environment with the predator through natural selection; it changed from a brightly coloured and spotted phenotype to a camouflage phenotype. This indicates directional selection because the guppies with camouflage (one extreme) were more fit than the brightly coloured ones (the other extreme), or the intermediates. Guppies in the non-predation environment propagated the brightly coloured phenotype because females preferred those males for mating: that is, non-random mating.

If these guppy populations remained separate and DNA mutations or migration and mixing of alleles from other populations occurred, and they had to continually adapt to different environments—such as the addition of more predators, or disease, or different food availability—eventually these populations would be so different that they would not be able to breed with each other and would be considered different species: an example of macroevolution.

CONCEPT REVIEW 7.4

1. What factors maintain Hardy-Weinberg equilibrium?

2. Consider the following: Hardy-Weinberg equilibrium occurs in a population of rats, and long-tailed rats are dominant over short-tailed rats, and the dominant allele is present in the population at a frequency of 0.5. How many short-tailed rats are in the population? How many long-tailed rats are in the population?

3. What factors cause allele frequency changes in populations?

4. What are three types of natural selection that can affect allele frequencies? Give an example of each.

5. What type of selection is occurring with the sickle cell anemia alleles in Africa?

6. What is a heterozygote advantage?

7. What is the selection pressure acting on the guppies?

8. Explain why the bottleneck effect does not select for the fittest individuals.

9. What is the source of new alleles in a population?

8 DNA Structure and Function

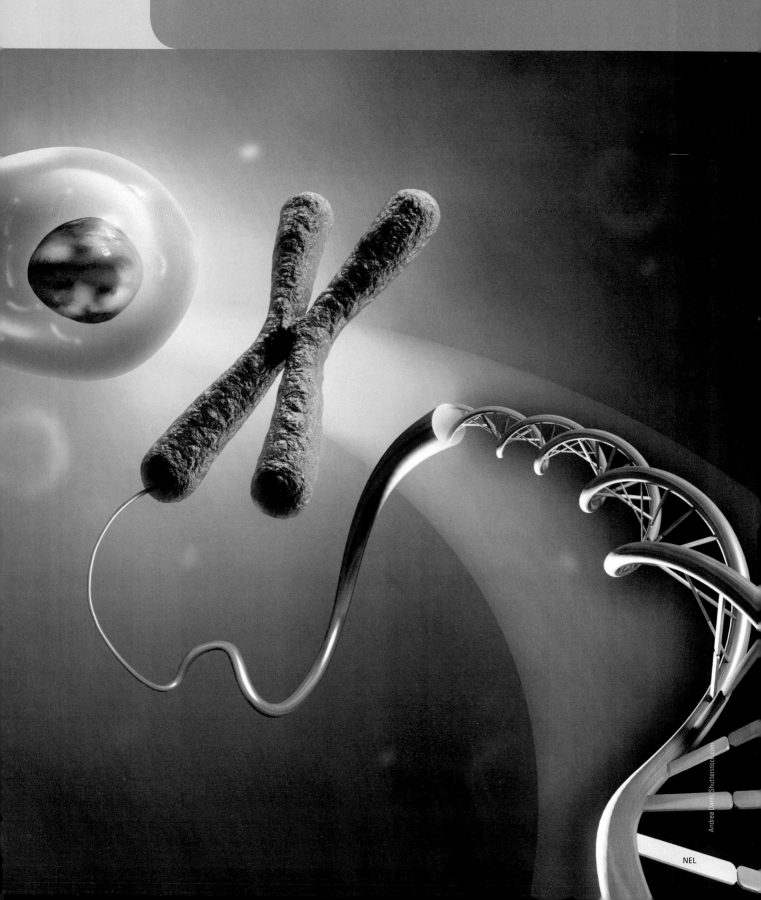

> *Learning is not attained by chance, it must be sought for with ardor and diligence.*
>
> Abigail Adams

8.1 DNA STRUCTURE

Composition of DNA

DNA is a macromolecule—a polymer—made up of building blocks—monomers— called **nucleotides** (Chapter 2). Every nucleotide consists of a phosphate molecule, a five-carbon sugar, and a nitrogenous base. DNA is made up of only four different nucleotides: adenine (A), guanine (G), thymine (T), and cytosine (C). A fifth nucleotide, called uracil, is found only in RNA. Adenine and guanine are purines, which are double-ringed molecules, and thymine, cytosine, and uracil are single-ringed molecules called pyrimidines (Figure 8.1). In the DNA molecule, a purine always binds with a pyrimidine, which is why DNA has a consistent symmetrical shape. In each nucleotide, the phosphate and sugar molecules are hydrophilic, and the bases are hydrophobic.

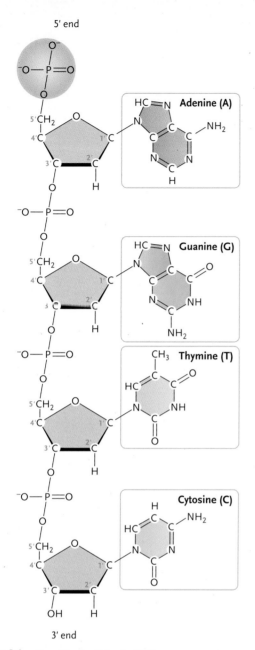

5′ end

3′ end

Figure 8.1   Four Nucleotides in DNA

DNA consists of sequences of four nucleotides: adenine, guanine, thymine, and cytosine.

Art source: From RUSSELL/HERTZ/STARR/FENTON. *Biology*, 2E. © 2013 Nelson Education Ltd. Reproduced by permission. www.cengage.com/permissions

Edwin Chargaff noted that DNA molecules always have *equal* amounts of purines and pyrimidines. This observation became known as Chargaff's rule: the amount of adenine always equals that of thymine, and the amount of cytosine equals that of guanine. We now

DID YOU KNOW?

Humans have 46 DNA macromolecules (chromosomes) in every cell, chimps have 48, chickens and dogs have 78, ants have two, and shrimp have 254.

know that adenine always binds with thymine, and cytosine always binds with guanine.

Comparison of sugar molecule in DNA and RNA

The sugar molecule in a DNA nucleotide is different from the sugar molecule in RNA nucleotides. DNA has the five-carbon sugar called **deoxyribose**, and RNA has the five-carbon sugar called **ribose** (Figure 8.2). The carbon atoms of each sugar are numbered 1, 2, 3, 4, 5 starting with the first carbon within the ring; this carbon atom binds to the nitrogenous base. Carbon 3 and 5 bind with the phosphate groups that make up the sugar-phosphate backbone. Deoxyribose has two hydrogen atoms attached to the second carbon, whereas ribose has a hydroxyl group attached to the second carbon.

Double-helix structure

DNA is composed of two strands that naturally twist into a helical shape, the DNA double helix (Figure 8.3). This structure was discovered in 1953 through the combined research efforts of Rosalind Franklin, Francis Crick, and James Watson at Cambridge University, England. Watson and Crick received a Nobel Prize for their work. The DNA backbone consists of alternating sugar-phosphate molecules that form covalent bonds called phosphodiester bonds. The sugar is connected to the purine or pyrimidine by a glycosidic bond, and the bases are connected in pairs by weak hydrogen bonds. DNA has two complementary strands, and they connect with each other by aligning in opposite directions; this means the two DNA strands are **antiparallel**. Phosphate molecules bind to either the fifth or third carbon on the deoxyribose which gives direction to the DNA strand. When the phosphate binds to the fifth carbon, which is called the 5′ (5 prime) end of one DNA strand, the other end has a free OH group on the third carbon, which becomes the 3′ end.

Nucleosides and nucleotides

Nucleosides are composed of a purine or pyrimidine base attached to either ribose or deoxyribose (Figure 8.4). A nucleotide is a nucleoside with a phosphate group. When DNA replicates, nucleoside triphosphates are added to the growing DNA strand, and two of the phosphates are removed. **ATP** (adenosine triphosphate) is

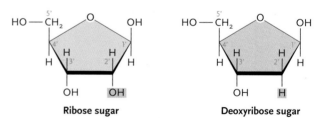

Figure 8.2   Ribose and Deoxyribose

Ribose and deoxyribose sugars in nucleic acids have five carbon atoms, numbered C_1–C_5.

Art source: From DIGIUSEPPE ET AL. *Nelson Biology 12*, 1E. © 2003 Nelson Education Ltd. Reproduced by permission. www.cengage.com/permissions

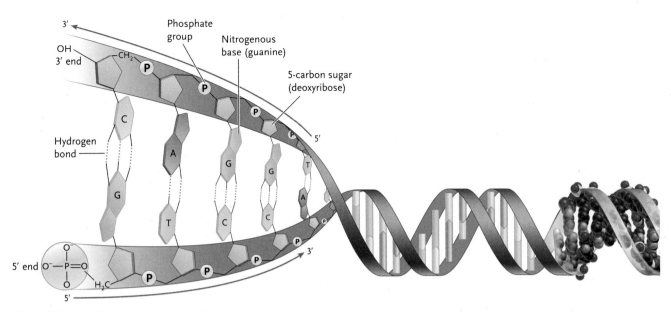

Figure 8.3 Double Helix Structure of DNA

Art source: From DIGIUSEPPE/FRASER. *Biology 12U*. © 2012 Nelson Education Ltd. Reproduced by permission. www.cengage.com/permissions

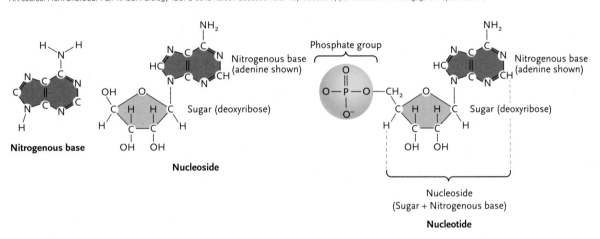

Figure 8.4 Base, Nucleoside, and Nucleotide

Nucleotides are built from nucleosides (sugar and nitrogenous base) and a phosphate group.

a nucleoside triphosphate. Recall from Chapter 4 that ATP is produced in all cells and is the primary source of energy for all cell functions. GTP can also be used as energy in cells. Both ATP and GTP provide energy through the loss of one or more phosphate groups.

Our diet is the source of the nucleosides that are made into nucleotides for DNA replication or the transcription of RNA molecules. The liver can synthesize nucleosides, but our main source is the foods we eat: anything that is plant or animal, not processed foods. All plant and animal cells contain DNA and RNA, which get broken down in our digestive tract with specific enzymes (Chapter 19) and then used to form nucleotides.

Telomeres

All eukaryotic chromosomes have special repeated DNA sequences at the ends of each chromosome: **telomeres** (Figure 8.5). The human telomere repeat is 5′ TTAGGG 3′,

and it is 5–10 kb (kilobases) long. Telomere sequences protect the genes on the ends of chromosomes from deteriorating during DNA replication. Each time the cell enters S-phase and DNA replication occurs, the very first 5′ end of each chromosome is not replicated. The beginning of the DNA replication process involves the addition of primers, which are later removed and replaced with DNA (see Section 8.2). All the primers along the chromosome can be replaced with DNA, except the first 5′ end. Therefore, with each successive replication, the chromosomes become slightly shorter.

The body's cells have a specific lifespan, mainly because of this shortening process, and this is part of the reason why we age as a whole organism. If telomeres were not present the genes at the ends of the chromosomes would degrade and cause cell dysfunction. When the telomeres shorten extensively after many cell divisions, the cell dies or becomes **senescent** and loses the ability to divide. New cells arise from the adult stem cells in many tissues.

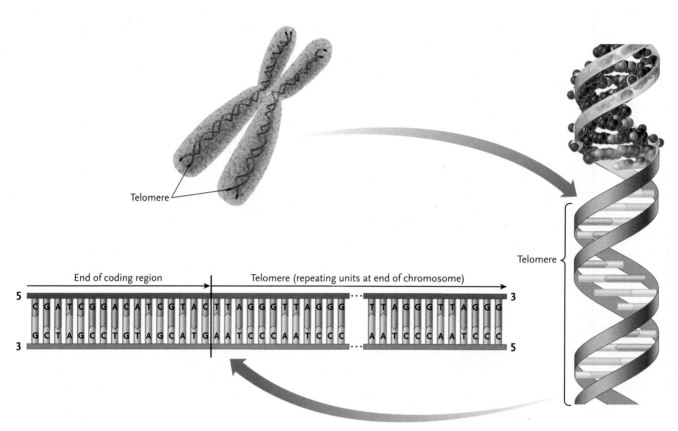

Telomere

End of coding region

Telomere (repeating units at end of chromosome)

5 CGATCGGACATCGTAC TTAGGGTTAGGG ··· TTAGGGTTAGGG 3
3 GCTAGCCTGTAGCATGAATCCCAATCCC ··· AATCCCAATCCC 5

Telomere

Figure 8.5 Telomeres
All eukaryotic chromosomes have telomere sequences at the end of each chromosome, and they shorten with each DNA replication.
Art source: From RUSSELL/HERTZ/STARR/FENTON. *Biology*, 2E. © 2013 Nelson Education Ltd. Reproduced by permission. www.cengage.com/permissions

DID YOU KNOW?

Different organisms have different rates of aging. A mouse lives for approximately three years; humans, 80 years; tortoises, 160 years; and black coral, 4000 years.

Lifespan of stem cells

Stem cells and germ cells that give rise to gametes contain an enzyme called **telomerase**, which functions to replace lost telomere sequences. If inducing telomerase expression in cells prevents telomere shortening and therefore cell death, why don't we use telomerase as the "fountain of youth"? Researchers have successfully prevented the shortening of telomeres in mice; however, this process is directly linked to the development of cancer. If cells do not die or decrease cell division after a period of time—depending on the cell type—many mutations would accumulate in that cell, depending on the amount of exposure to mutagens and carcinogens. Research has shown the risk of cancer development increases when cell death is prevented. In fact, cancer cells begin

to express telomerase, which explains why late stage cancer cells do not die; they become immortalized and can spread throughout the body.

DID YOU KNOW?

Recent research indicates that our health can affect telomere length. In a study of men with prostate cancer, telomere length was measured after five years during which the men increased their consumption of plant foods, exercised 30 minutes each day for six days a week, and added stress-reduction strategies, such as yoga or meditation. The group that incorporated the lifestyle changes had 10% longer telomeres, indicating that healthy lifestyle affected telomere length and lifespan by slowing the rate of telomere degradation.

Absence of telomeres in prokaryotes

Most bacterial species do not have telomeres because their DNA molecules are circular. Bacteria can also divide indefinitely and don't have a specified lifespan unless environmental factors cause them to die.

CONCEPT REVIEW 8.1

1. What are the five nucleotides that make up nucleic acid macromolecules?

2. Which of the nucleotides are purines, and which are pyrimidines?

3. Which nucleotides always combine in the DNA molecule?

4. What three types of bonds occur in the DNA molecule?

5. What is Chargaff's rule?

6. What did Franklin, Watson, and Crick discover?

7. What is the difference between a nucleoside and a nucleotide?

8. How do our cells acquire nucleosides?

9. What kind of structure does DNA form?

10. How are DNA and RNA different?

11. Explain how the carbons of the sugar molecules are named and why this is important.

12. What is the difference between ribose and deoxyribose?

13. What is the energy molecule of the cell?

14. What are telomeres? Why are they important?

15. Which cells produce telomerase? Why is this important?

16. Do prokaryotes have telomeres? Why or why not?

8.2 DNA REPLICATION

DNA replication

DNA replicates according to a method known as **semi-conservative replication**: the original strands of DNA separate and act as a template for two new strands of DNA (Figure 8.6). The DNA in every living organism replicates during the S-phase of the cell cycle before mitosis (see Chapter 5).

In prokaryotes, DNA replication begins at a specific location in the genome, called the **origin of replication** (Figure 8.7). Eukaryotes have a lot more DNA, and so each chromosome usually has multiple sites of origin of replication.

In all living organisms, several enzymes perform specific functions that enable the replication process and ensure that two complete copies of the DNA are produced. **Helicase** is the enzyme that breaks the hydrogen bonds between the bases of the complementary strands of DNA, starting at the origin of replication. Helicase gets its name because DNA has a double helix structure. When the DNA begins to separate and the helicase moves along the DNA molecule separating the two strands, the open portion is called the **replication fork**. **Single-strand binding proteins** stabilize the single-strand region of the DNA molecule, and the DNA polymerase enzyme adds the complementary nucleotides to each single strand. Adenine always pairs with thymine, and cytosine always pairs with guanine. It is possible for all living things to replicate DNA almost perfectly because a purine always binds with a pyrimidine. Some incorrect nucleotides may be added and lead to DNA mutations (discussed further in Section 8.3).

Polymerase can add **nucleoside triphosphates** only to the 3′ end of a nucleotide on the *new* strand. Then two phosphate groups are removed, which leaves a nucleotide: sugar, phosphate, and base (Figure 8.8). The problem arises when the replication fork begins and there is no 3′ hydroxyl group yet available. This problem is solved when the enzyme **primase** adds to the open strand of DNA a small RNA primer: a short sequence of nucleotides that has a free 3′ hydroxyl group.

DID YOU KNOW?

E. coli bacteria have 4.7 million nucleotides on their single chromosome, which can be replicated within 40 minutes. (1000 bases per second). An average human chromosome contains approximately 150 million nucleotides, but due to the multiple origins of replication, an entire genome can be replicated in about one hour.

After an RNA primer has been added, DNA polymerase copies the DNA by adding the complementary nucleotides to the template strand. Because the two strands of DNA are antiparallel, and helicase moves only in one direction, and polymerase can add only to the 3′ end, the two strands of DNA are copied in different ways. The two template strands are called a leading strand and a lagging strand. On the **leading strand**, only one RNA primer is required, and polymerase can follow the helicase and add complementary nucleotides in the 5′ to 3′ direction (on the new strand). The lagging strand, however, needs to replicate in fragments; these are called Okazaki fragments, named after the scientist who discovered them. These fragments

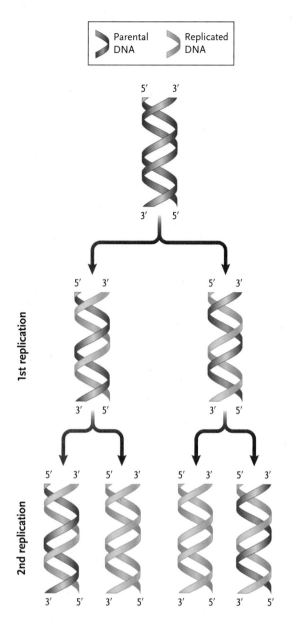

Figure 8.6   DNA Replication

Each original DNA strand separates during DNA replication and acts as a template for the new strand. The process is called semi-conservative replication because the new strands of DNA are made up of one original strand and one new strand.

Art source: From DIGIUSEPPE/FRASER. *Biology 12U.* © 2012 Nelson Education Ltd. Reproduced by permission. www.cengage.com/permissions

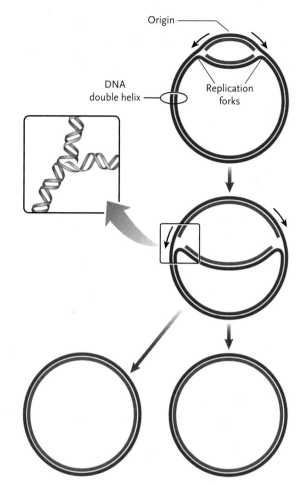

Figure 8.7   Bacterial DNA Replication

Bacterial DNA replication is also semi-conservative; the process starts at one point of origin, and the DNA is replicated in both directions at once.

Art source: From DIGIUSEPPE/FRASER. *Biology 12U.* © 2012 Nelson Education Ltd. Reproduced by permission. www.cengage.com/permissions

Supercoiling in prokaryotes

DNA replication enzymes in bacteria are the same as in eukaryotic cells: helicase unwinds the DNA, and polymerase adds nucleoside triphosphates in the 5′ to 3′ direction, beginning with a primer. However, because prokaryotic chromosomes are small and circular, only one primer is required on each strand initially; then polymerase replicates DNA continuously on both strands, without the need of Okazaki fragments. Furthermore, because bacterial DNA is circular, over-twisting occurs. This is due to the unwinding of the DNA, and an extra enzyme, **topoisomerase**, is required to relieve the supercoils. In viruses, topoisomerase is used to integrate viral DNA into the genome of the cell they are infecting. Some antibiotics, such as the fluoroquinolone antibiotics, prevent bacterial cell growth by inhibiting topoisomerase enzymes, with the result that the bacteria cannot replicate DNA and divide. In humans, topoisomerase regulates alternative splicing of pre-mRNA transcripts (Chapter 9).

require new RNA primers so that the polymerase can still add nucleotides in the 5′ to 3′ direction (Figure 8.9).

As the replicated DNA is produced, the RNA primers are removed by DNA polymerase, and DNA nucleotides replace the RNA primer. An enzyme called **ligase** forms phosphodiester bonds to seal the gaps. As the replication fork moves along the chromosome, the double helix structure reforms naturally. Eventually, in eukaryotic cells, the replication forks meet, and two new strands of DNA are produced. In prokaryotic cells, the polymerase adds approximately 1000 nucleotides per second. In eukaryotic cells, the polymerase adds approximately 50 nucleotides per second.

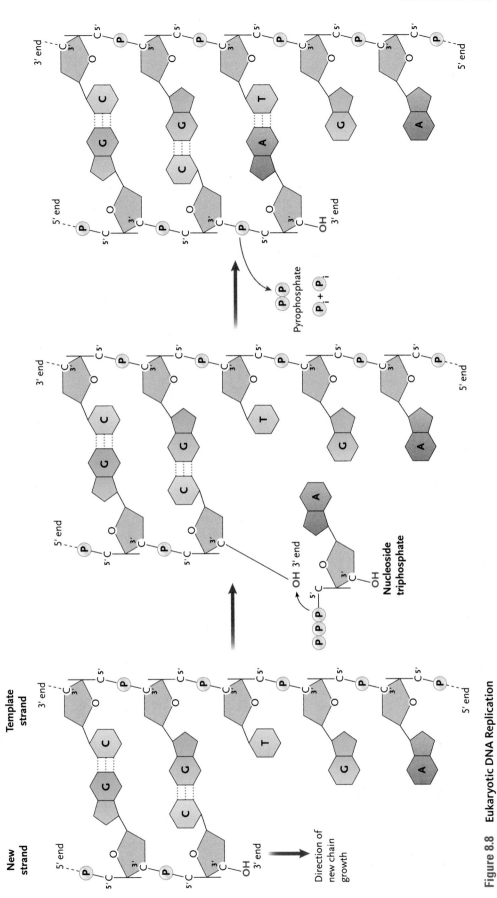

Figure 8.8 Eukaryotic DNA Replication
Nucleoside triphosphates are added by polymerase to the 3′ OH group of an existing nucleotide.

8

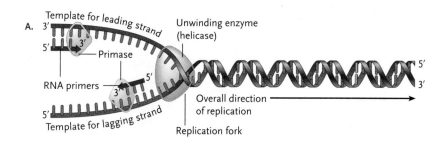

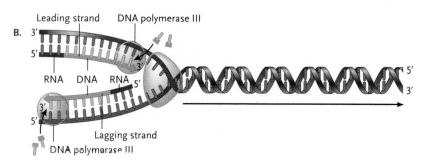

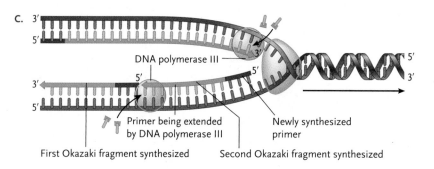

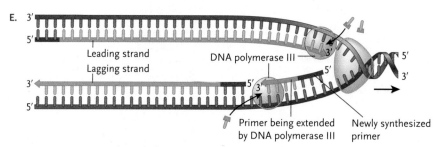

Figure 8.9 Leading Strand and Lagging Strand

DNA can replicate continuously on the leading strand. However, on the lagging strand DNA replicates by adding many Okazaki fragments, because polymerase can add nucleotides only in the 5′ to 3′ direction.

CONCEPT REVIEW 8.2

1. What is semi conservative replication?

2. Make a chart to show the main functions of the following enzymes and molecules involved in DNA replication: helicase, single-strand binding protein, primase, polymerase (two functions), and ligase.

3. Explain how DNA replication differs on the leading strand and the lagging strand.

4. Why do we need primase?

5. What is topoisomerase?

6. How does DNA replication differ between prokaryotes and eukaryotes?

8.3 DNA MUTATIONS

DNA mutations are the primary source of genetic variation and the creation of new alleles within a population. Some gene sequences are highly conserved between species because of their importance to survival. For example, the sodium-potassium membrane protein gene is found in every living organism, as are genes that produce proteins required for glycolysis—the production of ATP in the absence of oxygen. DNA mutations are produced through errors in replication, exposure to environmental chemicals, radiation, or some viruses. However they occur, mutations may be detrimental to the survival of the organism, beneficial, or have no effect at all (silent mutations).

Errors in replication

Errors in replication occur approximately once in every one million nucleotides. These mutations would accumulate very quickly if they were not fixed. Not all DNA mutations are fixed, and this leads to many illnesses (Figure 8.10). During replication, polymerase adds complementary nucleotides according to the sequence of nucleotides on the template strand. However, the following errors can result from the action of polymerase: (1) **substitution** or mismatch, when the wrong nucleotide is added, (2) **insertion** or addition when an extra nucleotide is added, or (3) **deletion** when a nucleotide is skipped. Errors in the DNA sequence lead to errors in the amino acid sequence of proteins, which can affect the function of the particular protein.

Categories of mutations

Point mutations involve either one or several nucleotides and can be substitutions, insertions, or deletions. A **frameshift mutation** occurs when an insertion or deletion mutation occurs, and the whole gene sequence is shifted. This type of mutation is very harmful because the amino acid sequence of a protein is determined by every three nucleotides of an RNA molecule, and the production of RNA is based on the DNA sequence; therefore, an entire protein is affected. **Recombination mutations** involve the moving of large sequences of DNA from one location to another within the genome.

Transposable elements (TEs), also called *transposons* or *jumping genes*, refer to sequences of DNA that move or transpose themselves to another chromosome within a cell. The mechanism of transposition can be either "copy and paste" or "cut and paste." All living organisms contain TEs, and these play a significant role in

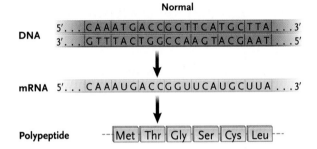

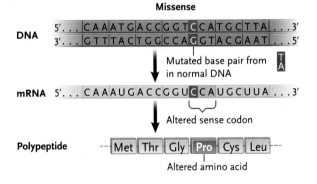

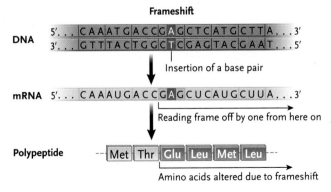

Figure 8.10   DNA Mutations

DNA mutations can lead to the production of non-functional proteins.

Art source: From DIGIUSEPPE/FRASER. *Biology 12U.* © 2012 Nelson Education Ltd. Reproduced by permission. www.cengage.com/permissions

phenotypic variation and evolution, but they can also cause harmful mutations if genes are damaged. When a mutagen causes a break in the sugar-phosphate backbone, the cell tries to repair this by adding the resulting fragment to another piece of DNA. Then some segments of DNA move to entirely different chromosomes. If this occurs in germ-line cells during gametogenesis, the homologous chromosomes may not line up properly, and meiosis cannot continue.

Insertional inactivation occurs when a transposable element inserts a sequence of DNA into the middle of a gene, and so that gene can no longer produce a functional protein (see Chapter 10). This can happen in a number of ways: (1) a TE may insert a sequence of DNA into the regulatory region upstream of the beginning of a gene sequence—sometimes hundreds of nucleotides before the beginning of the actual gene—and alter how often a particular protein is transcribed and translated, (2) a TE may turn on or off that gene in the wrong cell type, and (3) a TE that inserts into a non-coding region of the chromosome may have no noticeable effect at all. (See Figure 8.11.)

Mutations can occur in somatic cells or gametes (Chapters 5 and 6). Somatic cell mutations are not passed on to future generations but can be detrimental and possibly lethal to the individual. Mutations that occur in gametes are passed on to offspring, and they are a source of genetic variation between individuals (Figure 8.12). DNA mutations can affect morphological traits, development, biochemical processes, behaviour, or the regulation of expression of other genes. Mutations in gametes passed on to future generations can be dominant or recessive, autosomal, or X-linked (Chapter 6).

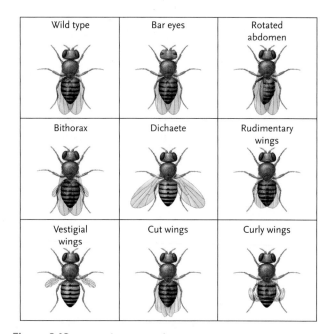

Figure 8.12 Non-detrimental DNA Mutations

Non-detrimental DNA mutations are the primary source of new alleles in a population, for example, morphological differences in bees.

Cellular response to DNA mutations

The enzyme polymerase recognizes some mistakes during replication through a process known as **proofreading**. When polymerase adds the wrong nucleotide, it can stop replicating, reverse direction (move in the 3′ to 5′ direction), and remove incorrect nucleotides using another enzyme function called **exonuclease activity**. Then replication continues in the 5′ to 3′ direction.

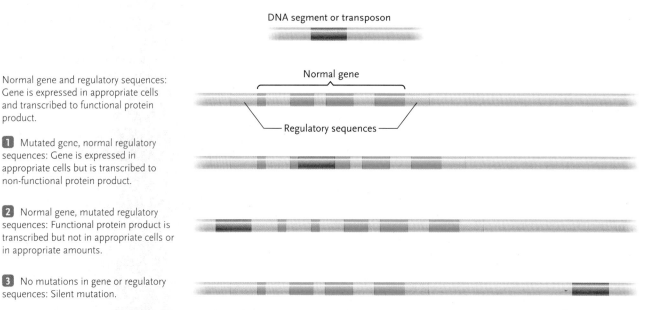

Normal gene and regulatory sequences: Gene is expressed in appropriate cells and transcribed to functional protein product.

1 Mutated gene, normal regulatory sequences: Gene is expressed in appropriate cells but is transcribed to non-functional protein product.

2 Normal gene, mutated regulatory sequences: Functional protein product is transcribed but not in appropriate cells or in appropriate amounts.

3 No mutations in gene or regulatory sequences: Silent mutation.

Figure 8.11 Transposition Mutations

Transposition or recombination can result in the production of a non-functional protein, an altered regulation of protein expression, or no noticeable effect.

Humans have approximately 130 different genes that code for **DNA repair enzymes**. DNA mutations within a DNA repair enzyme gene would be significantly damaging. DNA repair enzymes can locate mutations that polymerase missed during replication. Any purine-purine combinations create a bulge in the DNA molecule, and pyrimidine-pyrimidine combinations create an indent; repair enzymes can recognize both of these combinations. Also, during DNA replication some methyl groups ($CH3$) are added to some of the cysteine bases or adenine bases; in this way the new strand is distinguished from the strand of the original molecule. DNA repair enzymes, however, change only the non-methylated strand, which means they would fix the correct strand.

If mutations continue to persist in the DNA molecule, tumour suppressors function to prevent cell division (recall the cell-cycle checkpoints in G_1, G_2, and M). Also DNA repair enzymes can become recruited to fix mutations. When DNA mutations cannot be corrected, p53 induces apoptosis (programmed cell death) so that cells with mutations do not continue to divide and propagate those mutations. If apoptosis does not occur and the cell continues to divide with mutations, the body uses its last mechanism, the immune system, to kill damaged cells (Chapter 22). **Natural killer cells** are a type of white blood cell that recognize and kill cancer cells that are formed in our body (Figure 8.13). If mutations continue to exist, disease may occur. As we age, the accumulated amount of DNA mutations increases, as does the risk of cancer and other illnesses.

Causes and consequences of mutations

Mutagens and carcinogens, such as radiation and chemicals, are the most common causes of mutations (see Chapter 5 regarding cancer-causing mutations). The other causes of mutation are mistakes during DNA replication, transposition of genetic material, and viruses.

Mutations in viruses are responsible for diseases such as the following:

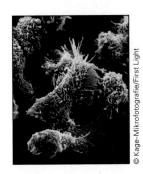

Figure 8.13 **Cancer and Natural Killer Cells**
Natural killer cells can kill cancer cells.

- Human papilloma virus (HPV)—genital warts, can cause cervical cancer
- Human immunodeficiency virus (HIV)—AIDS, can cause Kaposi sarcoma (abnormal growth of blood vessels)
- Hepatitis B or C—liver infection, can cause liver cancer
- Epstein Barr Virus (EBV)—mononucleosis, can cause lymphoma (rare)

Genetic diseases caused by inherited DNA mutations account for approximately 10% of all diseases (Chapter 6). The following list describes major genetic diseases:

- Cystic fibrosis—one mismatch base in the gene that codes for chloride channels
- Huntington's disease—insertion of multiple CAG repeats in a gene on chromosome 4
- Sickle cell anemia—one mismatch in the hemoglobin gene
- Cancer—usually due to two or more mutations in genes that code for repair enzymes, or genes that affect the cell cycle, oncogenes, or tumour-suppressor genes
- Phenylketonuria (PKU)—point mutation in liver enzyme gene, causes brain damage; babies tested for this at birth
- Nonpolyposis colorectal cancer—autosomal dominant, repeated CA mutation in DNA repair enzyme gene expressed in cells of the large intestine.

CONCEPT REVIEW 8.3

1. What is proofreading?
2. How do our cells cope with DNA mutations?
3. Explain the differences between the following types of DNA mutations: substitution (mismatch), addition (insertion), deletion, recombination (transposition).
4. What is a frameshift mutation?
5. What are the consequences of DNA mutations that occur in somatic cells or gametes?
6. How do mutations contribute to genetic variation?
7. What is insertional inactivation?
8. What are transposable elements? Do they always cause DNA mutations?
9. How do we acquire DNA mutations?
10. Make a list of the most common types of carcinogens.
11. What are some common viruses that can cause cancer through DNA mutations?
12. Make a list of some common genetic diseases caused by the inheritance of DNA mutations, and state what type of DNA mutation is involved.

9 Gene Expression and Regulation

The object of education is to prepare the young to educate themselves throughout their lives.

Robert M. Hutchins

9.1 CENTRAL DOGMA OF GENE EXPRESSION

Recall from Chapters 2 and 8 that each DNA macromolecule is a chromosome that's made up of sequences of the nucleotides: adenine, cytosine, guanine, and thymine (Figure 9.1). DNA is double stranded, and complementary nucleotides on each strand always bind so that adenine binds to thymine and cytosine binds to guanine; these are called **base pairs**. Sequences of nucleotides make up genes; for example, the human insulin gene sequence involves 4044 nucleotides on chromosome number 11. Only one strand of the DNA codes for the gene: known as the **coding strand** or **sense strand**. The complementary DNA strand is called the **template strand** or **antisense strand** (Figure 9.2). The coding strand has the *same* sequence as the mRNA molecule (except that uracil takes the place of thymine in the RNA). We have two copies of every gene—alleles—on each homologous chromosome (Chapter 6), one allele from each parent.

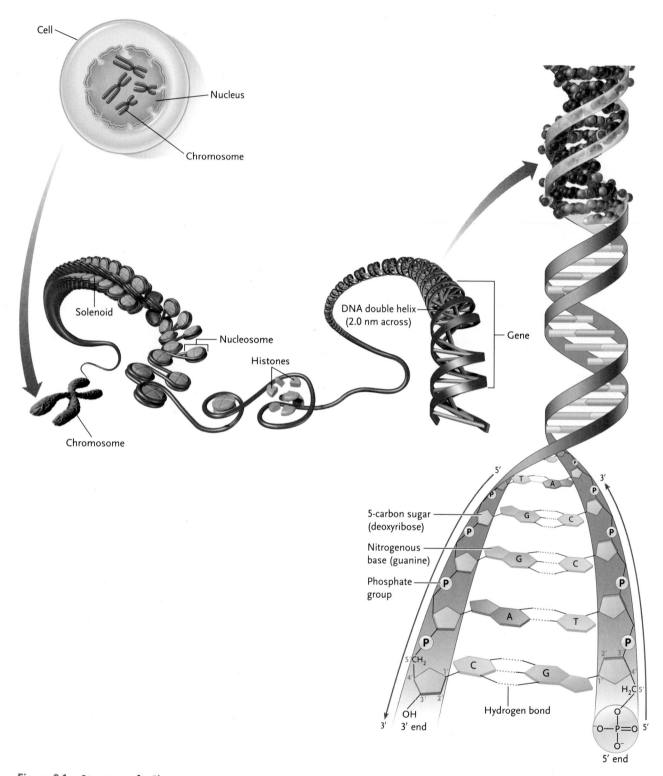

Figure 9.1 **Structure of a Chromosome**

Art source: From DIGIUSEPPE/FRASER. *Biology 12U.* © 2012 Nelson Education Ltd. Reproduced by permission. www.cengage.com/permissions

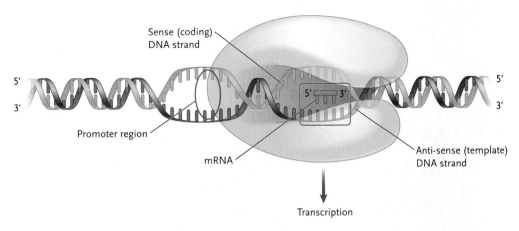

Figure 9.2 Sense and Antisense Strands of DNA

The sense strand is also called the *coding strand*, and the antisense strand is also called the *template stran*

Art source: From RUSSELL/HERTZ/STARR/FENTON. *Biology*, 2E. © 2013 Nelson Education Ltd. Reproduced by permission. www.cengage permissions

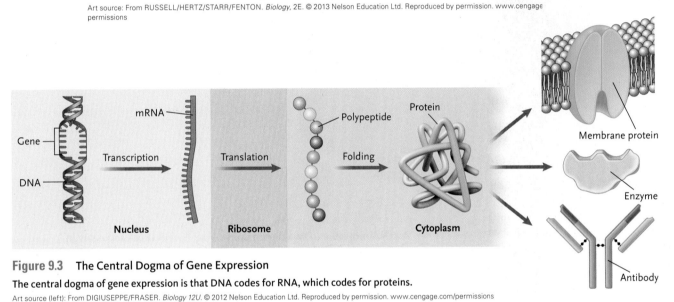

Figure 9.3 The Central Dogma of Gene Expression

The central dogma of gene expression is that DNA codes for RNA, which codes for proteins.

Art source (left): From DIGIUSEPPE/FRASER. *Biology 12U*. © 2012 Nelson Education Ltd. Reproduced by permission. www.cengage.com/permissions

Genes code for messenger RNA (mRNA) sequences that consist of nucleotides (Chapter 2), and this process is called **transcription**. A mature mRNA molecule moves out of the nucleus and through nuclear pores to a ribosome, where protein synthesis (translation) occurs (Chapter 3). The flow of information from gene to RNA to protein is what's known as the **central dogma** (Figure 9.3).

Although every cell nucleus contains all 46 chromosomes, which consist of the approximately 25,000 genes that make up the total human genome, only certain genes are transcribed and translated in certain cells, at specific times, and in specific amounts. This process is called **gene expression**. The regulation of gene expression is covered in Section 9.4.

CONCEPT REVIEW 9.1

1. Which nucleotides form complementary base pairs in DNA?

2. Explain what the central dogma is.

3. What is the difference between the sense strand and the antisense strand of DNA?

4. How many genes are found in the human genome?

9.2 TRANSCRIPTION

Transcription is the process of forming a **messenger RNA (mRNA)** molecule from a gene sequence in the nucleus of a cell. Three different types of RNA molecules play a role in the production of a functional protein (Figure 9.4).

1. **mRNA—messenger RNA—**is the nucleotide sequence that's complementary to the template sequence of the gene. The mRNA moves to the ribosome and codes for the amino acid sequence of the protein (translation). Each three-nucleotide segment, called a **codon,** determines the order of amino acids.

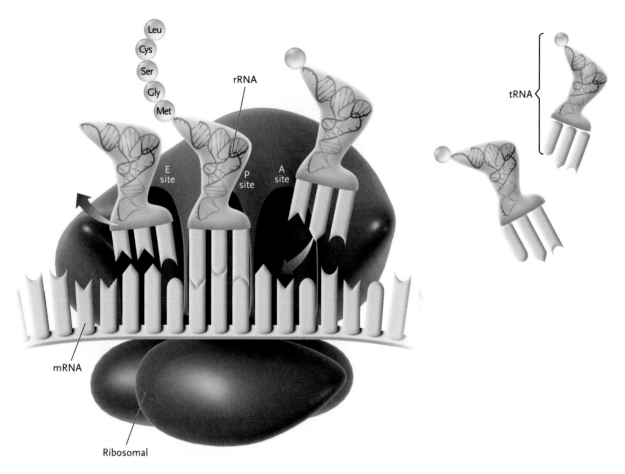

A.

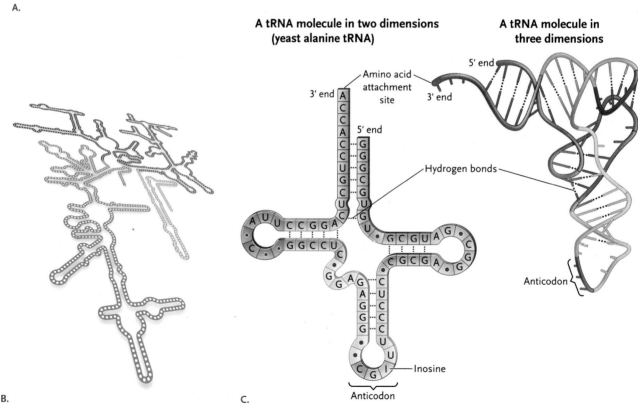

B.

C.

A tRNA molecule in two dimensions (yeast alanine tRNA)

A tRNA molecule in three dimensions

Figure 9.4 Types of RNA Molecules

(A) mRNA, (B) rRNA, and (C) tRNA

2. **rRNA—ribosomal RNA**—makes up part of the ribosome and plays an important role in ensuring the mRNA can bind to the ribosome. Recall from Chapter 3 that transcription of rRNA molecules occurs in the nucleolus.

3. **tRNA— transfer RNA**—carries amino acids to the ribosome during protein synthesis. Each tRNA carries a special nucleotide sequence, the **anti-codon**, that binds to the codons on the mRNA. The 3′ end of the tRNA carries a specific amino acid that is determined by the particular anticodon sequence.

Stages of Transcription

Transcription in prokaryotes and eukaryotes is very similar, but there are a few exceptions: (1) Prokaryotes can undergo transcription and translation at the same time because they do not have a nucleus, and the mRNA doesn't have to leave a nucleus to bind to a ribosome; (2) Transcription is a simpler process in prokaryotes because, unlike eukaryotes, they have no non-translated regions—**introns**. Prokaryotes and eukaryotes also have different mechanisms for regulating gene expression (Section 9.4). Transcription in both prokaryotes and eukaryotes has three stages: initiation, elongation, and termination.

Initiation

Transcription begins when the enzyme **RNA polymerase** binds to a **promoter region** of the DNA upstream from the gene that is transcribed (Figure 9.5). Promoters are nucleotide sequences that have many variations. The most common bacterial promoter sequences are TATTAA or CAAT, which are called a "TATA box" or a "CAT box." Promoters act as **regulatory regions** of the DNA; they are not transcribed but help the polymerase locate the beginning of a gene. The first three nucleotides (first codon) of almost all mRNA molecules is **AUG**. Therefore, the *start* site of the gene begins with ATG on the coding strand.

Elongation

Starting with the first nucleotide, RNA polymerase adds complementary nucleotides to the template strand of the gene so that the mRNA sequence occurs in the same order of nucleotides as the coding strand of the DNA—except that RNA contains uracil instead of thymine (see Figure 9.4). RNA polymerase can add nucleotides only in the 5′ to 3′ direction on the mRNA molecule: just like DNA polymerase functions during replication. RNA polymerase requires ATP.

Termination

Transcription of a gene is complete when the polymerase enzyme reaches a **terminator sequence**. This sequence causes the mRNA molecule to form a hairpin loop that makes the RNA polymerase dissociate from the DNA. Note that RNA polymerase causes the DNA strands to separate during transcription, and immediately after the formation of a small section of mRNA, the complementary DNA bases reform. Upon completion of transcription, the DNA is intact, and the mRNA molecule is ready for translation.

mRNA sequencing exercise

Examine the following DNA sequence, and figure out the corresponding mRNA sequence. Note that this sequence provides the gene between the start site and the terminator sequence; don't forget to label the 5′ and 3′ ends.

sense strand: 5′ ATGGCCTATGAATCG 3′
antisense strand: 3′ TACCGGATACTTAGC 5′
mRNA: _____

Answer: Polymerase uses the antisense (template) DNA strand to determine the correct and opposite nucleotides; therefore, the mRNA molecule has complementary bases to the antisense strand. These are the same nucleotides as the sense strand, except that uracil replaces thymine. Also, the mRNA is anti-parallel to the antisense strand. Therefore, the answer is

mRNA: 5′ AUGGCCUAUGAAUCG 3′

RNA processing

In eukaryotes, genes contain sequences of DNA, the introns, that do not translate into part of the amino acid sequence. The regions that do become the code for the protein are called **exons** (Figure 9.6). In eukaryotes, transcription produces a pre-mRNA molecule that must be processed so that a mature mRNA molecule is ready for translation. This process, called **splicing**, involves removing the introns and adding a **5′ cap** and a **3′ tail**. Splicing occurs with the help of **snRNPs** (pronounced *snurps*), which are small nuclear ribonucleoprotein particles. These particles consist of small nuclear RNA molecules, approximately 100–200 nucleotides long, associated with six to 10 proteins. Six different snRNAs plus associated proteins form a large, complex called a **spliceosome,** which removes the introns and forms new bonds between the exons. The 5′ cap is a methylated GTP (guanine triphosphate), and it enables the mRNA to efficiently begin translation at the ribosome. The 3′ polyA tail consists of several hundred adenine nucleotides attached to the end of the mRNA. This process, called **polyadenylation,** is required to protect mRNA from rapid enzymatic degradation. RNA processing occurs in the nucleus after transcription, and then the mature mRNA molecule moves out of the nucleus to a ribosome in the cytoplasm or onto the rough ER for translation.

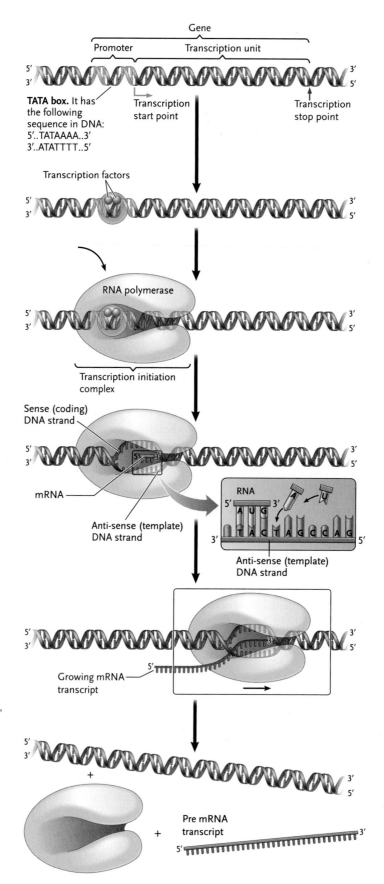

Figure 9.5 Transcription

Transcription occurs in the nucleus. RNA polymerase unwinds the DNA and adds complementary nucleotides to the template strand. After transcription, the RNA leaves the nucleus, and the DNA rewinds.

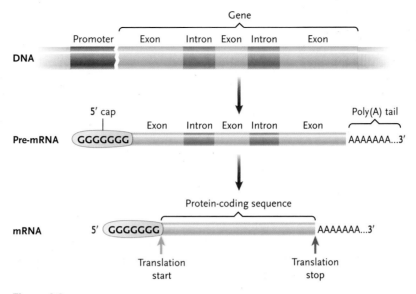

Figure 9.6 RNA Processing

RNA processing involves splicing out the introns and adding the 5′ cap and the 3′ tail.

Art source: From DIGIUSEPPE/FRASER. *Biology 12U.* © 2012 Nelson Education Ltd. Reproduced by permission. www. cengage.com/permissions

CONCEPT REVIEW 9.2

1. How does transcription differ in prokaryotes and eukaryotes?

2. Where does transcription take place in human cells?

3. What is the role of RNA polymerase?

4. How does RNA polymerase "know" where the beginning of a gene is located?

5. What is almost always the first codon in an mRNA molecule?

6. Which strand of DNA does the polymerase use to form the mRNA molecule? In which direction does polymerase transcribe mRNA?

7. What prevents the RNA polymerase from transcribing nucleotides that are not part of the gene sequence? (What stops transcription?)

8. Explain what occurs during RNA processing and why each process is important.

9. Extra study: compare the similarities and differences between DNA replication (Chapter 8) and transcription.

10. What mRNA sequence corresponds to the following sequences?

    a. sense strand: 5′ TACGGCTATC 3′
    b. antisense strand: 3′ ATGCCGATAG 5′

9.3 TRANSLATION

Translation is the process of synthesizing a protein from an mRNA sequence on a ribosome. Recall from Chapter 2 that the human body produces over 100,000 different proteins that have various functions in the body. This section describes how an mRNA molecule becomes a functional protein from a gene inherited from our parents.

Composition of ribosomes

Ribosomes are made up of a small subunit and a large subunit, both of which contain ribosomal RNA and proteins. The mRNA binds to the small subunit, and the tRNAs bind to the large subunit (Figure 9.7). The large subunit has three binding sites for tRNA molecules: the A, P, and E sites. The anticodon of the tRNA must match the codon of the mRNA that moves between the

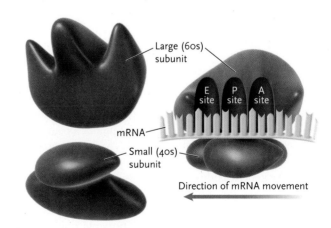

Figure 9.7 Structure of a Ribosome

Ribosomes are composed of ribosomal RNA and proteins.

small and large subunits. Since the first codon, located at the 5′ end of the mRNA, is **AUG**, the first amino acid is always **methionine** (Met). The first tRNA carrying the Met amino acid binds to the P site, where it pairs with the first codon. The second tRNA then binds to the A site, pairing with the second codon. The enzymatic activity of the RNA in the large subunit catalyzes the formation of the peptide bond between the amino acids in the P site. Because RNA in ribosomes has enzyme activity, ribosomes are also called **ribozymes**. The first tRNA, now with no amino acid attached, moves into the E site. The third tRNA binds to the third codon in the A site, and the second peptide bond forms between the second and third amino acid. This process continues—each new tRNA entering the A site, the peptide bond forming in the P site, and the empty tRNA moving into the E site (Figure 9.8)—until a *stop* codon is reached. Three codons (UAA, UAG, and UGA) do not code for any amino acid. Since no tRNA molecules bind to the A site when a stop codon is present, the ribosome dissociates, and translation ends. Similar to transcription, the process of translation requires energy. An empty tRNA becomes recycled when it leaves the ribosome, and **activating enzymes** in the cytoplasm attach the correct amino acid to the correct tRNA molecules (Figure 9.9).

As translation takes place, the primary protein structure begins to form (see Chapter 2), and immediately the secondary structures (alpha helices and beta sheets) begin to form due to hydrophobic and hydrophilic interactions between the functional groups of the amino acids. The full tertiary structure is complete when translation has finished. Proteins that have been translated on the ribosomes in the cytoplasm are released and can immediately function in the cell. Some examples of proteins and their functions include actin to make actin filaments, tubulin to make microtubules, and enzymes involved in glycolysis. Proteins produced on ribosomes on the rough ER are packaged into **vesicles** that pinch off and move to the **Golgi bodies** for further processing (see Chapter 3), called **post-translational modification**. Within the Golgi, certain proteins combine with lipids or sugars to produce **lipoproteins** (e.g., LDLs made in the liver), or **glycoproteins** (e.g., blood type markers, A, B, and AB).

Use of the genetic code

Each mRNA codon consists of three nucleotides. The sequence of codons determines the amino acid sequence. So we can predict the amino acid sequence of any protein from the gene, because each tRNA always carries a specific amino acid, depending on its anticodon. A chart called the **genetic code** can be used to figure out an amino acid based on the mRNA codon (Figure 9.10). There are 64 possible codons—each of the three nucleotides in a codon can be one of the four possible nucleotides. However, there are only 20 amino acids, so usually more than one codon matches each amino acid.

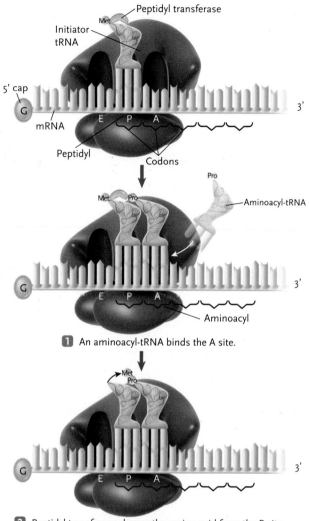

1 An aminoacyl-tRNA binds the A site.

2 Peptidyl transferase cleaves the amino acid from the P site tRNA and bonds it to the amino acid on the A site tRNA.

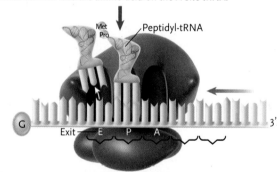

3 The ribosome moves along the mRNA to the next codon, thereby bringing the tRNA with the growing polypeptide to the P site and moving the empty tRNA to the E site.

Figure 9.8   Translation

Translation of mRNA occurs on ribosomes to synthesize proteins.

Example: Use the genetic code in Figure 9.10 to find the amino acid that coincides with the ATG codon. Find the *A* on the left hand side and follow that line across until you see the second letter as *U*. Then you'll see the AUG codon is Met: methionine. Now use the gene

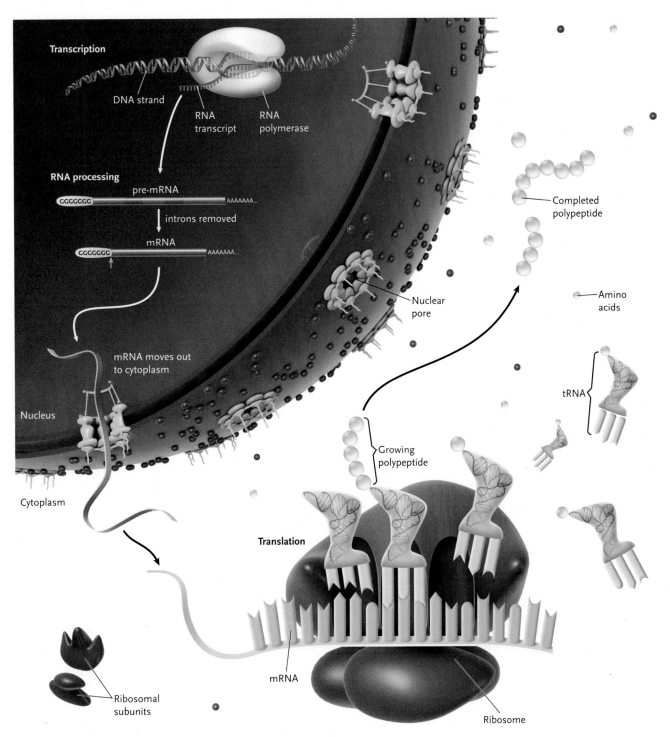

Figure 9.9   Overview of Transcription, RNA Processing, and Translation

Second base of codon

Figure 9.10   The Genetic Code
The genetic code is determined by the three-nucleotide sequences called codons that code for specific amino acids.

Art source: From DIGIUSEPPE/FRASER. *Biology 12U.* © 2012 Nelson Education Ltd. Reproduced by permission. www.cengage.com/permissions

sequence from Section 9.2, and figure out all the amino acids for that sequence. Remember that a normal gene codes for hundreds of amino acids. Note: a chart of the amino acid short forms can be found in Appendix A.

> sense strand: 5′ ATGGCCTATGAATCG 3′
> antisense strand: 3′ TACCGGATACTTAGC 5′
> mRNA: 5′ AUG GCC UAU GAA UCG 3′

Effect of mutations on the protein sequence

Recall from Chapters 5 and 6 that mutations in a gene sequence can alter a protein, making it non-functional: for example, alterations in the cystic fibrosis chloride channel protein or in the formation of oncogenes that cause unregulated cell division. Some changes in nucleotide sequence may have no effect on the amino acid sequence; they are called **silent mutations**. The following

examples show a single nucleotide mismatch; these can cause either a detrimental mutation or a silent mutation.

Example A
Detrimental mutation causing a truncated and non-functional protein

> sense strand: 5′ ATGGCCTAAGAATCG 3′
> antisense strand: 3′ TACCGGATTCTTAGC 5′
> mRNA: 5′ AUG GCC UAA GAA UCG 3′
> amino acid sequence: Met – Ala – Stop

Example B
Silent mutation causing no change in the amino acid sequence

> sense strand: 5′ ATGGCTTATGAATCG 3′
> antisense strand: 3′ TACCGAATACTTAGC 5′
> mRNA: 5′ AUG GCA UAU GAA UCG 3′
> amino acid sequence: Met – Ala – Tyr – Glu – Ser

CONCEPT REVIEW 9.3

1. What are ribosomes composed of?

2. How are mRNA, rRNA, and tRNA molecules involved in translation?

3. What happens in the A, P, and E sites of the ribosome?

4. Why are ribosomes also called ribozymes?

5. What is the role of activating enzymes?

6. Briefly explain how translation works.

7. What happens when a stop codon enters the A site?

8. When do proteins fold into tertiary structure?

9. How can proteins be modified after translation? Give two examples.

10. Use the genetic code to determine the protein sequence from the following mRNA:
    mRNA: 5′ CUC GAU ACC GAG UAG 3′

11. What is a silent mutation?

9.4 REGULATION OF GENE EXPRESSION

Every cell must be able to regulate when and how much protein is made from each gene that is expressed in its particular cell type. When we eat food, our pancreas beta islet cells (Chapter 16) detect an increase in blood sugar and then respond by producing insulin. Insulin regulates the transcription of the cell's glucose transporters (GLUT proteins) that then take up the blood sugar.

Although every cell that contains a nucleus contains all 46 chromosomes and all 25,000 genes, only certain genes are transcribed in certain cell types. Pancreas beta islet cells are the only cells that produce insulin, only red blood cells produce hemoglobin, and only muscle cells produce dystrophin. However, all cells produce the enzymes involved in ATP production. The regulation of gene expression is the essence of homeostasis (Chapter 1). When our body temperature decreases, receptors detect the temperature change and signal the hypothalamus integrating centre. Then the hypothalamus responds by signalling the skeletal muscles to shiver to create heat. All of this occurs because of regulated gene expression. Hundreds of genes are involved in this simple example: those involved in the production of neurotransmitters, ion channels, and synaptic vesicles and those involved in muscle contraction and the production of ATP and therefore heat. The regulation of gene expression can occur at the level of transcription, RNA processing, RNA degradation, or translation.

Prokaryotic gene expression

The **lac operon** in *Escherichia coli* (*E. coli*) bacteria was one of the first systems of gene regulation that scientists studied. The lac operon is a set of genes involved in lactose metabolism. The bacterial cell does not use energy to produce proteins that are not required, so the lac operon genes become transcribed and translated only when lactose is present and glucose is not. (*E. coli* preferentially metabolizes glucose rather than lactose if both are present.) In this case, gene regulation occurs at the level of transcription using a combination of an **activator** and a **repressor**.

It is important to understand the differences between regulatory DNA sequences and the proteins that bind to them (Figure 9.11). Figure 9.11 shows that in *E. coli* the three genes required for lactose metabolism are clustered in an **operon** so that all three are transcribed and translated together whenever needed; this clustering does not occur in eukaryotes. The regions of the regulatory DNA sequences are not transcribed, but they are binding sites for proteins that either initiate or prevent transcription. The **CAP binding protein** (catabolite activator protein) is the **activator** that initiates transcription by binding to the CAP binding sequence. The **repressor** prevents transcription by binding to the operator sequence, and the RNA polymerase binds to the promoter region.

Transcription will occur or not occur depending on the presence of lactose and glucose. Transcription occurs only if lactose *is* present and glucose is *not* present. When lactose is present, some molecules bind to the repressor and remove it from the operator sequence; therefore, the lactose acts as the **inducer**. In the absence of glucose, the CAP activator binds to the CAP binding site and directs the RNA polymerase to the promoter (Figure 9.12). So transcription occurs when the repressor is removed by a molecule of lactose and CAP binds and directs the RNA polymerase to the promoter. When glucose is present, the CAP activator does not bind, and the lactose metabolizing genes are not transcribed. Instead, genes involved in the utilizing of glucose will be transcribed.

Eukaryotic gene expression

In eukaryotes, the regulation of gene expression is much more complex than that in prokaryotes because of the condensed structure of their DNA (Figure 9.1). Different regions of each chromosome are condensed to a different degree, depending on the regions of DNA transcribed by a particular cell type. Four of the mechanisms involved in regulation of gene expression are described below: methylation, RNA processing, mRNA degradation, and transcription factors.

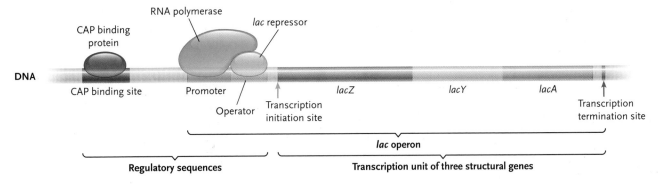

Figure 9.11 The Lac Operon

The lac operon is found in *E. coli* bacteria and contains three genes involved in lactose metabolism, all regulated by one promoter.

Art source: From RUSSELL/HERTZ/STARR/FENTON. *Biology*, 2E. © 2013 Nelson Education Ltd. Reproduced by permission. www.cengage.com/permissions

A. Glucose absent; Lactose absent; lac operon off

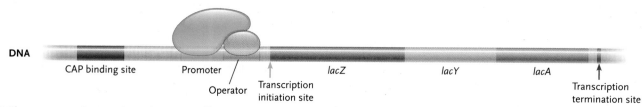

B. Glucose present: Lactose present; lac operon off

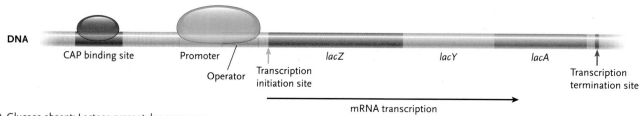

C. Glucose present; Lactose absent; lac operon off

D. Glucose absent; Lactose present; lac operon on

Figure 9.12 Overview of Regulation of the Transcription of the Lac Operon

Art source: From RUSSELL/HERTZ/STARR/FENTON. *Biology*, 2E. © 2013 Nelson Education Ltd. Reproduced by permission. www.cengage.com/permissions

1. **Methylation.** The prevention of gene transcription by the addition of methyl groups ($-CH_3$) is called **gene silencing.** Methylation of cytosine nucleotides most commonly occurs in the tightly condensed regions of DNA where the genes do not become transcribed in certain cell types. For example, all cells except the red blood cells have the hemoglobin gene region tightly condensed and methylated to prevent transcription. Methylation also occurs during DNA replication (Chapter 8, Section 3). Abnormal methylation is thought to play a role in cancer.

2. **RNA processing.** A mature mRNA molecule forms after introns have been removed and exons have been spliced together during RNA processing, but this does not have to occur in the exact same manner each time the gene is transcribed. In a process called **alternate splicing** or **exon shuffling,** certain exons become part of the mature mRNA at certain times, and for other exons this happens at other times. The particular function of a protein depends on what exons are involved in forming the mature mRNA. This mechanism makes it possible for over 100,000 different proteins to be produced from only 25,000 genes.

3. **mRNA degradation.** Various mRNA transcripts have a specified lifespan, and mRNA molecules can undergo translation as long as they are not broken

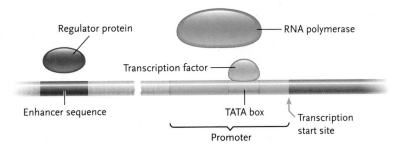

Figure 9.13 **Transcription Factors**

Transcription factors regulate eukaryotic gene expression.

Art source: From DIGIUSEPPE/FRASER. *Biology* 12U. © 2012 Nelson Education Ltd. Reproduced by permission. www.cengage.com/permissions

down. Some mRNA molecules are broken down immediately, but others may stay in the cytoplasm for several hours. The 3′ untranslated region (UTR) that follows the poly A tail has a repeated sequence of AUUUA: the greater the number of repeats in this region, the shorter the lifespan of the mRNA. The lifespan of each mRNA molecule is determined by the gene.

4. **Transcription factors.** Various transcription factors can bind to regulatory regions of the DNA to promote transcription of certain genes. This is the most common mechanism for regulating gene expression in eukaryotes. All hormones, both water soluble and fat soluble, are signalling molecules that regulate the transcription of other genes. However, only steroid hormones that directly bind to the DNA are considered transcription factors. Insulin is a water-soluble hormone that binds to a receptor on the cell surface, which then initiates a signalling cascade that eventually causes a transcription factor to bind

a regulatory sequence on the DNA; this then recruits RNA polymerase to the promoter upstream of the glucose transporter gene (Figure 9.13). Transcription factors are regulated by cellular signals as well as environmental stimuli.

DID YOU KNOW?

Thinking can affect gene expression! Close your eyes and think of something stressful or scary. Your thoughts just caused the transcription and translation of many molecules involved in the stress response. Conversely, meditation works by slowing thoughts and induces expression of genes involved in rest and relaxation. Meditation is becoming a widely popular adjunctive treatment for many illnesses. It has also been shown that optimistic people live longer on average than pessimistic people.

CONCEPT REVIEW 9.4

1. How is the regulation of gene expression related to homeostasis? Give one example.

2. In prokaryotes, what are two types of regulatory proteins that affect transcription?

3. What is the lac operon?

4. What is the regulatory region of DNA that the repressor, CAP activator, and RNA polymerase each bind to?

5. How is the repressor removed from the DNA?

6. Under what conditions is the lac operon transcribed?

7. If glucose and lactose are both present, are the activator and/or repressor bound or not? Does transcription occur? Why or why not?

8. List and briefly describe the main mechanisms of gene regulation in eukaryotes. Which is the most common?

9. What is gene silencing?

10. What is alternate splicing and why is it important?

10 Biotechnology

Education is the passport to the future, for tomorrow belongs to those who prepare for it today.

Malcom X

10.1 OVERVIEW OF BIOTECHNOLOGY

Biotechnology is the use of living organisms to produce biological products. For example, bacteria are used to produce cheese, yeast is used to make alcohol, and a fungus is used to produce penicillin. In the last few decades, it has become possible to move specific sequences of DNA from one organism to another; this process is called **recombinant DNA technology**, **genetic engineering**, or **gene splicing**. DNA sequences, or genes, can be cut at specific sites in the genome of any organism and can be inserted into the genome of any other organism. A *hybrid* or **recombinant DNA** molecule contains nucleotide sequences from two different organisms. For example, human DNA sequences, such as the insulin gene, can be inserted into a bacterial plasmid DNA molecule to produce a human-bacteria recombinant DNA molecule.

A **genome** consists of all of the genes in an organism. A **proteome** is all of the proteins produced from the DNA in an organism. **Genetic modification** refers to any kind of alteration in the genome, including the deleterious effects of mutations caused by radiation or chemicals. Humans have been altering DNA in plants and animals gradually for over 200,000 years by selective breeding (Chapter 7). It is now possible to insert specific DNA sequences into organisms such as plants or animals; the resulting organisms are called **genetically modified organisms (GMOs)**.

Overview of recombinant DNA technology

To create recombinant DNA molecules the following essential "ingredients" are required:

1. A gene of interest
2. A **restriction endonuclease** enzyme (a restriction enzyme) that cuts DNA at specific sequences
3. A **vector** that carries the inserted gene; most commonly a bacterial plasmid, but also other types of vectors (Section 10.3)
   a. Vectors must be able to replicate so that the inserted gene is copied.
   b. Vectors must be able to be transcribed and translated to produce a functional protein.
4. Ligase, an enzyme that's used to bind DNA together
5. A **host cell,** usually bacteria, to take up the recombinant plasmid; will replicate and copy the DNA, as well as transcribe and translate the inserted gene

Human insulin was the first genetically engineered medication, produced in 1982. The human insulin gene can be inserted into a bacterial plasmid; the bacteria replicate and transcribe the human insulin gene and then synthesize the protein, which can be extracted and used to treat people with type 1 diabetes.

DID YOU KNOW?

Papaya trees are genetically engineered to contain a ringspot virus coat protein so that the trees react by producing antiviral chemicals. This prevents the trees from being infected by actual ringspot viruses. The virus coat protein is not expressed in the fruit, so it doesn't harm humans.

CONCEPT REVIEW 10.1

1. What is recombinant DNA technology?
2. What is recombinant DNA?
3. What is a genetically modified organism? Give an example.

4. What are the primary ingredients required for recombinant DNA technology?

10.2 RESTRICTION ENZYMES

Restriction enzymes are special enzymes that cut DNA phosphodiester bonds at very specific symmetrical sequences called **palindromes,** usually four to eight nucleotides long. A palindrome sequence can be read the same way in both directions, such as the word RADAR. The following is an example of a DNA palindrome:

5′ GAATTC 3′
3′ CTTAAG 5′

On both strands of the DNA, the sequence is read the same in the 5′ to 3′ direction. Restriction enzymes do not require ATP to function.

Restriction enzymes have been isolated from many species of bacteria. Bacteria produce many types of restriction enzymes that function to prevent viral infections caused by **bacteriophages:** viruses that infect bacteria. The restriction enzymes cut the DNA molecules and, therefore, prevent viral replication. Each restriction enzyme always recognizes and cuts the DNA at the exact same sequence. Figure 10.1 shows the enzyme *Eco*RI, isolated from *E. coli* bacteria, always cuts 5′ GAATTC 3′ sequences. Bacteria protect their own DNA from these enzymes by using **methylase enzymes** to block those recognition sites on their chromosome with methyl groups (CH_3).

When restriction enzymes cut DNA, they don't necessarily cut it exactly in half (Figure 10.1). The DNA is usually cut so there are overhanging ends of non-paired nucleotides: called **sticky ends.** This is a very useful feature that allows for DNA cut with a

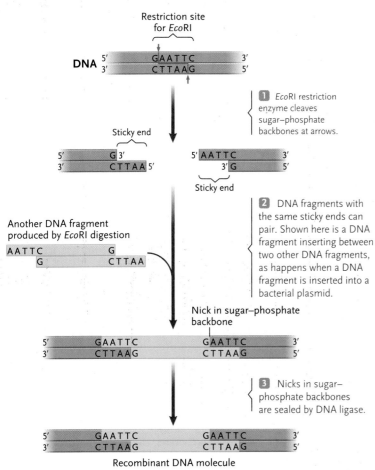

1 *Eco*RI restriction enzyme cleaves sugar–phosphate backbones at arrows.

2 DNA fragments with the same sticky ends can pair. Shown here is a DNA fragment inserting between two other DNA fragments, as happens when a DNA fragment is inserted into a bacterial plasmid.

3 Nicks in sugar–phosphate backbones are sealed by DNA ligase.

Figure 10.1   Restriction Enzymes

Restriction enzymes cut specific palindrome sequences.

Art source: From RUSSELL/HERTZ/STARR/FENTON. *Biology*, 2E. © 2013 Nelson Education Ltd. Reproduced by permission. www.cengage.com/permissions.

restriction enzyme to match up (**anneal**) with other DNA that has been cut with the same enzyme. The enzyme DNA ligase is used to re-form the phosphodiester bonds. Over 3000 restriction enzymes have been studied, and several hundred of these are available for genetic research. Each restriction enzyme cuts different palindrome sequences. The following shows the palindrome sequences related to three of the restriction enzymes:

*Eco*RI	5′.......GAATTC.......3′
	3′.......CTTAAG.......5′
*Bam*HI	5′......GGATCC.......3′
	3′......CCTAGG.......5′
*Hind*III	5′......AAGCTT.......3′
	3′......TTCGAA.......5′

Usefulness of restriction enzymes

Restriction enzymes can be used to cut DNA from two different organisms—such as bacterial DNA and human DNA—to produce recombinant DNA molecules. For example, the human insulin gene can be cut with a restriction enzyme and then inserted into a bacterial plasmid that has been cut with the *same* restriction enzyme. Once the human insulin gene anneals with the bacterial plasmid DNA, that recombinant DNA molecule can be taken up by bacterial cells; there the human insulin protein will be expressed by the bacteria. Many other examples are discussed throughout this chapter.

CONCEPT REVIEW 10.2

1. What are palindromes?
2. Where do restriction enzymes come from?
3. How do bacteria protect themselves from cutting their own DNA?

4. What bonds are broken by restriction enzymes?
5. Why are sticky ends important?
6. Why are restriction enzymes useful to genetic researchers?

10.3 VECTORS

A vector, also called a *vehicle*, is a DNA molecule that can carry extra pieces of DNA. Among the many different types of vectors, the most common ones are plasmids, bacteriophages, cosmids, and YACs. All vectors must have certain characteristics for them to be useful to genetic scientists.

- They must be able to **replicate** within the host cell.
- They must contain **restriction enzyme cut sites**.

- They should carry a **selectable marker** to allow for identification of the host cells that contain the recombinant DNA.
- The vector should be easy to recover from the host cells that contain it.
- The inserted gene must be expressed as a protein.

Plasmids

Plasmids are small pieces of bacterial cell DNA that are separate from the main chromosome (Figure 10.2). Plasmids usually contain around 2000 to 5000 nucleo-

10

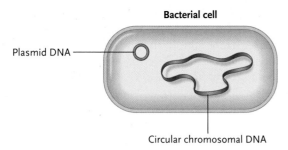

Bacterial cell

Plasmid DNA

Circular chromosomal DNA

Figure 10.2   Bacterial Plasmid

A bacterial plasmid is a small, circular, extra piece of DNA found in some bacteria.

tides and are self-replicating once inside bacterial host cells. Figure 10.3 shows an example of a plasmid. Plasmids are smaller than other vectors and can hold only small pieces of DNA, but they are usually sufficient for replicating single gene sequences. Bacteria are easy to grow and easily take up plasmids by a process called **transformation**. Many plasmids have been engineered to contain all of the required characteristics listed above. A plasmid contains at least one selectable marker that is usually an antibiotic resistance gene. It also contains many restriction enzyme cut sites, which are typically placed into a **multiple cloning site** (MCS) that is sometimes found within another selectable marker, such as the **lacZ gene**.

Use of selectable markers

Genetically modified bacteria are grown in the presence of the antibiotic that the plasmid is resistant to, such as ampicillin. It is very important for plasmids to have an antibiotic resistance gene for two reasons: (1) only bacteria that contain plasmids will grow, and (2) antibiotics in the growth media will kill any contaminating bacteria.

LacZ is one of the genes in the *E.coli* bacterial lac operon (Chapter 9), and it is used as a selectable marker. LacZ is one of the proteins produced by the *E.coli* that are involved in breaking down lactose (Chapter 9). When bacteria are grown in the presence of a molecule that is similar to lactose, called X-galactosidase (X-gal), the bacterial colonies turn blue, making this an excellent selectable marker (Figure 10.4). When bacteria are grown on a plate containing all nutrients required for bacterial growth, including an antibiotic, some of those bacteria may contain plasmids that do not contain a piece of recombinant DNA. In Figure 10.3, notice how the lacZ gene is located within the MCS. The MCS contains all of the restriction enzyme cut sites. When the plasmid is cut with a restriction enzyme and then another piece of DNA is inserted into that region, the lacZ gene becomes inactivated: a process known as insertional inactivation. If lacZ is inactivated, the bacteria do not turn blue in the presence of X-gal. Therefore, only the white bacterial colonies will contain recombinant DNA (Figure 10.4).

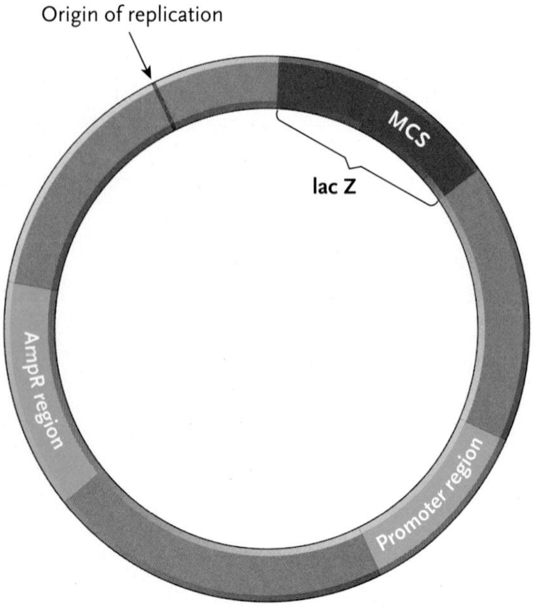

Origin of replication

MCS

lac Z

AmpR region

Promoter region

Figure 10.3   Important Features of a Plasmid

The multiple cloning site (MCS), a promoter, the origin of replication, and a selectable marker are important features of a plasmid.

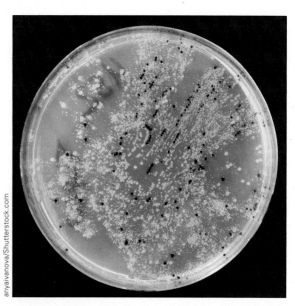

anyaivanova/Shutterstock.com

Figure 10.4   Blue and White Bacterial Colonies

Blue bacterial colonies do not contain recombinant DNA; white bacterial colonies do contain recombinant DNA.

Other useful vectors

Bacteriophages are viruses that infect bacteria. The lambda phage is widely used in recombinant DNA technologies. All of the lambda phage genes are known and sequenced. The genes involved in the lysogenic phase can be removed and replaced with foreign DNA without affecting the ability of the phage to infect bacteria and replicate. The phage containing recombinant DNA can then infect bacterial cells: a process called **transduction**.

Cosmids are laboratory-constructed vectors that contain the *cos* site from the lambda phage, which is required for packaging chromosomes into virus particles. They contain plasmid features such as antibiotic resistance and an origin of replication. They are used to carry much larger sequences of DNA than plasmids, and then they can be introduced to bacterial cells by transformation, in the same way as plasmids.

Yeast artificial chromosomes (YACs) are beneficial because these vectors can be grown in yeast cells, which are eukaryotic. Since prokaryotes lack the ability to process RNA, a eukaryotic system is very beneficial. YACs have an origin of replication; a centromere, so that DNA can be distributed during mitosis; restriction enzyme cut sites; and yeast-specific selectable markers. Yeast cells are easily grown and have been used to express many mammalian genes.

CONCEPT REVIEW 10.3

1. What are the important features of a vector?
2. Describe the important components of a plasmid.
3. What is the role of selectable markers?
4. How is the lacZ gene used as a selectable marker?
5. Do white or blue colonies contain recombinant DNA?
6. What are the main benefits of using a bacteriophage, a cosmid, and a YAC?

10.4 CLONING A GENE

Using a plasmid vector as an example, we can look at the process of **gene cloning**: making many copies of a specific gene with bacteria. If the plasmid is an expression vector then the gene can be transcribed and translated into a protein. A gene of interest, such as the human insulin gene, can be cut from the human chromosome and ligated into a bacterial plasmid that is cut with the same restriction enzyme (Figure 10.5).

Human DNA can be isolated from any human cell that contains a nucleus, because all 46 human chromosomes occur in every nucleated cell (mature red blood cells do not contain a nucleus). Since the human genome project was completed in 2003, all three billion nucleotides have been sequenced. Therefore, the human insulin gene sequence is known, along with the restriction enzyme cut sites that are within and outside the gene. The insulin gene can be found on the short arm of chromosome 11. Knowing the restriction enzymes that will cut the insulin gene out of chromosome 11 without cutting the gene into pieces is a very important first step. Once the restriction enzymes are chosen, they are used to cut the bacterial plasmid DNA. The cut plasmid DNA and the cut human DNA can be mixed together with DNA ligase, to form recombinant DNA.

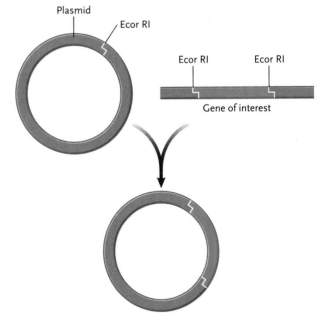

Figure 10.5 **Insertion of a Gene into a Plasmid**

Insertion of a gene into a plasmid involves cutting the gene and the plasmid with the same restriction enzyme, and then ligating the pieces of DNA together with the ligase enzyme.

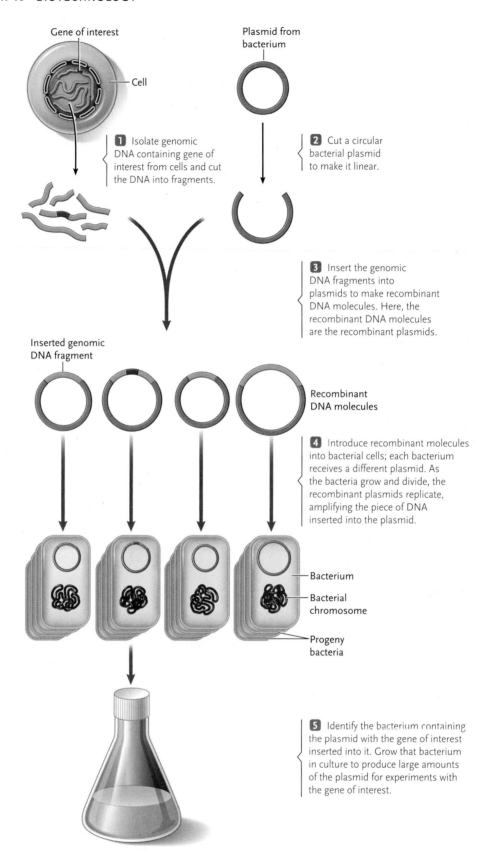

Gene of interest

Plasmid from
bacterium

Cell

1 Isolate genomic
DNA containing gene of
interest from cells and cut
the DNA into fragments.

2 Cut a circular
bacterial plasmid
to make it linear.

3 Insert the genomic
DNA fragments into
plasmids to make recombinant
DNA molecules. Here, the
recombinant DNA molecules
are the recombinant plasmids.

Inserted genomic
DNA fragment

Recombinant
DNA molecules

4 Introduce recombinant molecules
into bacterial cells; each bacterium
receives a different plasmid. As
the bacteria grow and divide, the
recombinant plasmids replicate,
amplifying the piece of DNA
inserted into the plasmid.

Bacterium

Bacterial
chromosome

Progeny
bacteria

5 Identify the bacterium containing
the plasmid with the gene of interest
inserted into it. Grow that bacterium
in culture to produce large amounts
of the plasmid for experiments with
the gene of interest.

Figure 10.6 Cloning a Gene

Cloning a gene involves inserting a gene into a plasmid and then transforming bacteria as they take up that plasmid
and grow and replicate the DNA.

Art source: From RUSSELL/HERTZ/STARR/FENTON. *Biology*, 2E. © 2013 Nelson Education Ltd. Reproduced by permission. www.cengage.com/
permissions.

The recombinant plasmid needs be taken up by bacteria through a process called transformation. Simply by mixing the plasmids and bacterial cells together and incubating them at 37°C for two minutes (using a specific strain of *E. coli* bacteria), the bacterial cells will take up the plasmid DNA. The transformed bacteria can be grown overnight to make millions of bacteria that contain a recombinant plasmid (Figure 10.6).

You can find a **restriction map** (Figure 10.7) for any gene through NCBI, the National Center for Biotechnology Information. You can also use restriction mapping software—that does virtual digests with various restriction enzymes—to determine possible outcomes. In the example shown in Figure 10.7, you can see that you would not choose *Sma*I or *Hind*II to cut out your insulin gene because those will cut your gene into pieces. *Eco*Ri or *Hind*III can be used with *Bam*HI to safely extract the gene.

Screening colonies

The transformed bacteria can be grown overnight on plates containing a growth medium, an antibiotic

that the plasmid has a resistance gene for, and X-gal. Any bacteria that have not taken up a plasmid cannot grow on the antibiotic medium because the antibiotic resistance gene is on the plasmid. Any bacteria that have not taken up the recombinant DNA will have an intact lacZ gene and turn blue in the presence of X-gal (Figure 10.8). Therefore, any white colonies growing on the plate should contain the insulin gene. Then the process of **hybridization** is used to determine if the colonies that have recombinant DNA actually contain your insulin gene.

Hybridization

Once you have a plate of bacterial cells with the white colonies containing recombinant DNA, the final step is to determine which of those white colonies contains the insulin gene. Remember that complementary nucleotides always bind to each other. Use that knowledge to create a **probe** that is complementary to your gene of interest. A probe in this case is approximately 20 nucleotides of the insulin gene sequence that are unique to insulin. Information can be found on the NCBI website for any gene. For the probe to be visible, it needs a tag: usually, **radioactive phosphate**, a colourimetric enzyme, or a fluorescent antibody-based protein (Figure 10.9).

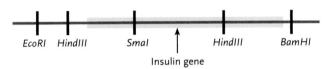

| EcoRI | HindIII | SmaI | HindIII | BamHI |

Insulin gene

Figure 10.7 Restriction Map

A restriction map is used to locate restriction enzyme cut sites.

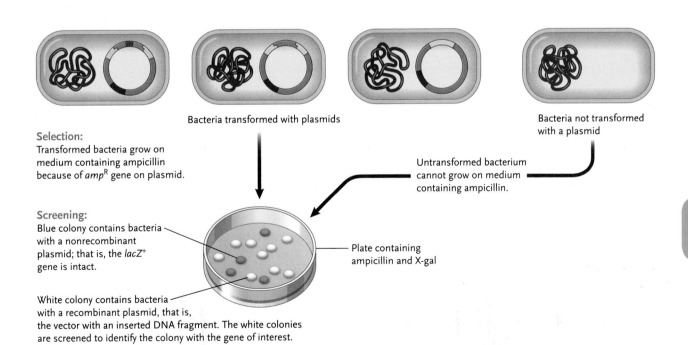

Bacteria transformed with plasmids

Bacteria not transformed with a plasmid

Selection:
Transformed bacteria grow on medium containing ampicillin because of *amp*^R gene on plasmid.

Untransformed bacterium cannot grow on medium containing ampicillin.

Screening:
Blue colony contains bacteria with a nonrecombinant plasmid; that is, the *lacZ*^+ gene is intact.

Plate containing ampicillin and X-gal

White colony contains bacteria with a recombinant plasmid, that is, the vector with an inserted DNA fragment. The white colonies are screened to identify the colony with the gene of interest.

Figure 10.8 Screening Bacterial Colonies

Screening bacterial colonies involves the use of selectable markers, such as an antibiotic resistance gene and the lacZ gene.

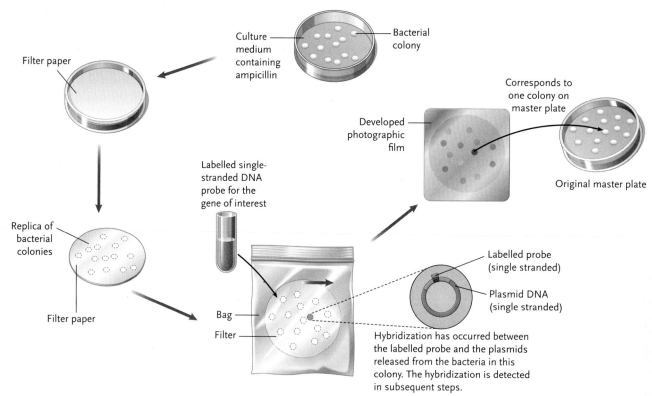

Figure 10.9 Hybridization

Hybridization involves the transfer of bacterial DNA to filter paper and the use of a probe to find the colony that contains the gene of interest.

Art source: From RUSSELL/HERTZ/STARR/FENTON. *Biology*, 2E. © 2013 Nelson Education Ltd. Reproduced by permission. www.cengage.com/permissions.

Hybridization steps

1. Grow DNA library bacterial cells on a plate containing growth media and an antibiotic.

2. Use special filter paper (with a positive charge to attract negatively charged DNA) to take a small sample of each of the bacterial colonies on the plate.

3. Treat the filter paper with a buffer solution that breaks open bacterial cells and denatures the DNA into single strands.

4. The single-stranded DNA sticks to the filter paper and all other cell contents are washed off.

5. Mix radioactive probe solution onto the filter paper and then rinse off excess—only the probe that matched the complementary sequence remains stuck to the DNA on the filter paper.

6. Overlay the filter paper onto photographic film to find the bacterial colony that contains your gene of interest.

Those specific bacterial colonies can now be grown in large flasks so that the bacterial cells can transcribe and translate the insulin gene. The hormone can then be isolated and used to treat diabetics. Note that not all medications are proteins.

Summary of steps in making recombinant protein medications

1. Isolate plasmid DNA and human DNA.

2. Look at the restriction map to choose the restriction enzymes that will not ruin the gene of interest, such as the insulin gene.

3. Cut both plasmid DNA and human DNA with the same restriction enzyme.

4. Mix the cut plasmid and human DNA, which contains the insulin gene, with ligase to form recombinant DNA.

5. Mix recombinant DNA with *E. coli* bacterial cells, and heat-shock the mixture to transform the bacteria.

6. Grow the transformed bacteria on a plate overnight.

7. Conduct DNA hybridization to screen colonies.

8. Grow specific bacterial colony containing the gene of interest and isolate the synthesized protein.

Growth hormones produced by genetically altered bacteria can be injected into animals such as cows and pigs to make them grow faster or produce more milk. The genetically altered bacterial plasmids contain the bovine (cow) somatotropin gene (growth hormone), and the result is **recombinant bovine somatotropin, rBST** (Figure 10.10). A lot of concern has been raised

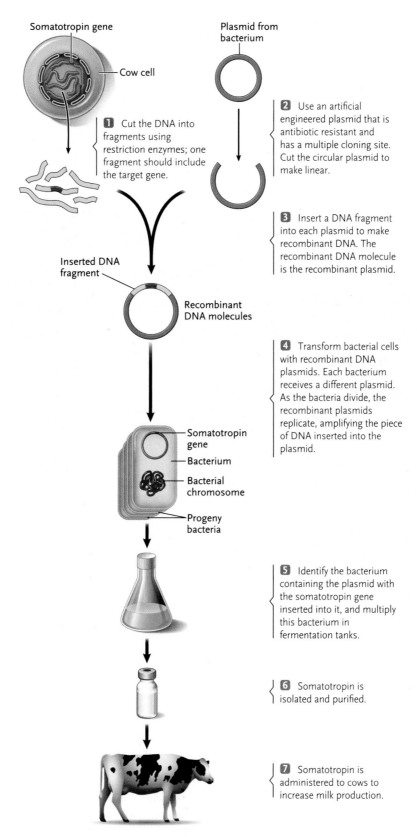

Figure 10.10 **Production of Recombinant DNA**

The production of recombinant DNA can be used to grow specific proteins within transformed bacteria; the protein can then be isolated and given to humans or animals.

Art source: From DIGIUSEPPE ET AL. Nelson Biology 12, 1E. © 2003 Nelson Education Ltd. Reproduced by permission. www.cengage.com/permissions

about our eating meat or drinking milk that contains growth hormones, and this has stimulated an increase in the purchase of organic traditionally raised meat and milk. Research results range from stating that genetically modified foods cause no harmful effects at all to stating that they cause massive tumour growth in animal studies. At this point, there is no conclusive evidence that proves GMOs are safe or not safe for human consumption. The rBST hormone is not used in dairy cows in Europe or Canada, but it is used in the United States. Cows that are given growth hormone have decreased fertility and increased infections that require antibiotic treatment. Milk containing growth hormones has been implicated in the earlier onset of puberty in girls since the 1980s. Drinking milk that contains rBST can stimulate other hormones in humans, such as **insulin-like growth hormone (IGF)**, which is another type of growth factor, and this is why it is not approved for use in Canadian dairy farming. See Table 10.1.

DID YOU KNOW?

Scientists have created bacteria that do not produce lactic acid, and these organisms could be used to colonize people's mouths and therefore prevent cavities. Cavities are formed by the lactic acid produced by *streptococcus mutans* bacteria in the mouth, which digest sugars and form lactic acid. If those bacteria didn't produce lactic acid we would not form cavities. In any case, we can also prevent cavities by eating less sugar.

DID YOU KNOW?

Genetically modified (transgenic) bacteria are used to produce toxic chemicals that kill insects, and therefore used as pesticides on crop foods. Other transgenic bacteria have been created to take up mercury and help decrease pollution.

TABLE 10.1

Protein Medications Made by Bacteria to Treat Human Conditions

Protein Medication	Condition Treated
Insulin	Type 1 diabetes
Erythropoietin	Anemia
Growth hormone	Dwarfism
Anticoagulants	Blood clotting after stroke or heart attack
Clotting factors	Hemophilia
Interferons	Some autoimmune diseases

CONCEPT REVIEW 10.4

1. What is gene cloning?
2. What is transformation?
3. Explain how a restriction map is useful.
4. How can you determine which bacterial colonies have the gene of interest?
5. Explain how hybridization is used to locate a specific colony.
6. List the steps required for bacteria to produce a medication.
7. What are some examples of medications made by bacteria? What conditions do they treat?

10.5 OTHER DNA TECHNIQUES

DNA libraries

A **genomic DNA library** is a collection of all DNA fragments that have been digested with restriction enzymes and ligated into plasmids. It contains all the possible DNA fragments of an organism or of a specific chromosome. For example, a human genomic DNA library contains all the possible gene sequences in the human genome within the plasmids within bacteria. Similar to the way information is contained in books on shelves in a library, pieces of DNA (information) are contained in plasmids in bacteria in a genomic library; DNA information is sorted, organized by species, and stored in a freezer. These libraries enable researchers to find any gene sequences they want by selecting bacterial samples and using the hybridization technique with a specific probe to find a specific gene. Since so many libraries have been made, it is now possible to order any sample from a catalogue from a company that makes and stores various types of DNA libraries.

cDNA

Recall that in eukaryotes, genes are encoded in both translated (exons) and non-translated (introns) segments (Chapter 9). In all eukaryotic organisms, the introns must be removed, by splicing, to form the processed mRNA molecule before translation can occur. Bacteria are prokaryotes and do not have introns or the mechanism for splicing out introns, so it is highly advantageous for genetic engineers to transfer DNA that has the introns removed already. Therefore, only the coding DNA sequence is ligated into bacterial plasmids.

It is possible to isolate mRNA that has already been processed from eukaryotic cytoplasm because splicing occurs in the nucleus. Recall that all mRNA has a **poly A tail**; therefore, a **polyT primer** would bind to all processed mRNA. We cannot insert RNA into a plasmid, so the mRNA has to be converted into DNA. This can be done using **reverse transcriptase**—an enzyme found in retroviruses such as HIV—that converts RNA into DNA before the retroviruses can replicate (Chapter 23). This DNA molecule, containing only the coding region of the gene, is called **complementary DNA (cDNA)**.

Gel electrophoresis

Gel electrophoresis is a method that uses an electrical current to separate strands of DNA based on the length of the DNA sequence. A type of plant carbohydrate, called agarose, is used to form a tiny mesh-like gel. The phosphate molecules in nucleotides are negatively charged, so DNA moves toward the positive end of the gel. The DNA molecules have to squeeze through the gel as the electrical current attracts the negative charges. The small fragments can move through the gel faster than the longer fragments (Figure 10.11). Gel electrophoresis is beneficial for determining if the correct length of fragment has been inserted into a plasmid.

The process involves the following steps: (1) a population of bacteria containing the gene of interest is grown; (2) the plasmid DNA from a small sample of cells is isolated and cut with the same restriction enzymes, separating the plasmid into two pieces, the original plasmid plus the inserted fragment; (3) the DNA is run on the gel along with a series of known fragments ranging in size from 300 to 2000 base pairs; this is called a *ladder*; and (4) the plasmid and the inserted fragment should

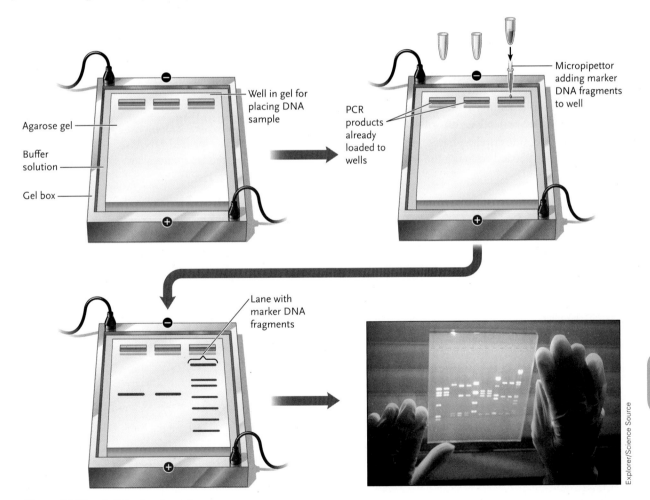

Figure 10.11 **Gel Electrophoresis**

Gel electrophoresis is used to separate strands of DNA based on the length of the DNA fragments. Small strands of DNA move through the gel faster than large strands.

Art source: From RUSSELL/HERTZ/STARR/FENTON. *Biology*, 2E. © 2013 Nelson Education Ltd. Reproduced by permission. www.cengage.com/permissions.

be seen in the gel beside the corresponding ladder fragment of the correct length. If the plasmid is visible but the fragment is not, then you know you don't have the gene of interest. Once you have a fragment that is the correct size, you can isolate that DNA from the gel and have it sequenced to ensure that it is the correct gene sequence and that no mutations occurred during the cloning process.

PCR

Polymerase chain reaction (PCR) is a technique to generate multiple copies of a specific DNA sequence. A specific primer must be used: a single-stranded fragment of approximately 20 nucleotides that match up with one strand of the gene of interest. Since DNA is double-stranded, a primer is needed for each end of the DNA, that is, a 5′ primer and a 3′ primer. The primers must match the correct end of the DNA since DNA polymerase can replicate DNA only in the 5′ to 3′ direction (Chapter 8). Using PCR to amplify a gene is a much faster and more reliable technique than using bacteria because the bacteria can produce point mutations during DNA replication.

Main ingredients required for PCR

- **Template DNA**—can come from a plasmid, or a blood sample or any DNA that contains the gene you want to amplify
- **Primers**—complementary DNA sequences that will bind to the DNA sequence you want to amplify
- **dNTPs**—nucleotide triphosphates—the (A, T, C, and G) nucleotides needed to make the new DNA

- **taq polymerase**—polymerase enzyme from a strain of bacteria (Thermus aquaticus) that lives at very high temperatures and therefore can withstand the increase in temperature needed to separate the DNA strands without itself being denatured.

Three steps involved in PCR

1. **Denaturation.** At increased temperature, DNA strands separate by breaking the hydrogen bonds that hold the bases together.
2. **Primer annealing.** 5′ and 3′ primers bind to the single-stranded DNA.
3. **Primer extension.** DNA polymerase replicates the DNA using the primers as a starting point.

Each cycle occurs simply by changing the temperature: **DNA denaturation** occurs at 91–95°C, annealing occurs at 50–55°C, and extension occurs at 72–75°C. This cycle is repeated 25 to 35 times to produce millions of copies of the template DNA.

Using PCR, researchers can amplify very small samples of DNA and then use them for testing: for example, when looking for an inherited genetic mutation, or for determining paternity, or for identifying a specific individual (Figure 10.12). PCR is used to amplify specific regions of DNA where known genetic markers are linked to hereditary diseases, and this makes it a valuable tool used for genetic testing.

DNA fingerprinting

Everyone has repeated DNA sequences in their introns. These repeated sequences are called **variable number**

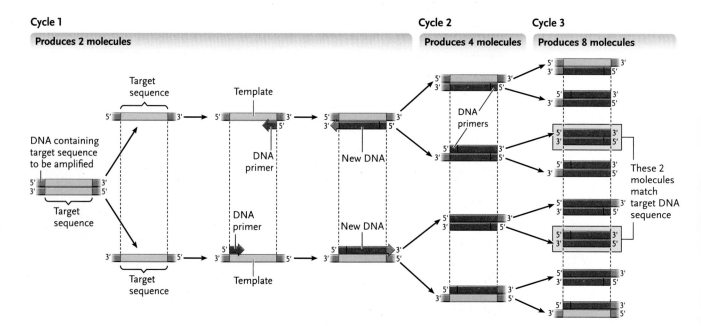

Figure 10.12 Polymerase Chain Reaction (PCR)

PCR is used to produce millions of copies of a DNA sequence.

tandem repeats (**VNTRs**). Everyone has the same sequences but different numbers of repeats of those sequences. Here is an example of a human VNTR:

GTCGGAATCG GTCGGAATCG GTCGGAATCG GTCGGAATCG

One person may have 10 copies of this sequence, and someone else may have 23. Consider how this information could be used. Imagine a crime scene where blood samples are taken from the victim, the crime scene, and the suspects. The DNA from the samples is digested with restriction enzymes, and then the DNA fragments are run on a gel to separate the varying *sizes* of DNA. Since each person has a different restriction pattern, the variation in lengths of their DNA shows up, except in identical twins. To identify specific people, the following techniques are used (Figure 10.13):

- PCR—to make many copies of a DNA sequence
- Restriction enzyme digestion—to cut DNA at specific sites
- Gel electrophoresis—to separate DNA strands based on size
- Hybridization—to use a specific probe to find VNTR DNA sequences

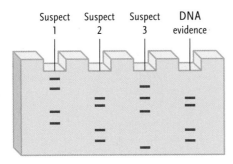

Figure 10.13 DNA Fingerprinting

DNA fingerprinting can be used to identify people because everyone except identical twins has their own unique DNA fingerprint.

Art source: From DIGIUSEPPE/FRASER. *Biology 11U* © 2011 Nelson Education Ltd. Reproduced by permission. www.cengage.com/permissions

An individual's DNA fingerprint is not 100% accurate. However, when multiple probes are used on multiple VNTRs, it is as reliable as traditional fingerprinting; everyone has their own specific pattern. DNA fingerprinting has been a reliable resource in criminal court since 1987.

Gene therapy

Gene therapy refers to the process of inserting a gene into an individual's cells to replace a mutated gene. The goal is to restore a cell's normal function with the healthy gene. So gene therapy can be used for single-gene inherited diseases such as cystic fibrosis or muscular dystrophy. Despite some great advances in this area, it remains difficult to insert a particular gene into the correct location in the genome in the correct cell type. Viruses are often used to target a specific cell type since viruses bind only to specific receptors (Chapter 23), and, therefore, only the targeted cell type will receive the gene. However, viruses inject cells with DNA that can be integrated randomly into the genome, not specifically; as a result this alters the normal regulation of transcription of the gene. A more common variation of gene therapy is called *antisense therapy*, where a homologous strand of DNA or RNA is used to bind to its complementary sequence and inactivate it. This technique is most commonly used to treat some cancers and has shown some success in blocking viral infections such as cytomegalovirus.

CONCEPT REVIEW 10.5

1. What is a genomic DNA library?
2. How can you find a specific sequence in a library?
3. What is cDNA and why is it useful?
4. What is gel electrophoresis? What is the main use for it?
5. Why is PCR useful?
6. What are the main ingredients and steps in PCR?
7. What are VNTRs? How can we make use of them?
8. What is gene therapy?
9. Why is gene therapy frequently not successful?

10

10.6 GENETICALLY MODIFIED ORGANISMS—GMOs

Genetically modified organisms (GMOs) include not only bacteria that contain human genes or genes from any other organism but also entire plants or animals that contain extra genes or are modified so that certain genes are turned off. Following are some examples of GMOs (see also Figure 10.14):

1. A gene that codes for a blue pigment can be added to plant cells, creating blue roses or blue strawberries.

2. A gene that causes fruit to produce ethylene gas, which causes fruit to ripen, can be inhibited with antisense RNA, and so the fruit doesn't ripen and rot during shipping; ethylene gas sprayed on the tomatoes at the store causes the tomatoes to ripen—known as *flavr savr* tomatoes.

3. Frogs engineered to be transparent are used to study the living frog.

4. Fluorescent genes from jellyfish can be injected into mouse and cat embryonic cells; Mr. Green Genes was the first glow-in-the-dark cat.

5. Chickens that don't grow feathers have been created so they don't have to be plucked.

The list of possible GMOs increases exponentially every year. Since all organisms are made of DNA—the same ATCG nucleotides—any DNA molecules can be combined to produce genetically modified organisms.

Figure 10.14 Examples of Genetically Modified Organisms

Genetically modified animals

Genetically modified (also called *transgenic*) animals can be produced by inserting a recombinant plasmid into a germ-line cell population (Figure 10.15). Instead of transforming bacteria with the recombinant plasmid to produce a protein, the plasmid can be introduced to

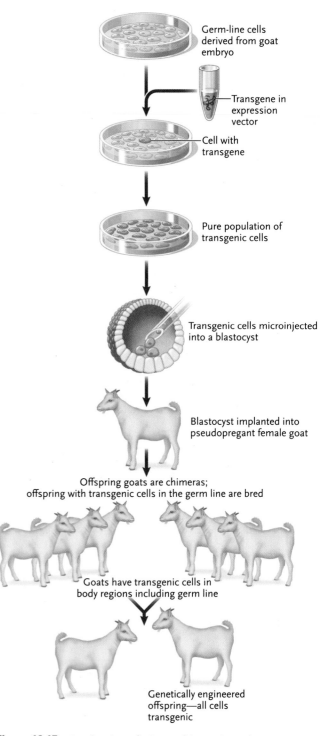

Germ-line cells derived from goat embryo

Transgene in expression vector

Cell with transgene

Pure population of transgenic cells

Transgenic cells microinjected into a blastocyst

Blastocyst implanted into pseudopregant female goat

Offspring goats are chimeras; offspring with transgenic cells in the germ line are bred

Goats have transgenic cells in body regions including germ line

Genetically engineered offspring—all cells transgenic

Figure 10.15   Production of a Recombinant Organism

The production of a recombinant organism involves the introduction of a new gene into embryonic stem cells.

Art source: From RUSSELL/HERTZ/STARR/FENTON. *Biology*, 2E. © 2013 Nelson Education Ltd. Reproduced by permission. www.cengage.com/permissions

stem cells that can be implanted into the blastocyst of a mouse, or cat or goat, etc. Once the new gene is incorporated into all of the embryonic cells, that gene will be expressed in the organism. *Where* the gene is expressed depends on the promoter the gene is attached to because that will affect how transcription is regulated.

Simplified steps to make a transgenic goat produce spider silk (BioSteel)

1. Isolate the spider silk gene.

2. Ligate the gene into a plasmid that contains the promoter sequence involved in the regulation of milk production in goats.

3. Microinject the recombinant plasmid into stem cells that can be injected into a goat blastocyst produced by *in vitro* fertilization.

4. Implant the blastocyst into a pseudopregnant goat—a female goat that has been given hormone injections to cause the uterus to be prepared for implantation.

5. The offspring of the goat will have the spider silk gene connected to the promoter region involved in lactation. When lactation occurs in that goat, the milk will contain spider silk that can be isolated and used to build many very strong substances. The first product ever made from BioSteel was a bullet-proof vest.

Since animals are eukaryotic and have the cellular machinery to process RNA and have Golgi bodies that modify proteins, they can be more reliable than bacteria for producing complete functional proteins. For example, ATryn, a drug produced in the milk of goats, is used to treat hereditary antithrombin deficiency.

Animals can also be genetically engineered to *not* produce a specific protein. For example, genes are deleted from the genomes of mice—known as **knockout mice**—so researchers can figure out the functions of those genes and what would happen to people with mutations in those genes. Several thousand single-gene or double-gene knockout mice exist for research purposes, and many are used to model human diseases, such as cancer, Parkinson's, diabetes, arthritis, and heart disease; these mice have been very valuable in finding treatments for many diseases.

Genetically modified plants

Many food crops are engineered to contain genes that give plants particular properties, such as the following:

- Resistance to herbicides called Roundup Ready®. Herbicides can be sprayed on whole crops, and only the weeds die.

- Ability to produce their own pesticides. This eliminates the need to spray for insects.

- Resistance to cold temperatures. This lengthens the growing season.

- Ability to prevent ripening. This enables long-distance transport without loss from rotting.

- Added nutrients. Golden rice contains a gene for beta-carotene, which is a precursor for vitamin A production.

One prevalent method of food plant engineering involves integrating a bacterial gene into a plant that codes for a protein that is toxic to insects, thereby producing insect-resistant crops. *Bacillus thuringiensis* (bt) is a natural soil bacterium that produces bt toxin. This toxin kills insects that eat it by creating pores in their digestive tracts, which leads to cell death and death of the insect. The crop plants that are now genetically engineered to contain either bt toxin or some other genetic modification include tobacco, potatoes, tomatoes, soybeans, corn, broccoli, canola, flax, sugar beet, rice, papaya, and cotton.

DID YOU KNOW?

Approximately 70–85% of processed foods contain genetically modified ingredients, mainly because many processed foods contain some form of corn starch or soybean oil.

DID YOU KNOW?

No research published to date has proven that consumption of genetically engineered foods cause harm. One study attempted to show that GMO corn caused large tumours in rats, but that study has major flaws, and the conclusions drawn are incorrect. See Forbes online references that explain why the rat tumour research is flawed: http://www.forbes.com/sites/stevensalzberg/2012/09/24/does-genetically-modified-corn-cause-cancer-a-flawed-study/

Implications of genetically engineered plants

- Insects become resistant to the bt toxin. Recall from Chapter 7 that microevolution occurs in every population. There are always some organisms that resist drugs, or antibiotics, or toxins. In fact, in 2009 the first resistant organism was found: a bollworm that infected cotton plants. As a result genetic engineers made cotton plants that contained multiple types of bt toxin-producing genes.

- Changes in ecosystems occur if even one population of organisms is altered. In many cases secondary pests become a problem, and the crops need to be sprayed with pesticides anyway.

- The engineered gene can be transferred into other closely related species. Many plants cross-pollinate with other similar species, and a wide variety of weed plants are similar to food plants. If the bt toxin gene crosses into other species it could have an effect on that ecology.

- The implications of genetically engineered foods on human health are not known. No reliable, long-term studies have been done with humans. The bt toxin

gene is expressed only by the parts of the plant that humans don't eat, such as the stem or leaves, but it is found in all of the plant's cells, including the parts used for food. Roundup Ready crops are sprayed with toxic herbicides that do not wash off the plants: a topic of considerable debate by health experts.

CONCEPT REVIEW 10.6

1. Give some examples of genetically engineered organisms that have been made.

2. In a genetically engineered animal, how is the expression of the gene controlled? For example, describe the difference between goats that produce spider silk in milk and mice that express GFP in their paws.

3. What is the advantage of using transgenic animals rather than bacteria to produce medications?

4. List the steps involved in making a transgenic goat that can produce spider silk in its milk.

5. Why are knockout mice so valuable to researchers?

6. What are the main ways that plants are genetically engineered?

7. What is bt toxin and how is it used to genetically engineer plants?

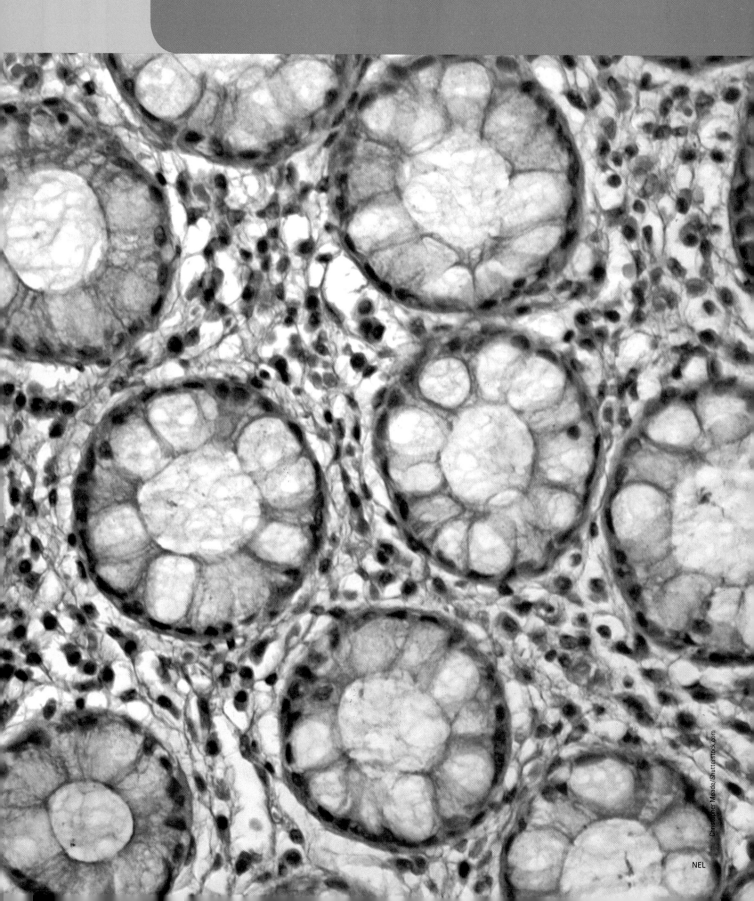

Any fool can know. The point is to understand.

Albert Einstein

A tissue is a combination of cells and proteins that function together in a specialized role. **Histology** is the study of tissues. Four main types of tissues are found throughout the body and support the function of our organ systems: **epithelial tissue**, connective tissue, muscular tissue, and nervous tissue.

11.1 EPITHELIAL TISSUE

Epithelial tissue has two major functions: (1) covering and lining the outside of the body and the inside of hollow organs, body cavities, and ducts; (2) forming glands. The first function protects organs and structures, and allows molecules and nutrients to move from one location to another by processes such as diffusion, filtration, secretion, and absorption. Recall from Chapter 3 that simple diffusion refers to the movement of small molecules across a cell membrane, and facilitated diffusion (facilitated transport) occurs when a membrane protein is required for an ion, such as a sodium ion, to cross the cell membrane. **Filtration** occurs in the glomerulus of the kidney when substances move from the blood into the Bowman's capsule (Chapter 20). **Secretion** is the movement of chemicals or molecules from a cell, usually into the bloodstream or into the lumen of a hollow organ, such as the small intestine or the kidney nephron. **Absorption** that involves epithelial cells is the movement of chemicals or molecules into the bloodstream, such as from the lumen of the digestive tract, or through the skin. In the kidney, molecules are reabsorbed through nephron epithelial cells from the lumen into the surrounding capillaries.

Glandular epithelial tissue is found in exocrine glands, such as sebaceous oil glands in the skin (Chapter 12), and endocrine glands such as the thyroid (Chapter 16). Glands are usually composed of several cell types: cells of the endocrine glands that secrete specific molecules, such as hormones, into the bloodstream; and cells of the exocrine glands that secrete sweat, saliva, milk, enzymes, or other substances to the surface of the body or through a duct into the lumen of a hollow organ.

Classification of epithelial cells

Epithelial cells are categorized based on their shape and the number of layers (Figure 11.1). They can have the following shapes: **squamous** (flattened), **cuboidal** (cube), **columnar** (rectangular, or column-like). Some epithelial cells that change shape are called **transitional epithelial cells**, such as bladder cells that change from cuboidal, when the bladder is empty, to squamous, when the bladder is full. *Simple* epithelial cells are a single layer, whereas those that are **stratified** occur in multiple layers. Some epithelial cells appear to be stratified when in fact they are a single layer: these cells are called **pseudostratified** because some cells do not reach the surface and so their nuclei appear layered. Some epithelial cells can have extra surface structures: such as **microvilli** to increase surface area for absorption of nutrients in the small intestine (Chapter 19), or **cilia** in the upper airways that help move mucus up and out of the respiratory tract (Chapter 18). See Table 11.1.

Characteristics of epithelial cells

Since epithelial cells cover and line surfaces and cavities of the body or glands, each side of the cell has different functions. Their specific function depends on whether the side of the cell is facing the surface

Classifying epithelium:
Cell shape

Transitional epithelial   Squamous   Cuboidal   Columnar

Arrangement of layers

Apical surface
Basal layer
Basement membrane

Simple   Pseudostratified   Stratified

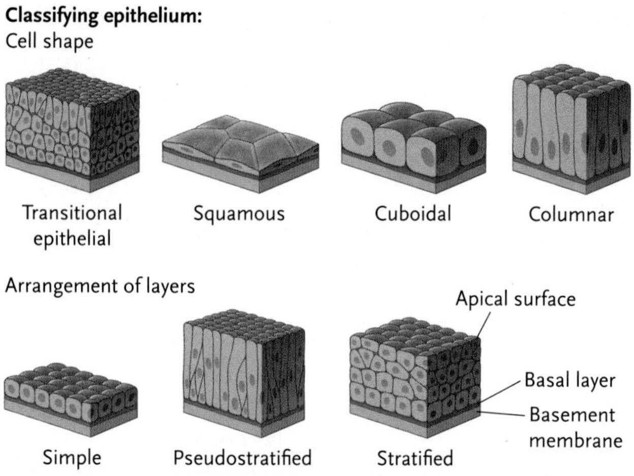

Figure 11.1   Epithelial Tissue
Epithelial tissue is classified by the shape of the cells and the arrangement of the layers.

or cavity (**apical** side), or if it is in contact with the **basement membrane** (**basalateral** side)—which is a protein region composed mainly of laminin and integrins that connect the epithelial cells to the connective tissue (Section 11.2). Epithelial tissue is **avascular**, meaning that it does not have any blood vessels, and so old cells can be shed without causing bleeding. Nutrients and oxygen diffuse into these cells from the underlying connective tissue. Epithelial cells are also highly connected by cellular junctions, such as gap junctions, tight junctions, adherens junctions, and desmosomes (Chapter 3). This means they are strongly held together and can transfer signalling molecules that allows the cells to act as a unit, that is, as a tissue. Because of the location of epithelial tissue and its role in protecting other structures, these cells undergo the most rapid cell cycle in the body and new cells are formed daily.

DID YOU KNOW?

A pap test is used to look at a sample of non-keratinized, stratified, squamous epithelial cells taken from the cervix to determine if cells are normal, precancerous, or cancerous. The cells are examined microscopically for shape or density changes. The most common type of cervical cancer is squamous cell carcinoma. Prognosis is very good if the cancer is detected early.

Classification of Epithelial Tissue

Simple Squamous

Squamous cell
Apical surface
Basal surface
Basement membrane

From RUSSELL/HERTZ/STARR/FENTON. Biology, 2E. © 2013 Nelson Education Ltd. Reproduced by permission. www.cengage.com/permissions

© Dr. Robert Calentine/Visuals Unlimited/Corbis

Functions

Single layer of flattened cells allows for diffusion, osmosis, filtration (glomeruli of kidneys), and secretion (surfactant in the alveoli).

Location

These cells line heart, blood vessels, lymphatic vessels, alveoli in lungs, and glomeruli in the kidney. It forms part of any serous membrane that covers internal organs, such as the heart (pericardium), lungs (pleura), and abdominal organs (peritoneum).

Simple Cuboidal

Apical surface
Squamous cell
Basal surface
Basement membrane

From RUSSELL/HERTZ/STARR/FENTON. Biology, 2E. © 2013 Nelson Education Ltd. Reproduced by permission. www.cengage.com/permissions

© Carolina Biological/Visuals Unlimited/Corbis

Functions

Single layer of cuboidal cells allows for secretion and absorption.

Location

These cells are found primarily in the ducts of exocrine glands, the lumen of the nephron in the kidney, the surface of the ovaries, and parts of the thyroid gland.

Simple Columnar

Cilia
Apical surface
Goblet cell
Ciliated columnar cell
Basolateral surface
Basement membrane
Underlying tissue

Jose Luis Calvo/Shutterstock.com

Functions

Ciliated, simple, columnar epithelial cells function to *move* substances, such as foreign particles trapped by mucus in the upper respiratory tract, or to move the egg along the fallopian tubes. Secretion of mucus.

Location

These ciliated cells are found in the upper respiratory tract, sinuses, fallopian tubes, and uterus.
(Cilia are hairlike projections that move in wavelike patterns.)

Apical surface (location of microvilli)
Non-ciliated columnar cell
Basal surface
Basement membrane

From RUSSELL/HERTZ/STARR/FENTON. Biology, 2E. © 2013 Nelson Education Ltd. Reproduced by permission. www.cengage.com/permissions

Jose Luis Calvo/Shutterstock.com

Functions

Non-ciliated, simple, columnar epithelial cells usually contain specialized extensions called microvilli that are cell extensions that allow for absorption of nutrients. This tissue contains goblet cells (considered unicellular glands) that secrete mucus.

Location

These non-ciliated cells are found primarily in digestive tract (stomach to anus), also found in the gallbladder. They secrete mucus that protects the lining of the digestive tract.

(Continued)

11

TABLE 11.1

Classification of Epithelial Tissue (*continued*)

Pseudostratified Columnar

Apical surface   Cilia
Goblet cell
Ciliated
pseudostratified
columnar cell
Basal cell
Basolateral
surface
Basement
membrane
Underlying
tissue

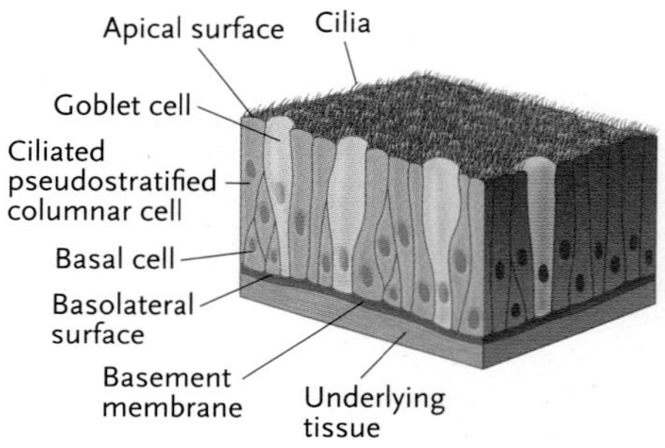

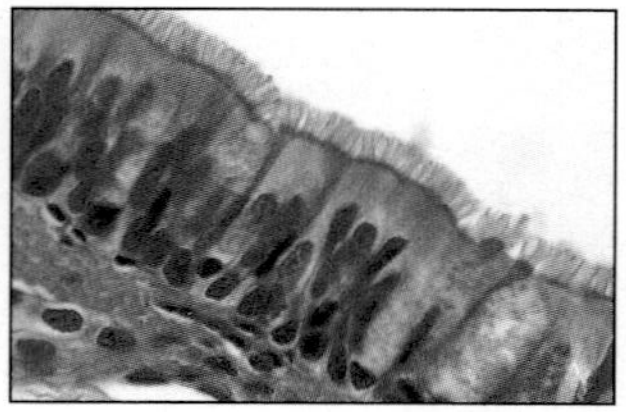

Jose Luis Calvo/Shutterstock.com

Functions	Location
Ciliated, pseudostratified, columnar epithelial cells function to secrete mucus, trap foreign particles, and move substances along the surface of the tissue. Notice how the cells appear to be stratified, but they are actually in a single layer with some cells not extending to the surface (*pseudo* means "false"). Non-ciliated cells function in absorption and secretion.	These cells are found in similar locations as ciliated, simple, columnar cells— primarily the respiratory tract. But non-ciliated, pseudostratified epithelial cells can be found in the epididymis and many glands.

Stratified Squamous

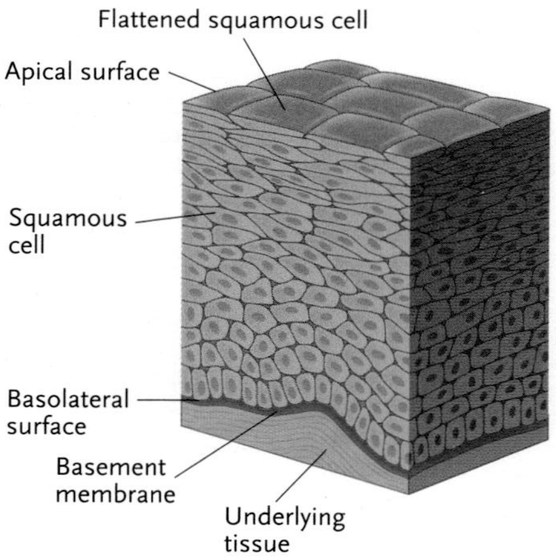

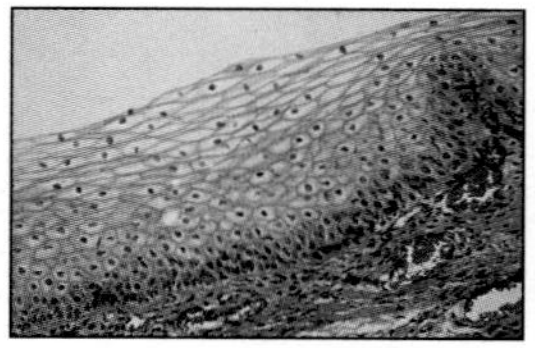

Flattened squamous cell
Apical surface

Squamous
cell

Basolateral
surface
Basement
membrane
Underlying
tissue

Biophoto Associates/
Science Source

Functions	Location
Stratified squamous is the most abundant of the stratified tissue types. The multiple layers function to protect the underlying tissues from abrasion and from microorganisms. Note that the apical cells are squamous, and the deeper basal cells are cuboidal.	**Keratinized,** stratified epithelial cells are found in the skin; the outermost layers contain keratin protein in dead cells to provide a tough barrier (Chapter 12). Non-keratinized, stratified epithelial cells are located in the mouth, tongue, esophagus, pharynx, and vagina.

11

Stratified Cuboidal

Apical surface
Cuboidal cells
Basolateral surface
Basement membrane
Underlying tissue

© Visuals Unlimited/Corbis

Functions

Stratified, cuboidal, epithelial cells can be two or more layers, and they function to protect underlying tissues and have some secretion and absorption properties.

Location

These cells are found mainly in esophageal ducts, sweat glands, mammary glands, and salivary glands.

Stratified Columnar

Apical surface
Columnar cells
Basolateral surface
Basement membrane
Underlying tissue

SCIENCE VU, VISUALS UNLIMITED/SCIENCE PHOTO LIBRARY

Functions

This less common cell type, functions in protection and secretion.

Location

These cells are found in the ducts of some exocrine glands, such as esophageal glands, and some parts of the urethra.

Transitional

Apical surface
Epithelial cells
Basolateral surface
Basement membrane
Underlying tissue

© M I (Spike) Walker/Alamy

Functions

Having characteristics of both squamous and cuboidal, transitional epithelial cells allow for a hollow organ to distend and stretch. Cell shape is squamous when stretched and cuboidal when relaxed.

Location

These cells are found in the bladder, ureters, and urethra.

11

1. What are the main functions of epithelial cells?
2. How are epithelial cells classified?
3. Briefly describe the location and function of each type of epithelial tissue.

11.2 CONNECTIVE TISSUE

Connective tissue is the most abundant and diverse tissue in the body. It is composed of combinations of different **cell types** and **extracellular matrix.** The matrix consists of fibrous proteins and molecules that make up the **ground substance,** which is a gel-like substance composed of water and non-fibrous proteins (Figure 11.2).

Overview of connective tissue functions

- **Loose connective tissue**—found in almost every body structure—connects epithelial tissue to muscle layers (skeletal or smooth), gives tissues strength and elasticity, stores fat (adipose tissue), and acts as a filter in spleen and lymph nodes (loose reticular tissue).

- **Dense connective tissue** forms strong attachments between structures, such as ligaments connecting bones within a joint, and allows organs to stretch and recoil, such as within arteries or the lungs.

- **Cartilage**—a very strong tissue—acts as a smooth surface for joint movement and gives particular structures their shape, such as the ear.

- **Bone** forms the skeleton, stores calcium and other minerals, produces red and white blood cells, stores some fat, and protects internal organs.

- **Blood** transports oxygen, nutrients, waste products, and hormones throughout the body.

Cell types in connective tissues

Fibroblasts make up the most abundant connective tissue cell type. They produce the collagen, elastin, and reticular fibres, as well as ground substance that make up the extracellular matrix. Fibroblasts are very important in the healing of wounds; when tissues are damaged, fibroblasts increase the synthesis of these substances to bind tissues together.

Adipocytes are cells that store fat as triglycerides. Any extra calories we eat—whether in the form of protein, carbohydrate, or fat—that are not used to build cell structures or to produce energy become converted into fat and stored in adipocytes in the subcutaneous layer of the skin, which is loose adipose connective tissue.

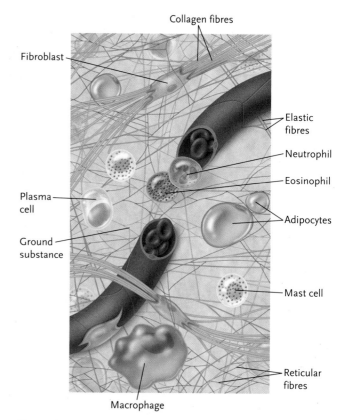

Figure 11.2 **Connective Tissues**
Connective tissues contain various cell types, protein fibres, and ground substance.

Mast cells, a type of immune cell, are located within loose and dense connective tissue along blood vessels. Mast cells produce **histamine** when there is tissue injury, which is part of the inflammatory response involved in healing tissue injuries. Mast cells also produce histamine during an allergic reaction.

Reticular cells are found in loose reticular connective tissue that produce the reticular fibres involved in filtering pathogens and old red blood cells in lymph nodes, the spleen, and the liver. **Chondrocytes** are the cells that make up cartilage connective tissue. **Osteocytes** are the cells that make up bone connective tissue. **Blood cells** include red blood cells that carry oxygen and some carbon dioxide, and the various white blood cells involved in the immune response (Chapter 22). See Table 11.2.

Types of protein fibres in connective tissue

Collagen, the most abundant protein in the body, makes tissues very strong. The more collagen fibres in a tissue, the stronger that tissue is. Tendons have a very high amount of collagen and are very strong because they connect muscles to bones and must withstand high pressure when muscles contract and pull on bones. Elastin is a protein that gives tissues the ability to stretch and recoil back to their original shape. The connective tissue layers beneath the epithelial cells of the skin have a high amount of elastin.

Reticular fibres are produced by reticular cells and are a type of thin collagen fibre surrounded by glycoproteins. These fibres form a framework for many tissues and act as a filter in the spleen and lymph nodes. Reticular fibres are also found in blood vessels; they surround fat cells, nerves, and muscles; and they make up one of the two layers of basement membranes (composed of basal lamina and the reticular lamina).

Composition of ground substance

Ground substance is the fluid part of the connective tissue that supports the functions of cells and fibres. Oxygen, nutrients, and hormones move from the blood vessels through ground substance to all the cells in the body. Ground substance contains water and combinations of proteins and carbohydrate chains, called **proteoglycans**. Some of the most important examples include glucosamine sulfate, hyaluronic acid, and chondroitin sulfate.

Glucosamine sulfate makes up a significant proportion of the ground substance in cartilage, tendons, and ligaments, and it is found in the fluid surrounding joints. **Chondroitin sulfate** is an important component of the fluid surrounding joints and within cartilage that allows the joints to withstand compression.

Hyaluronic acid is found in most types of connective tissue and plays a role in cell proliferation and migration. Immune cells and bacteria produce the **hyaluronidase** enzyme that breaks down hyaluronic acid, and this lets the immune cells and bacteria move easily through connective tissues. It is thought that cancer cells also produce this enzyme, which allows cancer cells to spread (metastasis).

11

TABLE 11.2

Classification of Connective Tissue

Loose Areolar

	Functions	Location
Patrick J. Lynch/Science Source	This is the most widely distributed connective tissue and it functions to strengthen, support, and provide elasticity to every body structure.	Found around all epithelial tissues, such as the subcutaneous layer and dermis layer of the skin, mucus membranes, blood vessels, nerves, and all organs.

Adipocyte Reticular fibres Fibroblast Ground substance Elastic fibres Mast cell Plasma cell Collagen fibres

Loose Adipose

	Functions	Location
Courtesy of Wendi Roscoe	Adipose tissue, which stores fat as triglycerides, can be used to produce energy (ATP), insulate the body, and to protect and cushion organs.	Found wherever there is areolar connective tissue: in the subcutaneous layer of the skin and around blood vessels, nerves, and organs.

Nucleus Adipocyte Fat storage Blood vessel

Loose Reticular

Reticular fibres

Reticular cells

Biophoto Associates/
Science Source

Functions

This type provides structural support, forms part of basement membranes that surround organs, blood vessels, muscles, and acts as a filter to remove old red blood cells in the spleen and lymph nodes.

Location

Found in liver, spleen, lymph nodes, bone marrow, blood vessels, all organs, and smooth muscle.

Dense Regular

Collagen fibre

Fibroblast cell

Ed Reschke/Photolibrary/
Getty Images

Functions

This type provides very strong attachments for connecting muscle to bone (tendons) and connecting bone to bone (ligaments). Dense, regular, connective tissue contains a high amount of parallel strands of collagen, making it very strong. This tissue does not contain blood vessels, so injuries heal very slowly.

Location

Found in tendons, ligaments, and aponeuroses (the sheetlike tendons that connect muscle to muscle) (Chapter 14).

(Continued)

11

TABLE 11.2

Classification of Connective Tissue (*continued*)

Dense Irregular

	Functions	Location
Collagen fibre Elastic fibre Fibroblast	This type consists of primarily collagen fibres, like dense, regular tissue, except that the fibres are not parallel; they are arranged randomly and form a strong sheetlike connective tissue that provides strength when pulled in many directions.	Makes up the fascia that surrounds muscles, organs— such as the heart (pericardium)—digestive tract, bones (periosteum), and that forms part of the lower dermis region of the skin.

Biophoto Associates/
Science Source

Dense Elastic

	Functions	Location
Elastic fibre Fibroblast	This tissue allows high amount of stretch and recoil.	Found in structures that undergo stretching forces, such as arteries, lungs, trachea, bronchial tubes, and vocal cords.

Biophoto Associates/
Science Source

Hyaline Cartilage

Lacuna with chondrocytes inside

Ground substance

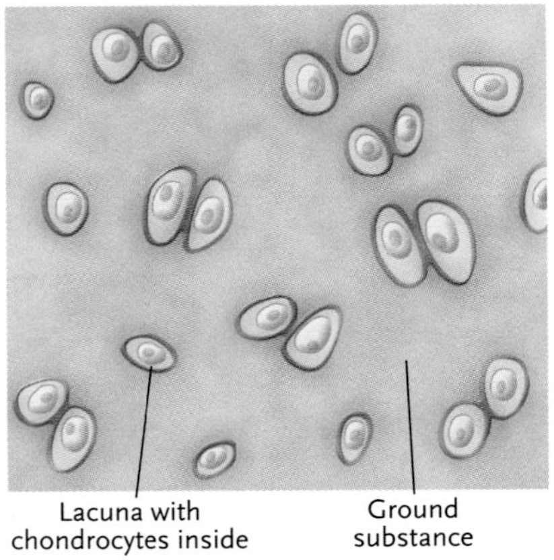

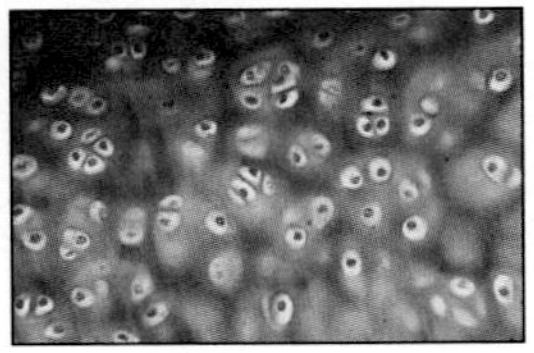

M. I. Walker/Science Source

Functions	Location
This tissue provides a strong, smooth surface for joint movement. Cartilage does not contain any blood vessels. Hyaline cartilage is the weakest of the three types and can fracture easily.	Found at the ends of long bones, ribs, nose, trachea, larynx, bronchi, and comprises most of the fetal skeleton.

Elastic Cartilage

Lacuna with chondrocytes inside

Elastic fibre

Ground substance

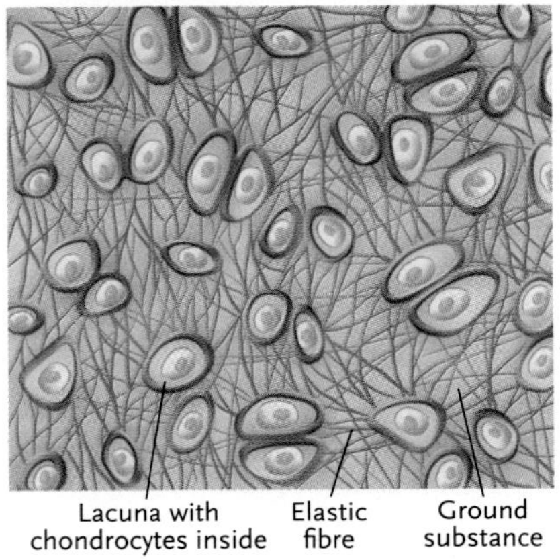

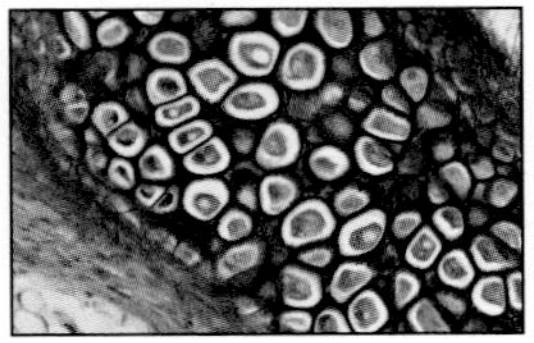

HERVE CONGE, ISM/ SCIENCE PHOTO LIBRARY

Functions	Location
This tissue provides strength as well as elasticity and shape of certain structures.	Found in the epiglottis, external ear, and the Eustachian tubes (tubes that connect the inner ear to the throat).

(Continued)

TABLE 11.2

Classification of Connective Tissue (continued)

Fibrocartilage		Functions	Location
 Lacuna Chondrocytes Collagen fibre	 Biophoto Associates/ Science Source	This is the strongest type of cartilage, containing thick collagen fibres that connect certain structures together and act as shock absorbers.	Found in the intervertebral discs, pubic symphysis, and knee joint.

Bone		Functions	Location
 Lacunae containing osteocytes Ground substance Vein Nerve Artery	 HERVE CONGE, ISM/ SCIENCE PHOTO LIBRARY	Bones move when muscles contract and produce movement. Bones are important in mineral homeostasis and provide a reservoir for calcium and phosphorus, and they protect internal organs. Red marrow produces red and white blood cells; yellow marrow stores some fat.	Includes long bones, skull, ribs, vertebrae, and pelvis (Chapter 13).

Blood		Functions	Location
 White blood cell Plasma Red blood cells Platelets	 BIOPHOTO ASSOCIATES/SCIENCE PHOTO LIBRARY	Blood transports oxygen, nutrients, hormones, and waste products throughout the body. Red blood cells transport oxygen and some carbon dioxide (Chapters 17 and 18). White blood cells include a variety of immune cells (Chapter 22). Platelets are important for blood clotting.	Found within blood vessels, but also where red and white blood cells and platelets are produced in red bone marrow.

CONCEPT REVIEW 11.2

1. Name the three main components of connective tissue.
2. What are the main functions of each type of connective tissue?
3. Describe the main functions of each cell type found in connective tissue.
4. What kinds of protein fibres are found in connective tissue, and why are they important?
5. Why is ground substance important?
6. Describe the main function and locations of each type of connective tissue.

11.3 MUSCLE TISSUE

The function of muscle tissue is to contract and generate force. When muscle fibres contract they produce movement and heat. Muscle fibres can contract without producing movement—called muscle tone—which is important for maintaining body posture and protecting internal organ structures. Specialized muscle tissue, **sphincters,** regulates the movement of substances within the body: for example, in diverting blood flow into certain capillary beds, and the movement of food through the digestive track.

Muscle tissue can be classified into three categories: skeletal, smooth, and cardiac. Each type of muscle tissue has distinctive properties that allow it to function optimally (see Figure 11.3).

Skeletal muscle

Skeletal muscle—the muscle tissue that is connected to our bones (skeleton)—allows our limbs to move. Muscles are connected to two different bones across a joint so that contraction will cause movement of the bones. For example, to bend the arm at the elbow, the brachialis muscle is connected to the humerus (upper arm bone), crosses the elbow, and connects to the ulna (one of the forearm bones). Mostly, skeletal muscle cells (also called fibres) contract as a result of something we decide to do. For example, the motor cortex of the brain's frontal lobe is primarily involved in contraction of skeletal muscles, for example, picking up something from the ground, and in this sense skeletal muscle contraction is considered **voluntary.** Sometimes certain skeletal muscles contract involuntarily; these are **reflexes,** such as pulling our hand away from a hot stove.

Skeletal muscle cells are **multinucleated** because during their development, many muscle cells merge together. Skeletal muscle fibres are **striated** due to a light and dark banding pattern that's visible under a microscope. The individual units, called sarcomeres, line up evenly, unlike smooth muscle. Skeletal muscle fibres can be subdivided into fast twitch, slow twitch, and intermediate, giving various muscle groups their different capacity for endurance or power movement (Chapter 14). (See Figure 11.3a.)

Smooth muscle

Smooth muscle is found within the walls of hollow structures, such as blood vessels, lymphatic vessels, stomach, intestines, airways, fallopian tubes, ureters, bladder, uterus, vas deferens, erector pili muscles in the skin, and the iris of the eye. Smooth muscle contraction and relaxation is controlled by the **autonomic nervous system** and is therefore considered **involuntary movement** that is unconsciously controlled. Different smooth muscles contract or relax when stimulated by either the sympathetic ("fight or flight") or parasympathetic ("rest and digest") components of the nervous system (see Chapter 15).

Smooth muscle cells each contain one single nucleus, and they are **non-striated** because they overlap unevenly; so there are no visible banding patterns. Smooth muscle cells are much smaller and contract very slowly, with much less force compared to skeletal and cardiac muscle cells. There are two kinds of smooth muscle cells, single unit and multiunit. **Single unit smooth muscle** occurs in sheets that surround visceral organs, where many cells contract simultaneously: for example, as found in the stomach, intestines, uterus, and bladder. **Multiunit smooth muscle** consists of many cells contracting individually, which gives more precise control: for example, as found in blood vessels, the eye, erector pili, and airways (see Figure 11.3b).

Cardiac muscle

Cardiac muscle, found only in the walls of the heart, contracts in order to circulate blood throughout the body. Cardiac muscle has some features similar to skeletal muscle and some features similar to smooth. Cardiac muscle fibres are involuntary in the sense that their contraction is not under our conscious control. Cardiac muscle is **autorhythmic,** meaning that specialized cells in the heart, called **pacemaker cells,** control the rate of contraction. However, the autonomic nervous system can influence the heart rate. Cardiac muscle fibres are striated like skeletal muscle, but have one to three nuclei per cell. Cardiac muscle fibres are branched and connected by **intercalated discs,** which contain gap junctions that allow signalling molecules to quickly pass between cells (see Chapter 14). This very unique feature of cardiac muscle is required for controlled and simultaneous contraction of the chambers of the heart so that blood circulates efficiently throughout the body (see Figure 11.3c).

11

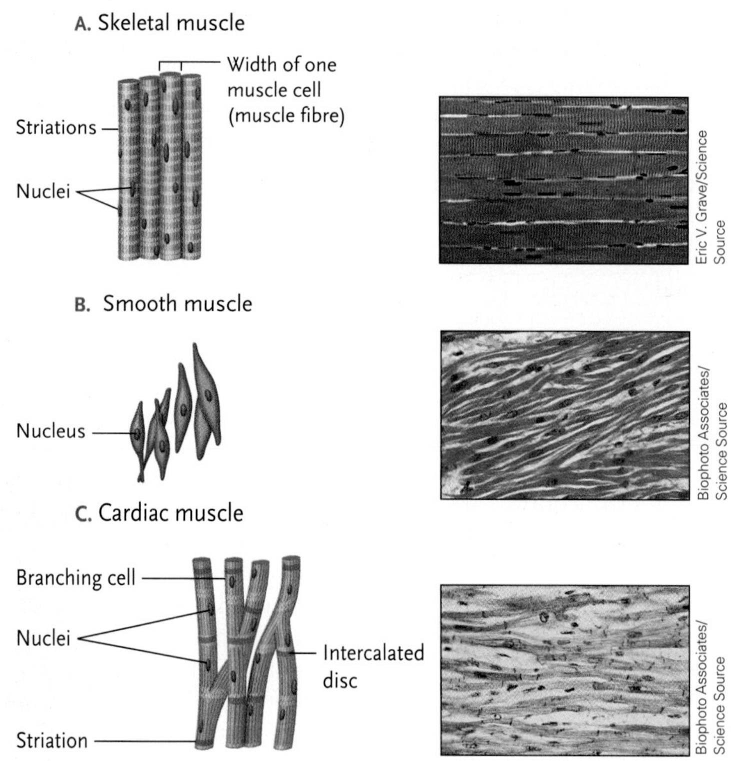

A. Skeletal muscle

Width of one
muscle cell
(muscle fibre)

Striations

Nuclei

B. Smooth muscle

Nucleus

C. Cardiac muscle

Branching cell

Nuclei

Intercalated
disc

Striation

Figure 11.3   Muscle Cell Types
There are three types of muscle cells: (A) skeletal, (B) smooth, and
(C) cardiac.

Art source: From RUSSELL/HERTZ/STARR/FENTON. Biology, 2E. © 2013 Nelson Education Ltd. Reproduced by permission. www.cengage.com/permissions

CONCEPT REVIEW 11.3

1. What are the functions of muscle tissue?

2. What are the three types of muscle tissue?

3. Make a study chart that summarizes the characteristics of each type of muscle.

4. Where is smooth muscle located?

5. What are the two types of smooth muscle and where are they located?

6. What are the special features of cardiac muscle tissue?

7. Why are intercalated discs an important component of cardiac muscle cells?

11.4 NERVOUS TISSUE

The function of nervous tissue is to detect and transmit sensations, interpret environmental stimuli, and respond to stimuli by signalling other **neurons**, muscles (skeletal, smooth, or cardiac), or glands (endocrine or exocrine). Nervous tissue is composed of the neurons that send and receive signals (Figure 11.4), and **neuroglia**, which are cells that support the neurons in various ways.

There are three general categories of neurons: multipolar, bipolar, and unipolar (Figure 11.5). Each type of neuron is found in different areas of the nervous system. **Sensory neurons**, also called **unipolar neurons**, transmit information from sensory receptors to the central nervous system (brain and spinal cord); **bipolar neurons** are found in the olfactory sensory system and the retina; and **motor neurons**, also called **multipolar neurons**, transmit information from the CNS to muscles or glands (Chapter 15).

Neuroglia are the cells that support the neurons and ensure proper functioning of the central (CNS) and peripheral (PNS) nervous systems. **Oligodendrocytes** (CNS) and **Schwann cells** (PNS) produce the protective myelin covering of neurons as found in CNS white matter and on long sensory and motor neurons. **Astrocytes** (CNS) and **satellite cells** (PNS) maintain ion concentrations surrounding neurons, and astrocytes form the blood-brain barrier. **Microglia** are resident immune cells in the brain and spinal cord. **Ependymal cells** produce cerebral spinal fluid.

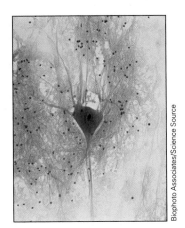

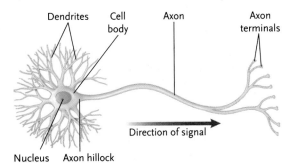

Figure 11.4 **Structure of a Typical Neuron**

A typical neuron is composed of dendrites, a cell body, an axon, and axon terminals.

Art source: From RUSSELL/HERTZ/STARR/FENTON. Biology, 2E. © 2013 Nelson Education Ltd. Reproduced by permission. www.cengage.com/permissions

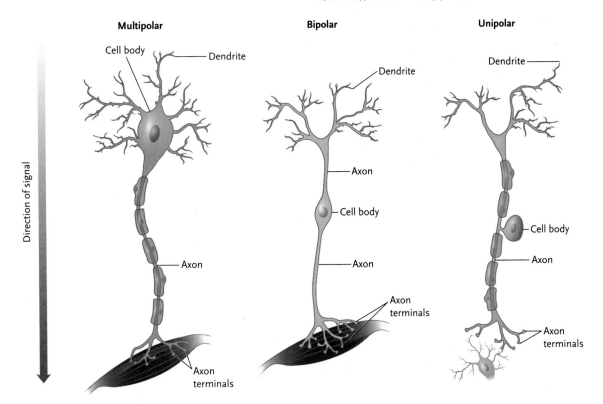

Figure 11.5 **Types of Neurons**

The three types of neurons are multipolar, bipolar, and unipolar.

Art source: From RUSSELL/HERTZ/STARR/FENTON. Biology, 2E. © 2013 Nelson Education Ltd. Reproduced by permission. www.cengage.com/permissions

CONCEPT REVIEW 11.4

1. What are the functions of nervous tissue?

2. What types of cells can neurons signal?

3. What cell types are found in nervous tissue?

4. Name the three types of neurons and where they are located.

5. Briefly, name the types of neuroglia and their main functions.

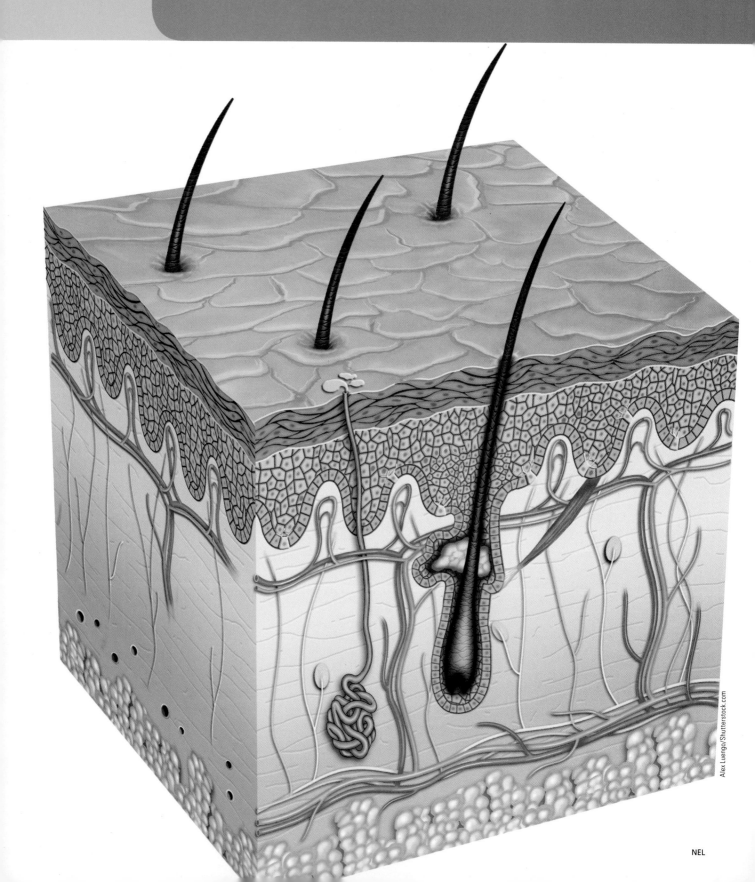

Live as if you were to die tomorrow. Learn as if you were to live forever.

Mahatma Gandhi

The **integumentary system** consists of skin, glands, hair, and nails. The **skin** is the largest organ in the body, making up around 12% of our total body weight. In other animals the integumentary system includes fur, feathers, scales, claws, and hooves.

12.1 FUNCTIONS OF THE INTEGUMENTARY SYSTEM

- Protection. Our skin protects our internal organs from infections, physical damage, dehydration, and environmental hazards, such as chemicals and UV light. The skin is the first barrier to pathogens (bacteria, viruses, fungi, or parasites) that we come in contact with. Sebaceous glands secrete oil, called sebum, which lubricates the outer layers of skin (epithelial cells), helps skin act as a waterproof barrier, and provides antimicrobial properties (Section 12.3).

- Regulation of body temperature. The deepest layer of the skin, the hypodermis (also called the subcutaneous layer), contains fat cells (adipose tissue) that help the body retain body heat. Some layers of the skin (dermis and hypodermis) have blood vessels that constrict in cold temperatures to keep heat within the body, or dilate in hot temperatures to promote heat loss. When blood vessels constrict and there is less blood flow near the cold body surface, less body heat escapes. By contrast, sweat glands in the dermis layer increase heat loss from the body through evaporation of sweat (water and ions). As water is pulled away from the body during evaporation, heat also leaves the body.

- Vitamin D synthesis. Cholesterol molecules in the lower layers of the epidermis of the skin are converted into a vitamin D precursor by sunlight, which the liver and kidneys then convert into active vitamin D, called calctriol (Chapter 20). Although vitamin D can also be acquired through the diet, some daily sunlight exposure is very important. Vitamin D plays many important roles in bone mineral regulation, hormone synthesis, and immune system regulation (see Chapters 16 and 22).

- Sensation. Our skin contains several different sensory receptors that allow us to be aware of changes in temperature, pain, touch, vibration, and pressure.

- Waste removal. Some waste products, including salts and urea, are removed from the body through the sweat glands. The skin, lungs, liver, and kidneys are all involved in breaking down or removing waste products from the body (Chapter 20).

- Protection provided by the hair and nails. Humans have hair on most surfaces of the skin, except on the palms of the hands and soles of the feet (thick skin regions). Having more hair on the head provides protection from UV damage, and eyebrows and eyelashes help protect the eyes from damage by foreign particles. For example, certain reflexes cause blinking when objects come close to our eyes, and the hairs of the eyebrows and lashes help trap small particles. Differences in hair distribution patterns are related to many factors, including genetics, hormones, and age. Nails are thick layers of dead keratinocytes that help us grip objects. Our nails are not as strong or sharp as claws in other animals, which need them for survival, such as in climbing to find food or escaping predators (Chapter 7).

DID YOU KNOW?

In animals that have fur or feathers, vitamin D is produced in an oily substance secreted by the skin. This substance covers the surface of the fur or feathers, and when the animal is grooming it ingests the vitamin D.

DID YOU KNOW?

Foods that are good sources of vitamin D include oily fish such as cod, herring, trout, sardines, tuna, and salmon, and also egg yolks and fortified dairy products. Food sources and dietary supplements of vitamin D are important for people living in regions that have limited sunshine during winter months.

CONCEPT REVIEW 12.1

1. List the components of the integumentary system.
2. List the main functions of the integumentary system.

3. Why is vitamin D important?
4. How does the skin help us regulate body temperature?

12.2 EPIDERMIS, DERMIS, AND HYPODERMIS

The skin is made up of an **epidermis** and a **dermis**, which is attached to a subcutaneous layer called the hypodermis (Figure 12.1). The epidermis is the outermost region and is made up of either four layers of cells (thin skin) or five layers of cells (thick skin), depending on the location in the body. Thick skin is located on the palms of the hands and soles of the feet. The dermis is divided into a **papillary layer** and a **reticular layer**, which is primarily connective tissue. The subcutaneous layer contains adipose tissue and is the main area for storing fat, which provides both stored energy and insulation for helping maintain body temperature.

Epidermis

The epidermis is made of stratified, squamous epithelial cells (mostly keratinocytes), and it is avascular (Chapter 11). Nutrients and oxygen must diffuse from capillaries in the dermis into the epidermis. The deepest layer of the epidermis where it contacts the dermis is the **stratum basale,** which consists of a single row of **keratinocytes** that divide very rapidly, some Merkel cells, and some melanocytes (Figure 12.2). This basale layer contains the most rapidly dividing cells in the whole

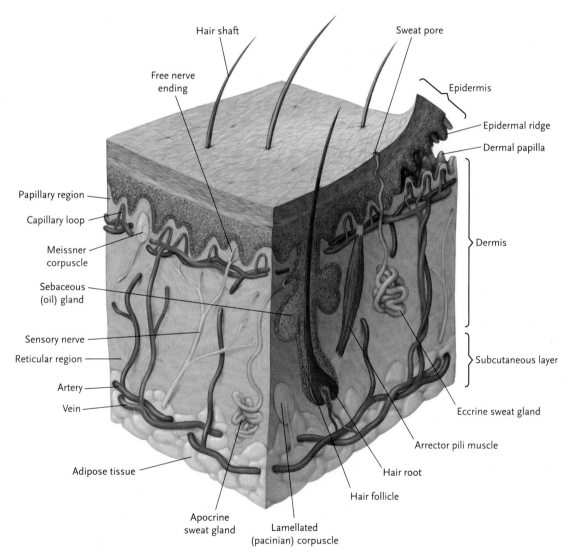

Figure 12.1   Integumentary System

The integumentary system is composed of the epidermis, dermis, and subcutaneous layers.

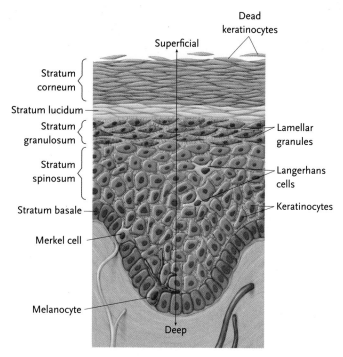

Figure 12.2   Epidermis

The epidermis is composed of layers of keratinocytes from the stratum basal to the stratum corneum. The epidermis also contains melano-cytes, Langerhans cells, and Merkel cells.

body, and it is where all epidermis cells are produced. **Merkel cells** are specialized mechanoreceptors that detect touch. They are closely associated with sensory neurons that send information to the somatosensory cortex of the brain where sensory stimuli are interpreted (Chapter 15). **Melanocytes** are the cells that produce **melanin** when exposed to UV light; they are the brown pigmentation associated with a tan. Melanin is produced by melanocytes from the amino acid tyrosine, and it acts as a photo-protectant because it can absorb UV energy, which then protects the DNA in skin cells from UV damage.

Newly replicated keratinocytes move up into the next layer and make up the greatest proportion of the epidermis: the **stratum spinosum**. This second layer is composed primarily of keratinocytes but also contains **Langerhans cells**. Langerhans cells are a type of immune cell that helps prevent infections by phagocytosis of infectious pathogens (Chapters 3 and 22). Some mitosis does occur in this layer, but not to the extent of the cell division that occurs in the basale layer. In this layer, keratinocytes begin to produce keratin, which is a strong, fibrous, structural protein that forms into intermediate filaments (Chapter 3) and makes the skin tough and strong; it also occurs in large amounts in hair and nails. Keratin is found in abundance in animal claws, scales, hooves, shells, and horns. A **callus** is formed when we use an area of skin repetitively. This happens because the keratinocytes increase the synthesis of the tough keratin protein that hardens the skin and protects it from excessive damage. A callus also has stratum corneum layers.

As keratinocytes continue to migrate to the next layer, the **stratum granulosum, lamellar granules** are released and keratin production increases. Lamellar granules are the organelles in keratinocytes. They merge with the cell membrane and release lipids, enzymes, and proteins into the extracellular space, which acts as a protective waterproof barrier, and they also help prevent infection by breaking down bacterial cell walls (Chapter 22). Lamellar granules are also found in certain cells of the lung, and they secrete surfactant (Chapter 18).

In areas of the body that have thicker skin, such as the soles of the feet and the palms of the hands, an extra layer of a few rows of dead keratinocytes form the **stratum lucidum**. This layer is composed of three to five layers of dead keratinocytes that have high amounts of keratin; this is necessary to protect these areas that endure much more physical stress compared with other skin areas.

The final layer of the epidermis is the **stratum corneum**, which consists of approximately 30 layers of dead keratinocytes that are completely filled with keratin. The stratum corneum cells are continually being sloughed off our bodies and replaced with cells migrating up from the deeper layers.

Dermis

The dermis is composed of strong and flexible connective tissue that allows the skin to move and stretch and return to its original shape. The dermis is divided into two sections, the papillary layer, which binds the dermis to the epidermis, and the deeper reticular layer, which connects to the underlying subcutaneous region.

The papillary layer is made up of loose areolar connective tissue that contains fibroblasts, collagen, reticular fibres, and elastin fibres (Chapter 11). The papillary layer of the dermis projects into the epidermal layer. This is called **dermal papillae**, which increases the surface area for the diffusion of oxygen and nutrients into the avascular epidermis. On the palms of the hands and soles of the feet, the dermal papillae lie on top of **dermal ridges** and cause the epidermis to also form **epidermal ridges**; these form our fingerprints and footprints (Figure 12.3). The pattern of the epidermal ridges is genetically deter-mined and everyone has a different pattern.

The reticular layer of the dermis consists of dense, irregular, connective tissue that contains many elastin and collagen fibres and fibroblasts in an irregular pattern that gives the skin strength and the ability to withstand pulling forces and to recoil from stretch.

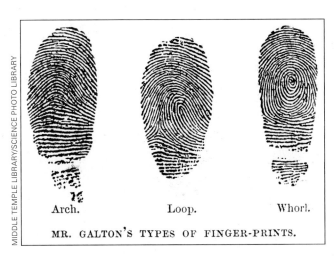

MR. GALTON'S TYPES OF FINGER-PRINTS.

Arch. Loop. Whorl.

Figure 12.3 Fingerprints
Fingerprints are formed from epidermal ridges.

Many structures are found within the dermis layer, including blood vessels, hair follicles, sebaceous glands, sweat glands, neurons, sensory receptors, and the arrector pili muscles attached to every hair follicle.

Merkel cells are touch receptors located in the epidermis, but there are many other sensory receptors in the dermis. **Pacinian corpuscles** detect pressure, nerve endings (nociceptors) detect pain, thermoreceptors detect temperature, and **Meissner corpuscles** detect touch and vibration. It is important to note that sensory receptors are not the same as membrane protein receptors (Chapter 3). Membrane receptors bind to specific molecules, such as a hormone or neurotransmitter, but a sensory receptor can be a whole cell, such as Merkel cells and pacinian corpuscles.

Arrector pili muscles contract and cause hairs to stand up (**piloerection**)—"goose bumps"—when the sympathetic nervous system is stimulated by fear or cold. The evolutionary significance of this response is much more useful in animals with fur; they appear larger to try to avoid confrontation and they can maintain body temperature. Since humans do not have fur, piloerection doesn't do much to prevent heat loss or scare off rivals and is therefore considered a vestigial structure (a remnant of evolution that is no longer useful).

DID YOU KNOW?

Tattoo ink is injected into the dermal layer of the skin. Cell division is much slower in this region and most of the dermis is connective tissue, so the ink will stay permanently. The epidermis is clear so the ink in the dermis is visible through the epidermis.

Hypodermis

The hypodermis, also called the subcutaneous layer, contains blood vessels, nerves, and adipose loose connective tissue. The main function of the hypodermis is to connect the dermis and epidermis to underlying muscle tissue and to store energy as fat. This fat can be broken down and used to make ATP when our blood sugar level decreases in between meals, when glucagon is produced by our pancreas (Chapters 4 and 16). It is also essential to have some fat storage to provide insulation in cold temperatures. People that are overweight have an overabundance of adipose tissue under their skin.

Skin colour

The colour of our skin is determined by several genes (polygenetic inheritance, Chapter 6). Variations in these genes give us a vast range of human skin tones. Melanin is the black-brown pigment secreted by melanocytes. Hemoglobin is the oxygen-carrying molecule in red blood cells and appears redder when more oxygen is bound and is visible through the skin. If blood vessels in the skin dilate, the skin appears more red or pink in colour. **Carotene** is the orange-yellowish pigment found in many foods we eat, such as carrots and egg yolk, and some is stored in the skin. Eating excessive amounts of carrots can cause the skin to look orange; this is called **carotenemia** and is not harmful. Carotene is an important pigment to have in our diet because it is the precursor for **vitamin A**, which is essential for healthy skin and vision.

CONCEPT REVIEW 12.2

1. Name the layers of the epidermis.
2. What types of cells are found in the different layers of the epidermis?
3. Describe the function of the various cell types.
4. Describe the function of melanin, keratin, and lamellar granules.
5. What are the layers and components of the dermis?
6. What types of sensory receptors are found in the epidermis and dermis? What type of sensation does each detect?
7. What are the main functions of the hypodermis?
8. What molecules affect skin colour?

MIDDLE TEMPLE LIBRARY/SCIENCE PHOTO LIBRARY

12

12.3 GLANDS

Several types of glands are located in the integumentary system: **sudoriferous glands** (sweat glands), **sebaceous glands** (oil glands), **ceruminous glands** (produce ear wax), and **mammary glands** (in females produce milk). The two types of sudoriferous glands are eccrine and apocrine (Figure 12.4).

Eccrine glands are found throughout the body, particularly on the forehead, palms, and soles. Eccrine glands are located in the deep reticular region of the dermis, and they have a duct leading to a pore on the surface of the epidermis. Eccrine sweat production is important for body temperature regulation.

Apocrine glands are located primarily in the axilla (armpit), the areola of the nipples, and the genital regions. These glands are located deep in the reticular dermis or in the hypodermis. Sweat secreted from these glands is thicker because it contains fatty acids in addition to water and ions. When the fatty acids are metabolized by the bacteria that reside on the skin, the fatty acids are broken down into **butyric acid**, which has an odour. Apocrine glands begin to function during puberty. Apocrine gland sweat is thought to function as **pheromones**, molecules that are important for attracting a mate in most animal species.

Sebaceous glands that produce sebum and surround each hair follicle in all regions of the body are not found on the palms or soles. Sebum is primarily composed of fatty acids, triglycerides, and **squalene**, which is similar in structure to long-chain fatty acids. Sebum lubricates the skin and hair, and it contains important molecules, such as antioxidants, pheromones, and antimicrobial substances that help protect us from infection. Acne is caused in part by an increased production of sebum that is not released from the gland and by an increase in bacteria that ingest sebum and initiate a small inflammatory response. During puberty, increased sebum production coincides with increased testosterone or estrogen production in males and females, respectively. Acne medication kills the bacteria that cause the inflammatory response.

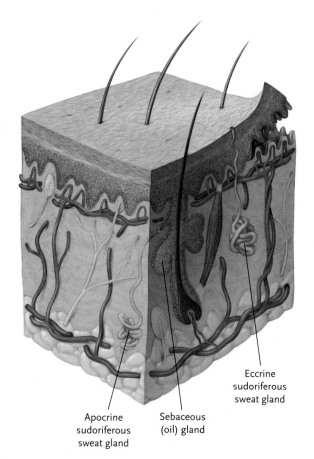

Apocrine sudoriferous sweat gland

Sebaceous (oil) gland

Eccrine sudoriferous sweat gland

Figure 12.4 **Glands of the Skin**
The skin contains sebaceous oil glands and two types of sudoriferous sweat glands: eccrine and apocrine.

DID YOU KNOW?

The only difference between a pimple and a blackhead is exposure to oxygen. Blackheads turn dark because the sebum is oxidized when exposed to air. Pimples are not dark because the sebum stays under the skin within the gland.

DID YOU KNOW?

A thick layer of sebaceous oil, called vernix, covers a baby during its embryonic development. This layer helps protect the baby's skin while surrounded by amniotic fluid, and sometimes babies are born with some vernix still on their skin.

Ceruminous glands located in the dermis of the ear canal produce ear wax. Ear wax is a substance called **cerumen**, which is very similar in composition to sebum, but it is thicker and coats the inside of the external ear canal to trap foreign particles. It also acts as water barrier and inhibits the growth of bacteria.

Mammary glands are located in breast skin tissue and are specialized to produce milk during lactation (Chapters 16 and 21).

CONCEPT REVIEW 12.3

1. What are the two types of sudoriferous glands? Where are they found? What is their function?
2. What is the location and function of the sebaceous glands?
3. What causes acne?
4. What is the function of ceruminous glands and mammary glands?

12

12.4 BURNS, SKIN CANCER, AND AGING

Burns

Burns can range from very mild to very severe depending on the layers of the skin that are affected. A **superficial burn** (first degree burn) involves the epidermis, such as sunburn. Peeling that occurs after sunburn involves the loss of epidermal tissue only. A **partial thickness burn** (second degree burn) damages the epidermis and the papillary region of the dermis, and blisters are visible. A **full thickness burn** (third and fourth degree burn) results in damage to both the epidermis and dermis of the skin, and often includes underlying tissues, such as the subcutaneous layer and possibly muscle or bone tissue (Figure 12.5). Full thickness burns require skin grafting in order for it to heal because there are no epithelial or dermal cells left for regeneration. Cells of the epidermis can regenerate quite easily through cell division of the undamaged healthy cells. The dermis can regenerate to some extent, but any burns that include the reticular dermis or the subcutaneous layer usually form scar tissue and possibly permanent damage.

The primary cause of death from extensive burns is loss of body fluids; the second major cause of death is infection. Both the severity of the burn and the extent of the body surface affected are important factors in determining the risk of fluid loss and infection. One method of determining the percentage of surface area of the skin affected by burns is to use the "rule of nines": certain body parts make up approximately 9% of the total surface area of the skin (Figure 12.6). If more than 15–20% of the body has deep or full thickness burns, the person may go into shock and require intravenous to replace body fluids. The percentages for each area differ in infants, children, and people who are obese because the surface area ratios differ. The surface areas in Figure 12.6 are based on an average weight adult.

DID YOU KNOW?

Burns are not only caused by heat or fire. People can have significant burns from electricity, chemicals, or radiation.

Skin cancer

The three most common types of skin cancer are basal cell carcinoma, squamous cell carcinoma, and melanoma (Figure 12.7). **Basal cell carcinoma** originates in the basal cells of the stratum basale layer of the epidermis. Recall that it is this layer of the epidermis that undergoes very rapid cell division. Basal cell carcinoma is the most common type of skin cancer and is caused by excessive sunlight exposure. It most commonly occurs on the face or areas exposed to excessive sunlight, and it appears as a raised, smooth bump that can sometimes ulcerate. This type of cancer rarely spreads and is easily treated.

Squamous cell carcinoma originates in the keratinocytes of the stratum spinosum layer. Recall that these cells make up the largest portion of the epidermis and have some cell division, but not as rapid as the basale layer. This type of cancer is also caused by excessive sun exposure and can spread quickly if it is not removed, but it is still easily treatable. Squamous cell carcinoma usually looks like a red, raised, scaly patch on the skin and can become quite large if not treated. These types of cancers can bleed and ulcerate.

Melanoma is the uncontrolled growth of the pigment-producing cells, melanocytes. Melanoma is the most dangerous type of skin cancer, often arising from moles; if left untreated they can be fatal. Melanoma is the rarest type of skin cancer but has the highest mortality rate because it can spread to many other tissues, making treatment much more difficult. A common method to determine possible melanoma is the mnemonic ABCDE, where A = asymmetrical, B = borders are irregular, C = colour is dark brown or black but not consistent, D = diameter is larger than 6 mm, and E = evolves over time. Harmless moles are generally round with smooth edges, all the same colour, and do not change over time. The cause of melanoma is actually 75% more related to ethnicity and pigmentation than to sun exposure.

DID YOU KNOW?

Sunlight is not the only cause of skin cancer. Other factors that play a significant role in skin cancer development include smoking; chronic inflammation (true of all cancers) or continual wound healing in the same location; being immunosuppressed or using immunosuppressant medication; chemicals found in cosmetics, lotions, and sunscreens that can contain harmful chemicals that induce production of free radicals. Warts caused by the human papilloma virus (HPV)—of which there are over 100 different types—can also increase the risk of squamous cell carcinoma.

DID YOU KNOW?

According to the Public Health Agency of Canada, the incidence of melanoma in Canada is 11.5 cases per 100,000 people in the population, and the five-year survival rate is 90%. As a comparison, breast cancer affects 96 people per 100,000 and has a five-year survival rate of 83%.

12

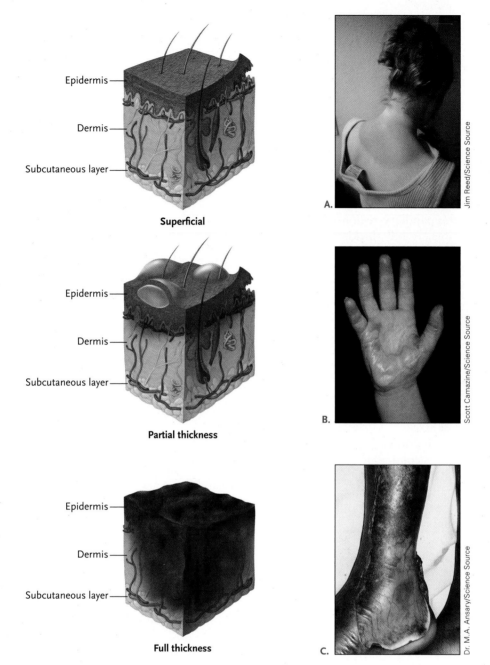

Superficial

Partial thickness

Full thickness

A.

B.

C.

Figure 12.5 Burns

Burns are categorized based on the layer of skin that is damaged: superficial affects the epidermis, partial thickness affects the dermis, and full thickness affects the subcutaneous layer.

Aging

The most common effects of aging on the skin include wrinkles, sagging skin, age spots, increased visible blood vessels, skin tags, and increased dryness. These effects are related to genetic and environmental factors and can begin to be noticeable in a person's late 30s and early 40s.

When we age, the skin becomes thinner and can sag and wrinkle because of decreased production of connective tissue proteins, mainly collagen and elastin. Collagen keeps the skin tough, and elastin allows the skin to go back to its normal position after being stretched. Sebaceous glands become less active, and therefore the skin becomes dry and more likely to crack, itch, and wrinkle. Wrinkles occur in areas where the skin is often creased from smiling or frowning. If you make a big smile or frown in the mirror, you can predict where you will end up with wrinkles later in life. Keeping skin moisturized can help slow the wrinkling process. With aging, the melanocytes decrease in number but increase in size, and they contribute to the dark age spots that become more noticeable with excessive sun exposure. The amount of subcutaneous fat decreases,

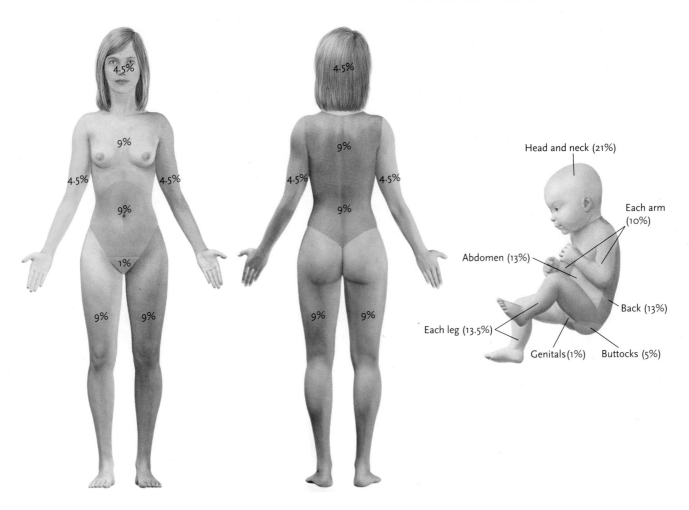

Figure 12.6 Rule of Nines

The rule of nines can be used to determine the approximate percentage of an area of skin that is burned.

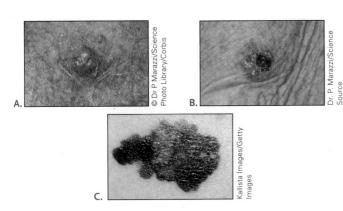

Figure 12.7 Types of Skin Cancer

(A) Basal cell, (B) squamous cell, and (C) melanoma

which leads to sagging skin especially in the face and neck region. Losing a lot of weight when older causes more sagging of the skin since the skin becomes less able to repair and regenerate.

Ways to slow the aging effects on skin

Certain factors can increase the rate of the effects of aging on the skin: excessive sun exposure, such as too much use of tanning beds; increased exposure to toxins, such as cigarette smoke and pollution; lack of sleep; lack of exercise; and poor nutrition. For example, smoking prevents proper collagen formation, adds toxins to the body that damage cells and DNA, and causes vasoconstriction and therefore decreased blood flow to the skin. Burning of the skin from excess sun exposure damages the epidermis, prevents sebum production, damages DNA, and prevents protein synthesis (of collagen and elastin), and causes inflammatory reactions that damage cells.

Some hormones help maintain the skin and the connective tissue proteins, including **growth hormone**, insulin-like growth factor (IGF), and the sex hormones (Chapter 16). When we get older, growth hormone production decreases. Exercise helps ensure the production of some growth hormone and other **androgens** in males and females, and this helps maintain cell regeneration processes and protein synthesis. When women go through menopause, the growth-promoting effects of **estrogen** are lost, and exercise becomes much more important to slow the effects of aging. Exercise also has many other health benefits (Chapter 14).

12

The most important vitamins that contribute to skin health are vitamins C, A, and E. Vitamin C (ascorbic acid) is required for collagen production in our connective tissues and is an important antioxidant. Therefore, eating a diet high in foods such as fresh fruits and vegetables, as well as fermented cabbage, is very beneficial. Vitamin E is a powerful antioxidant secreted in sebum, and it plays an important role in protection from UV light. Vitamin A has many functions and, with respect to the skin, it is important for epithelial cell growth. A form of vitamin A is *retinoic acid*, which is an important growth factor and is added to many anti-aging skin creams. Eating vitamin A-rich foods in addition to using good skin creams containing retinoic acid can reduce the signs of aging (fat-soluble vitamins can easily be absorbed through the skin). It is important to note that if you use a skin cream that contains retinoic acid you need to stay away from UV light because the retinoic acid will oxidize and can cause burning.

DID YOU KNOW?

Fat-soluble vitamins such as A and E are found in foods such as fish, shellfish (including oysters, snails, and clams), healthy oils such as cod liver oil and olive oil, eggs, nuts, and dairy products. Vitamin A can also be produced in our body by converting the beta-carotene found in orange foods, such as carrots, squash, pumpkin, and sweet potatoes.

CONCEPT REVIEW 12.4

1. Describe the three types of burns and which skin layers of the skin are affected for each.

2. Name the three main types of skin cancer and the cells that are affected.

3. What are the risks with each type of skin cancer? How could you determine possible melanoma?

4. What are the most common effects of aging on the skin?

5. What environmental factors can increase the rate of skin aging?

6. What are some things you can do to slow the skin aging process?

12

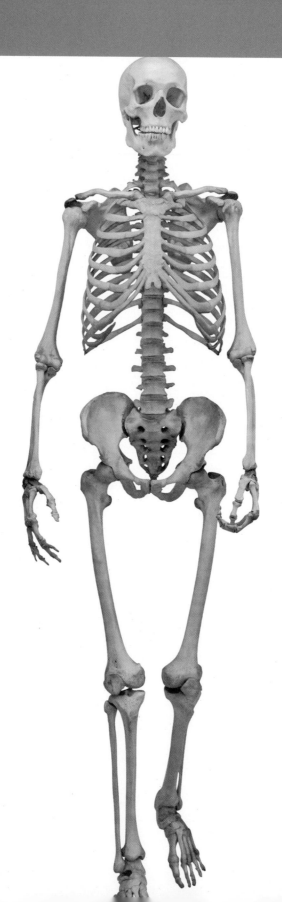

That is what learning is. You suddenly understand something you've understood all your life, but in a new way.

Doris Lessing

13.1 OVERVIEW OF THE SKELETAL SYSTEM

The skeletal system is made up of the axial skeleton and the appendicular skeleton (Figure 13.1). The axial skeleton includes our skull, vertebrae, sternum, and ribs. The appendicular skeleton includes the shoulders, pelvis, and limbs. The adult skeleton has 206 bones: 80 in the axial skeleton and 126 in the appendicular.

13

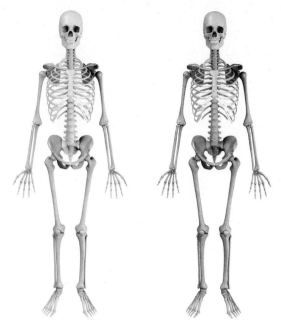

Figure 13.1 Axial Skeleton and Appendicular Skeleton

The axial skeleton consists of the skull, vertebrae, chest, and ribs; the appendicular skeleton consists of the shoulders, pelvis, and upper and lower limbs.

The skeletal system performs the following functions:

- Providing a structural framework for the body
- Providing movement because of attachments of skeletal muscles

- Protecting internal organs, such as the brain, spinal cord, heart, and lungs
- Storing minerals such as calcium, magnesium, and phosphorus, and being involved in the regulation of blood levels of those minerals
- Producing red and white blood cells and platelets in the red bone marrow by a process called hematopoiesis
- Storing fat as triglycerides in the yellow bone marrow
- Facilitating hearing by means of the very small bones in the inner ear that vibrate and transmit sound to the auditory nerve, which signal the brain and allow us to hear and interpret sounds

DID YOU KNOW?

Exercise can help prevent type-2 diabetes. Exercise that puts stress on the bones stimulates bone growth, specifically osteoblasts, which are cells that build up bone density. Stimulated osteoblasts secrete a hormone called **osteocalcin**, which signals fat cells to produce adiponectin, which then increases cellular sensitivity to insulin. (Type-2 diabetes is caused by insulin resistance; see Chapter 16.)

CONCEPT REVIEW 13.1

1. How many bones are in the adult body?
2. What major bones make up the axial skeleton and the appendicular skeleton?

3. What are the main functions of the skeletal system?

13.2 BONE TISSUE

Bones are an organ composed of *mineralized* bone connective tissue (Chapter 11), as well as bone marrow, epithelial tissue, blood vessels, and nerves. Bone tissue is made up of cells and extracellular matrix collagen and minerals (Figure 13.2). There are three main types of bone cells. **Osteoblasts** are the cells that build up the bone tissue. First osteoblasts produce the collagen scaffolding, and then they take up minerals from the blood and deposit them in the extracellular matrix, consisting mainly of calcium, phosphate, and some magnesium. Osteoblasts are very important for bone growth and repair. Once the bone tissue is formed

around the osteoblast, it becomes a mature osteocyte. Osteocytes are the mature "maintenance" cells that take up nutrients, get rid of waste, produce ATP, and communicate with other bone cells. **Osteoclasts** are the cells that break down the mineral portion of the extracellular matrix to release calcium, phosphate, and magnesium into the blood stream; then these minerals are used by other cells, such as muscles or neurons. Bone breakdown is a normal and very important process for bone remodelling and is regulated by hormones, such as **calcitonin (CT)** that's produced by the parathyroid gland (Chapter 16).

The extracellular matrix of bone tissue is composed of water, collagen protein fibres, and minerals. Although

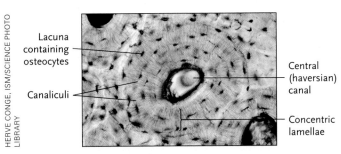

Figure 13.2 Bone Tissue

Bone tissue is made up of cells, collagen fibres, and minerals.

central canal (also called a haversian canal) containing blood vessels and lymphatic vessels. Each central canal is connected to other central canals by **perforating canals**, which also connect blood and lymph vessels with the larger vessels surrounding the outer regions of the bone. The rest of the osteon is bone tissue made up of osteocytes found within small spaces called **lacunae**, and concentric rings of mineralized extracellular matrix; each ring is called a **lamella**. Each lacuna is connected—by very small channels called **canaliculi**—to other lacunae so that cells within the bone can communicate and nutrients and oxygen can reach all cells.

Spongy bone (also called trabecular bone) is made up of **trabeculae** that are composed of osteocytes in lacunae surrounded by mineralized extracellular matrix; but it is not arranged into osteons like compact bone. The space around the trabeculae in spongy bone is filled with **red bone marrow**, where hematopoiesis occurs. Since spongy bone has more space between the trabeculae, it is lighter in weight. Spongy bone is found in short, flat, and irregular bones and also in the ends of long bones, and it surrounds the hollow medullary cavity (Figure 13.3).

bone is very dense tissue, water makes up approximately 25% of the extracellular matrix. Collagen is a very strong but flexible protein, composing another 25% of the bone tissue and allowing bones to have some *flexibility*. The minerals make up approximately half of the bone matrix and give the bones their density and *strength*.

There are two types of bone: compact bone and spongy bone (Figure 13.3). **Compact bone** is tightly arranged into units called **osteons**. Each osteon has a

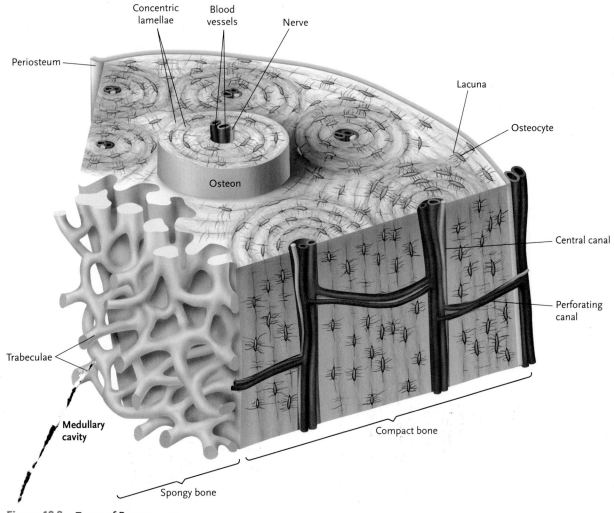

Figure 13.3 Types of Bones

There are two types of bone: compact and spongy.

Classification of bones

Every bone in the body can be generally classified into one of five categories: long, short, flat, irregular, and sesamoid (Figure 13.4). The **long bones** are found in the arms, legs, fingers, and toes. **Short bones** are in the wrists and ankles. **Flat bones** are in the sternum, skull, ribs, and shoulder blades. **Irregular bones** include the vertebrae, some bones of the face and skull (such as the ethmoid and sphenoid bones), and the pelvis. **Sesamoid bones** are embedded within a tendon and cross over a joint; they provide strength to the tendon. The sesamoid bones in the body are the patella, two bones in the hand, the pisiform bone in the wrist, and one bone in the foot that connects to the big toe.

Structure of a typical long bone

All bones are covered by a layer of dense, irregular, connective tissue called the **periosteum** (Chapter 11). Under the periosteum and surrounding the bone tissue are blood vessels, nerves, and lymphatic vessels that infiltrate the entire bone tissue through haversian and perforating canals. Bones are highly vascular. The periosteum also provides an attachment point for tendons and ligaments. The bone precursor cells that become osteoblasts lie directly under the periosteum and form the compact bone. These cells contribute only to bone width (increases with weight-bearing exercises), and they are important in repair of fractures and providing nutrition.

Each end of a long bone (Figure 13.5) is called an epiphysis; the superior end is the proximal epiphysis and the inferior end is the distal epiphysis. (See Chapter 1 for directional terms.) Each epiphysis is covered by a protective layer of articular cartilage. This type of cartilage articulates or connects to another bone. It is hyaline cartilage connective tissue (Chapter 11). The epiphyses contain spongy bone with red bone marrow where hematopoiesis occurs.

A region next to each epiphysis, called the metaphysis, is the region of the bone that grows during development. During bone development there is a layer of hyaline cartilage, called the epiphyseal plate, where chondrocytes produce extracellular matrix, and then mineralization occurs, which allows the bone to lengthen. Once growth is

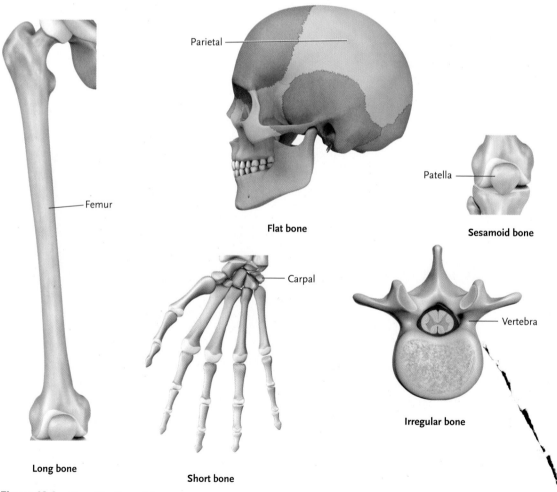

Figure 13.4   Classification of Bones
Bones can be classified as long, short, flat, irregular, and sesamoid.

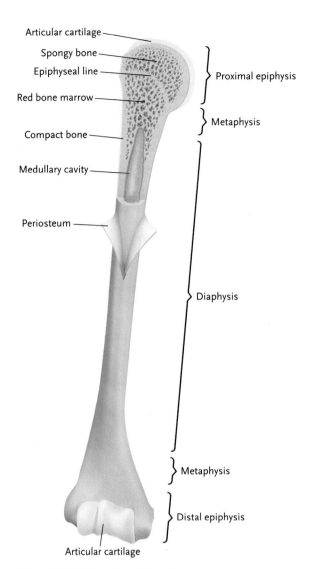

Articular cartilage

Spongy bone

Epiphyseal line

Red bone marrow

Compact bone

Medullary cavity

Periosteum

Proximal epiphysis

Metaphysis

Diaphysis

Metaphysis

Distal epiphysis

Articular cartilage

Figure 13.5 **Long Bone Structure**

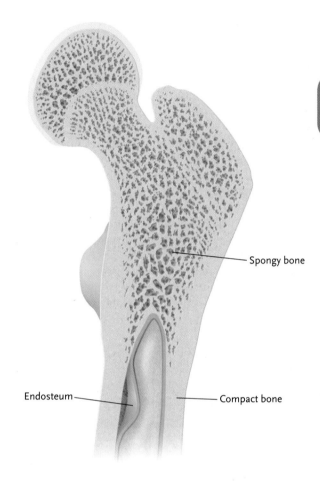

Spongy bone

Endosteum

Compact bone

complete, the epiphyseal plate is replaced by bone, and is called the epiphyseal line.

The mid-region or shaft of the long bone is called the diaphysis. In the diaphysis, under the periosteum, is the compact bone. Within the centre of the shaft is a hollow cavity called the medullary cavity. This cavity is lined with dense, irregular connective tissue, called the endosteum: like the periosteum, this tissue contains bone precursor cells. The medullary cavity is highly vascularized and contains yellow bone marrow, which is a storage site for triglycerides.

DID YOU KNOW?

Exercise increases bone strength and density because the mechanical stress stimulates osteoblasts to increase production of the extracellular matrix collagen proteins and mineralization. Without some weight-bearing exercise, the bone doesn't remodel and becomes weak, increasing the risk of injury. It has been shown that astronauts can lose as much as 1% of their bone mass (and muscle mass) per week when in space without gravity.

CONCEPT REVIEW 13.2

1. What is the difference between a bone and bone tissue?

2. What is the role of the three major cells types?

3. What components of bone tissue provide strength and flexibility?

4. Name the five general categories of bone types and an example of each.

5. Label a blank diagram of a long bone.

13.3 AXIAL SKELETON

The axial skeleton consists of the eight cranial bones, 14 facial bones, one hyoid, six auditory ossicles, 26 vertebrae, one sternum, and 24 ribs.

Cranial and facial bones

Twenty-two **craniofacial bones** provide significant protection for the brain, eyes, and inner ear. The cranial bones include the frontal bone, two temporal bones, two parietal bones, and the occipital bone: the names of all these bones correspond to the aspects of the cerebral cortex (Chapter 15). There are also the cranial bones behind the eye sockets, and the sphenoid bone and the ethmoid bone (Figures 13.6–13.11).

The cranial bones are fused together after birth and form **sutures**, which are a type of joint, categorized as fibrous or **synarthrosis** because they are not movable (Section 13.7). There are four main sutures: (1) the **coronal suture** between the frontal bone and the anterior portion of the parietal bones; (2) the s**agittal suture** between the medial edges of the parietal bones; (3) the **lambdoid suture** between the occipital bone and the posterior edges of the parietal bones;

(4) two **squamous sutures**, between the temporal bone and the lateral parietal bone and also at the edges of the occipital and sphenoid bones on each side of the skull.

The 14 facial bones are all in pairs, except the **mandible** and the **vomer** (Figures 13.6–13.11). Note that the vomer can be seen in Figure 13.10. The other 12 facial bones include two nasal bones, two **maxillae**, two **zygomatic bones**, two **lacrimal bones**, two **palatine**, and two **inferior nasal conchae**.

The **frontal bone** forms the forehead, superior eye orbits, and most of the cranial floor, and it contains two frontal sinuses. The **parietal bones** form the lateral and superior portions of the skull.

The **temporal bones** form the lateral portions of the skull and contain the **external auditory meatus**, which is where sound waves enter the inner ear. The temporal bone also forms the **zygomatic arch** and connects to the zygomatic bone. The **condylar process** of the mandible sits in a grove in the temporal bone, called the **mandibular fossa**; this forms the **temporomandibular joint (TMJ)**. The temporal bone has a process extending posteriorly— called the **styloid process**—and a small opening, called the **carotid foramen**, where the carotid artery passes through the bone (Chapter 17).

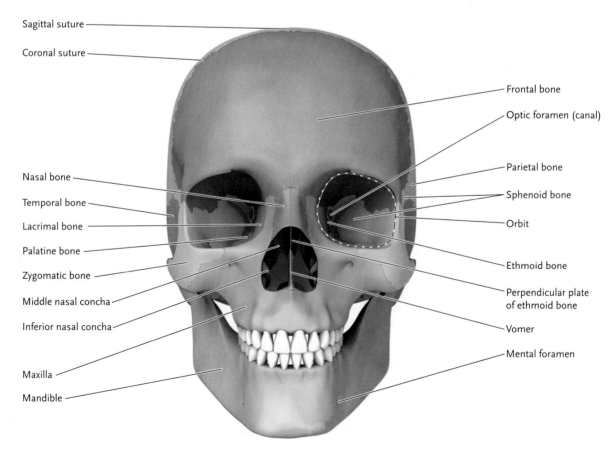

Figure 13.6   Craniofacial Bones: Frontal View

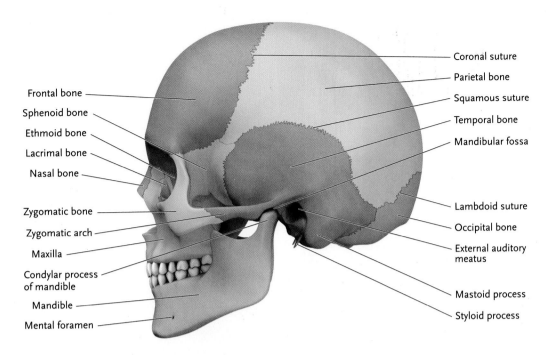

Figure 13.7 **Craniofacial Bones: Lateral View**

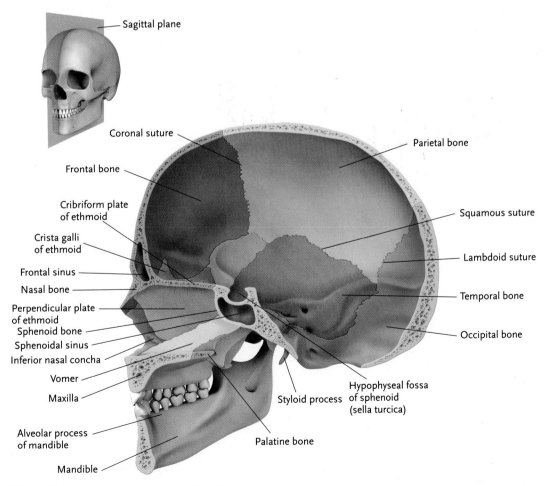

Figure 13.8 **Craniofacial Bones: Sagittal View**

13

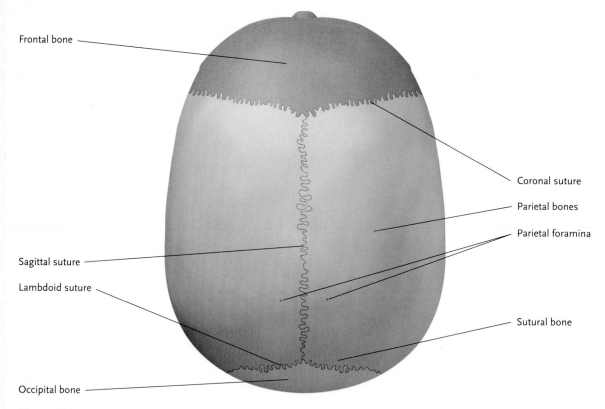

Frontal bone

Coronal suture

Parietal bones

Parietal foramina

Sagittal suture

Lambdoid suture

Sutural bone

Occipital bone

Figure 13.9 Craniofacial Bones: Superior View

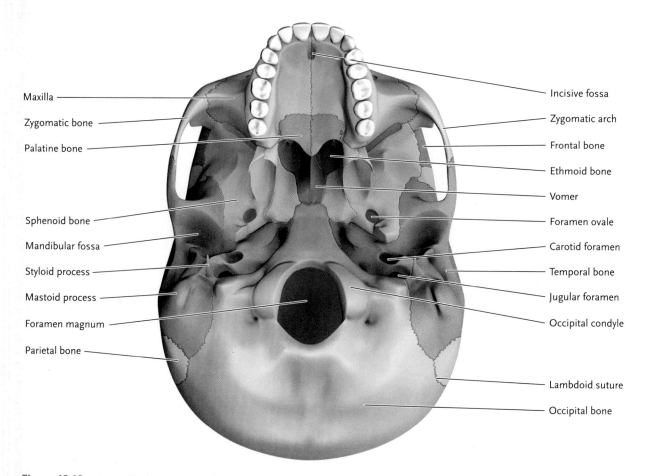

Maxilla

Zygomatic bone

Palatine bone

Sphenoid bone

Mandibular fossa

Styloid process

Mastoid process

Foramen magnum

Parietal bone

Incisive fossa

Zygomatic arch

Frontal bone

Ethmoid bone

Vomer

Foramen ovale

Carotid foramen

Temporal bone

Jugular foramen

Occipital condyle

Lambdoid suture

Occipital bone

Figure 13.10 Craniofacial Bones: Inferior View

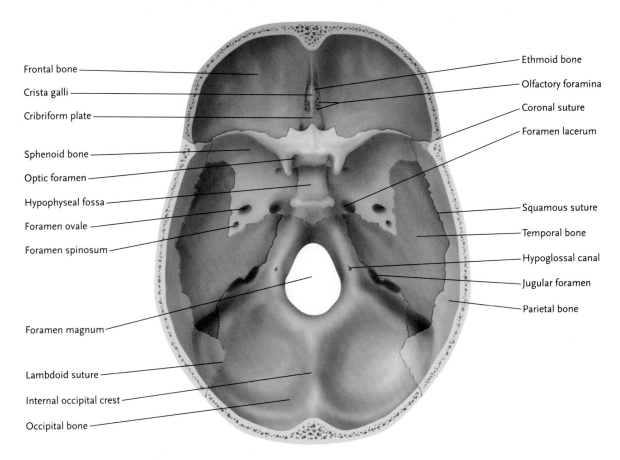

Frontal bone
Crista galli
Cribriform plate
Sphenoid bone
Optic foramen
Hypophyseal fossa
Foramen ovale
Foramen spinosum
Foramen magnum
Lambdoid suture
Internal occipital crest
Occipital bone

Ethmoid bone
Olfactory foramina
Coronal suture
Foramen lacerum
Squamous suture
Temporal bone
Hypoglossal canal
Jugular foramen
Parietal bone

13

Figure 13.11   Craniofacial Bones: Inferior-Transverse View

DID YOU KNOW?

Myofacial pain is most commonly caused by an improper align-ment of the condylar process within the mandibular fossa, which causes friction and inflammation, and is often accompanied by clicking or popping sounds when the jaw is opened. This is most often caused by imbalanced contraction of the chewing muscles because many people grind their teeth when they have stress. Both chiropractors and massage therapists can help relax the muscles and align the jaw to minimize the problem.

The **occipital bone** forms the posterior skull and most of the base of the skull. It contains the largest foramen, called the **foramen magnum,** and is the opening that the spinal cord passes through. If you feel the back and base of your skull, you will feel two large processes; those are called the **occipital condyles,** and they sit on the first vertebra and allow the head to move.

The **sphenoid bone** forms the middle base of the skull and is behind the eye sockets. The sphenoid bone contains sinuses, and also has a structure called the **sella turcica,** which protects the pituitary gland. The sphenoid bone contains the **foramen ovale,** which is the opening for the mandibular nerve. The **optic foramen** is located in the sphenoid bone behind each eye and is where the optic nerve passes through.

The **ethmoid bone** forms the anterior portion of the base of the skull (the entire base is composed of the occipital bone, the sphenoid bone, the vomer, and the ethmoid bone). The ethmoid bone forms the **nasal septum,** and the **cribiform plate** forms the roof of the nasal cavity. The olfactory nerves pass through the **olfactory foramen.** The ethmoid bone also forms the superior and middle nasal conchae; the inferior nasal conchae are facial bones.

All of the **paranasal sinuses** are within the frontal bone, sphenoid bone, ethmoid bone, and maxilla bones (Figure 13.12). The sinuses are lined with mucus

13

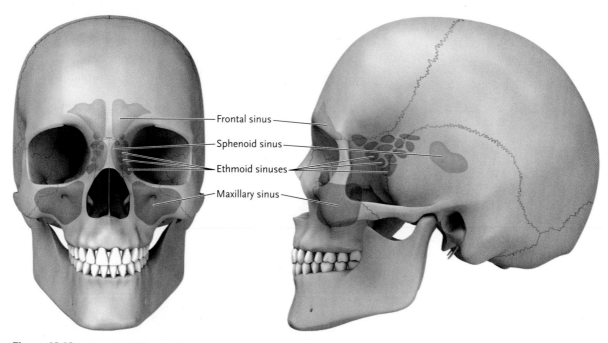

Frontal sinus

Sphenoid sinus

Ethmoid sinuses

Maxillary sinus

Figure 13.12 **Paranasal Sinuses**

Hollow cavities in the craniofacial bones are paranasal sinuses, located in the frontal, ethmoid, sphenoid, and maxilla bones.

membranes composed of epithelial tissue that is highly vascularized to warm the incoming air, and the mucus moisturizes the air before it enters the lungs. The sinuses also decrease the weight of the skull and give resonance to the voice. Since everyone has a unique structure and size of craniofacial bones and sinuses, everyone has a unique voice.

The vomer forms a small portion of the base of the skull articulates with the maxilla and palatine bones, and is one of the components of the nasal septum. The mandible, which forms the lower jaw bone, is the largest and strongest facial bone. The mandible articulates with the temporal bone, forming the TMJ joint, and the condylar process on the mandible articulates with the mandibular process of the temporal bone. The mandible contains **alveolar processes and sockets** that house the lower teeth.

The nasal bones form the top-centre portion of the nose, superior to the movable hyaline cartilage that forms most of the shape of our nose.

The maxilla bones form the upper jaw bone and, like the mandible, have alveolar processes and sockets that house the upper teeth. The maxilla bones articulate with every other facial bone, except the mandible, and they have sinuses that empty into the nasal cavity. They also form most of the upper hard palate of the mouth: with the **palatine bones** forming the posterior region of the hard palate.

The zygomatic bones form the cheekbones, and they articulate with the zygomatic arch of the temporal bones. The lacrimal bones—the smallest facial bones—form the inner part of the medial portion of the eye sockets. The inferior nasal conchae form the inferior portion of the nasal cavity (the ethmoid bone forms the middle and superior nasal conchae). Air circulates through the nasal cavity and the sinuses before it enters the lungs.

The **hyoid bone** is a unique U-shaped bone that is connected to the styloid processes of the temporal bone by ligaments, but it does not form a joint with any other bone. The hyoid is located just inferior to the mandible and is an attachment site for the tongue and neck muscles.

Infant skull

When babies are born, the bones of the skull are not fully fused; this allows the infant skull to squeeze through the mother's vaginal canal and allows space for the skull bones to grow. Newborns have two very important "soft spots" called fontanels. The **anterior fontanel,** located between the frontal bones and anterior

DID YOU KNOW?

The mandible has two openings called mental foramen, where a branch of the trigeminal nerve enters (Chapter 15). This nerve is used by dentists to "freeze" the lower jaw for dental work.

edge of the parietal bones, can be palpated for up to two years after birth; this fontanel takes the longest to close (Figure 13.13). The **posterior fontanel** is located between the occipital bone and the posterior edge of the parietal bones. The two **anterolateral fontanels** are located on each side of the head: superior and anterior to the temporal bone and posterior to the sphenoid bone. Two **posterolateral fontanels** are located on each side of the head, posterior to the temporal bone where the temporal, occipital, and parietal bones meet.

Vertebrae

Babies have 33 vertebrae, but adults have only 26 because the **sacrum** and **coccyx vertebrae** fuse. Specific types of vertebrae are located in specific regions of the spinal column: cervical (seven), thoracic (12), lumbar (five), sacrum (five fused bones), coccyx (four fused bones).

The vertebrae protect the spinal cord, articulate with thoracic ribs, and play a role in movement because of the many muscles that are attached to them (Figure 13.14). The curve of the spinal column allows for increased strength and movement, as well as absorption of shock during physical activities such as jumping or running.

DID YOU KNOW?

The cervical curve forms at approximately three months of age, when the baby begins to hold its head up, and the curve of the lower back develops once the child begins learning to walk, when gravity and muscle contractions pull the spinal column into the curved position.

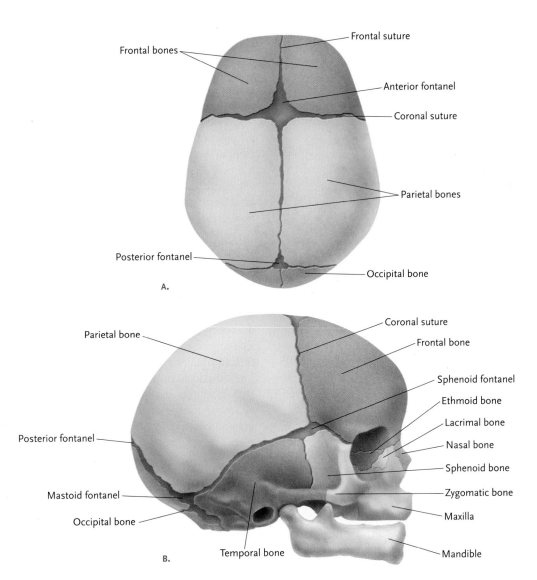

Figure 13.13 Infant Skull
The infant skull has fontanels where the skull bones have not yet fused.

13

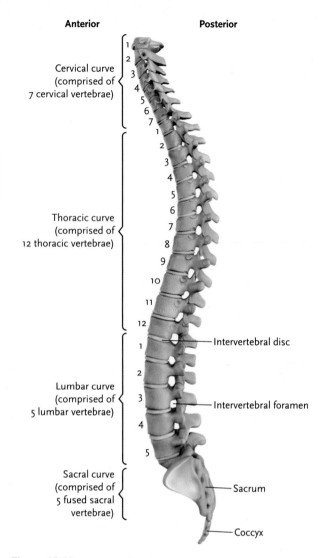

Anterior   **Posterior**

Cervical curve
(comprised of
7 cervical vertebrae)

Thoracic curve
(comprised of
12 thoracic vertebrae)

Lumbar curve
(comprised of
5 lumbar vertebrae)

Sacral curve
(comprised of
5 fused sacral
vertebrae)

Intervertebral disc

Intervertebral foramen

Sacrum

Coccyx

Figure 13.14   Vertebral Column

The vertebral column consists of seven cervical, twelve thoracic, five lumbar vertebrae, and also the sacrum and coccyx.

Vertebrae in each region of the spinal column differ in structure and size but have some similarities. The similar components include the body, foramen, articular facets, and processes. In between each vertebra is an **intervertebral disc** (Figure 13.15), which is composed of fibrocartilage (Chapter 11) that is very important for shock absorption and protection of the vertebrae. The body supports the weight of the spinal column, and the largest vertebrae, the **lumbar vertebrae,** bear the most body weight and attach to large back muscles.

Foramen are openings within the vertebrae. All vertebrae have a **vertebral foramen,** which is the opening for the spinal cord. The cervical vertebrae also have transverse foramen on either side of the body of the vertebrae, and these protect the blood vessels and nerves of the neck. All vertebrae have an intervertebral foramen, which is the opening for the spinal nerves.

Superior and inferior articular facets connect to vertebrae above and below and have a hyaline cartilage surface to prevent friction and allow smooth movement. The **spinous processes** extend out posteriorly and are the ridges you can feel along your spinal column; the **transverse processes** extend out transversely. Both spinous and transverse processes provide attachment sites for deep and superficial muscles of the back (Chapter 14).

There are seven cervical vertebrae (C1–C7). The C1, also called the **atlas,** has superior articular facets that connect with the occipital condyles of the occipital bone and bear the weight of the skull. Due to the shape of the articular facets, the only movement provided at this joint is nodding the head forward and backward. The atlas does not have a vertebral body or spinous process. Movement side to side is due to the C2, also called the **axis,** which contains a unique structure—the **dens**—that projects into the atlas where a vertebral body would be located, and allows for rotation of the head (Figure 13.16). Also unique to the cervical vertebrae are the bifurcated spinous processes (Figure 13.16c). The C7 is also called the vertebra prominens because of its long spinous process, which is the large palpable bone at the base of the neck.

Twelve **thoracic vertebrae** (T1–T12) connect with 12 pairs of ribs. Thoracic vertebrae have long, downward-sloping spinous processes that minimize movement in the thoracic region to protect the delicate lungs. Thoracic vertebrae also have two extra articular processes for rib attachment; all articular processes have hyaline cartilage to decrease friction (Figure 13.17).

The five lumbar vertebrae (L1–L5) have the largest vertebral bodies and bear more weight than the other vertebrae (Figure 13.18). The spinous processes are shorter and allow for greater movement of the back in the lumbar region.

The adult sacrum is composed of five fused bones, and the coccyx is composed of four fused bones (Figure 13.19). The superior articular facet of the sacrum connects with the fifth lumbar vertebra and also articulates with the pelvis. The fused sacrum provides a large, strong, supportive attachment for the spine and the pelvis and contains a pair of sacral foramen that accommodate nerves and blood vessels. The coccyx is known as the tailbone.

DID YOU KNOW?

Your sacral bones are mostly fused by the age of 16, but are not completely fused until approximately age 30.

13

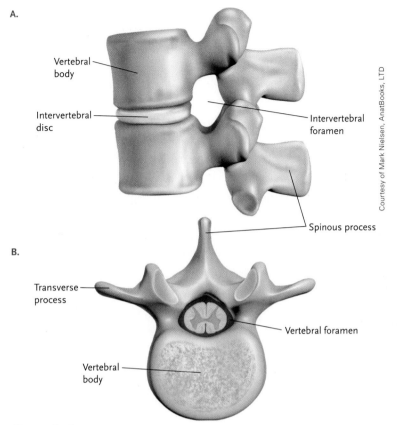

Courtesy of Mark Nielsen, AnatBooks, LTD

Figure 13.15   Anatomy of Vertebrae
(A) Lateral view and (B) superior view

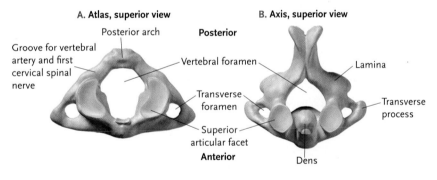

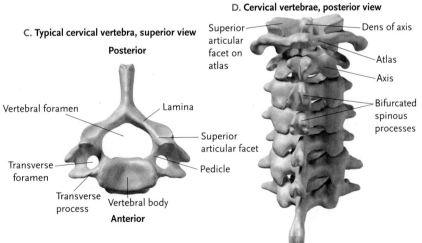

Figure 13.16   Cervical Vertebrae
(A) Atlas, (B) axis, (C) typical cervical vertebrae, and (D) posterior view

13

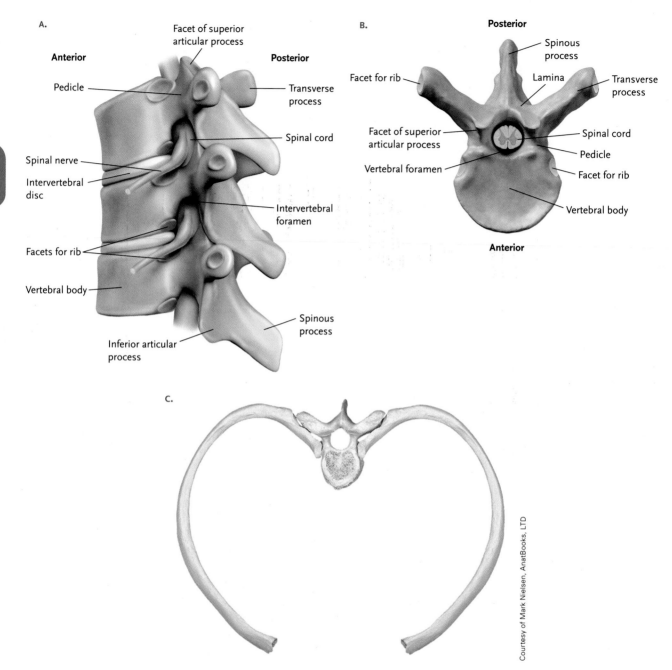

A.

Facet of superior
articular process

Anterior

Pedicle

Spinal nerve

Intervertebral
disc

Facets for rib

Vertebral body

Inferior articular
process

Posterior

Transverse
process

Spinal cord

Intervertebral
foramen

Spinous
process

B.

Posterior

Facet for rib

Spinous
process

Lamina

Transverse
process

Facet of superior
articular process

Vertebral foramen

Spinal cord

Pedicle

Facet for rib

Vertebral body

Anterior

C.

Courtesy of Mark Nielsen, AnatBooks, LTD

Figure 13.17   Thoracic Vertebrae

(A) Lateral view, (B) superior view, and (C) articulation with ribs

Alterations in the curve of the spinal column can occur for several reasons, the most common being genetic or developmental problems, or degenerative diseases such as arthritis or osteoporosis. **Scoliosis** is an S-shaped lateral curve of the spinal column; **kyphosis** is an over-curvature of the thoracic vertebrae, also called hunchback; and **lordosis** is an over-curvature of the lumbar region, also called sway back (Figure 13.20).

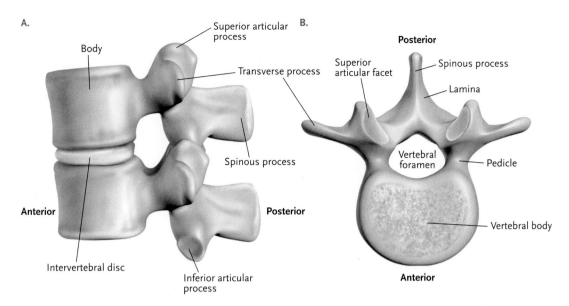

Figure 13.18 Lumbar Vertebrae
(A) Lateral view and (B) superior view

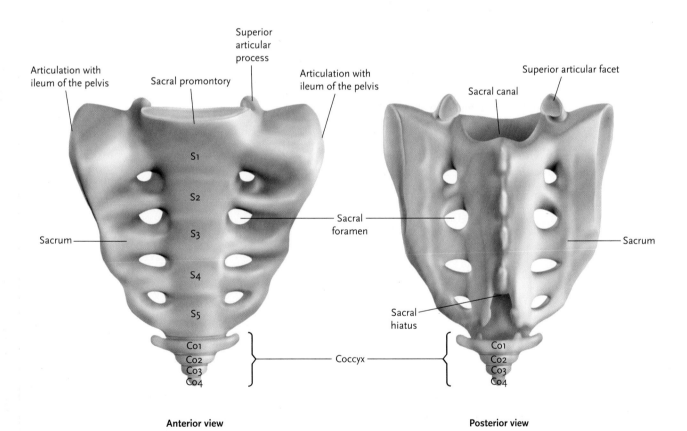

Figure 13.19 Sacrum and Coccyx

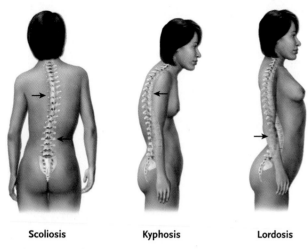

Figure 13.20   Abnormal Curvatures of the Spine
Abnormal curvatures of the spine include scoliosis, kyphosis, and lordosis.

Scoliosis   Kyphosis   Lordosis

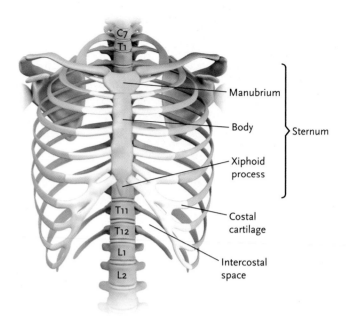

Figure 13.21   Sternum
The sternum consists of the manubrium, body, and xiphoid process. Hyaline cartilage connects the ribs to the sternum.

Sternum and ribs

The bones of the thorax include the 12 thoracic vertebrae that articulate with the 12 pairs of ribs (T1–T12), and the sternum. The bones of the thorax protect the lungs and heart. The **sternum** consists of three parts: the manubrium, the body, and the xiphoid process. The **manubrium** articulates with the first rib, part of the second rib, and the clavicles (Figure 13.21). The body of the sternum articulates with part of the second rib and with the third to seventh ribs by means of **costal (hyaline) cartilage**. The true ribs are ribs 1–7 that articulate directly with either the manubrium or the sternum. Ribs 8–12 are the false ribs because they either connect to the sternum indirectly through the cartilage of the seventh rib (rib 8–10) or they do not connect to the sternum at all (ribs 11–12): also called floating ribs. The **xiphoid process** is a small triangular piece of hyaline cartilage directly below and slightly deeper than

the **sternal notch**, and it doesn't ossify into bone tissue until approximately adulthood. The sternal notch can be palpated where the seventh ribs attach to the lower region of the sternum.

DID YOU KNOW?

The sternal notch is used as a landmark for ensuring that chest compressions during CPR are done above the xiphoid process. If chest compressions are given too low on the sternum, the xiphoid process can break off and its sharp edges can lacerate the lungs.

CONCEPT REVIEW 13.3

1. What are the sutures of the skull and where are they located?
2. Name the bones that form the temporomandibular joint.
3. What are fontanels?
4. Name three abnormal curvatures of the spine.

13.4 APPENDICULAR SKELETON

The appendicular skeleton has a total of 126 bones that include the upper and lower limbs, the shoulder blades (scapula and clavicles), and the hips (pelvis).

Upper limb

The upper arms are connected to the body through the scapula and clavicles, which articulate with the thorax (Figure 13.22). The **scapula** bones are the shoulder blades and are flat bones with a slightly concave anterior surface that articulates with the upper posterior ribs. The posterior scapula has a ridge called the spine that ends laterally at the **acromion**. Anteriorly, the scapula has a protrusion called the coracoid process; the acromion and the **coracoid process** form the edges of the "socket" of the shoulder joint, called the **glenoid cavity**, which is where the head of the humerus sits.

The **clavicles** are the collar bones, and they articulate laterally with the acromion and medially with the superior surface of the manubrium of the sternum. It is common for children to break their clavicles because falling sideways onto the upper arm transfers force to the body through the clavicle.

The shoulder joint is a ball-and-socket joint (Section 13.7) like the hip, and is the most highly movable

Anterior view

Clavicle

Acromion

Coracoid process

Glenoid cavity

Scapula

Humerus

Posterior view

Clavicle

Acromion of scapula

Spine

Body

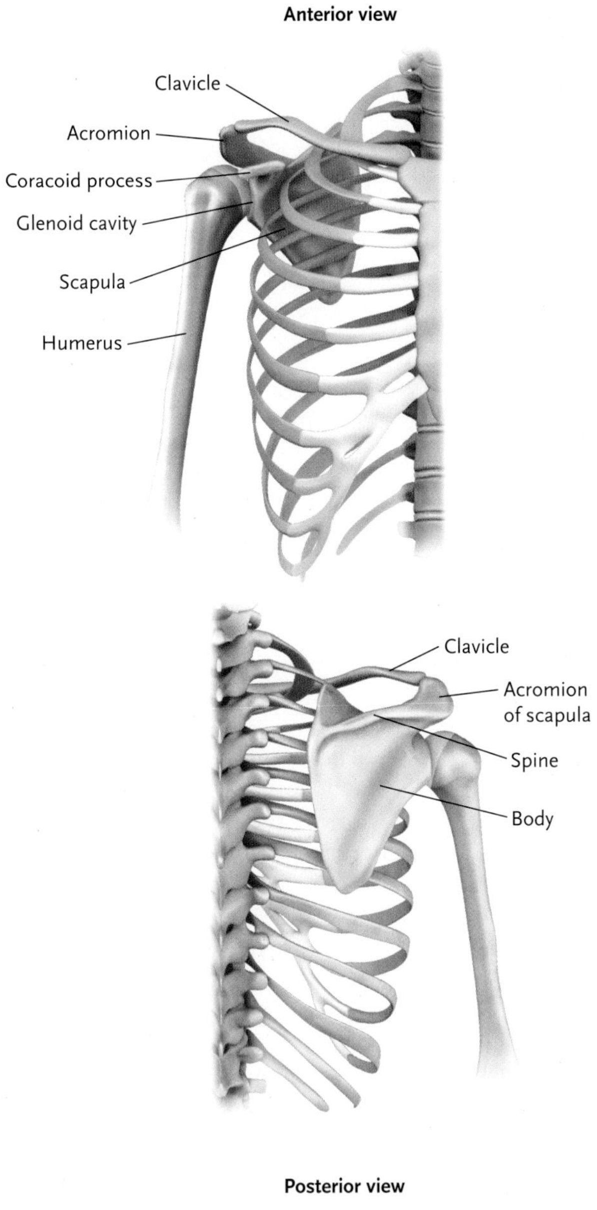

Figure 13.22 Shoulder
The shoulder girdle is made up of the scapula and clavicle bones.

joint in the body. The shoulder joint is formed by the head of the humerus in the shallow glenoid cavity.

The **humerus** is the only bone in the upper arm (Figure 13.23) and is the largest bone of the upper limb. The humerus has an **anatomical neck**, which is the area where the articulating cartilage meets the bone, and a **surgical neck**, which is the narrow region that is the most commonly broken part of the humerus. The **intertubercular groove** (bicipital groove) is where the biceps tendon lies. The **deltoid tuberosity** is the attachment site for the deltoid muscle. At the distal end, the **capitulum** articulates with the proximal **radius**, and the **trochlea** articulates with the proximal **ulna**. Just proximal to the capitulum and trochlea are the **radial fossa** and **coronoid fossa**. These indentations allow the head of the radius and the coronoid process of the ulna to move together, which increases the range of motion when the elbow is bent. On the posterior, distal region of the humerus there is a large indentation called the **olecranon fossa**, which allows room for the **olecranon** (elbow) of the ulna when the arm is straight. This fossa allows the arm to be straight, but prevents the elbow from hyperextending. The medial and later bone processes that can be felt at the elbow are the **medial and lateral epicondyles** of the distal humerus.

The forearm is made up of two bones, the radius and the ulna (Figure 13.24). The head of the radius and the coronoid process of the ulna articulate with the humerus. The head of the radius also articulates with the ulna at the **radial notch**, forming the **radioulnar joint**, which allows the forearm to rotate (Section 13.7). In between the radius and ulna is **interosseous membrane,** composed of dense, irregular connective tissue. At the distal end, the radius and ulna articulate with the carpals at the wrist. The medial and lateral palpable bones in the wrist are the styloid process of the radius (lateral in anatomical position, which is near the thumb), and the styloid process of the ulna (medial, near the little finger).

The wrist bones are formed from eight **carpals**, which are short bones that form a small concave region in the front of the wrist, called the **carpal tunnel**. All the flexor tendons that move the fingers—as well as the median nerve—enter through the carpal tunnel. The bones of the hand are formed by the five metacarpals. The fingers are formed by 14 phalanges. Each finger has a proximal, medial, and distal phalange, whereas the thumb has only two.

Lower limb

The pelvis is formed by three components of the hip bones: the ilium, pubis, and ischium on each side (Figure 13.25). The **ilium** is the large superior region, and the palpable hip bones are the **iliac crest**. The right and left pubis bones articulate anteriorly with the **pubic symphysis** made of fibrocartilage: a cartilaginous joint (Section 13.7). The ilium bones articulate with the sacrum posteriorly, forming the sacroiliac joint. When we sit down, the bones we sit on are the ischium bones, specifically the **ischial tuberosities**, which are the

13

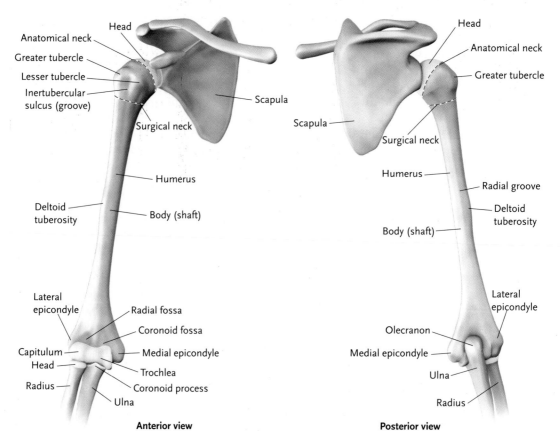

Anatomical neck
Greater tubercle
Lesser tubercle
Inertubercular
sulcus (groove)
Head

Surgical neck

Scapula

Head

Anatomical neck

Greater tubercle

Scapula

Surgical neck

Humerus

Deltoid
tuberosity

Body (shaft)

Humerus

Radial groove

Deltoid
tuberosity

Body (shaft)

Lateral
epicondyle

Radial fossa

Coronoid fossa

Medial epicondyle

Capitulum
Head
Radius

Trochlea

Coronoid process

Ulna

Anterior view

Olecranon

Medial epicondyle

Lateral
epicondyle

Ulna

Radius

Posterior view

Figure 13.23   Anatomy of the Upper Arm

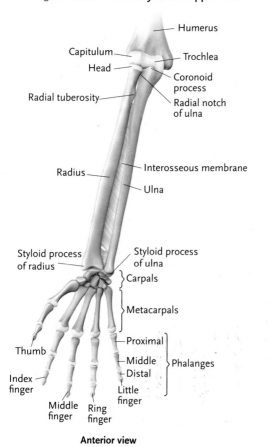

Humerus

Capitulum

Head

Radial tuberosity

Radius

Trochlea

Coronoid
process

Radial notch
of ulna

Interosseous membrane

Ulna

Styloid process
of radius

Styloid process
of ulna

Carpals

Metacarpals

Thumb

Index
finger

Middle
finger

Ring
finger

Proximal

Middle

Distal

Little
finger

Phalanges

Anterior view

Figure 13.24   Anatomy of Forearm and Hand

DID YOU KNOW?

Carpal tunnel syndrome occurs when there is inflammation in the median nerve that travels through the narrow carpal tunnel. Carpal tunnel syndrome is a repetitive use condition and causes pain and numbness. If left untreated, carpal tunnel syndrome can cause permanent nerve damage.

protruding processes. The opening in the pelvis formed by the sacrum, ilieum bones, and pubis bones is the **pelvic brim,** and this forms the top of the **true pelvis,** which protects the internal lower abdominal organs. The true pelvis is wider in females to allow for childbirth. The false pelvis is the region between the top of the iliac crests to the top of the pelvic brim. Within the interior of the true pelvis are medial protrusions from the ischium, called the **ischial spines;** these mark the narrowest part of the pelvic opening.

The openings in the ischium bones are the **obturator foramen,** which allow nerves and blood vessels and lymphatic vessels to pass through. These are the largest foramen in the body. All three pelvic bones meet in a deep socket called the acetabulum, which is where the head of the femur articulates and forms the hip joint. The iliac

13

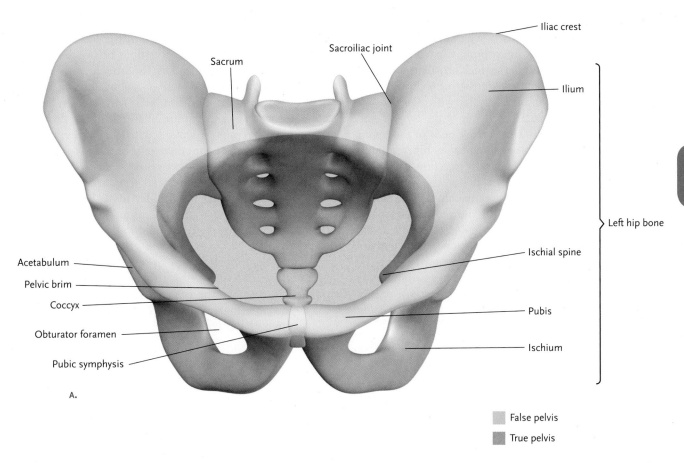

Iliac crest

Sacrum

Sacroiliac joint

Ilium

Left hip bone

Acetabulum

Pelvic brim

Coccyx

Obturator foramen

Pubic symphysis

Ischial spine

Pubis

Ischium

A.

False pelvis

True pelvis

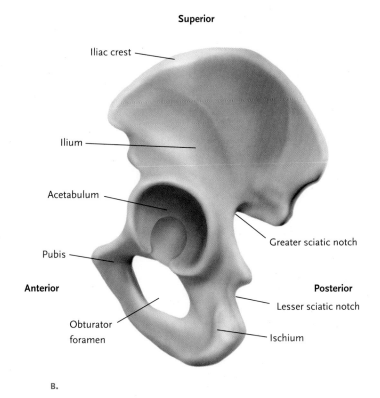

Superior

Iliac crest

Ilium

Acetabulum

Greater sciatic notch

Pubis

Anterior

Posterior

Lesser sciatic notch

Obturator
foramen

Ischium

B.

Figure 13.25 Pelvis

(A) Anterior view and (B) lateral view

bone has a large indentation posteriorly, and this is called the **greater sciatic notch**; this is where the **sciatic nerve** travels down the posterior lower limb (Chapter 15).

The head of the **femur** in the thigh articulates with the acetabulum in the pelvis, forming the hip joint. Like the shoulder, this is a ball-and-socket joint with a large range of motion (Figure 13.26). The femur is the longest and largest bone in the body. The narrow neck of the femur is the most commonly broken region in hip fractures. The femur has a palpable, lateral projection called the **greater trochanter** and a medial lesser trochanter, and these both provide sites for muscle attachment. The **gluteal tuberosity**, the attachment site for the gluteal muscles, is located on the proximal, posterior femur. The large medial and lateral knee bones are the medial and **lateral condyles** of the femur. Connecting to the distal anterior femur is the **patella**, also called the knee cap, which is a sesamoid bone within the tendon that attaches the quadriceps. The patella helps protect the knee joint. The **intercondylar fossa** is the indentation at the distal posterior femur.

The distal femur articulates with the superior **tibia**, which is the larger of the bones in the lower leg. The bump on the proximal anterior tibia below the patella is the **tibial tuberosity** and is the attachment site for the **patellar ligament**, which is a continuation of the quadriceps tendon. The **fibula** is the smaller, lateral bone in the lower leg that articulates with the proximal tibia, but does not articulate with the femur (Figure 13.27). Like the forearm, there is connective tissue—the interosseous membrane—between the tibia and fibula. At the distal end, the fibula and tibia articulate at the fibular notch, forming **tibiofibular joint** (Section 13.7). The protruding medial and lateral ankle bones are the lateral malleolus of the fibula and the medial malleolus of the tibia.

The ankle bones consist of seven tarsal bones. The foot consists of five metatarsals, and the toes are named phalanges, just like the fingers. All toes except the big toe have a proximal, medial, and distal phalange; the big toe has two phalanges.

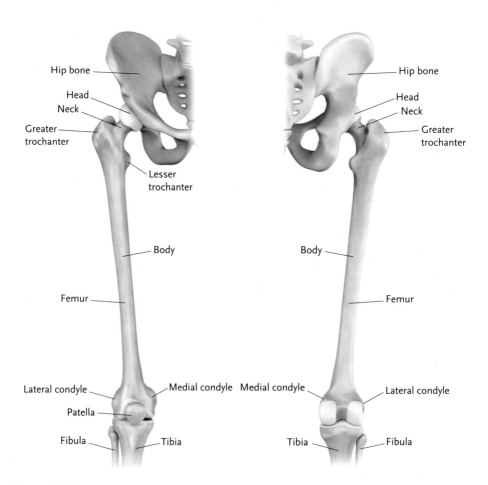

Figure 13.26 **Anatomy of the Femur**

13

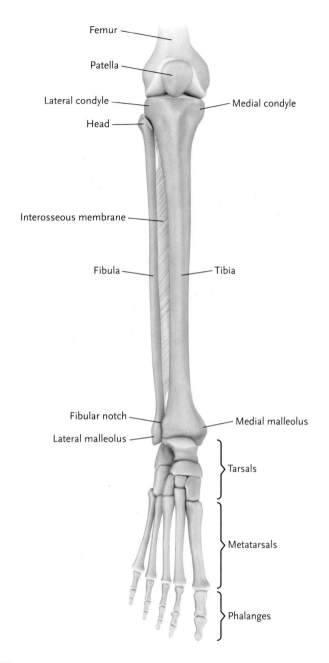

Femur

Patella

Lateral condyle — Medial condyle

Head

Interosseous membrane

Fibula — Tibia

Fibular notch — Medial malleolus

Lateral malleolus

Tarsals

Metatarsals

Phalanges

Figure 13.27   **Anatomy of Lower Leg and Foot**

CONCEPT REVIEW 13.4

1. List the bones in one upper limb.

2. List the bones in one lower limb.

3. Which pelvic bones do we sit on?

4. What region makes up the true pelvis?

5. What is the function of the patella?

13

13.5 MINERAL HOMEOSTASIS

The hormones involved in regulating bone growth and remodelling include **parathyroid hormone (PTH)**, calcitonin, growth hormone, insulin-like growth factors, thyroid hormones, insulin, and sex hormones (estrogen and testosterone) (Chapter 16).

PTH is released by the parathyroid gland in response to low blood calcium. PTH stimulates the osteoclasts to break down bone extracellular matrix and release calcium into the bloodstream. PTH also causes the kidneys to reabsorb Ca^{++} and Mg^{++} (Chapter 20), and stimulates the kidneys to produce **calcitriol** (active form of vitamin D) that causes the small intestine to absorb more calcium from the diet (Chapter 19).

Calcitonin is released from the thyroid gland and prevents the osteoclasts from breaking down bone extracellular matrix. This hormone plays only a small role in the regulation of calcium.

Growth hormone (GH) is produced by the anterior pituitary gland in response to stimulation from the hypothalamus during exercise in adults or growth spurts in adolescents. GH stimulates the growth of many tissues including bone and cartilage by stimulating the production of **insulin-like growth factors (IGFs)**.

IGFs are secreted mainly by the liver, bone, and muscle cells in response to growth hormone and, in bone, stimulate osteoblasts to increase bone mineralization and the production of bone extracellular matrix collagen fibres.

Thyroid hormones (T3 and T4) are produced by the thyroid gland and play an important role in metabolism in many tissues (Chapter 16), including bones where T3 and T4 stimulate osteoblasts.

Insulin, secreted by the pancreas when we eat food, is a general signal to many cells of the body to take up and use or store nutrients that are circulating in the bloodstream. This take up of nutrients also occurs in bones, where extra circulating calcium is taken up by osteoblasts and incorporated into the bone matrix.

Sex hormones are produced by the gonads (ovary or testes) and stimulate osteoblasts to increase bone growth. These hormones play a significant role through growth spurts during puberty and continue to stimulate bone remodelling during adulthood. Sex hormones also cause the epiphyseal plate to mineralize into the epiphyseal line when bone length is complete.

DID YOU KNOW?

Women in menopause have a higher risk of osteoporosis due to the loss of the bone growth, which promotes the stimulation of estrogen. It is very important for menopausal women to continue to exercise to ensure the production of growth hormone and adrenal androgens to prevent loss of bone density.

The key minerals required for bone health are calcium, magnesium, and phosphorus because they make up the majority of the mineral component of the extracellular matrix. Magnesium is required as a co-enzyme in every tissue in the body. Vitamin requirements for healthy bones include vitamins D, A, and C. **Vitamin D** is produced from cholesterol molecules in the skin when we are exposed to sunlight, and it is converted into calcitriol by the kidneys. Vitamin D is required for the absorption of calcium from food in the small intestine. Without vitamin D, any calcium in food stays in the small intestine and is excreted. Vitamin A is required for osteoblast function; however, an excessive amount of vitamin A has been shown to stimulate osteoclasts and have a detrimental effect on bone health. It is highly unlikely that anyone would take in too much vitamin A from dietary sources alone, but this can occur by frequent consumption of supplements. Vitamin C is required for the movement of collagen from cells where it is produced and into the extracellular matrix.

DID YOU KNOW?

Vitamin D deficiency in children causes **rickets**, which causes malformation of the bones, most commonly leading to bowing femurs, greenstick fractures, lordosis, and dental problems.

CONCEPT REVIEW 13.5

1. Which hormones increase bone growth, and which hormones cause bone to be broken down?

2. Which type of bone cells are stimulated by hormones that promote growth?

3. Which hormones target osteoclasts?

4. What are the key minerals required for bone growth?

5. What is the role of vitamins A, C, and D in bone health?

13.6 FRACTURES AND OSTEOPOROSIS

A fracture is any type of break in a bone (Figure 13.28). The most commonly broken bones are the long bones in the upper and lower limbs; this includes hip fractures because they usually involve the head of the femur, not the pelvis. Types of fractures include the following: greenstick, transverse, comminuted, spiral, stress, or open.

Fractures that do not break the skin are called closed, and if the bone protrudes through the skin the break is called open. **Greenstick fractures** are more common in children whose bones are very flexible. Like when you break a green tree branch, the bone cracks on one side but only bends on the other side. Adults tend to have fractures that completely break the bone. A **transverse fracture** is a simple break across the transverse section of the bone, a twisted break is called a **spiral fracture**, and crushed bone is called **comminuted fracture**. **Stress fractures** that cause the bone to crack but not break apart are common in athletes that jump or run on hard surfaces. Since bones are highly vascularized, a break can cause blood vessels to be broken and blood to fill up the area between the periosteum and the bone. This is called a hematoma, which is very painful.

When fractures heal, phagocytic immune cells (Chapter 22) move in to clean up the dead cells. Chondrocytes form **fibrocartilage** (Chapter 11) in the broken region and join the broken sections of bone. Then osteoblasts begin to lay down the new bone matrix of collagen and mineralize the newly forming bone: first spongy bone, which is then converted into compact bone. Osteoclasts are also highly involved in remodelling the new compact bone to replace the spongy bone in the right areas.

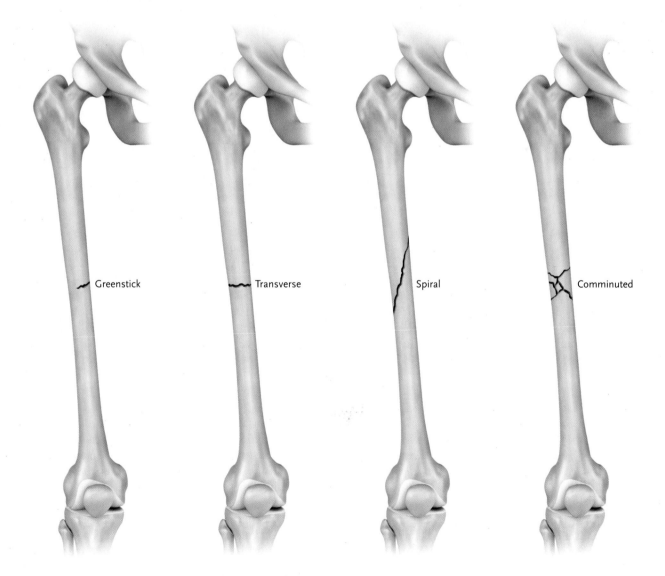

Figure 13.28 Bone Fractures

The categories of bone fractures are named according to how the bone breaks.

Osteoporosis

Bones need to be under mechanical stress to maintain density and strength. This is mostly accomplished by the hormonal influences mentioned above, including exercise. The combination of hormones (mainly thyroid hormones, growth hormones, IGFs, insulin, and sex hormones) as well as vitamins and minerals from the diet, ensure that osteoblasts continue to build up the extracellular matrix of bone tissue. When one or more of these hormones or nutrients are not available, the osteoclasts break down bone faster than the osteoblasts can build it up. A loss of bone density leads to a loss of tensile strength, and this process is called osteoporosis (Figure 13.29).

In adults, the primary factors are diet and exercise. Calcium levels in the blood must be maintained because calcium is required for the proper functioning of the nervous system (Chapter 15) and the muscles (Chapter 14). Exercise must include gravity and mechanical stress, such as jogging or running, jumping, weight lifting, or most other physical activities. Swimming does not produce enough force on the bones to increase bone density. Osteoporosis tends to affect primarily the bones of the wrist, shoulder, hip, and vertebrae. It mostly affects elderly people, women in menopause, and people who do not or cannot engage in physical activity. Among people age 50 and over, 80% of hip fractures are due to osteoporosis.

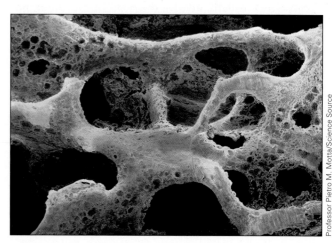

Professor Pietro M. Motta/Science Source

Figure 13.29 **Osteoporosis**
Osteoporosis is a progressive disease caused by a loss of mineralization of the bones, and it leads to an increased risk of bone fractures.

CONCEPT REVIEW 13.6

1. Name the different types of fractures. Which type is most common in children?
2. Which cell types are involved in healing fractures?
3. What are the primary factors involved in osteoporosis?
4. Why is osteoporosis more common in menopausal women?

13.7 JOINTS AND MOVEMENT

A **joint** is an articulation of either two or more bones, cartilage and bone, or teeth and bone. Different types of joints have different structures and functions. Structurally, joints are classified as fibrous, cartilaginous, or synovial. Functionally, they are classified as syndesmosis (no movement), amphiarthrosis (slight movement), and diarthrosis (highly movable) (Figure 13.30).

Fibrous joints can be either immovable (syndesmosis) or only slightly movable (amphiarthrosis) and are connected by dense, irregular connective tissue. Skull sutures are immovable fibrous joints and also **syndesmosis** joints because they do not move. The joint between a tooth and the socket of the alveolar process, the *dentoalveolar joint*, which is connected by the *periodontal ligament*, is immovable and also called a **gomphosis joint**. Although this joint is considered immovable, there is enough space between the tooth root and the alveolar bone process to allow some movement; this is why teeth can be straightened with orthodontic braces. The connection between the tibia and fibula by the interosseous ligament (also called the anterior tibiofibular ligament), as well as between the radius and ulna (also connected by an interosseous membrane), is **amphiarthrosis** because it is slightly movable.

Cartilaginous joints are connected by fibrocartilage or hyaline cartilage, rather than connective tissue, and are tightly connected with little or no movement. Examples of cartilaginous joints are the connection between the costal (hyaline) cartilage of ribs and the sternum, the connection between the pubic symphysis (fibrocartilage) and the pelvic bones, the epiphyseal plates in the ends of the long bones before they ossify, the connection between the manubrium and the body of the sternum, and the connection between the intervertebral (fibrocartilage) discs and the vertebrae.

A **synovial joint**, also called **diarthrosis**, is highly movable. Synovial joints have a joint cavity called the **synovial cavity**. The synovial cavity contains synovial fluid, and a synovial membrane encloses the ends of the connecting bones in a fibrous capsule (Figure 13.31).

Synovial fluid is secreted by the cells in the synovial membrane (made up of epithelial and connective tissue). The fluid consists of plasma from the nearby blood vessels as well as hyaluronic acid, which increases the viscosity of the fluid. The purpose of synovial fluid is to provide nutrients, prevent friction between the ends of the connecting bones, and to absorb shock. Exercise and movement stimulates the production of synovial fluid.

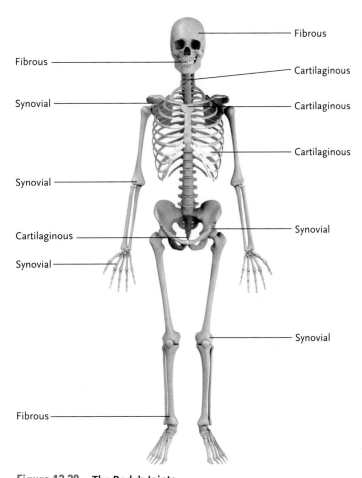

Figure 13.30 The Body's Joints

The joints in the body are categorized based on their structure—fibrous, cartilaginous, or synovial—or their function—syndesmosis, amphiarthrosis, or diarthrosis.

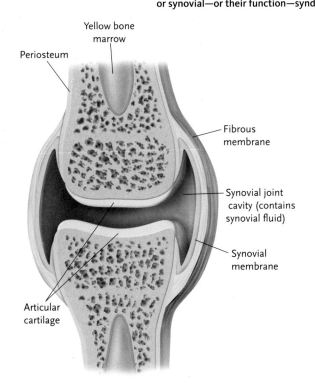

Figure 13.31 Synovial Joints

Synovial joints are highly movable and contain a capsule filled with synovial fluid between articulating bones to prevent friction during movement.

The bones in synovial joints are held together by **ligaments** (dense, irregular connective tissue) and muscles are connected to bones by **tendons**, which are also composed of dense, irregular connective tissue.

Synovial joints often contain **bursa sacs**, which are small sacs filled with synovial fluid and lined with a synovial membrane. Located outside the synovial cavity, bursa sacs add extra protection and cushion between bones and tendons or muscles.

DID YOU KNOW?

Being "double-jointed" does not mean that you have two joints; it means that you have more flexibility in the ligaments between the bones in that joint.

Types of synovial joints

Planar joint. The bones in the joint move in one plane back and forth or side to side and may allow rotation. The articulating surfaces are flat, and there is minimal movement. Examples of planar joints are the carpals in the wrists, the tarsals in the ankles, and the acromioclavicular joint (Figure 13.32a).

13

Hinge joint. The bones in the joint move only in one plane, like a door on a hinge. Hinge joints are connected by strong ligaments. Examples of hinge joints are the elbow, knee, ankle, and interphalangeal joints of the fingers and toes (Figure 13.32b).

Pivot joint. This joint occurs where one bone in the joint fits into a curve or ring of another bone and allows for rotational movement. Examples of pivot joints include the radioulnar joints that allow rotation of the forearm, and the atlanto-axial joint between the first and second cervical vertebrae, which allows the head to turn side to side (Figure 13.32c).

Condylar joint. Also called condyloid, this joint occurs where one bone of the joint has a concave surface that fits into the convex region of the other bone, allowing movement around two axes. One example is the metacarpophalangeal joint in our fingers, where the proximal phalanges connect with the metacarpals in the hand, which allows the fingers to rotate, as well as move back and forth and side to side. The wrist is also a condylar joint (Figure 13.32d).

Saddle joint. This is similar to the condylar joint, except that the bones fit together like a person sitting on a saddle, and it allows movement back and forth and side to side. The main example of a saddle joint is the thumb where the proximal phalange connects with a metacarpal (Figure 13.32e).

Ball-and-socket joint. One bone has a ball-shaped surface that fits into a concave depression on the other bone. These joints are the most highly movable of all synovial joints. Examples include the shoulder and hip joints (Figure 13.32f).

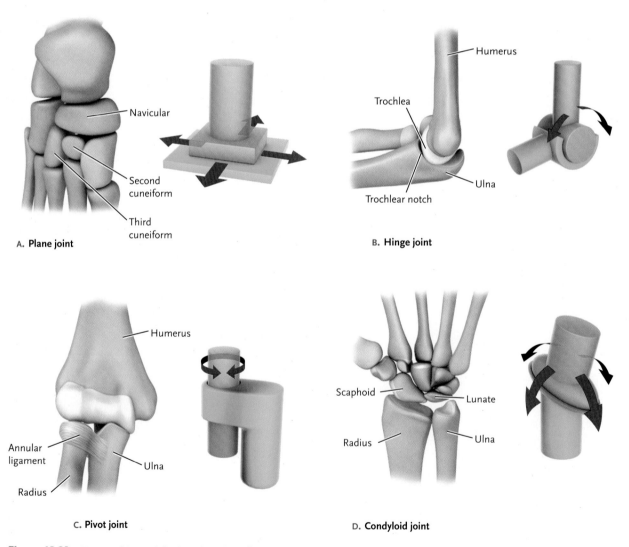

A. **Plane joint**

Navicular

Second cuneiform

Third cuneiform

B. **Hinge joint**

Humerus

Trochlea

Ulna

Trochlear notch

C. **Pivot joint**

Humerus

Annular ligament

Ulna

Radius

D. **Condyloid joint**

Scaphoid

Lunate

Radius

Ulna

Figure 13.32 **Types of Synovial Joints** *(continued)*

(A) Plane, (B) hinge, (C) pivot, (D) condylar, (E) saddle, and (F) ball-and-socket

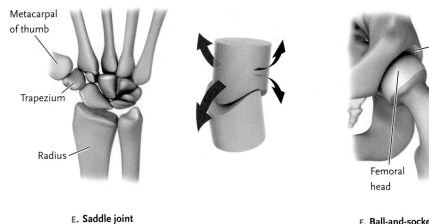

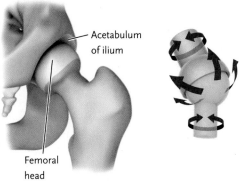

E. **Saddle joint**

F. **Ball-and-socket joint**

Figure 13.32   *(Continued)*

Types of movement

Gliding movement occurs when nearly flat surfaces of bones move back and forth and side to side. This type of movement occurs in planar joints (Figure 13.33). **Rotation** is a type of movement that allows a bone to move around its longitudinal axis; it occurs in pivot joints (Figure 13.33). **Angular** movement occurs when the angle between bones increases or decreases. It occurs in hinge, condylar, saddle, and ball-and-socket joints. Specific angular movements that occur in these joints include flexion, extension, hyperextension, abduction, adduction, and circumduction (Figure 13.33)

At certain specific joints, other types of movements occur, including elevation, depression, protraction, retraction, inversion, eversion, dorsiflexion, plantar flexion, supination, pronation, and opposition (Figure 13.33).

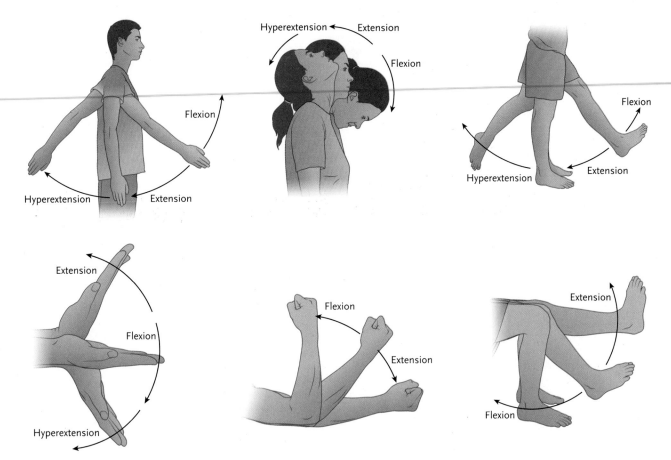

Figure 13.33   **Types of Synovial Joint Movement (***continued***)**

13

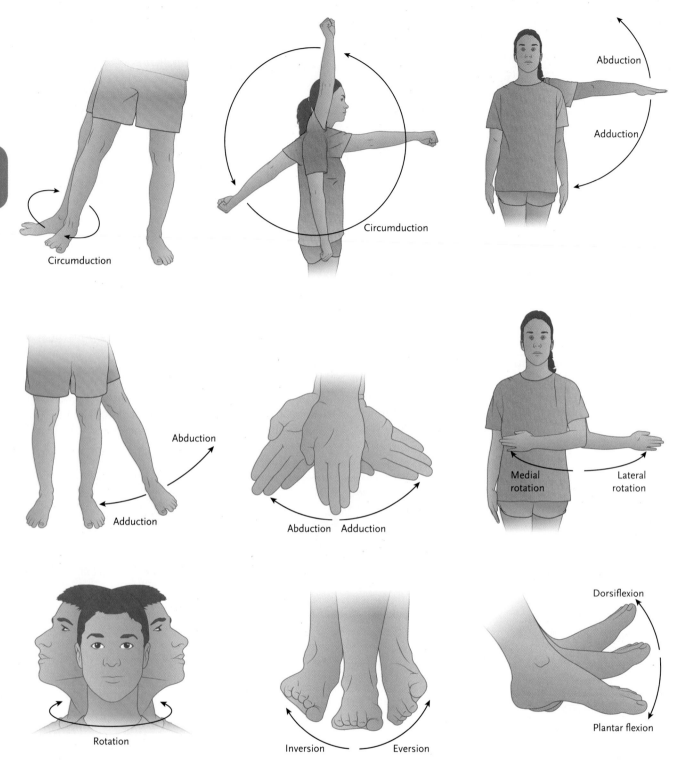

Figure 13.33 *(Continued)*

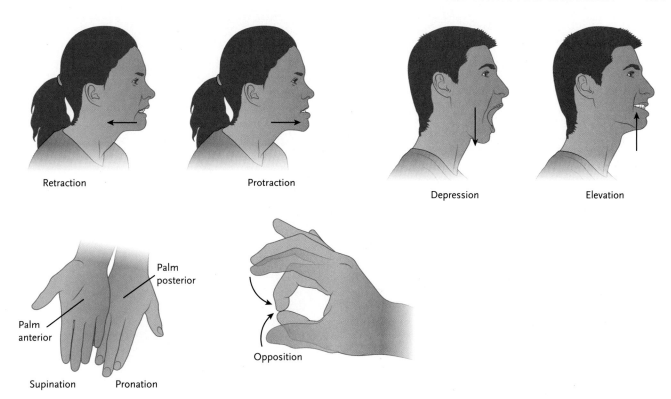

Figure 13.33 *(Continued)*

CONCEPT REVIEW 13.7

1. Explain how joints can be categorized by structure or function.
2. What are some examples of joints that are fibrous, cartilaginous, and synovial?
3. What are some examples of joints that are syndesmosis, amphiarthrosis, and diarthrosis?
4. What is the function of synovial fluid?
5. What are bursa sacs?
6. Label a diagram of a typical synovial joint.
7. What are the six types of synovial joints? Give an example of each.
8. What types of movements can occur in ball-and-socket joints?
9. What is an example of a joint that rotates?
10. What is the difference between pronation and supination?

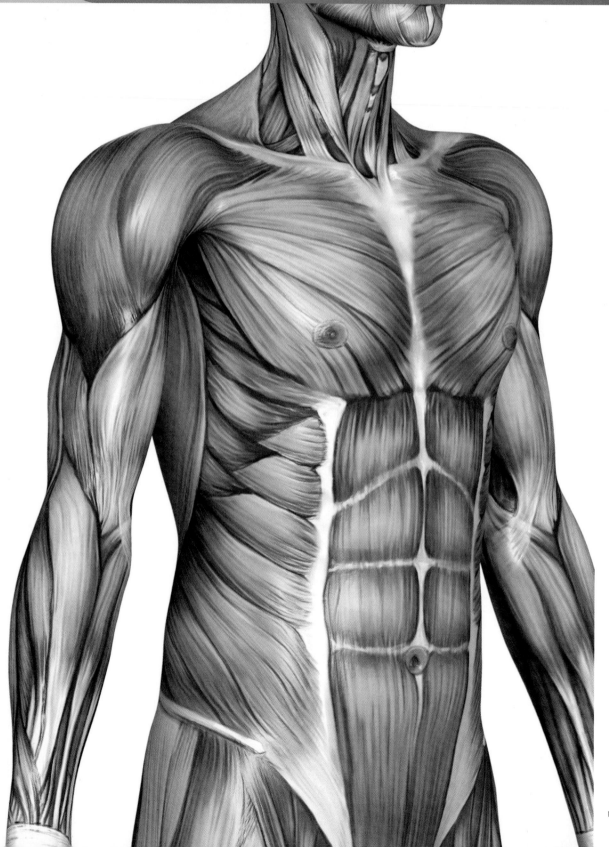

The beautiful thing about learning is that nobody can take it away from you.

B.B. King

14.1 OVERVIEW OF THE MUSCULAR SYSTEM

There are three types of muscle tissue: cardiac, smooth, and skeletal (Chapter 11). This chapter focuses on the superficial skeletal muscles and the major muscle groups. Each type of muscle tissue has different functions, including the following:

- Producing body movements
- Stabilizing body positions and posture
- Causing movement of internal organs, such as the smooth muscles of the blood vessels, digestive tract, fallopian tubes, bronchial tubes, and ureters

- Regulating capillary blood flow and movement of contents through the digestive tract by means of sphincters that are smooth or skeletal muscle
- Producing heat, as a by-product of ATP synthesis, with muscles producing the most ATP of any body tissue and playing an important role in body temperature homeostasis

DID YOU KNOW?

The strongest muscles in the body have the largest cross-sectional area and can exert the most force; these include the rectus femoris and the gluteus maximus. Shorter muscles with a shorter lever arm, such as the masseter (jaw muscle), can exert tremendous force. The heart muscle performs the largest quantity of contraction force in a lifetime.

CONCEPT REVIEW 14.1

1. What are the three types of muscle tissue?

2. What are the functions of muscles?

14.2 MUSCLE TISSUE

Table 14.1 summarizes the main features of each type of muscle tissue: cardiac, smooth, and skeletal.

Skeletal muscle tissue

There are approximately 650 skeletal muscles in the human body. Skeletal muscles are organs and are composed of all four tissue types: muscle fibres, nerves,

TABLE 14.1

Types of Muscle Tissue

Muscle Feature	Cardiac	Smooth	Skeletal
Appearance	Cardiac muscle — Branching cell, Nuclei, Intercalated disc, Striation (Biophoto Associates/Science Source)	Smooth muscle — Nucleus (Biophoto Associates/Science Source)	Skeletal muscle — Width of one muscle cell (muscle fibre), Striations, Nuclei (Eric V. Grave/Science Source)
Location	Heart	Blood vessels, digestive tract, bronchial tubes, some sphincters, iris	Connected to bones or skin, some sphincters
Nervous system control	Involuntary Autonomic nervous system	Involuntary Autonomic nervous system	Voluntary Somatic nervous system
Cell structure	Branched, connected with intercalated discs containing desmosomes and gap junctions	Small cells with tapered ends; can function as a single cell or multi-unit group of cells connected by gap junctions	Long, parallel cells that function as small or large motor units
Nuclei	One to three nuclei per cell	One nucleus per cell	Multinucleated
Striated	Yes	No	Yes
Diameter of fibres	Intermediate	Small	Large, intermediate, and small
Speed of contraction	Intermediate	Very slow	Fast, intermediate, and slow

epithelial tissue, and connective tissue. Muscles are highly vascular, so blood can carry oxygen and nutrients to the muscle cells, also called **muscle fibres,** for ATP production. Every muscle fibre is in contact with a neuron that stimulates contraction.

Skeletal muscle tissue is attached to bones by tendons and, when contracted, causes bones to move. Tendons are composed of dense, regular connective tissue. Three dense, irregular connective tissue layers surround muscles: the epimysium, perimysium, and endomysium (Figure 14.1). The epimysium is the outermost covering and below the **fascia,** which is also dense, irregular, connective tissue that supports and surrounds all muscles and connects with the subcutaneous layer of the skin (Chapter 12). Each muscle is composed of bundles of muscle fibres, called **fascicles,** and these are surrounded by the **perimysium.** Each individual muscle fibre is surrounded by the **endomysium.** All three layers of connective tissue meet beyond the muscle tissue at each end and are continuous with the tendons that connect the muscle to the bone. Within the fascicles

and surrounding every muscle fibre are blood vessels, lymphatic vessels, and neurons.

Skeletal muscle fibres (cells) contain all of the organelles of any typical cell type, except that these cells are very long, multi-nucleated cells. The prefix *sarco* means flesh and is associated with many terms regarding muscles. The cytoplasm is called **sarcoplasm.** The muscle fibre membrane is called the **sarcolemma,** and it is different from other cell membranes because it forms tunnel-like extensions throughout the cell: called **transverse tubules (T-tubules)** (Section 14.4). Muscle cells contain mitochondria for producing ATP, ribosomes for protein synthesis, and also rough and smooth endoplasmic reticulum (Chapter 3). The smooth endoplasmic reticulum in muscles cells is called **sarcoplasmic reticulum** and has the special feature of storing calcium ions that are involved in muscle contraction.

In addition to the cellular organelles, muscle cells have an extensive arrangement of contractile and **regulatory proteins,** which are arranged into small bundles called

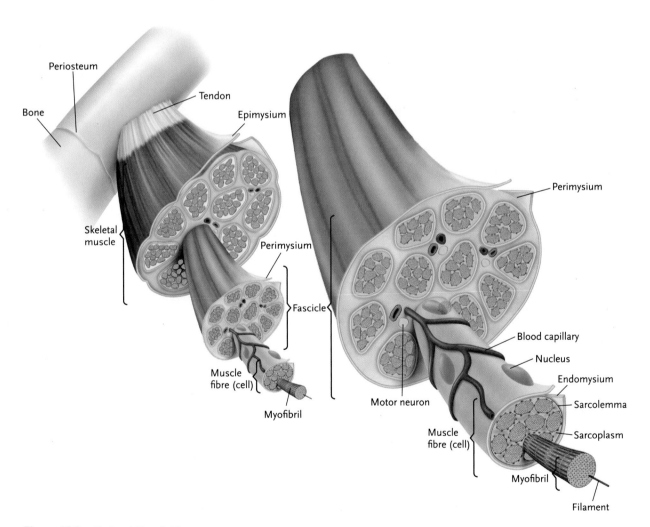

Figure 14.1 Skeletal Muscle Tissue
Skeletal muscle tissue is connected to bone by tendons. It is made up of bundles called fascicles, which are made up of muscle fibres that are made up of smaller bundles called myofibrils.

myofibrils (Figure 14.1). Each myofibril is composed of thick and thin filaments, which consist of the contractile and regulatory proteins. The **thick filaments** consist of **myosin** proteins that have a movable head region that causes the contraction. The **thin filaments** consist of **actin** proteins, which myosin binds to during contraction, as well as the two main regulatory proteins, **troponin** and **tropomyosin**. The thick and thin filaments overlap in highly organized structures called **sarcomeres**, which are considered the smallest single contractile unit; each muscle is composed of millions of sarcomeres. The details of how sarcomeres contract is covered in Section 14.4.

CONCEPT REVIEW 14.2

1. What structure connects muscle to bone?
2. What is fascia?
3. What are the three layers of connective tissue in muscles?
4. List the structures of a muscle from smallest to largest, starting with thick and thin filaments, and ending with epimysium.

14.3 MUSCLES OF THE BODY

Muscles are named based on their location, orientation to the body, size, shape, and/or action (see Table 14.2).

Movement produced by skeletal muscles

The end of the muscle that attaches to the stationary bone is called the **origin**. The end of the muscle attached to the movable bone is called the **insertion** (Figure 14.2). For example, when you contract your biceps muscle, the forearm moves toward the shoulder, so the end of the biceps that attaches to the scapula is the origin, and the insertion attaches to the radius (Chapter 13).

TABLE 14.2

Muscle Names and their Definition	
Muscle Name	**Definition**
Maximus	Largest
Minimus	Smallest
Major	Larger
Minor	Smaller
Longus	Longer
Rectus	Parallel to midline
Transverse	Perpendicular to midline
Oblique	Diagonal to midline
Flexor	Decreases joint angle
Extensor	Increases joint angle
Abductor	Moves away from midline
Adductor	Moves toward midline
Deltoid	Triangular shape
Trapezius	Trapezoid shape
Biceps	Has two origins
Triceps	Has three origins
Quadriceps	Has four origins

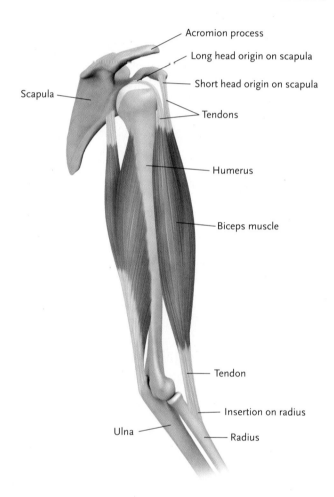

Figure 14.2   Muscle Insertion and Origin
The insertion of a muscle is connected to the moving bone, and the origin is connected to the unmoving bone.

The middle of the muscle is called the **belly,** and it is largest during a maximal contraction.

When we do complex movements, multiple muscles contract at the same time. The main muscle causing the contraction is called the **agonist**. In the example of doing a biceps curl, the biceps and underlying brachialis are the agonists. Since those two muscles work together during a curl, they are **synergistic**. The muscle on the opposite side of the upper arm, the triceps, must relax and stretch when the biceps are contracting and shortening. Since the triceps cause the arm to extend—which is *opposite* to flexion during the curl—the triceps are the **antagonist**. We also have muscles called **fixators** that stabilize the origin of the prime mover so that it can act more efficiently.

Muscle movements can also be categorized based on the force of contraction throughout the movement. **Isotonic contraction** causes the muscle to shorten with even tension throughout a particular movement, such as using a machine to do a bicep curl so that the level of difficulty is the same throughout the movement. Using a dumbbell to do a bicep curl is tougher when the bicep is extended and gradually becomes easier as the muscle shortens. When the velocity remains the same but the tension varies, this is called an **isokinetic contraction**. Isotonic and isokinetic contractions can occur in two ways. If the muscle is shortening during the contraction, it is a **concentric contraction**. If the muscle is lengthening, such as slowly lowering the dumbbell and straightening your arm, it is an **eccentric contraction**. It is also possible to have a contraction without any movement. For example, if you hang from a bar with biceps contracted so that your eyes stay at the level of the bar (called a flexed-arm hang), the biceps are contracted and hold you in place without shortening or lengthening; this is called **isometric contraction**.

The rest of this section describes individual muscles and some groups of muscles, specifically, origin and insertion points, and how certain muscles contract to produce movement. It is important to review Chapter 13, Section 13.7 (joint movements) before continuing with this chapter. See Figures 14.3—14.8 and Tables 14.3—14.5.

DID YOU KNOW?

Your muscles are approximately 40% stronger during eccentric contractions. If you add some eccentric contraction exercises (called negatives) into your regular concentric exercise routine, you will gain more strength if you do concentric contraction exercises alone.

Muscles of the face and neck are numerous and small, so we have the ability to make many facial expressions, chew, talk, swallow, squint, blink, sniff, move the eyes, and move the head. Facial expressions can be made because the insertion for many muscles is not on a bone but on the skin of the face. Depending on which facial muscles contract and how much they contract, humans can make over 10,000 different expressions.

DID YOU KNOW?

The limbic system, our primitive brain centre involved in emotions, is highly active when people are recognizing facial expressions and emotions of others. This capacity relates to our system of non-verbal communication, or body language.

The **pectoralis** major and minor muscles are the main chest muscles that allow pushing movements, such as push-ups, bench press, and anything involved in moving the arms forward. During pushing movements, other muscles contribute to the movement, including the triceps and anterior deltoids.

The abdominal muscles are considered **core muscles**, and their main role is to *stabilize* the body during movements involving multiple muscles, such as running, lifting, squatting, jumping, pushing, and pulling. The abdominals also help us bend forward, such as during sit-ups or raising legs up. The erector spinae muscles, the deep muscles of the back, are also core muscles.

The **diaphragm** is the primary muscle involved in breathing. It moves downward in the thoracic cavity during contraction, causing an increase in thoracic volume, and this causes a brief decrease in pressure, which pulls air into the lungs, quickly equalizing the pressure (Chapter 18). During exhalation, the diaphragm relaxes, moves upward, and lung volume decreases, which increases the pressure and causes the air to move out of the lungs. The synergistic **external intercostal muscles** help expand the rib cage during inspiration.

During exercise, more oxygen is required for muscle cells to produce ATP; therefore, breathing rate increases to match demand, and additional muscles play a role in breathing. During forced inspiration, the sternocleidomastoid contracts to help elevate the sternum and increase thoracic cavity volume. During a severe asthma attack, it is also possible to contract pectoralis muscles as well as many other deep muscles of the trunk. During forced expiration, the **internal intercostal muscles** and **abdominal muscles** contract to help push air out of the lungs.

The **rotator cuff** muscles include the supraspinatus, infratspinatus, teres minor, and subscapularis, which function to stabilize the shoulder joint and to produce movement. The shoulder joint is the most movable synovial joint in the whole body.

The **deltoid** muscle has an anterior, lateral, and posterior head that allows for different movements. Note that the insertion of the deltoid is the deltoid tuberosity, which is midway along the humerus, giving the deltoid muscles good leverage for movement; this makes the deltoids very strong muscles. Moving the arm up and forward (forward flexion) uses the anterior deltoid; moving the arm up sideways (lateral flexion) uses the lateral deltoid; and moving the

14

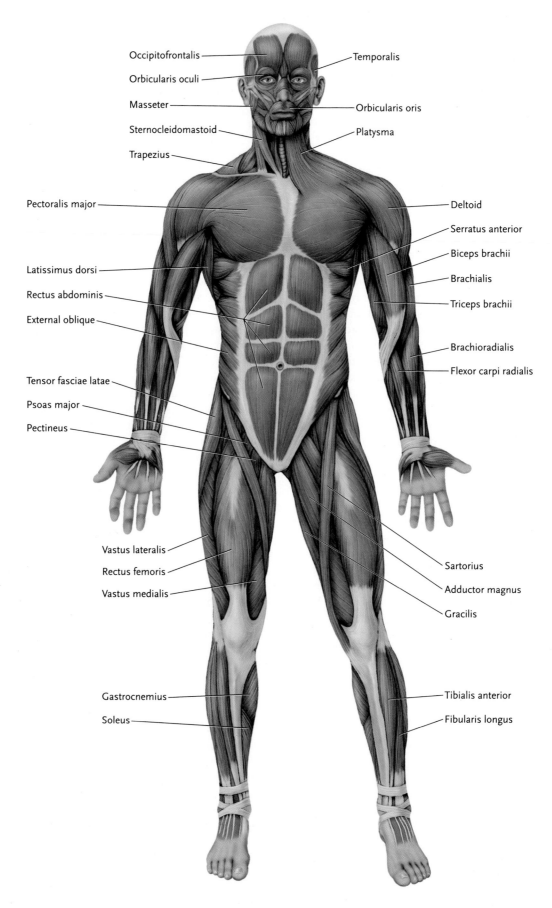

Occipitofrontalis

Temporalis

Orbicularis oculi

Masseter

Orbicularis oris

Sternocleidomastoid

Platysma

Trapezius

Pectoralis major

Deltoid

Serratus anterior

Biceps brachii

Latissimus dorsi

Brachialis

Rectus abdominis

Triceps brachii

External oblique

Brachioradialis

Flexor carpi radialis

Tensor fasciae latae

Psoas major

Pectineus

Vastus lateralis

Rectus femoris

Sartorius

Vastus medialis

Adductor magnus

Gracilis

Gastrocnemius

Tibialis anterior

Soleus

Fibularis longus

Figure 14.3 **Major Superficial Muscles of the Body: Anterior View**

14

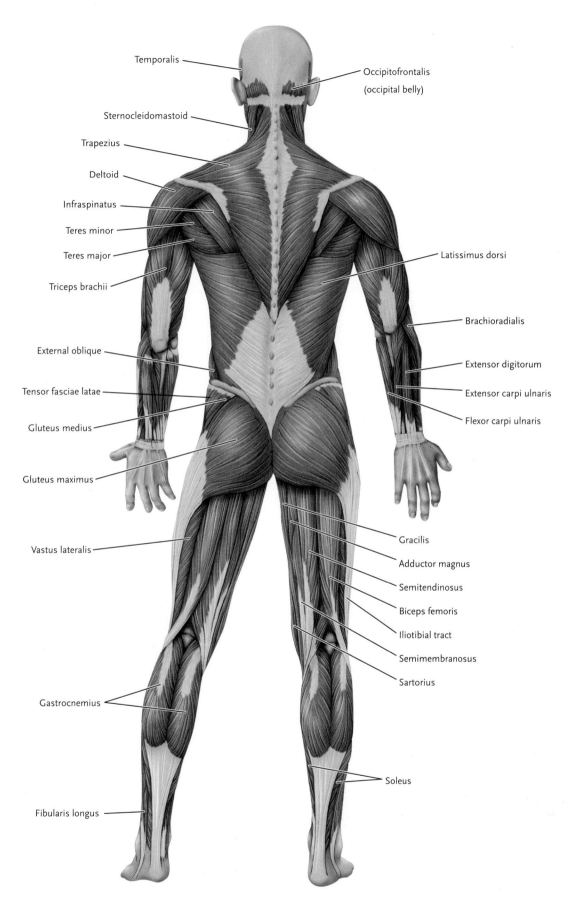

Temporalis

Occipitofrontalis
(occipital belly)

Sternocleidomastoid

Trapezius

Deltoid

Infraspinatus

Teres minor

Teres major

Latissimus dorsi

Triceps brachii

Brachioradialis

External oblique

Extensor digitorum

Tensor fasciae latae

Extensor carpi ulnaris

Gluteus medius

Flexor carpi ulnaris

Gluteus maximus

Vastus lateralis

Gracilis

Adductor magnus

Semitendinosus

Biceps femoris

Iliotibial tract

Semimembranosus

Sartorius

Gastrocnemius

Soleus

Fibularis longus

Figure 14.4 **Major Superficial Muscles of the Body: Posterior View**

TABLE 14.3

Muscles of the Head and Neck

Muscle	Origin	Insertion	Action
Occipitofrontalis Frontal belly	Epicranial aponeurosis	Skin superior to orbit	Raises eyebrows
Occipitofrontalis Occipital belly	Occipital and temporal bone	Epicranial aponeurosis	Draws scalp backward
Orbicularis oculi	Wall of orbit	Eyelid	Closes eye
Orbicularis oris	Muscles around mouth	Skin around mouth	Protrudes lips (as in kissing)
Zygomaticus major	Zygomatic bone	Skin at corners of mouth	Smiling
Platysma	Fascia over deltoid	Mandible, skin of lower face	Depresses mandible
Masseter	Zygomatic arch	Mandible	Elevates mandible (chewing)
Temporalis	Temporal bone	Mandible	Elevates and retracts mandible (chewing)
Sternocleidomastoid	Manubrium and clavicle	Occipital bone and mastoid process	Flexes and rotates head
Trapezius	Occipital bone and thoracic vertebrae	Clavicle and scapula (acromion anteriorly, and spine posteriorly)	Extends head and elevates scapula (shrugging) and pulls arms back, such as a seated row movment

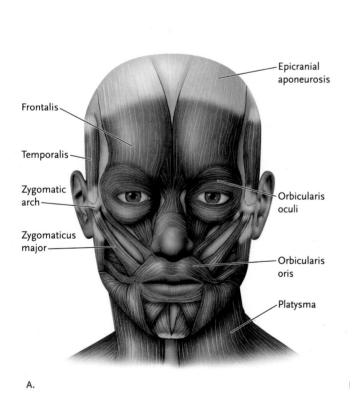

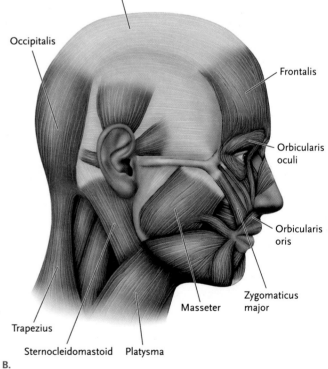

A.

B.

Figure 14.5   Head and Neck

(A) Frontal view and **(B)** lateral view

TABLE 14.4

Muscles of Anterior Trunk

Muscle	Origin	Insertion	Action
Pectoralis Major	Medial clavicle and sternum	Humerus	Adduction of the humerus, such as push-ups or moving straight arm toward midline; medial rotation of shoulder, such as arm wrestling
Pectoralis Minor	Ribs 3–5	Coracoid process of the scapula	Elevates ribs during forced inspiration, and pulls scapula forward and down
Rectus Abdominus	Pubic symphysis	Cartilage of ribs 5–7 and xiphoid process	Flexes vertebral column, aids in defecation, urination, forced expiration, and childbirth
External Oblique	Ribs 5–12	Linea alba	Flexion and rotation of vertebral column
Internal Oblique	Ilium	Cartilage of ribs 9–12	Flexion and rotation of vertebral column
Transverse Abdominus	Ilium, lumbar fascia, and ribs 6–12	Xiphoid process, linea alba, and pubis	Compresses abdomen and stabilizes trunk

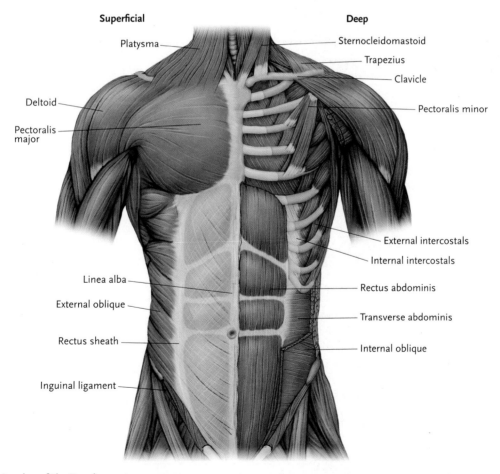

Figure 14.6   Muscles of the Trunk

14

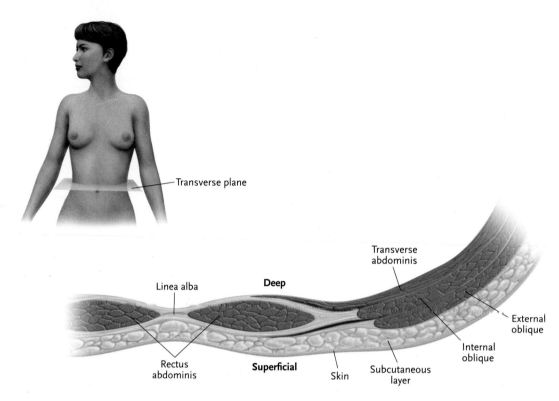

Transverse plane

Transverse abdominis

Linea alba

Deep

External oblique

Internal oblique

Rectus abdominis

Superficial

Skin

Subcutaneous layer

Figure 14.7   Abdominal Muscles

Transverse section

TABLE 14.5

Muscles Used for Breathing

Muscle	Origin	Insertion	Action
External intercostals	Rib above	Rib below	Increases thoracic volume during normal inspiration, involved in relaxation during normal expiration
Internal intercostals	Rib below	Rib above	Depresses ribs (forced expiration)
Diaphragm	Xiphoid process, cartilage of ribs 6–12, and lumbar vertebrae	Central tendon	Causes flattening during contraction, increases thoracic volume during normal inspiration

TABLE 14.6

Muscles of Posterior Trunk and Shoulder

Muscle	Origin	Insertion	Action
Supraspinatus	Supraspinous fossa of scapula	Superior humerus	Abduction of arm at shoulder
Infraspinatus	Infraspinous fossa of scapula	Superior, lateral humerus	Main external rotator
Subscapularis	Anterior scapula at medial border	Medial head of humerus	Medially rotates the humerus
Teres minor	Dorsal, lateral scapula	Lateral humerus	Mainly functions to hold humerus in the glenoid cavity
Teres major	Lower than the teres minor on the dorsal, lateral scapula	Lower than teres minor on anterior, lateral humerus	Extension and adduction of arm, such a chin-up, as well as medial rotation
Latissimus dorsi	Spines of ribs T7–L5, sacrum, ilium	Upper medial humerus	Extension and adduction of arm, such as wide grip, overhand pull-up, as well as internal rotation of the humerus
Deltoid	Clavicle and scapula (spine)	Humerus (deltoid tuberosity)	Abduction, lateral and anterior flexion and extension, rotation of arm, and circumduction

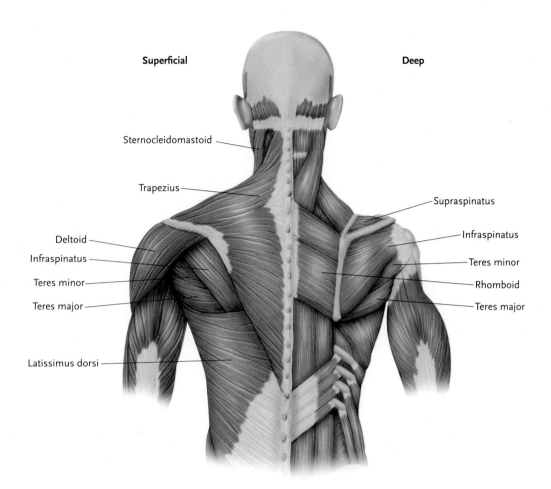

Superficial

Deep

Sternocleidomastoid

Trapezius

Deltoid

Infraspinatus

Teres minor

Teres major

Latissimus dorsi

Supraspinatus

Infraspinatus

Teres minor

Rhomboid

Teres major

Figure 14.8 **Posterior Trunk Muscles**

humerus backward, as in any kind of pulling motion, uses the posterior deltoid. Pulling movements also use many other muscles, including the lower trapezius, teres major, rhomboids (deep to trapezius), as well as accessory muscles such as biceps and muscles of the forearm (Table 14.6).

The **latissimus dorsi,** the largest back muscle, connects to many ribs as well as the pelvis. Note that it inserts lower and more medial on the humerus than the teres major. The teres major and latissimus dorsi are synergistic and are both involved in moving the humerus down from overhead, whether the arm is forward or out to the side.

Deep muscles of the back

The **erector spinae** are bundles of muscles that extend the length of the vertebral column. They are categorized into three main groups: spinalis, longissimus, and iliocostalis. The **spinalis** muscles originate and insert on the vertebral spinous processes. **Longissimus** muscles attach to the sacrum and the vertebral transverse processes. The **iliocostalis** muscles attach primarily to ribs (Figure 14.9). The function of the erector spinae muscles is primarily to maintain body postures and the extension of the vertebral column. We use our erector spinae muscles to lift objects up off of the floor, but because these are small muscles

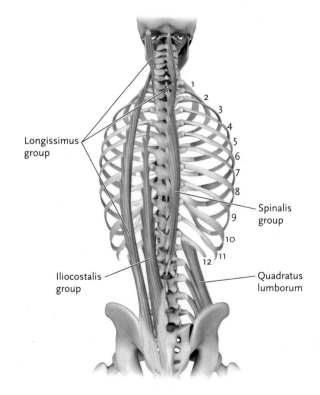

Longissimus group

Iliocostalis group

1
2
3
4
5
6
7
8
9
10
11
12

Spinalis group

Quadratus lumborum

Figure 14.9 **Deep Muscles of the Back**

bundles attached with small tendon insertions, they are easily injured with sudden heavy lifting. These muscles are part of the core muscles used for stabilization and also used synergistically with other large muscles during lifting movements: such as the gluteus maximus, hamstrings, and upper back muscles. It is important when lifting a heavy object to include these large muscles; this is done simply by bending the knees and using the legs as well as the back muscles. It is also important to keep erector spinae muscles strong with exercise (Section 14.5).

DID YOU KNOW?

Lower back pain affects 80% of the population at some point in a person's life. Most frequently, lower back pain is due to muscle sprains, strains, tears, or weakness of the erector spinae muscles; these conditions are completely preventable with strength exercises and proper technique when lifting.

The **biceps** and **brachialis** muscles work together to flex the forearm; however, only the biceps supinates the forearm, because it inserts on the radius (Table 14.7). Both the biceps and brachialis are involved in pulling movements that include bending the elbow, such as chin-ups. The **triceps** are the agonist for extension of the arms and the antagonist to flexion. The forearm muscles cross the wrist joint, and long tendons insert onto the bones of the fingers. In the anatomical position, the anterior forearm muscles are the **flexors**, and the posterior forearm muscles are the **extensors** (Figure 14.10).

The **hip flexors (iliopsoas)** are the **psoas major** and **iliacus** muscles. Both function to flex and laterally rotate the thigh: for example, when pulling your leg toward your body or your body toward your legs, as in doing leg raises or sit-ups (Table 14.8).

TABLE 14.7

Arm Muscles			
Muscle	**Origin**	**Insertion**	**Action**
Biceps Brachii	Long head—superior coracoid process Short head—anterior coracoid process	Radius	Flexion and supination of the forearm, flexion of arm at shoulder
Brachialis	Lower humerus, near the deltoid tuberosity	Ulna	Flexion of forearm
Triceps Brachii	Lateral posterior scapula	Posterior ulna	Extension of forearm

DID YOU KNOW?

Shortened hip flexors (iliopsoas) often result from sitting or riding a bicycle for extended periods. This causes a common form of lower back pain during sleep when lying prone. This occurs because the extended leg elongates the hip flexors, causing them to pull on the lumbar vertebrae—their point of origin—and thus causing the pelvis into an anterior tilt. The pulling action on the pelvis creates pain, which is relieved when the knees are raised and the feet are brought toward the buttocks, that is, as the hip flexors are shortened.

The quadriceps are a group of four muscles. The largest is the **rectus femoris** (Figure 14.11), which is anterior to the **vastus intermdius**. The **vastus medialis** is the muscle on the medial distal thigh near the knee, and the **vastus lateralis** is the large, long, lateral muscle of the thigh. The quadriceps all meet at the quadriceps tendon just above the knee; that tendon covers the patella and the patellar ligament attaches to the tibial tuberosity of the tibia. Since the quadriceps are all inserting on the tibia and crossing the knee, the primary movement is extension of the knee, like kicking a ball. The quadriceps are involved in every leg movement, including walking, running, jumping, squatting, kicking, as well as full body pushing or pulling. For example, during a deep knee bend or squat, as you slowly sink down, the quads are contracting eccentrically because the knee is bending and the quads are lengthening. When you stand up, the most difficult part of the movement, the quads are contracting concentrically, with the hamstrings and gluteal muscles acting synergistically. If you were to do a squat with weight, you could lift more weight on the eccentric contraction—the negative part of the movement—than on the concentric contraction, standing up. The hamstrings and gluteal muscles are also involved as synergystic muscles.

The **sartorius** muscle is important for the extension of the knee and lateral rotation of the leg. For example, when you are going to cross your legs, you lift your leg up, then move the knee out and the foot in. Several muscles, such as the **adductor longus,** help to adduct the leg in toward the midline.

DID YOU KNOW?

Pushing a sled is considered one of the best full body exercises for building strength and cardiovascular fitness. The muscles used include the forearm flexors; triceps; pectoralis major and minor; deltoids and rotator cuff muscles; lats, as stabilizers if arms are straight and shoulders down; other upper back muscles, also as stabilizers; abdominals and erector spinae, as core stabilizers; hip flexors; and all leg muscles.

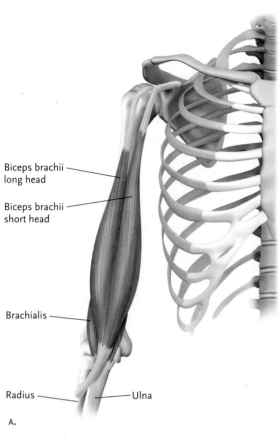

Biceps brachii
long head

Biceps brachii
short head

Brachialis

Radius

Ulna

A.

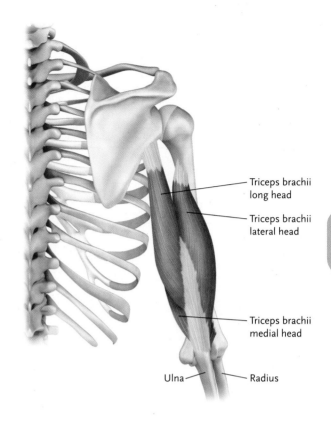

Triceps brachii
long head

Triceps brachii
lateral head

Triceps brachii
medial head

Ulna

Radius

B.

Flexors

C.

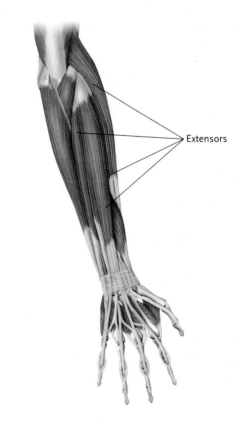

Extensors

D.

Figure 14.10   Muscles of the Arm

(A) Anterior arm, (B) posterior arm, (C) flexors, and (D) extensors

TABLE 14.8

Anterior Thigh Muscles

Muscle	Origin	Insertion	Action
Psoas major	Lumbar vertebrae	Femur	Hip flexion and lateral rotation
Iliacus	Ilium	Femur	Hip flexion and lateral rotation
Rectus femoris	Anterior iliac spine	Tibia (tibial tuberosity) via the patellar ligament	Extends the knee and flexes the hip
Vastus lateralis	Proximal femur	Tibia (tibial tuberosity) via the patellar ligament	Extends knee
Vastus Medialis	Proximal femur	Tibia (tibial tuberosity) via the patellar ligament	Extends knee and adducts thigh
Vastus Intermedius	Anterior lateral femur	Tibia (tibial tuberosity) via the patellar ligament	Extends knee
Sartorius	Ilium	Tibia	Flexes hip, abducts and laterally rotates thigh
Adductor longus	Pubis	Femur	Adducts and laterally rotates femur

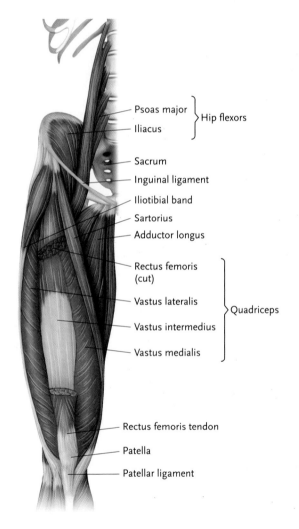

Psoas major ⎫
Iliacus ⎬ Hip flexors

Sacrum

Inguinal ligament

Iliotibial band

Sartorius

Adductor longus

Rectus femoris (cut)

Vastus lateralis ⎫
Vastus intermedius ⎬ Quadriceps
Vastus medialis ⎭

Rectus femoris tendon

Patella

Patellar ligament

Figure 14.11   Anterior Leg Muscles

The **gluteus maximus** is the largest and strongest muscle in the body and is involved in all lower body movements. The gluteus maximus causes the legs to extend and hyperextend (Figure 14.12 and Table 14.9).

The **hamstrings** primarily function to flex the knee, but since they attach to the pelvis, they also play a role in extending and hyperextending the thigh. During walking or running, as we push off the ground and the leg moves backwards, the gluteus maximus (agonist) and the hamstrings (synergists) contract to push the leg back. Bending the knee of the back leg primarily uses the hamstrings. The forward leg uses hip flexors to pull the knee up and the quadriceps when landing on the front leg and pushing off; the movement quickly switches to the glutes and hamstrings as that leg moves back.

DID YOU KNOW?

You do not have to use as much posterior muscle strength to run on a treadmill compared to running on a road or track. Since the treadmill is moving, it pulls your leg back; however, the quadriceps have to work as much as usual to pull the leg forward and extend the leg. So the strength of the anterior and posterior muscles can become imbalanced in people who run only on treadmills. For those who run on a treadmill during the winter it is also common for them to pull a hamstring in the spring when they start running outside again; this occurs because the hamstrings have lost some strength compared to the quadriceps.

The lower leg muscles are used mainly to move the foot. The two largest posterior muscles are the soleus and the gastrocnemius. The **soleus** is deep to the flatter gastrocnemius and is the main muscle used to plantar flex the foot, as in standing on your toes, or pointing your toes. The origin for the **gastrocnemius** is on the distal femur, so it is also involved in flexing the knee with the hamstrings. When the knee is bent, the gastrocnemius has less tension and is less involved

TABLE 14.9

Posterior Thigh Muscles

Muscle	Origin	Insertion	Action
Gluteus maximus	Posterior ilium, sacrum, and coccyx	Iliotibial tract and femur	Extends, hyperextends, and laterally rotates thigh
Gluteus medius	Ilium	Femur	Abducts and medially rotates thigh
Semitendinosus	Ishium	Medial tibia	Flexes knee and extends thigh
Biceps femoris	Ishium	Lateral tibia	Flexes knee and extends thigh
Semimembranosus	Ishium	Medial tibia	Flexes knee and extends thigh

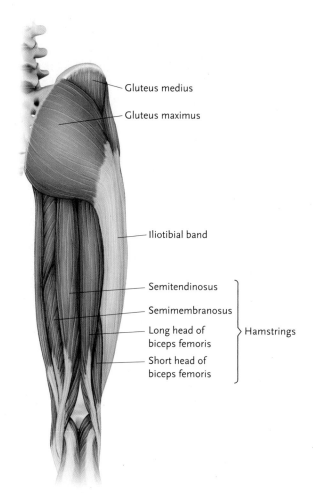

Figure 14.12 **Muscles of the Posterior Thigh**

TABLE 14.10

Lower Leg Muscles

Muscle	Origin	Insertion	Action
Soleus	Proximal tibia and fibula	Calcaneus via calcaneal (achilles) tendon	Plantar flexes foot
Gastrocnemius	Distal femur	Calcaneus via calcaneal (achilles) tendon	Plantar flexes foot and flexes knee
Tibialis anterior	Proximal tibia	First metatarsal	Muscle dorsiflexes foot

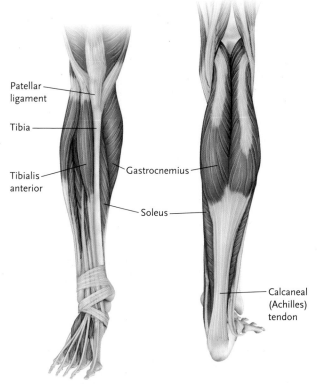

Figure 14.13 **Lower Leg**

in plantar flexion. A seated calf-raise exercise targets the soleus, whereas a standing calf-raise targets the gastrocnemius (Table 14.10).

The main muscle in the anterior leg that dorsiflexes the foot, as in lifting your toes when you walk, is the tibialis anterior. This muscle can cramp and cause pain and inflammation when someone first begins jogging or running; this condition is called shin splints (Figure 14.13).

CONCEPT REVIEW 14.3

1. Name the insertion, origin, and action of each major muscle.

2. Describe how the muscles of the thigh and leg work together during walking or running.

14.4 MUSCLE CONTRACTION AND THE NEUROMUSCULAR JUNCTION

Muscle cell structure

Skeletal muscles cells (fibres) are surrounded by a connective tissue covering called the endomysium. The cell membrane, called the **sarcolemma**, folds into the cell, forming the transverse tubules (T-tubules), so that action potentials from the nervous system can penetrate the interior of the cells to cause contraction. Each muscle cell contains bundles of filaments called myofibrils that contain thick and thin filaments that contract (Figure 14.14).

Myofibrils are surrounded by mitochondria that produce ATP, the T-tubules, and the sarcoplasmic reticulum, which contains calcium ions required for contraction.

A single sarcomere, the smallest contractile unit, is composed of thick and thin filaments (Figure 14.15). The thick filaments are composed of myosin proteins that look like bundles of golf clubs wrapped together. The thin filaments are composed of actin, which looks like two strings of beads twisted together. Myosin consists of **contractile proteins**. The thin filaments also include the regulatory proteins, troponin and tropomyosin. The protein of each myosin head (like the head of a golf club) has two binding sites: one binds to the actin proteins of the thin filament, and the other binds an ATP molecule. Energy

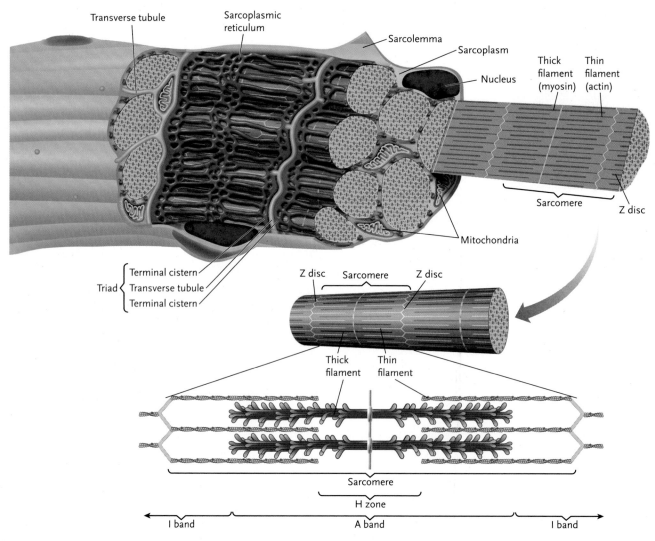

Figure 14.14 Skeletal Muscle Cells

Skeletal muscle cells are composed of bundles of myofibrils that are made up of thick and thin filaments that contract.

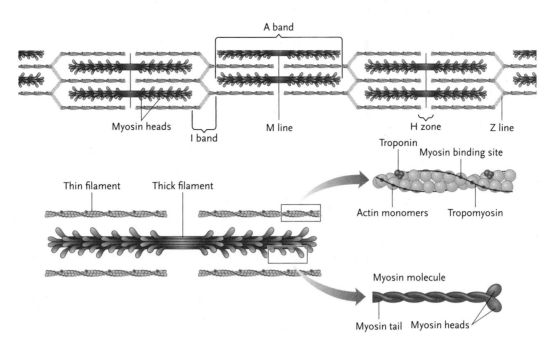

Figure 14.15 **Thick and Thin Filaments**
Thick and thin filaments are composed of actin and myosin.

in the form of ATP is required for every myosin head protein to bind to actin to form a cross-bridge. The thick and thin filaments overlap each other in a very organized pattern, which forms light and dark bands that are visible under a microscope; these bands are called **striations**.

Sarcomere

The banding pattern of each sarcomere has specific labels and contains specific thick and thin filament components (Figure 14.16). The ends of the sarcomere are the **Z discs**, which are structural proteins that hold the thick and thin filaments together and parallel to each other. In the very centre of the sarcomere is another structural protein that forms the **M line**. The central region that contains only myosin is called the **H zone**. The entire region of the myosin proteins, including the overlapping thin filaments, is the **A band**. Between the A band and the edge of the A band of the next sarcomere—containing only thin filaments and the Z disc—is the **I band**.

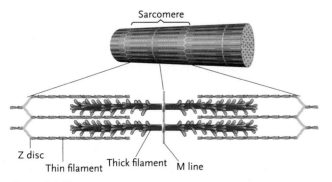

Figure 14.16 **Sarcomere**
A sarcomere is the smallest contractile unit.

Sliding filament theory

When the muscles contract, the myosin heads bind to actin, then pivot and pull the actin toward the M line. Each myosin head pivots toward the M line, pulling the actin of the thin filament toward the M line so that the thin filaments slide past the thick filaments; this is called the **sliding filament theory**. As the muscle contracts and the actin slides across the myosin, there are changes in the sarcomere (see Figure 14.17). The A band does not change because the myosin proteins are not contracting or shortening; they are pulling the thin filaments inward. The I bands and the H zone decrease and the Z discs move closer together.

Regulatory proteins

The main regulatory proteins are troponin and tropomyosin (Figure 14.18). The myosin-binding sites on the actin proteins are blocked by tropomyosin when there is no stimulation by the nervous system. So skeletal muscle contraction can only occur during voluntary movement: that is, when we choose to move and contract certain skeletal muscles. When a neuron stimulates a muscle fibre, it causes calcium to be released from the sarcoplasmic reticulum. The calcium ions then bind to troponin, which is bound to tropomyosin, which causes the tropomyosin to roll off of the myosin-binding sites on the actin (Figure 14.19). After the calcium is released, it stays in the cytoplasm only as long as there is stimulation by a neuron. Once the stimulation stops, the calcium is pumped back into the sarcoplasmic reticulum by an active transport membrane protein; therefore, ATP is required. As the cytoplasmic calcium level drops during relaxation, the tropomyosin moves back to cover the myosin-binding sites on actin.

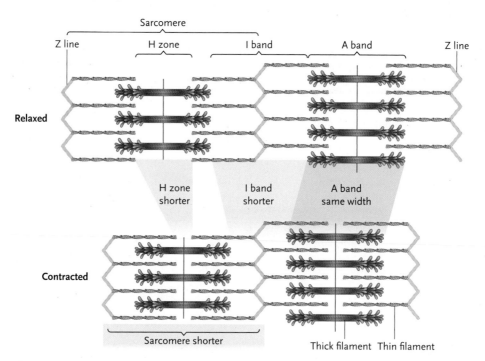

Figure 14.17 **Changes in the Sarcomere during Contraction**

During contraction the H zone and I bands get smaller, but the A band does not.

Art source: From SHERWOOD, Human Physiology, 8E. © 2013 Cengage Learning.

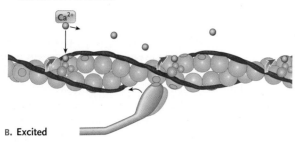

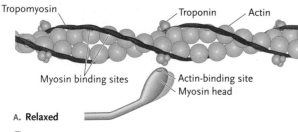

A. Relaxed

❶ No excitation

❷ No cross-bridge binding because myosin binding site on actin is physically covered by troponin–tropomyosin complex

❸ Muscle fibre is relaxed

B. Excited

❶ Muscle fibre is excited and Ca^{2+} is released

❷ Released Ca^{2+} binds with troponin, pulling troponin–tropomyosin complex aside to expose cross-bridge binding site

❸ Myosin binding site

❹ Binding of actin and myosin cross-bridge triggers power stroke that pulls thin filament inward during contraction

Figure 14.18 **Role of Regulatory Proteins**

Regulatory proteins bind calcium and allow contraction to occur only after an action potential has occurred.

Art source: From RUSSELL/HERTZ/STARR/FENTON. Biology, 2E. © 2013 Nelson Education Ltd. Reproduced by permission. www.cengage.com/permissions

DID YOU KNOW?

A decrease in magnesium (Mg^{2+}) is often a cause of muscle cramps or twitching. Mg^{2+} helps to balance the movement of calcium ions across the membrane in muscle cells. After a contraction, Mg^{2+} enters the cell as Ca^{2+} leaves. A deficiency in Mg^{2+} prevents the normal movement of Ca^{2+} out of the cell, which leads to continued binding of myosin to actin, even in the absence of an action potential.

Cross-bridge cycle

How the myosin binds to actin, pulls the thin filaments, and is released can be broken down into four simple steps (Figure 14.19).

1. Myosin is unbound from the actin and contains an ATP molecule in the ATP-binding site on the myosin head. The ATP hydrolyzes, the ADP and phosphate stay bound, and the action of the bond breaking in the ATP energizes the myosin by changing the myosin head to tilt from its normal conformational shape.

2. The phosphate molecule leaves the myosin, and the energized myosin binds to the myosin-binding site on the actin proteins, as long as calcium ions are present to bind troponin and move the tropomyosin off the binding sites. The binding of myosin to actin forms the **cross-bridge**.

3. The ADP molecule leaves the myosin, the myosin head tilts back to its normal shape, and because it is

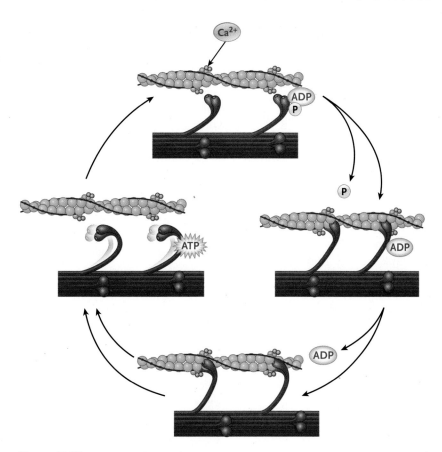

Figure 14.19   Cross-Bridge Cycle

The cross-bridge cycle occurs when ATP hydrolyzes and causes myosin to bind to actin and move the actin towards the centre of the sarcomere.

Art source: From RUSSELL/HERTZ/STARR/FENTON. Biology, 2E. © 2013 Nelson Education Ltd. Reproduced by permission. www.cengage.com/permissions

bound to actin, the myosin moves the actin toward the M line; this is called the power stroke. After the **power stroke**, the myosin remains bound to the actin.

4. A new ATP molecule binds to the myosin head, which allows the myosin to be released from the actin. Then the cycle repeats, starting with ATP hydrolysis in step 1.

DID YOU KNOW?

When a person dies, they go into a state where the body becomes completely rigid, called rigor mortis. This is caused because ATP is no longer produced after death, so the amount of ATP decreases until all the myosin heads remain bound to actin. Once the cells start to break down, calcium is released from the sarcoplasmic reticulum and floods the cells, binding troponin and allowing myosin to bind to actin. Furthermore, there is no ATP to pump the calcium back into the sarcoplasmic reticulum. It takes the body approximately 12 hours to be in full rigor mortis, and then gradually the stiffness decreases as the cells break down.

If the motor neuron continues to stimulate the muscle, calcium ions stay bound to troponin, leaving the myosin-binding sites unblocked by tropomyosin, and the cross-bridge cycle continues, if ATP is available, until muscle contraction is complete. You can visualize a millipede walking with different legs touching the ground at different times as the millipede moves itself forward. It is similar with myosin. Not all myosin heads bind at the same time, and they usually bind multiple times during a single muscle contraction.

During a single contraction, such as a bicep curl, the **load** determines the **force** of contraction required by the muscle. The heavier the load, the more force required and the more muscle fibres and myosin heads within each muscle fibre will be involved in the contraction. To lift a glass of water, not very many actin-myosin connections are required, but to lift a heavy dumbbell requires many more muscle fibres and myosin heads to be involved in the contraction.

During an eccentric contraction, such as lowering yourself slowly from a chin-up, the myosin heads bind to actin, but each time they are released they bind to an actin protein that is closer to the Z disc rather than to the M line; this is due to the sarcomeres being pulled into their lengthened state.

Muscle tone

When muscles are relaxed, there are always some myosin heads binding and releasing actin, so muscles are never 100% relaxed. This small amount of binding keeps muscles somewhat firm, and this is called **muscle tone**. Muscle tone is important for preventing our muscles from over-stretching, and it helps maintain posture and balance when we are moving.

Neuromuscular junction

Skeletal muscle contraction is voluntary, although many complex movements, such as walking or swimming, involve muscle contractions that we do not have to specifically think about; they become learned motor patterns that do not involve conscious thought. For the sake of simplicity, skeletal fibres are considered under voluntary control. Motor movements are initiated from the **motor cortex** of the brain, which connects to motor neurons in the spinal cord that allow action potentials to travel through spinal nerves to all skeletal muscle fibres (Chapter 15).

The neuromuscular junction (NMJ) is the **synapse**, which is the area where a motor neuron connects to a skeletal muscle fibre. A motor neuron may contact multiple muscle fibres (possibly up to 3000 fibres in a large motor unit) and multiple areas of the same muscle fibre because a neuron branches into **axon terminals**. Axon terminals contain **synaptic end bulbs**, which contain vesicles containing **neurotransmitters**. **Acetylcholine (Ach)** is the only neurotransmitter in motor neurons that leads to skeletal muscle contraction. The neurotransmitters cross the synaptic cleft—the space between the axon terminal and the sarcolemma—and bind to the specific receptors on the sarcolemma, thus stimulating an electrical signal in the muscle cell (Figure 14.20).

The electrical signal that travels down the motor neuron, from the cell body in the spinal cord, along the axon, to the axon terminals, is called an **action potential**, which is discussed in detail in Chapter 15. Simply, the action potential is the movement of sodium ions across the neuron membrane. As sodium ions cross the membrane in one section of axon, they signal sodium ions to cross the next section of axon, until sodium ions enter the axon at the axon terminal. At the axon terminal, the action potential (sodium ions moving into the cell) triggers the opening of **voltage-gated** calcium ion membrane channels. *Voltage-gated* means membrane channels open because of a change in the charge across the membrane, which is caused by the Na+ entering the cell. When the Ca^{2+} channels open, Ca^{2+} can enter the synaptic bulb and trigger the release of ACh neurotransmitters. The vesicles containing the ACh fuse with the axon membrane, and the neurotransmitters are released by exocytosis into the synaptic cleft. ACh then diffuses across the synaptic cleft and binds to specific receptors on the sarcolemma, causing **ligand-gated** sodium ion channels to open on the muscle membrane, which causes an action potential, just like in the neuron. Sodium ions enter the muscle cell sarcoplasm all along the T-tubules, which are located directly beside the sarcoplasmic reticulum. The action potential in the T-tubules causes the release of the calcium ions from the sarcoplasmic reticulum. The Ca^{2+} binds to troponin, moving the tropomyosin, and the cross-bridge cycle can begin. Back in the synaptic cleft, ACh is broken down by an enzyme called **acetylcholinesterase**.

DID YOU KNOW?

The choline used to make acetylcholine is acquired when we eat plant and animal cell membranes. Recall from Chapter 2 that the polar head of many phospholipids that make up cell membranes contains choline. The foods that have very high amounts of choline include seafood, eggs, chicken, beef, peanuts, flaxseeds, cauliflower, and spinach. Choline is also required to convert homocysteine to the amino acid methionine; a deficiency of choline leads to increased homocysteine levels, which are correlated with cardiovascular disease (Chapter 17, Section 12).

Steps of skeletal muscle contraction from nervous system stimulation to relaxation

1. Action potential travels down motor neuron to axon terminal (Figure 14.21).
2. Action potential triggers movement of Ca^{2+} into the axon terminal.
3. Ca^{2+} triggers release of ACh neurotransmitters into synaptic cleft.
4. ACh binds to specific receptors on the sarcolemma that triggers the opening of Na^+ channels.
5. Action potential occurs in muscle cell membrane as Na^+ crosses the membrane from the T-tubules and enters the sarcoplasm.
6. The action potential in the sarcolemma causes the sarcoplasmic reticulum to release Ca^{2+} into the sarcoplasm.
7. Ca^{2+} binds to troponin, which moves tropomyosin off the myosin-binding sites on the actin proteins.
8. ATP is bound to myosin in the relaxed state. Once an action potential occurs and the myosin-binding sites on the actin are available, ATP hydrolyzes,

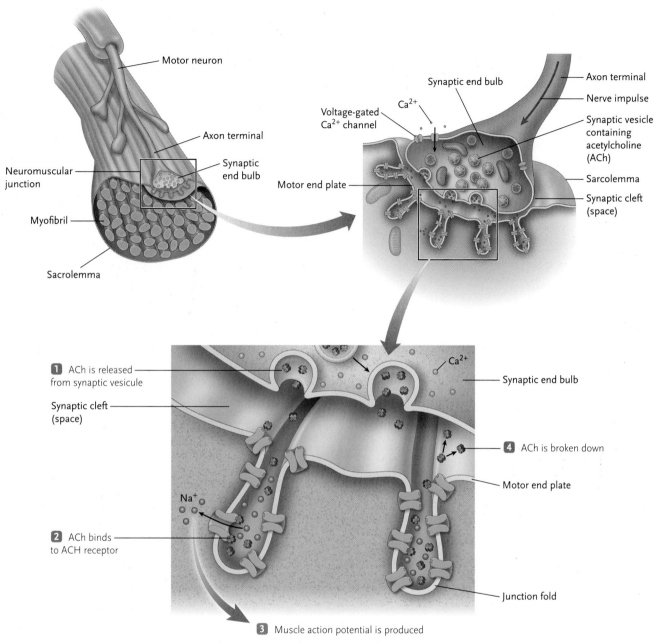

Figure 14.20   **Neuromuscular Junction**

The neuromuscular junction is the connection between a neuron and a muscle cell.

releasing the phosphate group, and myosin binds to actin (cross-bridge formation).

9. ADP leaves the myosin and causes myosin to tilt back into its normal position, which causes the action to slide across the myosin (power stroke), and sarcomeres shorten.

10. A new ATP molecule binds to myosin, causing the release of myosin from actin. The cycle continues as long as a motor neuron stimulates the muscle cell; the muscle relaxes when there are no further action potentials, and calcium has been pumped back into the sarcoplasmic reticulum.

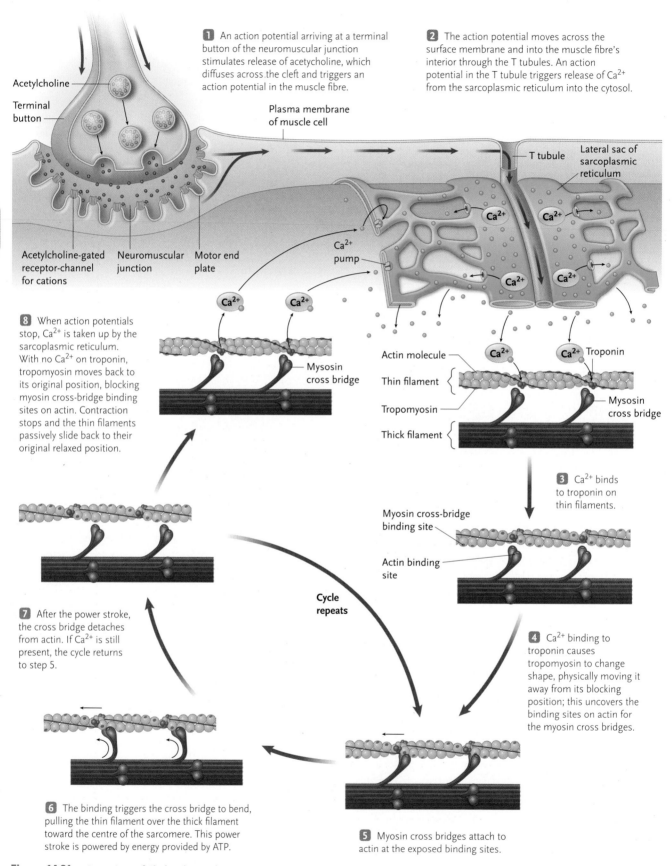

1 An action potential arriving at a terminal button of the neuromuscular junction stimulates release of acetycholine, which diffuses across the cleft and triggers an action potential in the muscle fibre.

2 The action potential moves across the surface membrane and into the muscle fibre's interior through the T tubules. An action potential in the T tubule triggers release of Ca^{2+} from the sarcoplasmic reticulum into the cytosol.

Acetylcholine

Terminal button

Plasma membrane of muscle cell

T tubule

Lateral sac of sarcoplasmic reticulum

Ca^{2+}

Ca^{2+}

Acetylcholine-gated receptor-channel for cations

Neuromuscular junction

Motor end plate

Ca^{2+} pump

Ca^{2+}

Ca^{2+}

8 When action potentials stop, Ca^{2+} is taken up by the sarcoplasmic reticulum. With no Ca^{2+} on troponin, tropomyosin moves back to its original position, blocking myosin cross-bridge binding sites on actin. Contraction stops and the thin filaments passively slide back to their original relaxed position.

Ca^{2+}

Ca^{2+}

Mysosin cross bridge

Actin molecule

Ca^{2+}

Ca^{2+} Troponin

Thin filament

Tropomyosin

Mysosin cross bridge

Thick filament

3 Ca^{2+} binds to troponin on thin filaments.

Myosin cross-bridge binding site

Actin binding site

Cycle repeats

7 After the power stroke, the cross bridge detaches from actin. If Ca^{2+} is still present, the cycle returns to step 5.

4 Ca^{2+} binding to troponin causes tropomyosin to change shape, physically moving it away from its blocking position; this uncovers the binding sites on actin for the myosin cross bridges.

6 The binding triggers the cross bridge to bend, pulling the thin filament over the thick filament toward the centre of the sarcomere. This power stroke is powered by energy provided by ATP.

5 Myosin cross bridges attach to actin at the exposed binding sites.

Figure 14.21 Overview of Skeletal Muscle Contraction

The process of skeletal muscle contraction, from neuron signalling to contraction

Art source: From SHERWOOD, Human Physiology, 8E. © 2013 Cengage Learning.

CONCEPT REVIEW 14.4

1. What are the two contractile proteins in muscle cells?

2. What are the two regulatory proteins? Which one binds calcium, and which one blocks the myosin-binding sites?

3. Describe the sliding filament theory.

4. Which regions of the sarcomere shorten during contraction?

5. Where is calcium stored in muscle cells?

6. What causes calcium to be released?

7. Name the four steps of the cross-bridge cycle.

8. Name two reasons why muscle cells use so much ATP.

9. What causes more muscle fibres to be involved in a contraction?

10. Name three functions of muscle tone.

11. What ion causes action potentials in neurons? In muscle cells?

12. Describe how calcium plays a role in the neuron.

13. What neurotransmitter triggers the action potential in the muscle fibre?

14. What happens to the neurotransmitter after the action potential?

15. List the order of events that occur during muscle contraction, beginning with stimulation by the motor neuron.

14.5 MUSCLE METABOLISM

Recall from Chapter 4 that ATP can be produced by aerobic respiration, with oxygen, or by anaerobic respiration, without oxygen. Muscles primarily use glucose as an initial fuel source, but they can also use lactic acid, ketones that are created from breaking down fat (Chapter 19), fatty acids, or amino acids. Muscle cells have a stored supply of the polysaccharide glycogen (Chapter 2) and an oxygen-binding molecule called **myoglobin**, which is similar to hemoglobin that binds oxygen in red blood cells. Glycogen can readily be broken down into free glucose molecules by a process called **glycogenolysis**, and it is the first supply of glucose that's used. Once the glycogen supply decreases, after approximately 20 to 30 minutes of moderate activity, the muscles take up glucose from the bloodstream. The liver maintains blood sugar levels by breaking down liver glycogen and fat, as well as fat stored in adipose tissue. (See Chapter 19 for details of liver function in metabolism.)

Muscle cells also use a molecule similar to ATP: **creatine phosphate** (Figure 14.22). Creatine is made from amino acids in the liver and is transported to the muscles. In the muscle cells, creatine is phosphorylated by an enzyme called **creatine kinase**. The phosphorylated creatine, which is stored in small amounts in the muscle, is used initially as a rapid supply of energy. Creatine phosphate can also transfer its phosphate group to ADP to form ATP, and this gives the cell enough energy for intense activity for approximately two to seven seconds, depending on the intensity of the activity. This brief period is enough time for the muscle cells to begin generating more ATP through aerobic or anaerobic mechanisms. When the intense exercise is complete, any excess ATP that was produced can be used to re-form phosphorylated creatine, which can be used during the next bout of exercise. This is a protective mechanism for muscle cells; it ensures there is always a supply of energy, without a lag period, while ATP is produced through glycolysis or oxidative phosphorylation (Chapter 4). Creatine phosphate can be taken as a supplement by athletes, but it only benefits athletes in high-intensity sports such as sprinting.

Muscle fibres use anaerobic respiration during high-intensity exercise: for example, during sprinting, which involves rapid contractions, or during weight lifting, which involves contraction under a heavy load. Anaerobic respiration does not require oxygen; it involves glycolysis where a net of two ATP molecules are produced from one glucose molecule. The pyruvate produced at the end of glycolysis is then converted into lactic acid (Figure 14.23). Lactic acid can be converted back into pyruvate—which can be used by the cell for aerobic respiration—once oxygen is available,

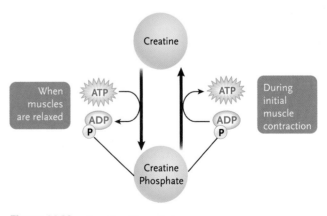

Figure 14.22 Creatine Phosphate

Creatine phosphate is the first energy used by muscle cells at the beginning of exercise.

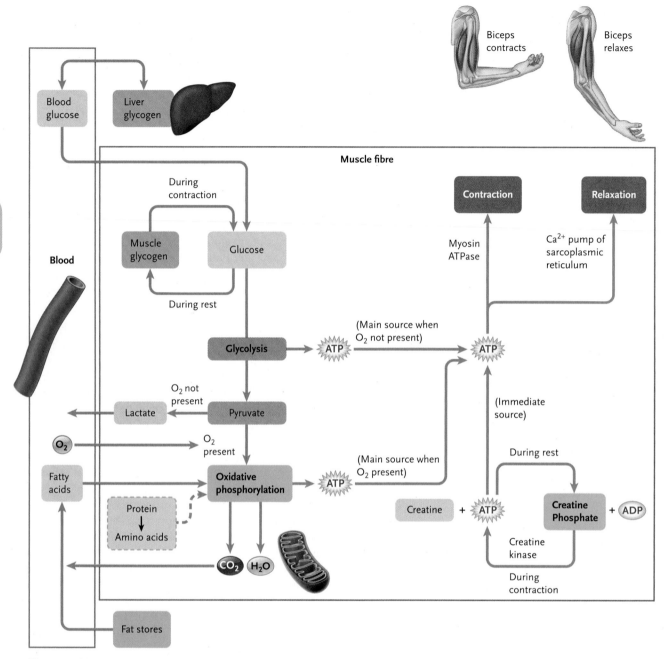

Figure 14.23 Aerobic and Anaerobic Respiration
Aerobic respiration utilizes oxygen, and anaerobic respiration involves the production of ATP in the absence of oxygen.

blood flow increases, or the intensity of the exercise decreases. The anaerobic pathway can produce enough ATP to sustain maximal effort for 30 to 40 seconds; after that, intensity must decrease or muscle fatigue will prevent further contraction. It is extremely difficult to maintain maximal effort for 30 to 40 seconds, and recovery time can be several minutes. It has been shown that sprinting or biking or some equivalent activity in bursts of even 20 seconds at maximal effort is very good for cardiovascular conditioning.

Aerobic respiration is used when plenty of oxygen is available so that pyruvate continues on to the Kreb's cycle,

and chemiosmosis can occur via the electron transport chain, producing a total of approximately 36 ATP molecules from one glucose molecule (Figure 14.23). Aerobic respiration produces CO_2 as a waste product and as heat. Muscle contractions are the main source of heat produced in our body because muscles produce the most ATP; as ATP is produced, heat is released. Aerobic respiration is involved in any kind of endurance activity that can be sustained for long periods of time. Most muscle cells have the capacity for both anaerobic and aerobic respiration; however, in Section 14.6, we see how different types of muscles primarily use one pathway over the other.

CONCEPT REVIEW 14.5

1. What is the difference between aerobic and anaerobic respiration?

2. How is creatine phosphate used as an energy source for muscle cells?

3. How much ATP is produced from anaerobic respiration?

4. How much ATP is produced from aerobic respiration?

5. Which method of ATP production would allow for longer-duration activities, such as jogging 5–10km?

14.6 TYPES OF SKELETAL MUSCLE FIBRES AND EXERCISE

There are three main types of skeletal muscle fibres: (1) slow oxidative, also called slow-twitch fibres; (2) fast oxidative-glycolytic, also called intermediate fibres; (3) fast glycolytic, also called fast-twitch fibres.

Each fibre type has special characteristics that allow muscles to function in either endurance activities, intermediate-intensity activities, or high-intensity, short-duration activities. All the specific muscles discussed in Section 14.3 contain some ratio of all three fibre types, and some muscles are more slow oxidative and others are more fast glycolytic.

Slow oxidative fibres

Slow-twitch fibres have the smallest diameter, so they contain fewer myofibrils per cell than other fibre types. They are adapted for long periods of continuous, low-intensity contractions and are the most resistant to fatigue. They are called oxidative because they primarily use aerobic respiration to produce ATP. Muscles with more slow-twitch fibres are a deep red colour because they contain many blood vessels that carry the required oxygen and glucose and also a lot of myoglobin, which is red like hemoglobin when carrying oxygen. Since these fibres use aerobic respiration, in comparison to other fibre types, they have the most mitochondria. Slow-twitch fibres store little glycogen and do not store creatine phosphate; therefore, they develop tension more slowly than the fast fibres. You can think of a chicken: the legs have the dark meat because leg muscles consist of the slow-twitch fibres; chickens walk around using their leg muscles for long periods of time.

Fast glycolytic fibres

Fast-twitch fibres have the largest diameter, contain the most myofibrils per cell, and are the strongest of all the fibre types. They are adapted for short bursts of high-intensity activity, such as sprinting or lifting weights, and they fatigue quickly. Fast-twitch fibres are adapted for rapid quick movements, with periods of rest in between burst of activity. For example, when bike riding, you can sprint uphill for approximately 30 seconds, and then you coast or peddle more slowly to catch your breath, and then you can sprint again once your muscles have recovered and produced more ATP. Or when weight lifting, you can do repetitions and sets so that the muscles recover between lifts. Fast-twitch fibres primarily use glycolysis for ATP production; therefore, they do not need as much oxygen. They have fewer blood vessels, no myoglobin, and appear white in colour. They store creatine phosphate that is used in the initial burst of activity, and they have a large glycogen supply. Picture the chicken again: the chicken breast is primarily fast glycolytic fibres because the chicken only uses the breast muscle to flap its wings in short bursts for very short periods of time. Exercising fast glycolytic fibres causes increased muscle mass because the muscle fibres build up more myofibrils in each cell. Having more myofibrils means that more ATP is required to fuel those muscles and therefore burn more calories.

Fast oxidative-glycolytic fibres

Intermediate fibres have qualities similar to both slow oxidative and fast glycolytic. They are adapted to use glycolysis as well as oxidative phosphorylation for ATP production and are involved in moderate to somewhat high-intensity activities and are moderately resistant to fatigue. We use intermediate fibres for the majority of our daily movements such as carrying groceries, climbing stairs, jogging, biking, dancing, and swimming. These fibres have an intermediate diameter, contain many mitochondria, blood vessels, and myoglobin, and appear red in colour.

Muscle types and their fibre speed

Most skeletal muscles have combinations of all three types of fibres. Approximately 50% of our muscle fibres are slow oxidative, and the other 50% is a combination of fast glycolytic, and fast oxidative-glycolytic. Different people have different proportions of each type based on their genetics and their primary types of activity. Many researchers now believe that fibres can be converted to slow- or fast-twitch fibres, either to increase blood vessel density and increase mitochondria or to increase fibre diameter and utilize glycogen, but it is not known to what extent this can occur. You can probably determine if you are dominant in slow fibres or fast fibres by the types of sports or activities you like to engage in. Endurance athletes have more slow-twitch fibres, and power-lifters and sprinters have more fast-twitch fibres.

Muscles that contain more slow oxidative fibres include postural muscles, erector spinae, abdominals, calves, forearm flexors and extensors, and deep upper back muscles. Also the muscles of the face used for facial expressions and chewing contain more slow fibres. Muscles that tend to have a large number of fast glycolytic fibres include the deltoids, gluteus maximus, quadriceps (depending on the person and type of exercise), chest, latissimus dorsi, teres major, biceps, and triceps. The quadriceps and hamstrings have a lot of intermediate fibres and are adapted for walking long distances for hours, as well as for sprinting; so the exact proportions depend on the individual.

Exercise and weight loss

The key to losing weight is making more ATP. Producing ATP, whether through aerobic or anaerobic means is how we burn calories (Chapter 2). All cells use ATP. To fuel the sodium-potassium pumps in all cells of the body uses close to one-third of our daily required calories. We also use ATP to think, to digest food, to breathe, to fight infections, to pump blood, and to move around using our skeletal muscles. The amount of calories we need to consume to maintain our cellular functions without gaining or losing weight is called our **basal metabolic rate (BMR)**. If we want to lose weight, stored as fat in our subcutaneous adipose tissue and in the abdominal cavity surrounding our organs, then we need to break down the stored fat and use it to make ATP; this can be accomplished by eating fewer calories or by increasing the need for ATP through exercise. If we consume more calories than required to maintain our BMR any excess nutrients will be stored as fat.

Since muscle cells use by far the most ATP compared to any other tissue, and increasing the amount of muscle contraction requires more ATP, we need to exercise. All three types of muscle fibres use ATP and burn calories. Slow-twitch fibres burn calories at a slower rate than fast-twitch fibres, but you can use slow-twitch fibres for a much longer period of time. Using fast-twitch fibres during high-intensity, short-duration exercises means more recovery time is required, and muscle cells need to repair. The repair process can last 24 hours or longer after a workout, and building new myofibrils requires a lot of ATP. It depends on which type of activities you prefer. If you enjoy the activity then you are more likely to stick with it. The most important thing is to be active all the time: some high-intensity activities to build speed and strength, and also many low-intensity activities, such as walking, carrying groceries, taking the stairs, play golf or tennis, or swimming. Any kind of activity causes muscles to contract; this requires ATP, so the more active you are, the more you will increase your BMR.

Factors affecting weight loss

Is weight loss really just that simple: eat less and be more active? The simple answer is yes, but there are several factors that play a significant role.

- People that weigh more burn more calories per day because they have more cells that require ATP.
- People with more muscle mass burn more calories because muscles use the most ATP.
- Males generally burn more calories than females, primarily because they usually have more muscle mass.
- Younger people burn more calories than older people. A teenage male that is very active in sports may require 3500 calories per day, whereas a small elderly woman that does not exercise may only require 1200 calories per day.

There are also physiological factors that affect BMR. **Lipoprotein lipase (LPL)** is a group of enzymes that break down triglyceride fats into components that can be utilized by cells to make ATP. LPLs are expressed primarily in adipose tissues, small intestine, heart, and skeletal muscles. Increased LPL is correlated with increased use of fats for ATP production. The hormones **glucagon** and **epinephrine** stimulate the production of LPL during fasting (Chapter 16), and it is also increased in growing children and teens. LPL expression decreases in sedentary people, making it more difficult to use fat for energy. Research has shown that people who exercise for 30 minutes, three times per week (the most common advice), but spend most of the rest of their time sitting, will have decreased LPL levels and will not use stored fat for energy. To summarize, yes, weight loss can be as simple as eating less and being active. Eating less means having periods of time between meals when you are not eating, and this stimulates hormones that break down fat. Being active means being active *throughout* the day, not just three times each week.

Health benefits of exercise

Exercise, especially exercise that uses all three fibre types increases bone density and prevents osteoporosis; strengthens muscles; increases balance and coordination, and therefore decreases the potential of falling and breaking bones; increases the production of good mood hormones; possibly prevents or reverses mild depression, and helps decrease stress. Exercise also helps decrease inflammation in the body, and helps prevent type-2 diabetes and cardiovascular disease.

CONCEPT REVIEW 14.6

1. Make a chart comparing the main features of slow oxidative fibres, fast glycolytic fibres, and fast oxidative-glycolytic fibres.

2. From the following list of activities, determine which muscle fibre type would be primarily involved. Note some activities may include more than one fibre type.

   a. squatting 50 lb
   b. jogging for 1 hour
   c. running 1500 m
   d. sitting in a chair with good posture
   e. typing
   f. eating
   g. downhill skiing
   h. walking up a flight of stairs
   i. running long jump

3. What are the two factors involved in maintaining a healthy body weight?

4. Name some benefits of exercise other than weight loss.

14

Learn everything you can, anytime you can, from anyone you can, there will always come a time when you will be grateful you did.

Sarah Caldwell

15.1 OVERVIEW OF THE NERVOUS SYSTEM

The nervous system is composed of the **central nervous system (CNS)** and the **peripheral nervous system (PNS)** (Figure 15.1). The CNS includes the brain and the spinal cord, and the PNS includes **cranial nerves**, spinal nerves, and all the neurons and sensory receptors outside the CNS. There are billions of neurons and supporting cells in the central and peripheral nervous systems. A **nerve** is a bundle of hundreds or thousands of neurons. The human body has 12 pairs of cranial nerves and 31 pairs of spinal nerves that innervate every tissue in the body.

The three major divisions of the PNS are the **somatic** nervous system, the autonomic nervous system (ANS), and the enteric nervous system (Figure 15.2).

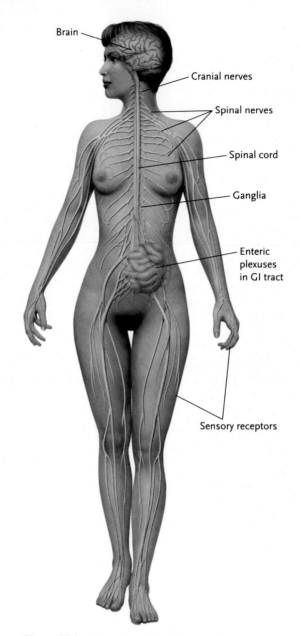

Brain

Cranial nerves

Spinal nerves

Spinal cord

Ganglia

Enteric
plexuses
in GI tract

Sensory receptors

Figure 15.1   Overview of the Nervous System
The nervous system is composed of the brain, spinal cord, and peripheral nerves.

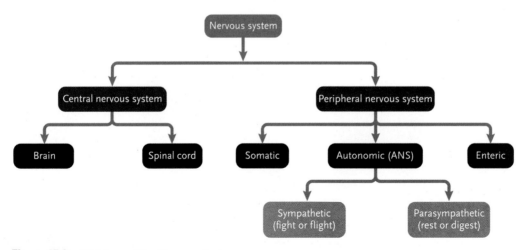

Nervous system

Central nervous system

Peripheral nervous system

Brain

Spinal cord

Somatic

Autonomic (ANS)

Enteric

Sympathetic
(fight or flight)

Parasympathetic
(rest or digest)

Figure 15.2   Divisions of the Nervous System
The nervous system is divided into the central nervous system and the peripheral nervous system.

The **somatic nervous system (SNS)** is involved in *voluntary* control of movements. The somatic nervous system includes receptors, sensory neurons, also called **afferent neurons**, and the motor neurons, also called **efferent neurons,** which stimulate skeletal muscle fibres to contract (Chapter 14) (Figure 15.3).

The autonomic nervous system (ANS) is involved in *involuntary* control of organs, blood vessels, endocrine glands, and exocrine glands. The ANS works with the endocrine system to regulate homeostasis, such as body temperature, blood pressure, and blood sugar levels. The components of the ANS, like that of the somatic system, include sensory receptors, afferent neurons, and efferent neurons. The efferent neurons in the ANS innervate smooth muscle, cardiac muscle, or glands (Figure 15.3). The ANS is discussed in detail in Section 15.7.

The **enteric nervous system** is a network of neurons that regulates the digestive system (Chapter 19). The enteric nervous system is made up of the **submucosal plexus,** which stimulates the secretion of enzymes, gastric

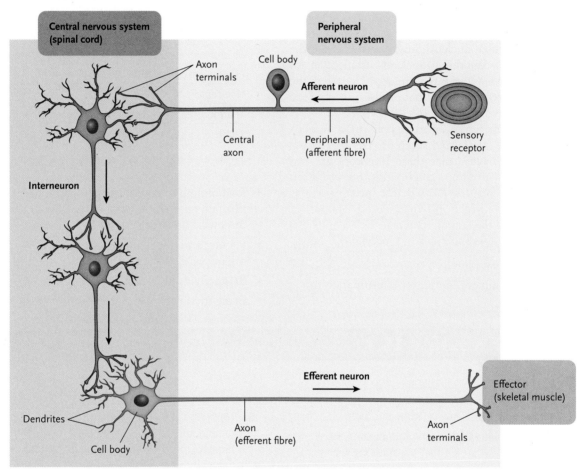

A. Somatic nervous system

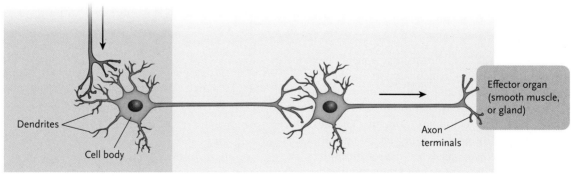

B. Autonomic nervous system

Figure 15.3 Afferent Neurons and Efferent Neurons

Afferent neurons are sensory neurons that bring information to the brain, and efferent neurons are motor neurons that bring information from the brain to muscles, glands, or other neurons.

acid, and other substances required for digestion, and the **myenteric plexus**, which stimulates the contraction of the smooth muscles involved in digestion. The enteric nervous system can be influenced by the autonomic nervous system.

Plexus is a term that defines a specific network of nerves and neurons. For example, the brachial plexus is the group of nerves in the brachial region that innervates the arm.

In the PNS, groups of cell bodies are called **ganglia.** For example, each dorsal root ganglion is part of every spinal nerve and contains sensory neuron cell bodies (Section 15.4). In the CNS, groups of cells bodies are called **nuclei** (singular, nucleus); however, this is not the same as the DNA-containing nucleus within cells. An example of the CNS nuclei is the supraoptic nuclei in the hypothalamus, a group of cells that produce antidiuretic hormone (ADH) (see Chapter 16).

Sensory receptors may be extensions of a sensory neuron or may be specialized receptor cells that signal a sensory neuron. Note that these receptors are not the same as the membrane protein receptors discussed in Chapter 2. Sensory receptors detect specific sensory information. Table 15.1 provides an overview of receptor types and the stimuli they detect.

The nervous system functions in the following ways:

- Detecting sensory stimuli
- Analyzing and integrating sensory information in the CNS

TABLE 15.1

Receptor and Stimulus

Receptor	Stimulus
Chemoreceptor	Chemical concentrations, such as hormones, neurotransmitters, and nutrients
Osmoreceptor	Osmolarity, concentration of water and ions
Baroreceptor	Blood pressure; found in certain blood vessels
Photoreceptor	Light; found in the retina
Mechanoreceptor	Stretch and physical movement
Proprioceptor	Body position
Nociceptor	Pain
Thermoreceptor	Temperature; found in the PNS and CNS
Tactile receptors	Touch, pressure, vibration; found in the skin

- Responding to environmental and internal stimuli through efferent neurons of the somatic, autonomic, or enteric nervous systems
- Playing a major role in the regulation of various organ systems (homeostasis)
- Being involved in reflexes
- Being involved in learning, memory, and emotions

CONCEPT REVIEW 15.1

1. Describe the major divisions of the nervous system.
2. What tissues are innervated by the ANS?
3. What are the two plexuses of the enteric nervous system, and what are their main functions?
4. Which division of the peripheral nervous system is involved in voluntary movements?
5. Name the stimuli detected by the following types of receptors: chemoreceptor, osmoreceptor, baroreceptor, photoreceptor, mechanoreceptor, proprioceptor, nociceptor, thermoreceptor, tactile receptors.

15.2 NERVOUS TISSUE

Neuron structure

Nervous tissue consists of neurons that transmit electrical signals and the **glial cells,** which support the functioning of the neurons. Neurons and glial cells are found within the CNS and the PNS. A typical motor neuron is shown in Figure 15.4. The cell body contains all the same organelles as any other cell type, including a nucleus with the DNA, where genes of neurotransmitters are transcribed; mitochondria for ATP production; and rough endoplasmic reticulum for protein synthesis. The **dendrites** are extensions of the cell body that detect stimuli.

The dendrites have membrane receptors (not the same as the sensory receptors discussed in Section 15.1); these are membrane proteins that are linked to ion channels that cause an action potential to be transmitted along the axon (Section 15.3). Action potentials occur in the *axon,* the main extension of the cell body, and move toward the axon terminals, where neurotransmitters are released and signal another neuron, a muscle, or gland. **Axon collaterals** are branches of axons that also end with axon terminals. Neurotransmitters are chemicals or proteins that can bind to the membranes of other cells and cause muscle contraction, stimulate a gland to secrete hormones, or signal another action potential in another neuron. Some neurotransmitters can also inhibit an action potential in other neurons (Section 15.3). The **axon hillock** is the region where the cell body connects to the axon, and it contains the **initial segment,** which is the starting point for an action potential. Many

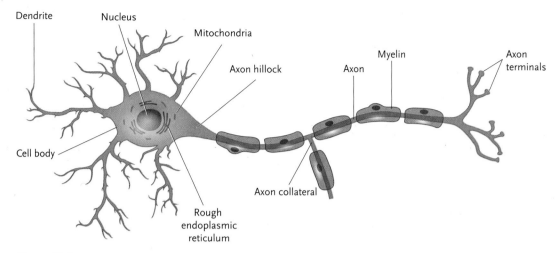

Figure 15.4 Structure of a Typical Motor Neuron

neuron axons contain a protective fatty sheath—called the **myelin sheath**—that helps speed up the transmission of action potentials along the axon. The junction of an axon terminal and the membrane of the neuron it innervates is called the synapse (Section 15.3).

Types of Neurons

There are three main types of neurons: multipolar, bipolar, and unipolar. Multipolar neurons have many dendrite extensions from the cell body (Figure 15.4). They are the common structure of the **motor neurons** (efferent) that innervate muscle tissue (skeletal, smooth, or cardiac) or glands, as well as of the much shorter **interneurons** in the CNS (Figure 15.3). Motor neurons signal the effector in a negative feedback response (Chapter 1). Motor neurons travel from the brain or spinal cord through nerves. Most motor neurons that innervate the skeletal muscles are extremely long. The cell bodies are located in the spinal cord, the axons extend the length of the nerve, and the axon terminals synapse onto the target tissue. One motor neuron can be over one metre long.

Sensory (afferent) neurons bring information to the brain or spinal cord. Those neurons are unipolar, and the cell body projects out to the side of the axon. Sensory neurons have dendrites in the tissue and receive sensory stimuli either directly or from a separate sensory receptor

cell (Section 15.1). The axon then travels through the nerve carrying signals to the spinal cord, where the cell bodies from each spinal nerve are grouped together in dorsal root ganglia (Section 15.4), and the axon terminals synapse onto interneurons in the spinal cord

Bipolar neurons have one main dendrite that extends from the cell body and then branches out (Figure 15.5). Bipolar neurons, the least common of the three types of neurons, are found mainly in the retina, inner ear, and the olfactory area of the brain.

Glial cells of the CNS

In the CNS there are four types of supporting glial cells: astrocytes, oligodendrocytes, microglia, and ependymal cells. Astrocytes are the most abundant glial cell in the CNS and are extremely important for regulating nervous system functions.

- Astrocytes help form the **blood-brain barrier** by producing extensions—called end foot processes—that wrap around the outside of capillary endothelial cells where tight junctions between those endothelial cells separate the circulating blood from the CNS extracellular fluid (Figure 15.6).

- Astrocytes regulate ion concentrations surrounding neurons, for example, mainly the concentrations of sodium, potassium, and calcium during action potentials (Section 15.3).

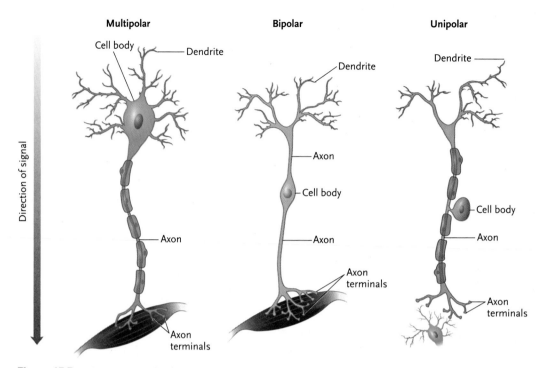

Figure 15.5   Three Types of Neurons

Neurons can be divided into three categories: multipolar, bipolar, and unipolar.

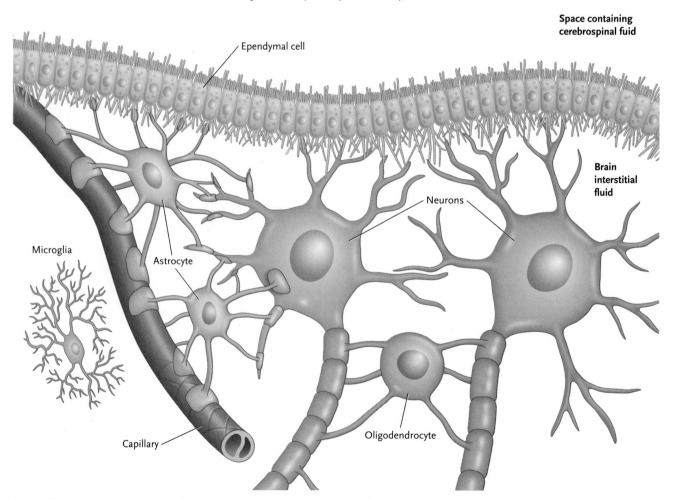

Figure 15.6   Glial Cells

Glial cells in the CNS include oligodendrocytes, astrocytes, microglia, and ependymal cells.

- Astrocytes can store glycogen and can break it down into glucose (glycogenolysis) and provide it to the neurons when needed, and they remove waste products such as ammonia.
- Astrocytes play a key role in the development of the nervous system by producing nerve growth factors that guide migrating neurons.
- Astrocytes are involved in producing scar tissue during healing of any CNS injuries.

Oligodendrocytes are the type of glial cells that forms the myelin sheath that surrounds the axons of many neurons in the CNS. Some small interneurons are not myelinated. Oligodendrocytes have branching fatty membrane extensions that can wrap around several axons. Myelin acts like an insulator so that electrical signals cross the neuron cell membrane only in the spaces between the myelin; these spaces are called the nodes of Ranvier. In these nodes are the ion channels that propagate the action potential. **Saltatory conduction** is

DID YOU KNOW?

Multiple sclerosis is an autoimmune disease where the immune system targets the myelin sheath as though it were fighting an infection. This process eventually destroys the myelin, and nerve impulses cannot travel to the correct locations. It often results in loss of sensations and paralysis.

the term that refers to action potentials travelling along myelinated neurons.

Microglia are resident immune cells that are phagocytic and fight infections that occur in the CNS. Microglia also engulf dead cells or cell debris.

Ependymal cells are found in the choroid plexus of each of the four ventricles of the brain, and they produce **cerebral spinal fluid (CSF)** from contents in the blood. The composition of CSF is similar to blood plasma, but not exactly the same because some of their ion concentrations differ. The CSF bathes the cells of the CNS with nutrients, ions, and oxygen. CSF is produced continuously, and it circulates through the ventricles and spinal cord and eventually re-enters the circulatory system at the superior sagittal sinus, which is the largest vein in the brain (Chapter 17).

Glial cells of the PNS

The peripheral nervous system has two kinds of glial cells: satellite cells and Schwann cells. Satellite cells are similar in function to astrocytes and regulate ions and nutrients in the interstitial fluid surrounding cell bodies in the PNS ganglia, such as the dorsal root ganglia that contain the cell bodies of sensory neurons. Schwann cells are similar in function to oligodendrocytes and produce myelin that surrounds the very long axons of afferent and efferent neurons. Schwann cells wrap around the axons of one neuron only and do not branch and wrap around several axons as do the oligodendrocytes in the CNS.

CONCEPT REVIEW 15.2

1. What two cell types are found in nervous tissue?
2. What is the main function of the dendrites? Cell body? Axon? Axon terminals?
3. Name three types of neurons.
4. Make a chart to summarize the main functions of each type of glial cell in the CNS and PNS.

15.3 ACTION POTENTIALS AND GRADED POTENTIALS

Action potentials

Recall from Chapter 3 that the ion concentrations in the cytoplasm and in the interstitial fluid are not the same. There are more sodium ions and chloride ions in the interstitial fluid, and there are more potassium ions in the cytoplasm. Along with the contribution of charges from proteins and other molecules inside and outside the cell, the charge inside the membrane of the cell is *negative* relative to outside the cell. Also recall that the gradient of sodium and potassium ions is maintained by the **Na$^+$/K$^+$ pump** (Chapter 3). The charge difference between the cytoplasm side and the interstitial fluid side of the cell membrane creates a **concentration gradient** as well as an electrical gradient, and the electrical gradient

makes the cell polarized. The charge inside a resting neuron is approximately –70mV, and this is called the resting membrane potential (Figure 15.7).

Ions cannot move across a cell membrane without the aid of a membrane protein ion channel. Ion channels in the neuron are regulated so that they are either open or closed depending on stimulation of the neuron. If we poke the back of our hand, tactile sensory receptors in the skin will stimulate a sensory neuron. The stimulation takes the form of opening sodium ion channels. Because sodium ions (Na$^+$) are positively charged, and in higher concentration outside the cell, the opening of Na$^+$ membrane channels causes Na$^+$ to move *into* the cell, down the concentration gradient, *and* down the electrical gradient. When Na$^+$ moves into the cell, the inside of the cell becomes positively charged, which is opposite to the negative charge of the resting membrane potential; this is called

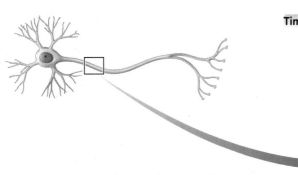

depolarization, and the cell is no longer polarized. The depolarization and subsequent repolarization of that region of the membrane is called an action potential. When the axon depolarizes in one area, it signals Na^+ channels in the next segment to open, and in this way depolarization occurs along the axon toward the axon terminals. The movement of depolarization along the axon propagates an action potential.

Ion channels open or close in response to a stimulus. When the stimulus is a chemical, such as a neurotransmitter from another neuron, the channel is called a **chemically gated channel**. When an ion channel opens or closes because of a change in the *charge* of the membrane, it is called a voltage-gated channel.

An action potential begins at the initial segment of an axon hillock in response to a change in the charge of the membrane that's caused by a graded potential (described in the next section). If the charge change reaches the threshold level of approximately $-55mV$, an action potential will occur and voltage-gated Na^+ channels open, allowing Na^+ to enter the cell and cause depolarization (Figure 15.8). Action potentials are considered "all-or-none" because they either occur or they do not occur, and this depends on the charge of the axon hillock reaching the threshold level. Action potentials always have the same magnitude; so a stronger stimulus causes an increase in the *frequency* of the action potentials, not a larger action potential.

Stages of an action potential

The resting membrane potential of a neuron is $-70mV$. A stimulus that causes the inside of the cell membrane to become more positive (early depolarization) will trigger an action potential *if* the cell charge reaches the threshold level of $-55mV$ (Figure 15.8).

1. **Depolarization.** Once the cell membrane potential reaches threshold, the voltage-gated sodium channels open, allowing Na^+ ions to move into the cell,

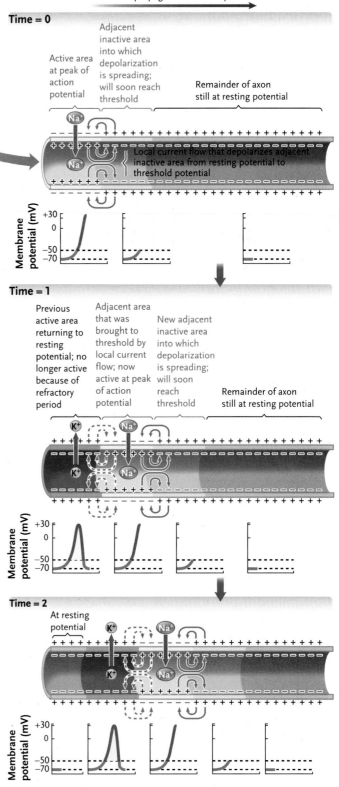

Figure 15.7 **Action Potential**

An action potential occurs in stages: depolarization, repolarization, and hyperpolarization.

Art source: From Russell/Hertz/Starr/Fenton. Biology, 2E. © 2013 Nelson Education Ltd. Reproduced by permission. www.cengage.com/permissions

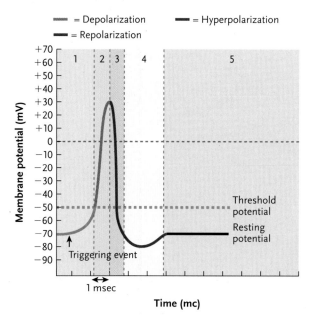

Figure 15.8 Depolarization and Repolarization

Depolarization is caused by Na⁺ moving into the cell, and repolarization is caused by K⁺ moving out of the cell.

further increasing the charge to about +30mV (and further depolarizing the membrane).

2. **Repolarization.** The cell membrane reaches +30mV, the Na⁺ channels become inactivated, so no more sodium ion can enter the cell, which stops further depolarization. Then voltage-gated potassium ion channels open. When K⁺ channels open, K⁺ moves out of the cell, causing the membrane potential to again become negative and the membrane to repolarize.

3. **Hyperpolarization.** The potassium channels have a slight delay in closing once the cell's charge becomes negative; this means more K⁺ leave the cell than are required to repolarize the membrane. So the membrane becomes more negatively charged than in the resting state, causing the membrane to be hyperpolarized. This is important because it means the charge of the cell is further from the threshold level, and it will prevent another immediate action potential unless the stimulus is strong enough to reach threshold. This refractory period ensures that the action potential moves in one direction down the axon toward the axon terminals.

4. Once the potassium ion channels close, the Na⁺/K⁺ pump helps restore the resting membrane potential by moving K⁺ in and Na⁺ out of the cell, and then another action potential can occur.

Note that one action potential in one section of the axon occurs within two to three milliseconds, that is, two to three 1000ths of a second. Also note that in a non-myelinated axon, action potentials continue along the entire length of the axon until it reaches the axon terminal.

Once an action potential reaches the axon terminal, neurotransmitters are released into the **synaptic cleft**. If the neurotransmitter binds to a muscle fibre contraction occurs; if it binds to another neuron it stimulates a **graded potential**.

Synapse

Recall from Chapter 14 that an axon terminal that connects with a skeletal muscle fibre is called a neuro-muscular junction, which is one kind of synapse. The release of neurotransmitters at the synapse is the same whether the axon terminal is synapsing with a smooth muscle, cardiac muscle, or another neuron. The following is the order of events that occur in the axon terminal after an action potential has been triggered (Figure 15.9):

1. The action potential causes the neuron membrane to depolarize, and the depolarization triggers voltage-gated calcium channels in the axon terminal to open and allow Ca⁺⁺ to enter the cell.

2. When calcium ions enter the axon terminal, the synaptic vesicles containing neurotransmitters fuse with the cell membrane, and exocytosis of the neurotransmitters occurs.

3. Once the neurotransmitters are released into the synaptic cleft, they bind to *specific* receptors on the next neuron, the **post-synaptic neuron**.

4. The receptors on the post-synaptic neuron are chemically gated ion channels that open when the neurotransmitter binds to them, and this allows ions to enter the next cell, which causes a graded potential.

5. If enough positively charged ions, such as Na⁺, enter the post-synaptic neuron, an action potential can be triggered in that neuron (Figure 15.9).

Graded potentials

A graded potential is a small change in the membrane potential, usually on the dendrites or cell body. Whereas action potentials always cause a depolarization of the axon membrane with sodium ions moving into the cell, a graded potential can be either depolarizing (becoming *positively* charged inside the cell) or hyperpolarizing (becoming more *negative* inside the cell). Also, graded potentials can be caused by positive or negative charged ions.

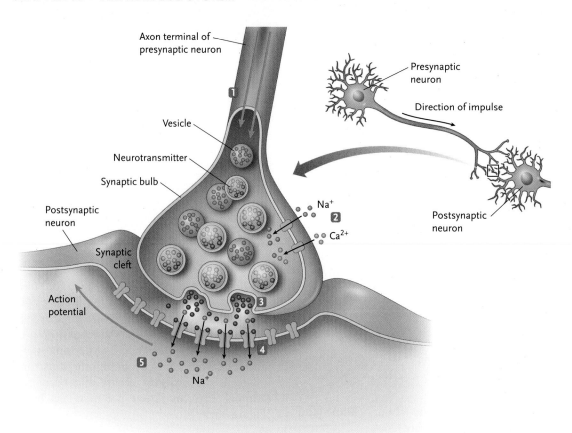

Figure 15.9   The Synapse
The synapse consists of the axon terminal, the synaptic cleft, and another neuron or a muscle cell.

When one neuron is signalling another neuron, the one that is releasing the neurotransmitter is the **pre-synaptic neuron,** because it is before the synapse. The neurotransmitters released by the pre-synaptic neuron act on the post-synaptic neuron (Figure 15.10).

For example, if a neurotransmitter from a pre-synaptic neuron stimulates chemically gated chloride (Cl⁻) channels to open in the post-synaptic neuron, Cl⁻ moves *into* the cell—because of the concentration gradient—and causes it to become more *negative*, hyperpolarized. Therefore, an action potential is *less* likely to occur at the axon hillock because the charge will be further from the threshold level. Neurotransmitters that cause hyperpolarization and decrease the likelihood of an action potential are **inhibitory neurotransmitters.** Inhibitory neurotransmitters cause graded potentials that are called **inhibitory post-synaptic potentials (IPSPs)** (Figure 15.10). If a neurotransmitter stimulates chemically gated sodium channels, Na+ enters the cell, making the charge more positive (depolarized), which brings the cell membrane closer to the threshold level at the axon hillock and increases the probability of an action potential. Neurotransmitters that cause depolarization and increase the likelihood of an action potential are excitatory, causing **excitatory post-synaptic**

potentials (EPSPs). When multiple pre-synaptic neurons are affecting a post-synaptic neuron, an action potential will occur only if the threshold level is reached.

Graded potentials can be summed, so multiple excitatory or inhibitor input from multiple connecting neurons can influence a single neuron. The single neuron will have an action potential only if the excitatory signals are stronger than the inhibitory signals and cause the cell to reach threshold level. Graded potentials can have **variable amplitude** depending on the amount of stimulation; more stimulation causes a greater depolarization (EPSP) or hyperpolarization (IPSP). Graded potentials do not have a refractory period. Chemically gated ion channels in the dendrites and cell body can be opened whenever neurotransmitters from the pre-synaptic neuron bind.

Stages of action potential from stimulation to transmission

1. Resting membrane potential in a neuron is –70mV.

2. EPSPs (graded potential) cause depolarization of the membrane, and if that increase in positive charge inside the cell is enough to reach the threshold level of –55mV, then an action potential can occur.

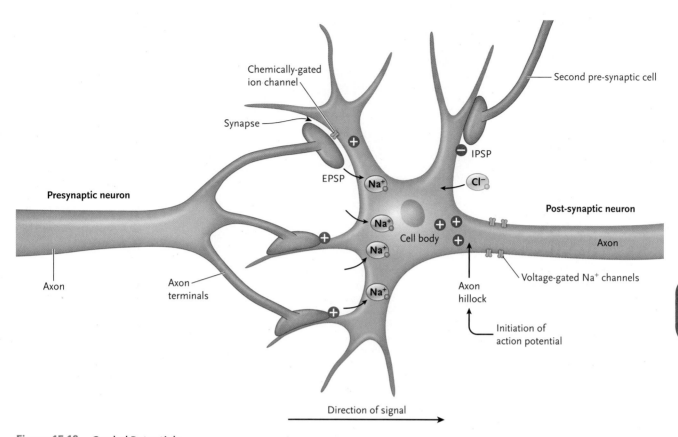

Figure 15.10 **Graded Potentials**

Graded potentials can be excitatory or inhibitory, and they determine whether or not the post-synaptic cell will have an action potential.

3. Voltage-gated sodium channels open in the axon, allowing Na+ to move *into* the cell, resulting in depolarization.

4. When the inside of the cell becomes positively charged, sodium channels close and voltage-gated potassium channels open, allowing K+ to move *out* of the cell, causing the cell to become negatively charged inside and resulting in repolarization.

5. The potassium channels are a bit slow to close again once the membrane is repolarized, so the cell becomes a little more negative than the resting membrane potential. This results in hyperpolarization, which is required to ensure unidirectional movement of action potentials.

6. The Na+/K+ pump helps keep the distribution of sodium and potassium ions in their correct concentrations; but notice that the Na+/K+ pump does *not* cause repolarization of the membrane.

7. Steps 3 to 6 repeat all along the axon. As depolarization occurs in one section, the positive charge inside the cell stimulates the next voltage-gated channels to open. This continues along the axon until the axon terminal is reached.

8. When an action potential reaches the axon terminal, it triggers voltage-gated Ca2+ channels to open and allow Ca2+ to move into the cell.

9. Calcium entering the cell triggers exocytosis of neurotransmitters into the synaptic cleft.

10. Neurotransmitters bind to chemically gated channels on the post-synaptic cell membrane (dendrites or cell body) and cause a graded potential, either an EPSP or an IPSP.

11. The combination of graded potentials from all connecting neurons determines whether the post-synaptic cell reaches threshold level.

Action potentials in myelinated axons

The propagation of action potentials in myelinated axons—saltatory conduction—occurs in exactly the same way as along non-myelinated axons, expect that voltage-gated sodium channels are only present in the nodes of Ranvier (Figure 15.11). When the membrane depolarizes at one node of Ranvier, the Na+ current flows within the cell near the membrane and alters the charge of the membrane at the next node, triggering voltage-gated channels to open at the next node. This causes the action potentials to occur only in the nodes, which is like taking big steps when walking instead of walking by placing one foot in front of the other. Action potentials move faster in myelinated axons and axons with a large diameter, such as motor neurons.

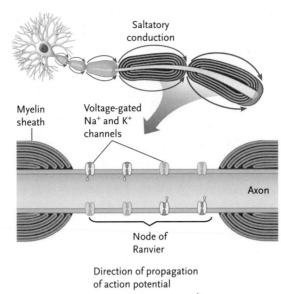

Figure 15.11   Saltatory Conduction

Saltatory conduction involves the transmission of an action potential along a myelinated axon.

Neurotransmitters

Over 100 different molecules are neurotransmitters that are released from neurons and affect other neurons or muscles. The following is a list of some of the most common. Notice that some of these are also hormones. The only difference, in epinephrine for example, is when it is released from a neuron it is a neurotransmitter, and when it is released by the adrenal medulla into the bloodstream, it is a hormone (Chapter 16). Neurotransmitters can be amino acids, derivatives of amino acids, peptides (several amino acids held together by peptide bonds), or proteins (Table 15.2).

TABLE 15.2

Neurotransmitters and Their Functions

Neurotransmitter	Function
Acetycholine (ACh)	Causes muscle contraction, excitatory only
Glutamate and Aspartate	Excitatory
Gamma Aminobutyric Acid (GABA)	Inhibitory
Epinephrine and norepinephrine (also hormones)	Involved in autonomic nervous system (Section 15.7)
Serotonin	Elevates mood
Dopamine	Increases ability to concentrate and plays a role in the "reward" pathway and addictive behaviours
Endorphins	Natural opiates, which decrease pain

DID YOU KNOW?

Alcohol can decrease the function of the calcium channels in the axon terminals, causing fewer neurotransmitters to be released in certain parts of the brain; when the neurons can't communicate, cognitive abilities are decreased. Alcohol increases the production of serotonin, making people feel more relaxed and sociable. Alcohol also decreases the effects of GABA (inhibitory neurotransmitter), which tends to make people feel more outgoing and spontaneous. Effects of alcohol are dose-dependent, causing symptoms ranging from feeling relaxed to having uncoordinated movements, to loss of memory, to engaging in high-risk activities, to unconsciousness. Blood alcohol concentrations of 0.4% can be fatal; when neurons in the brainstem cannot function, breathing cannot be controlled.

DID YOU KNOW?

Some anti-anxiety medications, such as valium (diazepam) or Ativan (lorazepam), act by increasing the effects of GABA. The medications bind to GABA receptors (mimicking GABA) and cause chemically gated chloride ion channels to open. Cl^- moves into the cells because of the strong concentration gradient, causing the membrane to become hyperpolarized, thus reducing the probability of an action potential occurring in the post-synaptic neuron.

CONCEPT REVIEW 15.3

1. Where is the higher concentration of sodium, potassium, and chloride in a resting neuron: intracellular or extracellular?

2. What is the difference between a chemically gated and a voltage-gated ion channel?

3. What ion causes depolarization and repolarization of the membrane during an action potential?

4. Why is the hyperpolarization phase of an action potential important?

5. Why is calcium important for proper neuron function?

6. What is saltatory conduction?

7. What two factors increase the speed of transmission of action potentials?

8. Which neurotransmitter is important for (1) addictive behaviour (2) mood (3) muscle contraction?

15.4 SPINAL CORD AND SPINAL NERVES

Sensory information travels to the CNS through afferent (sensory) neurons and from the CNS to effectors (muscles and glands) through efferent (motor) neurons. A sensory neuron carries information from a tissue all the way to the spinal cord before there is a synapse with another neuron. Multiple interneurons synapse with the axon terminals of the sensory neuron, which then communicate with various regions of the brain. Any thoughts or motor movements that originate in the brain must travel through the spinal cord before reaching the target tissue in the body. The spinal cord is a very important connection centre between the brain and our tissues (Figure 15.12).

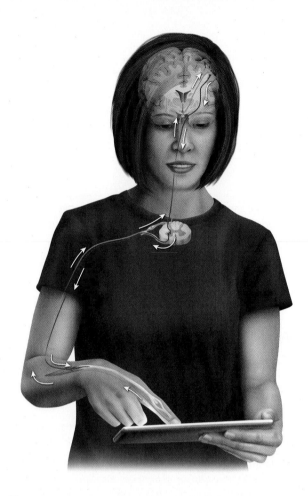

Figure 15.12 Neural Signalling Through the Spinal Cord
Signals travel to and from the brain through the spinal cord.

Anatomy of the spinal cord

The spinal cord is located within the vertebral foramen of each vertebra (Chapter 13). The spinal cord is highly protected by bone, the CSF produced by the ependymal cells, and protective connective tissue layers called **meninges** (Figure 15.13). Three layers of meninges protect the spinal cord and extend to cover the brain. The **dura mater** is the toughest dense, irregular, connective tissue outer layer. Between the dura mater and the vertebral bone is the **epidural space**, which contains blood and lymphatic vessels, the nerve roots, and fat that cushions the spinal cord. Having an "epidural" means that an anesthetic is injected into this space to block action potentials in the neurons and therefore decrease or block sensations in the body below the point of injection; this is often used during labour and delivery of babies. The middle layer of the meninges is the **arachnoid mater**, which looks like a spider web and is composed of collagen and elastin fibres. The innermost layer is called the **pia mater** and covers the outside of the spinal cord

and brain. The pia mater, like the arachnoid mater, is composed of collagen and elastin fibres, but it forms a sheet (Figure 15.13). The space between the pia mater and the arachnoid mater is called the **subarachnoid space** and contains blood vessels and cerebral spinal fluid. The CSF flows through the arachnoid space and surrounds the spinal cord with nutrients. Molecules in the arachnoid space can diffuse through the pia mater to reach the neurons in the spinal cord.

The spinal cord has a central canal that carries CSF. The posterior (dorsal) side of the spinal cord has a groove called the **posterior median sulcus**, and an anterior groove called the **anterior median fissure** (Figure 15.14).

The spinal cord is composed of grey matter and white matter. **Grey matter** includes neurons that are short interneurons with no myelin, and also the cell bodies, dendrites, and axon terminals of myelinated neurons. The **white matter** is made up of just the myelinated axons, which look white. The grey matter forms an H shape in the centre of the spinal cord. This is where sensory neurons entering the spinal cord synapse with interneurons that eventually signal the brain, and where motor information from the brain synapses with the cell bodies of motor neurons that signal the tissues (Figure 15.14). Spinal nerves are located between each vertebra and leave the CNS through the intervertebral foramen on the right and left side of the vertebral column. Each spinal nerve branches inside the vertebral column into anterior and posterior roots—called the dorsal and ventral root, respectively (Figure 15.13).

Sensory information is brought to the spinal cord through the **dorsal root** (posterior side of the spinal cord) to the **dorsal horn** (Figure 15.14). The unipolar sensory neuron cell bodies are located in the dorsal root ganglion, and the axon terminals synapse with interneurons in the dorsal horn. Those interneurons then synapse with the dendrites of somatic or autonomic motor neurons, as well as with long myelinated neurons, **ascending tracts**, that travel through the white matter to the brain.

Motor information is brought from the brain to the spinal cord through the white matter axons—the **descending tracts**—that synapse with interneurons and motor neurons in the **ventral horn**. The cell bodies of the motor neurons are in the ventral horn, and the axons extend to the cell bodies through the **ventral root** to the spinal nerve, where the axon terminals synapse with a target muscle or gland in the tissues. **Lateral horns** contain cell bodies of autonomic neurons, only in the thoracic region.

Spinal nerves

The spinal cord extends from the **medulla oblongata** of the brainstem, at the first cervical vertebra, to the **conus medullaris,** at the beginning of the lumbar vertebrae.

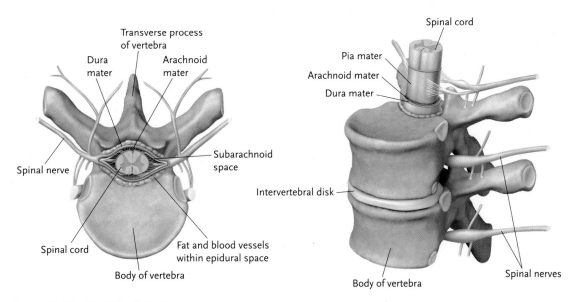

Figure 15.13   The Spinal Cord

The spinal cord contains the same meningeal layers as the brain: the outermost is the dura mater, the middle layer is the arachnoid mater, and the inner layer, the pia mater, covers the spinal cord.

There the spinal nerves branch out in a structure that looks like a horse's tail and is called the **cauda equina**. The spinal cord has two enlarged regions, the **cervical enlargement** that contains many neurons that innervate the upper limbs, and the **lumbar enlargement** that contains neurons that innervate the lower limbs (Figure 15.16).

Each spinal (and cranial) **nerve** is covered by layers of connective tissue similar to the way muscle and its bundles of muscle fibres are organized (Figure 15.15). Nerves contain groups of axons from sensory and motor neurons that may or may not be myelinated.

The outermost connective cover that protects the whole nerve is called the **epineurium**. Within the nerve are bundles of axons called fascicles (just like bundles of muscle fibres), and these are wrapped in the **perineurium**. Each fascicle contains bundles of individual axons that are covered by the **endometrium**; underneath is either the neuron cell membrane or the protective myelin sheath.

There are 31 pairs of spinal nerves: eight cervical nerves (note that there are only seven cervical vertebrae), 12 thoracic, five lumbar, five sacral, and one pair of coccygeal nerves. Spinal nerves from T1 to L2 contain

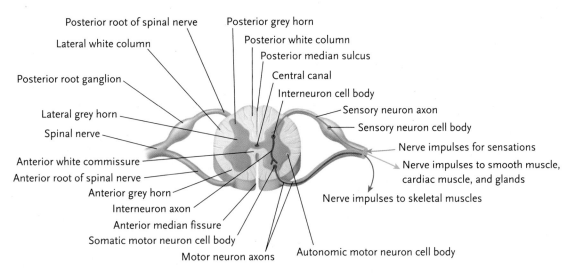

Figure 15.14   Structures in a Transverse Section of the Spinal Cord

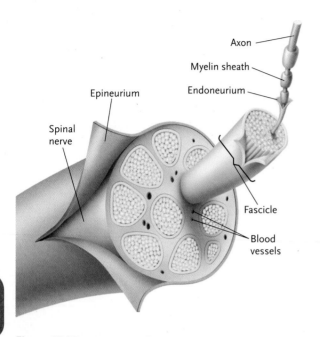

Figure 15.15 **Structure of a Nerve**

Nerves are made up of bundles of neurons and layers of protective connective tissue. The epineurium surrounds the nerve, the perineurium surrounds the fascicles, and the endoneurium surrounds each axon.

neurons of the sympathetic component of the autonomic nervous system.

Some of the spinal nerves form a network of nerves called a plexus (Figure 15.16 and Figure 15.17). The **cervical plexus** (C1–C5) contains neurons that innervate the head, neck, and shoulders and also the **phrenic nerve** that innervates the diaphragm. The **brachial plexus** contains nerves that innervate the upper limbs, neck, and shoulders. Nerves of the brachial plexus include the **axillary nerve** that innervates the deltoid and rotator cuff muscles, the **musculocutaneous nerve** that innervates the biceps and brachialis, the **median nerve** that innervates the forearm flexor muscles that affect the thumb and first three fingers, the **radial nerve** that innervates the thumb and first finger, and the ulnar nerve that innervates the last two fingers.

The **lumbar plexus** contains nerves that innervate the abdominal wall, genitals, and parts of the lower limbs. The **femoral nerve** innervates the quadriceps and hip flexors, and the **obturator nerve** innervates the hip joint and adductor muscles. The **sacral plexus** contains nerves that innervate the lower limbs, genitals,

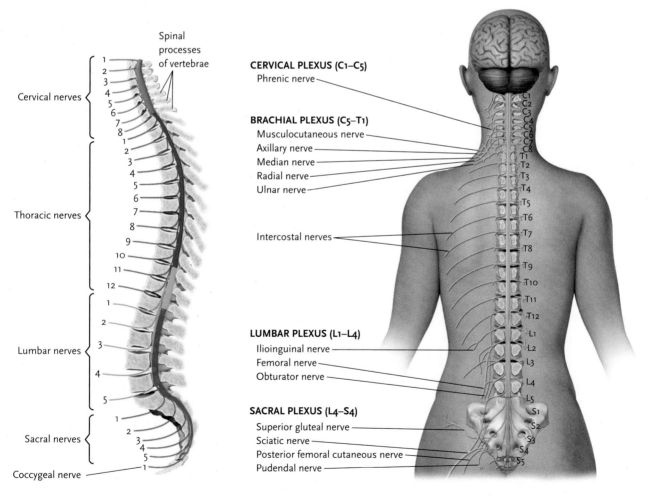

Figure 15.16 **Spinal Nerves**

There are 31 pairs of spinal nerves.

and gluteal muscles. The **superior and inferior gluteal nerves** innervate the gluteal muscles. The sciatic nerve innervates the hamstrings and then divides at the knee into the **tibial nerve,** which innervates the calves and feet, and the **common fibular nerve** that innervates the tibialis anterior.

DID YOU KNOW?

Approximately 50% of people who are diabetic will suffer from diabetic neuropathy, which is damage to any of the peripheral nerves, including sensory and motor nerves, as well as nerves of the autonomic nervous system. Depending on the nerves that are affected, symptoms can include numbness and tingling, burning pain sensations, trouble swallowing, facial drooping, urinary incontinence, and muscle twitches.

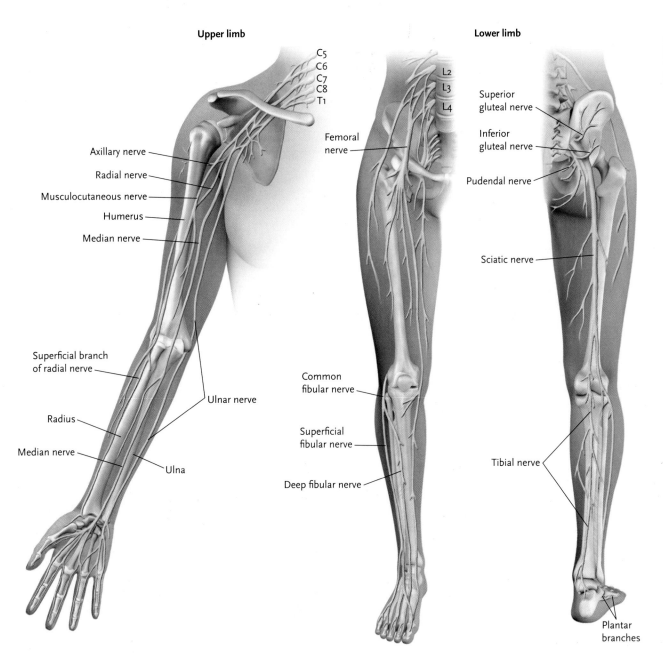

Figure 15.17   **Nerves of the Upper and Lower Limbs**

CONCEPT REVIEW 15.4

1. List the meninges from innermost layer to the outermost layer.

2. What is found within the epidural space and within the arachnoid space?

3. What neuron structures are found in grey matter and white matter?

4. Explain how sensory information enters the spinal cord and motor information leaves the spinal cord.

5. Describe the structure of a nerve, including the connective tissue coverings.

6. How many spinal nerves do humans have in the cervical, thoracic, lumbar, sacral, and coccygeal regions of the spinal column?

7. Fill in the following chart:

Nerve	Vertebral Region	Tissue Innervated
Phrenic	Cervical	Diaphragm
Axillary		
Musculocutaneous		
Median		
Radial		
Ulnar		
Obturator		
Pudendal		
Femoral		
Sciatic		
Tibial		
Fibular		
Superior and inferior gluteal		

15.5 REFLEXES

A **reflex** is a rapid and involuntary response to a stimulus, also called a **reflex arc**. Reflexes are protective mechanisms that prevent damage to any part of our body.

Monosynaptic reflexes involve a sensory neuron with *one* synapse onto a motor neuron in the spinal cord, and this causes a muscle to contract. For example, a **stretch reflex** involving the patellar tendon—called the knee-jerk response or **patellar reflex**—is monosynaptic. A common test in a doctor's office to determine the integrity of the peripheral nervous system is to tap the patellar ligament to cause the leg to extend. The quadriceps muscles are all connected by the quadriceps tendon, which encloses the patella and inserts on the tibial tuberosity on the proximal, anterior region of the tibia bone. Tapping the patellar ligament pulls on the quadriceps muscles making them *stretch*. The stretch is detected by sensory neurons in the quadriceps muscles; the signal is sent to the spinal cord where the axon terminals of the sensory neuron synapse directly onto a motor neuron that innervates the quadriceps muscle and causes contraction, which causes the leg to extend (Figure 15.18). In all types of reflexes, some neurons connect to interneurons in the spinal cord that send the sensory information to the brain, and so we become conscious of the movement.

Polysynaptic reflexes involve more than one synapse from sensation to response, such as the **withdrawal reflex**. Again, with any type of reflex, there are always neurons that synapse with neurons of the white matter tract and these carry the sensory information to the brain. A withdrawal reflex protects the body from damage, and

DID YOU KNOW?

We can acquire reflexes through learning. These acquired reflexes are different from the reflex arc, which involves just the spinal cord, because they involve the brain and complex movements. Acquired reflexes gradually develop over time, such as learning to ride a bike, drive a car, or play a guitar. Once we learn certain movements, these become a type of reflex because we no longer have to think about the movements; although they involve somatic motor neurons, they become unconsciously controlled. This is beneficial because as we learn, we can spend less conscious thought on those movements, which allows us to concentrate on other things at the same time. Imagine if we always had to concentrate on each muscle contraction when we walk, like a child first learning.

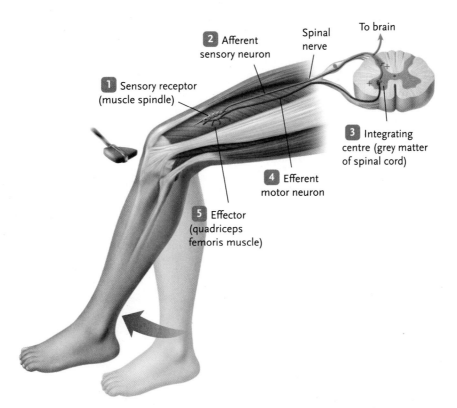

Figure 15.18 **Monosynaptic Reflex**

Monosynaptic reflexes involve the signalling of a sensory neuron directly to a motor neuron in the spinal cord.

the sensory neuron involved detects pain. For example, if you touch something hot, your hand will jerk away from the heat source before you are consciously aware of the pain because the sensory neuron synapses in the spinal cord with an interneuron, which immediately sends a motor response through a motor neuron, causing the biceps to contract (Figure 15.19). Often, **reciprocal innervation** occurs during a withdrawal reflex; this

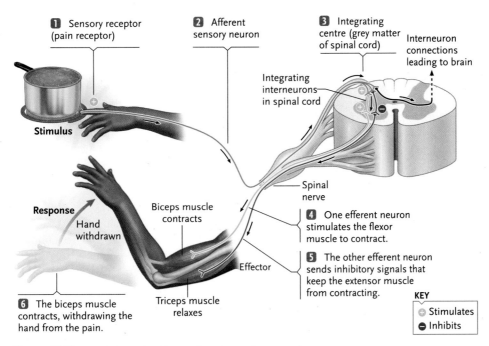

Figure 15.19 **Withdrawal Reflex and Reciprocal Innervation**

The withdrawal reflex involves the contraction of a muscle and the inhibition of contraction of its antagonist. Reciprocal innervation involves the opposite muscle contracting on the opposite side of the body.

means the antagonistic muscle is inhibited to ensure that the agonist can contract (Figure 15.15). Recall from Chapter 14 that at a neuromuscular junction, acetylcholine is released from the motor neuron and the neurotransmitter is always stimulatory, causing contraction. So a motor neuron cannot inhibit muscle contraction. The inhibition takes place in the spinal cord by the interneuron.

CONCEPT REVIEW 15.5

1. What is the difference between a monosynaptic and polysynaptic reflex?
2. Why do humans have reflexes?
3. Describe the events that occur in a patellar reflex.
4. Describe the events that occur in a withdrawal reflex.

15.6 BRAIN AND CRANIAL NERVES

The brain is protected by the bones of the skull, three layers of meninges, and cerebral spinal fluid (CSF). The bones of the skull are discussed in Chapter 13. The meninges are the same as those protecting the spinal cord (Figure 15.20). The outermost layer, the dura mater, is the thick, tough layer that lies next to the skull bones and contains blood vessels. The middle layer, the arachnoid mater, has the same spider web appearance as this layer surrounding the spinal cord. The innermost layer, the pia mater, lies directly next to the brain. The pia mater contains blood vessels that branch into and throughout the brain. In between the arachnoid mater and the pia mater is the subarachnoid space, which contains CSF that bathes the brain with nutrients and protects it by absorbing shock when we move around and jump.

DID YOU KNOW?

Inflammation in the meninges is called meningitis. This serious infection of the meninges can be caused by several different bacteria or viruses. Symptoms include severe headache accompanied by a stiff neck and sometimes vomiting and confusion. Meningitis can be life threatening, and immediate medical attention is required.

Recall from Section 15.2 that astrocytes are glial cells that have extensions that wrap around the capillaries in the brain and help form the blood-brain barrier (mainly due to the tight junctions between the capillary endothelial cells). The blood-brain barrier prevents the brain from exposure to any toxins that might be in the bloodstream. The CSF is produced by ependymal cells (Section 15.2) in a structure called the **choroid plexus** within each ventricle (Figure 15.21). The choroid plexus is a region where the ependymal cells lie close to blood vessels, and they extract water and nutrients from the blood plasma to produce CSF. As the CSF flows in the arachnoid space throughout the CNS, it eventually reaches the **arachnoid villi** where it re-enters the circulatory system through a

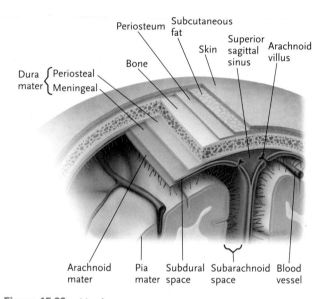

Figure 15.20 Meninges
There are three layers of meninges: the dura mater, arachnoid mater, and the pia mater.

vein called the **superior sagittal sinus**. The human body produces approximately 150 ml of CSF each day.

The brain can be divided into four main regions: the diencephalon, brainstem, cerebellum, and cerebrum (Figure 15.22). The structure and function of each region is described in the following sections.

Diencephalon

The **diencephalon** includes the thalamus, hypothalamus, and the pineal gland. The **thalamus** is well known as the "relay station" where all peripheral sensory signals are sent before being distributed to other regions of the brain, such as the cerebral cortex, for interpretation. The thalamus also plays an important role (along with the cerebellum) in controlling the coordination of skeletal muscle contractions during complex movements. Sensory information from the muscles, including amount of stretch, force of contraction, and body position, are sent to the thalamus, and that information is used to fine tune body movements.

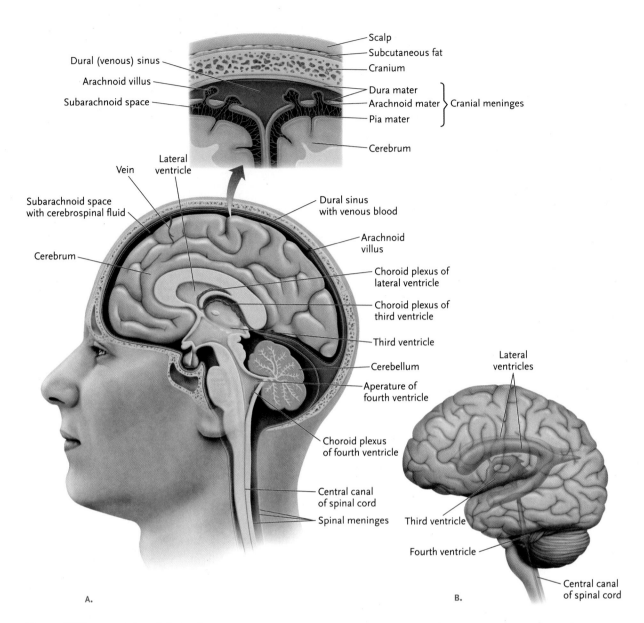

Figure 15.21   Ventricles of the Brain
Ventricles are hollow cavities within the brain. Ependymal cells within each choroid plexus in each ventricle produce cerebral spinal fluid.

The hypothalamus is an endocrine gland and produces releasing hormones that stimulate or inhibit the **pituitary gland**. The hormones of hypothalamus and pituitary gland are involved in essential homeostatic mechanisms—such as hunger, growth, temperature regulation, water and ion balance, regulation of the reproductive system—and are part of the autonomic nervous system (Section 15.7). The hypothalamus and pituitary gland are explained in Chapter 16.

The **pineal gland,** a small gland located in the posterior region of the diencephalon, produces a hormone called melatonin. **Melatonin** is produced in low light, like in the evening when the sun goes down, and it induces sleep. The pineal gland plays an important role in regulating our daily sleep-wake circadian rhythm.

Brainstem

The **brainstem** includes the midbrain, pons, and medulla oblongata (Figure 15.23). The brainstem is essential for life and is the most primitive brain region. All information travelling through neurons in the spinal cord passes through the brainstem before reaching any other part of the brain. The central core of the brainstem contains neurons that regulate **consciousness**: the *reticular activating system*. We wake up in the morning because of neuronal activity in the region of the brain known as the **reticular formation**. This is the region that is inhibited with general anesthetic during surgery. The reticular formation as well as the thalamus are also parts of the brain that allow us to focus attention on one specific stimulus while being surrounded by multiple

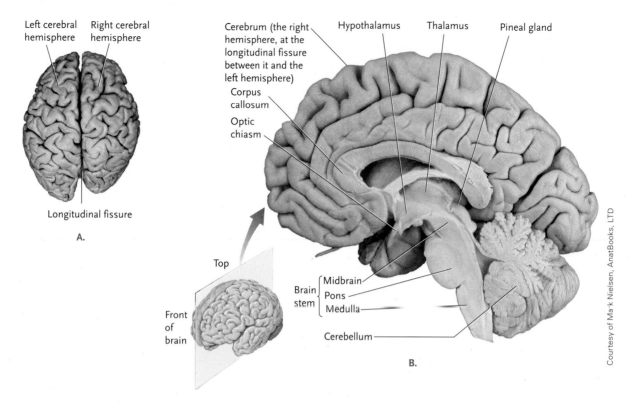

Figure 15.22   Regions of the Brain

The regions of the brain include the diencephalon, brainstem, cerebellum, and cerebrum.

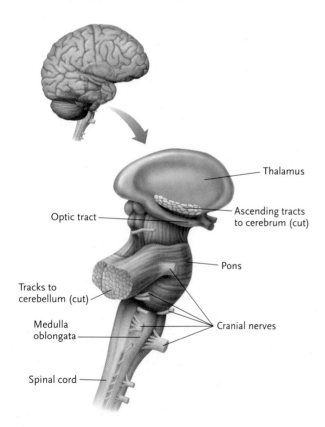

Figure 15.23   Brainstem

The brainstem is composed of the midbrain, pons, and medulla oblongata.

stimuli. The reticular formation interacts with many other regions of the brain.

At the top of the brainstem is the **midbrain,** which connects the pons to the diencephalon. The midbrain contains nuclei (groups of cell bodies with similar functions) called **substantia nigra.** In conjunction with the cerebellum and the thalamus, these nuclei regulate motor movements. The substantia nigra neurons produce the neurotransmitter **dopamine,** which acts on a region of the cerebrum called the **basal nuclei** (explained below) to control motor movements. **Parkinson's disease** is caused by a loss of neurons in the substantia nigra and results in a decreased ability to control motor movements.

The **pons** is located in the brainstem between the midbrain and the medulla oblongata. The pons receives signals from the motor cortex of the cerebrum and directs them to the cerebellum. It also helps in the relay of information that allows the cerebellum to send motor signals to the muscles in a smooth and coordinated way. Other nuclei in the pons play a secondary role in the regulation of breathing. The medulla oblongata contains the main respiratory centre but the pons helps smooth out respiratory movements and is involved when we voluntarily hold our breath (Chapter 18).

The medulla oblongata is the lowest region of the brainstem and contains nuclei that regulate the rhythm of breathing (Chapter 18) and the rate and force of heart contraction (Chapter 17). The medulla oblongata controls the reflexes for vomiting, coughing, and sneezing.

Cerebellum

The **cerebellum** is found at the base of the brain, posterior to the brainstem. The cerebellum plays a key role in coordinating, learning, and remembering complex motor movements, such as walking, running, swimming, and dancing. The cerebellum receives signals from many parts of the brain, including the thalamus, the pons, the motor cortex, the vestibular system, and the visual cortex. The signals received by the cerebellum serve to increase precision and accuracy of movements, maintain posture and balance during complex movements, and allow for fine motor control, such as in writing with a pen, or typing, or controlling facial muscles involved in talking. People with **cerebellar disease** have clumsy and uncoordinated movements and cannot learn new movements.

DID YOU KNOW?

A standard test for cerebellar function is also used to determine if a person is inebriated from alcohol. The test involves reaching the nose with the tip of the finger starting with the arm fully extended. A person who has a healthy cerebellum, and is sober, can touch the nose in one fluid motion; whereas a person with cerebellar damage, or someone intoxicated, will reach for the nose slowly and erratically. Alcohol consumption depresses the function of neurons throughout the nervous system, including the cerebellum, causing uncoordinated movements.

Cerebrum

The **cerebrum** is made up of the cerebral cortex, which is the outer surface of the brain, and the subcortical region, which lies beneath the cortex. The cerebrum is made up of white matter and grey matter, like the spinal cord. However, the grey matter in the brain is located in the cortex, and the white matter is in the subcortical region; this is opposite to their positions in the spinal cord.

The **subcortical region** of the brain contains several important areas, including the basal nuclei and the amygdala. The basal nuclei are sometimes called the *basal ganglia*. But recall that in the CNS groups of cell bodies are called nuclei, and in the PNS groups of cell bodies are called ganglia. The basal nuclei are composed of several groups of neurons that are primarily involved in regulating skeletal muscle movements, posture, and complex behaviour patterns (Figure 15.24). The basal nuclei control movements in conjunction with the midbrain, frontal lobe and motor cortex, thalamus, and cerebellum. One of the most well-known of the basal nuclei is the **nucleus accumbens**, which has been shown to play a role in reward, pleasure, addiction, compulsions, fear, and also the placebo effect. One of the main neurotransmitters released by neurons in this region is dopamine (Section 15.3). It is thought that addictive drugs such as nicotine, cocaine, and

amphetamines increase the amount of or response to dopamine. Another well-known basal nucleus is the **caudate nucleus**, which plays an important role in learning and memory. The **amygdala** is a subcortical nucleus that is involved in emotions, memory linked to emotions, and complex social and sexual interactions.

The **limbic system** consists of a group of structures, including regions of the thalamus, hypothalamus, amygdala, hippocampus (part of the cerebral cortex), olfactory bulb (detects sense of smell) fornix (fibre tracts that connect the limbic system structures), and the cingulate gyrus (forms connections with many regions of the brain) (Figure 15.25). In combination, these structures play a major role in our emotions, learning, memory, and social interactions. Because the olfactory bulb is part of the limbic system, the sense of smell is highly linked to memories; you may have noticed that a particular scent causes you to recall past events in great detail. This strong connection between scents and emotions is also the basis of the perfume industry. The limbic system is like a primitive brain. It functions to protect us by reacting rapidly to stimuli: for example, getting angry and becoming aggressive, or feeling sad and wanting to cry. Our limbic system reactions are highly modified by the cerebral cortex, which develops as we live in social environments. For example, we may feel angry but still respond in a socially appropriate way. A baby or a two-year-old uses primarily their limbic system with very little input from their cerebral cortex.

The **cerebral cortex** differentiates humans from other animals. It has a highly folded surface that increases the surface area of grey matter. An indentation is called a **sulcus**, and the protruding portion is a **gyrus** (Figure 15.26). A deep fissure—called the **longitudinal fissure**—runs along the centre of the brain and splits the brain into the right and left hemispheres. A groove along the centre of the brain divides the brain into anterior and posterior regions (separates the frontal lobe from the parietal lobes); this groove is called the **central sulcus**. The protruding region immediately anterior to the central sulcus is the **pre-central gyrus**, also known as the motor cortex. The protruding region immediately posterior to the central sulcus is the **post-central gyrus**, also known as the *sensory cortex*. The cerebral cortex is made up of the following lobes, which are named after the skull bones they are associated with: frontal lobe, left and right parietal lobe, left and right temporal lobe, and the occipital lobe. There is also a lobe called the insula, which is a region of the cortex located within folds deep to temporal lobe (Figures 15.22 and 15.26b).

The right and left hemispheres of the brain are connected by the **corpus callosum** in the subcortical region. The motor cortex and sensory cortex on one side of the brain control and receive input from the *opposite* side of the body. The right and left sides of the brain communicate through the corpus callosum, but sensory information crosses to the opposite side either in the spinal cord, the midbrain, or the thalamus.

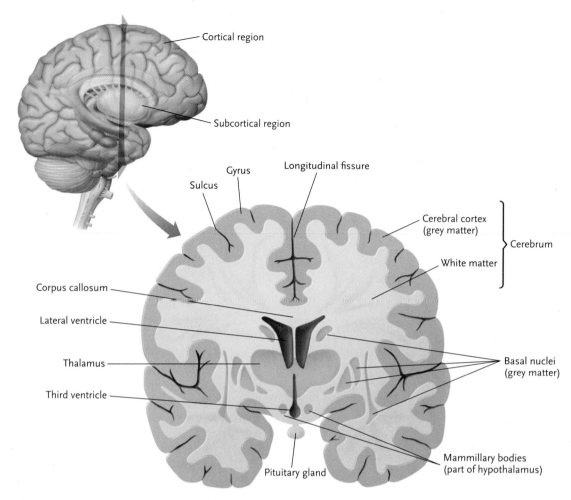

Figure 15.24   Cerebrum

The cerebrum is divided into the cortex and the subcortical regions.

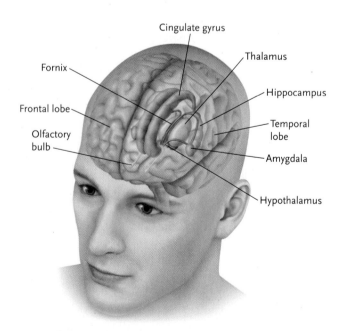

Figure 15.25   Limbic System

The limbic system is a group of structures that are involved in emotions.

The **frontal lobe** contains the motor cortex, which is where conscious control of skeletal muscles begins. The neurons in specific regions of the motor cortex as well as input from the cerebellum and basal nuclei control voluntary movements. The motor cortex has specific designated regions that associate with specific parts of the body (Figure 15.27). The **pre-frontal cortex**, located just anterior to the motor cortex, is involved in the planning of motor movements; it is the region of the brain where we make the decision to initiate a movement. The frontal lobe has many other important roles, including higher-thinking functions, such as setting goals, planning, reasoning, making decisions, concentrating on a task, and also short-term memory. The frontal lobe plays an important role in self-control and regulates the expression of emotions that are triggered by the limbic system. The frontal lobe also contains a very important region in the left hemisphere, called **Broca's area**, which is a language centre. The Broca's area controls the motor movements involved in speaking.

The **parietal lobe** contains the sensory cortex, also called the **somatosensory cortex** because it receives input from sensations coming from the skin, muscles, and joints, and not the eyes or ears or other special senses.

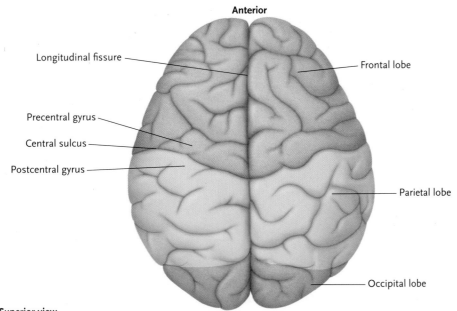

Anterior

Longitudinal fissure

Frontal lobe

Precentral gyrus

Central sulcus

Postcentral gyrus

Parietal lobe

Occipital lobe

A. **Superior view**

Posterior

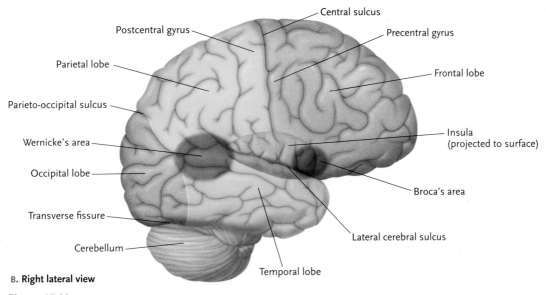

Central sulcus

Postcentral gyrus

Precentral gyrus

Parietal lobe

Frontal lobe

Parieto-occipital sulcus

Wernicke's area

Insula
(projected to surface)

Occipital lobe

Broca's area

Transverse fissure

Lateral cerebral sulcus

Cerebellum

Temporal lobe

B. **Right lateral view**

Figure 15.26 **Cerebral Cortex**
(A) Superior view and (B) lateral view

DID YOU KNOW?

People with attention deficit disorder have less dopamine released in neurons of the frontal lobe that are involved in concentration and thinking about a task. With decreased neurotransmitters in the frontal lobe, a person is easily distracted. Ritalin and other similar medications stimulate dopamine secretion in the frontal lobe, which helps increase concentration. Many children lack the ability to focus and are easily distracted; this is normal and usually decreases as they grow older and practise concentrating. The more a person thinks and concentrates, the more the brain adapts to doing this. For the most part, children grow out of ADD, usually by the time they reach the late teens.

The sensory cortex, like the motor cortex, has specific, designated regions that associate with specific regions of the body (Figure 15.27). The parietal lobe has many other important functions, such as *integrating* sensory information from the somatosensory cortex, and the **occipital lobe,** which receives visual information, and the **temporal lobe,** which receives auditory information. The parietal lobes use the incoming sensory information for visual-spatial processing: being able to recognize distance and changes in the environment that could impact our motor movements, such as avoiding a pothole when jogging or grasping an object. The right and left hemispheres of the parietal lobes have some differences. The left hemisphere is more involved in recognizing numbers and patterns and language, and the right side

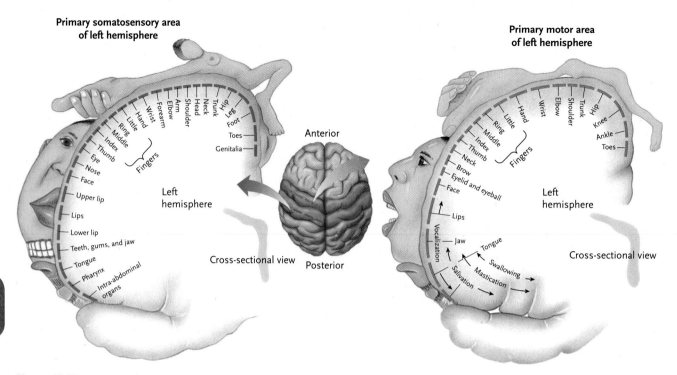

Figure 15.27 Motor Cortex and Somatosensory Cortex

The motor cortex and the somatosensory cortex each have a specific distribution of neurons that is linked with specific regions of the body.

Art source: From DIGIUSEPPE/FRASER. Biology 12 U. © 2012 Nelson Education Ltd. Reproduced by permission. www.cengage.com/permissions

is more involved in spatial relationships and interpreting images and maps. The parietal lobe is also the region of the brain involved in human self-awareness and an awareness of being part of a larger universe.

The temporal lobe contains the **auditory cortex,** which receives information about sound from the auditory nerve, and *association areas* that interpret sounds, including speech. The temporal lobe has a region called **Wernicke's area,** which is important for comprehending the meaning of spoken language; this is not to be confused with Broca's area in the frontal lobe that allows us to physically speak. The temporal lobe is the region of the brain that stores memories of names and different sounds. The **hippocampus** is a structure located within each temporal lobe that plays a key role in long-term memory. The hippocampus is one of the structures of the limbic system involved in emotions, and so long-term memory and learning is enhanced when it is linked with an emotional experience.

The occipital lobe contains the primary visual cortex and association areas that interpret light and colours and recognize objects. The occipital lobe also plays a role in interpreting motion. Information about our body position and movement occurs in both the occipital lobe and the parietal lobe, which receives sensory input from joints and muscles. When information about body position appears to the brain as different than usual, such as standing in a rocking boat, the visual information tells the occipital lobe that the body is *stable*, but the information from contracting muscles and moving joints, as well as the vestibular system in the

inner ear, tells the parietal lobe the body is *moving*; this incongruence causes motion sickness.

The final lobe of the brain is the **insula,** located in the folds deep to the temporal lobe. The insula is involved in recognizing pain, as is the somatosensory cortex. However, the insula seems to be more involved in how we perceive the pain, compared to the sensory cortex that interprets the location and intensity of the pain. The insula also interprets gastric and bladder distention.

The motor cortex and the somatosensory cortex have very similar brain maps that associate with specific areas of the body (Figure 15.27). The parts of the body that have the most corresponding brain area have the highest amount of motor movement and sensitivity. On the motor cortex, there is a large amount of brain tissue dedicated to the face, mouth and lips, and hands. This large area relates to our need for the very extensive and finely controlled muscle movements in facial expressions, moving the lips and mouth during chewing and talking, and in all the detailed movements of our hands. The same is true for the sensory cortex. There is a lot of brain tissue dedicated to receiving sensory information from highly sensitive areas—lips and hands.

Cranial nerves

There are 12 pairs of cranial nerves that originate in the brain but are part of the peripheral nervous system. Table 15.3 summarizes the main functions of each cranial nerve. Cranial nerve I, the olfactory nerve that is

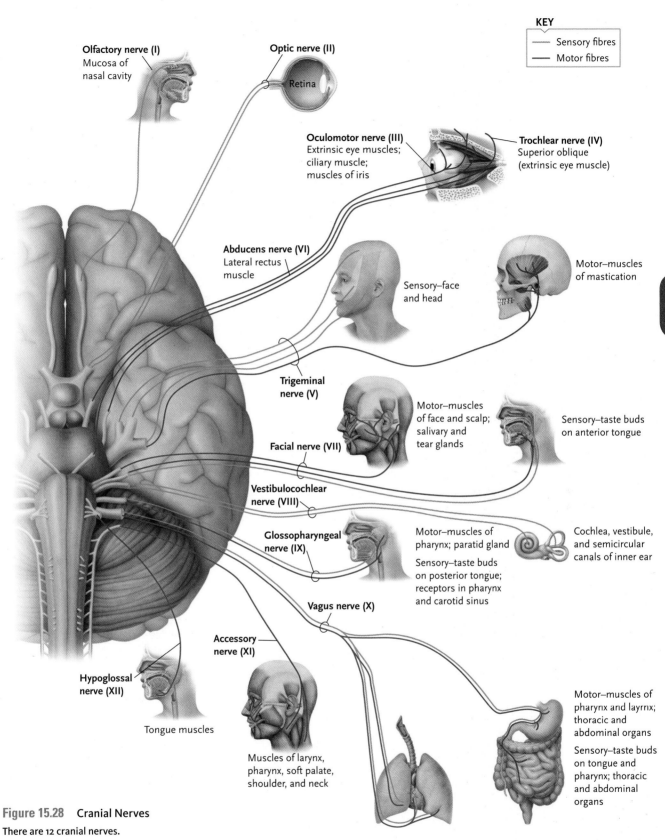

KEY
— Sensory fibres
— Motor fibres

Olfactory nerve (I)
Mucosa of
nasal cavity

Optic nerve (II)
Retina

Oculomotor nerve (III)
Extrinsic eye muscles;
ciliary muscle;
muscles of iris

Trochlear nerve (IV)
Superior oblique
(extrinsic eye muscle)

Abducens nerve (VI)
Lateral rectus
muscle

Sensory–face
and head

Motor–muscles
of mastication

**Trigeminal
nerve (V)**

Facial nerve (VII)

Motor–muscles
of face and scalp;
salivary and
tear glands

Sensory–taste buds
on anterior tongue

**Vestibulocochlear
nerve (VIII)**

**Glossopharyngeal
nerve (IX)**

Motor–muscles of
pharynx; paratid gland

Sensory–taste buds
on posterior tongue;
receptors in pharynx
and carotid sinus

Cochlea, vestibule,
and semicircular
canals of inner ear

Vagus nerve (X)

**Accessory
nerve (XI)**

**Hypoglossal
nerve (XII)**

Tongue muscles

Muscles of larynx,
pharynx, soft palate,
shoulder, and neck

Motor–muscles of
pharynx and layrnx;
thoracic and
abdominal organs

Sensory–taste buds
on tongue and
pharynx; thoracic
and abdominal
organs

Figure 15.28   Cranial Nerves
There are 12 cranial nerves.

not really a true nerve, stays within the CNS and carries information from the olfactory epithelial cells (sense of smell) to the olfactory bulb. Similarly, cranial nerve II is the optic nerve that carries information from the eyes to the occipital lobe and is not a true nerve. Cranial nerves III–XII originate from the brainstem and carry both motor and sensory information to specific tissues in the body (Figure 15.28).

TABLE 15.3

Function and Point of Origin of the Cranial Nerves

Nerve	Function	Origin
I Olfactory	Sense of smell	Cerebrum
II Optic	Vision	Cerebrum
III Oculomotor	Movement of eyelid, eyeball, lens, pupil constriction	Midbrain
IV Trochlear	Movement of eyeball	Midbrain
V Trigeminal (ophthalmic, maxillary, and mandibular branch)	Sensations in head, face, and jaw; motor control of chewing	Pons
VI Abducens	Movement of eyeball	Pons
VII Facial	Taste, facial expression, tears, salivation, and facial sensations	Pons
VIII Vestibulocochlear	Equilibrium and hearing	Lateral to facial nerve
IX Glossopharyngeal	Taste, swallowing, speech, and salivation	Medulla oblongata
X Vagus	Taste, sensations from pharynx, swallowing, coughing, voice, GI tract smooth muscle contraction, heart rate reduction, digestive secretion	Medulla oblongata
XI Accessory	Swallowing and movement of head and shoulders	Medulla oblongata
XII Hypoglossal	Speech and swallowing (tongue muscles)	Medulla oblongata

CONCEPT REVIEW 15.6

1. Describe the features that help protect the brain.
2. What cell type produces CSF, and where in the brain does this occur?
3. Where is CSF reabsorbed back into the bloodstream?
4. Name the four regions of the brain, and list the main structures found in each region.
5. What are the main functions of the thalamus and pineal gland?
6. What is the role of the reticular activating system? Where is it located?
7. What is the role of the substantia nigra in the midbrain?
8. Name two functions of the pons.
9. What two major organ systems are regulated by the medulla oblongata?
10. What is the function of the cerebellum?
11. Describe the main function of the basal nuclei and amygdala.
12. What is the function of the limbic system?
13. Name the lobe of the brain, and any specific region, if applicable, that is primarily involved in each of the following functions:
    a. recognizing the sound of someone's voice
    b. being aware of your body position
    c. having the physical ability to speak
    d. deciding to walk across a room
    e. calculating the cost of your school supplies
    f. recognizing the difference between red and blue paint
    g. remembering a childhood experience
    h. concentrating on a difficult task
    i. understanding language
    j. interpreting the pain of a bee sting as being more severe because of a fear of bees
14. Make a chart to list the main functions of each cranial nerve.

15.7 AUTONOMIC NERVOUS SYSTEM

The autonomic nervous system is the involuntary, unconsciously controlled component of the nervous system that regulates the smooth muscle, cardiac muscle, and some glands. The ANS has two divisions: the **sympathetic nervous system**, also known as the fight-or-flight response; and the **parasympathetic nervous system**, also known as the rest-and-digest response. Most tissues receive input from both systems; usually one division is stimulatory and the other is inhibitory. Table 15.4 describes the functions of each system in relation to various tissues.

The ANS neural pathway from the CNS to the tissues involves two neurons. The first neuron leaves the CNS through a cranial or spinal nerve and synapses with a second neuron in an **autonomic ganglion** (Figure 15.29). The first neuron is the **preganglionic** and the second neuron

TABLE 15.4

Comparison of the Somatic and Autonomic Nervous Systems

Somatic	Autonomic
Voluntary	Involuntary
Innervates skeletal muscle	Innervates smooth and cardiac muscle and some glands
Always excitatory	Can be excitatory or inhibitory
Neurotransmitter is acetylcholine (ACh)	Neurotransmitter can be acetylcholine or norepinephrine (NE)
One motor neuron connects the spinal cord with the muscle fibre	Two neurons—one preganglionic, the other postganglionic—connect the CNS to the muscle or gland

15

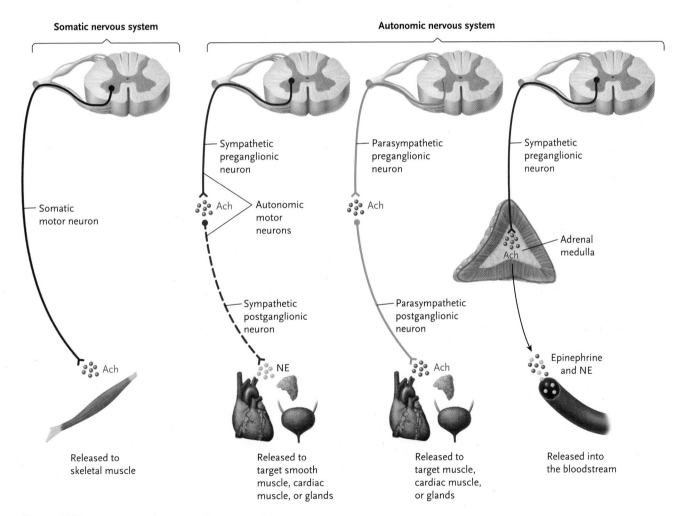

Figure 15.29 Somatic and Autonomic Nervous Systems

Somatic and autonomic nervous systems release specific neurotransmitters onto specific cell types to cause various effects.

15

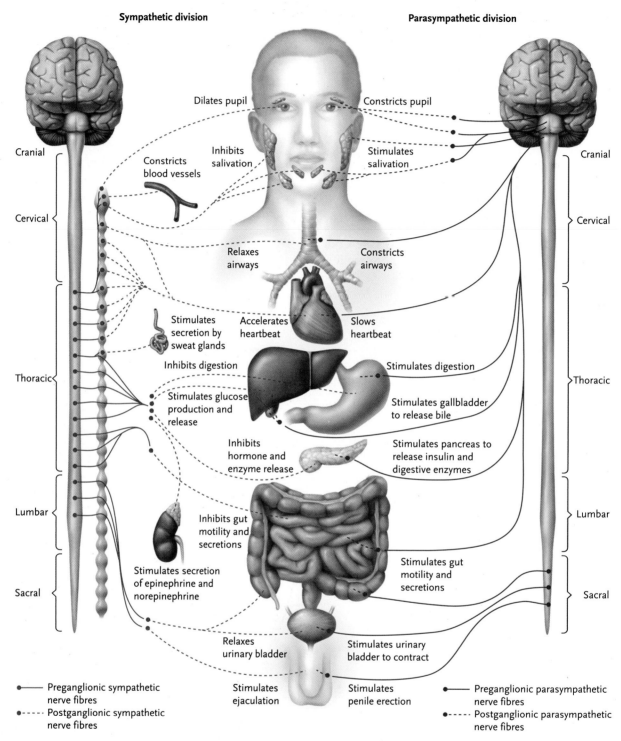

Sympathetic division

Parasympathetic division

Cranial

Cervical

Thoracic

Lumbar

Sacral

Dilates pupil

Constricts pupil

Inhibits salivation

Stimulates salivation

Constricts blood vessels

Relaxes airways

Constricts airways

Stimulates secretion by sweat glands

Accelerates heartbeat

Slows heartbeat

Inhibits digestion

Stimulates digestion

Stimulates glucose production and release

Stimulates gallbladder to release bile

Inhibits hormone and enzyme release

Stimulates pancreas to release insulin and digestive enzymes

Inhibits gut motility and secretions

Stimulates secretion of epinephrine and norepinephrine

Stimulates gut motility and secretions

Relaxes urinary bladder

Stimulates urinary bladder to contract

Stimulates ejaculation

Stimulates penile erection

Cranial

Cervical

Thoracic

Lumbar

Sacral

⎯•⎯ Preganglionic sympathetic nerve fibres
•---- Postganglionic sympathetic nerve fibres

⎯•⎯ Preganglionic parasympathetic nerve fibres
•---- Postganglionic parasympathetic nerve fibres

Figure 15.30 **Sympathetic and Parasympathetic Nervous Systems**
The sympathetic nervous system causes the fight-or-flight response, and the parasympathetic nervous system causes the rest-and-digest response.

is the **postganglionic**. One preganglionic neuron may synapse with many postganglionic neurons. Sympathetic preganglionic neurons are shorter than parasympathetic neurons, and the ganglia are often located close to the spinal cord. By contrast, parasympathetic ganglia are located very close to or within the target tissue. All ANS preganglionic neurons release acetylcholine (ACh)

at the first synapse. Parasympathetic postganglionic neurons release acetylcholine on the target muscle or gland, whereas sympathetic neurons release either **norepinephrine** or epinephrine. The sympathetic nervous system preganglionic neurons can signal the **adrenal medulla** to release epinephrine as a hormone into the bloodstream, which causes the same but longer lasting effects as compared to the postganglionic neuron releasing norepinephrine directly onto target tissues.

Neurons of the sympathetic nervous system leave the CNS from the thoracic and upper lumbar spinal nerves, and the parasympathetic neurons leave the CNS through cranial or sacral spinal nerves (Figures 15.28 and 15.29). The functions of the sympathetic nervous system include effects on the tissues required for a stress response, such as increasing heart rate; dilating bronchial tubes to increase the amount of oxygen that enters the blood for energy production; and increased blood flow to skeletal muscles (Figure 15.30). The stress response is triggered when we need to "fight or flee" a dangerous situation, and through evolution this has become very beneficial for a rapid physical response. However, in a modern technological world where we no longer need to hunt or gather food, or fight for territory, our stress response tends to be triggered primarily during exercise and emotional stress. It is healthy and beneficial for the sympathetic response to be activated during exercise, when the increased blood sugar is used to contract muscles, but chronically activating the sympathetic response for emotional, non-physical stress can be detrimental in the long term since the increased blood sugar is not used to make ATP.

The functions of the parasympathetic nervous system involve stimulating digestion and relaxation, such as increasing blood flow and smooth muscle contraction in the gastrointestinal tract, slowing heart rate, and promoting the storage of nutrients. The two divisions of the ANS are always functioning, and one system is more or less dominant at any point in time (Figure 15.30). Table 15.5 lists the effects of the sympathetic and parasympathetic nervous systems.

TABLE 15.5

Effects of the Autonomic Nervous System

Tissue	Sympathetic Response	Parasympathetic Response
Eyes	Dilates pupils and increases far vision	Constricts pupils and increases near vision
Lacrimal glands	No known effect	Production of tears
Salivary glands	Decreases production of saliva	Increases production of saliva
Heart	Increases heart rate and force of contraction	Decreases heart rate and force of contraction
Airways	Dilates bronchial tubes	Constricts bronchial tubes.
Blood vessels	Dilates vessels in skeletal muscles, heart, lungs, and brain; constricts them in skin, kidneys, and GI tract	No known effect
Smooth muscle in GI tract	Decreases contraction and motility	Increases contraction and motility
GI tract	Inhibits secretions	Stimulates secretions
Gallbladder	Stores bile	Contracts and releases bile
Sweat glands	Increases sweat production	No known effect
Erector pili muscles in skin	Contraction—goosebumps	No known effect
Pancreas	Inhibition of secretion of digestive enzymes and insulin; production of glucagon	Secretion of digestive enzymes and insulin
Liver	Breakdown of glycogen and fat	Synthesis of glycogen and fat
Adipose tissue	Breakdown of fat	No known effect
Kidney	Increases production of renin and therefore water retention and increased blood pressure	No known effect
Bladder	Relaxes, no urination*	Contracts and causes urination
Uterus	Contraction during pregnancy; inhibits contraction in non-pregnant women	No known effect
Reproductive tract	Ejaculation	Erection, vasodilation of blood vessels in penis/ clitoris

*Note that during a sympathetic nervous system response, the blood pressure increases, which can cause increased urine production and the urge to urinate; however, the parasympathetic nervous system is what stimulates urination.

15

DID YOU KNOW?

One of the healthiest ways to deal with emotional stress is to engage in regular, vigorous exercise. Here is a list of some of the benefits of exercise to relieve emotional stress.

- Muscles take up and use the increased level of blood sugar produced by the liver to make ATP.

- The brain releases endorphins that make you feel good—known as "runner's high."

- Dopamine increases, which helps you to concentrate and focus your attention. Try a 30 second sprint or do 10 push-ups the next time you need to take a study break,

- Increases other molecules involved in elevating mood such as serotonin and oxytocin

- Exercise results in a burst of sympathetic nervous system and cortisol release that quickly dissipates and leads you to feeling calm but energized.

CONCEPT REVIEW 15.7

1. What types of tissues are innervated by the somatic and autonomic nervous system?

2. Which neurotransmitter is released by the preganglionic neurons in the ANS?

3. What neurotransmitters are released by the sympathetic and parasympathetic postganglionic neurons?

4. Name any five tissues that are stimulated by the sympathetic nervous system.

5. What is the primary role of the parasympathetic nervous system?

6. What happens in the liver during a fight-or-flight response compared to a rest-and-digest response?

7. Where do the preganglionic neurons of each ANS division originate in the CNS?

8. How is the adrenal gland involved with the autonomic nervous system?

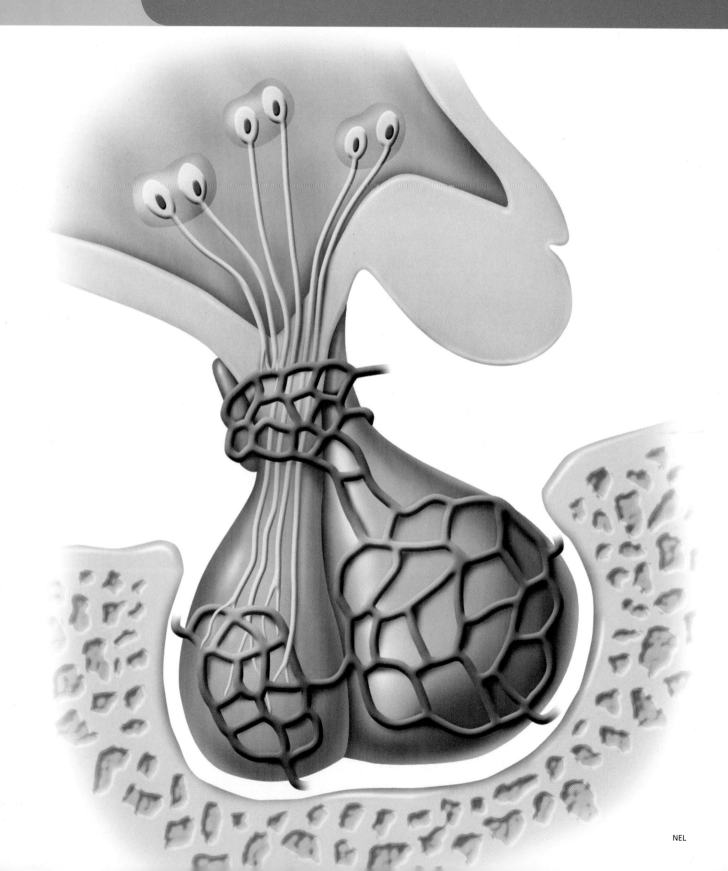

NEL

I am always ready to learn although I do not always like being taught.

Winston Churchill

16.1 FUNCTIONS OF THE ENDOCRINE SYSTEM

The endocrine system is one of the two major communication systems of the body. The nervous system communicates via neurotransmitters released by neurons, and the endocrine system communicates via **hormones** released by particular cells into the bloodstream. Both the nervous system and the endocrine system are key components of homeostasis. The nervous system responds to stimuli in the environment through various receptors (such as those for pain, temperature, pH, pressure), and neurotransmitters target other neurons, muscle cells (skeletal, smooth, or cardiac), or glands (including adrenal glands, salivary glands). The nervous system response time from stimulus to target cells is extremely rapid and only lasts for a brief period of time. Think of a time when something startled you and you felt that instant electrical feeling flood your body; that is a nervous system response. The lingering effects are from the epinephrine released from the adrenal gland into the bloodstream as a hormone.

The endocrine system involves many glands that produce hormones that can target any cell type in the body. Generally, the hormone response is initiated more slowly than the nervous system response, but the effects last much longer. It is essential that the nervous system and the endocrine system cooperate to ensure proper reactions to changes in the external environment. The major endocrine glands of the body include the hypothalamus, pituitary gland, pineal gland, thyroid and parathyroid glands, thymus, pancreas, adrenal glands, testes and ovaries (Figure 16.1). Many cells that are not considered endocrine glands also produce hormones; these are discussed in Section 16.7.

As a point of comparison, **exocrine glands** are part of the epithelial tissues that secrete substances (enzymes, sweat, oil, mucus) through a duct into a hollow cavity or to the surface of the body (Chapter 11). For example, digestive enzymes are released into the small intestine, and sweat glands secrete sweat onto the surface of the skin.

Functions of hormones

Hormones have many functions in the body, depending on the cell type they bind to and what signalling processes occur within the particular cell. The specific functions

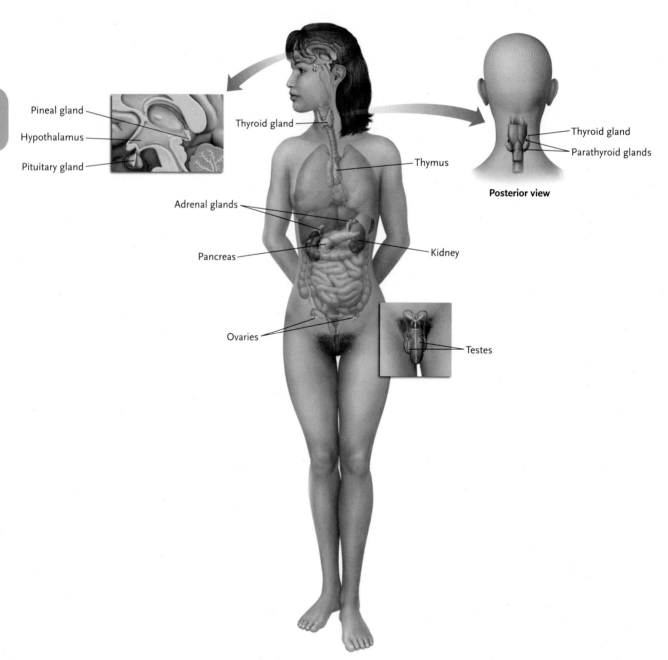

Figure 16.1 Organs of the Endocrine System

of particular hormones are covered in Sections 16.3–16.7, but the general functions of hormones include the following:

- Homeostatic regulation of blood sugar levels, water balance, blood calcium levels, blood pressure, thirst, hunger, blood cell production, as well as the regulation of other hormones in the blood

- Growth, metabolism, and energy production

- Reproductive functions such as secondary sex characteristics, lactation, childbirth, and development of sperm or eggs

- Stress response

- Regulation of digestion

- Regulation of circadian rhythms such as sleep-wake cycles

CONCEPT REVIEW 16.1

1. What are the similarities and differences between the nervous system and the endocrine system?

2. What is the difference between an endocrine gland and an exocrine gland?

3. List the major organs of the endocrine system and where they are located.

4. List the general functions of the endocrine system.

16.2 TYPES OF HORMONES

For hormones to exert an effect they must bind to a *specific* receptor either on the cell membrane or within the target cell. All hormones initiate a signalling pathway that either directly or indirectly affects transcription or translation of a gene or multiple genes (Chapter 9). For example, insulin is a hormone that is secreted by the pancreas in response to an increase in blood sugar after we eat food. Insulin binds to very specific receptors on all cells to initiate the transcription and translation of the glucose-transporter protein, which then embeds into cell membranes so that glucose can enter the cells.

There are two categories of hormones, water soluble and fat soluble. **Water-soluble hormones** cannot diffuse across membranes, so the specific receptor is located on the extracellular side of the membrane: only on cells that the hormone is meant to target. Water-soluble hormones elicit their response indirectly through a second messenger system (Figure 16.2). A second messenger relays a signal from the membrane receptor to a target molecule inside the cell, and this alters the activity of the cell. **Fat-soluble hormones** can easily cross cell membranes to bind to their receptor inside the cell, and usually they directly affect transcription and function as **transcription factors** (Chapter 9) (Figure 16.3).

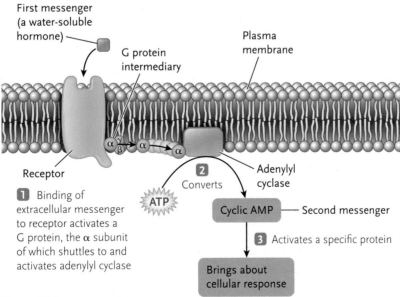

Figure 16.2 **Water-Soluble Hormones**

Water-soluble hormones indirectly affect transcription and translation by binding to a membrane surface receptor that triggers a second messenger signalling pathway.

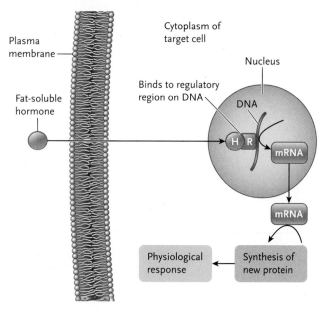

Figure 16.3   Fat-Soluble Hormones

Fat-soluble hormones directly affect transcription and translation by binding to receptors within the cell nucleus.

Water-soluble hormones are modified amino acids, small peptides, or proteins. The four most common water-soluble hormones are the following:

1. catecholamines—dopamine, epinephrine, and norepinephrine
2. pancreatic hormones—insulin and glucagon
3. pituitary hormones
4. hypothalamic hormones

Fat-soluble hormones are **steroids** produced from cholesterol (Figure 16.4), except thyroid hormones, which are made from **tyrosine**. Recall from Chapter 2 the many important functions of cholesterol. The five most common fat-soluble hormones are the following:

1. thyroid hormones—T_3 and T_4
2. cortisol
3. aldosterone
4. androgens—testosterone, estrogen, and progesterone
5. calcitriol (active Vitamin D_3)

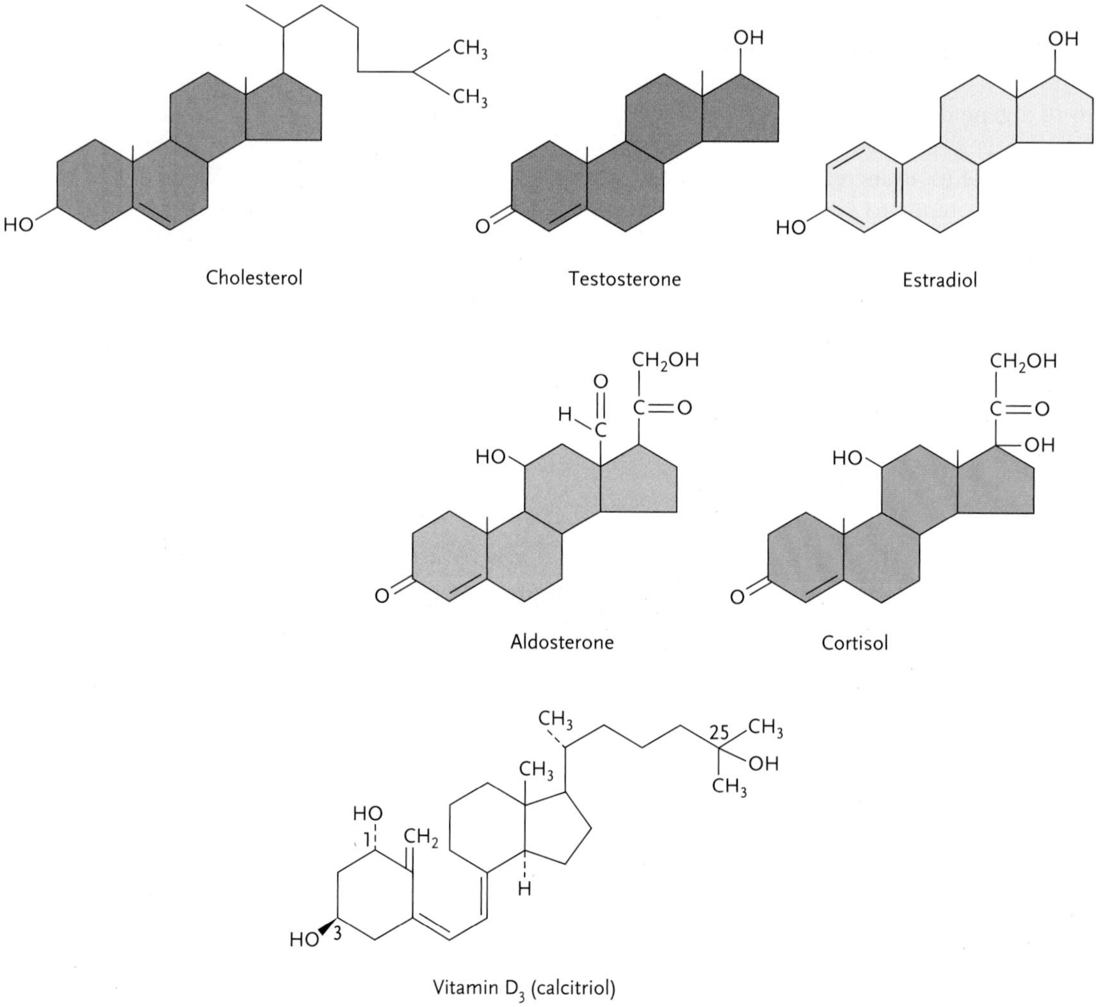

Cholesterol

Testosterone

Estradiol

Aldosterone

Cortisol

Vitamin D_3 (calcitriol)

Figure 16.4   Steroid Hormones Derived from Cholesterol

CONCEPT REVIEW 16.2

1. How does a hormone target a specific cell type?

2. What is the main difference in how water-soluble and fat-soluble hormones affect their target cell?

3. What is a second messenger?

4. What is the general effect of all hormones on their target cell?

5. What types of molecules do our cells use to produce water-soluble hormones?

6. Why is cholesterol important?

7. Give examples of important water-soluble and fat-soluble hormones in our body.

8. How does the structure and solubility of thyroid hormones differ from the other hormones?

DID YOU KNOW?

We make some very important molecules from amino acids other than proteins. Tryptophan and tyrosine are the amino acids used to produce hormones. Tryptophan is an essential amino acid, so it has to be acquired through the diet (mostly poultry, eggs, and dairy, but also fish and beef), and it is used to produce serotonin, which is both a hormone and a neurotransmitter and involved in feeling happy and content. Tyrosine is the amino acid used to produce dopamine, thyroid hormones, and the catecholamines. Tyrosine is not essential, so our cells can produce it from other amino acids.

16.3 HYPOTHALAMUS AND PITUITARY

The hypothalamus is the control centre of the brain and is involved in many aspects of homeostasis by being part of both the nervous system (composed of neurons in the brain) and the endocrine system (secretes hormones into the bloodstream). The hypothalamus is directly connected to the pituitary gland via neurons (posterior pituitary) and the *hypophyseal portal blood system* (anterior pituitary) (Figure 16.5). The hypothalamus produces **releasing hormones** or inhibiting hormones that travel through the hypophyseal portal system to the **anterior pituitary gland,** and these hormones

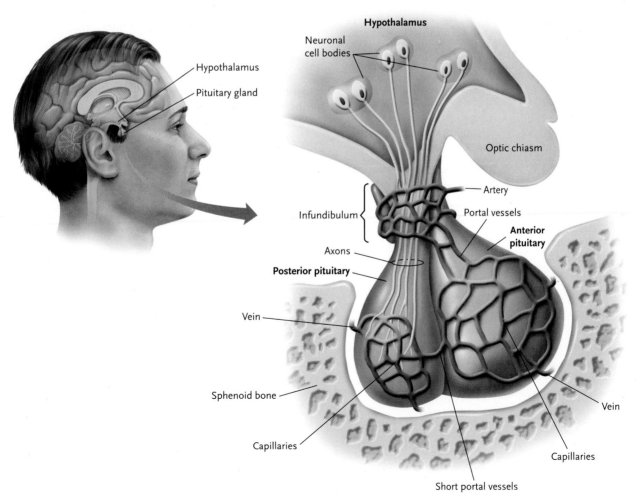

Figure 16.5   **Anatomy of the Hypothalamus–Pituitary Gland Interactions**

then signal the pituitary to produce specific hormones (Table 16.1). The hypothalamus also produces two hormones, **oxytocin** and **anti-diuretic hormone (ADH)**, also called vasopressin, which it transports to the **posterior pituitary gland**. The neuronal cell bodies are located in the hypothalamus where the hormones are produced. These hormones are then transported along the axons and released and stored in the axon terminals in the posterior pituitary until they need to be released into the bloodstream. Behind the hypothalamus is the pineal gland that produces the hormone melatonin, which is involved in inducing sleep (Chapter 15).

Posterior pituitary hormones

Oxytocin

Oxytocin is produced in the hypothalamus, stored in the posterior pituitary gland, and released during childbirth. The main functions of oxytocin are to cause uterine contractions and the ejection of breast milk from the mammary glands during breast-feeding. Uterine contractions and oxytocin release is an example of a positive feedback mechanism since the contracting uterus further stimulates the release of more oxytocin, which causes the uterus to contract, and so on. Oxytocin also acts as neuromodulator in the brain. It plays a role in maternal bonding with the infant, bonding and feelings of love for another person, and feelings for another person such as empathy. It also plays a role in orgasm, the regulation of immune response, and decreasing anxiety. Oxytocin's role in decreasing inflammation and promoting wound healing supports the idea that positive social interactions are good for our health.

> **DID YOU KNOW?**
>
> The drug ecstasy is called the "love drug" because it causes the hypothalamus to release large amounts of oxytocin into the blood.

> **DID YOU KNOW?**
>
> Oxytocin is used to induce labour in pregnant women that are passed full term. It can also be given to control bleeding after delivery because it causes the uterus to contract.

Anti-diuretic hormone (ADH)

Anti-diuretic hormone (ADH), also called vasopressin, is produced in the hypothalamus and stored in the posterior pituitary gland. ADH is released from the posterior pituitary gland when blood volume is low, or when ion concentrations are high. Both low blood volume and high ion concentrations indicate that water needs to be retained. An anti-diuretic has the opposite effect of a diuretic. Diuretics are drugs that cause water to be excreted by the kidneys. ADH acts on the **distal tubule** and **collecting ducts** of the kidneys to reabsorb water, which restores blood volume and therefore blood pressure (Chapter 20). When blood volume decreases, blood pressure also decreases, and ADH causes blood vessels to constrict in order to maintain blood pressure during periods of lower blood volume. Anything that causes a drop in blood volume, such as excessive sweating, hemorrhage, or diarrhea, will stimulate ADH secretion.

> **DID YOU KNOW?**
>
> Drinking alcohol and caffeine decreases the amount of ADH produced (dose-dependent response) and can lead to increased excretion of water. One of the main causes of a hangover is due to the dehydration effects of alcohol from inhibition of ADH.

Anterior pituitary hormones

Growth hormone

Human growth hormone (GH) (also called somatotropin) is a protein hormone that is released from the anterior pituitary gland in response to GHRH produced by the

TABLE 16.1

Hormone Action on the Pituitary Gland	
Hypothalamus Hormone	**Action on the Anterior Pituitary Gland**
Growth hormone-releasing hormone (GHRH)	Secretion of human growth hormone (GH)
Growth hormone-inhibiting hormone (GHIH)	Inhibition of GH
Thyrotropin-releasing hormone (TRH)	Secretion of thyroid stimulating hormone (TSH)
Corticotropin-releasing hormone (CRH)	Secretion of adrenocroticotropic hormone (ATCH)
Gonadotropin-releasing hormone (GnRH)	Secretion of follicle stimulating hormone (FSH) Secretion of luteinizing hormone (LH)
Prolactin-releasing hormone (PRH)	Secretion of prolactin (PRL)
Prolactin-inhibiting hormone (PIH)	Inhibition of PRL

hypothalamus. The main stimuli for the production of GH is deep sleep, high-intensity exercise, periods of increased androgen production (growth periods during childhood and puberty), and fasting (period between meals). Adolescents produce twice as much growth hormone as an adult. Even during adulthood, GH is very important for maintaining healthy cell division to replace old and damaged cells; a lack of GH results in increased signs of aging. As people age, less GH is secreted, and so it is very important to include exercise in a daily routine.

Growth hormone performs the following functions in the body:

- Promoting bone growth (height) in children and adolescents—GH causes calcium to be incorporated into bones, and decreases osteoporosis in adults.

- Stimulating protein synthesis, particularly in skeletal muscle where it is required to increase muscle strength.

- Aiding metabolism—GH causes the liver to break down glycogen and fat and thereby increases the level of blood sugar so fuel is available for cells.

- Stimulating the immune system—stress hormones inhibit growth hormone and inhibit the immune system

Excessive GH production during childhood and puberty can cause **gigantism**, too much GH produced after puberty causes **acromegaly**, and too little GH production causes **dwarfism**. With early diagnosis possible, these conditions are much less common. Dwarfism can now be prevented in children with exogenous growth hormone produced through genetic engineering.

DID YOU KNOW?

If you have a mild cold without a fever, moderate- to high-intensity exercise can help you get rid of the cold faster because you will stimulate the production of growth hormone and give your immune system a boost. Exercising while you have an infection that causes a fever will divert too much energy from your immune system and can make you feel worse.

Thyroid stimulating hormone

The anterior pituitary gland produces **thyroid stimulating hormone (TSH)** in response to the secretion of thyroid-releasing hormone (TRH) from the hypothalamus. The hypothalamus releases TRH when the blood levels of thyroid hormones T_3 and T_4 are low. TSH acts on the thyroid gland and is regulated—through negative feedback—by the concentration of thyroid hormones in the blood; a high level of circulating thyroid hormones inhibits the release of TSH from the anterior pituitary gland.

DID YOU KNOW?

Having low blood levels of thyroid hormones can contribute to depression because T_3 and T_4 play a role in the release of serotonin: known as the good mood molecule.

Adrenocorticotropic hormone

The anterior pituitary gland releases adrenocorticotropic hormone (ACTH) in response to the secretion of corticotrophin-releasing hormone (CRH) from the hypothalamus. ACTH acts on the **adrenal cortex** and regulates the production of glucocorticoids, mainly cortisol, in response to stress. ACTH causes cortisol to be produced by the **zona fasciculata** region of the adrenal cortex from cholesterol, and it stimulates adrenal cells to take up cholesterol from low-density lipoproteins (LDLs) from the blood.

Follicle stimulating hormone (FSH) and luteinizing hormone (LH)

FSH and LH have important roles in the stimulation of the gonads during puberty and reproduction. FSH is the primary hormone that stimulates development of the sperm in the testes and oocytes in the ovaries. LH stimulates the gonads to produce the hormones testosterone, estrogen, and progesterone. Males and females secrete all three hormones; however, females have much more of the **aromatase** enzyme that converts testosterone into estrogen.

Prolactin

Prolactin (PRL) is produced in the anterior pituitary gland and released in response to the secretion of prolactin-releasing hormone (PRH), which is produced in the hypothalamus. Prolactin is blocked by prolactin-inhibiting hormone (PIH), also produced in the hypothalamus. The primary role of prolactin is in **lactation**. It stimulates the growth during pregnancy of the female mammary glands and their production of milk after delivery of the baby. Recall that the posterior pituitary hormone oxytocin causes the ejection of the milk. As progesterone levels decrease at the end of pregnancy, the hypothalamus begins to produce PRH. Prolactin also plays a role in surfactant production in the fetus at the end of the pregnancy, and this helps the mother's immune system not reject the infant.

Prolactin is produced in both men and women after orgasm, causing a satisfied, relaxed feeling. Inappropriately high prolactin levels can cause impotence in males. In females, prolactin inhibits the

16

production of FSH and LH during lactation after a baby is born. This biological mechanism is meant to prevent another pregnancy before the newborn has developed enough to no longer need the mother's breast milk for nutrition. However, this inhibition is weak, so some FSH and LH is still produced, and many pregnancies occur while mothers are still nursing their newborns.

CONCEPT REVIEW 16.3

1. Which hormones are made in the hypothalamus and stored in the posterior pituitary?

2. List the hypothalamus hormones and their effect on the anterior pituitary gland.

3. List the main functions of oxytocin and ADH.

4. What are the functions of growth hormone? Does GH have any function in adults?

5. What are the effects of too much or too little growth hormone?

6. What endocrine glands are stimulated by the anterior pituitary hormones TSH, ACTH, FSH, and LH?

7. What is the function of prolactin?

16.4 THYROID AND PARATHYROID

The **thyroid gland** is located below the larynx. It consists of **thyroid follicles** composed of **follicular cells** that secrete the hormones **thyroxine** (T_4) and **triiodothyronine** (T_3), and **parafollicular cells** that secrete **calcitonin** (Figure 16.6). Follicular cells surround thyroid follicles, and the smaller parafollicular cells are located between the follicles.

Thyroid hormones

Thyroid hormones are produced by the thyroid gland in response to stimulation by the anterior pituitary hormone TSH, and they are extremely important regulators of **metabolism.** They act on every cell in the body because all cells must make ATP. Triiodothyronine (T_3) and thyronine (T_4) are produced from the amino acid tyrosine and the trace mineral **iodine** (Figure 16.7). T_3 contains three iodine atoms and T_4 contains four iodine atoms. The thyroid gland is the only tissue in the body that requires iodine. **Selenium** is also a very important micronutrient required for thyroid hormone synthesis, and selenium deficiency can cause severe thyroid malfunction. This micronutrient is found in foods such as nuts (especially Brazil nuts), fish, wheat bran, pork, shellfish, eggs, and mushrooms. The follicular cells of the thyroid gland produce both T_3 and T_4. However, once these hormones circulate in the blood and are taken up by cells throughout the body, most T_4 is converted into T_3, which is the more active form. Although thyroid hormones are produced from the amino acid tyrosine, they are not water soluble and must be transported through the blood on

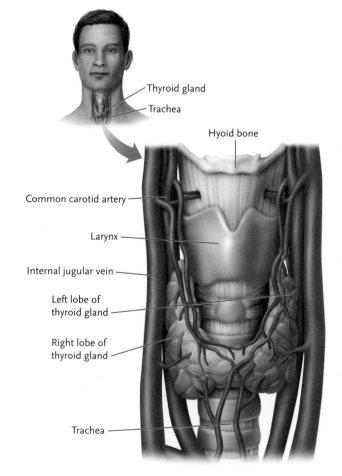

Anterior view

Figure 16.6 Thyroid Gland

The thyroid gland is made up of follicular cells that secrete the thyroid hormones T_3 and T_4, and parafollicular cells that secrete calcitonin.

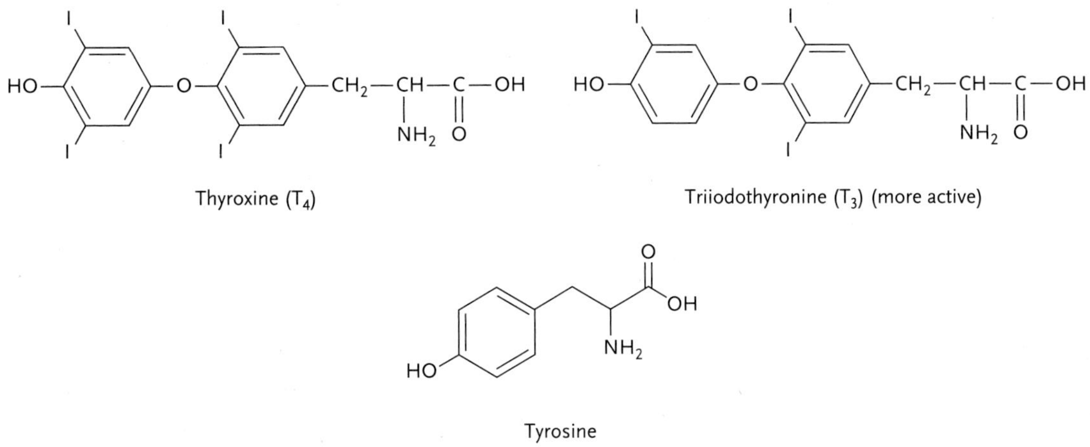

Figure 16.7 Thyroid Hormones

Thyroxine (T$_4$) and triiodothyronine (T$_3$) are thyroid hormones produced from the amino acid tyrosine.

plasma proteins. Once they reach a target cell, they enter the cell through a membrane transporter protein and act on receptors inside the cell that affect gene transcription.

DID YOU KNOW?

Table salt is "iodized" because it contains iodine in order to prevent deficiencies that can lead to hypothyroidism. Before iodization of salt, millions of people in the world had severe iodine deficiency that caused goiter. Iodine is common in soils located near oceans, but many inland regions lack iodine. In the early 1920s, iodine was routinely added to table salt in North America, and thyroid disease dramatically decreased.

The thyroid hormones have the following functions in the body:

- Regulating oxygen use by all cells during cellular respiration
- Increasing basal metabolic rate (BMR) by increasing the production of ATP from carbohydrates and fats
- Increasing the **synergistic effects** of epinephrine and norepinephrine in the breakdown of glycogen and fat for energy production
- Stimulating the production of growth hormone, especially important during fetal development
- Increasing heart rate, breathing rate, and cardiac output so that more oxygen can be delivered to cells

Abnormal functioning of the thyroid gland

Hypothyroidism is a condition caused by insufficient production of thyroid hormones. People with hypothyroidism are not able to utilize nutrients to produce energy, and so they feel very tired, have low energy, often gain weight, feel cold (since body heat is produced when ATP is produced), often have dry skin and brittle nails, and sometimes experience depression. Thyroid hormones play a role in the production of serotonin, and a lack of serotonin leads to depression. The most common causes of hypothyroidism are iodine deficiency, selenium deficiency, any malfunction of the hypothalamus or anterior pituitary gland, extreme stress, and anorexia. Note that very low-calorie diets cause the body to stop producing thyroid hormones; if nutrients aren't available for energy production, the body won't produce the necessary hormones to break down nutrients to produce energy. Eating properly and exercising regularly helps ensure sufficient thyroid hormone production.

Hashimoto's disease is an autoimmune disease in which the immune cells destroy the thyroid cells. The lack of thyroid hormone stimulates an increase in TSH secretion from the anterior pituitary gland and this can lead to goiter. Goiter can occur in cases of both hypothyroidism and hyperthyroidism.

Hyperthyroidism is a condition caused by excessive production of thyroid hormones. People with hyperthyroidism feel nervous, irritable, hot; have increased heart rate and difficulty sleeping; usually have weight loss; and sometimes have protruding eyes (**exophthalmos**) from the swelling of the fat and muscle tissue behind the eyes. The most common cause of hyperthyroidism is **Graves' disease**, which is an

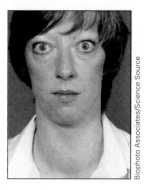

Figure 16.8 Graves' Disease

Typical appearance of an individual with Graves' disease experiencing exophthalmos.

autoimmune disease (Figure 16.8). Antibodies that act against the thyroid stimulating hormone (TSH) bind to the TSH receptors instead of TSH, and they continually stimulate the thyroid to produce thyroid hormones, which causes the thyroid gland to enlarge (goiter), and further increases thyroid hormone production.

DID YOU KNOW?

Radioactive iodine can be used to specifically destroy cancerous or overactive thyroid cells because no other cells in the body take up iodine.

Calcitonin

Calcitonin (CT) is the hormone produced by the parafollicular cells of the thyroid gland. The main function of calcitonin is to prevent calcium (Ca^{++}) blood levels from being too high; it balances the effects of parathyroid hormone, which increases blood calcium levels. Calcitonin secretion from the parafollicular cells is regulated by calcium blood levels. When levels increase, calcitonin inhibits **osteoclast** activity. Osteoclasts are the bone cells that break down the mineral matrix and release stored calcium and phosphate into the blood (Chapter 13).

The functions of calcitonin include the following: inhibiting osteoclast activity, inhibiting calcium absorption in the small intestine, and inhibiting reabsorption of calcium in the kidney. Because it prevents bones from losing calcium, calcitonin can be used to treat **osteoporosis.**

Parathyroid hormone

Parathyroid hormone (PTH) is produced by the **chief cells** in the parathyroid glands, which are located on the posterior side of the thyroid gland, and is the major

regulator of blood calcium (Ca^{++}), magnesium (Mg^{++}), and phosphate ions (HPO_4^{2-}) in the blood (Figures 16.9 and 16.10).

The functions of PTH include (1) stimulating osteoclasts to release calcium and phosphate into the bloodstream; (2) stimulating kidney cells to increase reabsorption of Ca^{++} and Mg^{++}; (3) increasing the excretion of HPO_4^{2-} by the kidneys; and (4) increasing the production of active vitamin D (calcitriol) by the kidneys, which then increases the absorption of calcium in the small intestines.

PTH is stimulated by low blood-calcium levels. Very low calcium levels caused by calcium deficiency can cause hyper-secretion of PTH, which causes excessive bone demineralization and osteoporosis. Calcium deficiency can be caused by insufficient dietary intake of calcium, vitamin D deficiency, or renal failure. Foods high in calcium include dairy products, such as milk, cheeses and yogurt; leafy green vegetables; fish; beans; and nuts.

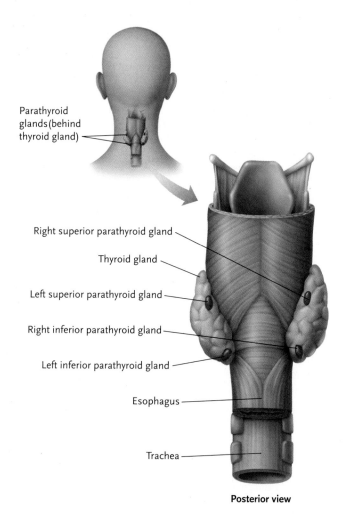

Parathyroid glands (behind thyroid gland)

Right superior parathyroid gland

Thyroid gland

Left superior parathyroid gland

Right inferior parathyroid gland

Left inferior parathyroid gland

Esophagus

Trachea

Posterior view

Figure 16.9 Parathyroid Glands

Four small parathyroid glands are located on the posterior side of the thyroid gland. PTH is the major hormone that regulates blood calcium, phosphate, and magnesium ion levels.

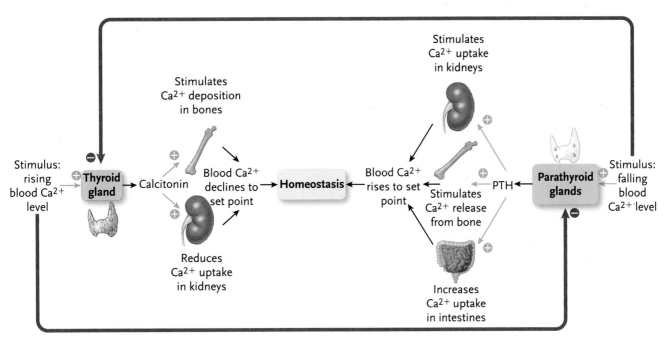

Figure 16.10 Regulation of Blood Calcium Levels

Blood calcium levels are regulated by calcitonin and parathyroid hormone.

Art source: From DIGIUSEPPE/FRASER. Biology 12 U. © 2012 Nelson Education Ltd. Reproduced by permission. www.cengage.com/permissions

CONCEPT REVIEW 16.4

1. Which hormones are produced by the follicular and parafollicular cells of the thyroid gland?

2. What is the difference between T_3 and T_4? Which is the more active hormone?

3. What amino acid is used by the thyroid to produce T_3 and T_4?

4. What two dietary minerals are required for the production of thyroid hormones? What foods contain them?

5. Are thyroid hormones fat soluble or water soluble? How are they transported through the blood?

6. List the functions of thyroid hormones.

7. What are the symptoms of hypothyroidism and hyperthyroidism?

8. What are the most common causes of hypothyroidism and hyperthyroidism?

9. What is the function of calcitonin? What cell types does it act on?

10. What are the functions of parathyroid hormone? What cell types does it act on?

11. How can low calcium cause osteoporosis?

12. What foods are high in calcium?

16.5 ADRENAL GLANDS AND THE STRESS RESPONSE

The **adrenal glands** are located above the kidneys. They consist of the outer adrenal cortex and the inner adrenal medulla (Figure 16.11). The adrenal cortex is divided into three zones. The outer **zona glomerulosa** secretes mineralcorticoids, mainly aldosterone; the middle zona fasciculata secretes **glucocorticoids**, mainly cortisol; and the inner **zona reticularis** secretes androgens, mainly dehydroepiandrosterone (DHEA) (Figure 16.12). The adrenal medulla is controlled by the autonomic nervous system, and stimulation from sympathetic preganglionic

neurons causes the medulla to secrete epinephrine and a small amount of norepinephrine.

Aldosterone

Aldosterone is the main mineralcorticoid produced by the adrenal cortex. A steroid hormone made from cholesterol (Figure 16.4), aldersterone is produced when blood pressure is low and water needs to be reabsorbed. Its production is under the control of the **renin-angiotensin system** (see Chapter 20). It causes the distal tubule and collecting duct of the nephrons in the kidney to produce sodium ion channels; this causes more sodium to be reabsorbed back into the bloodstream.

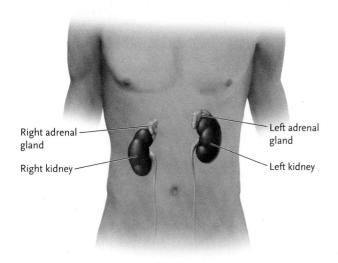

Figure 16.11 Adrenal Glands

The adrenal glands are located above each kidney.

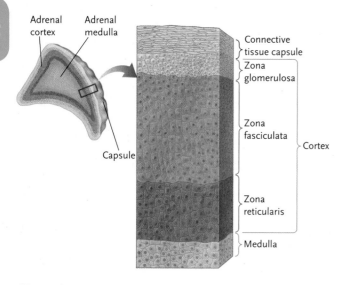

Figure 16.12 Adrenal Cortex

The adrenal cortex is composed of three layers that produce different hormones. The zona glomerulosa produces mineralcorticoids, the zona fasciculate produces glucocorticoids, and the zona reticularis produces androgens.

Art source: From SHERWOOD/KELL. Human Physiology, 2E. © 2013 Nelson Education Ltd. Reproduced by permission. www.cengage.com/permissions

When more sodium is reabsorbed, more water is also absorbed because of osmotic pressure gradients. And increase in water reabsorption brings blood volume, and therefore blood pressure, back to normal. Aldosterone also causes the nephrons to secrete potassium ions.

Cortisol

Cortisol is the main glucocorticoid produced in the adrenal cortex in response to the anterior pituitary hormone **adrenocorticotropic hormone (ACTH)**. Cortisol is a steroid hormone produced from cholesterol

(Figure 16.4). It is released during a 24-hour daily cycle, as well as during stress. The functions of cortisol include helping the body access stored nutrients so that cells can make energy (ATP). By stimulating glycogen and fat breakdown in the liver and in fat cells, cortisol increases blood sugar. It also inhibits the immune response in relation to the amount of cortisol secreted; that is, immune suppression is dose dependent. Normal cortisol secretions do not prevent the immune system from functioning, but they do act as a brake so the immune system doesn't over-react.

In people who have a regular sleep schedule, their daily cortisol level begins to rise just before they wake up in the morning. This has a beneficial effect because it helps increase our blood sugar level—which lowered during the period of sleeping and not eating—so we have energy in the morning until we eat breakfast (Figure 16.13). Cortisol secretion increases during a stress response, along with epinephrine and norepinephrine.

Cushing's syndrome

Cushing's syndrome results from an excess of cortisol secretion. An excess secretion can be caused by conditions that increase the production of corticotropin-releasing hormone (CRH) by the hypothalamus or increase the release of adrenocorticotropic hormone (ACTH) by the anterior pituitary gland. It can also be caused by long-term glucocorticoid drug treatment, which is used to treat **chronic inflammatory diseases** because of its immune suppression abilities. People with Cushing's syndrome have symptoms that result from excessive breakdown of fats, glycogen, and bone minerals; redistribution of stored body fat to the abdomen; and decreased production of new molecules, such as muscle proteins. This condition leads to hyperglycemia, muscle weakness, osteoporosis, type-2 diabetes, immune suppression, poor wound healing, thin skin, bruising, depression, moodiness, fat redistribution from limbs to trunk and face, "moon face," "buffalo hump," hair

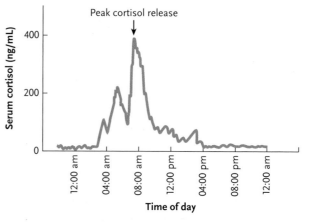

Figure 16.13 Daily Cortisol Secretions

Daily cortisol secretions are highly linked to circadian rhythm.

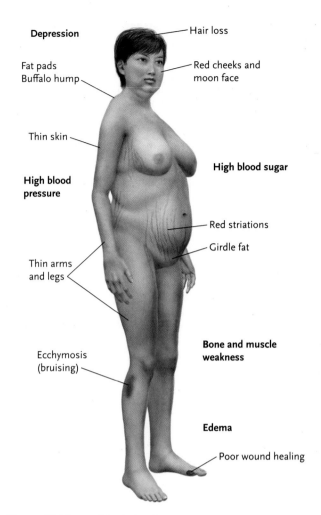

Depression

Hair loss

Fat pads
Buffalo hump

Red cheeks and
moon face

Thin skin

High blood sugar

High blood
pressure

Red striations

Girdle fat

Thin arms
and legs

Bone and muscle
weakness

Ecchymosis
(bruising)

Edema

Poor wound healing

Figure 16.14   Cushing's Syndrome
Over-secretion of cortisol or long-term treatment with glucocorticoid medication can cause Cushing's syndrome.

loss, decreased fertility, fatigue, insomnia, menstrual irregularities, water retention, and high blood pressure (Figure 16.14).

Androgens

Androgens are secreted by the adrenal cortex in both males and females. In males, the testosterone produced by the testes during puberty plays a much more significant role in the development of secondary sex characteristics and sperm development than does dehydroepiandrosterone (DHEA). In females, the adrenal cortex androgens contribute to libido and are converted into estrogens; after menopause, this is the only source of estrogen. Overstimulation of the adrenal glands can cause increased androgen production in some women, and, although rare, can lead to masculinization.

Adrenal medulla and the stress response

The adrenal medullae secrete epinephrine and norepinephrine (NE)—also called adrenaline and noradrenaline—that are released due to stimulation

of the preganglionic neurons of the sympathetic nervous system (Chapter 15). Each adrenal medulla secretes primarily epinephrine, but also small amounts of norepinephrine; the neurons of the sympathetic system release primarily norepinephrine. Note that epinephrine and norepinephrine are hormones when released into the bloodstream by the adrenal medulla, and they are neurotransmitters when released by neurons of the autonomic nervous system. Recall from Chapter 15 that the sympathetic nervous system controls the fight-or-flight response that's activated during physical or emotional stress. When stressed, we need energy available to fight or flee from a dangerous situation. Stressors that stimulate this response include fear, intense exercise, surgery, infection, illness, injury, pain, sleep deprivation, low blood sugar, cold temperatures, and any strong emotional response.

The effects of epinephrine—the main catecholamine that is secreted—include the following: increased heart rate; bronchiole dilation; pupil dilation; increased water retention and therefore increased blood pressure; vasodilation and increased blood flow in liver, skeletal muscles, cardiac muscle, and adipose tissue; vasoconstriction in digestive tract; breakdown of glycogen and fat and therefore increased blood sugar since epinephrine increases effects of glucagon and thyroid hormones.

Cortisol is also secreted during a stress response and further increases blood-sugar levels by breaking down fat and glycogen. This response is beneficial and adaptive for short-term physical stress: for example, when the blood sugar would be used and the immune response wouldn't over-react during infection, when healing could take place, or when muscles could contract during exercise or fleeing.

Long-term emotional stress, however, is very damaging to the body because the level of blood sugar remains high since it is not used by cells; protein synthesis cannot occur properly, so cells are not replaced; the immune system becomes overly suppressed, which leads to increased infections; and muscle-wasting and weakness can occur.

DID YOU KNOW?

Chronic cortisol secretion from emotional stress inhibits the production of proteins, including collagen and elastin in the connective tissue layers of the skin. This leads to premature aging of the skin, wrinkles, and loss of elasticity. Cortisol secretion also inhibits the immune response, so chronic emotional stress results in an increased risk of infection such as the common cold or influenza.

CONCEPT REVIEW 16.5

1. What hormones are secreted from the adrenal cortex and the adrenal medulla?

2. What are the three zones of the cortex, and what types of hormones are secreted from each?

3. What is the most common mineralcorticoid, glucocorticoid, and androgen?

4. How does the autonomic nervous system affect the adrenal gland?

5. What are the main functions of aldosterone?

6. What are the main functions of cortisol, and when does your body secrete it?

7. What condition can occur if too much cortisol is secreted? Describe the symptoms.

8. What is the function of cortical androgens in males and females?

9. What hormones are released during a stress response?

10. Describe how epinephrine acts on various parts of the body during a stress response.

16.6 PANCREAS

The pancreas is both an endocrine and an exocrine gland. The exocrine functions of the pancreas include secreting enzymes and bicarbonate through the **pancreatic duct** into the small intestine during digestion. The endocrine functions include secreting the hormones insulin and glucagon into the blood to regulate blood glucose levels. The pancreas is located in the abdominopelvic cavity, inferior and posterior to the stomach, within the curve of the first portion of the small intestine, the duodenum (Figure 16.15). The endocrine portion consists of

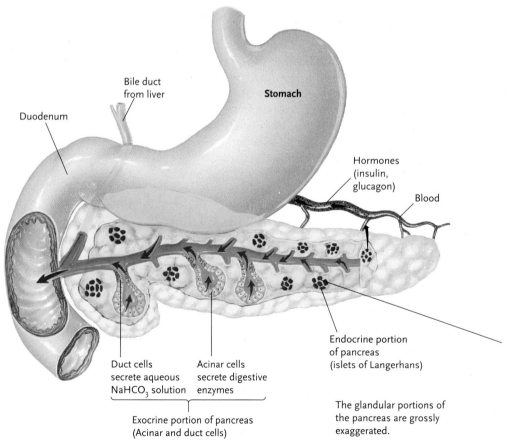

Bile duct from liver

Stomach

Duodenum

Hormones (insulin, glucagon)

Blood

Duct cells secrete aqueous NaHCO$_3$ solution

Acinar cells secrete digestive enzymes

Endocrine portion of pancreas (islets of Langerhans)

Exocrine portion of pancreas (Acinar and duct cells)

The glandular portions of the pancreas are grossly exaggerated.

Robert Markus/Science Source

Figure 16.15 Pancreas

The pancreas is composed of exocrine cells that produce digestive enzymes and endocrine cells that produce hormones. The alpha islet cells produce glucagon, and the beta islet cells produce insulin.

Art source: From SHERWOOD/KELL. Human Physiology, 2E. © 2013 Nelson Education Ltd. Reproduced by permission. www.cengage.com/permissions

pancreatic islets called the *islets of Langerhans*, which are made up of alpha cells that secrete glucagon and beta cells that secrete insulin.

Insulin is produced after we eat food, during what's called the **absorptive phase**. Different foods will stimulate different amounts of insulin secretion; this variability between foods is called the **glycemic index**. Foods that increase insulin secretion the most are sugars and starch, but eating protein and fat also cause insulin to be released. Insulin signals the body's cells to take up available nutrients from the bloodstream and store anything that isn't needed immediately. All cells respond to insulin by taking in glucose from the blood to use for ATP production. Excess nutrients are taken up by the liver, muscle, and fat cells. The liver and muscle cells store glucose as glycogen, the liver and fat cells store excess glucose, amino acids, and fatty acids as triglycerides (Figure 16.16).

Glucagon is produced in between meals when blood sugar levels decrease: known as the **fasting phase**. Glucagon causes cells to break down the stored glycogen and triglycerides in the liver, adipose tissue, and muscle cells so that glucose becomes available for ATP production. Normal blood sugar levels remain quite stable after eating, as well as during exercise or fasting: between 4.5 and 5.5 mmol/L. When blood sugar levels are low, some epinephrine is also secreted, and this also causes cells to break down stored nutrients to increase blood sugar. Notice that several hormones are involved in breaking down stored nutrients to increase blood sugar in order for our cells to have a constant supply of glucose for ATP production: thyroid hormone, epinephrine, cortisol, growth hormone, and glucagon.

Diabetes

Type-1 diabetes is an **autoimmune disease**; immune cells destroy the beta islet cells, and insulin cannot be produced by the pancreas. This is also known as insulin-dependent diabetes mellitus (IDDM), and careful monitoring of blood sugar levels and insulin injections are required to maintain blood glucose.

Type-2 diabetes is also known as non-insulin dependent diabetes mellitus (NIDDM), or **insulin resistance**. Eating a lot of sugar and starch, and therefore having constant high insulin levels, leads to down-regulation of insulin receptors on target cells. This causes glucose to stay in the bloodstream instead of being taken up and stored by cells. Type-2 diabetes progresses gradually and worsens as insulin receptors down-regulate further, until eventually the person becomes fully insulin resistant. This condition can be determined by a blood test that measures glucose tolerance. People with impaired glucose tolerance

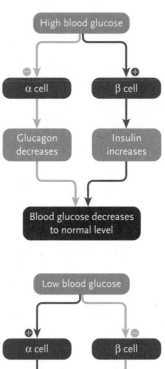

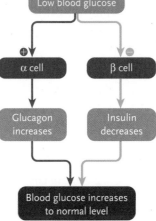

Figure 16.16 **Pancreatic Hormones**

Pancreatic hormones regulate blood glucose levels. Insulin is released when blood glucose levels increase, and it causes cells to take up and store nutrients. The glucagon that's released when blood glucose levels decrease causes cells to break down and release nutrients.

excrete glucose in the urine, which leads to increased water excretion and therefore excessive thirst; weakness and fatigue; impaired wound healing; increased tooth, skin, or bladder infections; and possibly tingling and numbness of extremities.

Prevention of type-2 diabetes

Insulin resistance will not occur if insulin secretions remain in the normal physiological range. This requires a moderate intake of sugar and starch (that breaks down into glucose), balanced meals that include healthy fats and fibre, which slow digestion and absorption and therefore prevents large insulin secretions; and exercise to use up blood glucose for ATP production.

CONCEPT REVIEW 16.6

1. What are the exocrine functions of the pancreas?

2. What hormones are secreted by the alpha and beta islet cells of the pancreas?

3. When is insulin secreted, and what are the effects on cells in the body? Which cells can insulin act on?

4. What is the glycemic index?

5. When is glucagon secreted, and what are the effects on cells in the body?

6. How and where are excess glucose, fat, and protein stored in the body?

7. Explain the difference between type-1 and type-2 diabetes.

8. Can type-2 diabetes be prevented?

16.7 OVARIES AND TESTES

Hormones produced in the gonads are stimulated by the anterior pituitary hormone, **luteinizing hormone (LH)** (**Figure 16.17**). LH causes the testes to produce **testosterone** and the ovaries to produce **estrogen** and **progesterone**. Sperm and egg development, **spermatogenesis** and **oogenesis**, respectively, are stimulated by the anterior pituitary hormone **follicle stimulating hormone (FSH)**.

Testosterone is produced by the **leydig cells** in the testes in response to LH and is the primary androgen in males (**Figure 16.18**). Testosterone is considered an **anabolic steroid** hormone because it causes growth and is involved in building up tissues. The main functions of testosterone are increasing protein synthesis; increasing

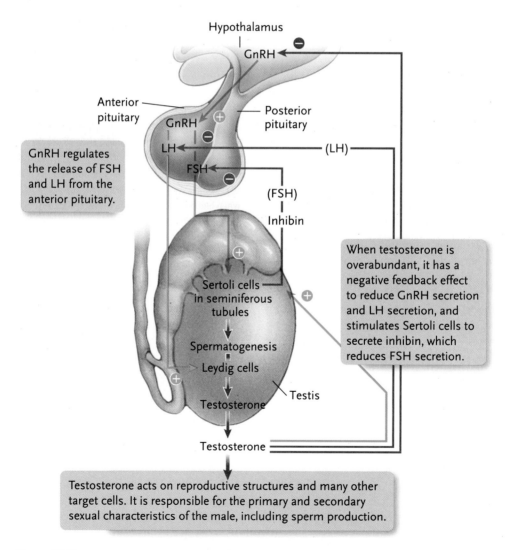

GnRH regulates the release of FSH and LH from the anterior pituitary.

When testosterone is overabundant, it has a negative feedback effect to reduce GnRH secretion and LH secretion, and stimulates Sertoli cells to secrete inhibin, which reduces FSH secretion.

Testosterone acts on reproductive structures and many other target cells. It is responsible for the primary and secondary sexual characteristics of the male, including sperm production.

Figure 16.17   Male Hormone Production

Art source: From DIGIUSEPPE/FRASER. Biology 12 U. © 2012 Nelson Education Ltd. Reproduced by permission. www.cengage.com/permissions

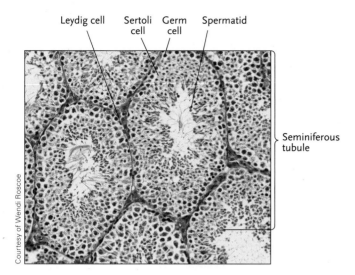

Leydig cell   Sertoli   Germ   Spermatid
               cell   cell

Seminiferous
tubule

Courtesy of Wendi Roscoe

Figure 16.18   Anatomy of the Testis

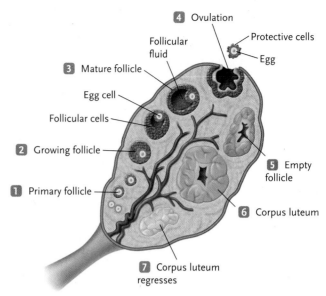

4 Ovulation

Follicular
fluid   Protective cells

3 Mature follicle   Egg

Egg cell

Follicular cells

2 Growing follicle   5 Empty
                                       follicle

1 Primary follicle   6 Corpus luteum

7 Corpus luteum
  regresses

Figure 16.19   Anatomy of the Ovary

muscle mass; increasing bone density and bone maturation; development of sex organs and secondary sex characteristics, such as facial and axillary hair growth and deepening voice.

Estrogen is produced by the granulose cells that surround the developing oocyte in the ovary because of stimulation by LH (Figure 16.19). Each month as the follicle grows, the surrounding granulose cells produce more and more estrogen, causing a surge of LH at approximately day 14 of the cycle, and this prompts **ovulation**. After ovulation, the corpus luteum secretes progesterone, which increases the growth of the uterine lining in preparation for a pregnancy. If pregnancy occurs, the placenta continues to produce progesterone until labour begins. If pregnancy does not occur, the corpus luteum breaks down and stops producing progesterone, which then stimulates menses—when the uterine lining is shed (Figure 16.20). This cycle occurs approximately every 28 days in females starting from puberty until menopause. Note that males and females both produce testosterone, estrogen, and progesterone, but in different amounts. See Figure 16.20.

Functions of estrogen and progesterone

The main functions of estrogen include the development of secondary sex characteristics during puberty; the LH

surge when estrogen increases during the menstrual cycle, which leads to ovulation and growth of uterine lining; increased bone formation—explains why osteoporosis is more common in menopausal women; and increased blood clotting—explains why women on birth control have higher risk of stroke. Estrogen affects the transcription of more than 100 different genes. The main function of progesterone is to help increase the uterine lining before fertilization, but its primary role is to support pregnancy and support fetal development in many ways.

DID YOU KNOW?

Birth control pills prevent pregnancy because the estrogen and progesterone in the birth control pills cause the blood level of those hormones to be similar to that during pregnancy; so ovulation is inhibited. Different types of birth control pills have varying levels of estrogen and progesterone.

16

16

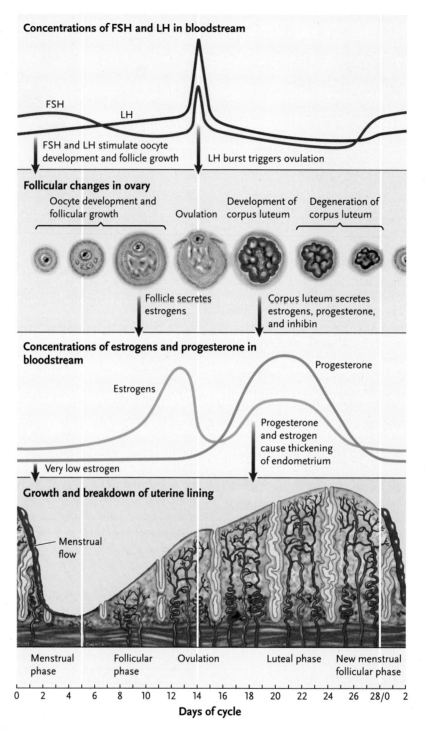

Figure 16.20   Female Hormone Cycle

Art source: From DIGIUSEPPE/FRASER. Biology 12 U. © 2012 Nelson Education Ltd. Reproduced by permission. www.cengage.com/permissions

CONCEPT REVIEW 16.7

1. What stimulates the production of testosterone, estrogen, and progesterone?

2. What enzyme, found in higher abundance in females than in males, causes females to have more estrogen than males?

3. What are the functions of testosterone, estrogen, and progesterone?

4. What is the role of FSH?

5. What causes ovulation?

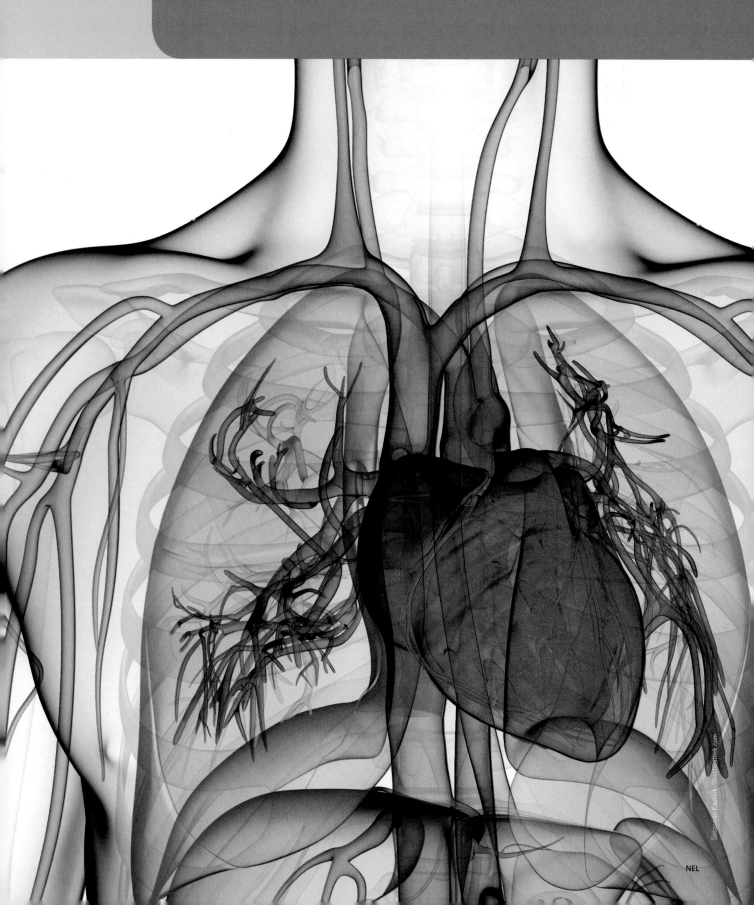

I had discovered that learning something, no matter how complex, wasn't hard when I had a reason to want to know it.

Homer Hickam

17.1 OVERVIEW OF THE CARDIOVASCULAR SYSTEM

The cardiovascular system is composed of the heart that pumps the blood; the blood vessels that transport the blood; and the blood, which contains red and white blood cells, platelets, and plasma. This **circulatory system** brings the blood very close to each and every cell in the body and this allows an optimal exchange of oxygen, carbon dioxide, nutrients, and waste products across cell membranes. The circulatory system is arranged into two circuits: the **pulmonary circulation** brings blood to the lungs to pick up oxygen and remove carbon dioxide, and the **systemic circulation** brings oxygenated blood to all the body cells that take up the oxygen and get rid of carbon dioxide (Figure 17.1).

The cardiovascular system (CVS) performs many functions in the body, including the following:

- Transporting oxygen from the lungs to the tissues
- Transporting carbon dioxide from the tissues to the lungs

- Distributing nutrients from the digestive system to the tissues
- Distributing waste products from the tissues to the urinary system
- Delivering immune cells and signalling molecules to the tissues
- Regulating **blood pressure** through hormonal and local vascular control mechanisms
- Regulating body temperature through the heat-absorbing properties of plasma and vascular shunting

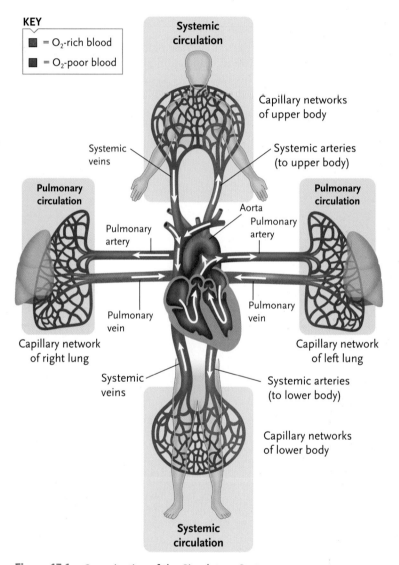

Figure 17.1   Organization of the Circulatory System

Art source: From RUSSELL/HERTZ/STARR/FENTON. *Biology*, 2E. © 2013 Nelson Education Ltd. Reproduced by permission. www.cengage.com/permissions

CONCEPT REVIEW 17.1

1. What are the functions of the cardiovascular system?

2. What are the main components of the cardiovascular system?

3. What is the difference between the systems of pulmonary circulation and systemic circulation?

17.2 CARDIAC MUSCLE TISSUE

The heart muscle, or myocardium (cardiac muscle tissue), is one of the most unique tissues in the body. This is because it has features of both skeletal muscle and smooth muscle tissue (Chapters 11 and 14). **Cardiac myocytes** (cardiac muscle cells, also called muscle fibres) are shorter than skeletal muscle fibres. They contain many mitochondria and one to three nuclei. They are branched and tightly connected at specialized intercellular structures known as intercalated discs (Figure 17.2). Two types of structures are found within intercalated discs: desmosomes and gap junctions (Figure 17.3).

Desmosomes consist of interlocking filaments of extracellular proteins that mechanically connect adjacent cardiac myocytes: much like velcro. This intercellular anchoring arrangement stabilizes and strengthens the muscle cells; this is an important feature because muscle fibres must continually contract in unison to develop force.

Gap junctions are membrane proteins that can form **chemical synapses,** which are channels that chemically connect cells together and allow the *direct* transfer of ions between cells. This direct transfer of ions makes it possible for action potentials to rapidly spread throughout the mass of myocardial cells so that large groups of muscle cells contract simultaneously. The two upper chambers, the right and left atria, beat as one coordinated unit. The two lower chambers, the right and left ventricles, beat as another distinct coordinated unit.

Cardiac muscle fibres are involuntary as their contraction is not under conscious control. Cardiac muscle is autorhythmic, which means that specialized cells in a particular region of the right atrium—the **sinoatrial node (SA node)**—act as a pacemaker to control the rate of contraction. However, the autonomic nervous system can also influence the heart rate.

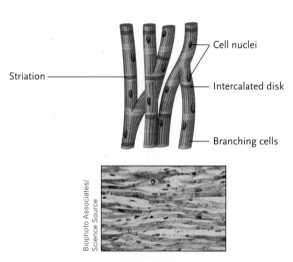

Figure 17.2   Cardiac Muscle Cells

Cardiac muscle cells are branched, striated, and contain many mitochondria.

Art source (top): From RUSSELL/HERTZ/STARR/FENTON. *Biology,* 2E. © 2013 Nelson Education Ltd. Reproduced by permission. www.cengage.com/permissions

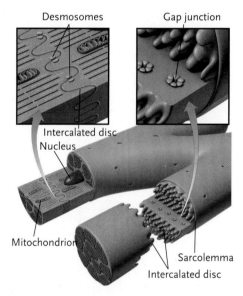

Figure 17.3   Intercalated Discs

Intercalated discs contain desmosomes and gap junctions that are important for holding cardiac muscle cells together and for transmitting electrical signals during contraction.

CONCEPT REVIEW 17.2

1. What are the features of cardiac muscle tissue?

2. How do desmosomes and gap junctions contribute to the function of the intercalated discs?

3. Why are intercalated discs important in cardiac tissue?

4. Explain how the heart is autorhythmic.

17.3 ANATOMY OF THE HEART

The heart is located superior to the diaphragm and near the midline of the thoracic cavity. The heart is enclosed in the mediastinum, which is the area between the lungs that extends from the sternum to the vertebral column and between the first rib and the diaphragm (Figure 17.4). Approximately two-thirds of the heart is left of the midline.

The heart has four chambers. The thin-walled right and left atria are separated from the thick-walled right

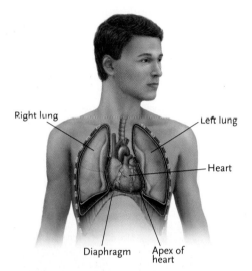

Figure 17.4   Position of the Heart in the Body

The heart is located in the mediastinum of the thoracic cavity.

and left ventricles by a fat-filled depression called the **coronary sulcus**: also known as the **interventricular sulcus** (Figure 17.5). This deep groove contains several blood vessels that supply the heart; these are the **coronary arteries**. The superior part of the heart is known as the base; the narrower, inferior part of the heart is the apex.

Blood vessels are named according to the direction of blood flow. **Veins** always transport blood *toward* the heart, and **arteries** always transport blood *away* from the heart. The *great vessels* include the inferior and superior vena cavae, the pulmonary arteries and veins, and the aorta. The **superior vena cava** returns blood to the right atrium from regions superior to the heart, such as the head, neck, face, and upper limbs. The **inferior vena cava** returns blood to the right atrium from regions inferior to the heart. These regions include the trunk, abdominal and pelvic organs, and the lower limbs. Blood returning from the coronary veins is collected in a thin-walled sac called the **coronary sinus**, located on the posterior wall of the right atrium. The coronary sinus has an opening that returns most of the coronary blood directly into the right atrium. The right and left **pulmonary arteries** receive blood from the right ventricle to bring deoxygenated blood to the right and left lungs, respectively. Two **pulmonary veins** from each lung return oxygenated blood to the left atrium. The **aorta** is the largest artery in the body, and it leaves the left ventricle as a short segment called the **aortic arch**. It then descends through the posterior trunk of the body

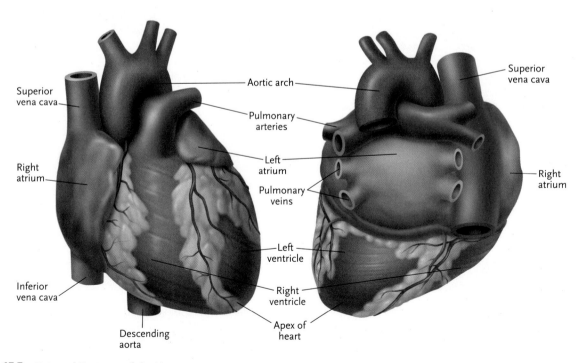

Figure 17.5   External Features of the Heart

as the **descending aorta** and supplies blood to all parts of the body, except the lungs.

Layers of the heart

The heart is contained within a double-walled membrane structure called the **pericardium**. The pericardium attaches to the diaphragm and the great vessels, and these attachments secure the heart in the mediastinum. The **parietal pericardium** forms the outer wall surrounding the actual cardiac muscle, and the **visceral pericardium** form the inner wall (Figure 17.6). Because the visceral part of the membrane forms the outer layer of the wall of the heart, it is also known as the **epicardium**. The space between the two layers is called the pericardial cavity, which is filled with a small volume of lubricating **serous fluid** that prevents friction between the two membranes when the heart is pumping. Next to the epicardium is the heart muscle: the **myocardium**. The myocardium is the thickest where it surrounds the left ventricle; this is because that region must generate the most force to move blood out of the heart to the systemic circulation. The right ventricle moves blood to the pulmonary circulation, and so it is under significantly less pressure than the systemic circulation. The atria have a very thin myocardial layer because blood moves from the atria into the ventricles mostly by the force of gravity. Atrial contraction contributes only a small amount of blood volume. Inside each of the chambers—the two atria and the two ventricles—is the final layer of the heart: the **endocardium**. The endocardium is composed of simple, squamous epithelial tissue since it is covering and lining a hollow organ. The myocardium is muscle tissue, and the pericardium is a combination of epithelial tissue and connective tissue. The endocardium is continuous with the inside of the blood vessels that are connected with the heart: the vena cavae, aorta, and the pulmonary arteries and veins. This inner layer is called the **endothelium** and is the innermost layer of all blood vessels (Figure 17.7).

Internal features of the heart

The **interatrial septum** separates the two atria, and the **interventricular septum** separates the two ventricles. The interatrial septum contains a shallow depression called the **fossa ovalis,** which is a remnant of a fetal structure known as the **foramen ovale** that allows fetal blood to flow from the right atrium to the left atrium.

Valves maintain unidirectional blood flow in veins so that blood does not flow backwards. These are found in large- and medium-diameter veins and in the heart. There are four valves in the heart: two atrioventricular (AV valves) between the atria and ventricles, and two semilunar (SL) valves between the ventricles and either the aorta or the pulmonary arteries. The AV valves prevent the backflow of blood from the ventricles into the atria when the ventricles contract. The right AV valve is the **tricuspid** and consists of three cusps (Figure 17.8).

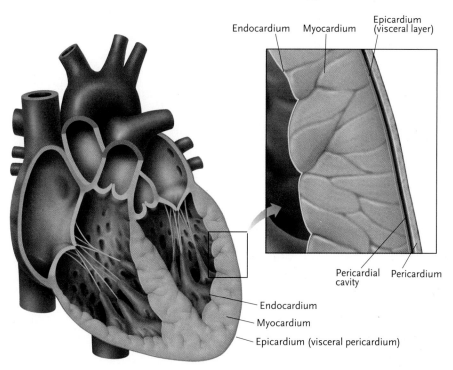

Figure 17.6   **Layers of the Heart**

The layers of the heart include the endocardium, which covers the inside of the atria and ventricles; the myocardium, which is the heart muscle tissue; and the epicardium, which is the outer covering of the heart.

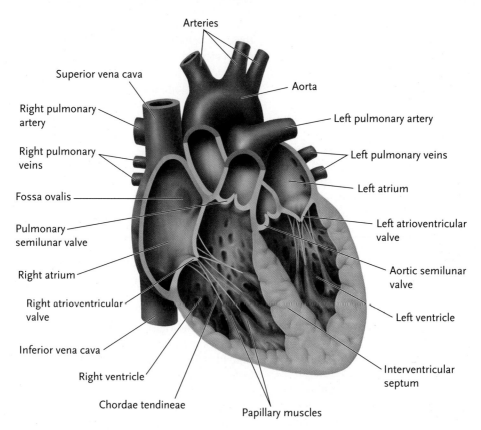

Figure 17.7 **Internal Structures of the Heart**

Art source: From DIGIUSEPPE/FRASER. *Biology 11 U.* © 2011 Nelson Education Ltd. Reproduced by permission. www.cengage.com/permissions

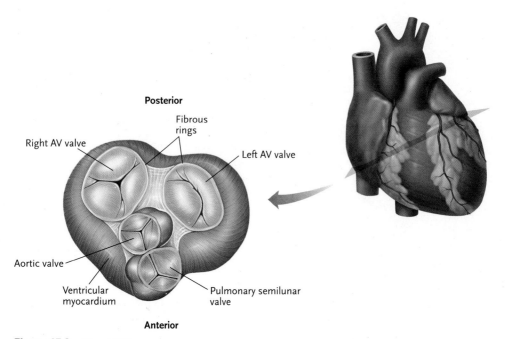

Figure 17.8 **Heart Valves**

The heart valves include the atrioventricular valves (tricuspid and bicuspid) and the semilunar valves (aortic and pulmonary).

Art source: From SHERWOOD/KELL. *Human Physiology*, 2E. © 2013 Nelson Education Ltd. Reproduced by permission. www.cengage.com/permissions

The left AV valve is the bicuspid, also called the **mitral valve**, and consists of two cusps. The cusps of the atrioventricular valves are connected to the chordae tendineae, which are connected to papillary muscles in the ventricles. When the papillary muscles contract, they pull on the chordae tendineae to close the valve and prevent the backflow of blood. The contraction of the ventricles produces a lot of force to move the blood into the aorta or pulmonary arteries, and the chordae tendineae prevent the valves from flipping inside the atria. The **aortic semilunar (SL) valve** is located between the left ventricle and the aortic arch, and the **pulmonary semilunar valve** is located between the right ventricle and the pulmonary arteries. The semilunar valves do not have chordae tendineae. Instead, they each have three concave, half-moon-shaped valves that are attached to the wall of the aorta and to the pulmonary trunk just before it branches into the pulmonary arteries.

Heart valves open and close based on the pressure gradient across the valve. When the atria contract and blood moves from atria to ventricles, the atrioventricular valves open and then close once the blood has moved into the ventricles. When the ventricles contract, the semilunar valves open and allow blood to move into the aorta and pulmonary arteries.

Heartbeat sounds are caused by the closing of the valves. The first heart sound—lub—is from the closing of the atrioventricular valves, and the second heart sound—dub—is from the closing of the semilunar valves. If a valve does not fully close, blood can leak back into the atria or ventricles, depending on which valve is damaged, causing a sloshing sound called a **heart murmur**.

DID YOU KNOW?

Leaking heart valves can be caused by high blood pressure (hypertension) because of the excess stress put on the heart and valves. People with heart murmurs may have no symptoms, or may feel tired (due to inefficient circulation of oxygen), or may have a fluttering feeling in the chest.

CONCEPT REVIEW 17.3

1. Describe the external structures of the heart.
2. Where are the coronary arteries located?
3. Briefly describe the heart chambers and large vessels.
4. What is the role of the pericardium?
5. Where are the myocardium and endocardium located?

6. Name the four heart valves.
7. Which valves have chordae tendineae? Why are these chordae important?
8. Explain how the heart valves create the heartbeat sounds?
9. What causes the valves to open and close?

17.4 CARDIAC CYCLE

Blood always flows from areas of higher pressure toward areas of lower pressure—through the heart and the blood vessels and back to the heart. The pressure is due to the heart muscle contracting and pushing blood from the atria to the ventricles and then to the blood vessels.

Blood flow circuit

The right atrium is usually considered the starting point for studying blood flow throughout the body. Blood comes from the superior and inferior vena cavae, which are the largest veins in the blood flow circuit and have the lowest pressure. Blood flows into the right atrium, through the tricuspid (right AV valves) to the right ventricle. When the ventricles contract, the force of the contraction causes significant pressure. Blood is pumped from the right ventricle through the pulmonary semilunar valve to the pulmonary arteries (region of high pressure), to the **arterioles** (where there is lower pressure), to the **capillaries** in the lungs (region of low pressure). In the lungs, the blood picks up oxygen and expels carbon dioxide, making the blood oxygenated. Blood continues to move from higher pressure areas to lower pressure areas: from the capillaries into the **venules**, into the pulmonary veins and then into the left atrium. Blood moves from the left atrium, through the bicuspid (left AV valve) to the left ventricle. Contraction of the left ventricle generates the most force, causing the highest blood pressure in the body to occur in the aortic arch. As blood flows through the aortic arch, two small openings supply blood to the coronary arteries (Figure 17.8). Blood flows from the aorta to vessels of the systemic circulation: arteries, arterioles, and capillaries. In the capillary beds, oxygen and nutrients move to the tissues, and carbon dioxide and other waste products move into the blood. From the capillaries, blood flows back to the heart through small venules, leading to small and then larger veins, until the blood again reaches the superior and inferior vena cavae (Figure 17.9).

The mechanical events associated with one heartbeat make up the **cardiac cycle**. These events include the contraction and relaxation of the different chambers of the heart, the opening and closing of heart valves,

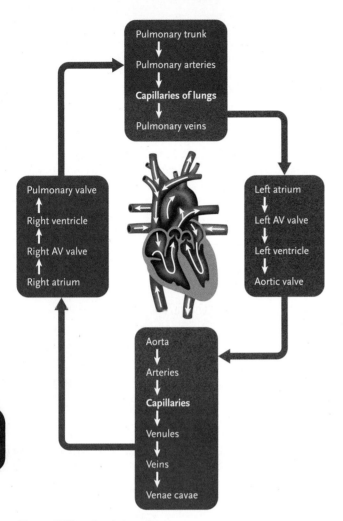

Figure 17.9   Blood Flow Circuit

Blood flows from the right side of the heart to the lungs and from the left side of the heart to the rest of the body through arteries, arterioles, and capillaries, where gas exchange occurs. Blood flows from the tissues back to the heart through venules, and then through veins to the inferior and superior vena cavae.

and the entry and exit of blood. When the heart contracts, the atria contract together, and the ventricles contract together; so blood moves to the pulmonary and systemic circulation simultaneously. The cardiac cycle consists of the following distinct stages:

1. When the atria contract, the AV valves are open, and blood moves into the ventricles. This stage is called **atrial systole.** Note that at this point the ventricles are almost full from blood flow due to gravity; atrial contraction adds the last 10–20 ml.

2. Atrioventricular valves close due to the rising pressure in the ventricles as they begin to contract. The closing of the AV valves prevents backflow of blood into the atria and makes the first heart sound: lub.

3. When the atria relax, blood flows into the right atrium from the superior and inferior vena cavae, and into the left atrium from the pulmonary veins. This stage is called **atrial diastole.**

4. At the same time as the atria relax, the ventricles contract. The AV valves are closed so blood cannot move back into atria, and the semilunar valves open when the pressure in the contracting ventricles increases to a threshold level that forces the high pressure blood flow into the aorta (from the left ventricle) and the pulmonary arteries (from the right ventricle). This stage is called **ventricular systole.**

5. Semilunar valves close. After ventricular contraction, a small amount of blood in the aorta and pulmonary arteries moves back toward the heart because of gravity, and this causes the SL valves to close. This closing prevents backflow of blood into the ventricles and makes the second heart sound: dub.

6. When the ventricles relax, the pressure in the ventricles decreases, which causes the AV valves to open. This stage is called **ventricular diastole.** Blood can begin flowing from the atria into the ventricles *before* the atria contract (back to step 1).

CONCEPT REVIEW 17.4

1. Trace the path of a red blood cell as it travels from the right atrium throughout the body and back to the right atrium.

2. How much of the final ventricular blood volume is due to atrial systole?

3. What causes the AV valves to open and close, and when does this occur?

4. What causes the SL valves to open and close, and when does this occur?

5. What does systole and diastole refer to?

DID YOU KNOW?

The human body has about 100,000 km of blood vessels, enough to wrap around Earth two and a half times!

17.5 ELECTRICAL CONDUCTION

The **conducting system** of the heart begins at the SA node located in the posterior region of the right atrium, inferior to the opening for the superior vena cava. The

SA node can spontaneously depolarize, a trait called **autorhythmicity**. These specialized cells do not have a steady resting membrane potential as do other cells in the body; however, they slowly depolarize, and this is called the **pacemaker potential**, which leads to an action potential, which leads to cardiac muscle contraction.

Recall from Chapter 15 that in a neuron, an action potential occurs if there is a stimulus that causes the inside of the cell to depolarize, which usually occurs when sodium ions enter the cell. Once a threshold level is reached, an action potential occurs as sodium ions move into the cell, then repolarization occurs when potassium leaves the cell. Action potentials in the SA node are slightly different. In the SA node, sodium ions begin to move into the cell as soon as the cell is repolarized to approximately –60 mV. As sodium ions slowly move *in*, the cell becomes more positively charged (pacemaker potential), and once it reaches approximately –40 mV, calcium channels open, allowing calcium ions to move *into* the cell, further depolarizing the cell (action potential) to approximately +10 mV. Once the inside of the cell is positively charged, potassium channels open and potassium ions move *out* of the cell, causing repolarization. Then the cycle continues without the need for any external stimulation (Figure 17.10).

The nervous system can influence heart rate, but there is no nervous tissue within the heart itself. Without any external stimulation, the SA node has a firing rate of approximately 100 beats per minute (bpm), but our regular heart rate is usually lower than that due to nervous system and endocrine stimulation. For example, the sympathetic nervous system will increase heart rate by releasing norepinephrine, which increases the opening of sodium channels in the SA node, causing the pacemaker depolarization phase to occur faster. The parasympathetic nervous system will release acetylcholine that slows the opening of sodium channels, so it takes longer for the action potential to occur. A typical resting heart rate is 60–75 bpm.

DID YOU KNOW?

If the SA node is damaged, the AV node can become the pacemaker. However, the heart rate would drop to 40–60 bpm, which may not be adequate to oxygenate the brain. In that case, an artificial pacemaker can be surgically implanted into the heart to deliver the electrical stimulation required to cause the heart to contract.

DID YOU KNOW?

The heart can beat for hours after being removed from the body if it is kept in oxygenated saline solution. This factor is very beneficial for people needing a heart transplant.

Once the SA node depolarizes, the electrical signal travels throughout the atrial muscle fibres and then to the **atrioventricular node (AV node)**. At the AV node, there is a delay in the transmission to the bundle of His, and this allows the atria to fully contract before the ventricles begin to contract. Once the signal reaches the **bundle of His** (pronounced Hiss)—also called **atrioventricular bundle**—it travels along the right and left **bundle branches** to the **purkinje fibres** at the apex of the heart. Action potentials travel quickly in the large-diameter purkinje fibres and cause the ventricular muscle cells to contract from the bottom up, which pushes blood from the ventricles out to the aorta and pulmonary arteries.

ECG

Action potentials in the conducting cells of the heart stimulate the contraction of the heart muscle fibres. An **electrocardiogram (ECG)** is a method of measuring the electrical currents caused by the movement of ions throughout the heart during the cardiac cycle (Figure 17.11). There are three distinct waves associated with the heart contractions.

1. The **P wave** represents the atrial depolarization that causes contraction of the atria.

2. The **QRS** represents ventricular depolarization that causes contraction of the ventricles. Although not visible on the ECG, at this time the atria are repolarizing and relaxing.

3. The **T wave** represents ventricular repolarization and relaxation.

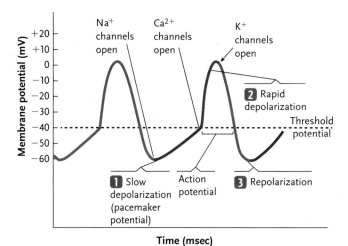

Figure 17.10 Pacemaker Potential

The pacemaker potential involves (1) slow depolarization caused by sodium ions moving into the cell; (2) rapid depolarization, caused by calcium ions moving into the cell; (3) repolarization, caused by potassium ions moving out of the cell.

Art source: From SHERWOOD/KELL. *Human Physiology*, 2E. © 2013 Nelson Education Ltd. Reproduced by permission. www.cengage.com/permissions

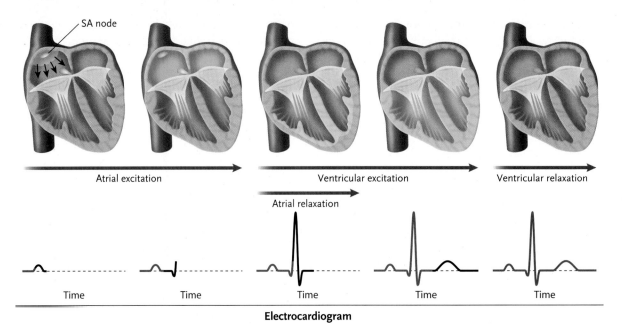

Figure 17.11 Conduction System of the Heart

(A) Anatomical structures and (B) ECG

Art source: From DIGIUSEPPE/FRASER. *Biology 11 U.* © 2011 Nelson Education Ltd. Reproduced by permission. www.cengage.com/permissions

Variations in ECG patterns can be used to diagnose malfunctions of the cardiac conduction system such as **arrhythmias,** which are irregular heart contractions.

An overview of the major events occurring during one cardiac cycle is shown in Figure 17.12 and summarized as follows:

1. P wave occurs when depolarization of the atria leads to contraction of the atria.

2. Slight pressure increases in ventricle due to increased blood volume.

3. Approximately 10–20 ml of blood volume is added to the ventricles due to atrial contraction.

4. **End diastolic volume (EDV)** occurs when the ventricles are full; it is the volume of blood in the ventricles after ventricular diastole (relaxation).

5. The first heart sound occurs from the closing of the AV valves after the atria have contracted. Closing the AV valves prevents the backflow of blood from the ventricles to the atria once the ventricles begin to contract.

6. The first isovolumetric phase occurs when the AV valves are closed and the semilunar valves have not yet opened, and so the volume of blood in the ventricles does not change.

7. QRS phase occurs when the ventricles depolarize, which leads to contraction of the ventricles.

8. Contraction of the ventricles causes the pressure in the ventricles to increase.

9. Increased ventricular pressure reaches a threshold level and causes the semilunar valves to open.

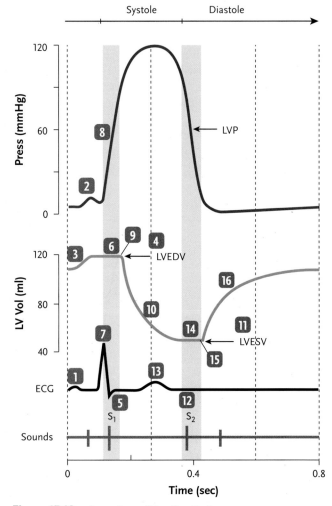

Figure 17.12 Overview of Cardiac Cycle

The cardiac cycle is shown in terms of ECG, heart sounds, and blood volume and pressure in the left ventricle.

10. Blood is pumped out of the left ventricle into the aorta and out of the right ventricle into the pulmonary arteries, and the volume of blood in the ventricles decreases.

11. End systolic volume (ESV) occurs when the ventricles are almost empty; it is the volume of blood left in the ventricles after systole (contraction). Note, as shown in Figure 17.12, approximately 50 ml of blood volume remains in the left ventricle after contraction.

12. Semilunar valves close once the pressure in the ventricles decreases, causing the second heart sound.

13. T wave occurs when the ventricles repolarize, the muscle relaxes, and the pressure inside the ventricle decreases.

14. The second isovolumetric phase occurs when the volume in the ventricles is not changing because the semilunar valves have just closed and the AV valves have not yet opened.

15. Atrioventricular valves open because the pressure is now higher in the atria than in the ventricles.

16. Ventricles begin filling with blood as soon as the AV valves open, and before the atria contract at the next P wave.

CONCEPT REVIEW 17.5

1. What structures make up the conducting system of the heart?

2. What is autorhythmicity?

3. What three ions play an important role in the pacemaker potential?

4. How is the pacemaker potential different from an action potential in a neuron?

5. What is happening in an ECG during the P wave? QRS? T wave?

6. What causes the first and second heart sounds?

7. What is an arrhythmia?

8. What is happening during the first and second isovolumetric phases of the cardiac cycle?

9. When does most of the blood enter the ventricle?

10. What is end diastolic volume and end systolic volume?

17.6 CARDIAC OUTPUT

Cardiac output refers to the total amount of blood ejected from the left ventricle per minute. It is determined by the number of beats per minute (heart rate) and the amount of blood ejected with each contraction (*stroke volume*).

cardiac output (CO) = heart rate (HR) x stroke volume (SV)

For an adult the average resting heart rate is approximately 70 bpm. An adult's average stroke volume is approximately 70 ml. Stroke volume is determined by the amount of blood in the ventricle after diastole (relaxation): that is, the end diastolic volume (EDV) minus the end systolic volume (ESV), which is the amount of blood left in the ventricle after systole (contraction).

stroke volume = EDV – ESV

Given an average EDV of 120 ml and an average ESV of 50 ml,

stroke volume = 120 ml – 50 ml = 70 ml

If we assume a resting heart rate of 70 beats per minute and a stroke volume of 70 ml per beat, the cardiac output at rest is calculated as follows:

cardiac output = 70 ~~beats~~/ 1 min x 70 ~~mL~~/1 ~~beat~~ x 1L/1000 ~~mL~~

There are approximately 5–6 L of blood in the body (depending on body weight). Therefore, at rest the heart is pumping nearly the entire blood volume each and every minute.

Cardiac output during exercise

With moderate exercise, we might have an increase in heart rate from 70 bmp to 100 bpm, and an increase in stroke volume from 70 ml to 100 ml. That would increase cardiac output to 10 L/min. This is known as **cardiac reserve**. Most people have a cardiac reserve of two to three times their resting value (approximately 10–15 L/min). Endurance-trained elite athletes may have a cardiac reserve of more than four times their resting value (20+ L/min).

Factors that affect cardiac output

Any factor that increases heart rate or stroke volume increases cardiac output up to a maximum (Figure 17.13). Once the heart rate gets very high (190–200 bpm) the heart becomes inefficient and the stroke volume decreases significantly. Heart rate varies on average between 60 and 180 bpm. Athletic individuals tend to have a lower resting heart rate because their heart is stronger and, therefore, more blood volume is pumped with each contraction. Cardiac output can be maintained with a lower heart rate if the stroke volume is higher. Stroke volume varies between 70 and 120 ml.

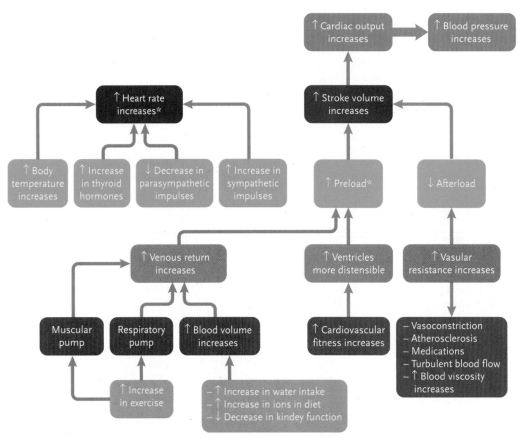

*An increase in heart rate and/or preload while stroke volume decreases from inefficient contraction force = Blood pooling in lungs, congestive heart failure, and/or deep-vein thrombosis.

Figure 17.13 Factors Affecting Cardiac Output

Factors that affect heart rate

When heart rate increases, the QRS depolarization time and the time required for the ventricles to contract does not change relative to a resting state; rather the relaxation period decreases when the heart rate increases. Heart rate is primarily affected by the autonomic nervous system (ANS). During exercise, stress, or relaxation, the ANS affects the rate of depolarization of the SA node. Epinephrine released by adrenal glands, or norepinephrine released by the sympathetic neurons increase the heart rate as well as the force of contraction. Acetylcholine from parasympathetic neurons slows heart rate and decreases the force of contraction (Section 17.5). The medulla oblongata in the brainstem is the cardiovascular centre that regulates heart rate based on information it receives from the body. Peripheral baroreceptors detect blood pressure in the aortic arch and the carotid arteries. If blood pressure increases, the medulla oblongata stimulates parasympathetic neurons to decrease heart rate and therefore cardiac output, which then decreases blood pressure.

Other factors that can increase heart rate include thyroid hormones; increased blood Na^+, Ca^{++}, and K^+; and increased body temperature. Females tend to have slightly higher resting heart rates than males, and the elderly tend to have higher heart rates than younger adults.

Factors that affect stroke volume

The strength of contraction is directly proportional to the amount of stretch in the myocardial fibres: a relationship known as the **Frank-Starling law.** A larger blood volume increases the amount the muscle fibre stretch, which increases the strength of the contraction. Any increase in the end diastolic volume will increase ventricular contraction, which leads to an increase in stroke volume. The ANS can increase the force of contraction, which also increases the stroke volume.

The volume of blood returning to the atria from the body and from the pulmonary circulation is called the **preload**. An increase in preload increases EDV and, therefore, increases stroke volume. The stroke volume is also dependent on the systemic vascular resistance, which is called **afterload.** People who have constricted blood vessels due to atherosclerosis (Section 17.11) have greater vascular resistance and therefore reduced stroke volume and cardiac output.

Preload is affected primarily by **distensibility** of the ventricles and by **venous return**, which is the amount of blood returning to the heart. Distensibility refers to how readily the heart wall expands to accommodate a returning volume of blood. A heart that is rigid, or one that has extensive scarring as a result of a heart attack, does not easily expand. As a result, a limited amount of blood can enter the heart, and this reduces stroke volume and cardiac output (at a given heart rate).

The heart can only pump the blood that returns to it. Venous return is affected by the total blood volume in the body and by the skeletal and respiratory pumps that help pull blood back to the heart from the systemic circulation. **Blood volume** can have a dramatic effect on venous return, stroke volume, and cardiac output. Low blood volume, as occurs with severe hemorrhage, reduces the returning volume of blood and decreases the stroke volume. To try to compensate for the decreased blood volume, the heart rate increases in an effort to maintain an optimal cardiac output. Raising the legs can help overcome the force of gravity and help to return blood to the heart.

Venous return increases with the contraction of skeletal muscles. The veins that carry blood back to the heart are located more superficially than the arteries, which are deeper within tissues. When skeletal muscles contract, they squeeze regions of the veins and push blood through; this is called the skeletal pump. Since the blood in the veins is under the lowest pressure of any of the other blood vessels, blood moves forward in the veins during a muscle contraction and back in between muscle contractions. The valves within veins prevent the backflow from moving too far, so the blood moves forward from one valve to another in small bursts (Figure 17.14). Exercise can greatly increase the venous return because the increased skeletal muscle contractions force the blood through the veins and back to the heart. This is one reason why cardiac output increases during exercise.

Venous return also increases with increased respiration. This is due to the pressure gradient created between the thoracic and abdominal cavities with each breath, when

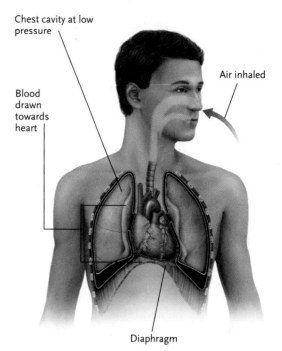

Figure 17.15 **Respiratory Pump**

The respiratory pump helps to move blood through veins back towards the heart, with each inhalation when the pressure in the thoracic cavity decreases.

the diaphragm contracts and relaxes (Figure 17.15). When we inhale, the diaphragm contracts and moves downward, increasing the volume of the thoracic cavity and *decreasing* intra-thoracic pressure, this is called the respiratory pump. Recall that blood flows through vessels from areas of higher to lower pressure. When the pressure in the thoracic cavity decreases with each inhalation, blood moves from the higher pressure area in the large abdominal veins toward the inferior vena cava in the thoracic cavity.

Factors that increase cardiac output

Heart rate increase, which is affected by the following:

- an increase in sympathetic stimulation by epinephrine and norepinephrine
- a decrease in parasympathetic stimulation
- exercise
- an increase in sodium, potassium, and calcium levels
- an increase in body temperature

Stroke volume increase, which is affected by the following:

- an increase of venous return (preload)
- exercise, by means of skeletal and respiratory pumps
- an increase in blood volume, for example, due to water retention when sodium level increases
- an increase in ventricular contractility and distensibility
- a decrease in vascular resistance (afterload)

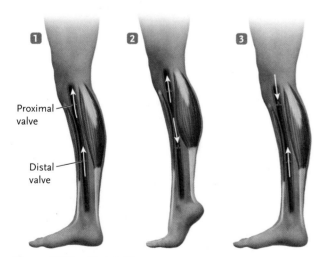

Figure 17.14 **Skeletal Pump**

The skeletal pump refers to how the contraction of lower leg muscles helps push blood through the veins and back toward the heart.

CONCEPT REVIEW 17.6

1. What is cardiac reserve?

2. Bill has an EDV of 130 ml, an ESV of 50 ml, and a heart rate of 75 bpm. What is his cardiac output?

3. How does the sympathetic and parasympathetic nervous system affect heart rate?

4. How do epinephrine and thyroid hormones affect cardiac output?

5. How would an increase in vascular resistance affect stroke volume?

6. What is preload and afterload?

7. Explain the Frank-Starling law.

8. What factors affect venous return?

9. How do the skeletal and respiratory pumps increase venous return?

10. What factors increase cardiac output?

17.7 ANATOMY OF THE BLOOD AND LYMPHATIC VESSELS

Arteries always move blood *away* from the heart, and the largest artery in the body is the aorta. Blood flows from the aorta, to the main arteries, to arterioles, and then to capillaries where oxygen and nutrients can move into the tissues. Blood flows back from the capillaries to venules, to veins, and then to the largest veins, which are the superior and inferior vena cavae (Figure 17.16). Our total blood volume at any one time is mostly found in the veins. The capillaries have the least total amount of blood, although they have the largest cross sectional area compared to any other type of blood vessel (Figure 17.17).

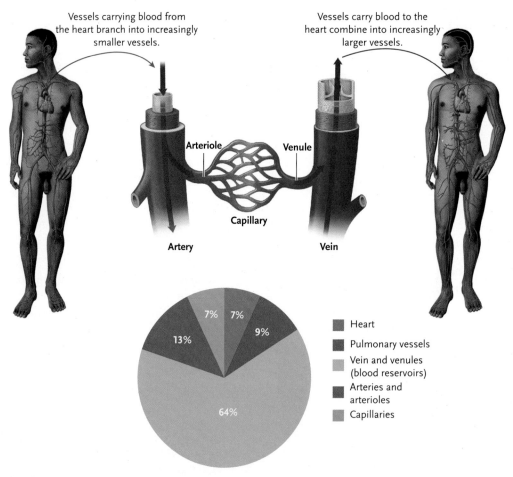

Figure 17.16 Overview of Blood Vessels

Most of the blood volume is in the veins at any given time.

Art source (centre): From RUSSEL/HERTZ/STARR/FENTON. *Biology*, 2E. © 2013 Nelson Education Ltd. Reproduced by permission. www.cengage.com/permissions

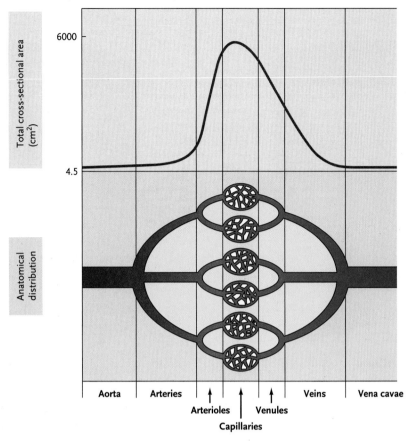

Figure 17.17   Cross-Section of Blood Vessels

Capillaries have the greatest cross-sectional area compared to all other blood vessels.

Art source: From SHERWOOD/KELL. *Human Physiology*, 2E. © 2013 Nelson Education Ltd. Reproduced by permission.
www.cengage.com/permissions

Arteries and veins

Arteries and veins are the largest blood vessels, and they consist of three layers of tissue: tunica intima, tunica media, and tunica externa (Figure 17.18). The innermost layer, the **tunica intima,** is composed of simple squamous epithelial tissue. This tissue, the endothelium, is a basement membrane that surrounds all epithelial tissue; it is composed of proteins that connect the epithelial layer to connective tissue. Next to the basement membrane in arteries, but not in veins, is a connective tissue layer that is high in elastic proteins. This layer is necessary in arteries, because the arteries must stretch with each heartbeat and the elastic layer helps the artery move back to its original shape. The middle layer, the **tunica media,** is composed of smooth muscle tissue, and arteries also have another layer of elastic connective tissue. The smooth muscle layer is larger in arteries because they must contract to help move the blood through the vessels with each heartbeat. The outermost layer, the **tunica externa,** is composed of a thick layer of loose connective tissue (areolar and reticular).

Of all the blood vessels, arteries must withstand the highest pressure. Veins withstand relatively low blood pressure but have much more blood volume. The elastic layer and larger smooth muscle layers are needed for arteries to contract and move blood to the arterioles. As the bolus of blood moves

through the blood vessels with each heartbeat, we can feel a pulse in the large arteries that are close to the surface of the body, but veins do not have a pulse. Veins have a much larger internal diameter because they hold a much larger volume of blood, and the veins have valves that prevent the backflow of blood as it moves back toward the heart.

DID YOU KNOW?

When the valves inside the veins do not work properly, blood can move backward, swirl around in the same area for a long period of time, and pool there. Since many veins are close to the surface of the skin, the pooled blood is visible, and this is called **varicose veins.**

Arterioles and venules

Arterioles and venules are like small arteries and veins. The larger arterioles and venules contain the same layers as arteries and veins, but each of the layers is smaller. The smallest arterioles are composed only of inner endothelium and basement membrane, a very small smooth muscle layer or just scattered smooth muscle cells in some regions, and an outer loose connective tissue layer. The smallest venules

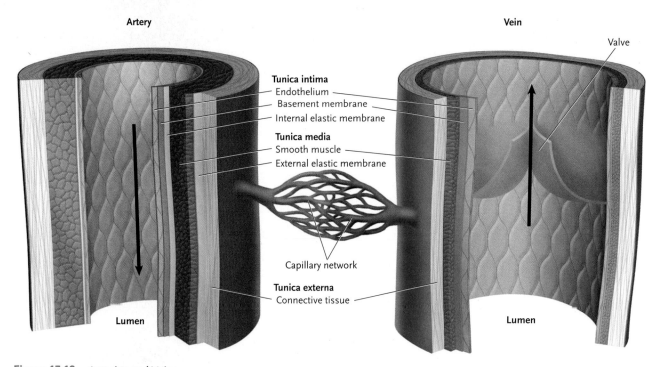

Figure 17.18 Arteries and Veins

Arteries and veins are made up of three layers: the tunica intima, tunica media, and tunica externa.

are composed only of inner endothelial cells and basement membrane, and outer connective tissue (Figure 17.19).

Arterioles play a very important role in the regulation of blood flow to specific regions of the body. They do this by responding to signals from the autonomic nervous system and the endocrine system. For example, when we exercise, more blood is required in the skeletal muscles and is diverted away from the digestive system, mostly because

of **vasodilation** of arterioles in the skeletal muscles and **vasoconstriction** of arterioles in the gastrointestinal tract.

Capillaries

Capillaries are composed only of inner endothelium cells and are surrounded by the basement membrane (Figure 17.20). Capillaries have the most cross-sectional

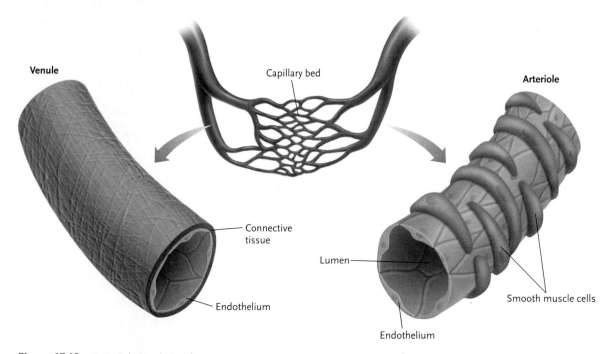

Figure 17.19 Arterioles and Venules

Arterioles and venules contain endothelial cells, a basement membrane, a layer of smooth muscle, and connective tissue.

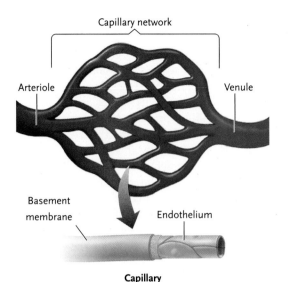

Figure 17.20 Capillaries

Capillaries are the site of exchange of gases and nutrients.

Art source: From RUSSELL/HERTZ/STARR/FENTON. *Biology*, 2E. © 2013 Nelson Education Ltd. Reproduced by permission. www.cengage.com/permissions

area of any blood vessel, and come within 100 µm of almost every single cell in our body (except the cells of avascular tissues). Capillaries have the important function of exchanging materials with the cells. Oxygen, nutrients, hormones, and other signalling molecules leave the bloodstream through the capillaries, diffuse across the interstitial space, and then enter the cells. Recall from Chapter 16 that hormones bind only to cell types that contain a specific receptor. Carbon dioxide and waste products leave the cells, enter the capillaries, and are expelled or broken down once they circulate to the lungs, liver, or kidneys. The movement of oxygen from the capillaries to the cells and carbon dioxide from the

cells to the capillaries is called **gas exchange**. Blood flow regulation can occur at the level of the capillaries as well as the arterioles. Not every capillary bed in our body contains the same amount of blood all the time. Tissues that need more blood oxygen and nutrients will receive more blood flow. Each capillary bed has precapillary sphincters, which are small circular smooth muscles that block blood flow when they contract (Figure 17.21). Precapillary sphincters contract or relax depending on the amount of blood flow required. For example, when the body temperature decreases, the precapillary sphincters in the deeper layers of the skin contract and decrease blood flow to prevent excessive heat loss.

In capillary beds, fluid leaves the blood vessels at the arteriole end, and some of it returns to circulation at the venule end. This loss of fluid is a normal process that helps move nutrients from the bloodstream to the cells. Depending on where the fluid is located, it has different names: (1) When the fluid is inside the capillaries, it is called **plasma**, which is the liquid portion of the blood that contains dissolved nutrients, ions, vitamins, and minerals; (2) when some of this fluid moves out of the capillaries and into the interstitial (intercellular) space, the fluid is called interstitial fluid, which continually bathes the cells in nutrients and dissolved oxygen so that each cell can readily take up whatever it needs; (3) when excess fluid either returns to the venule end of the capillary bed or is taken up by lymphatic vessels, the fluid is called **lymph**.

Why fluid leaks from the circulatory system

The movement of fluid out of the capillaries into the interstitial space increases as blood pressure increases (Figure 17.22a). This is called **hydrostatic pressure**, which is the physical pressure of the fluid moving through the blood

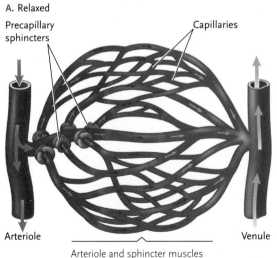

Figure 17.21 Precapillary Sphincters

Precapillary sphincters regulate blood flow into various capillary beds.

Art source: From RUSSELL/HERTZ/STARR/FENTON. *Biology*, 2E. © 2013 Nelson Education Ltd. Reproduced by permission. www.cengage.com/permissions

vessels (see Figure 17.22b, large arrow at arteriole end). During exercise, when blood pressure increases, more fluid is pushed from the capillaries into the interstitial space. All the fluid from the blood does not leave the vessels because certain proteins (mostly **albumin**) cause an **osmotic pressure** that keeps fluid inside the capillaries (see Figure 17.22b, small arrow at arteriole end). As blood moves through the capillary network toward the venule end, the osmotic pressure in the capillaries increases due to the loss of fluid along the way, but the proteins stay inside the blood vessels. The increased osmotic pressure at the venule end causes some interstitial fluid to re-enter the capillaries (see Figure 17.22b, larger arrow at venule end). Also at the venule end, fluid re-enters the capillary because of the increased hydrostatic pressure of fluid in the interstitial space (see Figure 17.22b, smaller arrow on venule end). In Chapter 20, the movement of fluid out of capillaries is discussed in relation to the urinary system.

About 20 L of fluid moves from the cardiovascular system into the interstitial spaces every day. All but about three or four litres of it re-enters the capillaries, and the rest must be taken up by the lymphatic vessels. If fluid builds up in the interstitial space, tissues swell; this is called **edema**.

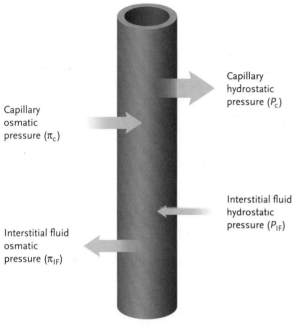

Capillary hydrostatic pressure (P_c)

Capillary osmatic pressure (π_c)

Interstitial fluid hydrostatic pressure (P_{IF})

Interstitial fluid osmatic pressure (π_{IF})

Net filtration pressure = $P_c + \pi_c - P_{IF} - \pi_{IF}$

A.

DID YOU KNOW?

Eating a lot of salt can cause edema because sodium ions are in higher concentrations just outside of cells, in the interstitial space. This causes an osmotic gradient that pulls water into the interstitial space. In a healthy person, the kidneys filter and excrete excess sodium, so persistent edema could be an indication that the kidneys are not functioning optimally.

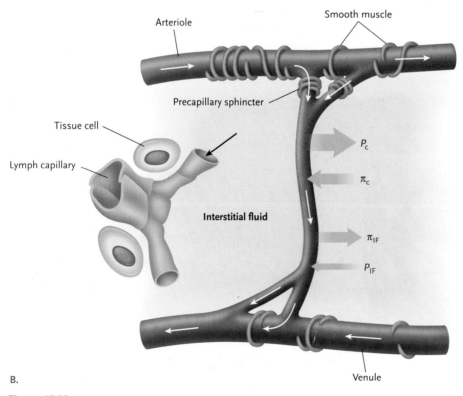

Arteriole

Smooth muscle

Precapillary sphincter

Tissue cell

Lymph capillary

Interstitial fluid

P_c

π_c

π_{IF}

P_{IF}

Venule

B.

Figure 17.22 Movement of Fluid

Movement of fluid into and out of capillaries is due to hydrostatic and osmotic pressures.

Lymphatic vessels

The lymphatic system plays three important roles: (1) returning fluid from the interstitial tissue back to the circulatory system, (2) transporting the fats absorbed in the digestive tract to the circulatory system, and (3) making up the majority of the immune system (Chapter 22). Lymphatic vessels are composed of endothelium and a basement membrane (as in capillaries), but they also contain some smooth muscle cells (as in small arterioles) and valves (as in veins) to prevent backflow of lymph fluid as it moves through the vessels (Figure 17.23). Lymph is moved through the lymphatic system by contraction of the skeletal muscle and also by contraction of smooth muscle cells that surround the lymph vessels. Fluid moves from regions of higher pressure to lower pressure—just like blood flow—through the vessels until it reaches the lymphatic ducts that drain into the venous system. The lymphatic system of the right, upper side of the body drains into the **right lymphatic duct**, then into the **right subclavian vein**. The lymphatic system of the rest of the body drains into the **thoracic lymphatic duct,** then into the **left subclavian vein** (Figure 17.24).

Summary of lymph flow

plasma—interstitial space—lymphatic vessel—lymph ducts—subclavian veins of the circulatory system

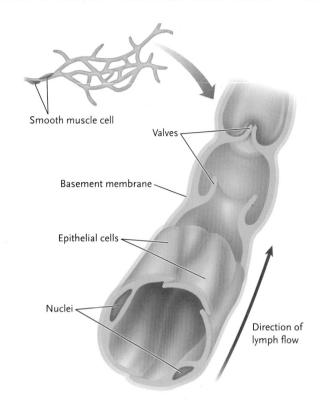

Figure 17.23 **Anatomy of a Lymph Vessel**

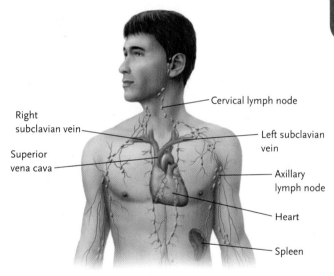

Figure 17.24 **Drainage Paths for the Lymphatic System**
Lymph drains back into the circulatory system via the right and left subclavian veins.

DID YOU KNOW?

Massage therapists can help with lymphatic drainage by massaging and pushing lymph through the vessels toward the subclavian veins; this process is called manual lymphatic drainage. Increasing lymphatic drainage increases blood circulation because blood volume and venous return increases. It is also believed that manual lymphatic drainage can help fight colds and flus. This is because lymph nodes filter the fluid that moves through lymph vessels, and this is where immune cells can fight infections.

CONCEPT REVIEW 17.7

1. Describe each of the three layers of an artery.
2. How are arteries and veins different?
3. Which blood vessels withstand the most blood pressure?
4. Which blood vessels have valves?
5. Where does gas exchange occur?
6. What is the importance of precapillary sphincters?
7. What is the difference between plasma, interstitial fluid, and lymph?
8. What is edema?
9. How is lymph moved through the lymphatic vessels back to the circulatory system?

17.8 CIRCULATORY ROUTES

Blood vessels are organized in terms of their specific routes to all the tissues of the body (Figure 17.25). The right atrium moves *deoxygenated* blood through the pulmonary semilunar valve into the **pulmonary trunk**, which divides into left and right pulmonary arteries, and then to the arterioles and to the capillary beds in the lungs (Figure 17.26). *Oxygenated* blood flows back through the pulmonary venules into the right and left pulmonary veins.

Blood is then pumped from the left ventricle to the aorta, where all other systemic arteries branch off (Figure 17.27). Blood flows from the aorta, which is the largest artery in the body. The main regions of the aorta are the **ascending aorta**, the aortic arch, the thoracic aorta, and the abdominal aorta—also called the descending aorta. From the aorta, major arteries branch to the upper and lower body. The right and left coronary arteries branch off from the ascending aorta and supply the heart muscle.

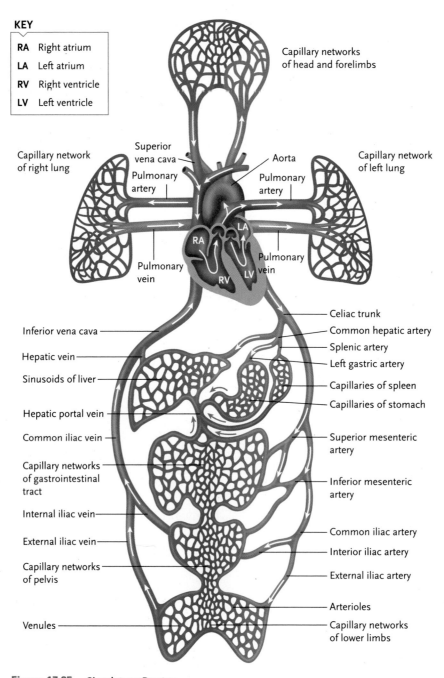

KEY

RA	Right atrium
LA	Left atrium
RV	Right ventricle
LV	Left ventricle

Capillary networks of head and forelimbs

Capillary network of right lung

Superior vena cava

Pulmonary artery

Aorta

Pulmonary artery

Capillary network of left lung

Pulmonary vein

Pulmonary vein

Inferior vena cava

Hepatic vein

Sinusoids of liver

Hepatic portal vein

Common iliac vein

Capillary networks of gastrointestinal tract

Internal iliac vein

External iliac vein

Capillary networks of pelvis

Venules

Celiac trunk

Common hepatic artery

Splenic artery

Left gastric artery

Capillaries of spleen

Capillaries of stomach

Superior mesenteric artery

Inferior mesenteric artery

Common iliac artery

Interior iliac artery

External iliac artery

Arterioles

Capillary networks of lower limbs

Figure 17.25   **Circulatory Routes**

Blood vessels are organized into various circulatory routes throughout the body.

Art source: From RUSSELL/HERTZ/STARR/FENTON. *Biology*, 2E. © 2013 Nelson Education Ltd. Reproduced by permission. www.cengage.com/permissions

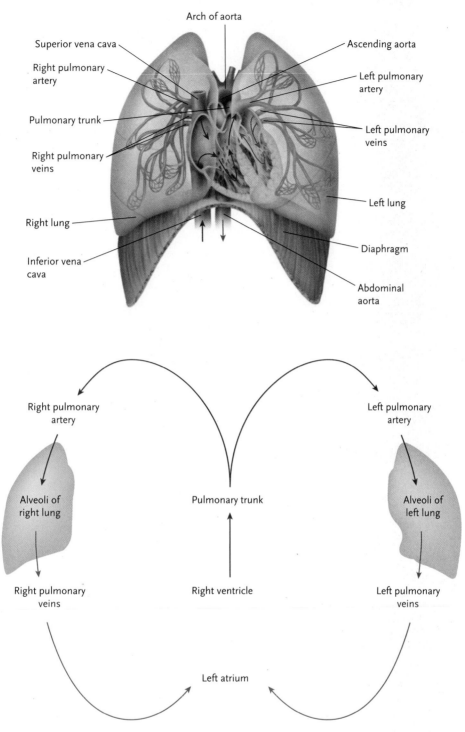

Figure 17.26 Pulmonary Circulation

Pulmonary circulation allows the blood to pick up oxygen and remove carbon dioxide in the lungs.

From the aortic arch, the **brachiocephalic trunk** branches to the **right subclavian artery** and the **right carotid artery**. The **left subclavian artery** and the **left carotid artery** branch directly from the aortic arch. The subclavian arteries supply the arms, and the carotid arteries supply the head and neck.

From the thoracic aorta, the bronchial arteries supply the bronchi in the lungs. The **esophageal arteries** supply the esophagus. The **intercostal arteries** supply the chest and intercostal muscles. The **superior phrenic arteries** supply the superior region of the diaphragm.

From the **abdominal aorta**, the inferior phrenic arteries supply the inferior region of the diaphragm. Blood flows to the digestive organs through the **celiac trunk**, and the **superior and inferior mesenteric arteries**. The celiac trunk branches into the **common hepatic**

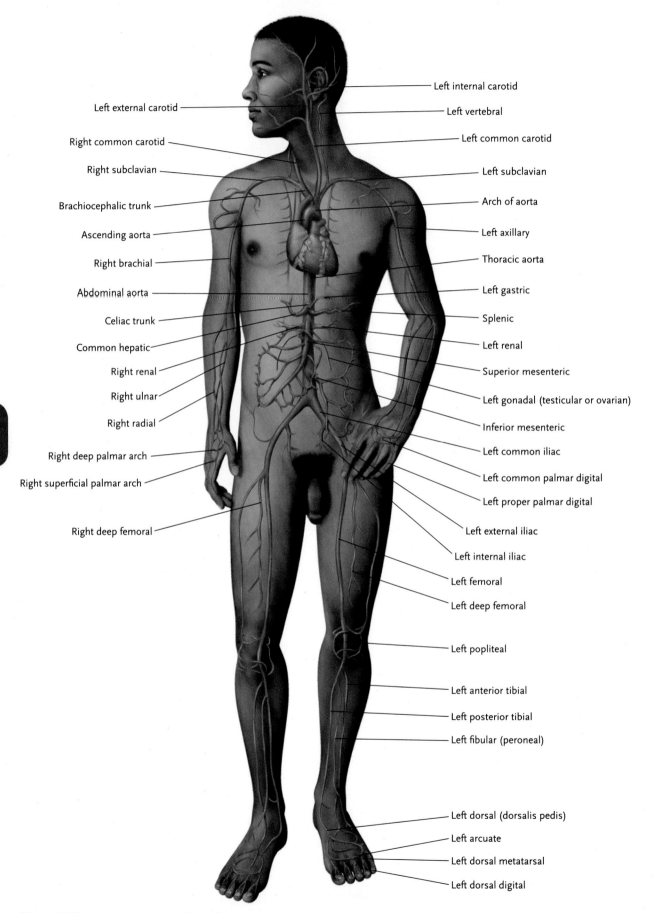

Left internal carotid

Left external carotid

Left vertebral

Right common carotid

Left common carotid

Right subclavian

Left subclavian

Brachiocephalic trunk

Arch of aorta

Ascending aorta

Left axillary

Right brachial

Thoracic aorta

Abdominal aorta

Left gastric

Celiac trunk

Splenic

Common hepatic

Left renal

Right renal

Superior mesenteric

Right ulnar

Left gonadal (testicular or ovarian)

Right radial

Inferior mesenteric

Right deep palmar arch

Left common iliac

Right superficial palmar arch

Left common palmar digital

Left proper palmar digital

Left external iliac

Right deep femoral

Left internal iliac

Left femoral

Left deep femoral

Left popliteal

Left anterior tibial

Left posterior tibial

Left fibular (peroneal)

Left dorsal (dorsalis pedis)

Left arcuate

Left dorsal metatarsal

Left dorsal digital

Figure 17.27 Major Arteries of the Body

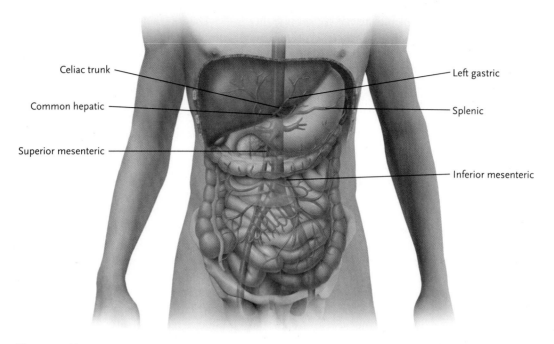

Figure 17.28   Abdominal Arteries

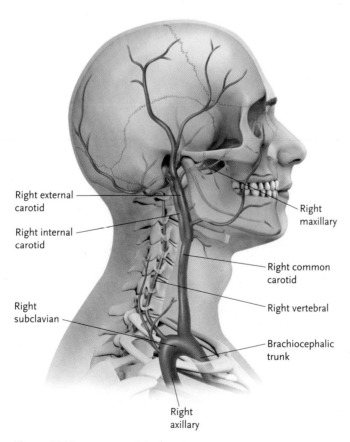

Figure 17.29   Arteries of the Head and Neck

artery to supply the liver, stomach, duodenum, and pancreas; the **left gastric artery** to supply the stomach and esophagus; and the splenic artery to supply the **spleen**, stomach, and pancreas (Figure 17.28). The superior mesenteric arteries supply the small intestine, pancreas, and the ascending and transverse colon. The inferior mesenteric arteries supply the transverse and descending colon, as well as the rectum. All blood that flows through the gastrointestinal tract circulates to the liver through the **hepatic portal vein** before it flows to the inferior vena cava.

Also branching from the abdominal aorta are the **renal arteries** that supply the kidneys, the **gonadal arteries** that supply the ovaries or testes, and the **common iliac arteries**. The common iliac arteries branch into the internal and external arteries. The **internal iliac artery** supplies the uterus or prostate, gluteal muscles, and the bladder. The **external iliac artery** supplies the lower limbs.

The right common carotid artery branches from the brachiocephalic trunk, and the left common carotid branches directly from the aortic arch. The common carotid arteries branch into the internal and external carotid arteries that supply the brain,

and these branch off to arteries that supply the face (Figure 17.28).

The internal carotid arteries supply structures inside the skull, brain, eyes, and ears. The inferior surface of the brain has an arrangement of blood vessels called the **circle of Willis** (cerebral arterial circle). These vessels include the internal carotid arteries, the anterior and posterior cerebral arteries, and the anterior communicating artery (Figure 17.29). The circle of Willis regulates the blood pressure inside the brain, and it also provides redundant or excess circulatory routes so that the brain can still receive blood flow if some vessels are damaged (Figure 17.30).

The right brachiocephalic trunk branches to the right subclavian artery, and the left subclavian artery branches directly from the aortic arch. The subclavian arteries pass through the **axillary area,** and at this point the name changes to **axillary artery.** The axillary artery continues into the upper arm and becomes the **brachial artery.** This is the artery that is used to determine blood pressure using a sphygmomanometer. At the elbow, the brachial artery branches into the **radial artery** and the **ulnar artery** (Figure 17.31).

The external iliac artery becomes the femoral artery in the thigh; this becomes the **popliteal artery** once it

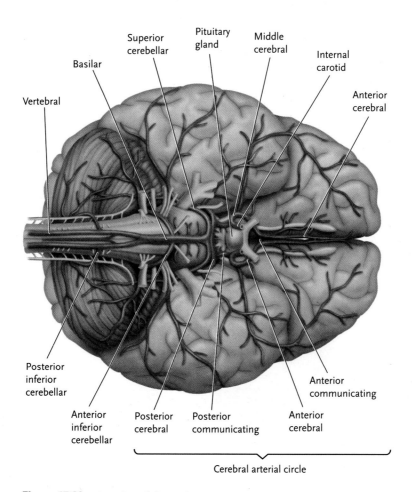

Figure 17.30 **Arteries of the Brain**

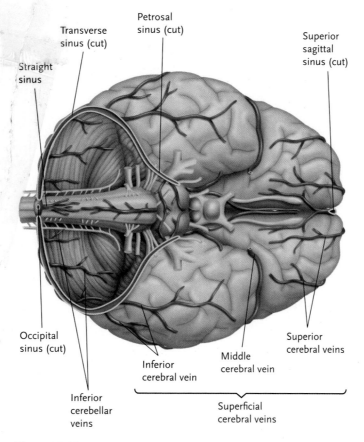

Figure 17.34   Veins of the Brain

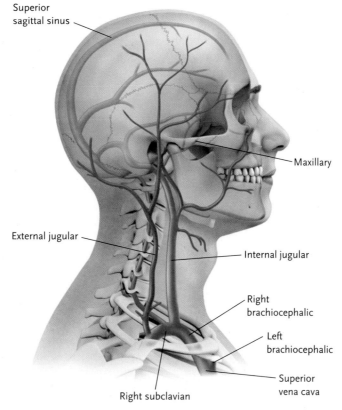

Figure 17.35   Veins of the Head and Neck

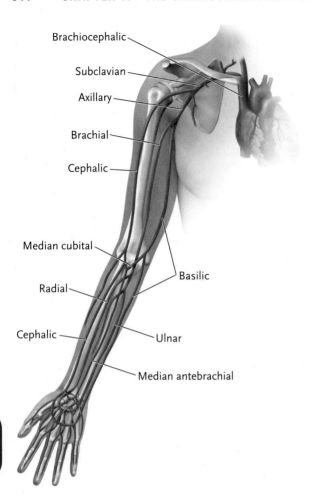

Brachiocephalic
Subclavian
Axillary
Brachial
Cephalic
Median cubital
Basilic
Radial
Cephalic
Ulnar
Median antebrachial

Figure 17.36 Veins of the Upper Limb

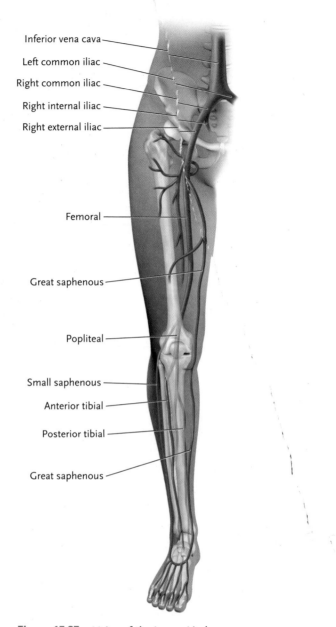

Inferior vena cava
Left common iliac
Right common iliac
Right internal iliac
Right external iliac
Femoral
Great saphenous
Popliteal
Small saphenous
Anterior tibial
Posterior tibial
Great saphenous

Figure 17.37 Veins of the Lower Limb

and the **basilica vein** in the forearm, to the **median cubital vein** in the elbow region. This is the most common vein used for inserting an IV or drawing blood, but any of the superficial veins can be used. The cephalic vein continues through the upper arm and along with the median cubital vein, and drains into the axillary vein.

The lower limb also carries blood through deep and superficial veins. The deep veins include the anterior and posterior tibial veins, and the popliteal and femoral veins—continuous with popliteal veins but located in the thigh. The femoral veins drain into the external iliac veins, then the common iliac, and then the inferior vena cava. The superficial veins include the **small saphenous vein** and the **great saphenous vein**. The small saphenous vein drains into the popliteal vein. The great saphenous vein is one of the longest in the body, extending the length of the lower limb and draining into the femoral veins (Figure 17.37).

Blood that flows through the organs of the digestive tract pick up water and nutrients that have been absorbed mainly by the small and large intestines. This nutrient-rich blood first circulates to the liver before continuing to the inferior vena cava and back to the heart. It is very important that the blood flows from the digestive organs through the hepatic portal vein to

DID YOU KNOW?

The great saphenous vein can be removed and used to replace coronary arteries during coronary bypass surgery.

the liver so that the liver can process and store certain nutrients (Chapter 19). Blood flows from the tissue capillaries to the splenic and mesenteric veins, to the hepatic portal vein, and then to the liver. Once nutrients and molecules have been processed by the liver, blood flows through the hepatic vein to the inferior vena cava (Figure 17.38).

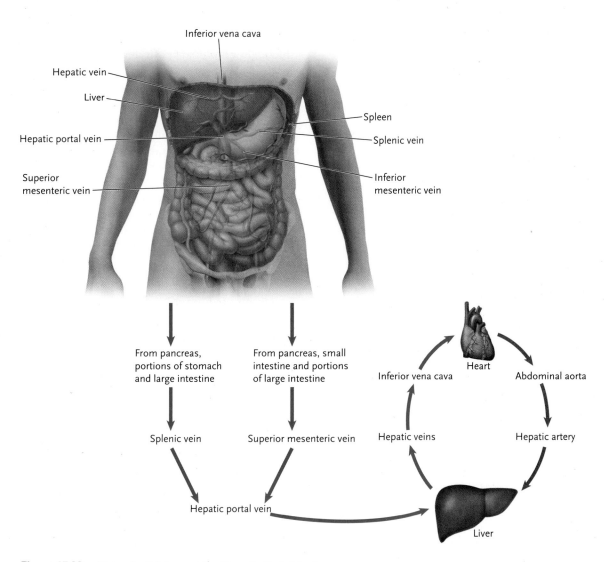

Figure 17.38 Abdominal Veins and the Hepatic Portal System

CONCEPT REVIEW 17.8

1. Why is the circle of Willis important?

2. How does the CSF drain from the ventricles of the brain into the circulatory system?

3. Why is the hepatic portal system important?

17.9 VASCULAR RESISTANCE AND BLOOD PRESSURE

Three factors relate to the movement of blood through our blood vessels: flow, pressure, and resistance. **Flow** is the amount of blood volume; **pressure** is the force of that blood volume on the walls of the blood vessels; and resistance is the force that opposes blood flow. The following equation is used to determine how these three factors relate:

flow = change in pressure / resistance

Flow increases if the pressure increases or if the resistance decreases. When the heart contracts strongly, pressure and flow increase. When the blood vessels in a tissue dilate, more blood flows into that tissue.

Blood always flows from areas of higher to lower pressure. The highest pressure is in the aorta because of the contraction of the heart. As blood flows from arteries to arterioles to capillaries, the pressure and flow continue to decrease (Figure 17.39).

Resistance is largely determined by the *diameter* of the blood vessels. As blood moves from arteries to arterioles, the vessel diameter decreases; therefore,

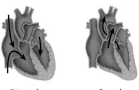

Diastole   Systole

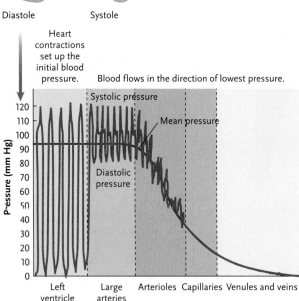

Figure 17.39   Blood Pressure Changes in the Vascular System

Blood always flows from areas of higher pressure to areas of lower pressure, which ensures that blood always flows from arteries to arterioles to capillaries, and then to venules, veins, and finally the vena cavae, which have the lowest pressure in the body.

Art source (top): From DIGIUSEPPE/FRASER. *Biology* 11 U. © 2011 Nelson Education Ltd. Reproduced by permission. www.cengage.com/permissions

resistance increases, causing flow to decrease. Recall that arteries and arterioles branch into many vessels that are progressively smaller, and the capillaries have the largest cross-sectional area. As blood flows from the aorta to the capillaries, the resistance increases in the smaller vessels (more vessel wall surface area to oppose blood flow), the pressure decreases, and the flow decreases in each individual vessel (Figure 17.40).

DID YOU KNOW?

When a person gains a lot of body weight, their blood vessel growth increases to supply oxygen and nutrients to the new tissue. This increase in total blood vessel length increases vascular resistance, and this is one reason why obesity contributes to cardiovascular disease.

DID YOU KNOW?

Gaining one extra pound of body fat leads to the growth of approximately 15 km of new blood vessels.

Factors that affect blood flow

Blood flow is primarily affected by changes in blood vessel resistance, which can occur due to vasoconstriction or vasodilation of arteries and arterioles. Vasoconstriction increases resistance and therefore decreases blood flow. Vasodilation decreases resistance and therefore increases blood flow. Changes in resistance, specifically due to constriction or dilation of blood vessels, are the primary means of regulating blood flow to specific tissues.

Other factors that influence blood flow include (1) metabolic requirements of tissues, (2) body temperature, (3) blood volume, (4) **blood viscosity**, (5) cardiovascular disease, and (6) medications.

1. Metabolic requirements. Tissues that need an increase in oxygen and nutrients require greater blood flow. For example, during exercise the skeletal muscles require more blood flow than the digestive system. Blood vessels carrying blood to the skeletal muscle dilate (decreasing resistance), and blood vessels leading to the digestive tract constrict (increasing resistance).

2. Body temperature. When our body temperature increases above 37°C, a negative feedback mechanism takes effect (Chapter 1). Thermoreceptors detect an increase in temperature; this information is carried to the hypothalamus (integrating centre); and the hypothalamus signals effectors, such as sweat glands and blood vessels. When our body needs to lose heat, the blood vessels in the skin dilate so more blood flows to the surface of the body and heat can be removed. When our body temperature decreases, blood vessels constrict and heat is retained within the body.

3. Blood volume. An increase in blood volume increases blood flow to all areas of the body because the increased blood volume increases cardiac output. Anything that increases cardiac output increases blood flow (Section 17.6). Blood volume fluctuates normally within a small range due to the dietary intake of fluids and ions, mainly salt. In healthy people the kidneys quickly regulate blood volume by excreting or retaining excess water and ions. People with high blood pressure need to closely monitor the amount of salt in their diet (see Section 17.10 regarding hypertension).

4. Blood viscosity. Viscosity refers to the thickness of the blood. Increased viscosity leads to increased resistance; thicker blood takes more pressure to move through the blood vessels. Our blood is composed of cells and plasma—the water portion that contains dissolved ions and nutrients. The viscosity is determined primarily by the amount of red blood cells (hematocrit level). Under normal circumstances the hematocrit level does not change, and this is why blood vessel diameter is the primary factor of

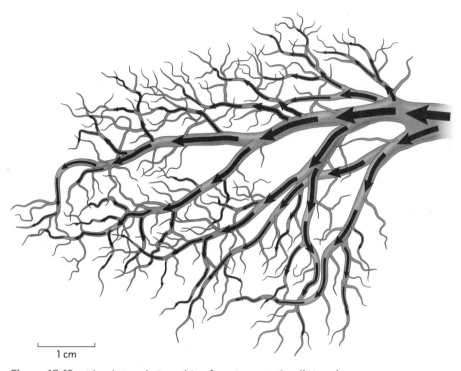

1 cm

Figure 17.40 Blood Vessels Branching from Large to Small Vessels

Resistance increases as the vessels become smaller, but pressure and flow decrease as blood moves from arteries to arterioles to capillaries because of the increasing cross-sectional area of the smaller vessels.

DID YOU KNOW?

Athletes have been known to engage in "blood doping," which means they take the hormone erythropoietin to stimulate the bone marrow to increase production of red blood cells (Chapter 13). This lets them carry more oxygen and, therefore, increase their endurance ability. The problem with blood doping is that excessive production of red blood cells increases the viscosity of the blood, which increases resistance and puts a lot of strain on the cardiovascular system. Many athletes have died from heart attack or stroke because the blood is simply too thick to circulate properly.

A. Laminar flow

B. Turbulent flow

Figure 17.41 Laminar Flow and Turbulent Flow

Laminar flow is the smooth healthy flow of blood through vessels. Turbulent flow occurs when vessel abnormalities, such as accumulation of plaque, cause blood to flow in a chaotic manner.

resistance. Anemia (decreased number of red blood cells) can also decrease resistance.

5. Cardiovascular disease. Atherosclerosis is the primary cause of cardiovascular disease, which is the buildup of plaque (mostly fat and immune cells) within the walls of the blood vessels. When this happens, the inside of the blood vessel walls are no longer smooth and uniform, and this affects how blood flows through a vessel. When the vessel is healthy and smooth inside the lumen, blood flows evenly: laminar flow (Figure 17.41). With atherosclerosis, the uneven surface of the lumen of a blood vessel causes the blood to flow in a random pattern, turbulent flow, which increases resistance and

decreases flow to that tissue. Atherosclerosis can also cause clot formation, leading to heart attack or stroke because turbulent blood flow activates platelets.

6. Medications that affect blood flow in turn affect blood volume, vascular resistance, or blood pressure. Diuretics are a class of medications that decrease blood volume and therefore decrease blood pressure by increasing water excretion by the kidneys. Vasodilators decrease vascular resistance by causing vasodilation and therefore decrease blood pressure.

CONCEPT REVIEW 17.9

1. How is blood flow related to blood pressure and blood vessel resistance?

2. What are the main factors that increase resistance?

3. What are six important factors that affect blood flow?

4. How does temperature affect blood flow?

5. How is resistance affected by an increased hematocrit level?

6. Explain laminar and turbulent blood flow and how cardiovascular disease can cause turbulent flow.

7. What is the effect of diuretics and vasodilators on blood pressure?

17.10 BLOOD PRESSURE REGULATION

Blood pressure in the arteries is higher during **systole** (contraction) and lower during **diastole** (relaxation). A typical, healthy blood pressure is 120/80 mmHg, where 120 is the systolic pressure and 80 is the diastolic pressure. Blood pressure and blood flow in the body is regulated by changes in heart rate, stroke volume, blood volume, and vascular resistance (vasoconstriction and vasodilation). Blood flow and pressure in various tissues depends on the metabolic needs of that tissue; during exercise, more blood is diverted to the skeletal muscles; after eating a meal, more blood is diverted to the gastrointestinal system. All the factors that increase cardiac output also increase blood pressure (Figure 17.13).

The body uses negative feedback mechanisms to regulate blood pressure. The cardiovascular centre is in the medulla oblongata in the inferior region of the brainstem, and it regulates heart rate and stroke volume through the autonomic nervous system (ANS). Recall from Chapter 1 that a negative feedback loop involves sensory receptors, sensory (afferent) neurons, integrating centre (in this case, the medulla oblongata), motor (efferent) neurons, and effectors.

Receptors that detect blood pressure are called baroreceptors, which are located in the aortic arch and the carotid arteries. These are key locations since the coronary arteries branch from the aorta, and the carotid arteries supply the delicate brain tissue. Monitoring blood pressure in these locations is very important. Baroreceptors are a type of mechanoreceptor that detects how much the arteries stretch when blood pressure increases. When the stretch of these receptors increases, the rate of action potentials sent to the medulla oblongata increases. Chemoreceptors in the aortic arch and carotid arteries detect blood levels of oxygen, carbon dioxide, and hydrogen ions. When oxygen is low, carbon dioxide is high or H^+ concentration is high, and more action potentials reach the medulla oblongata (Chapter 18). Proprioceptors located in the muscles and joints also play a role in blood pressure regulation by detecting

and providing information about body position. When we stand up from a laying down position, the blood vessels in our lower body constrict to prevent blood from being pulled into the lower body and away from the brain. When we begin exercising, the movement of the body signals the medulla oblongata to increase our heart rate before that's actually needed; this is known as an **anticipatory response** that prepares the body for exercise.

The medulla oblongata sends signals through sympathetic neurons when the heart rate or force of heart contraction needs to increase and signals through the parasympathetic neurons when the heart rate needs to decrease (Figure 17.42).

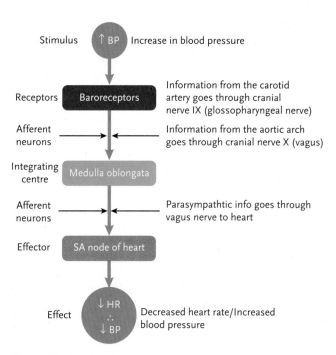

Figure 17.42 Blood Pressure Regulation

Baroreceptors detect blood pressure in the aortic arch and carotid arties. Changes in pressure are regulated by the cardiovascular centre: the medulla oblongata in the brainstem.

Information from the baroreceptors and chemoreceptors in the carotid artery travels to the medulla oblongata through afferent neurons in the **glossopharyngeal nerve**, and information from the aortic arch travels through afferent neurons in the **vagus nerve**. Information from proprioceptors travels through the peripheral sensory nervous system to the spinal cord, at which point there are synapses to neurons that signal the medulla oblongata. The cardiovascular centre in the medulla oblongata interprets the information (action potential frequency) from the sensory input. The CVS centre then sends signals through efferent neurons to the effectors.

Efferent information can travel from the medulla oblongata through sympathetic or parasympathetic neurons to the target tissues. Parasympathetic neural signals travel to the SA node of the heart through the vagus nerve to decrease heart rate. Sympathetic system information travels through the sympathetic spinal nerves to the SA node and ventricles of the heart to increase heart rate and increase the force of contraction of the ventricles. Recall from Chapters 15 and 16 that preganglionic neurons of the sympathetic nervous system can target the adrenal medulla to produce epinephrine hormones that enter the bloodstream. The hormonal response is slightly slower than the response of the nervous system, but it targets many different cell types. Efferent signals also target many different blood vessels, primarily arterioles, to constrict or dilate.

The primary effectors involved in blood pressure regulation include the SA node and ventricle myocardium of the heart, blood vessels, and the kidneys (targeted by hormones). The following factors are involved in blood pressure regulation:

When blood pressure increases,

- parasympathetic stimulation decreases heart rate,
- a lack of sympathetic stimulation decreases heart rate and the force of ventricular contraction,
- parasympathetic stimulation causes vasodilation, and
- the kidneys excrete more water.

When blood pressure decreases,

- sympathetic stimulation increases heart rate and the force of ventricular contraction;
- sympathetic stimulation causes veins to constrict, which increases venous return, cardiac output, and blood pressure;
- sympathetic stimulation causes vasoconstriction; and
- the kidneys reabsorb more water.

Hormones involved in regulating blood pressure

Antidiuretic hormone, released by the hypothalamus, increases blood volume and therefore blood pressure by signalling the distal tubule and collecting duct of the kidney to increase water reabsorption.

Aldosterone, released by adrenal cortex, increases blood volume through the reabsorption of sodium, and therefore water by osmosis, in the distal tubule.

Epinephrine and norepinephrine, released by adrenal medulla, increases blood pressure by increasing vascular resistance through vasoconstriction.

Angiotensin II, formed by the renin-angiotensin system (RAS), increases blood pressure by stimulating the production of aldosterone and by causing vasoconstriction.

Atrial natriuretic peptide (ANP), released by the atria of the heart when blood pressure is too high, causes the kidneys to excrete sodium and, therefore, lowers blood volume and blood pressure; it also lowers blood pressure by causing vasodilation.

CONCEPT REVIEW 17.10

1. What is a normal, healthy, blood pressure?
2. Name the major components of a negative feedback loop.
3. What is the function of the three main types of receptors involved in detecting changes in blood pressure?
4. Explain how sensory information travels from the carotid arteries and aortic arch to the brain.
5. Where is the cardiovascular centre?
6. How is the sympathetic nervous system involved in increasing blood pressure?
7. Name each important hormone that is involved in regulating blood pressure. Make a chart to show where each hormone is produced and how it affects blood pressure.

17.11 CARDIOVASCULAR DISEASE

Cardiovascular disease is the leading cause of death in North America, and it is almost completely preventable. Cardiovascular disease is caused by a buildup of fats, minerals, and immune cells (**plaque**) inside the walls of arteries. This buildup is called **atherosclerosis**. Because the immune cells are involved, cardiovascular disease is also considered a **chronic inflammatory disease**. Recall the layers of an artery wall: plaque builds up underneath the inner endothelial cell layer, not inside the lumen of the blood vessel (Figure 17.43). Plaque is primarily made up of **low-density lipoprotein (LDL)**, which is a cholesterol-protein molecule produced by the liver to transport dietary triglycerides from the liver to tissues. LDLs are called "bad" blood cholesterol, as compared to **high-density lipoprotein (HDL)**, known as the "good" blood cholesterol. HDLs also transport fats through the bloodstream, but they carry fat from the adipose tissue back to the liver to be broken down for energy. HDLs can remove fat from blood vessel walls when the fat is needed to produce energy; therefore, HDL levels can be increased by exercising.

Recall from Chapter 3 that molecules that have both a hydrophilic and hydrophobic component are called

TABLE 17.1

Blood Pressure and Hypertension

Category	Systolic (mmHg)	Diastolic (mmHg)
Healthy	less than 120	less than 80
Prehypertension	121–139	81–89
Hypertension	140–159	90–99
Severe hypertension	160 or higher	100 or higher

amphipathic. HDLs and LDLs are **amphipathic** so that they can bind to the triglycerides as well as circulate through the water-based blood. Once LDLs migrate into the artery wall they stay there and begin to oxidize. This causes tissue damage, which stimulates an immune response. The immune cells, mostly phagocytic macrophages, then migrate into the wall of the artery. As plaque increases, the diameter of the artery lumen decreases, which increases resistance for two reasons: (1) the lumen becomes narrowed, (2) an uneven endothelial layer causes turbulent blood flow (Figure 17.41). As resistance increases, overall blood pressure increases, resulting in high blood pressure, known as **hypertension (Table 17.1)**. Atherosclerosis is the underlying cause of most cardiovascular disease, including heart attack (myocardial infarction), stroke, congestive heart failure, pulmonary embolism, deep vein thrombosis, and kidney failure. Atherosclerosis can progress for 10 to 20 years without symptoms.

Risk factors for cardiovascular disease

The exact causes of cardiovascular disease are not known, but the following risk factors are highly linked to the condition:

- High blood LDL and triglyceride levels
- Low blood HDL levels
- Obesity, particularly abdominal fat, which is associated with high LDL levels
- Diet, particularly trans fats in many processed foods, which increase blood LDLs more than any other type of fat
- Lack of physical activity, which contributes to obesity and low HDLs
- Type-1 or type-2 diabetes, which doubles the risk of heart attack or stroke
- High blood levels of **c-reactive protein**, which is a strong indicator of chronic inflammation
- High blood homocysteine levels
- Smoking, which increases blood pressure
- Stress, possibly linked to behaviours such as smoking, drinking, and overeating, although no evidence that stress alone causes cardiovascular disease
- Menopausal women, related to decreased estrogen, although no higher risk for women with a healthy lifestyle

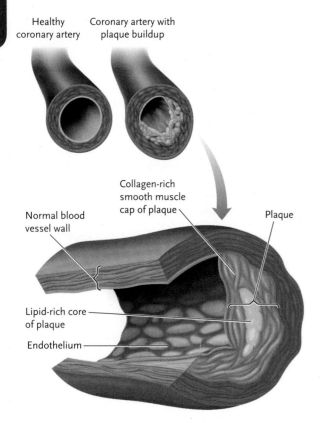

Healthy coronary artery   Coronary artery with plaque buildup

Collagen-rich smooth muscle cap of plaque

Normal blood vessel wall

Plaque

Lipid-rich core of plaque

Endothelium

Figure 17.43   **Atherosclerosis**

Atherosclerosis is caused by the buildup of plaque within the wall of the large vessels.

Art sources: (Top) From DIGIUSEPPE/FRASER. *Biology* 11 U. © 2011 Nelson Education Ltd. Reproduced by permission. www.cengage.com/permissions; (bottom) from SHERWOOD/ KELL. *Human Physiology*, 2E. © 2013 Nelson Education Ltd. Reproduced by permission. www.cengage.com/permissions

- Excessive alcohol consumption, more than two drinks per day for women, more than three drinks per day for men
- Genetic predisposition
- Ethnicity, with greater risk of high blood pressure and diabetes among First Nations people and those of African or South Asian descent

Metabolic syndrome

According to the International Diabetes Federation, people with **metabolic syndrome** have significantly increased risk of death from heart attack or stroke. A person is considered to have metabolic syndrome if they have central obesity (abdominal fat)—in females, more than 35 inch waist; in males, more than 40 inch waist—as well as two of the following:

- high blood triglycerides
- low HDLs
- high resting blood pressure—systolic > 130 mmHg, or diastolic > 85 mmHg
- high fasting blood glucose—greater than 5.6 mmol/L, or diagnosed diabetes

Complications of atherosclerosis

Myocardial infarction

Atherosclerosis or a blood clot in the coronary arteries that supply the heart muscle leads to **ischemia**, which is decreased blood flow to the myocardium. If one or more coronary arteries become fully blocked, areas of the heart muscle do not receive oxygen (**hypoxia**), can't make ATP, and so cannot contract. Typical symptoms of a heart attack include severe squeezing chest pain that may radiate down the left arm, nausea, weakness, and difficulty breathing. Often people with advanced atherosclerosis have **angina** for a period of time before they have a heart attack. Angina involves bouts of chest pain that diminish after several minutes, and it is a sign that significant artery blockage exists.

Stroke

There are two types of strokes: (1) **ischemic**, where the blood supply to areas of the brain has been decreased by a blockage or a clot that has travelled from another area of the body (Figure 17.44a); (2) **hemorrhagic**, where capillaries have broken and blood flows out of the circulatory system into the brain, putting pressure on the brain tissue, and preventing normal blood circulation from reaching other brain areas (Figure 17.44b).

Symptoms of stroke are often noticed on one side of the body since the left side of the brain communicates with the right side of the body and the right side of the brain communicates with the left side of the body. A person's symptoms depend on the location in the brain where the damage has occurred, but common symptoms include

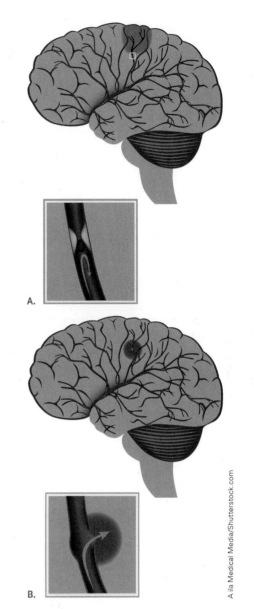

Figure 17.44 Types of Stroke
(A) An ischemic stroke, usually caused by a clot, occurs because there is a lack of blood flow and, therefore, oxygen to a part of the brain. (B) A hemorrhagic stroke is caused by damaged blood vessels.

one-sided weakness, numbness, or paralysis, slurred speech, or confusion. Sometimes people experience **transient ischemic attacks (TIAs)**, sometimes referred to as "mini strokes," which are similar to angina in that the symptoms resolve on their own; this condition indicates that significant blood vessel damage exists.

Kidney failure

The most common cause of kidney failure is chronic hypertension, also called **end stage renal disease**. Constant elevated blood pressure causes damage to the glomerular capillaries that filter blood, until eventually the kidneys cannot function to excrete waste, which builds up in the bloodstream and can quickly cause death.

Treatment for atherosclerosis

Atherosclerosis can be significantly decreased with lifestyle changes, such as eliminating trans fats from the diet and strictly limiting processed foods that contain trans fats, daily exercise, monitoring blood sugar, quitting smoking, losing weight, and managing stress. When the disease is advanced, the following types of medication may be used to decrease blood sugar level and the stress on the heart:

- **Diuretics** that increase water excretion and decrease blood volume
- **Beta blockers** that decrease the effects of epinephrine and therefore decrease heart rate and force of contraction

- **Alpha blockers** that decrease the effects of norepinephrine, which affects vascular tone and therefore decreases preload
- **Calcium channel blockers** that decrease the force of heart contraction, because calcium causes depolarization of the SA node as well as contraction in muscle fibres
- **Vasodilators** that dilate blood vessels, such as nitroglycerin that's taken to provide rapid relief from angina
- **ACE inhibitors**, which block the effects of angiotensin converting enzyme that converts angiotensin I into angiotensin II, and which leads to the production of aldosterone

CONCEPT REVIEW 17.11

1. What is atherosclerosis?
2. Explain the difference between LDLs and HDLs, and how HDLs can be increased and LDLs decreased.
3. Make a chart of risk factors and how each contributes to cardiovascular disease.
4. What are some of the major complications of long-term atherosclerosis?
5. What is metabolic syndrome?

6. What is the difference between angina and a heart attack?
7. What are TIAs?
8. What can you do to prevent cardiovascular disease?
9. Make a chart that lists the main function of the major classes of medications used to treat cardiovascular disease.

17.12 COMPONENTS OF THE BLOOD

The blood is composed of cells and plasma. The average total blood volume is 5–6 L, depending on body size. The main functions of the blood are to transport nutrients, oxygen, ions, and hormones to the body's cells and to transport waste to the liver and kidneys for excretion. The blood is also involved in regulating body temperature. As more blood flows to the surface of the body, more heat can be removed. Blood is also involved in protecting the body from infections because it contains white blood cells and antibodies that fight bacterial, viral, parasitic, and fungal infections (Chapter 23). Blood contains platelets that form clots when there is tissue injury.

Plasma

The plasma consists of water, ions, vitamins, nutrients (glucose, fatty acids, amino acids), hormones, plasma proteins, some dissolved oxygen and carbon dioxide, and waste products. Plasma makes up approximately 55% of the total blood volume (Figure 17.45). Plasma is mostly water—approximately 92% of the plasma. The most important ions dissolved in the blood plasma include Na^+, K^+, Ca^{+2}, HCO_3^-, H^+, Mg^{+2}, and Cl^-. These ions are collectively called **electrolytes,** and they must

be in balance to maintain the functioning of all cells, specifically the nervous system.

Some important plasma proteins include albumin, which is the main protein that creates the osmotic gradient that keeps fluid in the circulatory system; it

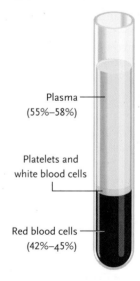

Plasma
(55%–58%)

Platelets and
white blood cells

Red blood cells
(42%–45%)

Figure 17.45   Components of Blood

Blood is composed of cells and plasma, and red blood cells make up 45% of the total blood volume.

makes up approximately 54% of all plasma proteins. **Antibodies,** also called *immunoglobulins*, are produced by specific white blood cells—called B cells—that can bind to specific pathogens during an infection. Complement is a plasma protein produced by the liver and also an important component of the immune response. **Fibrinogen** is essential for the formation of blood clots. Blood **serum** is a term that describes blood plasma without fibrinogen, and it is removed by allowing clotting. **Lipoproteins** are produced by the liver and are composed of proteins and cholesterol; they function to transport most fats through the bloodstream. Hormones make up approximately 1% of the plasma proteins (some hormones are lipid soluble). They are transported by means of the circulatory system from the organ where it is produced to its target cell type (Chapter 16).

Important blood values

Blood testing is an important way for doctors to determine our state of health. The following are some of the most common blood values:

> **Blood pH**—7.36–7.41. Venous blood is slightly more acidic than this because of higher carbon dioxide levels. Blood pH is extremely tightly regulated by the respiratory system and kidneys.
> **Blood glucose**—4.5–6.5 mmol/L. Blood sugar fluctuates during the day depending on when and what we eat.
> Blood cholesterol (healthy levels)
> - LDLs (low density lipoproteins, "bad" cholesterol)—lower than 3.3 mmol/L
> - HDLs (high density lipoproteins, "good" cholesterol)—higher than 1.0 mmol/L
> - Triglycerides—lower than 2.2 mmol/L
>
> **Homocysteine**—lower than 7.2 μmol/L. High levels of this amino acid are associated with increased risk of heart disease.

Cells

Platelets, also called thrombocytes, are *fragments* of cells, and they are produced in the bone marrow. **Red blood cells,** also called **erythrocytes,** make up most of the remaining 45% of the total blood volume, and this is called the hematocrit level. Red blood cells contain a protein called hemoglobin that consists of four proteins plus iron and carries oxygen from the lungs to the tissues. Red blood cells are produced in the bone marrow by a process called hematopoiesis. During **hypoxia,** when oxygenation of cells decreases, the kidneys secrete the hormone **erythropoietin,** which stimulates hematopoiesis.

As red blood cells mature they lose their nucleus, so there is more room for hemoglobin-iron complexes to carry oxygen. A red blood cell lives for approximately 120 days and then is destroyed by the spleen or the liver.

When the red blood cells are broken down, the amino acids are recycled and used to make other proteins, and much of the iron is recycled back to the bone marrow to make new red blood cells. The **heme** portion of the red blood cells is broken down, and **bilirubin** is formed. Bilirubin circulates to the liver and is added to the **bile** and then excreted into the small intestine when the gallbladder releases bile. Because of the bright yellow colour of bilirubin, if the liver is not functioning properly, the bilirubin stays in the body, giving the skin a yellow appearance: called **jaundice.**

DID YOU KNOW?

Vitamin B_{12} is essential for the production of red blood cells. A molecule called *intrinsic factor* is produced by the stomach and allows B_{12} to be absorbed from the intestine into the blood. B_{12} is found mostly in animal products (meat, fish, dairy, and eggs), and it is common for inexperienced vegetarians to become B_{12} deficient, causing anemia. People who cannot produce intrinsic factor have pernicious anemia. Also, excessive alcohol intake inhibits the absorption of B_{12} and can cause anemia, fatigue, depression, weakness, and poor memory.

DID YOU KNOW?

Alcohol, caffeine, the tannins in red wine and grapes, and phytic acid in grains inhibit the absorption of iron. Vitamin C increases the absorption of iron. A lack of iron causes iron-deficiency anemia.

White blood cells, also called **leukocytes,** are cells involved in the immune response. Like red blood cells, white blood cells are produced in the bone marrow through the process of hematopoiesis. White blood cells can be divided into two categories: **granular leukocytes** and **agranular leukocytes.** When stained and viewed under a microscope, granular leukocytes have visible cytoplasmic vesicles that contain enzymes that digest microorganisms.

Granular leukocytes

> **Neutrophils,** the most abundant type of white blood cells, are most often the first cell type at an area of infection. They secrete chemicals that neutralize the infected area, primarily killing bacteria with enzymes and oxidants, such as hydrogen peroxide.
> **Basophils,** the least common of the white blood cells, release histamine during allergic reactions.

Eosinophils, are involved in allergic reactions, as well as fighting multicellular parasites, such as worms.

Natural killer cells are large granular lymphocytes, that directly kill virus-infected body cells and cancer cells, and they play important roles in both the innate and adaptive immune response.

Agranular leukocytes

Monocytes, the largest of the white blood cells, migrate from the bloodstream to tissues and then differentiate into mature **macrophages** and **dendritic cells,** which are very important antigen-presenting cells.

B cells are lymphocytes that are involved in the adaptive immune response, respond to specific antigens, and differentiate into plasma cells that secrete antibodies.

T cells are lymphocytes that are involved in the adaptive immune response, respond to specific antigens, and differentiate into helper T cells and cytotoxic T cells.

CONCEPT REVIEW 17.12

1. What are the four main functions of the blood?

2. What are electrolytes?

3. What are the main functions of plasma proteins?

4. What is the role of platelets?

5. What is a healthy hematocrit level? How does a person feel if they have a low hematocrit level?

6. What is hematopoiesis, and what hormone stimulates this process?

7. What mineral is important for the production of red blood cells?

8. What happens to the cell components in old, worn out erythrocytes?

9. Describe the main functions of each type of leukocyte.

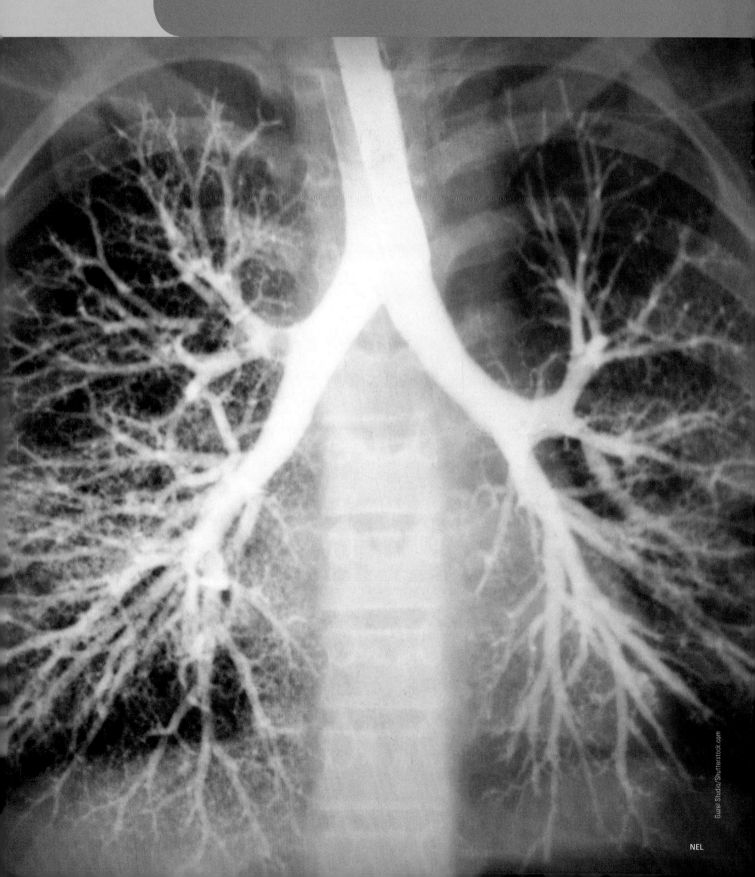

I have not failed. I've just found 10,000 ways that won't work.

Thomas A. Edison

18.1 OVERVIEW OF THE RESPIRATORY SYSTEM

The respiratory system is closely connected with the cardiovascular system and in combination these systems take in and transport oxygen to all tissues in the body and remove carbon dioxide. Structures in the respiratory system include the nose, mouth, pharynx, larynx, trachea,

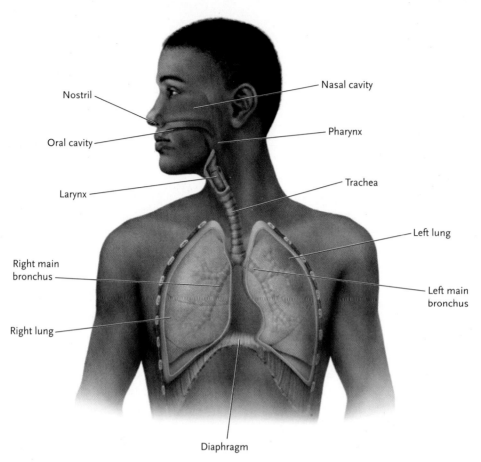

Figure 18.1 Structures of the Respiratory System

Structures in the respiratory system include the nose, mouth, pharynx, larynx, trachea, bronchi, bronchioles, and alveoli (air sacs).

bronchi, bronchioles, and alveoli (air sacs) (Figure 18.1). The movement of air into and out of the lungs is called **ventilation**. Exchange of gases in the alveoli is **respiration**. Breathing in is called **inhalation** or inspiration, and breathing out is called **exhalation** or expiration. Once oxygen has circulated to the tissues and is used by cells to make ATP, that process is called cellular respiration (Chapter 4).

The respiratory system has the following functions in the body:

- Taking in oxygen and removing carbon dioxide
- Regulating blood pH by removing carbon dioxide
- Warming and moistening inhaled air
- Filtering particles from inhaled air
- Providing a sense of smell
- Producing sound by moving air past the vocal cords

CONCEPT REVIEW 18.1

1. What are the functions of the respiratory system?

18.2 ANATOMY OF THE RESPIRATORY SYSTEM

The respiratory system can be divided *structurally* into the upper and lower respiratory system. The upper structures include the nose, mouth, pharynx, and larynx, and the lower structures include the trachea, bronchi, bronchioles, and alveoli. The respiratory system can be divided *functionally* into the **conducting zone** and the **respiratory zone** (Figure 18.1). The nose, mouth, pharynx, larynx, trachea, bronchi, bronchioles, and terminal bronchioles make up the conducting zone; these structures are involved in the movement of air

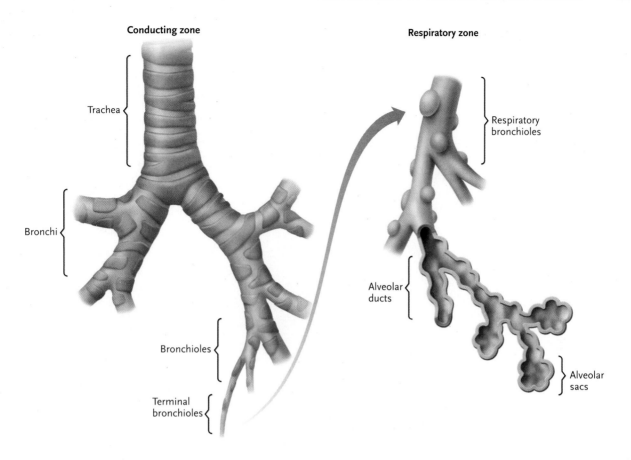

Conducting zone

Trachea

Bronchi

Bronchioles

Terminal bronchioles

Respiratory zone

Respiratory bronchioles

Alveolar ducts

Alveolar sacs

Figure 18.2   **Functional Divisions of the Respiratory System**
The conducting zone includes structures that transport air to the respiratory zone, which is where gas exchange occurs.

18

into and out of the lungs. The respiratory bronchioles, alveolar ducts, and alveoli make up the respiratory zone, which is where gas exchange occurs.

The nose and mouth are the entrances to the respiratory system (Figure 18.2). The external nose is made up of the nasal bones, hyaline cartilage, and skin, and its two openings, the external nares (nostrils), lead into the nasal cavity. The nasal cavity is separated by a septum, which is a perpendicular plate of the ethmoid bone. Air flows through the nasal cavity, which consists of the superior, middle, and inferior **nasal conchae**—also called **turbinates**—and through the frontal, ethmoid, maxillary, and sphenoid sinuses. As air flows through these passageways, which are lined with mucus membranes, it is warmed by heat transferred from the blood vessels and filtered by the hairs and cilia that line the upper airways. The mucus membrane of the superior conchae contains receptors for detecting scents; therefore, this membrane is also called the olfactory epithelium.

Air travels from the nasal cavity to the **pharynx** (throat). The pharynx is divided into an upper portion, called the **nasopharynx**; the middle portion where food also travels, called the **oropharynx**; and the lower portion called the **laryngopharynx**. Within the nasopharynx is the opening of the auditory tube, also called the **Eustachian tube**. This tube leads to the middle ear and is important for equalizing air pressure within the inner ear. When swallowing food, the **epiglottis** covers the opening to the larynx so that food does not enter the lower respiratory structures (Figure 18.3).

The **larynx**, also called the voice box, contains the **vocal cords** and connects the pharynx to the trachea. The mucus membrane of the larynx forms the vocal cords, and it contains elastic connective tissue that's connected to cartilage and also muscles that are under voluntary control. When these muscles contract, the vocal cords become pulled to a certain degree and produce sounds. As air flows over the vocal cords during exhalation, it causes them to vibrate, in a way similar to plucking strings on a guitar. When males produce an increased amount of testosterone during puberty, their vocal cords become thicker and longer and produce a lower vocal sound. Everyone's voice is slightly different because of tiny variations in the

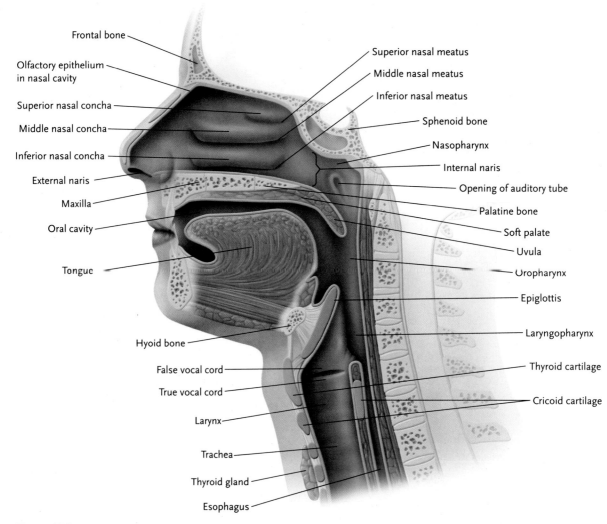

Figure 18.3 **Upper Respiratory Structures**

thickness and length of the vocal cords, as well as the shape and size of the nasal cavity, pharynx, and larynx. The larynx is protected by hyaline cartilage (also called thyroid cartilage); known as the Adam's apple, it can be palpated in the neck.

The **trachea**—or windpipe—is located anterior to the esophagus and is protected by rings of *hyaline cartilage* (Figure 18.4). The cartilage keeps the airway open and prevents the soft tissues from collapsing inward so that breathing is not obstructed. The trachea is lined with a mucus membrane composed of pseudostratified, ciliated, columnar, epithelial tissue (Chapter 11). The ciliated epithelial tissue contains **goblet cells** that produce **mucus** that protects the lungs by trapping inhaled particles and bacteria, and the cilia continually move the mucus up toward the pharynx

so it can be swallowed or coughed out of the lower airways. The lower airways also contain immune cells that fight infectious organisms that enter the lungs.

The trachea branches into the left and right **bronchi**, which enter the lungs (Figure 18.4). The trachea and bronchi have supportive cartilage rings, whereas the smaller conducting zone and respiratory zone structures do not have. The bronchi branch into many smaller and smaller **bronchioles**. The smallest conducting zone bronchioles are called **terminal bronchioles**, which lead to the first of the respiratory structures. Gas exchange occurs in the respiratory structures, which include the **respiratory bronchioles**, the **alveolar ducts**, and the **alveoli**.

The lungs are divided into the right and left side, and each side is divided into **lobes**. The right lung contains

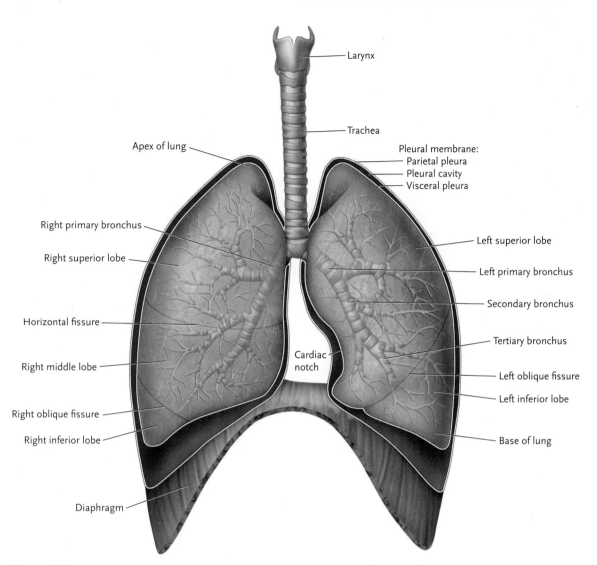

Larynx

Trachea

Apex of lung

Pleural membrane:
Parietal pleura
Pleural cavity
Visceral pleura

Right primary bronchus

Right superior lobe

Left superior lobe

Left primary bronchus

Secondary bronchus

Horizontal fissure

Right middle lobe

Cardiac notch

Tertiary bronchus

Left oblique fissure

Left inferior lobe

Right oblique fissure

Right inferior lobe

Base of lung

Diaphragm

Figure 18.4 **Lower Respiratory Structures**

three lobes (Figure 18.3). The left lung has two lobes and the **cardiac notch,** where the heart is located. On the right side, the superior and middle lobes are separated by the **horizontal fissure,** and the middle and inferior lobes are separated by the **oblique fissure**. On the left side the left oblique fissure separates the superior and inferior lobes.

The lungs are surrounded by mucus membranes called **pleura**. The membrane that directly surrounds the lungs is the **visceral pleura**. The **parietal pleura** covers the inside of the thoracic cavity. The space between the two pleural membranes is very small and is called the pleural cavity, which is filled with fluid that prevents friction during breathing when the lungs expand and contract.

The bronchioles branch into smaller and narrower bronchioles that lead to the smaller terminal and respiratory bronchioles, which lead to the alveoli. Each lobe of the lungs has **lobules** that contain respiratory structures, arterioles, venules, and lymphatic vessels (Figure 18.4). The respiratory structures are surrounded by capillaries so that gas exchange can occur efficiently (Figure 18.5). Bronchioles are surrounded by smooth muscle cells that can contract to cause constriction of the bronchiole tubes, or relax to cause dilation. The smooth muscles cells are controlled by the autonomic nervous system.

The alveoli are composed of two types of epithelial cells. **Type I alveolar cells** are simple squamous epithelial cells that make up the majority of the alveoli and are the site of gas exchange. **Type II alveolar cells** are specialized epithelial cells that produce and release **surfactant** into the alveoli. Surfactant is an amphi-

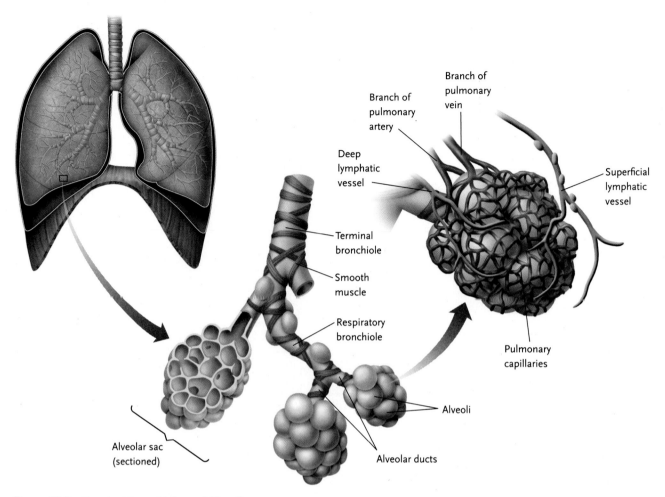

Figure 18.5   **Terminal Bronchioles and Alveoli**

Art source (centre): From DIGIUSEPPE/FRASER. *Biology* 12 U. © 2012 Nelson Education Ltd. Reproduced by permission. www.cengage.com/permissions

pathic fluid composed of phospholipids and lipoproteins that have both polar and nonpolar regions. The surfactant relieves the surface tension caused by the polarity of water molecules inside the alveoli. The alveoli are only one cell thick with minimal connective tissue between them and the blood capillaries. Without the surfactant, they would collapse from the pull of the hydrogen bonds in the water molecules even though there is just a very small film of water in the lungs (see Figure 2.1 from Chapter 2). Oxygen that is inhaled must dissolve in water before it can diffuse across cell membranes into the red blood cells. A molecule of oxygen must cross several alveolar epithelial cell membranes, pass through a small layer of connective tissue, cross the capillary endothelial cell membranes, and finally cross the membrane of the red blood cell (Figure 18.6). The total distance from the air inside the alveoli to the inside of a red blood cell is approximately 0.2 μm (the diameter of a typical cell is approximately 10 μm).

DID YOU KNOW?

Taking a deep breath causes the lungs to stretch and this stimulates the type II alveolar cells to produce surfactant, which will decrease surface tension, and will make breathing easier.

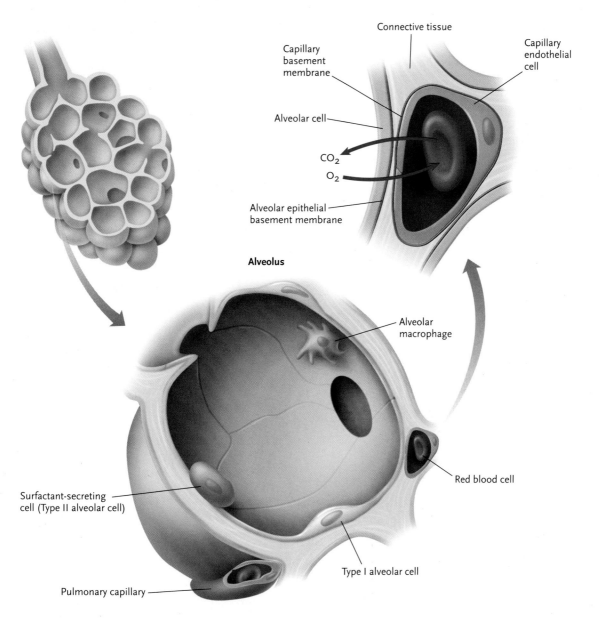

Figure 18.6   Gas Exchange

Gas exchange occurs in the alveoli.

Art source (top left): From DIGIUSEPPE/FRASER. *Biology* 12 U. © 2012 Nelson Education Ltd. Reproduced by permission. www.cengage.com/permissions

CONCEPT REVIEW 18.2

1. What is the difference in function between the conducting zone and the respiratory zone?

2. Why is it important for air to flow through the nasal conchae?

3. What are the three regions of the pharynx?

4. Why are the Eustachian tubes important?

5. How do the vocal cords produce sounds?

6. Name two things that prevent particulate matter such as dust from entering the lungs.

7. What is the difference between the visceral pleura and the parietal pleura?

8. Why is the pleural cavity filled with fluid?

9. Which alveolar cell type produces surfactant?

10. Why is surfactant necessary?

18.3 PULMONARY VENTILATION

Pulmonary ventilation occurs because of changes in the volume of the thoracic cavity that affect the air pressure and cause air to move in or out. The diaphragm is the primary muscle involved in relaxed breathing. During diaphragm contraction, thoracic volume increases, which causes a brief decrease in pressure, and this pulls air into the lungs, quickly equalizing the pressure. During diaphragm relaxation, a decrease in thoracic volume causes the air to move out, also quickly equalizing the pressure.

Muscles involved in relaxed breathing

During *relaxed* inspiration, the muscles that increase the thoracic cavity volume are the diaphragm and the **external intercostal muscles** (Figure 18.7). When stimulated by neurons in the phrenic nerve, the diaphragm contracts, causing it to move down, thus increasing the thoracic volume. The external intercostal muscles are located between the ribs, and their contraction causes the ribs to move upward and outward, which also increases the thoracic volume.

During *forced* inhalation, such as during exercise, more oxygen is required for muscle cells to produce

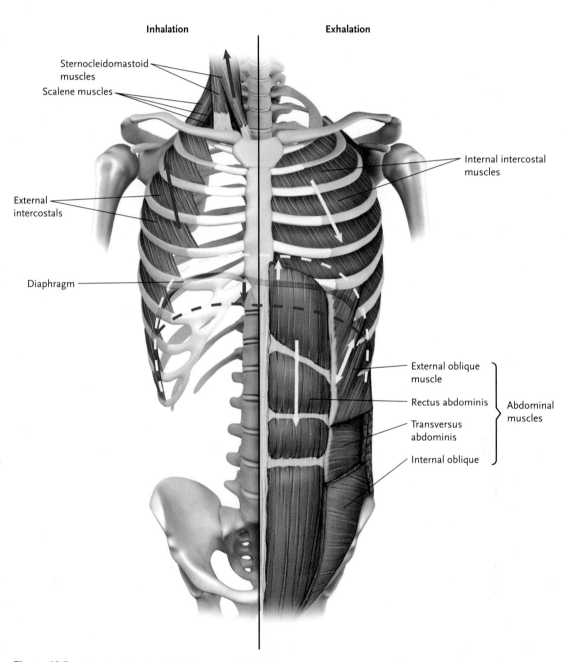

Figure 18.7 Muscles Involved in Pulmonary Ventilation

ATP. Therefore, breathing rate and depth increases to match demand, and additional muscles play a role in breathing. During forced inspiration, the **sternocleidomastoid** contracts to help elevate the sternum and clavicles, and the **scalenes** elevate the upper ribs to increase thoracic cavity volume (Figure 18.7). During a severe asthma attack, it is also possible to contract pectoralis muscles, which are connected to the sternum and the humerus. A person enduring an asthma attack also raises their shoulders to try to further increase lung volume.

DID YOU KNOW?

An inability to catch your breath for a long time after exercise can be an indication of early heart disease. The heart contracts to circulate oxygenated blood to all the body's cells so they can have enough oxygen to produce ATP. When the cells do not receive the oxygen, your breathing centre (medulla oblongata and pons) signals the respiratory muscles to increase your rate and depth of breathing, making you feel short of breath. Often the only symptom of a heart attack, particularly in females, is not being able to catch one's breath for a long time after exercise. A good measure of cardiovascular fitness is how easily you can regain normal breathing after intense exercise. Although shortness of breath is the most common symptom of a heart attack, many pulmonary problems can also cause shortness of breath; in either case, medical attention is required.

During *relaxed* exhalation, there is no muscle contraction; the diaphragm and the external intercostals simply relax, which causes the thoracic volume to decrease, which causes air to leave the lungs. The lung tissues have a high amount of elastic proteins that recoil and aid the movement of air out of the lungs.

During *forced* exhalation, such as exercise, sneezing, or coughing, other muscles contract to increase the rate of exhalation by causing the thoracic volume to decrease faster. The muscles involved are the internal intercostals and abdominal muscles (Figure 18.7). When contracted, the internal intercostal muscles—which lie in an oblique angle opposite to the external intercostals—cause the ribs to move down and inward, which decreases thoracic volume. The **external obliques** also move the ribs down and inward. When the **rectus abdominis**, **transverse abdominis**, and **internal obliques** contract, abdominal pressure increases, which forces the diaphragm to move upward and thereby decreases thoracic volume.

Changes in thoracic pressure during breathing

Boyle's law is a gas law that states that in a closed container with a constant number of molecules at a constant temperature, the pressure is inversely

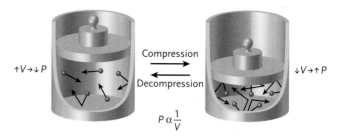

Figure 18.8 Boyle's Law
Boyle's law states that volume is inversely proportional to pressure.

proportional to the volume. If the volume increases, the pressure decreases; if the volume decreases, the pressure increases (Figure 18.8). The amount of pressure is determined by the amount of force of the molecules on the container. In our lungs, the molecules are the components of air (see Table 18.1), and the container is the lungs, which are in close contact with the pleural membranes of the thoracic cavity. When we breathe, a pressure change of 2 mmHg (millimeters of mercury) is equivalent to approximately 500 ml of air.

As we breathe, the pressure is not constant because as soon as the pressure drops, air moves into our lungs, and therefore the pressure quickly becomes equal to that of the environment (Figure 18.9). The air pressure in the environment is the **atmospheric pressure**. At sea level atmospheric pressure is approximately 760 mmHg; this measurement changes with altitude and various weather patterns. The pressure inside the lungs is the **alveolar pressure**, and the pressure between the visceral and parietal pleura is the **intrapleural pressure**. The changes in volume and pressure during breathing are described in three stages:

1. No change in volume. Any time you are not changing the volume of air in your lungs, the alveolar pressure is equal to the atmospheric pressure, whether you are holding a lung full of air or if your lungs are empty. If the atmospheric pressure is 760 mmHg, and the thoracic volume is constant, the alveolar pressure will also be 760 mmHg. The intrapleural pressure must always be *lower* than the alveolar pressure—also called negative pressure—so the lungs do not collapse. At rest, the intrapleural pressure is −4 mmHg compared to the alveolar and atmospheric pressure.

2. Inhalation. During inhalation, muscles contract, causing the thoracic volume to *increase*, which causes the alveolar pressure to *decrease* relative to the atmospheric pressure (Figure 18.9). Since the lungs are not a closed container, the decrease in pressure causes more air molecules to move into the alveoli and equalize the pressure in the larger volume. During inhalation there is a corresponding change in the intrapleural pressure, so a difference of 4 mmHg ensures that the lungs remain open as the pressure decreases inside the alveoli (very briefly since the incoming air immediately equalizes the pressure).

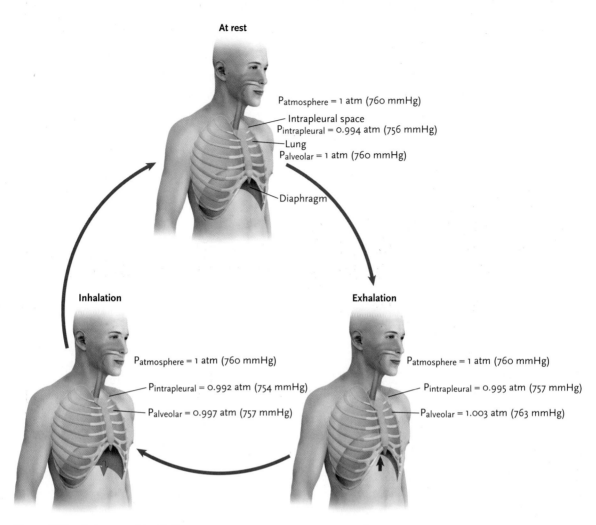

At rest

P_{atmosphere} = 1 atm (760 mmHg)

Intrapleural space
P_{intrapleural} = 0.994 atm (756 mmHg)

Lung
P_{alveolar} = 1 atm (760 mmHg)

Diaphragm

Inhalation

P_{atmosphere} = 1 atm (760 mmHg)

P_{intrapleural} = 0.992 atm (754 mmHg)

P_{alveolar} = 0.997 atm (757 mmHg)

Exhalation

P_{atmosphere} = 1 atm (760 mmHg)

P_{intrapleural} = 0.995 atm (757 mmHg)

P_{alveolar} = 1.003 atm (763 mmHg)

Figure 18.9 **Pulmonary Ventilation**
Inhalation occurs when the volume in the thoracic cavity increases, causing the pressure inside the lungs to decrease.

3. Exhalation. During exhalation, muscles relax, and this causes the thoracic volume to *decrease*, which makes the alveolar pressure *increase* relative to the atmospheric pressure. The increase in pressure causes air molecules to move out of the lungs, equalizing the pressure. During exhalation, the intrapleural pressure remains negative compared to the alveolar pressure to ensure that the lungs do not collapse.

The ability of the lungs to expand with inhalation and retract with exhalation is dependent on two important factors, **compliance** and **recoil**, which are determined by the health of the extracellular matrix proteins (connective tissue), such as collagen and elastin, as well as the amount of surfactant required to decrease surface tension inside the alveoli. In healthy lungs compliance is high, and the lungs can stretch and expand to allow a large volume of air to enter the lungs if needed, such as during exercise. Someone with decreased compliance has decreased lung capacity and cannot take in as much air. Healthy recoil allows air to easily move out of the lungs during exhalation, without the need for internal

intercostal or abdominal muscles to contract to force the air out of the lungs.

Lung volume

When we are sitting at rest, we breathe a certain volume of air in and out. However, we can modify that amount by taking an extra deep breath in or by forcing out extra air after a regular exhalation. Each of these volumes of air can be labelled (Figure 18.10).

1. The volume of air in one inhaled breath or exhaled breath is the tidal volume, which is equal to approximately 500 ml.

2. After taking a normal breath in and continuing to breathe in as much as possible, the lung volume is approximately 3 L, and this is called the **inspiratory reserve volume**. The inspiratory reserve volume is about six times greater than the resting tidal volume.

3. The sum of the tidal volume and the inspiratory reserve volume is called the **inspiratory capacity**.

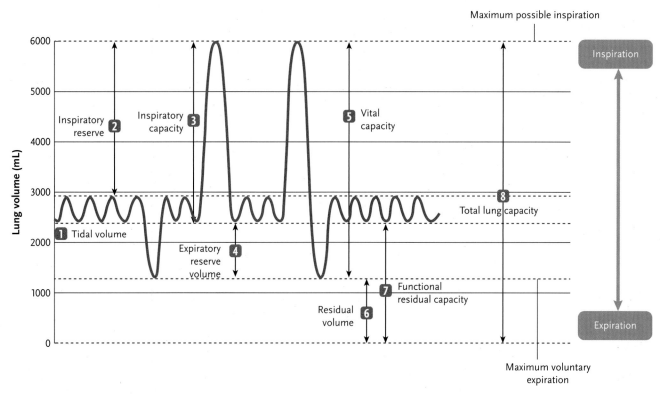

Figure 18.10   Lung Volumes

4. After taking a normal breath out and continuing to breathe out as much as possible— which requires contraction of the intercostal and abdominal muscles, the lung volume is approximately 1.2 L, and this is called the **expiratory reserve volume.**

5. The amount of air that we can move into and out of our lungs by breathing in and out as much as possible is approximately 4.7 L, but this varies with age, body size, and level of health; this amount is called the **vital capacity.** Vital capacity is equal to the sum of tidal volume, inspiratory reserve, and expiratory reserve.

6. There is always some air left in our lungs even after we have exhaled as much as possible. Approximately 1.2 L remains within the lungs that we cannot exhale, and this amount is called the **residual volume.**

7. The sum of the expiratory reserve and the residual volume is called the **residual capacity,** which is approximately 2.4 L; this amount is significantly higher than the fresh air coming into the lungs with each 500 ml tidal volume.

8. The total maximum amount of air that the lungs can hold is approximately 5.8–6 L, and this is called the **total lung capacity:** the sum of the total amount of air we can breathe in (vital capacity) plus the residual volume.

DID YOU KNOW?

One pulmonary function test is called the forced expiratory volume (FEV). It involves breathing in the inspiratory reserve volume and then breathing out as fast and as much as possible into a spirometer that measures lung volumes. It is used as a diagnostic tool to determine if someone has an obstructive pulmonary disease that involves bronchoconstriction, such as asthma. A healthy FEV is 80% of your vital capacity in one second; whereas someone with an obstructive pulmonary disease cannot breathe as much air out as fast, so their FEV percentage is lower.

Certain categories of individuals have larger lung volumes than others. People with larger lung volumes include those who are tall, those living at high altitudes, those with normal weight compared to those who are obese, men more than women, athletes more than non-athletes, non-pregnant women compared to pregnant women, and those who are healthy compared to those with restrictive lung diseases.

Dead space

The body's respiratory structures are within either the conducting zone or the respiratory zone. The conducting zone only transports air; gas exchange occurs only in the respiratory zone, which includes the terminal bronchioles, alveolar ducts, and alveoli. The air in the conducting zone does not contribute to gas exchange and is therefore called **anatomical dead space**. Approximately 30% of each tidal volume stays within the conducting zone. After a normal inhalation of 500 ml of air, only 350 ml is new air from outside; 150 ml is already in the conducting zone from the previous breath. This is why we can't swim underwater by breathing through a garden hose, which would simply create a very long dead space where eventually there would be no oxygen available for gas exchange. This also explains why it is important to take a deep breath every once in a while: to exchange the air in the dead space to ensure there is always enough oxygen for gas exchange. During exercise, deeper and longer breaths deliver the extra oxygen that the muscles need for ATP production, and this can make a very significant difference in your ability to continue to contract muscles and resist fatigue.

Alveolar ventilation refers to the amount of air that reaches the alveoli in one minute. It is similar to cardiac output, which is amount of blood ejected by the heart per minute. To calculate alveolar ventilation use the following equation:

$$\text{alveolar ventilation} = [\text{tidal volume (ml)} - \text{dead space (ml)}] \times \text{respiration rate (breaths/min)}$$

A typical resting alveolar ventilation = $(500 - 150) \times 12 = 4.2$ L/ min)

CONCEPT REVIEW 18.3

1. Which muscles are involved in relaxed inhalation and relaxed exhalation?

2. What muscles are involved in inhalation during exercise or an asthma attack?

3. Which muscles would contract when you have forced exhalation such as coughing?

4. What is Boyle's law?

5. How are lung volume and pressure related, and what are the changes in volume and pressure that occur during inhalation?

6. Suppose you take a very deep breath in and hold it. Assuming you are at sea level, what is the atmospheric pressure, the alveolar pressure, and the intrapleural pressure?

7. Why does the intrapleural pressure have to be negative in relation to the alveolar pressure?

8. What is the difference between lung compliance and recoil?

9. Provide the name of a particular lung volume that matches the following definitions:

   a. amount of air inhaled in one normal breath
   b. total amount of air in lungs after a full deep breath in
   c. amount of air further exhaled after a normal exhalation
   d. amount of air in the lungs after exhaling as much air as possible

10. Name three factors that affect lung volume.

11. What is dead space?

12. What is the alveolar ventilation if the tidal volume is 600 ml, dead space is 180 ml, and respiratory rate is 14 breaths per minute?

18.4 GAS EXCHANGE AND TRANSPORT OF OXYGEN AND CARBON DIOXIDE

Gas exchange

When the oxygen in the inhaled air reaches the respiratory zone, oxygen dissolves in the small amount of water that's in the alveoli and then moves by diffusion through the alveolar epithelial cell membranes into the bloodstream. Inhaled air is a mixture of gasses of which oxygen makes up approximately 21% (Table 18.1).

Dalton's law states that the pressure of each gas in a mixture is independent of the pressure exerted by the other gases. If atmospheric pressure at sea level is 760 mmHg, oxygen will exert 21% of that pressure:

TABLE 18.1

Composition of Air	
Molecule	**Percentage**
Nitrogen	78.08
Oxygen	20.9
Argon	0.9
Carbon dioxide	0.03
Ozone	0.01
Other trace elements	~0.5

this percentage is called a **partial pressure**. The partial pressures of all the gases in the mixture, add up to the total pressure.

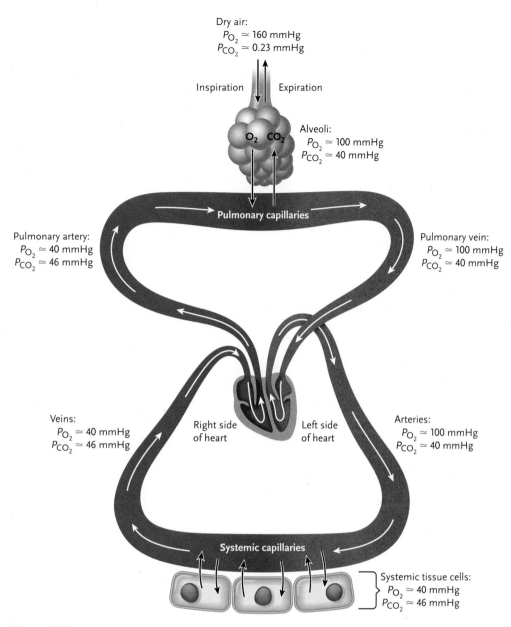

Figure 18.11   **Changes in Partial Pressures during Breathing**

Partial pressure of oxygen in the atmosphere can be calculated as follows:

$$P_{O2} = 760 \, (0.209) = 158.8 \text{ mmHg}$$

During respiration, oxygen and carbon dioxide move down their partial pressure gradients from an area of higher concentration to an area of lower concentration (Figure 18.11). As we breathe in, the concentration of oxygen is higher in the inhaled air compared to that in the bloodstream, so oxygen moves into the blood in the alveoli. The partial pressure of oxygen in the air is approximately 159 mmHg, but as it enters the lungs and mixes with the air in the dead space, the concentration of oxygen decreases, and the partial pressure of oxygen in the alveoli is approximately 105 mmHg (Figure 18.9).

The blood flowing through the capillaries is *deoxygenated* until oxygen begins to diffuse across the alveolar membranes into the capillaries. The deoxygenated blood in the alveolar capillaries has a partial pressure of approximately 40 mmHg, which is lower than the alveolar partial pressure of 105 mmHg. As oxygen moves into the blood vessels, the partial pressure increases to approximately 100 mmHg, and at this point the blood is *oxygenated*. Oxygenated blood flows from the lungs back to the left atrium of the heart and is pumped from the left ventricle to the rest of the body. Once the oxygenated blood reaches any cells with low oxygen, the oxygen in the blood again moves toward the area of lower concentration, and into the cells. As the oxygen moves from the capillaries in the tissues into the cells, the blood becomes deoxygenated, and the partial pressure gradually becomes 40 mmHg as the oxygen moves into the cells. The deoxygenated blood circulates back to the lungs, and the cycle continues such that oxygen from inhaled air continuously moves from the alveoli into the blood and then into the body's cells to be used for the production of ATP.

Carbon dioxide gas also moves from areas of higher concentration to areas of lower concentration. As cells take up and consume oxygen, carbon dioxide is produced as a waste product that diffuses into the capillaries within the tissues. As carbon dioxide moves into the capillaries, the partial pressure of carbon dioxide becomes approximately 46 mmHg in the deoxygenated blood, which is then circulated to the right side of the heart. Blood then circulates to the lungs where the CO_2 moves down the concentration gradient into the alveoli. The air in the alveoli contains a mixture of inhaled air and air recycled within the dead space, and so the partial pressure of the carbon dioxide is approximately 40 mmHg. As we exhale, the carbon dioxide mixes with the atmospheric air—where the concentration of CO_2 is very low and its partial pressure is approximately 0.3 mmHg.

If you were to breathe air that has less than 21% oxygen, such as at high altitudes, the partial pressure of oxygen would be lower, and therefore the partial pressure of oxygen in the alveoli would be lower, and less oxygen would diffuse into the bloodstream. The body compensates for this reduction in oxygen initially by increasing the rate and depth of ventilation. If you stay at a high altitude for a long period of time, your body will also compensate by increasing the production of red blood cells to carry the oxygen in the blood.

Conditions that decrease the partial pressure of oxygen in the alveoli

Certain conditions cause the partial pressure of oxygen in the alveoli to decrease, including (1) decreased atmospheric pressure at high altitude; (2) decreased ventilation; and (3) increased metabolism. Ventilation has to increase if metabolism increases to ensure that enough oxygen molecules diffuse into the blood.

DID YOU KNOW?

Yawning is stimulated by an increase in blood CO_2 levels, which occurs more when we are tired and do not breathe deeply. The carbon dioxide levels increase and trigger the yawning reflex, causing you to take a deep breath in, much deeper than the amount of dead space volume, which decreases the alveolar carbon dioxide partial pressure. While yawning, you also hold your breath for a second after the deep inhalation, which allows more time for CO_2 to diffuse from the blood into the alveoli before you exhale. Yawning allows for more carbon dioxide to be removed from your blood.

Hyperventilation occurs when the ventilation rate increases beyond what is necessary for removing CO_2 and acquiring O_2. Hyperventilation can cause the blood CO_2 level to drop below the normal concentration, causing hypocapnia (low CO_2), which can lead to an abnormal increase in blood pH, called respiratory alkalosis. Symptoms include dizziness; tingling in fingers, toes, and face; weakness; and possibly fainting.

Hypoventilation occurs when the ventilation rate decreases below what is necessary for acquiring O_2 and removing CO_2. Hypoventilation can cause the blood CO_2 levels to be too high, which is called hypercapnia, and that condition can lead to respiratory acidosis. Certain drugs, alcohol, stroke, or tumour affecting the brainstem, or altitude sickness can cause hypoventilation and can be life threatening.

DID YOU KNOW?

Sleep apnea refers to a periodic absence of breathing during sleep, which can last several seconds or up to one or two minutes, and may occur many times a night without the person being aware. It is more common in people who are overweight, smokers, or borderline diabetics. Symptoms often include loud snoring, gasping, or choking sensations during sleep, daytime fatigue, difficulty concentrating, and impaired memory.

Transport of gasses in the blood

Oxygen and carbon dioxide cross all membranes by diffusion; no membrane proteins are required in either the lungs or the tissues. Most oxygen is transported in the bloodstream by being bound to hemoglobin proteins in red blood cells. The rest, approximately 1.5%, is carried dissolved in the plasma. Each hemoglobin molecule is composed of four globin proteins containing heme and an iron ion that binds the oxygen (Figure 18.12). Heme is a large heterocyclic molecule containing iron, and it is a red pigment that turns bright red when oxygen is bound. The function of heme and oxygen binding is regulated by both the partial pressure gradient of oxygen and the pH of the blood (Figure 18.13).

In the lungs when we inhale, the partial pressure of oxygen becomes lower in the blood, and as we exhale carbon dioxide, the pH of the blood in the lungs becomes slightly more alkaline (carbon dioxide is acidic). Therefore, oxygen binds to hemoglobin forming oxyhemoglobin (Figure 18.14). As the blood circulates to the tissues, the blood pH decreases because the cells produce CO_2 that enters the blood, and the partial pressure of oxygen in the cells becomes lower than in the blood. Consequently, oxygen leaves the hemoglobin and enters the cells. Body temperature also affects the binding of oxygen to hemoglobin.

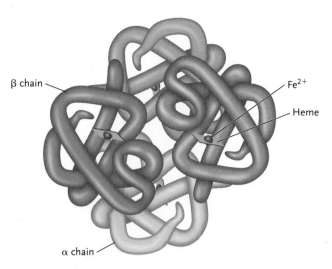

β chain

Fe²⁺

Heme

α chain

Figure 18.12   Hemoglobin

Hemoglobin is the protein in red blood cells that binds oxygen.

During exercise, when the temperature increases, more oxygen is released from hemoglobin in the tissues. During exercise blood pH decreases because more CO_2 is produced, and this increases the amount of oxygen that dissociates from the hemoglobin in the muscles.

Exercise also increases the amount of oxygen consumed, which increases the partial pressure gradient of oxygen, causing more oxygen to be released from hemoglobin in the tissues. Note that in the tissues when oxygen dissociates from hemoglobin and enters the cells, the hemoglobin in the red blood cells is still approximately 78% saturated.

Unlike oxygen, carbon dioxide is transported in the following three ways (Figure 18.15):

- 7–10% is transported dissolved in plasma
- 25–30% is transported bound to hemoglobin
- 60–65% is transported in the form of bicarbonate ions

Carbon dioxide dissolves in water much easier than oxygen does, and therefore some carbon dioxide can be transported dissolved in the plasma. About 25% of the carbon dioxide can be transported on hemoglobin. In this case, CO_2 binds to amino groups that make up the hemoglobin molecules—not to the iron ion—and forms carbaminohemoglobin ($HbCO_2$). The majority of CO_2 is carried within the red blood cell in the form of bicarbonate ions (HCO_3^-), which are formed according to the following chemical equation:

$$CO_2 + H_2O \rightarrow H_2CO_3 \rightarrow HCO_3^- + H^+$$

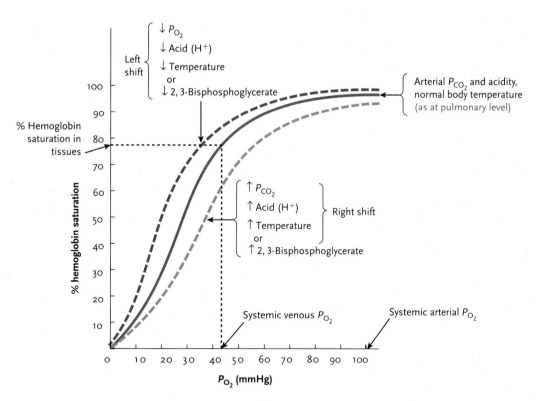

Figure 18.13   Hemoglobin Saturation Curve

The hemoglobin saturation curve is an important tool for understanding how our blood carries and releases oxygen.

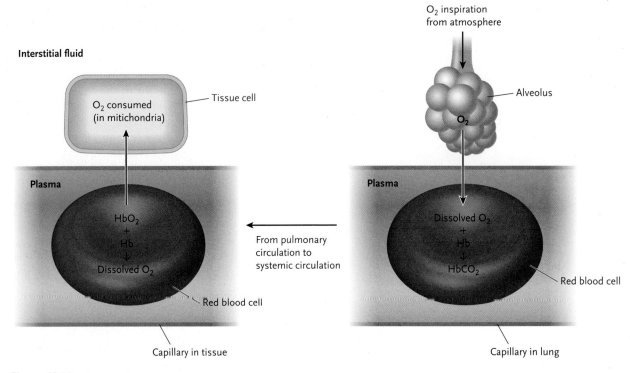

Figure 18.14 Oxygen Transport

Oxygen is transported by being bound to hemoglobin or dissolved in the blood plasma.

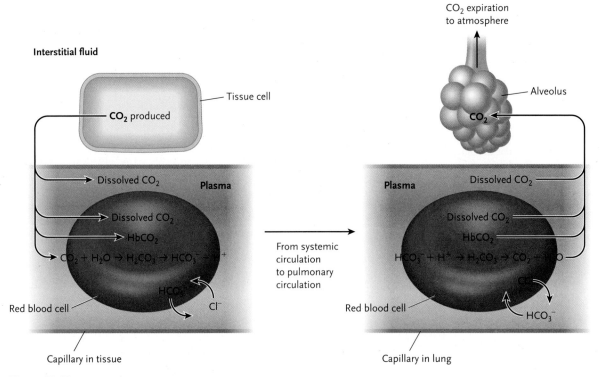

Figure 18.15 Carbon Dioxide Transport

Carbon dioxide is primarily transported in the form of carbonate.

In the tissues, carbon dioxide and water form carbonic acid (H_2CO_3) in the presence of an enzyme, called carbonic anhydrase, that's found inside red blood cells. Carbonic acid dissociates into bicarbonate (HCO_3^-) and hydrogen ions (H^+). This is why an increase in CO_2 in the blood causes a decrease in blood pH, and this plays an important role in the regulation of breathing. Bicarbonate ions diffuse out of the red blood cells and are transported in the plasma. Because bicarbonate ions are negatively charged and leave the red blood cell, the charge of the red blood cell (RBC) has to be balanced. This is accomplished by Cl– ions moving into the red blood cells, which is called the **chloride shift** (Figure 18.15).

When blood circulates to the lungs, the CO_2 dissolved in the plasma diffuses across the RBC and alveolar membranes to reach the inside of the alveoli and is exhaled. Any CO_2 bound to the hemoglobin amino acids is released and also exhaled. Lastly, the reverse bicarbonate reaction occurs. The bicarbonate ions move back into the red blood cells, chloride ions move back out and into the plasma, and the bicarbonate and H^+ ions combine to form carbonic acid. Carbonic anhydrase then converts carbonic acid back into CO_2 and H_2O, which is exhaled.

CONCEPT REVIEW 18.4

1. How is Dalton's law different from Boyle's law?

2. Using Table 18.1, calculate the partial pressure of each gas in air at an atmospheric pressure of 760 mmHg.

3. Describe the changes in the partial pressure of oxygen in the inhaled air, the alveoli, and the blood within the lungs compared to the blood within body tissues.

4. Name three factors that can affect the partial pressure of oxygen in the alveoli and therefore affect the amount of oxygen that diffuses into the blood.

5. Describe the changes in the partial pressure of carbon dioxide in the blood within the tissues compared to the blood within the lungs, the alveolar air, and the atmospheric air.

6. Which of the following causes an increase in the amount of oxygen that dissociates from hemoglobin?

   a. an increase or decrease in pH
   b. an increase or decrease in partial pressure gradient
   c. an increase or decrease in body temperature

7. Describe how carbon dioxide is transported from the tissues to the lungs.

8. Why is the chloride shift important?

18.5 REGULATION OF BREATHING

The medulla oblongata and the pons located in the mid-to-lower part of the brainstem are both involved in the regulation of breathing. The medulla oblongata is the primary respiratory control centre and contains two regions of neurons: the **dorsal respiratory group (DRG)** and the **ventral respiratory group (VRG)**. The DRG controls inspiration by signalling the diaphragm and external intercostal muscles to contract, causing inhalation. Motor neurons that lead from the medulla oblongata to the diaphragm are located in the phrenic nerve. The VRG is composed of pacemaker cells that receive signals from and send signals to the DRG, and together these cells regulate the *rhythm* of breathing. The VRG can also signal the muscles involved in expiration, such as abdominals and the internal intercostals. The dorsal and ventral neurons in the medulla oblongata are involved in both quiet and forceful breathing.

The pons is located above the medulla oblongata and contains two regions that influence the medulla oblongata. The upper pons is called the **pneumotaxic centre.** It smooths the transitions between inhalation and exhalation and can affect the rate of breathing by sending inhibitory signals to the medulla oblongata. The pneumotaxic area also influences the lower pons, which is called the **apneustic centre.** The apneustic centre also helps to regulate the rate of breathing by stimulating the inspiratory neurons in the DRG of the medulla oblongata, which can cause inhalation to be deeper and longer.

Stretch receptors, located within the smooth muscles of the airways, stretch as the lungs expand. When stretch increases sensory signals are sent to the medulla oblongata to inhibit the DRG. Then further depolarization of the motor neurons stops, which stops contraction of the inspiratory muscles. If you inhale as much as you possibly can, once your lungs reach the inspiratory capacity, the stretch receptors inhibit the dorsal respiratory neurons of your medulla oblongata.

Summary of events during inhalation and exhalation

During inhalation the following processes occur:

1. DRG depolarization of neurons signals *inspiratory* muscles to contract; this may include scalenes and sternocleidomastoid if breathing is forceful.

2. VRG does not depolarize motor neurons that signal the *expiratory* muscles.

3. The inspiratory and expiratory muscles do not contract at the same time.

4. The pons helps smooth the transition to exhalation.

5. The apneustic centre of the pons can stimulate an increase in the rate and depth of inhalation.

6. Stretch receptors prevent over-inflation of the lungs by inhibiting the DRG when the lungs are full.

During exhalation the following processes occur:

1. The pneumotaxic centre of the pons can inhibit the inspiratory muscles and prevent inhalation.

2. DRG of the medulla oblongata does not depolarize, so inspiratory muscles do not contract.

3. No motor neurons are stimulated during quiet breathing, and so inspiratory muscles relax and lung recoil pushes air out of lungs.

4. During forced breathing, the VRG stimulates expiratory muscles to contract and increase the rate and force of air moving out of the lungs.

Factors that influence the medulla oblongata and the pons

- Signals from the cerebral cortex can stimulate or inhibit the neurons in both the dorsal and ventral cells of the medulla oblongata. As a result, we can voluntarily hold our breath, take a deep breath, or change the rate of our breathing.

- The limbic system, also known as the emotional part of our brain, is involved in emotions that impact breathing: for example laughing, or crying, or expressions of pain.

- The autonomic nervous system is involved because a sympathetic nervous system response increases the rate and depth of breathing, and a parasympathetic nervous system response decreases the rate and depth of breathing.

- Irritation to airway structures, such as chemical irritants or blowing air, can signal a reflex that inhibits the medulla oblongata and prevents contraction of inspiratory neurons, which prevents the inhaling of damaging chemicals into the delicate alveoli. Have you ever walked outside into a large gust of wind and lost your breath for a second? Your medulla oblongata was inhibited to prevent the possibility of over-inflation of the lungs in very high wind.

- Temperature is a factor because increased body temperature, such as during a fever, increases the rate of breathing.

- Medications such as general anaesthetic, morphine, and alcohol, inhibit the medulla oblongata and depress respiration, but amphetamines stimulate respiration.

- Blood levels of O_2, CO_2, and H^+ are involved because breathing rate and depth increases if oxygen decreases, or if carbon dioxide or hydrogen ions increase.

Peripheral and central chemoreceptors

Peripheral chemoreceptors detect blood levels of O_2, CO_2, and H^+, which fluctuate based on metabolic rate. We need oxygen to produce ATP, and carbon dioxide is produced as a waste product. Recall that as carbon dioxide levels increase, blood pH decreases; it becomes more acidic as CO_2 is converted into carbonic acid. Whenever metabolism increases, ATP production increases, and so does the amount of CO_2 and H^+, which then enters the blood. Exercise increases the need for ATP production, and therefore our rate and depth of breathing increases during exercise; high-intensity exercise that requires more ATP increases breathing more than low-intensity exercise. Peripheral chemoreceptors are located in key areas: the aortic arch and the carotid arteries (Figure 18.16) to detect arterial blood levels in the blood that is leaving the heart and circulating to the body and the brain. The peripheral chemoreceptors are located very close to the peripheral baroreceptors that detect blood pressure (Chapter 17). The medulla oblongata also detects blood levels of O_2, CO_2, and H^+, which contain the **central chemoreceptors**.

Central chemoreceptors in the medulla oblongata primarily respond to increased H^+ in the extracellular fluid of the brain, which is caused by an increase in CO_2 due to increased energy requirements. The peripheral receptors respond to increased H^+, as well as directly detecting increased CO_2 and reduced O_2. The receptors that detect these molecules are directly connected to sensory neurons that send information to the brainstem. The overall strongest stimulus for increasing breathing is the increase in carbon dioxide. If you hold your breath as long as possible, the buildup of CO_2 stimulates the medulla oblongata to signal inspiratory muscle contraction and eventually forces you to breathe.

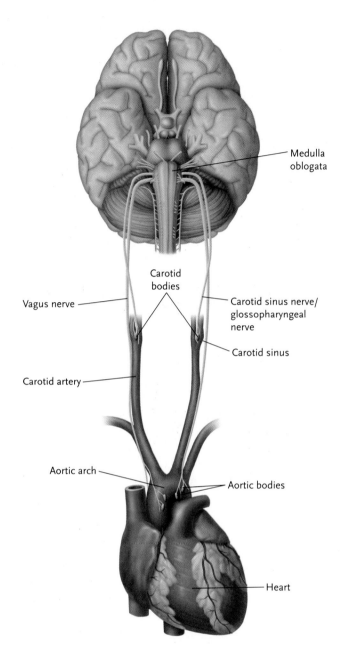

Figure 18.16 **Peripheral Chemoreceptors**

Peripheral chemoreceptors detect blood oxygen and carbon dioxide levels.

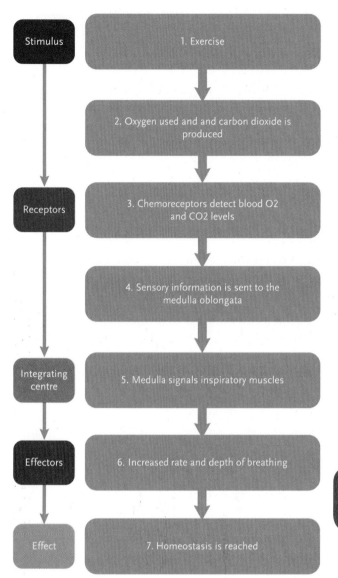

Figure 18.17 **Negative Feedback Control of Breathing**

Regulation of breathing during exercise

Stimulus → receptors → integrating centre → effectors → effect

1. Exercise requires an increase in ATP for skeletal muscle contraction (Figure 18.17).

2. ATP production requires oxygen from the blood to be taken up by muscle cells and produces carbon dioxide.

3. Peripheral receptors detect a decrease in blood O_2 (hypoxia), an increase in CO_2, and a decrease in pH (increased formation of H^+). Central receptors detect increased CO_2 in brain extracellular fluid.

4. Sensory information from the receptors is sent through afferent neurons to the medulla oblongata.

5. The medulla oblongata signals motor neurons to cause the alternating contraction of inspiratory muscles and expiratory muscles (depending on the intensity of the exercise).

6. Increased contraction of inspiratory muscles (and possibly expiratory muscles during high-intensity exercise) causes increased rate and depth of breathing.

7. Increased breathing causes blood oxygen levels to increase, and blood carbon dioxide levels to decrease. Decreased carbon dioxide causes a decrease in free hydrogen ions and therefore increases blood pH.

CONCEPT REVIEW 18.5

1. Describe the function of the dorsal and ventral neurons of the medulla oblongata.

2. How does the pons influence breathing?

3. How is over-inflation of the lungs prevented?

4. Make a list of some factors that can affect the rate or depth of breathing.

5. Where are the central and peripheral chemoreceptors located?

6. What is detected by the central and peripheral chemoreceptors?

7. Identify the stimulus, receptor, integrating centre, and effectors involved in the increased breathing during exercise.

18.6 COPD AND ASTHMA

Chronic obstructive pulmonary disease (COPD) is a chronic, progressive, lung disease caused primarily by smoking or long-term breathing of pollutants or chemical irritants (Figure 18.18). Cigarette smoking is the leading cause of COPD, and the two main types of COPD are **emphysema** and **chronic bronchitis**. People with COPD have damage in both the conducting and respiratory zones of the respiratory system, and this is characterized by progressive difficulty breathing. In emphysema, the elastic recoil of the air sacs is lost, and exhaling becomes very difficult. Damage to the air sacs leads to fewer and larger alveoli, and this eventually causes a dramatic decrease in gas exchange. In chronic bronchitis, inflammation causes swollen, irritated airways with significantly increased mucus production, often causing a chronic cough. Most people with COPD have both emphysema and chronic bronchitis. There is no cure for COPD, but medications may help decrease symptoms.

Asthma is a chronic disease caused by inflammation and narrowing of the airway conducting zones. It is also known as **bronchoconstriction** and results in difficulty breathing (Figure 18.19). In people who have asthma, the immune system responds to stimuli that normally it should not react to. In a way similar to an allergic reaction, common particles stimulate an inappropriate immune reaction. The exact cause is not known, but the triggers can include stress, change in air temperature, or exercise. The inflammatory response causes swelling of the lining of the airways and mucus production, which can obstruct airflow. Also smooth muscle contractions in the bronchi and bronchioles narrow the diameter of the conducting zone structures. Symptoms include

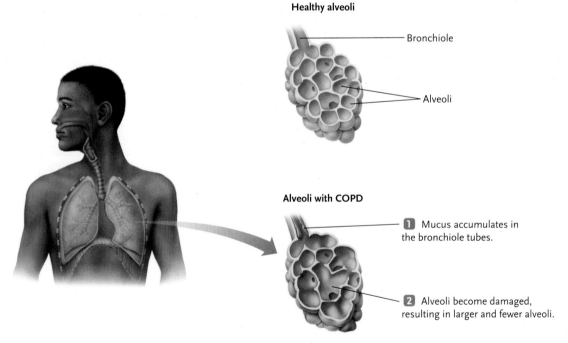

Healthy alveoli

— Bronchiole

— Alveoli

Alveoli with COPD

1 Mucus accumulates in the bronchiole tubes.

2 Alveoli become damaged, resulting in larger and fewer alveoli.

Figure 18.18 Chronic Obstructive Pulmonary Diseases (COPD)

COPD includes emphysema and bronchitis.

Art sources (right): From DIGIUSEPPE/FRASER. *Biology 12 U.* © 2012 Nelson Education Ltd. Reproduced by permission. www.cengage.com/permissions

difficulty breathing, which requires the use of the additional inspiratory muscles—such as the scalenes and sternocleidomastoid muscles—to increase lung volume; coughing, wheezing, and chest tightness due to increased mucus and narrow airways; tachycardia, which means the heart rate increases to compensate for the decreased oxygen in the blood and the effort to increase cardiac output to meet the demands of the cells; and fatigue due to decreased oxygen reaching the cells and less ATP production. A severe asthma attack can cause anxiety because of the suffocating feeling of the lack of oxygen. Asthma can be treated with **bronchodilators** and anti-inflammatory medications. People with asthma must learn what their specific triggers are and then try to avoid or minimize exposure.

There are four types of hypoxia, and each is characterized by inadequate oxygen delivery to the tissues:

1. Anemic hypoxia involves poor delivery of oxygen because of either too few red blood cells or abnormal hemoglobin.
2. Ischemic hypoxia involves impaired blood circulation.
3. Histotoxic hypoxia involves the inability of the body's cells to use O_2 (cyanide poisoning causes this).
4. Hypoxemic hypoxia involves reduced arterial O_2, which can result from a lack of ventilation due to diseases such as COPD or asthma.

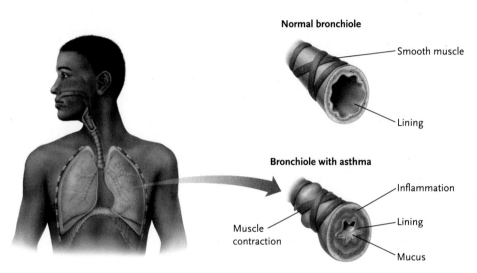

Normal bronchiole
— Smooth muscle
— Lining

Bronchiole with asthma
— Inflammation
Muscle contraction —
— Lining
— Mucus

Figure 18.19 Asthma

Asthma is an inflammatory disease of the airways.

Art sources (right): From DIGIUSEPPE/FRASER. *Biology 12 U.* © 2012 Nelson Education Ltd. Reproduced by permission. www.cengage.com/permissions

CONCEPT REVIEW 18.6

1. What two main diseases are considered chronic obstructive pulmonary diseases?
2. What is different between chronic bronchitis and emphysema?
3. What is the primary cause of COPD?
4. What are the main symptoms of asthma?
5. Name two types of medications that can be used to treat the symptoms of asthma.
6. What are the main causes of hypoxia?

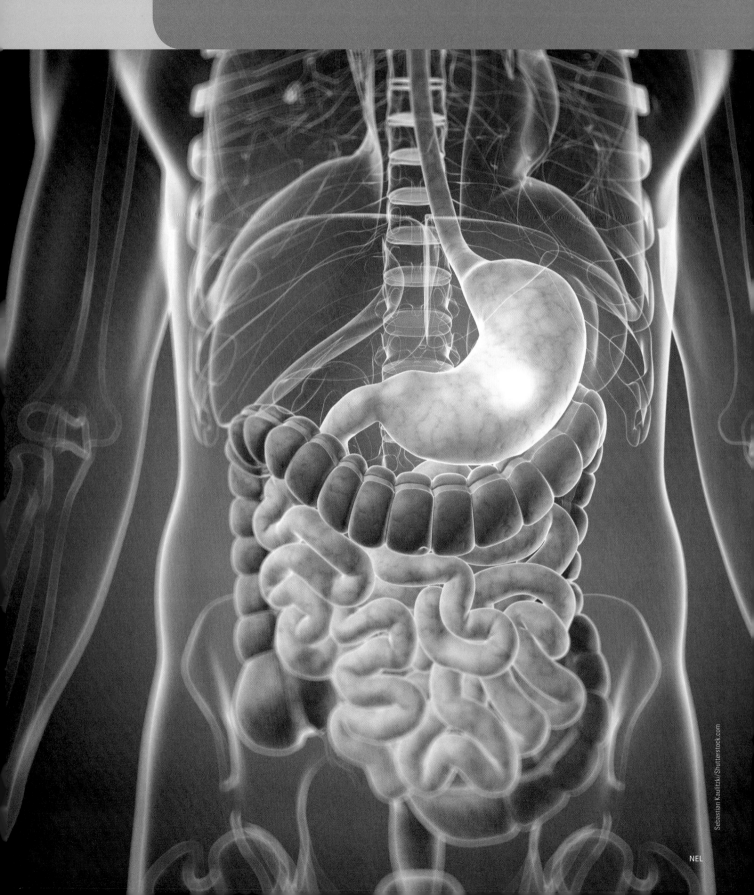

Whenever you find yourself on the side of the majority, it is time to pause and reflect.

Mark Twain

19.1 OVERVIEW OF THE DIGESTIVE SYSTEM

The digestive system consists of the gastrointestinal tract and the accessory organs. The **gastrointestinal tract** (GI tract) is the tube that food travels through: the mouth, esophagus, stomach, **small intestine**, **large intestine**, and the anus (Figure 19.1). The **accessory organs** help the digestive process by producing enzymes, bile, or other molecules; they include the salivary glands, liver, gallbladder, and pancreas.

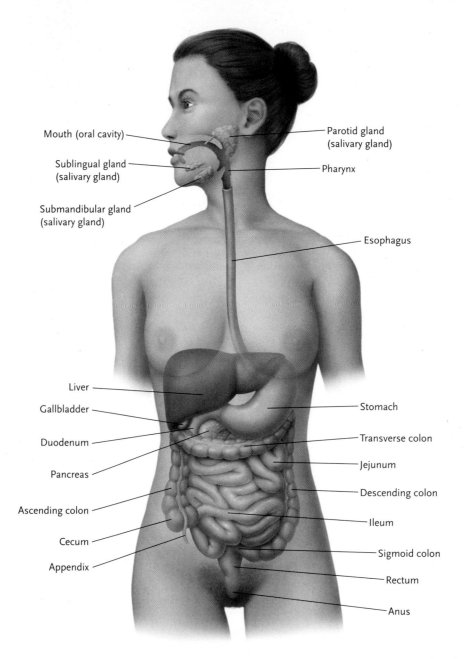

Figure 19.1 Overview of the Digestive System

The organs of the GI tract include the mouth, esophagus, stomach, small intestine, large intestine, and the anus. The accessory organs include the salivary glands, liver, gallbladder, and pancreas.

Six basic processes occur in the GI tract: ingestion, mechanical and chemical digestion, secretion, absorption, and defecation (Figure 19.2). **Ingestion** occurs when we take in and swallow food. **Mechanical digestion** refers to the physical mixing and breaking down of food into smaller pieces. This occurs when we chew, as well as during muscular contractions in the stomach and small intestine. Muscular contractions also propel food through the digestive tract. **Chemical digestion** involves digestive enzymes that chemically break bonds between molecules—such as the peptide bonds in proteins. In this way macromolecules are broken down into monomers (amino acids, glucose, fatty acids, and nucleotides). Secretion involves the release of water, ions (mostly sodium), gastric acid, enzymes, bile, and bicarbonate into various regions of the gastrointestinal tract to aid the mechanical and chemical digestion process. Most of the water and ions become reabsorbed during the absorption process. Absorption involves the process of moving digested nutrients and water and ions from the lumen of the digestive tract into the bloodstream. **Defecation** is the removal of undigested matter, called feces, from the body through the anus.

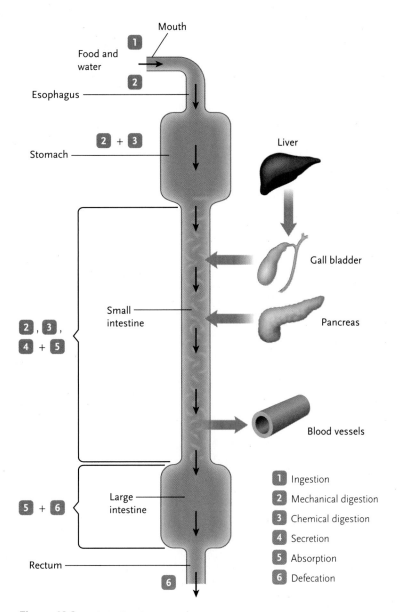

Figure 19.2   Digestive System Processes

Processes of the digestive system include ingestion, digestion, secretion, absorption, and defecation.

The overall functions of the digestive system include the following:

- Physically and chemically breaking down ingested food

- Absorbing digested nutrients into the bloodstream

- Removing undigested waste

- Producing intrinsic factor, which is required for the absorption of vitamin B12

- Housing symbiotic bacteria—the normal flora—mostly in the large intestine

DID YOU KNOW?

The human digestive tract is approximately 5 m long. If removed after death when it has no muscle tone, it can stretch to 9 m (30 ft).

CONCEPT REVIEW 19.1

1. What are the organs of the gastrointestinal tract?

2. What are the accessory organs?

3. Describe the six main processes of the digestive system.

4. List the main functions of the digestive system.

19.2 TISSUES OF THE DIGESTIVE SYSTEM

The digestive tract is an organ system composed of the organs described in Section 19.1. These organs consist of all four types of connective tissue: epithelial, connective, muscle, and nervous. From its innermost layer to the outside layer, the GI tract includes the mucosa, submucosa, muscularis, and serosa (Figure 19.3).

The **mucosa,** the inner lining of the digestive tract, is a mucus membrane composed of epithelial cells—mostly simple columnar epithelial cells in the small intestine for absorption. The mucosa is connected to a layer of areolar connective tissue, called the **lamina propria**. This layer is surrounded by a thin layer of smooth muscle tissue, called the **muscularis mucosa**. The mucosa is highly vascularized since nutrients need to be absorbed from the GI tract lumen into the bloodstream. The mucosa also contains many lymphatic vessels; capillaries and small lymph vessels extend into every villus (Figure 19.3).

The **submucosa** is the layer next to the mucosa, and it contains areolar connective tissue, larger blood and lymphatic vessels, and a network of neurons called the submucosal plexus. The neurons that innervate the GI tract are collectively called the enteric nervous system, which is part of the autonomic nervous system. A plexus is a branching network of neurons. The digestive tract has two: the submucosal plexus and the myenteric plexus in the muscularis layer. The submucosal plexus detects stimuli such as stretch in the GI tract, and it controls the secretion of substances into the digestive tract.

DID YOU KNOW?

Approximately 70% of the immune system lymph nodes are located in the region of the digestive tract because of the amount of exposure to infectious organisms in food. This part of the immune system is called the gut associated lymphatic tissue (GALT).

The **muscularis** layer contains two thick layers of primarily smooth muscle: an inner circular layer that contracts to mix food contents within the GI tract, and an outer longitudinal layer that propels food through the GI tract. Certain regions of the GI tract contain skeletal muscle in this layer: the tongue, pharynx, and

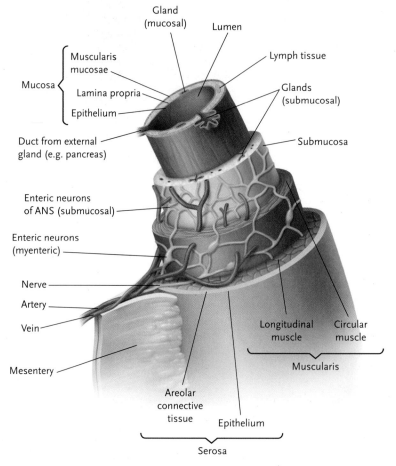

Figure 19.3   **Layers of the Digestive Tract**

The layers of the digestive tract include connective tissue, smooth muscle, nerves, and the epithelial cell lining.

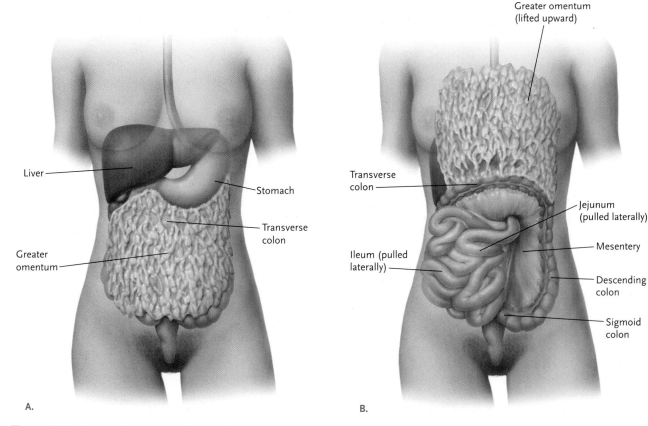

Figure 19.4   Greater Omentum and Mesentery

(A) The greater omentum, is composed of adipose tissue. (B) The mesentery holds the abdominal organs in place.

upper esophagus, which allows for voluntary control of swallowing. The external sphincter of the anus is also skeletal muscle, which allows voluntary control of defecation. Between the circular and longitudinal muscle layers lies the myenteric plexus of neurons that controls the contraction of the muscle layers.

The **serosa** is the outermost layer of the gastrointestinal tract and is the **visceral peritoneal membrane** (peritoneum) made up of areolar connective tissue covered by simple squamous epithelial tissue. The **parietal peritoneum** covers the abdominal cavity wall. To prevent friction, there is fluid between the visceral and parietal membranes (similar to the visceral and parietal membranes of the heart and lungs). Some organs in the abdominal cavity lie *behind* the parietal peritoneum, including the kidneys, which are located in the posterior portion of the abdomen and are therefore located in the *retroperitoneal cavity*.

Within the abdominal cavity are numerous nerves, blood vessels, and lymphatic vessels that supply the digestive organs. The digestive organs, nerves, and vessels are held in place by an extension of the peritoneum—the **mesentery**—and cushioned with adipose tissue called the **greater omentum** (Figure 19.4). This fat in addition to the subcutaneous skin fat contributes to "belly fat."

DID YOU KNOW?

Increased abdominal fat within the abdominal cavity is a significant risk factor for heart disease and type-2 diabetes.

CONCEPT REVIEW 19.2

1. What are the four major layers of the gastrointestinal tract?

2. Name the three components of the mucosa.

3. What are the functions of the submucosal plexus and the myenteric plexus?

4. What are the functions of the circular and longitudinal muscle layers of the muscularis?

5. What makes up the peritoneum?

6. What is the difference between the visceral peritoneum and parietal peritoneum?

7. What are the functions of the greater omentum and the mesentery?

19.3 ORGANS OF THE DIGESTIVE SYSTEM

Mouth, pharynx, and esophagus

The **mouth** is formed by the lower jaw bone (mandible) and upper jaw bone (maxilla), the hard palate (anterior maxilla and posterior palatine bones), the muscular tongue, cheeks, gums, and lips. The gums, also called the **gingiva,** cover the joint between the tooth sockets and the teeth; when inflamed, the condition is known as **gingivitis**. The front incisors and the canines tear and cut food; the premolars and molars grind food, giving us the ability to eat a wide range of foods. Adults have a total of 36 teeth: eight incisors, four canines, eight premolars, 12 molars, and four wisdom teeth (Figure 19.5).

The **tongue** is a strong muscle that aids in chewing and swallowing by moving food between the teeth and then rolling it into a ball—called a **bolus**—that can be easily passed through the pharynx and into the esophagus. The tongue contains many taste receptors that detect food flavours such as salty, sweet, bitter, sour, and savory. Savory is also called umami since it detects the amino acid glutamate, for example, in savory meats. Sensory taste information is sent to the brain via cranial nerves VII, IX, and X. When we swallow, the uvula moves up and covers the opening to the nasal passages, directing food into the pharynx.

The pharynx is the space posterior to the mouth and superior to the opening of the esophagus and trachea. When food moves through the pharynx, the epiglottis covers the opening to the trachea, and so food is directed into the **esophagus**, located behind the trachea (Figure 19.6). Once the food is in the esophagus, the muscular contractions are no longer voluntary; from this point on they are controlled by the autonomic nervous system. As the circular and longitudinal muscle layers of the esophagus contract, the bolus of food is propelled downward to the stomach. The wavelike muscle contractions that move food through the entire digestive tract is known as **peristalsis**. Before food enters the stomach it moves through the lower esophageal sphincter. This is a ring of muscle that relaxes to let food into the stomach and then contracts to prevent the movement of food back into the esophagus.

DID YOU KNOW?

Many animals lick their wounds to promote healing because their saliva contains a molecule called NGF (nerve growth factor) that helps wounds heal twice as fast. Humans do not have NGF in their saliva.

Stomach

The divisions of the stomach include the **cardia,** which is the opening of the stomach that's connected to the lower esophageal sphincter; the **fundus,** which is the top portion; the body, which is the middle portion; and **pylorus,** the inferior region (Figure 19.7). The stomach is a strong muscular organ with three layers of muscle that contract in different directions to mix and mechanically break down food: the inner oblique layer, the middle circular layer, and the outer longitudinal layer. The muscle layers are smooth muscle controlled by the enteric nervous system. Two sphincters contract to close off each opening to the stomach: the lower esophageal sphincter and the **pyloric sphincter**. The inner mucosal layer of the stomach—the **rugae**—is highly folded to allow for the stomach to expand. The volume of an empty stomach is approximately 75–100 ml, but it can easily expand to hold 1–2 L of food.

Once the bolus of food enters the stomach, the stomach adds secretions. At this point the food that breaks down is called **chyme**. Different cell types in the mucosal lining of the stomach produce different secretions that aid digestion. Food is mainly broken down by mechanical digestion in the stomach from the contractions of the muscular layers, but proteins undergo chemical digestion in the stomach's acid environment. Goblet cells produce mucus to protect the lining of the stomach from being digested and from the low pH. Chief cells produce **pepsinogen**, which is

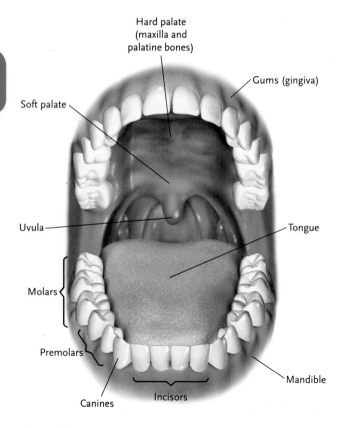

Figure 19.5   Mouth

Hard palate (maxilla and palatine bones)

Gums (gingiva)

Soft palate

Uvula

Tongue

Molars

Premolars

Mandible

Canines

Incisors

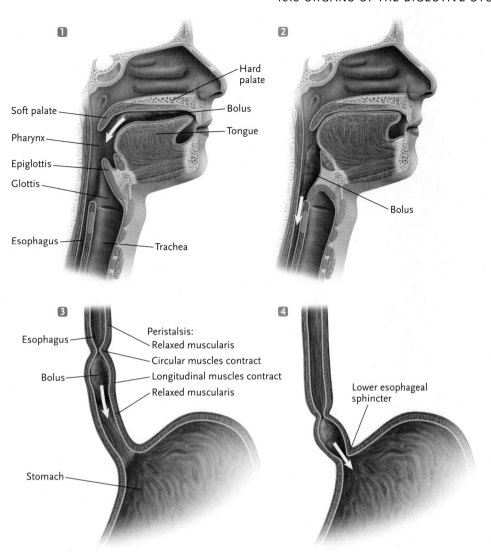

1

Soft palate
Pharynx
Epiglottis
Glottis
Esophagus

Hard palate
Bolus
Tongue
Trachea

2

Bolus

3

Esophagus
Bolus
Stomach

Peristalsis:
Relaxed muscularis
Circular muscles contract
Longitudinal muscles contract
Relaxed muscularis

4

Lower esophageal sphincter

Figure 19.6   Swallowing

Swallowing involves wavelike contractions that move food through the esophagus.

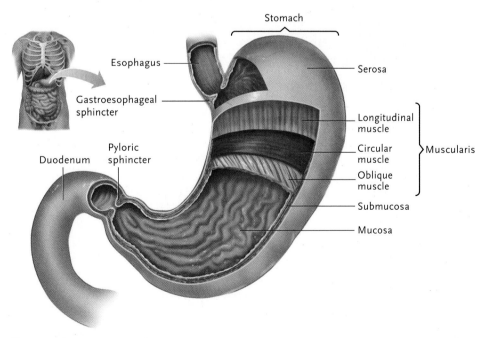

Stomach

Esophagus
Gastroesophageal sphincter

Duodenum
Pyloric sphincter

Serosa

Longitudinal muscle
Circular muscle   } Muscularis
Oblique muscle

Submucosa
Mucosa

Figure 19.7   Stomach

19

activated to form the enzyme **pepsin,** which digests proteins. Parietal cells secrete **hydrochloric acid (HCl)** that kills any pathogens that have been swallowed and converts pepsinogen into pepsin and a substance called **intrinsic factor,** which is needed to absorb vitamin B12 in the small intestine. Note that mucus, pepsinogen, intrinsic factor, and hydrochloric acid are *exocrine* secretions. *Endocrine* secretions (hormones) are discussed in Sections 19.5 and 19.6. During digestion in the stomach, small amounts of chyme are released into the small intestine in intervals through the pyloric sphincter so that only a small amount enters the small intestine at a time. Only water, ions, and some drugs and alcohol can be absorbed into the bloodstream from the stomach (Section 19.5).

Accessory digestive organs

Before considering what happens to the food once it reaches the intestines, it is important to understand the functions of the digestive system's accessory organs: the salivary glands, liver, gallbladder, and pancreas.

Salivary glands

Three pairs of salivary glands—the **parotid, submandibular,** and **sublingual** (Figure 19.6)—produce mucus, antibacterial lysozyme, enzymes, water, and salt.

In the mouth, salivary secretions moisten and soften food, which aids both mechanical and chemical digestion. Chewing breaks food into smaller pieces physically. Then two enzymes secreted by the salivary glands begin chemical digestion: **amylase** breaks down complex carbohydrates, and **lipase** begins the breaking down of fats. Lipase also breaks down bacterial cells walls and plays a role in role in decreasing tooth decay by controlling the growth of oral microorganisms. **Salivation** is stimulated by the parasympathetic division of the autonomic nervous system.

Liver and gallbladder

The **liver** is the second largest organ in the body after the skin. It is located in the upper right abdominopelvic quadrant. The liver is connected to the **gallbladder** through the common hepatic duct and the cystic duct (Figure 19.8). The liver produces bile that is stored in the gallbladder. The cells of the liver, called hepatocytes, perform all the major functions of the liver. The liver is highly vascularized, containing permeable capillaries called **hepatic sinusoids** between rows of hepatocytes. The nutrients absorbed into the bloodstream throughout the digestive tract are carried to the liver through the hepatic portal vein before they enter the main circulation to the rest of the body. The liver also

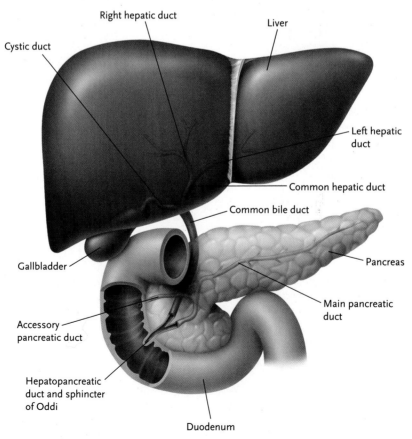

Right hepatic duct

Liver

Cystic duct

Left hepatic duct

Common hepatic duct

Common bile duct

Pancreas

Gallbladder

Main pancreatic duct

Accessory pancreatic duct

Hepatopancreatic duct and sphincter of Oddi

Duodenum

Figure 19.8 Accessory Digestive Organs
Accessory organs include the liver, gallbladder, and pancreas.

contains resident immune cells—**Kupffer cells**—that are important for breaking down old or damaged cells, including red blood cells, and any bacteria that may have entered the bloodstream from the GI tract.

During digestion, the bile is released into the duodenum of the small intestine via the **common bile duct**, which merges with a pancreatic duct that is called the **hepatopancreatic duct**. The duodenum controls entry of bile and pancreatic secretions by the **sphincter of Oddi** (Figure 19.8).

The liver performs the following functions:

- Synthesizing bile that is required for fat digestion
- Storing nutrients during the absorptive phase (stimulated by insulin)

  converts glucose into glycogen—**glycogenesis**
  converts glucose and amino acids into triglycerides—**lipogenesis**
  produces lipoproteins to transport fats from liver to adipose tissue
  stores excess fat-soluble vitamins—Vitamin A, D, E, and K
  stores minerals and trace metals

  - Breaking down nutrients during the fasting phase (stimulated by glucagon, epinephrine, thyroid hormones, and growth hormone) (Chapter 16)

  converts glycogen into glucose—**glycogenolysis**
  converts triglycerides into glucose—**gluconeogenesis**
  converts fatty acids into **ketones**, which can be converted into acetyl-CoA and can be used to produce ATP (Chapter 4)
  produces urea from breakdown products of proteins (Chapter 20)

- Breaking down old red blood cells (the spleen also does this); the heme component of hemoglobin becoming bilirubin, which is added to the bile and then excreted with other waste from digestion through defecation
- Breaking down many organic and inorganic toxins, such as drugs, alcohol, hormones, and foreign toxins from environmental exposure; and adding the breakdown products to the bile for excretion
- Excreting excess trace metals
- Synthesizing plasma proteins such as albumin, which is important for osmotic regulation as well as binding proteins for the transport of fat-soluble hormones in the bloodstream
- Synthesizing fibrinogen, the protein required for blood clotting (Chapter 17)
- Synthesizing **angiotensinogen,** involved in the renin-angiotensin system, to regulate blood pressure (Chapter 20)

- Activating vitamin D—the precursor produced in the skin during sun exposure— that's required for calcium homeostasis and immune regulation (also activated by kidneys)
- Producing insulin-like growth factors (IGFs) in response to growth hormone (Chapter 16)
- Secreting **complement proteins** involved in the immune response (Chapter 22)

DID YOU KNOW?

Bilirubin is a bright yellow pigment that contributes to the greenish colour of bile. If the liver is not functioning properly, such as during a hepatitis infection or cirrhosis of the liver, the bilirubin is not excreted, and it builds up in the tissues as red blood cells are broken down. This causes the yellow pigment to become visible in the skin: a condition called **jaundice**. Bilirubin also causes the yellowish colour of bruises as clotted red blood cells are gradually broken down.

The gallbladder stores bile that is an amphipathic molecule containing polar and nonpolar regions that act like soap to help *mechanically* break down ingested fats (Figure 19.9). Bile is produced by hepatocytes from cholesterol. Bile is composed of the bile salts, as well as many breakdown products produced by the liver, water, mucus, and other inorganic salts.

Pancreas

The pancreas is located behind and slightly below the stomach in the abdominopelvic cavity. The pancreas consists of glandular epithelial cells: a small number

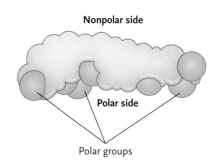

Figure 19.9 **Bile**
Bile is required for the digestion of fat.

of **islet cells** and some arranged in clusters called **acini**. The pancreas has endocrine and **exocrine** functions. Endocrine functions include the production of the hormone insulin (beta islet cells) during the absorptive phase (after eating) and the production of **glucagon** (alpha islet cells) during the fasting phase. The exocrine functions aid digestion. The pancreatic acini cells secrete **pancreatic juice** that contains enzymes, water, salt, and bicarbonate ions. Passing through the sphincter of Oddi, pancreatic juice enters the duodenum via the pancreatic duct, which merges with the hepatopancreatic duct. The enzymes break down proteins, fats, and complex carbohydrates. The bicarbonate ions neutralize the acidic chyme coming from the stomach.

Small intestine

The small intestine has three parts: the **duodenum,** the first and shortest section; the **jejunum,** the middle region; and the **ileum,** the final and longest portion. The small intestine is approximately three metres long from the pyloric sphincter to the **ileocecal sphincter,** which is the first part of the large intestine. Most digestion and absorption of nutrients occurs in the duodenum. The small intestine receives bile and pancreatic enzymes and also produces some of its own digestive enzymes. The intestinal mucosal epithelium contains goblet cells that produce alkaline mucus. This mucus protects the intestinal lining, helps neutralize the pH, and helps food move through the intestines. The circular and longitudinal muscles of the small intestines contract to mechanically digest food as well as propel it through the GI tract. The intestinal mucosa is highly folded to allow for an extensive surface area for the absorption of nutrients. In addition, the mucosal layer contains **villi** that increase the surface area and microvilli—also called the **brush border**—that further increase the surface area (Figure 19.3). Within each villus are capillaries and a lymphatic vessel, called a **lacteal,** that take up the absorbed nutrients (Figure 19.10). The ileum also has the

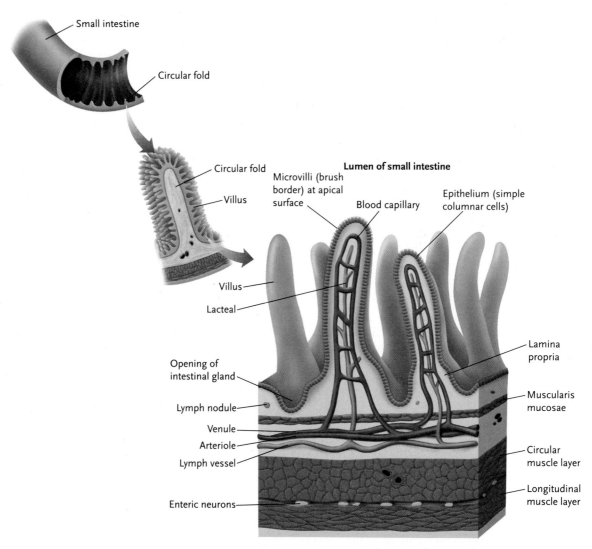

Figure 19.10   Small Intestine
The small intestine has folds, villi, and microvilli to increase the surface area for the absorption of nutrients.

important role of absorbing vitamin B12 from the diet along with the intrinsic factor that is produced by the parietal cells in the stomach.

Large intestine

The large intestine is larger in diameter than the small intestine, but shorter in length; it is approximately 1.5 m long. It begins with the ileocecal sphincter that joins the **cecum** in the lower right quadrant of the abdominopelvic cavity: also the location of the **appendix**. The **ascending colon** travels up the right side of the lower abdominopelvic cavity. The **transverse colon** sits approximately 5 cm above the umbilicus and across the lower part of the **epigastric region** (Figure 1.24). The **descending colon** travels down to the **sigmoid colon,** which connects to the **rectum, anal canal,** and the **anus** (Figure 19.11). The internal anal sphincter is regulated by the autonomic nervous system, but the external sphincter consists of skeletal muscle and can be voluntarily controlled.

The large intestine is composed of the same tissue layers as described in Section 19.2. As in the small intestine, the epithelial cells are primarily simple columnar cells that absorb water and ions. There are also goblet cells that secrete mucus to lubricate the large intestine and assist the movement of undigested and increasingly solid matter

though the final portion of the digestive tract. As the circular muscles in the large intestine contract, it takes on a pouched appearance (Figure 19.11). The functions of the large intestine include (1) absorbing water and ions, although the most water and ions are absorbed in the small intestine; (2) storing waste material and excreting it through the process of defecation; (3) absorbing vitamins created by the beneficial bacteria—the normal flora—that reside in the large bowel, specifically, B12, thiamine, riboflavin, and vitamin K.

Digestive waste, called feces, consists mostly of undigested fibre, water, dead intestinal cells, and bacteria. The bacteria in the large intestines play several very important roles in our overall health:

- Fermenting undigested substrates, mostly soluble fibre
- Regulating the immune system to tolerate food molecules
- Preventing growth of harmful bacteria
- Producing vitamin K and biotin
- Increasing growth of intestinal cells and helping maintain a healthy mucosal layer
- Producing lactic acid during fermentation, which lowers pH and prevents growth of pathogenic species

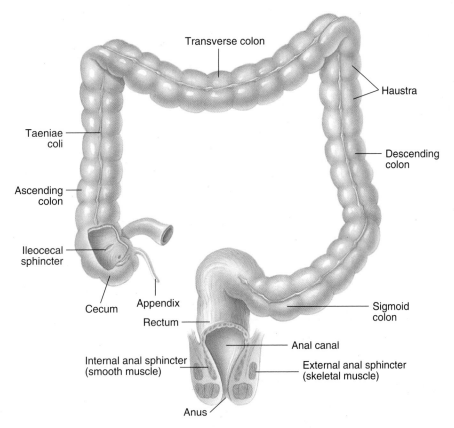

Figure 19.11   Large Intestine

The large intestine stores digestive waste, reabsorbs water, and houses beneficial bacteria.

- Converting carbohydrates into short-chain fatty acids that can be absorbed and used for energy
- Releasing important minerals from food, such as magnesium, calcium, zinc, and iron

- Metabolizing carcinogens, thereby having a significant role in preventing colon, breast, and prostate cancers

DID YOU KNOW?

Mice grown without gut bacteria in a sterile environment need to eat 30% more food to weigh the same as a mouse with normal gut bacteria.

DID YOU KNOW?

Eating fermented foods that contain live bacteria such as sauerkraut or yogurt is very beneficial for the overall health of the digestive system, immune system, and nervous system, and has been shown to improve conditions such as anxiety and depression.

CONCEPT REVIEW 19.3

1. Which bones make up the mouth?
2. What is the gingiva?
3. How many teeth do adults have if they have all of their wisdom teeth?
4. Name the functions of the tongue.
5. What is peristalsis?
6. What is the role of the uvula? Epiglottis?
7. What are the muscle layers of the stomach?
8. What is the role of the pyloric sphincter?
9. What exocrine secretions are produced by the parietal cells, chief cells, and goblet cells?
10. List the functions of the liver.
11. What is bile made from?
12. Describe the endocrine and exocrine functions of the pancreas.
13. Where does the most absorption of nutrients take place?
14. What special features are found in the small intestine to increase surface area for absorption?
15. Name the regions of the large intestine from the ileocecal sphincter to the anus.
16. What are the functions of the large intestine?
17. What are some important functions of gut bacteria?

19.4 DIGESTIVE ENZYMES

Recall from Chapter 2 that enzymes are a group of proteins that catalyze chemical reactions in every cell of the body. The enzymes involved in digestion bind to *specific* substrates and allow hydrolysis reactions to break the chemical bonds that hold together the monomers within the macromolecules (chemical digestion). Table 19.1 lists the macromolecules and their monomers.

TABLE 19.1

Macromolecules and Their Monomers	
Macromolecules	**Monomers**
proteins	amino acids
triglycerides	glycerol and fatty acids
nucleic acids (DNA and RNA)	nucleotides
complex carbohydrates (starch)	glucose only
disaccharides	
sucrose	glucose and fructose
lactose	glucose and galactose
maltose	glucose and glucose

Enzymes are secreted in the saliva, stomach, small intestine, and pancreas. Chemical digestion begins in the mouth, with amylase breaking starch into smaller chains of glucose molecules and some maltose (disaccharides), and lipase breaking triglycerides into glycerol and fatty acids. Because the food is in the mouth for only a short time, this process of breaking down macromolecules is incomplete at this stage.

In the stomach, the low pH denatures proteins, which become unfolded from their tertiary structure. Pepsin begins breaking the peptide bonds in proteins into peptides, which are short chains of amino acids. No other nutrients are chemically digested in the stomach.

The pancreas secretes several enzymes into the small intestine. In addition, the pancreas secretes bicarbonate ions into the small intestine to neutralize the acidic stomach contents. Enzymes can only function in the small intestine in a neutral pH. Pancreatic **amylase** breaks starch into short chains of glucose and maltose. After bile has emulsified the large fat droplets into small droplets, pancreatic **lipase** breaks down triglycerides into two individual fatty acids and monoglycerides (glycerol with one fatty acid attached). Lipase can act only on the triglycerides on the outside of the droplet, and so it breaks down all of the fats gradually. Fats take the longest of all macromolecules to chemically digest. Therefore, the rate of this step is limited: the more fat in the meal the longer it takes for the meal to be digested. For this reason, it is beneficial to have some healthy fats in a meal so the feeling of fullness lasts longer but not so much fat that the food stays in the stomach for a long period of time.

DID YOU KNOW?

The most common cause of heartburn is a large meal with a high fat content. Food leaves the stomach and enters the small intestine in small amounts, which must be almost completely digested before chyme from the stomach can enter the intestine. Meals with high-fat content take a very long time to digest. So the stomach continues to mix food for a long time, causing the acidic content to be pushed up into the esophagus through the lower esophageal sphincter, especially after a large meal, and causing damage to the lower esophagus. Heartburn is also more common in people with a large amount of abdominal fat, which increases pressure in the abdominal cavity, forcing stomach contents into the lower esophagus.

The pancreas secretes two enzymes that break down the nucleic acids DNA and RNA that are found in all plant and animal cells. **Ribonuclease** breaks RNA into nucleotides, and **deoxyribonuclease** breaks DNA into nucleotides. Nucleotides can also be broken down into nucleosides, which are just the sugar and base of the nucleotide (Chapter 8). The pancreas produces several enzymes that break down proteins. Pepsin from the stomach does not function once it reaches the neutral small intestine. Pancreatic **trypsin** and **chymotrypsin** break proteins into peptides. Then **carboxypeptidase** breaks the peptides into individual amino acids. It is possible to absorb individual amino acids as well as small peptides, whereas carbohydrates have to be broken down completely into individual monosaccharides to be absorbed.

The small intestine produces enzymes that break down disaccharides: **maltase** breaks down maltose, **sucrase** breaks down sucrose, and lactase breaks down lactose. People who are lactose intolerant do not produce lactase, so this sugar is not absorbed; eventually it reaches the large intestine where the residing bacteria population break it down and produce gas as a by-product. The small intestine also produces **peptidases** that work with pancreatic carboxypeptidase to break peptides into amino acids.

CONCEPT REVIEW 19.4

1. What are the breakdown products of proteins, carbohydrates, triglycerides, and nucleic acids?

2. Make a chart summarizing the enzymes involved in chemical digestion. Include where they are made and what food molecules they break down.

3. Which type of food molecule takes the longest to digest?

4. How does the acidic content of the stomach become neutralized once it reaches the small intestine?

19.5 ABSORPTION

Absorption refers to the movement of nutrients across the simple columnar epithelial cells of the intestine into the bloodstream. Nutrients must move across the apical and basolateral membranes of the epithelial cells into the lamina propria connective tissue, and then into either a blood capillary or the lacteal (Figure 19.12). Different nutrients are absorbed by different mechanisms, including simple diffusion, osmosis, facilitated diffusion, coupled transport, and active transport (Chapter 3). Anything that is not absorbed continues through the gastrointestinal tract and is eliminated as feces.

Water, ions, and water-soluble vitamins

Water can be absorbed into the bloodstream by moving through the epithelial cell membranes (osmosis), or between the epithelial cells. Just like the reabsorption of water from the kidney tubule into the bloodstream, as more ions are absorbed more water is absorbed. During digestion in the small intestine, water and ions are

secreted into the lumen to aid digestion. This secreted water and ions as well as the ingested water and ions are absorbed mostly in the duodenum. However, water is also absorbed in the stomach and all sections of the small intestine and the large intestine. All but approximately 200 ml of water is reabsorbed at some point along the digestive tract.

DID YOU KNOW?

Chicken soup broth helps you to feel better when you are sick simply because it is high in sodium. As we absorb the sodium in the intestine, we absorb more water, which minimizes water loss through diarrhea.

Ions and minerals can only move across cell membranes through protein ion channels (Figure 19.13). Each ion moves through a specific ion channel. Calcium can be absorbed only if vitamin D is present. Recall

19

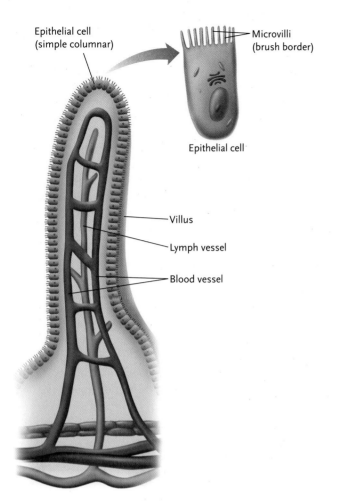

Epithelial cell
(simple columnar)

Microvilli
(brush border)

Epithelial cell

Villus

Lymph vessel

Blood vessel

Figure 19.12   Absorption in the Small Intestine

from Chapter 16 that vitamin D is required to signal the GI tract to absorb calcium; this occurs because vitamin D signals the epithelial cells to express the membrane protein that allows calcium to enter the cell. Other minerals, such as iron, zinc, potassium, chloride, magnesium, are all absorbed as ions by facilitated transport through their own specific membrane protein.

Water-soluble vitamins, the B vitamins, and vitamin C are co-transported with sodium ions across the apical membrane; the exception is vitamin B12, which requires intrinsic factor for it to be absorbed. Vitamin B12 is absorbed in the last part of the small intestine, the ilium. Some people cannot produce intrinsic factor and will develop pernicious anemia; which can be treated with injections of B12.

Monosaccharides

Glucose and galactose are absorbed on the apical membrane by co-transport with sodium ions (Figure 19.13) and through **GLUT** (glucose transporter) membrane carrier proteins on the basolateral side. Fructose is transported through both membranes by GLUT carrier proteins. The sodium that enters the

cell through co-transport with glucose or galactose is pumped out of the cell by the sodium-potassium pump (Chapter 3). Once monosaccharides pass through the basolateral membrane, they enter the capillary in the villus. All blood from the digestive tract flows via the hepatic portal vein to the liver where the nutrients are processed and stored. Once sugars enter the blood and circulate to the pancreas, the increased blood sugar stimulates the pancreas to produce insulin. Insulin circulates to the cells and signals the production of more GLUT membrane carrier proteins so that more sugar can be absorbed in the intestine as well as taken up by other body cells for ATP production.

Peptides and amino acids

Amino acids are absorbed through the apical membrane by co-transport with sodium ions (Figure 19.13), and they move through the basolateral membrane via amino acid carrier proteins. We also absorb small peptides of two or three amino acids by co-transport with hydrogen ions. Epithelial cells have peptidases that break down the small peptides into amino acids inside the cell before they are transported through the basolateral membrane. It is important to note that we also break down and absorb amino acids from our body's dead intestinal cells and the enzymes that were used during digestion; so we do not lose the amino acids that were used to make digestive enzymes. Amino acids move into the capillaries and circulate to the liver.

Nucleotides

Nucleotides are absorbed by co-transport with sodium ions on the apical membrane, and then transported through nucleotide carrier proteins on the basolateral membrane. Like monosaccharides and amino acids, nucleotides are absorbed into the bloodstream, unlike fats that are absorbed into the lacteal.

Fat and fat-soluble vitamins

Fats are mechanically broken down into small fat droplets by bile, and then lipase chemically breaks down triglycerides into monoglycerides and free fatty acids. Once the triglycerides are enzymatically broken down, they form very small droplets called **micelles**. Micelles continually form from newly digested fat droplets. Then they break down into individual fatty acids and monoglycerides and move into the epithelial cells (Figure 19.14). Unlike monosaccharides, amino acids, and nucleotides, fats can be absorbed into the epithelial cells by simple diffusion, rather than by facilitated transport, because cell membranes are composed of fats.

Once inside the epithelial cell, fatty acids and monoglycerides re-form into triglycerides. This occurs on the smooth endoplasmic reticulum in the epithelial cell. The triglycerides are then surrounded by proteins,

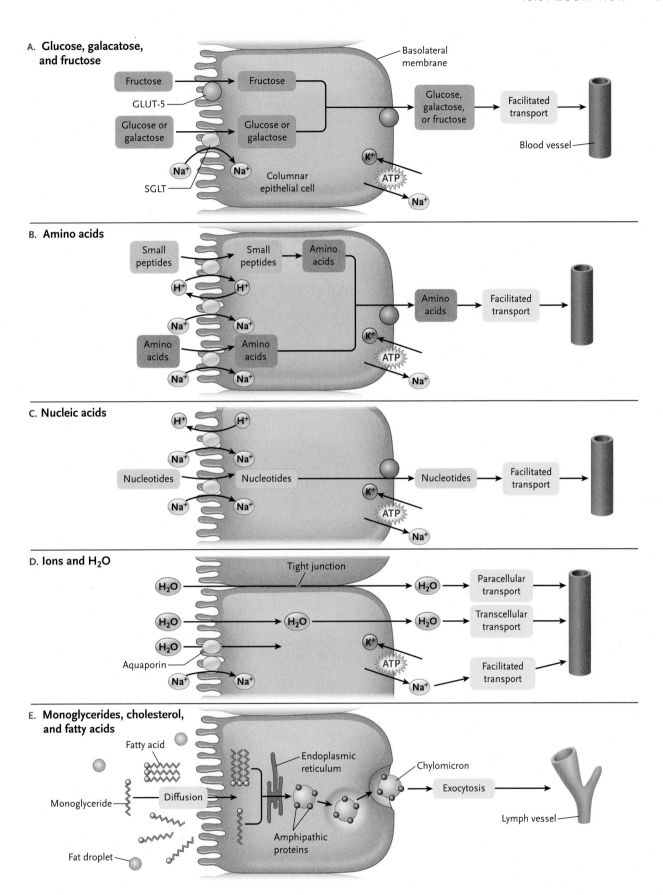

Figure 19.13   Absorption of Nutrients

Absorption of nutrients involves the movement of amino acids, monosaccharides, and fats across the epithelial cell membrane and then into the bloodstream.

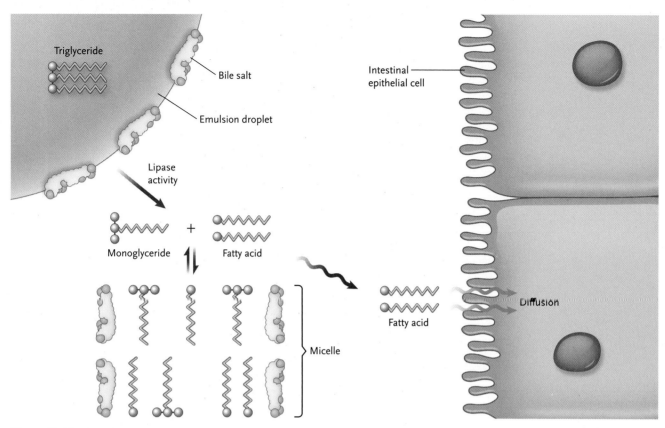

Figure 19.14   Fat Absorption

Fat absorption occurs by diffusion into epithelial cells.

forming a **chylomicron**, which then leaves the basolateral side of the cell and enters the lacteal. Chylomicrons are too large to enter the bloodstream, but lacteals have larger pores that allow fats to enter. Once in the lymph vessels, the fats flow with the lymph through the lymphatic system until they reach the thoracic duct and enter the left subclavian vein along with the lymph (Figure 19.15). After entering the bloodstream, chylomicrons circulate to the adipose tissue and the liver, where they are either directly taken up by fat cells or bound to lipoproteins in the liver and transported to cells for ATP production or to adipose tissue to be stored.

Short-chain fatty acids with less than 12 carbons are small enough to diffuse directly into the bloodstream. Some cholesterol and some triglycerides can also directly enter the blood capillaries in the villi. When blood cholesterol levels are tested, free triglycerides are also measured and should be lower than 2.2 mmol/L. When the epithelial mucosa is damaged, triglycerides are more likely to enter the bloodstream than the lymph vessels. The epithelial cells are held together by tight junctions, but these junctions can be damaged by inflammation, causing an increase in permeability of the intestinal lining (sometimes called leaky gut syndrome). Many things can affect the integrity of the epithelial barrier,

including excessive alcohol intake, inflammatory diseases, medications such as ibuprofen, and high blood sugar. This is one reason why diabetes is a risk factor for heart disease.

The cholesterol part of the composition of bile is also reabsorbed in the small intestine. A diet high in fibre decreases the amount of bile or dietary cholesterol absorbed.

Fat-soluble vitamins are absorbed along with fatty acids, monoglycerides, and cholesterol by simple diffusion across the apical membrane, and they are packaged into chylomicrons with other fats and enter the lacteal.

Gut bacteria can break down fibre that we cannot break down, such as cellulose, and convert it into short-chain fatty acids that can be absorbed by simple diffusion into the blood in the large intestine.

ATP is required during digestion for smooth muscle contraction, enzyme production, and moving molecules across membranes (the basolateral sodium-potassium pumps must maintain sodium balance). So the process of digesting food uses a certain amount of calories; this is called the **thermogenic effect**. It takes more energy to digest protein than it does to digest fats or carbohydrates.

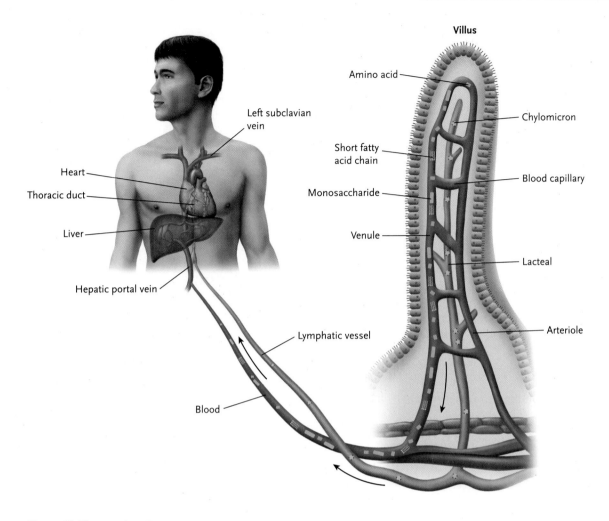

Figure 19.15 Nutrient Transport

The transport of nutrients occurs through the lymphatic and circulatory systems.

CONCEPT REVIEW 19.5

1. How are ions and water absorbed in the small intestine?

2. What vitamin is required for calcium absorption?

3. Which vitamin requires intrinsic factor for absorption?

4. Which nutrients enter the epithelial cells by co-transport with sodium ions?

5. Why are sodium-potassium pumps required in the basolateral membrane?

6. Explain how fats are absorbed into the lymphatic vessels.

7. How are water-soluble and fat-soluble vitamins absorbed?

8. Explain how the intestinal mucosa can be damaged and how this impacts the absorption of nutrients.

9. What is the thermogenic effect?

19.6 REGULATION OF DIGESTION

Digestion occurs in three phases: cephalic phase, gastric phase, and the intestinal phase. The **cephalic phase** is initiated by the sight, smell, or thought of food. It triggers parasympathetic signalling to the submucosal and myenteric plexus to begin secretion of gastric juices and to start muscle contractions in anticipation of ingesting food—causing the stomach to "growl." Salivation also begins during the cephalic stage, and once food is ingested the taste of food further increases secretion and motility of the GI tract. Motility refers to the contractions of the smooth muscle in the GI tract. The **gastric phase** begins when food enters the stomach. The physical presence of food contacting the walls of the stomach lining signals the gastric secretions and muscle contractions. The **intestinal phase** begins once the first bolus of chyme enters the duodenum and triggers intestinal motility and the release of pancreatic secretions, bile, and intestinal enzyme secretion. Once the intestinal phase begins, gastric motility begins to decrease to ensure that the intestine is not overloaded.

Foods that contain a large amount of fat require more time to completely digest; whereas simple carbohydrates, such as fruit, are digested very quickly. Certain hormones play a role in the regulation of the rate of digestion. In the stomach, food enters and stretches the stomach lining, causing the release of the hormone **gastrin**—secreted by *enteroendocrine* cells in the gastric glands in the pyloric region of the stomach (Figure 19.16). Gastrin acts on the parietal cells to produce hydrochloric acid and intrinsic factor. It also stimulates muscle contractions and stimulates the chief cells and mucus cells to produce pepsinogen and mucus, respectively.

As food enters the small intestine, the acidic chyme stimulates the intestinal mucosa to produce the hormone **secretin**. Secretin stimulates the pancreas to release bicarbonate ions that neutralize the acid and also the release of pancreatic enzymes. As food is chemically broken down, and proteins are broken down into amino acids, and triglycerides are broken down into fatty acids and monoglycerides, the presence of these building blocks stimulates the small intestine to produce the hormone **cholecystokinin (CCK)**. CCK acts on the stomach to slow gastric emptying, causes the gallbladder to contract and the sphincter of Oddi to relax, allowing the release of bile into the small intestine. CCK is the hormone that signals the satiety centre in the hypothalamus and causes the feeling of fullness. CCK is not produced until food starts entering the small intestine: during the intestinal phase. Eating quickly can lead to eating much more food than needed, so it is important to eat slowly to avoid overeating.

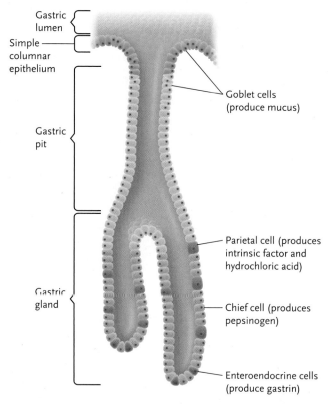

Figure 19.16 Gastric glands
Gastric glands contain cells that secrete hormones.

DID YOU KNOW?

Since CCK is triggered by the digestion of fat and protein, drinking sugary soft drinks or candy does not cause a feeling of fullness. A doughnut is sugar and fat, so a doughnut can cause a sense of fullness. Research shows that people who add soft drinks to their diet do not decrease their total calorie intake; one reason is because drinking a Coke does not stimulate CCK.

CONCEPT REVIEW 19.6

1. Describe the three phases of digestion.
2. Create a chart to summarize the hormones involved in digestion. Include where the hormone is produced, the stimulus for hormone production, and its function.
3. Which hormone causes us to feel full after eating?

19.7 DISORDERS OF THE DIGESTIVE TRACT

Many disorders can affect the digestive system, including certain infections, autoimmune diseases, cancer, allergies, and inflammation. The following list highlights some of the most common.

- Gastroesophageal reflux disease (GERD). This disease occurs when acidic contents of the stomach are forced up into the lower esophagus, causing irritation and a burning sensation. Acid in the lower esophagus can damage the sensitive tissue and cause ulcerations, inflammation, and possibly cancer. The most common causes are obesity, hiatal hernia, and a diet high in fat. Eating foods very high in fat slows digestion and keeps food in the stomach longer.

- Hiatal hernia. The lower esophageal sphincter is a ring of muscle located at the lower end of the esophagus and just above the stomach. It is meant to contract and keep the stomach contents in the stomach during digestion. When a hernia occurs, the sphincter and a portion of the upper stomach protrude through the diaphragm into the chest cavity. Hernias may not always cause symptoms, but symptoms can include a dull ache in the chest, difficulty swallowing as though food is staying in the esophagus, and possibly heart palpitations. The most common cause is excessive force in the abdominal cavity, such as from violent vomiting, coughing, lifting heavy weight, or straining with constipation. (See Figure 19.17a.)

- Ulcers. Located along the digestive tract, these sores occur most commonly in the duodenum, but they can occur within the stomach, esophagus, or any region of the small or large intestine. The most common cause is an overgrowth of the bacterium *Helicobacter pylori* (*H. pylori*), and therefore the most common treatment is a course of antibiotics. Other factors can contribute, such as overuse of pain killers, excessive alcohol consumption, and smoking. Smokers are more prone to duodenal ulcers, and alcoholics are more prone to esophageal ulcers. (See Figure 19.17b.)

- Crohn's disease. This inflammatory disease primarily affects the small intestine but can occur in any region of the digestive tract. Crohn's is a chronic disease with no known cure. The most common symptoms include abdominal pain, cramping, diarrhea, nausea and vomiting, fatigue, and weight loss. Chronic inflammation in the bowel is associated with an increased risk of bowel cancer.

- Ulcerative colitis. This chronic autoimmune disease causes chronic inflammation and ulcerations in the large intestine and often occurs in a relapsing-remitting cycle. There is no known cause or cure for any autoimmune disease (Chapter 22). Symptoms often include cramping, bloody diarrhea, and fatigue. As with Crohn's disease, people with ulcerative colitis have an increased risk of colon cancer.

- Gastritis. This is a general term related to inflammation of the stomach. However, gastritis most often refers to an acute infection causing nausea, vomiting, and diarrhea. The most common cause is an infection by a norovirus (Norwalk virus)—which is often called the stomach flu. But gastritis is not caused by influenza. In healthy people, the immune system controls these infections within 24 hours to three days.

- Diverticulitis. Small pouches called diverticula can form in regions of the bowel where the wall is weakened and inflamed. Symptoms include abdominal pain, nausea, diarrhea, and fever. The cause is unknown, but it is possible that increased pressure from straining with constipation is a risk factor and that a diet high in fibre can help to prevent the disease. It was previously thought that people with diverticulitis should avoid nuts and seeds, but research has shown there is no correlation, and, in fact, a higher intake of nuts may help to avoid the development of this disease. (See Figure 19.17c.)

- Irritable bowel syndrome. Symptoms include cramping, abdominal pain, gas and bloating, and possibly constipation or diarrhea. Unlike Crohn's disease and ulcerative colitis, irritable bowel syndrome is not an inflammatory disease. The exact cause is not known, but it may be related to food allergies or sensitivities, infections, malabsorption of nutrients, or abnormalities in the gut flora.

- Hemorrhoids. This condition involves clusters of swollen veins in the anus and rectum. It is caused by increased pressure, most commonly by straining with constipation. A high fibre diet is very important for preventing constipation.

DID YOU KNOW?

The most common food allergies are peanuts, eggs, and shellfish. Although there has been a dramatic increase in the number of people with food allergies in the last 20 years, no specific reason for this is known.

19

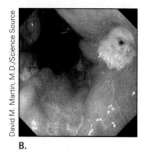

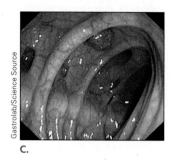

Figure 19.17 **Disorders of the Digestive Tract**
(A) Hiatal hernia, (B) ulcer, and (C) diverticulitis

CONCEPT REVIEW 19.7

1. Which of the disorders of the digestive tract explained in this section are chronic inflammatory diseases?

2. Which digestive disorders can be prevented with a high-fibre diet?

3. Which digestive disorders are most often caused by infectious organisms?

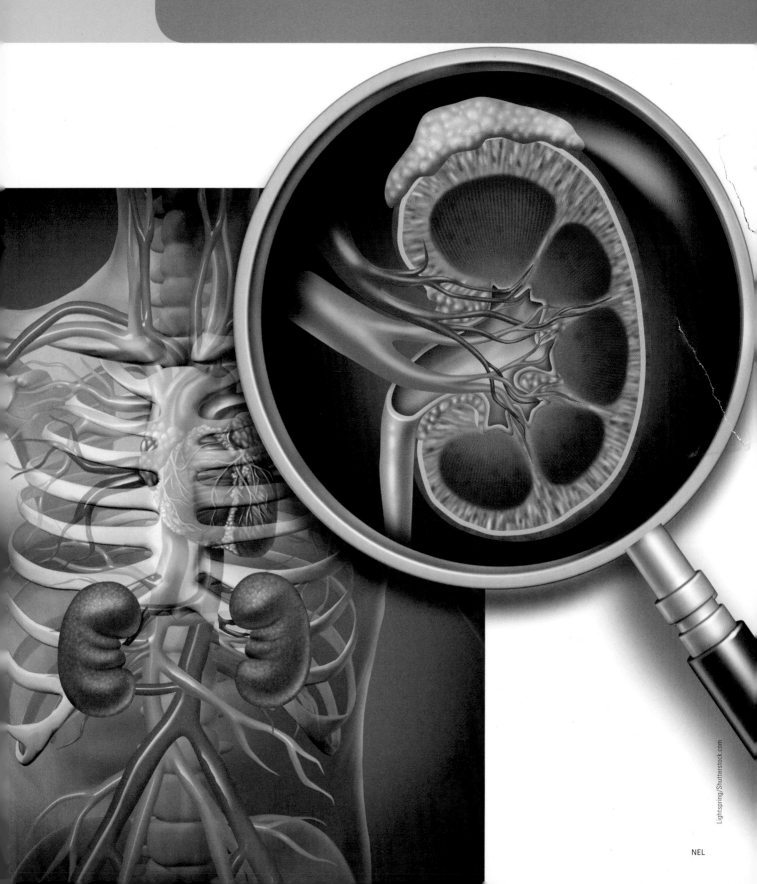

Live as if you were to die tomorrow. Learn as if you were to live forever.

Mahatma Gandhi

20.1 OVERVIEW OF THE URINARY SYSTEM

The urinary system is located in the abdominal cavity and consists of the kidneys, ureters, bladder, and urethra. The kidneys are located in the posterior abdominal cavity behind the digestive organs (Figure 20.1). The urinary system regulates the body's **osmotic composition**, water, and ions. The kidneys filter our entire blood volume more than 30 times every day excreting extra water, ions, and waste products as **urine**.

The urinary system performs the following functions:

- Regulating the blood's water and ion composition
- Excreting the waste products of metabolism
- Regulating blood pressure—the most important organ to do so
- Regulating blood pH by excreting excess H^+ ions
- Releasing two hormones—**calcitriol,** the active form of vitamin D, and **erythropoietin,** which stimulates the bone marrow to produce red blood cells

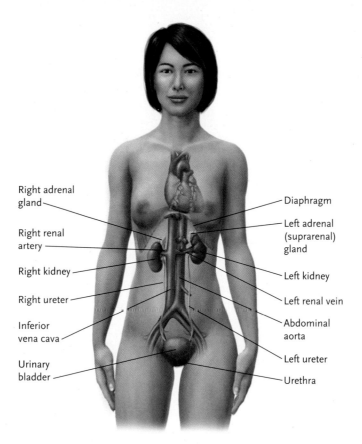

Figure 20.1 Location of the Kidneys
The kidneys are located in the posterior abdominal cavity.

CONCEPT REVIEW 20.1

1. What are the functions of the urinary system?

2. What structures make up the urinary system?

20.2 ANATOMY OF THE URINARY SYSTEM

Each kidney is shaped like a kidney bean, approximately three inches long. The adrenal glands are attached to the superior region of each kidney. The kidneys receive their blood supply from the **renal arteries** that branch from the abdominal aorta. The renal arteries enter each kidney from the medial side at a region called the **renal hilum;** this is the same region where the renal vein, lymphatic vessels, nerves, and the ureter leave the kidney (Figure 20.2a). The kidney is surrounded by a **renal capsule,** a protective covering for the kidneys that's composed of connective tissue and connected to adipose tissue that also helps to protect and cushion the kidneys.

The kidney consists of an outer cortex and an inner medulla. The functional unit of the kidney is the **nephron,** and each kidney contains about 1 million nephrons. Each nephron has a glomerulus located in the cortex region, which is where the initial filtering of the blood occurs. As the filtered fluid moves through the nephron, water and ions are reabsorbed back into the bloodstream, specifically,

into the peritubular capillaries that surround the nephron in both the cortex and medulla regions. The fluid left in the nephron will be excreted as urine. The medulla region of the kidney has a higher **osmotic gradient** compared to the cortex and is important for increased water reabsorption back into the bloodstream. The renal medulla contains areas called **renal pyramids,** which is where many nephrons converge onto **collecting ducts** and where the final reabsorption of water and ions occurs before the urine is fully formed. The regions of cortex that extend toward the middle of the kidney, in between the pyramids, are called the **renal columns.** Each section of kidney from capsule to renal hilum that encompasses one renal pyramid is called a **renal lobe** (Figure 20.2b).

Urine drains from the collecting ducts in the pyramids, through the **minor calyx** and then the **major calyx** to the large **renal pelvis** in the centre of the kidney. The renal pelvis is connected to the **ureters** that carry the urine to the bladder for storage until the bladder is full. The urinary bladder contains an internal layer of **transitional epithelial cells** (Chapter 11). It is surrounded by connective tissue and a muscle layer called the **detrusor muscle** (Figure 20.3). The **external urethral sphincter** is skeletal

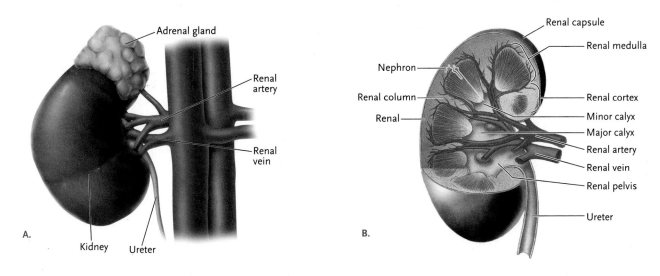

A.
- Adrenal gland
- Renal artery
- Renal vein
- Kidney
- Ureter

B.
- Renal capsule
- Renal medulla
- Nephron
- Renal column
- Renal
- Renal cortex
- Minor calyx
- Major calyx
- Renal artery
- Renal vein
- Renal pelvis
- Ureter

Figure 20.2   Anatomy of the Kidney

(A) External anatomy and (B) internal anatomy

Art source (B): From DIGIUSEPPE/FRASER. Biology 12 U. © 2012 Nelson Education Ltd. Reproduced by permission. www.cengage.com/permissions

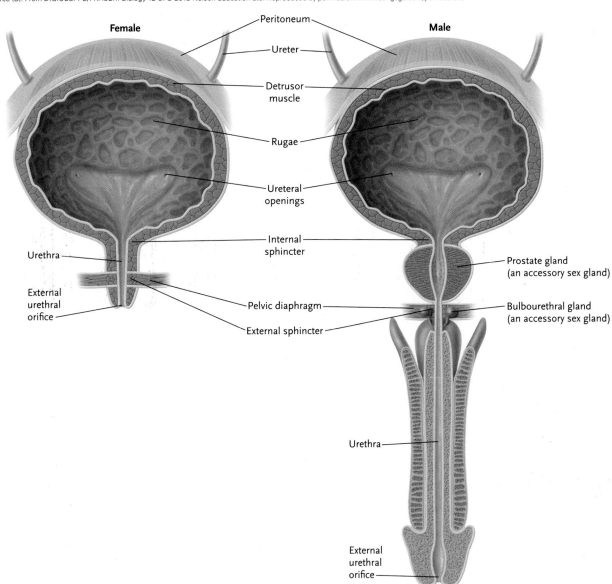

Female   **Male**

- Peritoneum
- Ureter
- Detrusor muscle
- Rugae
- Ureteral openings
- Internal sphincter
- Urethra
- External urethral orifice
- Pelvic diaphragm
- External sphincter
- Prostate gland (an accessory sex gland)
- Bulbourethral gland (an accessory sex gland)
- Urethra
- External urethral orifice

Figure 20.3   Anatomy of the Bladder

20

muscle and is voluntarily controlled. The **internal urethral sphincter** is controlled by the autonomic nervous system. Urine is excreted from the body through the **urethra**.

When the bladder is full, stretch receptors (mechano-receptors) trigger a spinal cord reflex, and information travelling through parasympathetic nerves causes the detrusor muscle to contract and the internal sphincter to relax, leading to urination: also called **micturition**.

Blood enters the renal hilum through the renal artery, which branches inside the kidney to the **segmental arteries** and then to the **interlobar arteries** that surround the pyramids. The blood flows toward the cortex of the kidney, where the glomeruli are located (Figure 20.4).

From the interlobar arteries, blood flows to the cortex through the **cortical radiate arteries** and finally to the **afferent arteriole**, which branches into the capillaries of each **glomerulus**. This is where filtration occurs. From the glomerulus, the portion of blood that was not filtered continues to the efferent arteriole, which branches to the **peritubular capillaries**. In the peritubular capillaries water and nutrients are reabsorbed from the nephron. The blood flows from the capillaries to the **cortical radiate veins**, to the **arcuate veins** that flow back from the renal pyramids to the **interlobar veins,** and finally back to the large **renal vein** that connects to the inferior vena cava.

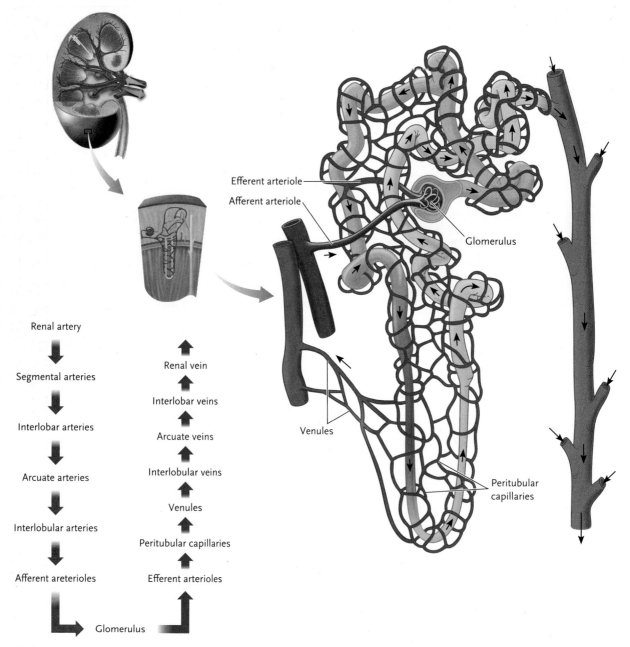

Figure 20.4   **Blood Flow to the Kidneys**

Blood flows into each kidney through the renal artery and leaves through the renal veins.

CONCEPT REVIEW 20.2

1. What is the difference between the cortex and the medulla of the kidney?

2. List the structures that fully formed urine flows through, from the collecting ducts to the urethra.

3. List the path of blood flow through the kidney from the renal artery to the renal vein.

20.3 FUNCTIONS OF THE NEPHRON

The nephron is the smallest functional unit of the kidney (Figure 20.5). A nephron consists of several specific regions that have specific functions. Fluid moves through the following kidney structures from the time it leaves the bloodstream until urine is produced: renal corpuscle, which is made up of the glomerulus and the Bowman's capsule; proximal convoluted tubule; descending loop of Henle; ascending loop of Henle; distal tubule; and collecting ducts.

The kidney performs three important tasks: filtration, reabsorption, and secretion of certain substances in certain regions of the nephron (Figure 20.5). **Excretion** is the physical removal of the urine from the bladder through the urethra. Filtration occurs when fluid (including ions and other molecules) move from the glomerulus into the Bowman's capsule. **Reabsorption**

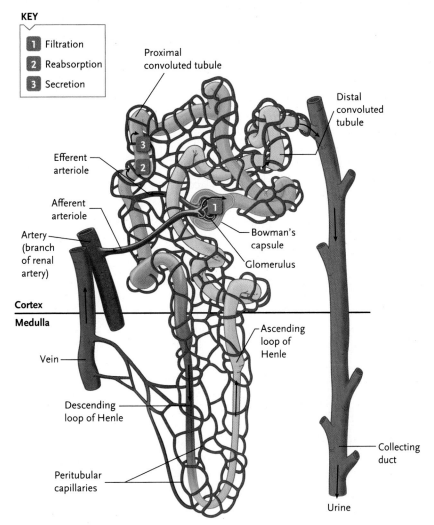

KEY

1 Filtration
2 Reabsorption
3 Secretion

Proximal convoluted tubule

Distal convoluted tubule

Efferent arteriole

Afferent arteriole

Artery (branch of renal artery)

Bowman's capsule

Glomerulus

Cortex

Medulla

Vein

Ascending loop of Henle

Descending loop of Henle

Peritubular capillaries

Collecting duct

Urine

Figure 20.5   Nephron

The nephron is the smallest functional unit of the kidney. The processes of filtration (1), reabsorption (2), and secretion (3) occur in specific regions of the nephron.

involves the movement of fluid from the tubule back into the bloodstream, and secretion involves the movement of fluid from the peritubular capillaries into the tubule.

amount of substance excreted = amount filtered –
amount reabsorbed + amount secreted

Some substances are completely reabsorbed after filtration: such as glucose and amino acids. Some substances are filtered and then a proportion is reabsorbed, and the rest is excreted. Some substances are completely excreted by filtration and secretion with no reabsorption, such as the breakdown products of medications and other toxins.

Filtration

Filtration occurs in the renal corpuscles. **Podocytes** surround the glomerular endothelium and form a leaky filtration membrane that permits plasma containing water, ions, nutrients, and waste products to move into the **Bowman's space** (Figure 20.6). Blood cells and most plasma proteins remain in the blood because they are too large to pass through the membrane. Once the fluid has moved from the blood to the Bowman's capsule, it is called **filtrate**. Recall from Chapter 17 that when plasma moves from the tissue capillaries to the interstitial space, it becomes interstitial fluid. The same thing occurs in the kidneys, except that the capillaries allow much more fluid to leave the circulatory system and enter the kidney tubules.

The **macula densa** and **juxtaglomerular cells** (Figure 20.6) play a role in the regulation of filtration and excretion. Autonomic neurons innervate the blood vessels in the kidney to regulate blood pressure via vasodilation and vasoconstriction of the arterioles, which have a smooth muscle layer that can contract and relax.

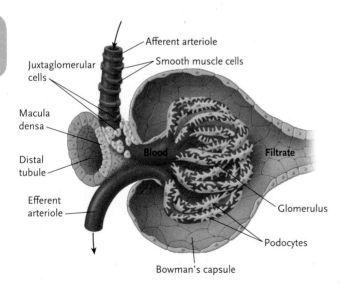

Figure 20.6 **Filtration**

Filtration occurs in the renal corpuscle.

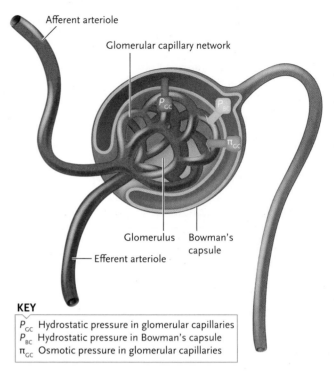

KEY

P_{GC}	Hydrostatic pressure in glomerular capillaries
P_{BC}	Hydrostatic pressure in Bowman's capsule
π_{GC}	Osmotic pressure in glomerular capillaries

Figure 20.7 **Pressures That Affect Filtration**

Hydrostatic and osmotic pressures affect filtration.

The pressure that causes filtration is the blood pressure in the glomerular capillaries, otherwise known as the **hydrostatic pressure** (P_{GC}) in the glomerular capillaries (Chapter 17). Although blood pressure accounts for some fluid leaving the capillaries in every tissue, there are also forces that oppose filtration (Figure 20.7). Because the plasma proteins are not filtered and they remain in the blood vessels, they exert an **osmotic pressure** (π_{GC}) that prevents too much fluid from leaving the blood circulation. Once the fluid moves into the Bowman's capsule, it exerts a hydrostatic pressure (P_{BC}), which prevents more fluid from entering the capsule. Once the capsule is filled with fluid, no more fluid can enter until the filtrate has moved into the proximal tubule. The filtrate is isotonic (Chapter 3) to the blood plasma since it is the plasma that enters the Bowman's capsule. So there is no opposing osmotic pressure in the Bowman's capsule. If blood pressure increases, filtration will increase. The total amount of fluid filtered per day is called the **glomerular filtration rate (GFR)**, which amounts to approximately 180 L/day.

Filtration is not selective. A fraction of all blood plasma that circulates through the kidneys is filtered into the nephron, including molecules that we do not necessarily want to excrete, such as glucose and amino acids, and most of the water and ions. Therefore, most substances need to be reabsorbed back into the bloodstream.

TABLE 20.1

Substances Reabsorbed in the Proximal Tubule	
Substance	**Percentage**
Water	65
Glucose and amino acids	100
Na^+ and K^+	65
HCO_3^-	80–90
Cl^-	50
Urea	50
Ca^{2+} and Mg^{2+}	variable*

*Calcium and magnesium are reabsorbed when blood levels of those minerals are low and parathyroid hormone is released.

Reabsorption

Reabsorption involves the movement of water, ions, or other nutrients back into the bloodstream (Table 20.1). The efferent arteriole leaves the glomerulus and then branches into many capillaries that surround the rest of the nephron and reabsorb molecules. The proximal convoluted tubule is the first portion of the nephron where reabsorption takes place. Most reabsorption occurs in the proximal tubule. All glucose and amino acids are reabsorbed back into the bloodstream in the proximal tubule. There should not be any glucose or amino acids in the urine of healthy people. When someone is diabetic, the very high blood sugar levels that enter the filtrate cannot be 100% reabsorbed; therefore, glucose can be detected in the urine. Recall from Chapter 3 that membrane carrier proteins can become saturated, and the rate of movement across the membrane cannot increase with a further increase in solute concentration, which is what happens in the proximal tubule of a diabetic.

The descending loop of Henle is permeable to the passage of water, but impermeable to ions and urea. Approximately 15% of the water that was filtered in the glomerulus is reabsorbed along the descending loop of Henle. As the filtrate moves down the descending loop and more water is reabsorbed, the filtrate becomes more concentrated: hypertonic compared to initial filtrate. The loops of Henle vary in length throughout the kidney; some are short and stay mostly within the cortical region, and others are much longer, reaching into the medulla region. The longer loops reabsorb more water than the cortical nephrons because of the increased osmolarity of the medulla. Due to the osmotic gradient, a higher concentration of ions in the medulla causes more water to be reabsorbed there.

As the filtrate moves up along the ascending loop of Henle, the epithelial cells of the nephron become *impermeable* to water and permeable to ions because of the expression of ion channel membrane proteins. Approximately 20–30% of sodium, potassium, and chloride ions are reabsorbed along the ascending loop.

In the distal tubule and collecting duct, hormonal regulation affects the remaining reabsorption of ions and water. The release of parathyroid hormone increases calcium reabsorption; ADH increases water reabsorption, and aldosterone increases sodium reabsorption, which then causes more water to be reabsorbed by osmosis. Aldosterone also increases potassium secretion. The last 9% of the water that was initially filtered is reabsorbed in the collecting ducts. This leaves the remaining 1% of water in the urine that is excreted, which amounts to approximately 1–2 L per day. The distal tubules and the collecting ducts also reabsorb a small amount of urea; although it is a waste product, we excrete only about 40% of the amount filtered.

DID YOU KNOW?

Animals living in very dry desert regions need to highly conserve their body water. Compared to other animals, they have the longest loops of Henle for maximal water reabsorption. A camel can last about 15 days in the desert without water, and a kangaroo rat can live on just the water in its food without drinking any extra water for its whole life.

Secretion involves active transport of **nitrogenous wastes, creatinine,** excess hydrogen and potassium ions, and some medications out from the peritubular capillaries and into the tubules. Secretion occurs mainly in the proximal tubule, distal tubule, and collecting duct. Nitrogenous waste is produced from breaking down proteins or nucleic acids that contain nitrogen because nitrogen is not used to make ATP and is not recycled by our cells. Nitrogenous wastes include **ammonia, urea,** and **uric acid.** Most ammonia is converted into urea. Ammonia is extremely alkaline and toxic, so only very small amounts remain in the blood. Most pyrimidines—the single-ring nucleotides of nucleic acids—are also converted into urea. The purines—adenine and guanine—are converted into uric acid. Two ammonia molecules combine with a molecule of carbon dioxide to produce urea (Figure 20.8). Uric acid is also secreted into the tubules to be excreted in the urine. An excess of uric acid can cause gout or kidney stones. Note that urea is filtered, reabsorbed, and secreted.

Creatinine is a waste product produced when muscle cells use creatine phosphate as an energy source (Chapter 14). H^+ ions are secreted depending on the pH of the blood. The respiratory system and the kidneys are

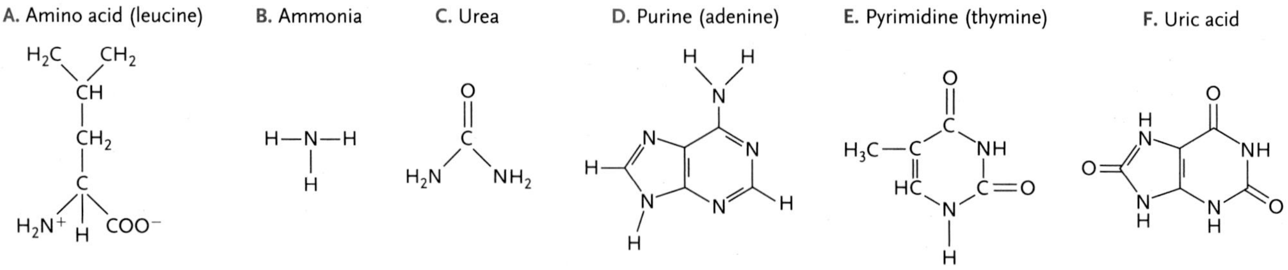

A. Amino acid (leucine)　　B. Ammonia　　C. Urea　　D. Purine (adenine)　　E. Pyrimidine (thymine)　　F. Uric acid

Figure 20.8　Waste Products from the Breakdown of Protein and Nucleic Acids
(A) Amino acid—leucine, (B) ammonia, (C) urea, (D) purine—adenine, (E) pyrimidine—thymine, and (F) uric acid

the regulators of blood pH, which must be maintained at a very stable 7.41–7.45 in the arteriole blood entering the renal corpuscles. When the blood becomes too acidic, more H+ ions are secreted into the tubule. Excess K+ ions are also secreted, mainly in the distal tubules; this is increased when aldosterone is present. Excretion of potassium depends on the blood levels, which are determined from dietary intake. Many medications, such as penicillin, digoxin, and tetracycline, are secreted into the tubule for excretion.

DID YOU KNOW?

Gout is an inflammation of the joints and is considered a type of arthritis, it is caused from an accumulation of uric acid in the joints. This very painful inflammatory disorder can be exacerbated by eating foods that tend to cause high uric acid production, such as herring, tuna, and sardines. So people with gout need to watch out for omega 3 fatty acid supplements that are extracted from small fish. Instead they can increase omega 3 fatty acids by eating flaxseeds. Omega 3 fatty acids are key to preventing chronic inflammation.

CONCEPT REVIEW 20.3

1. List the order of the structures in the nephron that filtrate flows through.

2. Where does most water and nutrients get reabsorbed?

3. What is significantly different between the ascending and descending loop of Henle?

4. Which nutrients are 100% reabsorbed?

5. What is reabsorbed when the following hormones are present: PTH, ADH, or aldosterone?

6. What are the three types of nitrogenous waste products? Why are they formed?

7. Where does secretion occur?

8. Which molecules are secreted?

20.4 TRANSPORT ACROSS MEMBRANES

Molecules move into or out of the nephron across cell membranes by three different processes in the kidney. Filtration is the movement into the Bowman's capsule; reabsorption is the movement from the nephron tubules into the peritubular capillaries and back into the circulatory system; and secretion is the movement from the peritubular capillaries into the tubules to be excreted (Figure 20.9a). Any movement into or out of the nephron tubules requires passage through the simple cuboidal epithelial cells that line the tubule. The focus of this section is the reabsorption of important nutrients from the lumen back into circulation (Figure 20.9b). The side of the epithelial membrane that is next to the lumen of the tubule is the **apical**, or luminal, membrane. The opposite side, close to the peritubular capillaries, is the basolateral membrane. In between, the epithelial cells are tight junctions that hold the cells together tightly so that molecules cannot pass between the cells but must be regulated by membrane proteins. In this way the nephron regulates which molecules and how many of them move back into the bloodstream. The water and molecules left in the tubules are the filtrate that is excreted as urine.

Different nutrients are absorbed by different mechanisms, including simple diffusion, osmosis, facilitated diffusion, coupled transport, and active transport. Water can be reabsorbed into the bloodstream by moving (1) through the epithelial cell membranes (osmosis); (2) through **aquaporin** membrane proteins; or (3) between

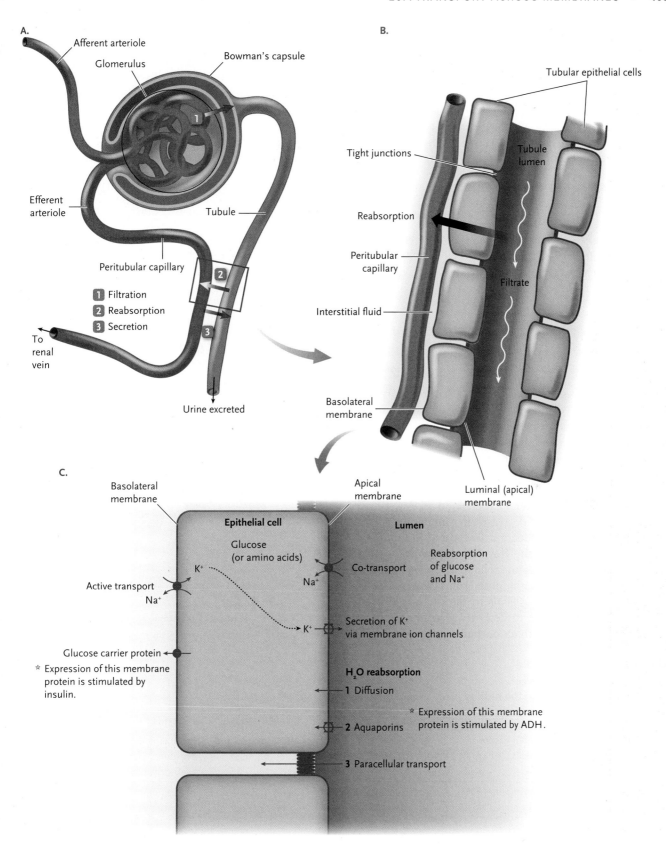

A.

Afferent arteriole

Glomerulus

Bowman's capsule

Efferent arteriole

Tubule

Peritubular capillary

1 Filtration
2 Reabsorption
3 Secretion

To renal vein

Urine excreted

B.

Tubular epithelial cells

Tight junctions

Tubule lumen

Reabsorption

Peritubular capillary

Filtrate

Interstitial fluid

Basolateral membrane

Luminal (apical) membrane

C.

Basolateral membrane

Apical membrane

Epithelial cell

Glucose (or amino acids)

Lumen

Co-transport

Na⁺

Reabsorption of glucose and Na⁺

Active transport

K⁺

Na⁺

Secretion of K⁺ via membrane ion channels

Glucose carrier protein

* Expression of this membrane protein is stimulated by insulin.

H_2O reabsorption

1 Diffusion

* Expression of this membrane protein is stimulated by ADH.

2 Aquaporins

3 Paracellular transport

20

Figure 20.9 **Movement across the Tubule Epithelial Cell Membranes**

(A) Filtration, reabsorption, and secretion. (B) Substances that are reabsorbed must move from the tubule lumen, through the epithelial cell, and into the bloodstream. (C) Substances cross the tubule epithelial cell membrane through various membrane transport proteins, carriers, or via diffusion.

the epithelial cells by a process called **paracellular transport**. Just like the absorption of water in the small intestine more water is absorbed in the kidneys as more ions are absorbed (Figure 20.9c).

In healthy people glucose and amino acids are 100% reabsorbed in the proximal tubule by co-transport with sodium on the apical membrane, and then through membrane carrier proteins (GLUT, glucose transporters) on the basolateral membrane (Figure 20.9c).

The sodium brought into the cell through co-transport is removed from the cell on the basolateral membrane by the sodium-potassium pump. Potassium ions are moved into the epithelial cells by the Na+/K+ pump and must be *secreted* into the tubule lumen. Aldosterone causes an increase in sodium reabsorption, while at the same time increasing the secretion of potassium. Hormones regulate reabsorption and secretion by affecting the transcription of membrane proteins.

CONCEPT REVIEW 20.4

1. What direction are molecules moving, into or out of the tubule, during filtration, reabsorption, and secretion?

2. Name three ways that water can be reabsorbed.

3. How are glucose and amino acids transported across the apical membrane and the basolateral membrane?

4. Where is the sodium-potassium pump located on tubule epithelial cells?

20.5 REGULATION OF THE URINARY SYSTEM

Hormones that regulate the kidneys affect reabsorption or secretion in the distal tubules and the collecting ducts. Regulation of kidney function is related to the regulation of blood pressure and blood mineral concentrations, such as that of calcium, sodium, potassium, and magnesium. See Table 20.2 for a summary of the hormones involved in regulation blood volume, ions, and minerals.

When sodium or water in the blood is low, blood pressure decreases. A decrease in sodium level is detected by the macula densa cells located in the distal convoluted tubule, which—through the folding of the tubules—is located beside the renal corpuscle where the afferent arteriole enters the glomerulus (Figure 20.6). Macula densa cells send signals to the smooth muscles of the afferent arteriole to decrease glomerular filtration, which decreases urinary output (Figure 20.10).

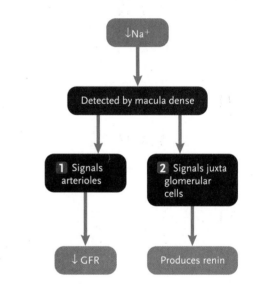

Figure 20.10 **Effect of Decreased Blood Sodium Level**
Blood sodium levels are detected by the macula densa cells.

TABLE 20.2

Hormones That Act on the Kidney

Hormone	Stimulus for Production	Production Site	Target Tissue	Function
ADH	Low blood pressure	Hypothalamus, released from posterior pituitary gland	Distal tubules and collecting duct	Producing aquaporin membrane proteins to increase water reabsorption
Aldosterone	Increased angiotensin II or high blood potassium	Adrenal cortex (zona glomerulosa)	Distal tubule and collecting duct	Increasing sodium reabsorption, therefore water reabsorption; increasing potassium secretion
PTH	Low blood calcium or magnesium	Parathyroid gland	Distal tubule	Increasing calcium and magnesium reabsorption
ANP	High blood pressure	Atria	Distal tubule and collecting duct	Decreasing sodium reabsorption, resulting in more water being excreted

Renin-Angiotensin System (RAS)

When blood pressure decreases, there is decreased stretch in the baroreceptors (stretch receptors) in the afferent arterioles. This stimulates the secretion of the enzyme **renin** from the juxtaglomerular cells (Figure 20.6). Renin moves into the bloodstream where it cleaves angiotensinogen, produced by the liver, into angiotensin I. Angiotensin I is then cleaved into angiotensin II (Figure 20.11). Angiotensin II has several functions that help increase blood pressure: (1) stimulating the adrenal cortex to produce aldosterone, (2) causing vasoconstriction; (3) stimulating the posterior pituitary to release ADH, (4) stimulating the sympathetic nervous system to increase blood pressure, (5) stimulating the reabsorption of sodium and chloride and therefore water, and (6) stimulating thirst and cravings for salt.

DID YOU KNOW?

There is a class of diuretics called ACE inhibitors that decrease blood pressure by preventing the formation of angiotensin II. Without angiotensin II, aldosterone production is not stimulated and therefore more water is excreted, causing blood pressure to decrease.

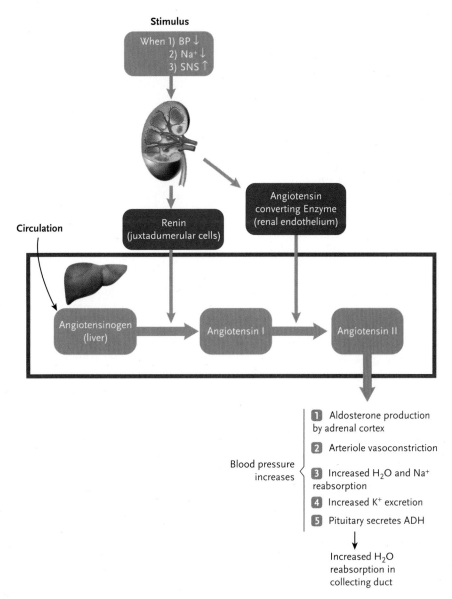

Figure 20.11 Renin Angiotensin System (RAS)

The renin-angiotensin system (RAS) involves the production of hormones that regulate water and ion concentrations.

CONCEPT REVIEW 20.5

1. List each hormone that plays a role in regulating blood pressure, where it is made, what stimulates its production, what tissue it acts on, and its main functions.

2. What is the function of renin?

3. What is the function of angiotensin-converting enzyme?

4. List all the functions of angiotensin II.

20.6 KIDNEY DISEASES

Chronic kidney disease

Chronic kidney disease (CKD), the most common kidney disease, is characterized by a gradual loss of kidney function and leads to kidney failure. Approximately 26 million people in North America have CKD. The primary causes are hypertension and diabetes. Kidney function can be determined by the glomerular filtration rate (GFR). As chronic kidney disease progresses, GFR decreases until eventually **renal failure** occurs, and there is very little or no urine output. Kidney function can also be determined by testing for **urine albumin** content. Recall from Chapter 17 that albumin is the main plasma protein involved in maintaining osmotic pressure in the blood vessels and preventing excess fluid from leaking into the interstitial tissues. Proteins are too large to be filtered at the renal corpuscle, and all proteins, cells, and large molecules should not be found in the urine of healthy people. If urine contains protein, kidney damage is present. Chronic hypertension can damage the delicate capillaries of the glomeruli, and high blood sugar from diabetes also damages blood vessels.

When the kidneys cannot properly remove waste products from the blood, regulate blood pH and ion levels, and properly regulate blood pressure, symptoms can become very severe very quickly. Symptoms of advanced kidney disease include nausea; weakness; fatigue; inability to concentrate; decreased appetite; muscle cramping (from lack of ion regulation); increased blood pressure; itching of the skin (from lack of histamine excretion); and edema in the tissues—commonly in the feet and around the eyes—from decreased control of blood pressure. Chronic kidney disease is often treated with blood pressure medication and by controlling diabetes. Once kidney failure occurs, the only treatment options are **dialysis** or a kidney transplant.

DID YOU KNOW?

Without dialysis, renal failure, with no urine output at all, would cause death within one week. People can live with kidney failure and dialysis treatments for many years.

Kidney stones

Kidney stones affect approximately 10% of the population and are most commonly caused from excess uric acid or calcium-oxalate precipitation. Recall that uric acid is a waste product produced from the breakdown of the purine nucleotides of DNA and RNA molecules. We eat nucleic acids any time we eat plants (fruit and vegetables) or meat because all cells contain these molecules. Uric acid buildup in the joints causes gout, which is becoming an increasingly common disorder. The buildup of uric acid as either kidney stones or gout is highly linked to metabolic syndrome and also has a strong genetic component.

Kidney stones are said to be one of the most painful health problems to endure. The buildup of uric acid or calcium crystals often form in the renal pelvis and then travel through the very small ureters to the bladder (Figure 20.12). The symptoms depend on the size of the stone. As the stone moves through the ureters, the smooth muscles in the ureters contract to try to help move the stone through; in addition to the inflammation caused from the stone scraping the delicate epithelial cells, this causes severe pain. Pain can radiate from the lower back toward the front lower abdominal groin area. Other symptoms include nausea, blood in urine, fever (from inflammatory response), and painful urination. Small stones often pass without treatment other than pain medication and drinking large amounts of water. Large kidney stones may need to be removed with surgery. Drinking enough water on a regular basis can help prevent the development of large kidney stones. A person with a family history of kidney stones needs to take dietary precautions.

DID YOU KNOW?

People who are prone to calcium-oxalate stones can decrease the amount of oxalate in their diet by rinsing and cooking such green vegetables as spinach, Swiss chard, and broccoli. The vegetables should be cooked in boiling water for at least two minutes to allow most of the oxalates to leave the food and diffuse into the water. Also, by eating enough calcium, any precipitation with oxalates will occur in the digestive tract and be removed with the feces; this will prevent the oxalates from being absorbed into the bloodstream and precipitating in the kidney.

Figure 20.12 Kidney Stones

Kidney stones are caused by excess precipitation of (A) uric acid and (B) calcium-oxalate.

CONCEPT REVIEW 20.6

1. What are the primary causes of chronic kidney disease?

2. How can chronic kidney disease be prevented?

3. Why would someone with hypertension need to decrease salt intake?

4. What are the symptoms of advanced chronic kidney disease?

5. What are the treatment options?

6. What are the two most common types of kidney stones?

7. What other disease can be caused from a buildup of uric acid?

The difference between a successful person and others is not a lack of strength, not a lack of knowledge, but rather a lack of will.

Vince Lombardi

21.1 OVERVIEW OF THE REPRODUCTIVE SYSTEM

The primary role of the reproductive system in both males and females is to produce **haploid gametes** in order to produce offspring. The production of gametes is called **gametogenesis,** which involves meiosis where **diploid** germ cells go through two cell divisions to create the haploid gametes with 23 chromosomes: **spermatozoa (sperm)** in males and **ova (eggs)** in females. Gametogenesis occurs within the **gonads:** the **testes** in males and the **ovaries** in females. Gametogenesis in males is called spermatogenesis and in females is called oogenesis. Both males and females produce steroid hormones that regulate the reproductive process; these are testosterone, estrogen, and progesterone.

The functions of the male reproductive system include spermatogenesis; producing steroid hormones, mainly testosterone; storing and transporting sperm; adding secretions from accessory glands to the sperm to produce semen.

The functions of the female reproductive system include oogenesis; producing steroid hormones, mainly estrogen and progesterone; storing and ovulation of ova; being the site of fertilization in the fallopian tubes; being the site of implantation and growth of the fetus in the uterus; producing milk in mammary glands to nourish the newborn.

This chapter outlines the anatomy of the male and female reproductive systems and reviews the hormonal regulation of gametogenesis. Further details related to the reproductive system are discussed in other chapters, specifically, meiotic cell division in Chapter 5, hormones in Chapter 16, and the most common sexually transmitted infections in Chapter 23.

CONCEPT REVIEW 21.1

1. What are the functions of the male and female reproductive systems?

21.2 MALE REPRODUCTIVE SYSTEM

The testes are the male gonads. They are located within the **scrotum,** which holds the testes away from the body and regulates their temperature to be 35–36°C; this is one to two degrees lower than body temperature, which is essential for normal sperm development. The testes are composed of **seminiferous tubules,** and this is where spermatogenesis occurs beginning in puberty. The testes are connected to a duct system that transports the mature sperm from the seminiferous tubules (Figure 21.1).

DID YOU KNOW?

If all the seminiferous tubules from each testis were removed and stretched out they would reach the length of two football fields. The epididymis alone measures approximately 6 m. Males can produce millions of sperm each day, and approximately 1 million in each ejaculation can survive three to five days inside a female.

The seminiferous tubules converge to a section of tubule called the **rete testis** that leads to the **efferent ductules,** and then to the coiled **epididymis** where sperm mature. The fully developed sperm travel through the epididymis to two **vas deferens.** The vas deferens are lined with smooth muscle tissue that contracts and moves sperm through the ducts during sexual arousal. Sperm can survive within the vas deferens for several months. The sperm move through the vas deferens, which loops around the anterior and then superior surface of the urinary bladder to the posterior side where the **accessory glands** are located.

The **seminal vesicles** join the vas deferens at the **ejaculatory duct** (Figure 21.1). The seminal vesicles secrete an alkaline viscous fluid that contains fructose, prostaglandins, mucus, proteins, enzymes, and vitamin C; contributes nutrients for the sperm; and makes up approximately 60–70% of the volume of the **semen** ejaculated. Sperm use fructose to make ATP, and the prostaglandins contribute to muscle contractions in the female reproductive tract. The alkalinity is important for neutralizing the acidic environment of the vagina. The ejaculatory duct leads to the **prostate gland,** located below the urinary bladder where the ejaculatory duct merges with the urethra. The prostate gland secretes citric acid, used by the sperm to produce ATP, and prostate-specific antigens (PSA), which are important for helping the sperm enter the cervix. Semen and urine cannot move through the urethra at the same time.

The **bulbourethral glands,** located below the prostate gland and on either side of the urethra, secrete alkaline mucus into the urethra during arousal to neutralize any acid residue from urine that may be in the urethra. Secretions from the bulbourethral glands begin before ejaculation. The urethra travels through the penis, which is composed of cylindrical masses of tissue called the **corpora cavernosa** and the **corpus spongiosum.** These highly vascularized tissues fill with blood, causing erection through stimulation of the autonomic nervous system.

DID YOU KNOW?

Pregnancy can occur without ejaculation. Smooth muscle contractions in the vas deferens that accompany arousal prompt the sperm that's stored in the vas deferens to begin to travel to the urethra. Sperm can move through the ducts and join the bulbourethral secretions that occur before ejaculation.

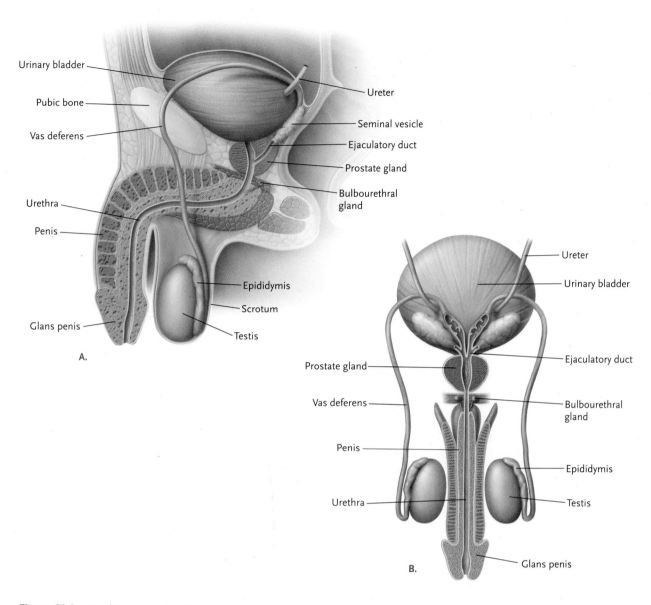

Figure 21.1 Male Reproductive Organs

From DIGIUSEPPE/FRASER. *Biology 12 U.* © 2012 Nelson Education Ltd. Reproduced by permission. www.cengage.com/permissions

Spermatogenesis

The seminiferous tubules consist of an outer basement membrane, Sertoli cells, germ cells called **spermatogonia,** and cells in various stages of meiotic cells division (Figure 21.2). Outside the seminiferous tubules are the Leydig cells that produce testosterone. Spermatogonia are the diploid (2n) cells that undergo both mitosis—to continually replenish the spermatogonia population—and meiosis. The spermatogonia that enter meiosis become differentiated: **primary spermatocytes** undergo the first meiotic cell division and become haploid **secondary spermatocytes.** Secondary spermatocytes then undergo a second cell division;

sister chromatids separate to become **spermatids,** each containing 23 chromosomes. Spermatids then differentiate into spermatozoa (sperm), but full and complete maturation continues and is completed in the epididymis.

The **Sertoli cells** play an integral role in the development of sperm and have several specific functions:

• They extend from the basement membrane to the lumen of the seminiferous tubules and are held together by tight junctions, forming a closed environment called the **blood-testis barrier,** not unlike the blood-brain barrier formed by astrocytes in the central nervous system. It is important that toxins

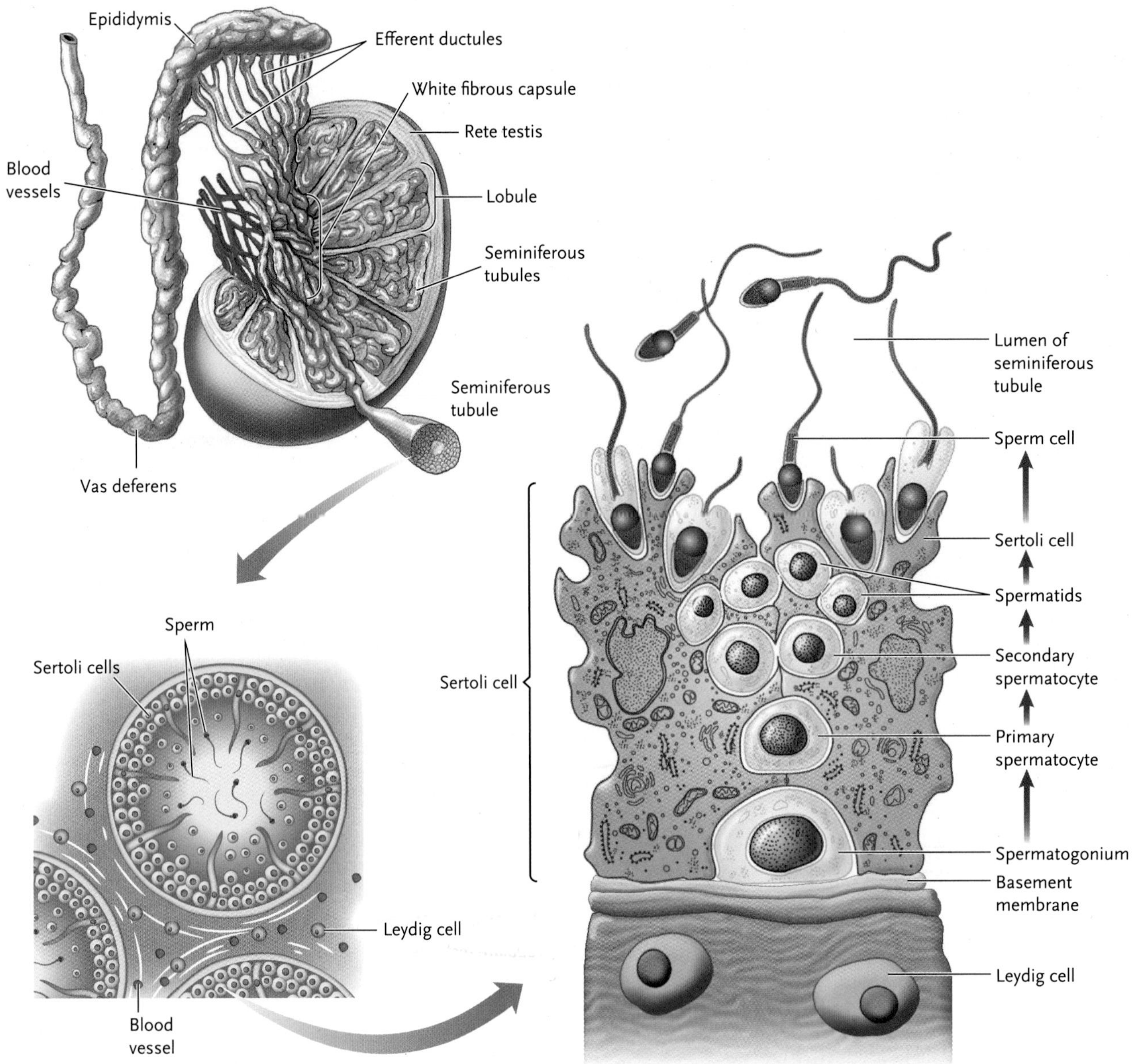

Figure 21.2 Spermatogenesis

Spermatogenesis is the production of sperm through meiosis.

Art source (top left and right): From SHERWOOD/KELL. *Human Physiology*, 2E. © 2013 Nelson Education Ltd. Reproduced by permission. www.cengage.com/permissions

from the bloodstream do not come in contact with the developing sperm. Spermatocytes move through the tight junctions between the Sertoli cells just before meiosis begins.

- They produce **androgen-binding protein (ABP)** that binds to the testosterone secreted by the Leydig cells in the interstitium outside the seminiferous tubules. Leydig cells produce testosterone in response to **luteinizing hormone (LH)** that's secreted by the anterior pituitary gland in response to the **gonadotropin-releasing hormone (GnRH)** produced by the hypothalamus.

- They are stimulated by follicle stimulating hormone (FSH) produced by the anterior pituitary gland in response to GnRH produced by the hypothalamus. FSH and testosterone stimulate the Sertoli cells to produce the many signalling molecules that regulate spermatogenesis. Testosterone also stimulates development of the male secondary sex characteristics.

- They produce the hormone **inhibin,** which signals the anterior pituitary gland to decrease FSH secretion (Figure 21.3).

- They engulf improperly developing sperm through phagocytosis.

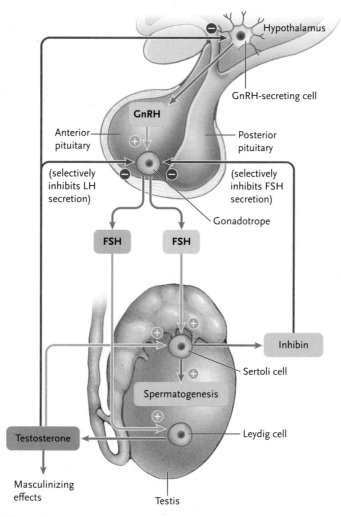

Figure 21.3 **Hormonal Regulation of Spermatogenesis**

Hormones of the anterior pituitary gland regulate testosterone production in the testes, which regulate spermatogenesis.

Structure of spermatozoa

Sperm have three sections: the head, mid-piece, and tail (Figure 21.4). The head of the sperm contains the nucleus that has 23 chromosomes and minimal cytoplasm. The head, surrounded by an **acrosome,** contains important enzymes that break down the outer coat of the egg and lead to fertilization. The mid-piece contains mitochondria that produce ATP from the fructose present in the secretions of the accessory glands. The tail is a flagellum composed of microtubules that allow the tail to move and propel the sperm through the female reproductive tract.

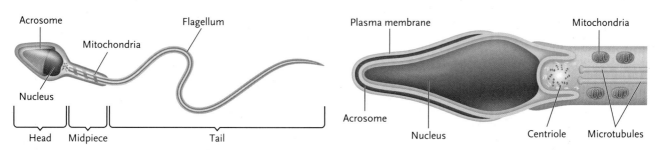

Figure 21.4 **Structure of a Spermatozoan**

CONCEPT REVIEW 21.2

1. List in order the structures that the sperm travel through from the seminiferous tubules to the urethra.

2. What is the function of the seminal vesicles, the prostate gland, and the bulbourethral glands?

3. Name the cell types that arise during spermatogenesis.

4. Which cell types are diploid and which are haploid?

5. How do Sertoli cells support spermatogenesis?

6. What is the function of the Leydig cells?

7. What is the role of FSH and LH in spermatogenesis?

8. Describe the structure of a spermatozoan.

21.3 FEMALE REPRODUCTIVE SYSTEM

The ovaries are the female gonads. They are located within the lower abdominopelvic region and protected by the sacrum, coccyx, and pelvic bones. The ovaries are located on each side of the uterus (Figure 21.5), where the oocytes develop through oogenesis. An **ovarian ligament** connects each ovary to the outside of the uterus and holds the ovary close to the opening of the fallopian tubes. Oogenesis begins during the embryonic development of the female fetus. As a result, newborn baby girls have already produced all the **oocytes** they will have for their lifetime. The development of the oocytes is halted in mid-meiosis and completed only if fertilization takes place. Ovulation begins during puberty, when approximately every 28 days one egg is released into the **fallopian tube**. The egg is pulled into the **infundibulum** of the fallopian tube by the **fimbriae** where it remains viable for 24 to 48 hours; if no fertilization occurs in that time period, the egg will break down. The fallopian tubes extend from the ovaries to the superior edges of the uterus and are lined with ciliated epithelial tissue that helps move the oocyte or zygote through the tubes to the uterus.

If fertilization does occur the haploid nucleus of the sperm and the egg combine to form the first single diploid cell: the zygote. Successive mitotic divisions occur in the fallopian tube, and after approximately 10 days a hollow ball of cells called a **blastocyst** is formed and can implant in the lining of the **uterus**. The uterus, also called the womb, is located between the urinary bladder and the rectum. The larger superior region is called the fundus, which narrows to the central region called the body, which leads to the inferior opening to the vaginal canal, called the **cervix** (Figure 21.5). The uterus, a thick-walled muscular organ, is made up of three layers: the endometrium, myometrium, and perimetrium. The endometrium is the inner layer of the uterine wall where the blastocyst implants, and it is the mucus membrane that sheds each month when fertilization does not occur. The **myometrium** is the smooth muscle layer that contracts during childbirth. The **perimetrium** is the outer visceral layer composed of simple squamous epithelial and areolar connective tissue.

The area between the thighs and buttocks of males and females is called the **perineum**. The external female genitalia—in combination called the **vulva**—include the **mons pubis,** an adipose layer covering the pubic symphysis; the **labia majora,** which are the outer skin folds; the **labia minora,** the inner skin folds; and the **clitoris.** The labia major and minor surround the opening to the urethra and the vagina; the vagina is posterior to the urethra. The region within the labia minora is called the **vestibule** (Figure 21.6). The labia are analogous to the male scrotum—that is, they come from the same embryonic tissue origin—and contain **sudoriferous glands** (sweat) and sebaceous glands (oil). The clitoris, analogous to the penis, contains erectile tissue and is located anterior to the urethra at the anterior junction of the labia minora.

DID YOU KNOW?

If a fertilized egg does move from the fallopian tubes to the uterus, it can implant within the wall of the fallopian tube. This is called an ectopic pregnancy or a tubal pregnancy, and it occurs in approximately 5% of all pregnancies. An ectopic pregnancy can cause hemorrhaging that would be fatal to both mother and fetus without surgical intervention.

Oogenesis

Oogenesis begins in the fetus during embryonic development. By birth the ovaries contain anywhere from 2 to 4 million primary oocytes, and only about 400 become ovulated during a lifetime. In females, the diploid germ cells are called *oogonia* and are analogous to the male spermatogonia. Oogonia stop mitotic cell division around the seventh month of gestation and enter the S phase of the cell cycle (Chapter 5) when DNA replicates, and are then called **primary oocytes.** Primary oocytes grow during childhood and divide just before ovulation to produce a **secondary oocyte** (Figure 21.7). The secondary oocytes are haploid, but still contain the sister chromatids from DNA replication before the first meiotic division. The last cell division, to separate the sister chromatids, does not occur unless that secondary oocyte is fertilized. After the first cell division, most of the cytoplasm and organelles move into only one of the

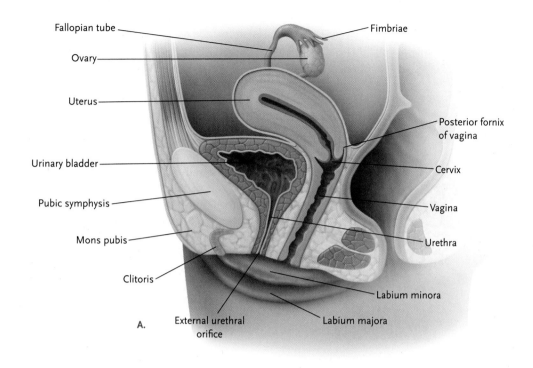

A.

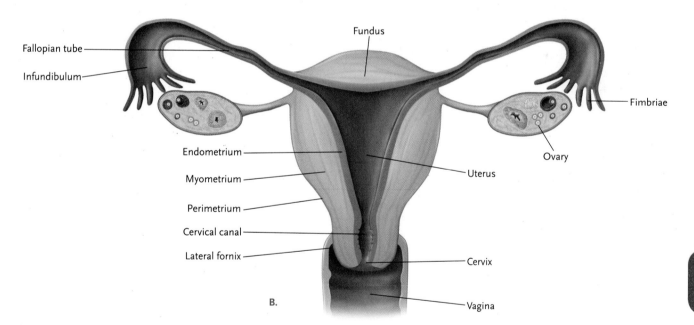

B.

Figure 21.5 Female Reproductive Organs

(A) Lateral view and (B) anterior view

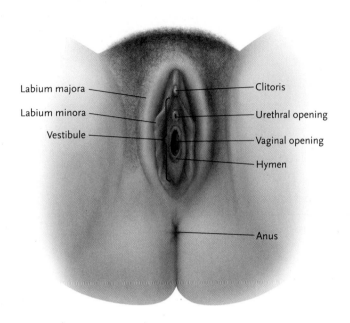

Labium majora — Clitoris

Labium minora — Urethral opening

Vestibule — Vaginal opening

— Hymen

— Anus

Figure 21.6 Structures of the Vulva

daughter cells with the chromosomes, leaving the other cell to become a **polar body** that degenerates. A polar body is also formed after the second cell division, so one primary oocyte gives rise to only one ovum; unlike the male counterpart where one primary spermatocyte gives rise to four spermatids.

Follicular development

The primary oocytes begin to develop within **follicles** from infancy and then continue through the meiotic stages during the menstrual cycle. Once a month, a mature follicle releases the primary oocyte, which undergoes the first meiotic cell division just before ovulation. If the secondary follicle is fertilized in the fallopian tube, the second meiotic division will occur, and the 23 chromosomes in the ovum combine with the 23 chromosomes in the sperm to form the first cell, the zygote.

A primordial follicle consists of a primary oocyte that contains the nucleus with 46 replicated chromosomes (92 total chromosomes). It is surrounded by a layer of glycoproteins, called the **zona pellucida**, and

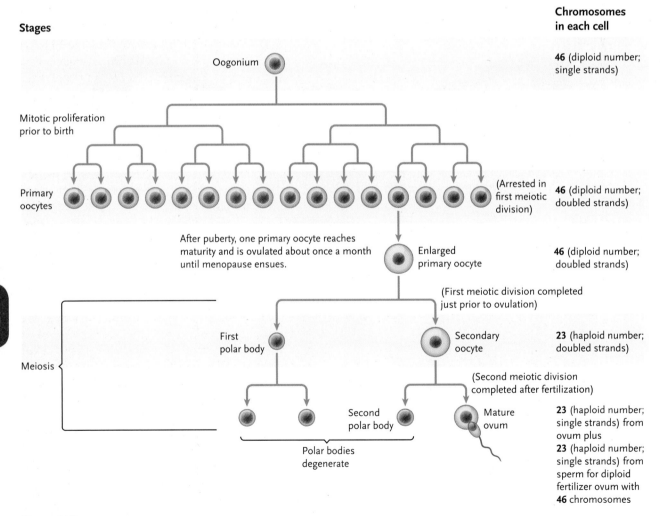

Stages

Chromosomes in each cell

Oogonium — **46** (diploid number; single strands)

Mitotic proliferation prior to birth

Primary oocytes — (Arrested in first meiotic division) — **46** (diploid number; doubled strands)

After puberty, one primary oocyte reaches maturity and is ovulated about once a month until menopause ensues.

Enlarged primary oocyte — **46** (diploid number; doubled strands)

(First meiotic division completed just prior to ovulation)

Meiosis

First polar body

Secondary oocyte — **23** (haploid number; doubled strands)

(Second meiotic division completed after fertilization)

Second polar body

Mature ovum — **23** (haploid number; single strands) from ovum plus **23** (haploid number; single strands) from sperm for diploid fertilizer ovum with **46** chromosomes

Polar bodies degenerate

Figure 21.7 Oogenesis

Oogenesis is the production of oocytes through meiosis.

21

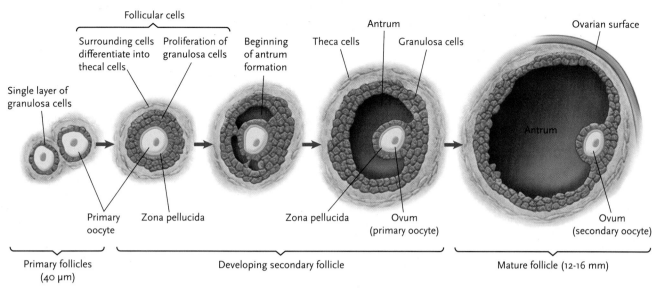

Figure 21.8 Follicular Development

also a single layer of **granulosa cells** (Figure 21.8). As the follicle grows during the follicular phase of the monthly cycle, the oocyte increases in size, the granulosa cells proliferate, and the outer connective tissue cells form the **theca layer**, which generates the preantral follicle. The theca layer continues to increase, and a fluid-filled central cavity begins to form. The cavity is called the antrum, and the follicle at this stage is called the **early antral follicle**. The antrum increases in size, and mitosis of granulosa cells and theca cells continues until a mature follicle is formed. Whichever antral follicle is dominant at the time of ovulation will release the oocyte. Antral follicles that do not release an oocyte regress and undergo apoptosis (programmed cell death).

DID YOU KNOW?

There are two types of twins—fraternal and identical. Fraternal twins (also called dizygotic because they come from two zygotes) occur when two eggs are ovulated at the same time and are fertilized by different sperm and have different combinations of the parents' chromosomes. Identical twins (also called monozygotic) occur when the zygote divides and becomes two separate cells (zygotes), and each then grows into a separate blastocyst that implants and develops. Identical twins have exactly the same DNA.

Hormonal regulation

The 28-day reproductive cycle (menstrual cycle) can be divided into two phases. The phase just described when the follicle develops is called the follicular phase, and the period between ovulation and menses is the luteal phase (Figure 21.9).

The **follicular phase,** which begins on the first day of the cycle, is the first day bleeding begins and the endometrial layer is shed. Early in the cycle, estrogen levels are low; therefore, the hypothalamus stimulates the anterior pituitary gland to release FSH and LH. LH stimulates the theca cells to produce androgens that diffuse into the granulosa cells, and FSH stimulates the granulosa cells to produce estrogen from the androgens produced by the theca cells (Figure 21.8). As follicles develop and granulosa cells proliferate, more estrogen is produced. One follicle becomes the dominant follicle, and the high level of estrogen has a negative feedback effect on the anterior pituitary gland. This regulation mechanism decreases FSH production, and the nondominant follicles undergo apoptosis. Estrogen levels reach high levels close to day 14 of the cycle, causing a positive feedback effect on the hypothalamus, which leads to a small increase in FSH just before ovulation and an **LH surge** that causes the primary oocyte to go through the first meiotic division, and then ovulation occurs (Figure 21.9).

The **luteal phase** begins when the oocyte is ovulated and the remaining granulosa and theca cells remain within the ovary and transform into a structure called a **corpus luteum**. The corpus luteum functions as a gland producing the hormones progesterone and some estrogen that support pregnancy if the secondary oocyte is fertilized. Progesterone acts on the endometrium to make it suitable for implantation, and it stimulates breast growth. If fertilization occurs the placenta takes over the role of producing progesterone during the pregnancy. If fertilization does not occur within 24 to 48 hours of ovulation, the oocyte will break down, and within 10 days the corpus luteum will break down. Once the corpus luteum breaks down, the levels of estrogen and progesterone decrease and cause menstruation. The low

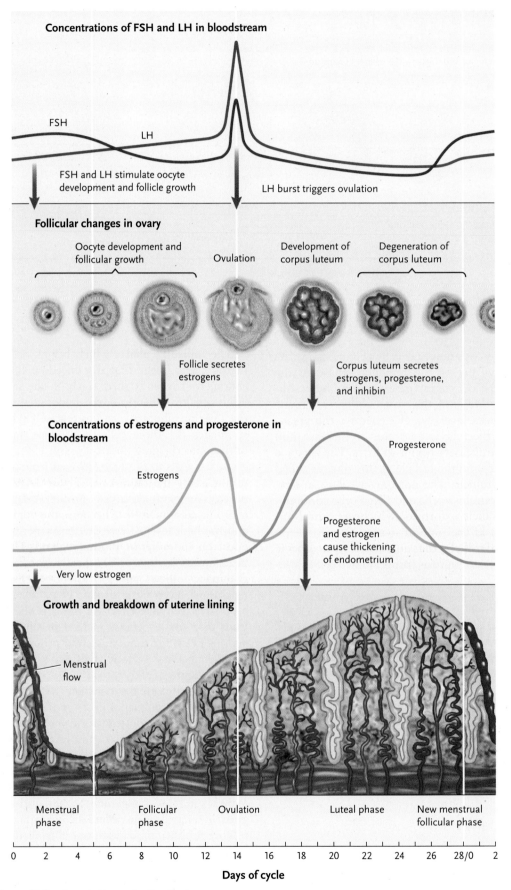

Figure 21.9 Female Reproductive Cycle

The reproductive cycle has a follicular phase and a luteal phase.

estrogen levels signal the release of FSH and LH, and a new follicular phase begins.

Granulosa cells have the following important supportive roles in oogenesis:

- They nourish the oocyte; note there is no blood-ovary barrier.
- They produce paracrine-signalling molecules that prevent the oocyte from continuing meiosis. The first meiotic division is stimulated by the LH surge.
- They use androgens produced by theca cells to produce estrogen when stimulated by FSH.
- They produce antral fluid.
- They produce inhibin, which signals the anterior pituitary gland to decrease FSH secretion.

DID YOU KNOW?

If fertilization occurs, and the blastocyst implants into the uterine wall (approximately six days after fertilization), the outer blastocyst cells that become the placenta produce a hormone called human chorionic gonadotropin (hCG). hCG stimulates the corpus luteum—which normally degenerates around day 10—to continue to produce progesterone and maintain the endometrium so that implantation and development of the placenta can occur. This hormone is the one that is measured during a pregnancy test and can be detected in the urine as early as the first day a period is missed: approximately 14 days after fertilization occurred. hCG is also the most likely cause of morning sickness early in pregnancy.

CONCEPT REVIEW 21.3

1. Where does fertilization occur?
2. Describe the three layers of the uterus.
3. What structures make up the vulva?
4. When does meiosis begin in females?
5. When does the first and second meiotic cell division occur?
6. Briefly describe the stages of follicle development.
7. Which hormone causes ovulation?
8. What hormones are produced by the corpus luteum?
9. What causes menstruation?
10. What cell type is stimulated by LH and by FSH?

21.4 DISEASES OF THE REPRODUCTIVE SYSTEM

Endometriosis is involved in up to 45% of cases of female infertility and affects approximately 10% of women of reproductive age. Endometriosis is the growth of endometrial tissue outside the uterus. It can occur on the ovaries, outside of the uterus wall, fallopian tubes, or on other structures in the abdominal cavity. The endometrial tissue breaks down and bleeds during menstruation just like the endometrial tissue inside the uterus, except that it cannot be shed through the vagina. The trapped tissue and blood causes severe pain and inflammation during the menstrual cycle. The cause of endometriosis is not known.

Pelvic inflammatory disease (PID) is an inflammatory condition in females that affects the uterus, fallopian tubes, or uterus and is a leading cause of infertility. PID is most often caused by a bacterial infection, such as chlamydia or gonorrhea, but it can also be caused by fungal, viral, or parasitic infection; this mostly occurs through sexually transmitted infections (STIs), but not all cases. More than

700,000 women are affected by PID in North America each year, and this causes over 100,000 cases of infertility annually. Gonorrhea, the cause of 60% of the cases, is problematic because it can be asymptomatic in women. As a bacterial infection, it can be treated in the acute phase with antibiotics. However, left untreated—since there are no symptoms—gonorrhea can lead to inflammation that requires anti-inflammatory medication to relieve symptoms.

Prostate cancer is the sixth leading cause of death from cancer in males in North America. Prostate cancer can be detected by measuring blood levels of prostate-specific antigen (PSA), which indicates an enlarged prostate that is producing more proteins than normal. The risk of prostate cancer increases with age. However, many men who have prostate cancer do not have symptoms, and they die from something unrelated. Prostate cancer that is left untreated has been shown to spread to other tissues, such as the bones and lymphatic tissues. Prostate cancer can be treated by surgical removal of the prostate, but this frequently leads to erectile dysfunction. Because of the very slow growth of most prostate cancers, surgery may not be required.

CONCEPT REVIEW 21.4

1. What is one of the leading causes of infertility in females?
2. What are the most common causes of PID?
3. What blood test can help to determine a person's risk of prostate cancer?

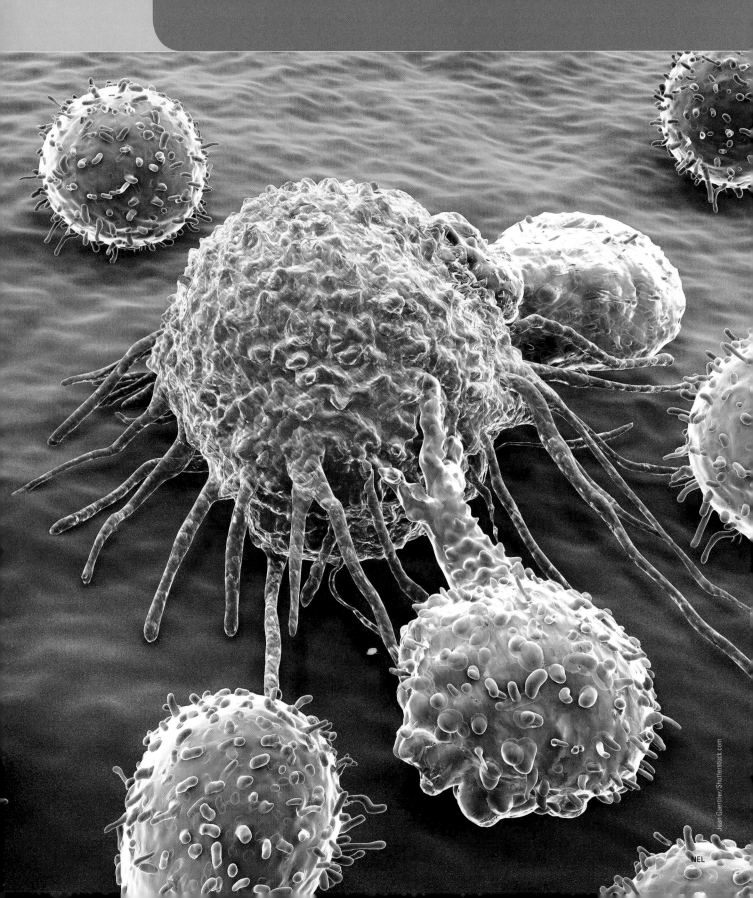

It is our choices that show what we truly are, far more than our abilities.

J. K. Rowling

22.1 OVERVIEW OF THE IMMUNE SYSTEM

The immune system helps the body maintain homeostasis by protecting the body against harmful pathogens. The immune system has mechanisms for recognizing infectious organisms, preventing infection, and killing many different infectious bacteria, viruses, fungi, and parasites. The immune system also develops tolerance for foreign molecules that do not cause harm, such as those in food, and environmental molecules, as in pollen. When the system of tolerance does not develop properly and the immune system reacts to non-harmful molecules, **allergic reactions** occur. The immune system also builds tolerance for "self" molecules, so it recognizes the difference between our own cells and that of infectious organisms. During embryonic development, our immune cells must undergo a selection process to remove any cells that would react against the self-molecules. When the immune system reacts to our own cells, this produces an autoimmune disease. The immune system can produce **memory cells** that recognize specific infections. The memory cells remain in the lymphatic tissue, and future exposure to the same microorganism will stimulate a rapid immune response.

Structures of the immune system include the lymphatic vessels, lymph nodes, **thymus**, spleen, tonsils, Peyer's patch of small intestine, and bone marrow (Figure 22.1).

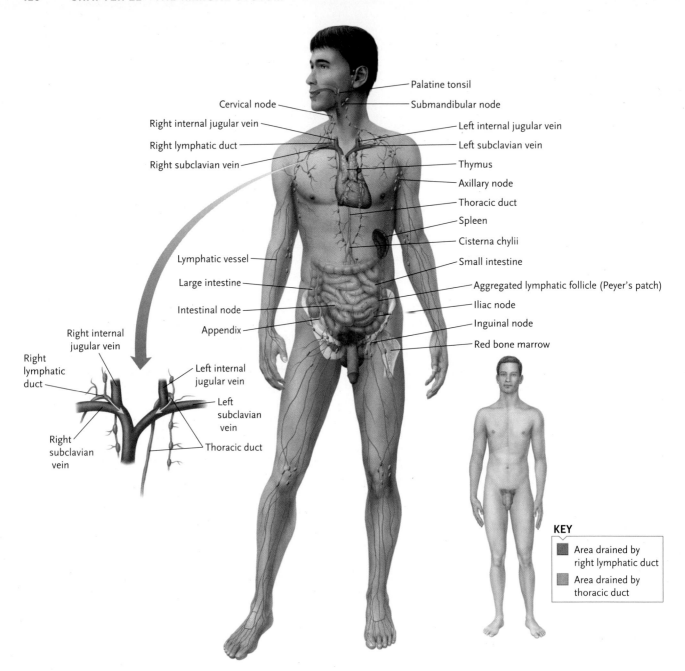

Figure 22.1   Structures of the Immune System

The immune and lymphatic systems perform the following functions in the body:

- Draining excess interstitial fluid from interstitial spaces into lymphatic vessels and then returning it to the bloodstream

- Transporting dietary fats and fat-soluble vitamins from the small intestine to the bloodstream

- Recognizing and killing infectious organisms, including bacteria, viruses, fungi, and parasites

- Recognizing and tolerating our own cells, as well as non-harmful foreign molecules such as food and environmental substances

- Producing immunological memory cells that prevent infection from the same organism in the future

1. List the structures of the immune system.

2. What are the functions of the lymphatic and immune system?

3. What are the main types of infectious organisms?

22.2 ANATOMY OF THE LYMPHATIC SYSTEM

Fluid in the bloodstream naturally moves out of capillaries into the interstitial space between cells; this is the interstitial fluid. Interstitial fluid either re-enters the capillary at the venule end or is taken up by lymphatic vessels (Figure 22.2). Lymphatic vessels are similar to capillaries in that they consist of a single layer of endothelial cells. However, lymphatic vessels have closed ends, and fluid is taken up through a one-way opening between the endothelial cells. Fluid does not move out of the vessel because increased fluid volume inside the vessel causes an increase in pressure, which closes the openings between the slightly overlapping endothelial cells (Figure 22.2). Once the interstitial fluid enters the lymphatic vessel, it is called lymph. The components of blood plasma, interstitial fluid, and lymph are the same: such as the ions, vitamins, nutrients, oxygen, CO_2. The only difference in these three fluids is their location. As fluid circulates through blood vessels to the interstitial space and back to the lymphatic system, cells are constantly bathed in water, nutrients, and ions.

Lymph travels through the lymphatic vessels and is returned to the bloodstream through the subclavian veins (Figure 22.1). Lymph moves through lymphatic vessels from the periphery toward the heart. Smaller vessels merge into larger vessels as they progress toward the heart. Lymphatic vessels contain one-way **valves**, as veins do, to prevent the backflow of fluid, and the larger vessels also contain a layer of smooth muscle that contracts and pushes lymph through the vessels, as do the larger blood vessels. In addition, the same mechanisms that help blood

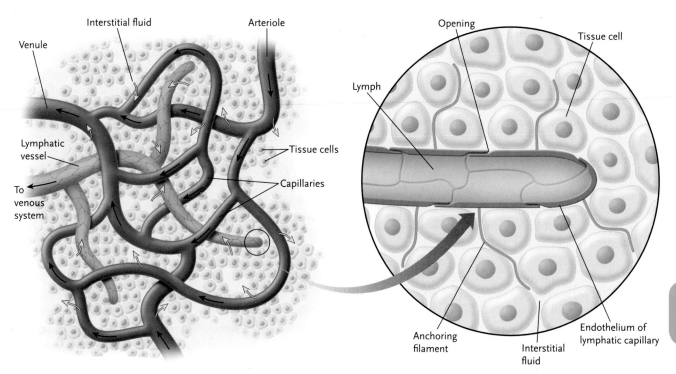

Figure 22.2   Capillary Exchange

Fluid that leaves the capillaries can be taken up by lymphatic vessels.

flow back to the heart also help lymph flow toward the heart. These mechanisms are the skeletal pump and the respiratory pump. When skeletal muscles contract, they squeeze the lymphatic vessels, altering the pressure, and causing fluid to move toward the heart. While breathing, inhalation causes the diaphragm to move down, decreasing the pressure within the thoracic cavity, which causes air to move into the lungs, but the lower pressure also causes fluid in lymphatic vessels and blood in the blood vessels to move forward toward the heart. So with each breath, blood and lymph move forward.

The right upper body lymphatic vessels drain through the right lymphatic duct into the junction of the right internal jugular vein and right subclavian vein to re-enter the circulatory system. Lymph drains from left upper body as well as the right and left lower body through the thoracic duct into the junction of the left internal jugular vein and the left subclavian vein (Figure 22.1).

Fluid accumulation within the interstitial space is called edema (Figure 22.2). Many different factors cause edema, including the following:

- Increased **capillary pressure** (hypertension) causes an increase in the amount of fluid, ions, and proteins that leave the vessel and enter the interstitial space, and this increase in proteins or ions causes greater osmotic pressure, which draws more fluid from the capillaries.

- Kidneys do not properly filter or excrete salt and water, as occurs in kidney failure.

- Valve failure, obstruction, and heart failure causes high pressure in the veins and therefore lack of venous return.

- Lack of skeletal muscle pump action due to immobility or paralysis results in more fluid retention in the veins and the lymphatic vessels.

- Medications can affect blood pressure, vascular resistance, the autonomic nervous system, kidney function, and osmotic balance.

- Lack of plasma proteins that help retain fluid in the circulatory system can be caused by wounds, burns, liver disease, or malnutrition.

- Lymphatic vessels can become blocked due to cancer, infection, inflammation, and surgical removal of tissue, such as mastectomy; an inflammatory response also increases capillary permeability.

Structures of the lymphatic system

Bone marrow

Many cells types are derived from stem cells in the bone marrow. The majority of bone marrow is composed of **yellow bone marrow**, which primarily stores fats in the medullary cavity. The red bone marrow is located primarily in flat bones, such as the pelvis, scapula, vertebrae,

ribs, and cranium, as well as the epiphyseal ends of long bones. The red bone marrow contains stem cells called **mesenchymal cells** or **stromal cells** that can differentiate into cartilage, osteoblasts, or adipose cells. The **hematopoietic stem cells** of the red bone marrow differentiate into red blood cells, white blood cells, and platelets. B cells and T cells are the major cell types involved in the adaptive immune response. B cells become mature cells in the bone marrow and then migrate to the spleen or lymph nodes. Immature T cells migrate to the thymus to differentiate into mature T cells, and then migrate to the spleen or lymph nodes.

Thymus

The thymus is located posterior to the sternum and superior to the heart (Figure 22.1). In the thymus, immature T cells—called **thymocytes**—that migrated from the bone marrow go through a maturation process where any cells that recognize self-proteins are destroyed through apoptosis. The thymus also contains macrophages that are white blood cells with phagocytic activity that engulf and remove dead cells.

T cells have specific receptors that recognize specific antigens. Every T cell develops a different receptor; theoretically, our entire set of T cells should be able to recognize any possible infectious organism. However, any T cells with a receptor that recognizes one of our own cellular molecules must be eliminated so they do not attack our own tissues. The epithelial cells in the thymus express proteins that are found in other regions of the body, and any thymocytes that bind strongly to these proteins are destroyed; this includes approximately 98% of the T cells that migrate to the thymus. This process of developing a T cell repertoire that does not attack self-molecules is called negative selection. The thymus is most active in newborns and children up to puberty. Then the thymus gradually atrophies, and cells are replaced with fat. Mature T cells leave the thymus and migrate to the spleen and lymph nodes. B cells also have specific receptors and undergo a similar selection process in the bone marrow before they migrate to the spleen and lymph nodes.

Lymph nodes

As lymph travels through the lymphatic vessels, it also passes through **lymph nodes**. The lymph nodes contain white blood cells that can recognize foreign molecules and initiate an immune response. Lymph nodes are located in larger numbers in areas where microorganisms can enter the body, such as the ears, mouth, lungs, digestive tract, urogenital region, and the mammary glands, to prevent transfer of infections from mother to infant (Figure 22.3). Fluid is constantly circulating from the capillaries, to the interstitial space, to the lymphatic system, and back to the circulatory system, ensuring that any pathogen can be detected by the immune system.

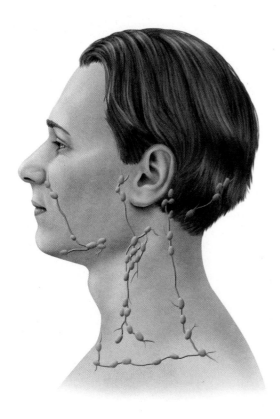

Figure 22.3   Lymph Nodes in the Head and Neck Region

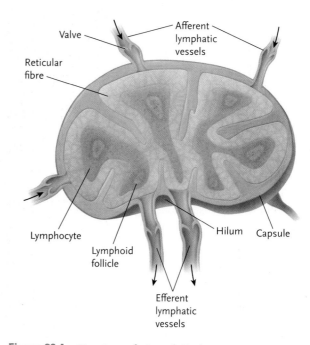

Figure 22.4   **Structure of a Lymph Node**
Lymph nodes contain immune cells and are located throughout the body.

Interstitial fluid containing water, ions, proteins, and some types of immune cells enter the lymph vessels and travel to the lymph nodes via the afferent lymphatic vessel. Inside the lymph node are many **reticular fibres,** which act as a filtering screen to slow the movement of lymph and to trap particles that should not re-enter the circulatory system (Figure 22.4); these fibres are surrounded by a fibrous capsule. Within the lymph nodes are the mature B cells and T cells, and also phagocytic macrophages and dendritic cells that engulf dead cells and any debris. All T cells and B cells have specific receptors that recognize specific proteins, and as the reticular fibres slow the flow of lymph through the lymph node, a specific immune response—the **adaptive immune response**, also called the acquired immune response—may occur. Fluid may flow through multiple lymph nodes and the spleen before it re-enters the circulatory system.

Spleen

The spleen is the largest lymphatic organ in the body, located in the superior, lateral abdominal cavity above the stomach and below the diaphragm. The spleen contains two distinct regions, the red pulp and the white pulp (Figure 22.5). The **red pulp,** which contains macrophages, red blood cells, and platelets, is the location for breakdown of old red blood cells. The liver also breaks down red blood cells. The **white pulp** contains lymphocytes, macrophages, and dendritic cells that are involved

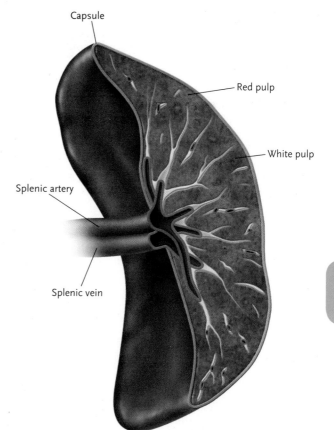

Figure 22.5   **Structure of the Spleen**

22

in the adaptive immune response. The spleen contains reticular fibres, as do the lymph nodes, and it slows the flow of molecules so that T cells or B cells can initiate an immune response. During embryonic development, the spleen is also involved in hematopoiesis.

Lymphatic nodules

Lymphatic nodules are very similar to lymph nodes, except they are not surrounded by a capsule like the lymph nodes. Lymphatic nodules are found throughout the body and are in higher numbers in regions where infectious organisms can enter the body. Some examples of lymphatic nodules include the **tonsils**, located in the mouth region, and **Peyer's patches**, located around the small intestines. Tonsils, adenoids, appendix, and Peyer's patches are also called the **gut-associated lymphoid tissue (GALT)**. See Figure 22.6.

DID YOU KNOW?

When infants are born they leave the sterile environment of the uterus and begin to colonize **beneficial bacteria**, called the normal flora, during delivery through the vaginal canal, breast-feeding, and any contact with the resident bacteria on the mother's skin. Babies that are breast-fed take in prebiotic oligosaccharides that feed beneficial bacteria that begin to populate the digestive tract. Those bacteria convert oligosaccharides into short chain fatty acids that can be used by the infant for energy. Breast-fed infants develop a normal flora that is higher in **bifidobacteria** and **lactobacilli bacteria**, which are very beneficial for the stimulation of immune defenses in the gut-associated lymphoid tissue that protects the infant from infectious gastrointestinal organisms, and also stimulates the proper regulation of the T cell response. Imbalanced T cell responses have been shown to be involved in Crohn's disease, ulcerative colitis, and inflammation of the stomach caused by *Helicobacter pylori* (a major contributor to gastric ulcers).

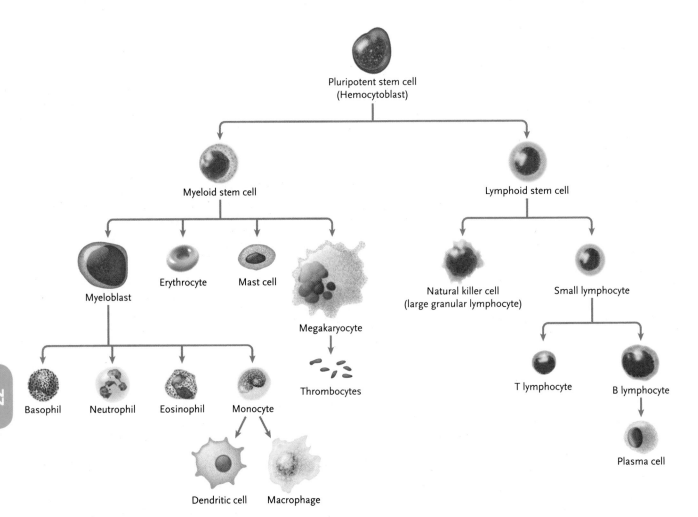

Figure 22.6   Cells of the Immune System

Cells of the immune system originate from pluripotent stem cells in the bone marrow.

Art source: From RUSSELL/HERTZ/STARR/FENTON. Biology, 2E. © 2013 Nelson Education Ltd. Reproduced by permission. www.cengage.com/permissions

CONCEPT REVIEW 22.2

1. Describe the difference between a blood capillary and a lymphatic vessel.

2. How are plasma, interstitial fluid, and lymph the same or different?

3. Where does lymph from the right and left sides of the body re-enter the circulatory system?

4. List reasons why edema can occur.

5. Where are new red and white blood cells produced? From what type of stem cell?

6. What cell types are produced from mesenchymal cells?

7. What is the function of the thymus?

8. What is the role of reticular fibres in the lymph node?

9. What major cell types are found in lymph nodes?

10. What type of immune response can occur in the lymph node?

11. What are the major functions of the spleen?

12. Give two examples of lymphatic nodules.

13. List the cells of the immune system and their main function.

22.3 INNATE IMMUNE RESPONSE

The **innate immune response** includes mechanisms that prevent infection and non-specific cellular responses to infectious organisms; it does not provide long-term protection like the adaptive immune response. Cells involved in innate immune responses trigger the cells involved in the adaptive immune response. Note that it will be important for you to continually refer to Table 22.1 where the functions of each type of immune cell are explained.

First line of defense—prevention

The skin, certain epithelial tissues, and secretions are the most important protective barriers to prevent infection:

- The skin, a physical barrier covered with beneficial microorganisms called the normal flora, helps fight off infectious organisms, for example, by producing lactic acid.

- Hairs and mucus in the nose block large particles, and movement of the hairs can stimulate the sneezing reflex.

- Cilia and mucus in the upper respiratory tract trap particles and propel them up toward the throat to be swallowed and destroyed by hydrochloric acid in the stomach, or removed by coughing.

- Antimicrobial chemicals such as **lysozyme** are secreted into tears, saliva, and sweat and break down bacterial cell walls.

- Sebaceous glands produce acidic oil that prevents the growth of certain bacteria.

- The stomach produces hydrochloric acid, a very strong acid that kills many organisms taken in with food.

- Resident beneficial bacteria in the digestive tract play an important role in preventing infection.

- The flushing actions of urine, tears, and saliva remove microorganisms from the urogenital region, eyes, and mouth, respectively.

Second line of defense—innate response

If microorganisms pass through the body's initial physical barriers and their numbers increase in the body, the cells of the innate immune system become activated. Infected cells secrete signalling molecules that alert nearby immune cells by a process called **chemotaxis**. The white blood cells (WBCs) are attracted by the increasing concentration of chemicals in the vicinity of the infection or damaged tissue. **Cytokines** are a large group of chemicals secreted by a variety of cells— including immune cells, epithelial cells, and connective tissue cells—and that have numerous effects. Cytokines mostly function as **autocrine** or **paracrine** signalling molecules that can act as chemo-attractants, stimulate proliferation or differentiation of other cells types, or cause cells to secrete other cytokines. Often during an immune response, cytokines released by one cell stimulate a cascade of reactions by many other cells, and these result in a positive feedback loop that increases immune cell stimulation until the infection is controlled. The whole system of cytokine function is complex, so this section discusses their function in general and a few cytokines, specifically.

The first cells to arrive at an area of infection are most often neutrophils that migrate from the bloodstream into the tissues. These cells are followed closely by other immune cells, such as monocytes, dendritic cells, macrophages, and natural killer cells. Neutrophils secrete cytokines that attract other immune cells, secrete inflammatory molecules that

TABLE 22.1

Immune System Cell Types and Functions

Cell Type	Function
Neutrophil	Most numerous of all white blood cells, and predominant cell type in pus
	First-responders that produce chemicals involved in the inflammatory response, such as histamine, vasodilators, reactive oxygen species, and cytokines
	Phagocytic cells that engulf many types of microbes, mostly bacteria and opsonized pathogens (have antibodies from B cell response)
	Migrates from blood vessels to infection area within tissues; attracted to damaged cells by process of chemotaxis (migration of immune cells toward an area of infection)
Basophil	Least common type of white blood cell
	Releases histamine and several cytokines involved in the inflammatory response
	Plays a role in allergic reactions and parasitic infections, specifically, ectoparasites such as ticks or lice
Eosinophil	Involved in parasitic infections such as helminths and worms
	Produces many chemicals, including histamine that is involved in allergies, asthma, and inflammation
Mast cell	Primarily involved in allergic reactions and anaphylaxis by releasing histamine if interaction with IgE antibodies occurs
	Plays an important protective role in wound healing
	Prevalent in tissues near openings to the body, such as mouth, nose, skin, lungs, and GI tract
Monocyte	Circulates in the blood and responds to inflammatory signals, migrates into tissues and differentiates into macrophages or dendritic cells
Dendritic cell	Important antigen-presenting cells (APCs) that stimulate lymphocytes (B cells and T cells) in the spleen, lymph nodes, or lymphoid nodules and initiate the adaptive immune response
	Phagocytic cells that can engulf pathogens or opsonized pathogens (when antibodies or complement proteins are bound to the pathogen)
	Present in tissues that are in contact with the environment, such as the skin (Langerhans cells) lungs, and GI tract
Macrophage	Important APCs that stimulate lymphocytes in the lymphatic tissues and initiate the adaptive immune response
	Phagocytic cells that can engulf pathogens or opsonized pathogens (when antibodies or complement proteins are bound to the pathogen)
	Secretes toxic chemicals that directly kill invading organisms
	Secretes cytokines and chemicals involved in the inflammatory response
Natural killer cell	Specifically kills virus-infected cells and cancer cells by lysis with perforin or induction of apoptosis
	Kills cells opsonized with antibodies
	Can produce antigen-specific memory cells important for detecting same infection a second time
B cell	Contains a specific receptor for a specific antigen that is a membrane-bound immunoglobulin (antibody)
	Circulates between blood and lymphatic tissues
	Phagocytic when antigen matches receptor
	Antigen-presenting cell
	Produces cytokines
	Differentiates into plasma cells when activated by helper T cell and then secretes antibodies
	Can produce memory cells
Helper T cell	Do not directly kill any invading microorganisms
	Can only respond to antigen presented by APC (macrophage, dendritic cell, or B cell)
	Produces cytokines that activate B cells, cytotoxic T cells, macrophages, dendritic cells, and NK cells
	Can produce memory cells
Cytotoxic T cell	Directly kills virus-infected cells, cancer cells, and transplant tissue
	Binds to antigen presentation on MHC class I molecules
	Can produce memory cells
Regulatory T cell	Inhibits other immune cells at the end of an infection and prevents auto-reactive immune cells that escaped the selection process in the bone marrow or thymus from damaging tissues

22

increase vascular permeability, and they engulf bacteria by phagocytosis. Monocytes attracted to the area also engulf and kill invading organisms, and they differentiate into phagocytic macrophages and **dendritic cells**. Macrophages and dendritic cells engulf any infectious organisms that are extracellular, including dead cells and cell debris. Once the adaptive immune response has been activated, phagocytic immune cells engulf cells that have antibodies bound to them.

Natural killer cells target virus-infected cells and cancer cells by puncturing the cell membrane with a molecule called **perforin**. Once perforin forms a hole in the membrane formed, interstitial fluid moves into the cell and cause it to burst (Figure 22.7). Then macrophages and dendritic cells engulf the dead cell.

Some cells infected with a virus produce a signalling molecule called **interferon (INF)**. Interferon binds to nearby uninfected cells, triggering the production of antiviral proteins that degrade RNA (Figure 22.8). Viruses replicate inside cells by using the host cell's organelles, building blocks, and ATP; however, if all RNA is degraded the virus cannot replicate. Unfortunately, this causes a temporary loss of normal cell transcription and translation. Interferons can also stimulate apoptosis by means of the p53 pathway in cells that are already infected.

The liver produces complement proteins that circulate in the blood in an inactive form until an infection occurs. The complement cascade can be activated either directly by bacteria or **fungi** (extracellular organisms), or by any microbe that has antibodies bound to it (Figure 22.9).

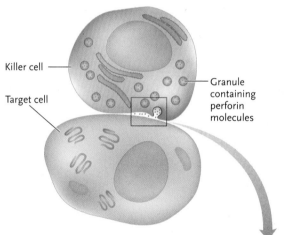

1 Killer cell binds to its target.

2 As a result of binding, killer cell's perforin-containing granules fuse with plasma membrane.

3 Granules disgorge their perforin by exocytosis into a small pocket of intercellular space between killer cell and its target.

4 On exposure to Ca²⁺ in this space, individual perforin molecules change from spherical to cylindrical shape.

5 Remodeled perforin molecules bind to target cell membrane and insert into it.

6 Individual perforin molecules group together like staves of a barrel to form pores.

7 Pores admit salt and H₂O, causing target cell to swell and burst.

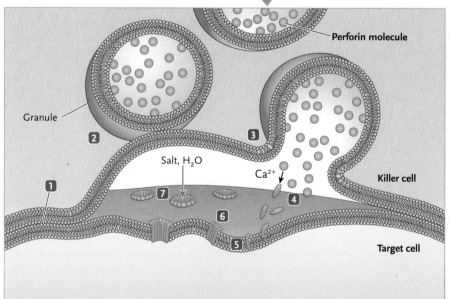

A. Details of the killing process for cytotoxic T cells and NK cells

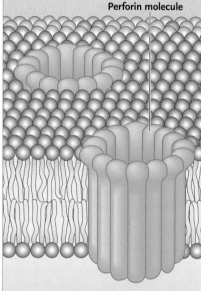

B. Enlargement of perforin-formed pores in a target cell

Figure 22.7   Perforin

Perforin is a secreted by natural killer cells to kill virus-infected cells or cancer cells.

Art source: From SHERWOOD, Human Physiology, 8E. © 2013 Cengage Learning.

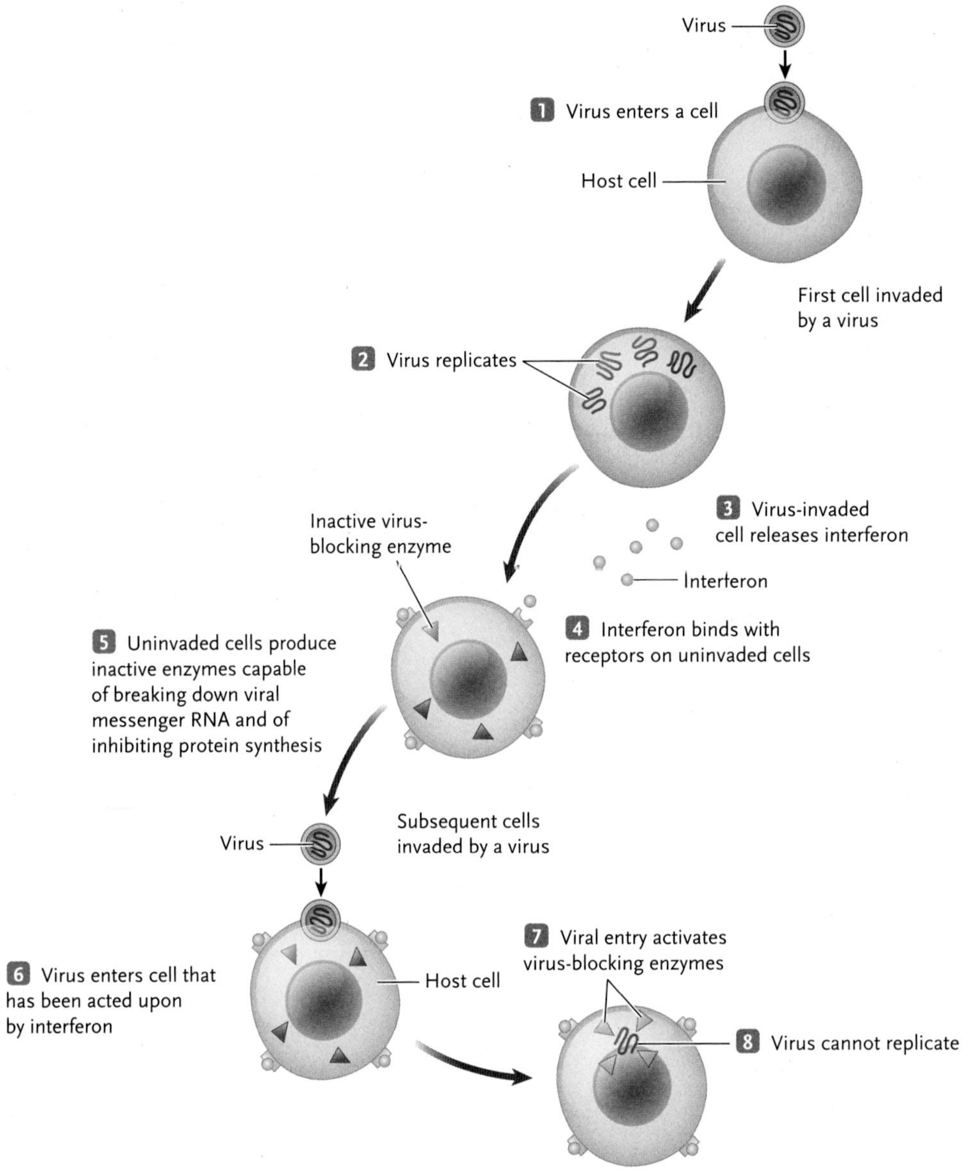

Figure 22.8 Interferon

Interferon acts as an alarm system to warn nearby cells of a viral infection.

Art source: From SHERWOOD/KELL. *Human Physiology*, 2E. © 2013 Nelson Education Ltd. Reproduced by permission. www.cengage.com/permission

This activation of complement proteins causes a cascade of reactions that lead to three main events:

1. Complement proteins bind to microbe and increase phagocytosis.

2. Complement proteins stimulate mast cells to produce histamine, which causes vasodilation, increases vascular permeability, and attracts phagocytic cells (chemotaxis).

3. Complement proteins combine to form a **membrane attack** complex that kills bacteria and fungi, similar to how perforin kills virus-infected cells or cancer cells.

Inflammatory response

The **inflammatory response** is triggered any time tissues have physical injury, such as a burn, puncture, cut, sprain, break, or are affected by toxins or infectious organisms. Tissue-specific inflammation is often named by the tissue that is undergoing the inflammation: for example, **tendonitis** is inflammation in a tendon, and **appendicitis** is inflammation of the appendix. The function of inflammation is to remove infectious organisms and heal damaged tissue. The classic signs of inflammation are redness, swelling, pain, and heat, which are due to molecules released by cells in the damaged tissue.

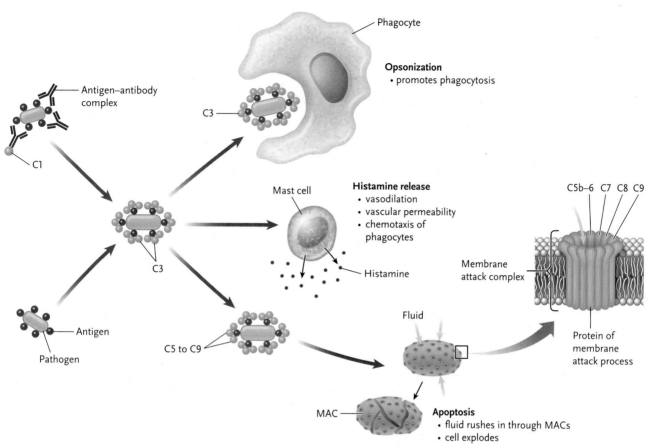

Figure 22.9 Complement Proteins

Complement proteins can kill bacteria and fungi.

Art source (right): From SHERWOOD, *Human Physiology*, 8E. © 2013 Cengage Learning.

The primary cells involved in initiating inflammation are cells already in the area that has been damaged; these include phagocytes, basophils, mast cells, and platelets. The main molecules released by these cells are histamine, cytokines, and eicosanoid. Histamine causes vasodilation to increase blood flow to the region and increased **vascular permeability** so that more fluid, proteins, and immune cells enter the tissues. Cytokines signal and activate other immune cells. **Eicosanoids** are cytokines that are produced from the essential fatty acids, omega 3 and omega 6, that we need to include in our diet. Omega 6 fatty acids promote the production of pro-inflammatory eicosanoids, and omega 3 fatty acids promote the production of anti-inflammatory ones; therefore, it is important to ensure a balance of these fats in our diet. One of the major types of eicosanoids is **prostaglandin,** which has the same functions as histamine and also signals pain sensations. Prostaglandin production is inhibited by drugs, such as aspirin and ibuprofen, that decrease the pain and swelling from inflammation.

If an injury has damaged blood vessels, platelets convert the fibrinogen protein that is always circulating in the blood into *fibrin* threads to clot blood, and it also traps pathogens in the tissue to prevent them from spreading and infecting nearby tissue. The combined effects of the chemicals produced during inflammation cause the following classic signs of an inflammatory response (Figure 22.10):

- Redness from increased blood flow
- Swelling from increased vascular permeability and therefore edema because fluid moves into the interstitial space
- Pain from prostaglandins, from swelling that puts pressure on pain sensing neurons, and from direct injury to neurons
- Heat from increased blood flow as well as increased heat production by immune cells producing ATP

In some cases an infection causes a fever to develop. A **fever** is a change in the set point of our normal body

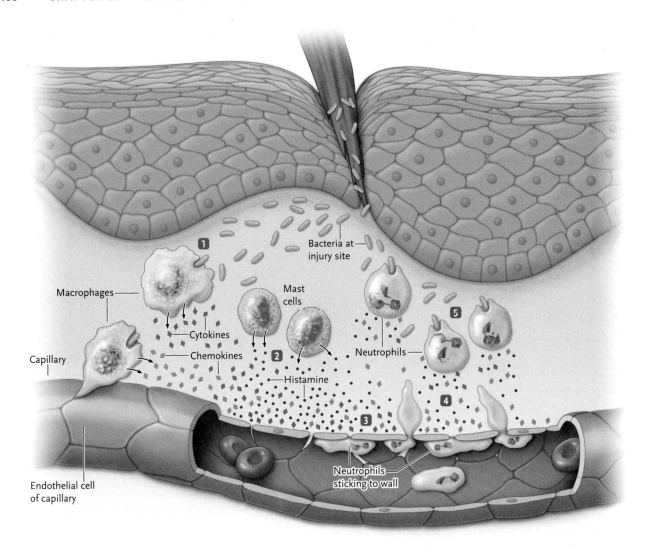

1 A break in the skin introduces bacteria, which reproduce at the wound site. Activated macrophages engulf the pathogens and secrete cytokines and chemokines.

2 Activated mast cells release histamine.

3 Histamine and cytokines dilate local blood vessels and increase their permeability. The cytokines also make the blood vessel wall sticky, causing neutrophils to attach.

4 Chemokines attract neutrophils, which pass between cells of the blood vessel wall and migrate to the infection site.

5 Neutrophils engulf the pathogens and destroy them.

Figure 22.10   Inflammatory Response

The inflammatory response is an important process for killing infections.

Art source: From RUSSELL/HERTZ/STARR/FENTON. *Biology*, 2E. © 2013 Nelson Education Ltd. Reproduced by permission. www.cengage.com/permissions

temperature of 37°C. The body's temperature set point is determined by the hypothalamus. An increase in body temperature by one or two degrees will increase the rate of chemical reactions and inhibit microorganisms that are adapted to reproduce at 37°C, giving our immune system an advantage. However, a fever of 42°C can be fatal because the body's proteins can denature. Molecules that cause the hypothalamus to increase the body's temperature set point are called **pyrogens**; these can be cytokines or prostaglandins produced by immune cells, or toxins produced by the infecting organism. When a fever reaches 40°C an **antipyretic**—a medication that reduces fever—should be taken.

Phagocytic cells continue to clean up old dead cells and pathogens, and the damaged tissue heals as new cells are formed through mitosis.

DID YOU KNOW?

People tend to have lots of omega 6 fatty acids in their diet and not enough omega 3. Therefore, many people have chronic inflammatory conditions because they are not getting enough omega 3 fatty acids. Omega 3 fatty acids are high in flax seeds, fish, and fish oils.

1. Describe the mechanisms involved in the first line of defense.

2. Which immune cells are phagocytic?

3. Which immune cell will target virus-infected cells?

4. Describe the function of the following chemicals involved in the innate immune response: lysozyme, perforin, interferon, histamine, complement, and prostaglandins.

5. Describe the causes of redness, pain, swelling, and heat involved in an inflammatory response.

6. What are pyrogens?

22.4 ADAPTIVE IMMUNE RESPONSE

The adaptive immune response involves long-term protection from infectious organisms. During an infection, immune cells can leave the cardiovascular system and enter the tissues to kill invading organisms, signal other immune cells, or carry pathogen particles to the lymph nodes to stimulate activation of **lymphocytes** involved in the adaptive immune response. The lymphocytes are the B cells and T cells. The innate and adaptive immune responses work together.

An **antigen** is any molecule on the surface of an infectious organism that can be recognized by any immune cell. Antigens can be proteins, glycoproteins, lipoproteins, lipopolysaccharides, or saccharides. Cells of the innate immune system can recognize general molecular features associated with pathogens, whereas T cells and B cells have specific receptors that only recognize specific antigens. There are millions of possible B cell receptors (BCRs) and T cell receptors (TCRs), and some will never be activated if they never come in contact with a matching antigen. T cells cannot be activated directly by a pathogen; they only recognize the antigens presented by **antigen-presenting cells** (**APCs**), such as macrophages, dendritic cells, or B cells. B cells can directly recognize antigens; however, further activation by a T cells increases their response.

Antigen recognition and presentation

APCs are phagocytic cells that engulf pathogens and dead cells. Phagocytic cells identify pathogens by recognizing that their general surface pattern of molecules is not the same as any found in our cells. These are called **pathogen-associated molecule patterns** (**PAMPs**), and they bind to receptors on the phagocyte: the **pattern recognition receptors** (**PRRs**). Binding of a PAMP to a PRR stimulates the phagocytic cells to secrete cytokines that act as autocrine and paracrine signalling molecules, and then the pathogen is engulfed. Once inside the phagocytic cell, lysosomes digest the microbe into many fragments. Those fragments are attached to **major histocompatibility** (**MHCII**) molecules and then brought to the surface of the membrane (Figure 22.11). All APCs are phagocytic, engulf microbes in the tissues where the infection originates, present antigen on their cell surface on MHCII molecules, and then circulate to the lymphatic tissues. It is this presented antigen that is recognized by T cells in the lymph nodes or spleen.

MHC molecules are encoded by the largest family of genes in humans. Expression of these genes in the receptor is co-dominant (both maternal and paternal genes are involved) and polygenic (more than one gene is involved). Every person has their own MHC molecules that their immune system recognizes. It is the particular combination of MHC and antigen that tells the T cells to react, or not. All nucleated cells in the body, including the immune APCs, express **MHCI molecules**. Since mature red blood cells do not have a nucleus, they do not express MHC. Cells continually present self- or foreign antigens on the surface of the cell. This is a constant process of presenting molecular fragments and then breaking them down and presenting other fragments. In cells infected with a virus or bacterial toxin, the antigens present microbial fragments that cytotoxic T cells will recognize and destroy. By contrast, normal cells that express self-proteins will not be harmed by T cells: due to the selection process that occurred in the thymus. When someone has a tissue transplant, the MHCI molecules on the transplanted tissue will be different from their own MHCI molecules, and, therefore, cytotoxic T cells will attack the new tissue. This explains why it is important to receive a transplant from a closely related person who has very similar MHC molecules.

The adaptive immune response can be divided into two different pathways: the **cell-mediated immune response** that uses T cells, and the **humoral immune response** that uses B cells and antibodies.

Cell-mediated immune response

T cells can be **helper T cells** or **cytotoxic T cells**. Helper T cells do not directly kill invading organisms but do produce many important cytokines that direct the other immune cells. Cytotoxic T cells directly kill cells infected with a virus, cancer cells, and transplant tissue with foreign MHCI molecules.

Helper T cells only recognize antigens that are presented by APCs on MHCII. Helper T cells also have a co-receptor called CD4, and are therefore sometimes called **CD4$^+$ T cells**. **HIV** (**human immunodeficiency virus**)

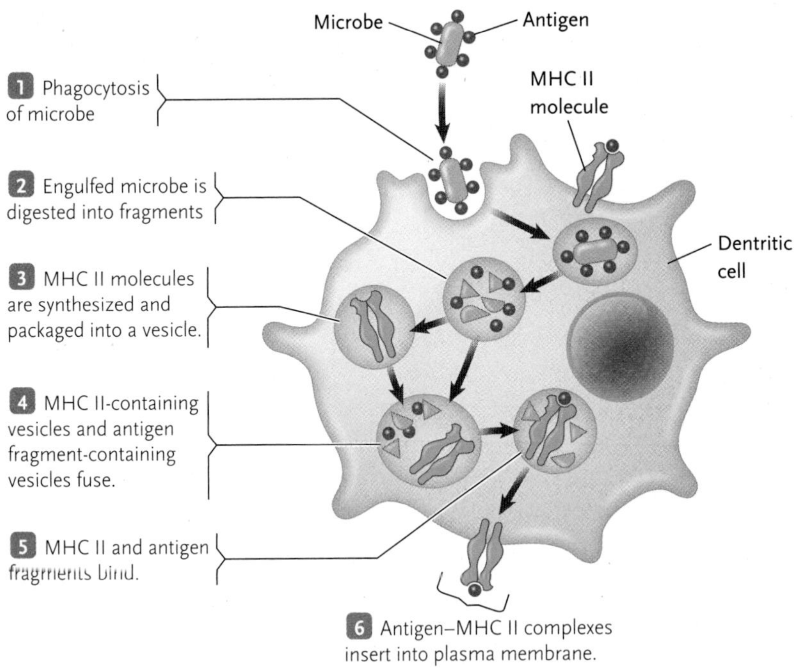

Microbe — Antigen

1 Phagocytosis of microbe

MHC II molecule

2 Engulfed microbe is digested into fragments

Dentritic cell

3 MHC II molecules are synthesized and packaged into a vesicle.

4 MHC II-containing vesicles and antigen fragment-containing vesicles fuse.

5 MHC II and antigen fragments bind.

6 Antigen–MHC II complexes insert into plasma membrane.

Figure 22.11   Antigen Presentation

Antigens are presented on MHC molecules on the surface of cells.

Art source: From RUSSELL/HERTZ/STARR/FENTON. *Biology*, 2E. © 2013 Nelson Education Ltd. Reproduced by permission. www.cengage.com/permissions

infects cells that have a CD4 receptor. When an APC engulfs a pathogen in the tissues, which could be bacteria, virus, fungus, or small parasite, it will digest the pathogen and present some fragments on its cell surface on MHCII. The APC then travels from the tissue to the nearest lymphatic vessel and travels with the lymph through the vessels and lymph nodes until it comes in contact with a T cell that has a matching T cell receptor (TCR) (Figure 22.12).

The binding of the APC–antigen–MHCII complex to the TCR–CD4 stimulates the APC to produce the cytokine **interleukin-1 (IL-1)**, which acts as a paracrine signal on the T cell and stimulates the T cell to secrete **interleukin-2 (IL-2)**, which acts as an autocrine signal that causes that T cell with that specific TCR to proliferate.

Helper T cells can differentiate into different subtypes based on which cytokines are produced by the

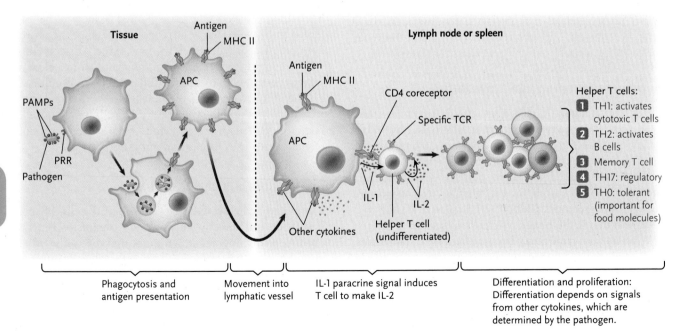

Figure 22.12   Helper T Cell Activation

Art source (right panel): From RUSSELL/HERTZ/STARR/FENTON. *Biology*, 2E. © 2013 Nelson Education Ltd. Reproduced by permission. www.cengage.com/permissions

APC. When APCs such as macrophages and dendritic cells encounter pathogens in the tissues, they recognize PAMPs on the pathogen that tell the APC generally what type of organism it is. Although both cell-mediated and humoral immune responses are stimulated for any type of infection, one or the other is stimulated *more* based on the type of infectious organism. Infections that are mostly **intracellular**, such as viruses or bacterial toxins, require cytotoxic T cells to kill infected host cells. **Extracellular** organisms, such as bacteria, fungi, and viruses when they are outside host cells, require antibodies and other extracellular mechanisms such as complement proteins and phagocytosis.

Helper T cells can stimulate one pathway or another by differentiating into subtypes of helper T cells. **TH1** cells increase activation of cytotoxic T cells in the case of viral infection, and **TH2** cells increase activation of B cells in the case of a bacterial infection. Differentiation depends on the types of cytokines produced by the antigen-presenting cell. Some helper T cells become memory cells that are capable of fighting the same infection in the future. Depending on the cytokines present in the environment, helper T cells can become **regulatory T cells (TH17)** because of the co-receptor they express. Regulatory T cells slow down the immune response at the end of an infection, and in other cases they prevent immune cells from reacting against self-antigens. Helper T cells can also differentiate into non-reactive, *tolerant* cells for certain antigens: called **TH0** cells (Figure 22.12).

This happens in the case of a high exposure to an antigen, especially through exposure along the GI tract. Molecules that we are constantly exposed to, such as normal flora, food molecules, and environmental molecules, do not stimulate an immune response unless this immune system regulation does not occur properly and allergies develop. In animal experiments, oral exposure to a pathogenic antigen can cause immune tolerance, and the immune system will not react to future exposure to that pathogen.

Cytotoxic T cell activation is a little different. Cytotoxic T cells only recognize foreign antigens presented on MHCI molecules that would be found on normal body cells if they were infected with a virus or if they produced cancer proteins. Cytotoxic T cells (and B cells) can migrate between the lymphatic system and the tissues; whereas helper T cells tend to stay in the lymphatic tissue until activated. Cytotoxic T cells have the CD8 co-receptor and are therefore also called **CD8+ T cells**.

Cytotoxic T cells each have a specific TCR that recognizes only one antigen. If a cytotoxic T cell recognizes and binds to an infected cell, it becomes fully activated only if a helper cell produces IL-2 to stimulate the cytotoxic T cell to proliferate. TH1 helper cells produce many cytokines that direct the immune response, but IL-2 is the main cytokine that causes the cytotoxic T cells to proliferate—producing many clones expressing the same TCR that can then kill other cells expressing the same foreign antigen on MHCI (Figure 22.13). Once activated, some cytotoxic T cells become memory cells.

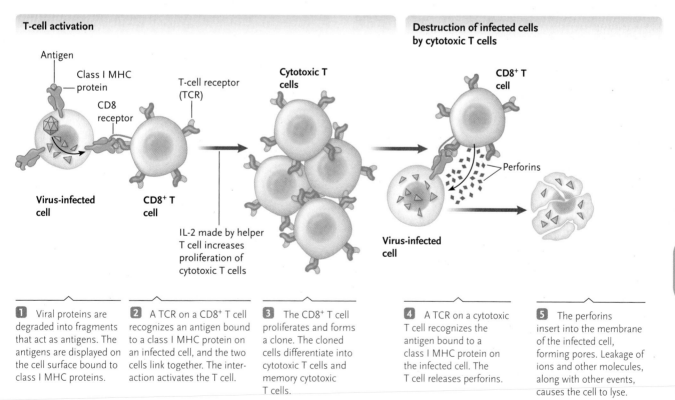

T-cell activation

Antigen
Class I MHC protein
CD8 receptor
T-cell receptor (TCR)
Cytotoxic T cells
Virus-infected cell
CD8+ T cell
IL-2 made by helper T cell increases proliferation of cytotoxic T cells

Destruction of infected cells by cytotoxic T cells

CD8+ T cell
Perforins
Virus-infected cell

1. Viral proteins are degraded into fragments that act as antigens. The antigens are displayed on the cell surface bound to class I MHC proteins.

2. A TCR on a CD8+ T cell recognizes an antigen bound to a class I MHC protein on an infected cell, and the two cells link together. The interaction activates the T cell.

3. The CD8+ T cell proliferates and forms a clone. The cloned cells differentiate into cytotoxic T cells and memory cytotoxic T cells.

4. A TCR on a cytotoxic T cell recognizes the antigen bound to a class I MHC protein on the infected cell. The T cell releases perforins.

5. The perforins insert into the membrane of the infected cell, forming pores. Leakage of ions and other molecules, along with other events, causes the cell to lyse.

Figure 22.13 **Cytotoxic T-Cell Activation**

22

Cytotoxic T cells kill infected body cells in two different ways: (1) by producing perforin (like natural killer cells) and other toxic molecules that enter the cell and cause the cell to die; (2) by producing molecules that trigger apoptosis in the infected cell. Phagocytic cells engulf and clean up any cells debris. As the infection diminishes, regulatory T cells inhibit the immune response and prevent excessive damage to tissues.

Humoral immune response

In the same way as T cells, B cells have specific receptors that recognize specific antigens. B cell receptors are **immunoglobulins** that are attached to the cell membrane. B cells can migrate between the tissues and the lymphatic system and the circulatory system, and when they come in contact with a matching antigen, they can bind directly with that pathogen. B cells are phagocytic cells, and they are also antigen-presenting cells. When an inactivated B cell encounters a matching antigen, it engulfs the organism, digests it, and presents fragments on MHCII. Then the B cell can migrate back to the lymphatic system. Upon contact with a helper T cell that recognizes the antigen presented on MHCII, further activation of the B cell will occur (Figure 22.14). Helper T cells that bind B cells presenting an antigen, secrete IL-2 and cause B cell to proliferate, producing many more B cells expressing the same receptor. The B cell also differentiates into a plasma cell that secretes antibodies that match the antigen. Antibodies can circulate throughout the body and bind to the infecting organism, called opsonization. Some of the B cell population differentiates into memory cells.

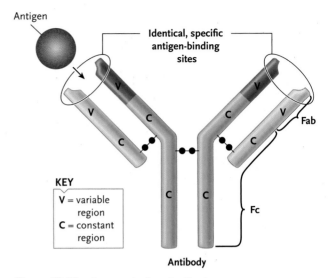

Figure 22.15 Structure of an Antibody

Art source: From SHERWOOD, *Human Physiology*, 8E. © 2013 Cengage Learning.

Functions of antibodies

When secreted antibodies bind to infecting organisms, they are **opsonized**; this is like they are tagged for destruction by other immune cells (Figure 22.15). When microbes are opsonized with an antibody, the following occurs:

1. Phagocytosis by macrophages and dendritic cells increases.
2. Targeting by complement proteins increases, and the bacterial or fungal organism is killed with the membrane attack complex.

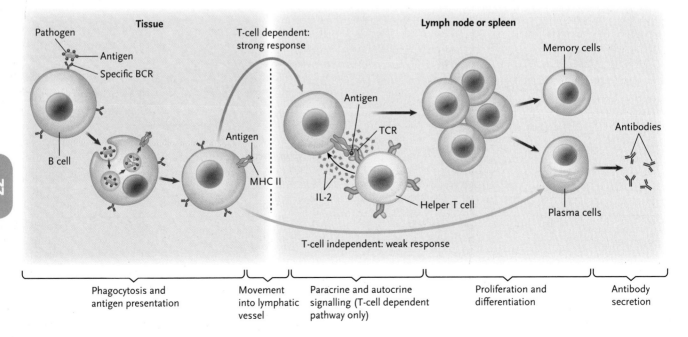

Figure 22.14 B-Cell Activation

Art source: From RUSSELL/HERTZ/STARR/FENTON. *Biology*, 2E. © 2013 Nelson Education Ltd. Reproduced by permission. www.cengage.com/permissions

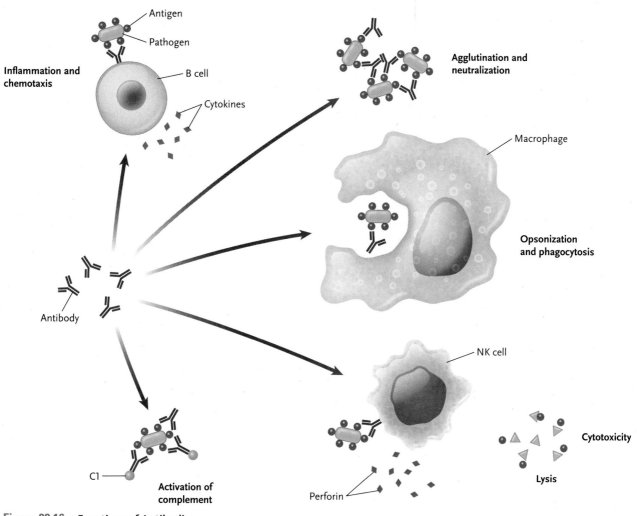

Figure 22.16 Functions of Antibodies

Antibodies are immunoglobulins that identify and neutralize infectious organisms.

3. Microbes become neutralized, so they cannot produce toxins or bind to the body's cells.

4. Bacteria are immobilized, so they cannot spread easily.

5. Agglutination occurs, which means many antibodies and microbes clump together and so they cannot function (Figure 22.16).

Role of memory cells

Memory cells can be formed from the activation of helper T cells, cytotoxic T cells, and B cells. Memory cells are a subpopulation of the activated cells with the same receptor that recognize the antigens from a specific infection. If we were ever infected with exactly the same, or a very similar, infection again, memory cells would be stimulated to activate, proliferate, and kill the infecting organisms very rapidly, and most often we wouldn't have any symptoms. This is called a **secondary immune response**.

If we cannot be infected by the same organism twice, why do we get sick from the common cold every year? Many viruses, such as those causing the common cold and influenza, change their surface antigens frequently and thereby evade our immune system. Every time the infectious organism changes its surface proteins (micro-evolution), the cells of the immune system recognize that the antigens are different and respond as if the infection is new. Since some surface proteins stay the same and some change, infectious organisms that have significantly different antigens cause a stronger immune response than organisms that have changed very little. This explains why having a common cold or the flu sometimes causes severe symptoms and other times only minor symptoms. Other factors also affect the severity of symptoms.

22

TABLE 22.2

Types of Adaptive Immunity

Type of Immunity	How Acquired
Naturally acquired active immunity	Acquired after an infection where memory cells are produced
Naturally acquired passive immunity	Acquired through transfer of antibodies from mother to baby during breast feeding, which temporarily prevents infection in the baby due to any infection the mother has made antibodies against
Artificially acquired active immunity	Acquired through vaccination—transfer of an antigen into the body from an infectious organism, which stimulates the immune system to produce memory cells related to the infection but without causing disease
Artificially acquired passive immunity	Acquired by transfer of antibodies into the body by injection, which provides temporary immunity to the organism specifically related to the injected antibodies

DID YOU KNOW?

The following factors can alter the effectiveness of an immune response and therefore prolong infection:

- Lack of protein, either from malnutrition or lack of available muscle protein that can be broken down into amino acids. The immune system requires a much higher amount of amino acids during illness in order to produce all the cytokines and signalling molecules involved in the innate and adaptive immune responses.

- Lack of sleep, although this connection to immune response is not simple. People suffering from sleep disorders have a significantly increased risk of getting colds and flu.

- Excess iron. Iron supplementation may not be beneficial during an infection since microorganisms require iron for replication and virulence. The liver decreases the level of blood iron during an infection, which decreases the ability of micro organisms to replicate.

- Excess stress. Stress causes the release of cortisol, which is normally secreted during infections to act as a negative feedback control system that prevents overstimulation of immune cells. Stress increases cortisol above normal levels and thereby increases the amount of immune inhibition.

CONCEPT REVIEW 22.4

1. What is an antigen?
2. Which cell types are antigen-presenting cells?
3. Why is it important for antigens to be presented on the surface of a cell?
4. Which cell types express MHCI? MHCII?
5. What is the function of IL-1 and IL-2 in activation of helper T cells?
6. What are the various types of cells that helper T cells can differentiate into during activation?
7. What is the main role of each type of helper T cell?
8. Which cells express CD4 and CD8?
9. What is the function of cytotoxic T cells?
10. How are helper T cells activated by B cells?
11. What types of cells can B cells differentiate into?
12. What is opsonization, and how does it help the immune response?
13. What is the difference between the cell-mediated immune response and the humoral immune response?
14. Which cell type is more involved in a viral infection? A bacterial infection?
15. Why are memory cells important?
16. Why do we get colds every year even though we produce memory cells?
17. What is the difference between active and passive acquired immunity?

22.5 VACCINES

As discussed in the previous section, **naturally acquired active immunity** involves the production during an infection of memory cells that can protect the body from future infection from the same organism. A different type of adaptive immunity, **artificially acquired active immunity** involves the use of **vaccines**. Edward Jenner in 1796 produced the first vaccine—a vaccine to fight **smallpox** infection.

Sometimes one microorganism is similar enough to another one that cross-reactivity can occur such that the memory cells for one infection protect against another similar infection. This was the case with smallpox (a deadly viral infection) and **cowpox** (a mild skin infection transferred between cows and humans). Edward Jenner determined that people infected with cowpox did not get infected with smallpox. The word *vaccine* that we use today comes from the Latin *vaccinus*, which means "from cows."

Although becoming infected with organisms creates an immune response that produces memory cells that protect us in the future, some infections we would not survive to enjoy the benefits of the memory cells: for example, smallpox, tetanus, tuberculosis, and hepatitis. It would be ideal to have a vaccine that could stimulate the immune system to produce memory cells without actually causing the disease, and then we could be protected for the rest of our lives from the infection. Different methods have been used to produce vaccines and each has different advantages and disadvantages.

Live attenuated vaccines

Viruses can be grown over several generations in a lab until an *attenuated* (less virulent) strain is produced by random DNA mutations. Similarly, cowpox is less virulent than smallpox, but similar enough to make the memory cells it produces be cross-reactive. A **live attenuated vaccine** stimulates the immune system to mount an adaptive immune response that leads to the production of memory cells, but with very low risk of causing the disease. The vaccine for measles, mumps, and rubella is known as the MMR vaccine. The benefits of a live attenuated vaccine is that long-term memory cells are produced even with a very low dose of the vaccine, and immune stimulants called **adjuvant** do not need to be added to the vaccine. The risk with this type of vaccine is possible contamination with other viruses, or the virus

may revert back to its virulent form and cause disease. This has not happened with the MMR vaccine; however, this is not a method of choice for the production of a vaccine for deadly viruses such as HIV or Ebola.

Killed virus vaccines

If the virus used in the vaccine is killed, it cannot cause infection. However, the immune system is difficult to trick because it can recognize the difference in pathogenicity between organisms, which is why various helper T cells are produced depending on the type of infection: for example, TH1 vs TH2. A killed virus still has intact surface antigens that can be recognized by the immune system, and it cannot cause infection. However, an adjuvant needs to be added to the vaccine to increase the immune system's response, and in many cases killed virus vaccines cannot cause the induction of memory cells. Killed virus vaccines have been produced for influenza and cholera, for example.

Toxoids

Some bacteria are only pathogenic when they produce toxins that damage our cells. Vaccines can be made from **inactivated toxins** so that the antigens are present to stimulate the immune system, but the bacteria are not present to cause infection. These vaccines are beneficial because they can't cause illness, and they can promote the production of memory cells, maybe not life-long, but for approximately 10 years. The childhood DPT vaccine for diphtheria, pertussis, and tetanus is an example of this kind of vaccine.

Vaccines produced using DNA technology

New vaccine research is moving into the field of biotechnology to try to produce safer and more effective vaccines against deadly infections. Recall from Chapter 10 that genes from one organism can be moved to another, such as moving the insulin gene into bacteria to have bacteria produce human insulin proteins. Likewise, virus surface proteins have been expressed in bacteria, yeast, mammalian cells, and low virulence viruses. These proteins can then be used to make the vaccine without concern that infection from the whole pathogen will result (Figure 22.17). Examples of this type of vaccine include the hepatitis B vaccine and a vaccine against West Nile virus for horses. Many other clinical trials are ongoing.

22

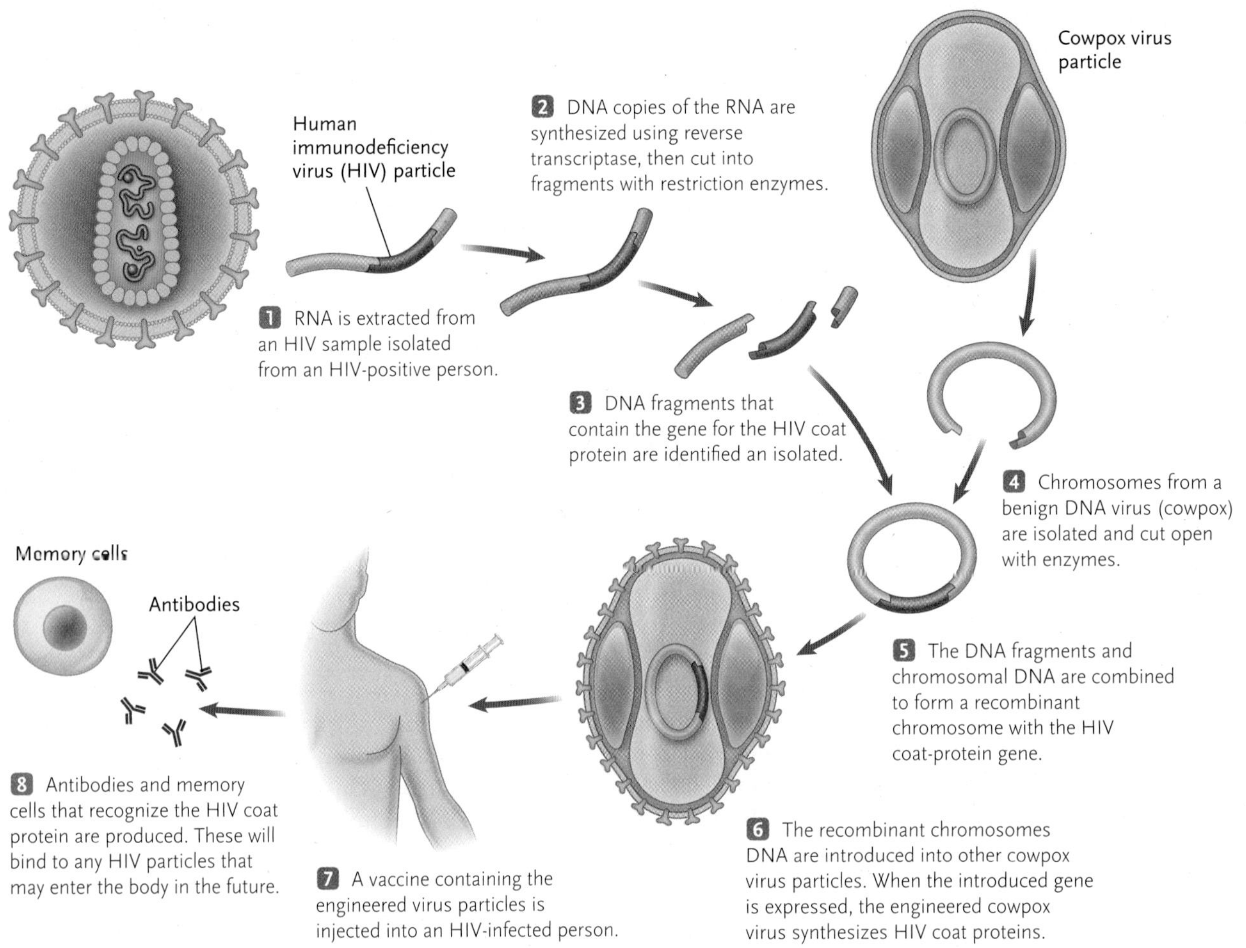

Cowpox virus particle

Human immunodeficiency virus (HIV) particle

2 DNA copies of the RNA are synthesized using reverse transcriptase, then cut into fragments with restriction enzymes.

1 RNA is extracted from an HIV sample isolated from an HIV-positive person.

3 DNA fragments that contain the gene for the HIV coat protein are identified an isolated.

4 Chromosomes from a benign DNA virus (cowpox) are isolated and cut open with enzymes.

Memory cells

Antibodies

5 The DNA fragments and chromosomal DNA are combined to form a recombinant chromosome with the HIV coat-protein gene.

8 Antibodies and memory cells that recognize the HIV coat protein are produced. These will bind to any HIV particles that may enter the body in the future.

7 A vaccine containing the engineered virus particles is injected into an HIV-infected person.

6 The recombinant chromosomes DNA are introduced into other cowpox virus particles. When the introduced gene is expressed, the engineered cowpox virus synthesizes HIV coat proteins.

Figure 22.17   Recombinant DNA Vaccines

Recombinant DNA vaccines, produced from genes of the infectious organism, stimulate an immune response without causing an infection.

CONCEPT REVIEW 22.5

1. Describe how Edward Jenner produced the first vaccine.

2. Give three examples to describe how live attenuated vaccines are produced.

3. Give three examples to describe how killed virus vaccines are produced.

4. How are vaccines produced for diphtheria, pertussis, and tetanus?

5. Explain how DNA technology has been used to produce the hepatitis vaccine.

22.6 AUTOIMMUNE DISEASES AND ALLERGIES

Autoimmune diseases occur when the immune cells attack normal tissues as though they were fighting an infection. During embryonic development, our lymphocytes go through a process called negative selection (self-tolerance), where any receptors that recognize self-proteins are eliminated by apoptosis. Negative selection of B cells occurs in the bone marrow and for T cells occurs in the thymus. By this process, any lymphocytes that would react with our normal tissues should be eliminated, but this process is not always perfect. Furthermore, there are regulatory T cells that inhibit the immune response and prevent overstimulation of the innate and adaptive immune responses. If the balance of immune regulation is not maintained this overstimulation can occur and damage our tissues.

The exact cause of any autoimmune disease is not known, but there are many theories. Researchers believe that once a mechanism of action is determined for one of the autoimmune diseases, it will open the door to

TABLE 22.3

Common Autoimmune Diseases

Disease	Tissue Affected
Multiple sclerosis	Myelin on axons in the CNS
Type 1 diabetes	Beta islets cells of pancreas
Celiac disease	Small intestine microvilli in cross-reaction with gluten
Addison's disease	Adrenal cortex
Aplastic anemia	Bone marrow
Crohn's disease	Gastrointestinal tract
Ulcerative colitis	Large intestine
Grave's disease	Thyroid gland
Lupus	Many tissues
Myasthenia gravis	Skeletal muscles
Rheumatoid arthritis	Joints

understanding and effectively treating others. Current treatments for many autoimmune diseases involve **immune-suppression medications** that are not sufficiently effective and do not cure the disease. The incidence of autoimmune disease has been increasing at an alarming rate in developed countries in the last several decades, and currently one-third of the population of North America lives with an autoimmune disease.

Theories of causes of autoimmune diseases

Molecular mimicry. Because the immune system sometimes cross-reacts with very similar antigens (memory cells against cowpox protect against smallpox), it is possible that the immune system will cross-react with a self-tissue after fighting a certain infection.

Hygiene theory. There is an inverse correlation between the incidence of infections and autoimmune diseases in certain populations. In countries and regions where infection rates are high, the incidence of autoimmune disease is very low; this is particularly the case for populations with parasitic worm infections (Figure 22.18).

Lack of exposure to beneficial bacteria in childhood. The role of normal flora organisms on our skin and GI tract is becoming more understood in its relation to immune regulation. With an increased human population, food must travel long distances and remain in stores for long periods of time, so foods are sterilized and pasteurized to prevent rotting and food poisoning. However, with the loss of traditional preparation of foods, such as bacterial fermentation, people have much less contact with the beneficial bacteria that play an important role in immune regulation. Children who require antibiotics (that also

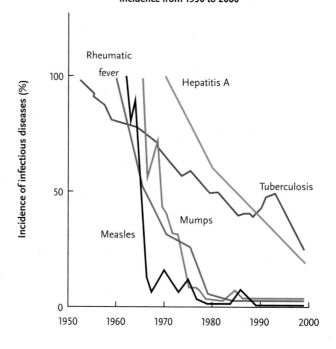

Prototypical infectious disease incidence from 1950 to 2000

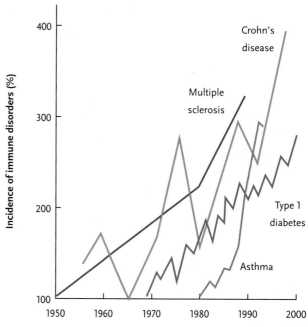

Autoimmune disorder incidence from 1950 to 2000

Figure 22.18   Hygiene Theory

Autoimmune diseases are prevalent in regions that have a low incidence of infections.

Source: Courtesy of Dr. Joel Weinstock

kill normal flora) in the first year of life have an increased risk of developing asthma and allergies.

Genetic factors. Research has shown that some variations of certain MHC genes affect antigen presentation.

Vitamin D. People living in equatorial regions with more exposure to UV light have a significantly lower incidence of autoimmune diseases than those in northern or southern regions, and this lower incidence also applies to those who lived in equatorial regions in childhood and then moved to northern or southern regions after puberty. Vitamin D does play an important role in immune regulation, but supplementation alone has not reversed the incidence of autoimmune disease in northern or southern regions. Multiple factors are likely involved.

Allergies

Allergic reactions are caused by the immune system reacting to non-harmful environmental antigens, such as pollen, dust, mites, bee venom, or foods. The immune system should develop tolerance for the antigens it comes into contact with in very high amounts and on a regular basis. Recall from Section 22.4 that helper T cells can differentiate to become TH1, TH2, TH17, or TH0 helper T cells, depending on the cytokines produced; when this process fails, allergies and asthma can develop. The exact cause is not known. Allergic reactions are also called hypersensitivity reactions. Asthma is considered a form of hypersensitivity reaction due to the release of histamine.

Allergic reactions are primarily humoral immune responses, where B cells are activated to become plasma cells that secrete antibodies against a harmless antigen.

The antibody type that is specific to allergic reactions is immunoglobulin E (IgE). B cells circulate throughout the tissues and lymphatic system. The activation of a B cell with a receptor that matches an environmental or food antigen can occur without IL-2 stimulation from a helper T cell (Figure 22.14). The B cell differentiates into a plasma cell that secretes IgE antibodies, which bind to mast cells or basophils (Figure 22.19). During the very first exposure to an antigen, there may not be any symptoms. The second exposure to the same antigen will bind the IgE antibodies already on mast cells or basophils and cause those cells to release large amounts of histamine.

Histamine is the main chemical responsible for allergy symptoms, which include the following:

- localized vasodilation that causes redness around the eyes and nose
- increased vascular permeability that causes watery eyes and runny nose
- itching
- bronchoconstriction that can cause wheezing

The severity of the allergic reaction depends on how the immune system has been sensitized, the type of allergen, and the mode of entry into body. Very severe reactions can cause increased swelling and bronchoconstriction that can be fatal: called **anaphylaxis (anaphylactic shock)**. Generally, the most severe allergic reactions are caused by foods or venoms such as a bee sting. The most common food allergies are to peanuts, eggs, and shellfish.

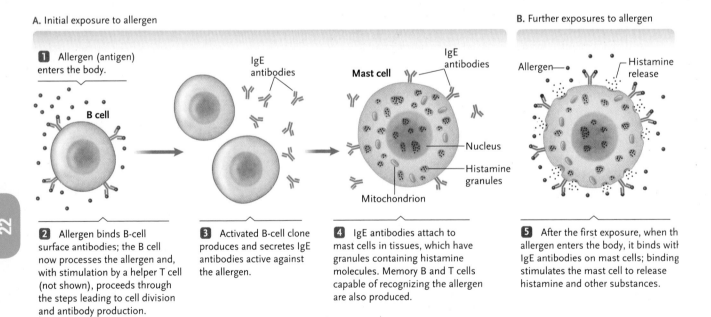

A. Initial exposure to allergen

1 Allergen (antigen) enters the body.

IgE antibodies

Mast cell

IgE antibodies

B cell

Nucleus

Histamine granules

Mitochondrion

2 Allergen binds B-cell surface antibodies; the B cell now processes the allergen and, with stimulation by a helper T cell (not shown), proceeds through the steps leading to cell division and antibody production.

3 Activated B-cell clone produces and secretes IgE antibodies active against the allergen.

4 IgE antibodies attach to mast cells in tissues, which have granules containing histamine molecules. Memory B and T cells capable of recognizing the allergen are also produced.

B. Further exposures to allergen

Allergen— Histamine release

5 After the first exposure, when th allergen enters the body, it binds witl IgE antibodies on mast cells; binding stimulates the mast cell to release histamine and other substances.

Figure 22.19 Allergic Reaction

Allergic reactions are primarily humoral immune responses involving IgE, as well as mast cells, basophils, and histamine.

Art source: From RUSSELL/HERTZ/STARR/FENTON. *Biology*, 2E. © 2013 Nelson Education Ltd. Reproduced by permission. www.cengage.com/permissions

CONCEPT REVIEW 22.6

1. Describe how autoimmune diseases cause disease.

2. List the main theories of the causes of auto-immune disease.

3. Where in the world are autoimmune diseases the most prevalent?

4. Describe what tissue is targeted in the following diseases:
   a. type-1 diabetes
   b. multiple sclerosis
   c. myesthenia gravis
   d. Grave's disease

5. What cell types are involved in allergic reactions?

6. Why do no symptoms occur at the time of the very first exposure to a potential environmental antigen?

7. Which type of antigens cause the most severe allergic reactions?

8. What is the main chemical involved in allergic reactions and asthma?

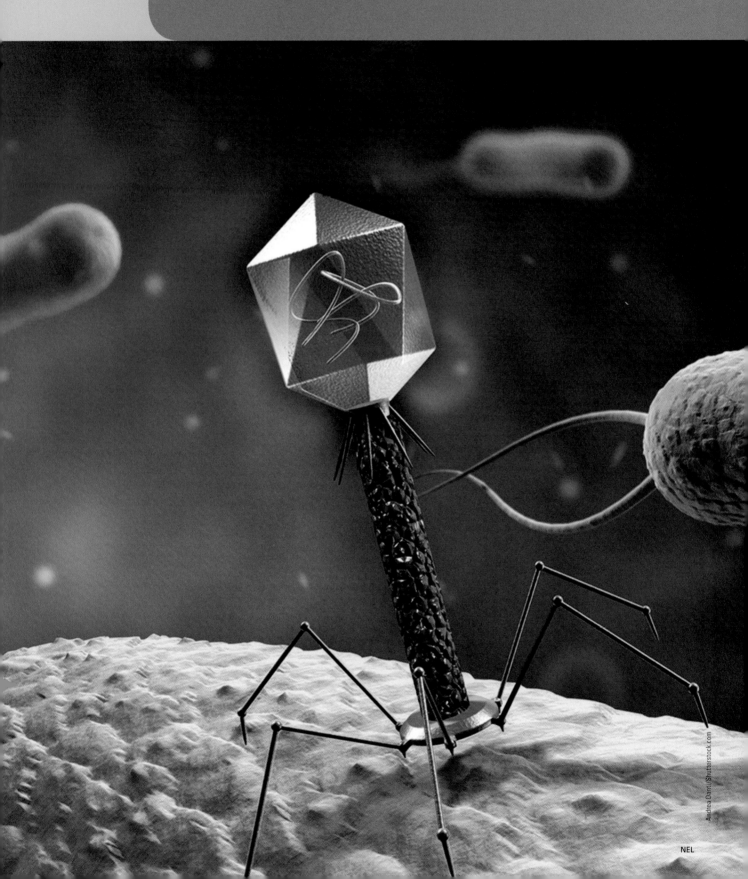

Every great dream begins with a dreamer. Always remember, you have within you the strength, the patience, and the passion to reach for the stars to change the world.

Harriet Tubman

23.1 BACTERIA

Bacteria are prokaryotes that make up two of the kingdoms: archaebacteria and eubacteria (Chapter 1). Prokaryotes, the first living organisms on Earth—beginning approximately 3.5 billion years ago—are still the most abundant life form on the planet. Of all bacterial species, 99% are beneficial and necessary for survival, and only 1% are disease-causing organisms. Photosynthetic bacteria, called **cyanobacteria,** were responsible for introducing oxygen into Earth's atmosphere and are still a primary source of oxygen, along with plants. Many species of bacteria live in the soil and recycle minerals by breaking down dead, organic, plant material and converting it into inorganic minerals that are used by growing plants, including food crops. Humans have learned to use bacterial organisms to make foods, including fermented vegetables, such as sauerkraut or kimchi, and also sourdough bread, yogurt, and cheese. Eating foods fermented by bacteria is very healthy and contributes to the bacteria population that lives inside our digestive tract and on the surface of our body; these are our normal flora. Without bacteria, no other organisms on Earth would be able to survive.

Normal flora

In the human body, there are more bacterial cells than human cells, making up approximately 5% of our body weight. Bacteria live on our skin and throughout our digestive tract and play a substantial role in our health. Six square centimetres (or one square inch) of skin contains more than 500,000 microorganisms: called normal flora. Normal flora contains fungal organisms such as various species of yeast, but bacteria make up the majority of microorganisms. Approximately 1000 species of bacteria exist in our digestive tract, and about 300 of these are the most abundant. The following list describes the most important functions of normal flora (also listed in Chapter 19):

1. Fermenting undigested substrates, mostly soluble fibre

2. Regulating the immune system to tolerate food molecules

3. Preventing growth of harmful bacteria

4. Producing vitamin K and biotin

5. Increasing the growth of intestinal cells and helping to maintain a healthy mucosal layer

6. Producing lactic acid during fermentation, which lowers pH and prevents growth of pathogenic species

7. Converting carbohydrates into short chain fatty acids that can be absorbed and used for energy

8. Releasing important minerals from food, such as magnesium, calcium, zinc, and iron

9. Metabolizing carcinogens by playing a role in preventing colon, breast, and prostate cancers

DID YOU KNOW?

It is very simple to make your own probiotic fermented sauerkraut at home. Thinly slice cabbage, which naturally contains the lactobacillus bacteria that makes up a large proportion of our gut bacterial population, and add 1 tablespoon of sea salt, chopped garlic, one half of a thinly sliced onion, and any other spices you like, such as oregano or chili peppers. Use your hands to mash up the cabbage and spices until the juice of the cabbage is released, then stuff into a 1L jar; you can top up the jar with tap water. Then simply leave the jar, with the lid on, on the counter at room temperature for three to four weeks and allow the bacteria to ferment the cabbage (without oxygen). Keep it in the fridge and it will last for more than a year. The salt is an important ingredient because it prevents the growth of any bacteria that we don't want to grow. Once the lactobacillus produces lactic acid, no harmful bacteria can grow.

Bacterial cell structure

Bacterial plasma membranes are composed of phospholipids, cholesterol, and integral membrane proteins similar to the cell membranes of every living organism (Chapter 3). Bacterial membranes have additional protection in the form of a cell wall that is composed of protein and carbohydrates called peptidoglycan: not to be confused with the cellulose structure of plant cell walls. Some bacteria have an extra outer layer on top of the cell wall—called **lipopolysaccharide (LPS)**—that's composed of lipids and sugar molecules. Bacteria that have an LPS layer are called gram negative because the outer fatty layer does not absorb crystal violet dye, in contrast to the thick peptidoglycan layer of gram positive bacteria that lack LPS and do absorb crystal violet dye (Figure 23.1).

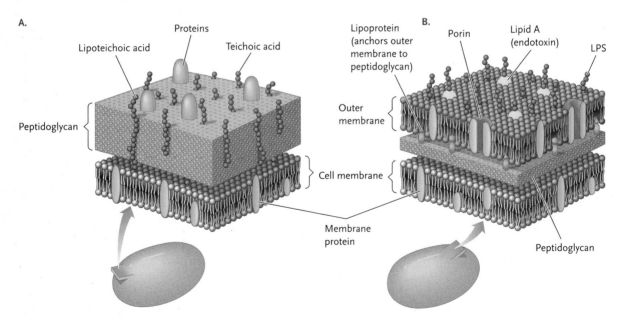

Figure 23.1 (A) Gram Positive and (B) Negative Bacteria

Gram positive and negative bacteria have different cell wall structures.

Although some species of bacteria do not fit into either category, the distinction between gram positive and gram negative helps in determining what is the best antibiotic to treat a bacterial infection. Gram negative infections are more resistant to treatment with penicillin. Various species of bacteria may have additional structures, such as a protein capsule for protection from environmental stress, flagella for mobility, or pili for transfer of genetic material (Figure 23.2) (see also Chapter 3).

Some bacteria have a mechanism—the formation of **endospores**—that allow them to endure extreme changes in their environment, such as lack of oxygen, extreme heat or freezing temperatures, detergents, radiation, or UV light. An endospore has a core of cytoplasm that contains the DNA and ribosomes; it has a surrounding, protective rigid coat, and it does not undergo metabolism (Figure 23.3). Endospores that are millions of years old have been found. Infectious endospore-forming bacteria include species such as tetanus and anthrax.

Bacteria have one of three general shapes, and they can exist as a single cell or in groups (Figure 23.4). Spherical bacteria are **cocci**; single spherical bacteria are **monococci**, and pairs are called **diplococci**. When they are grouped together in long chains, they are **streptococci**, and clusters are called **staphylococci**. Strep throat is caused by an overgrowth of streptococci bacteria. The hospital-acquired staph superbug infection is caused by *Staphylococcus aureus*, which is resistant to one of the strongest antibiotics—called methicillin—and has become known as **MRSA** (**methicillin-resistant staphylococcus aureus**). Many bacteria take on a rod shape, called **bacilli**, and can live as single cells or in long chains called **streptobacilli**. A common intestinal bacterium, *Escherichia coli*, *E. coli*, is rod-shaped. *E. coli* is also present in the intestinal tract of animals and can contaminate food, especially ground beef, which is the last meat pulled off of the cow carcass. *E. coli* infection can cause very serious illness and is the reason why all ground beef must be cooked thoroughly to ensure

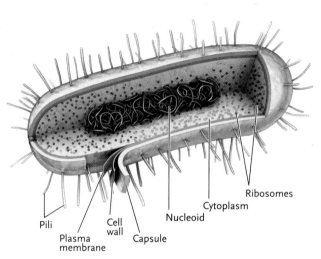

Figure 23.2 **Typical Structures Found in Prokaryotic Cells**

Art source: From RUSSELL/HERTZ/STARR/FENTON. Biology, 2E. © 2013 Nelson Education Ltd. Reproduced by permission. www.cengage.com/permissions

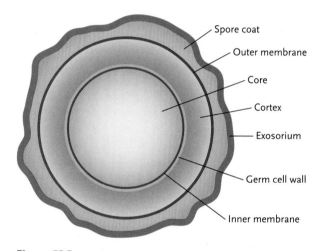

Figure 23.3 **Endospores**

Endospores allow bacteria to survive extreme environmental conditions.

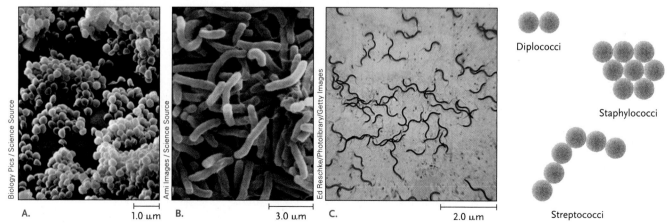

Figure 23.4 **Shapes of Bacterial Cells**

(A) Cocci, (B) bacillus, and (C) spirochete are the most common bacterial cell shapes. (D) Diplococci are a pair of cocci, staphylococci are cocci grouped in clusters, and streptococci are cocci joined in strands.

Art source: From DIGIUSEPPE/FRASER. Biology 11 U. © 2011 Nelson Education Ltd. Reproduced by permission. www.cengage.com/permissions

that bacteria are killed. The third common bacterial cell shape is a spiral shape, called **spirochete**: for example, the bacterium that causes syphilis or Lyme disease.

Recall from Chapter 5 that bacteria reproduce asexually by a process called binary fission, and they have three unique mechanisms for acquiring new DNA that contribute to their ability to have extensive genetic variation and evolutionary success. Bacteria can acquire new DNA directly from other bacteria by conjugation, or from viral infection by transduction, or from their surroundings by transformation. The quick generation time and genetic variation makes infectious bacteria a constant threat to human health; our immune system must be able to adapt to change equally rapidly (Chapter 22) to prevent the replication of pathogenic organisms inside our body.

Antibiotic resistance

An example of bacterial evolution that affects human health is the development of antibiotic resistance by some bacteria. Antibiotics are a selection pressure (Chapter 7). They kill many but not all bacteria; the bacteria that have the ability to survive the antibiotic will proliferate, leading to a new population that is resistant to the drug—an example of microevolution. Antibiotics function in two ways: by preventing replication (**bacteriostatic**) or by directly killing the bacteria (**bacteriocidal**).

Since bacteria can acquire new DNA through conjugation, transduction, or transformation, there are countless opportunities for antibiotic resistance genes to be shared between bacteria and **viruses**. Resistance genes most commonly reside on the small extra-chromosomal DNA plasmid within the cell (Figure 23.5). Bacteria have small genomes and keep only the genes that are beneficial. The more exposure bacterial populations have with antibiotics, the more important those resistance genes become.

Why antibiotic resistance occurs

The main cause of antibiotic resistance is the **prophylactic** treatment of food animals. Prophylactic treatment occurs when farmers give a constant low dose of antibiotics to all animals, whether they are sick or not, in order to try to *prevent* infections. This means that all the normal bacteria in the animals are constantly exposed to a selection pressure, and they protect themselves by retaining the resistance genes. Prophylactic treatment, over-treatment, or mistreatment of viral infections using antibiotics causes bacteria in humans to be exposed to a selection pressure. Antibiotics have no effect on viruses that lack a cell membrane, so when a person with a cold or flu is given antibiotics, much of the normal flora bacterial population is killed, which removes the protective healthy bacterial barrier and also promotes antibiotic resistance. Resistance genes can be transferred between healthy and **pathogenic bacteria**.

How resistance genes work

Resistance genes when transcribed and translated produce proteins that prevent the antibiotic from killing the bacterium (Figure 23.5). There are three main mechanisms: (1) genes can code for enzymes that *degrade* the antibiotic, (2) membrane protein pumps can *remove* the antibiotic from the cell, (3) proteins may *modify* the antibiotic so that it can no longer function. Bacteria that contain genes that allow them to be resistant to many different antibiotics are called multi-resistant, or superbugs, such as MRSA.

Opportunistic infections

Many bacterial species can live within the human body without causing any illness until the immune system becomes compromised. For example, staphylococcus bacteria are part of our normal flora and are not usually harmful, but an overgrowth can cause serious disease when the immune system does not keep the growth level in check. Sometimes bacteria cause disease only when the immune system is occupied fighting another infection; for example, bacterial pneumonia can occur after a viral infection such as influenza. Organisms that cause infection only when our immune system is suppressed or when our normal flora is damaged (after antibiotic treatment) are considered opportunistic, which means they cause disease only when they have the "opportunity." **Opportunistic infections** can be caused by bacteria,

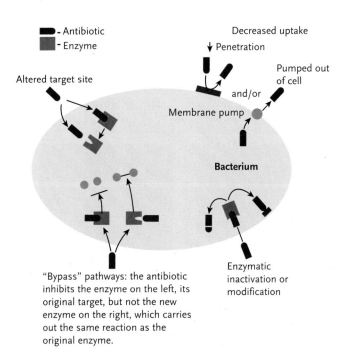

Figure 23.5 Antibiotic Resistance Genes

Antibiotic resistance genes code for proteins that prevent an antibiotic from killing the bacterium.

viruses, fungi, or parasites. Opportunistic infections can occur in the following situations:

1. Suppressed immune system, such as during stress when cortisol levels are high (Chapter 16)

2. Altered normal flora, most often caused by antibiotic treatment, which leaves physical space for other organisms to attach to the body's cells, particularly in the digestive tract.

   a. Note: After taking antibiotics, it is important to replenish the normal flora with probiotics, either supplements or probiotic food such as fermented vegetables or yogurt.

3. Malnutrition due to lack of nutrients, which exacerbates many infections and why infection rates are dramatically higher in regions of poverty

4. Cancer patients, who have a greater risk of opportunistic infections partly because the immune system is occupied with the cancer cells and also because treatment such as chemotherapy or radiation damages healthy cells including immune cells

5. Immunosuppressant medication, for example, those used when patients have had stem cell therapy (requires irradiating their bone marrow), or corticosteroids used to treat autoimmune disease

6. Immunodeficiency such as occurs in people with HIV

7. Damaged skin barrier, such as punctures, surgery, or burns, which allows organisms to enter the bloodstream where they would normally not be found

How bacteria make us sick

Many bacteria produce **toxins** that either prevent cells from functioning normally, or they kill the cells. Tetanus is caused by a bacterial toxin that prevents the inhibition of muscle contraction, causing severe and painful muscle contractions. **Botulism** is caused by a toxin produced only when the common soil bacterium *Clostridium botulinum* is forced to live in anaerobic conditions (such as in a jar or can); when consumed that toxin blocks neurotransmitters and causes paralysis. Certain strains of *E. coli* produce a toxin that causes cells to lyse and leads to severe breakdown of tissues. Other bacteria cause disease by invading the cells to reproduce and, ultimately, killing the host cell—a process similar to viruses. Examples include chlamydia, salmonella, and tuberculosis (Table 23.1).

DID YOU KNOW?

Botulism is caused by a normal soil bacterium that is toxic only if it lives in an anaerobic environment, for example, in canned vegetables that have not been properly sterilized. You eat a carrot out of the garden even though there are botulism bacteria growing on it. However, if you put a carrot in a jar or can with no oxygen, and without sterilizing, the botulism bacteria make a neurotoxin that causes paralysis. This is the same botulinum toxin that is used to make botox. It prevents wrinkles by paralyzing the muscles that make you frown.

TABLE 23.1

Pathogenic Bacteria

Bacterial Species	Tissue Infected	Characteristics
Methycillin-resistant *Staphylococcus aureus* (MRSA)	Skin, wounds, blood, respiratory tract, or urinary tract in immunosuppressed people	Staphylococcus, a normal flora bacterium, overgrows in immunosuppressed people and is commonly acquired in hospitals. Often localized to the skin, but can cause necrotizing fasciitis ("flesh-eating" disease), pneumonia, or blood poisoning.
Vancomyocin-resistant *Enterococcus* (VRE)	Diarrhea infection of the small intestine and urinary tract	Most commonly associated with hospitalized patients taking antibiotics that kill protective normal flora and allow enterococcus species to proliferate.
Chlamydia trachomatis	Genitals and eyes	Chlamydia is one of the most common sexually transmitted infections (STI), known as "silent" because many people do not have symptoms. It is transmitted through oral, vaginal, or anal sexual contact and can be transmitted to babies during childbirth. Chlamydia can cause a serious infection in women, called pelvic inflammatory disease (PID) that can cause infertility.
Neisseria gonorrhoeae	Genitals, mouth, throat	Gonorrhea, an STI, can cause symptoms in men, such as burning during urination and discharge, but many women have no symptoms. Like chlamydia, gonorrhea can cause PID and can be transmitted to a newborn.

(Continued)

TABLE 23.1 (*Continued*)

Bacterial Species	Tissue Infected	Characteristics
Streptococcus pyogenes	Most commonly infects the throat, but can infect skin and cause necrotizing fasciitis; can also cause serious blood infection where bacteria damage red blood cells.	Strep throat is caused by an overgrowth of several strains of strep bacteria, usually when immune system is suppressed; more likely to occur during high amounts of stress. Symptoms include very sore, red throat with visible white exudate on the tonsils, and fever. Usually easily treated with antibiotics. Untreated strep pyogenes infections can cause rheumatic fever that can cause damage to heart valves.
Escherichia coli (E. coli)	Gastrointestinal tract	*E. coli* is a normal flora organism that resides in the large intestine of humans and animals. If eaten, *E. coli* in the stomach and small intestine cause food poisoning, causing diarrhea and vomiting. It most commonly results from uncooked ground beef or chicken, or raw fruits and vegetables that have been exposed to farm runoff fertilized with manure.
Species of *Salmonella*	Gastrointestinal tract	Salmonella can cause food poisoning if ingested in large amounts, but usually the acidic stomach environment kills the bacteria, such as the tiny amount that may be in raw eggs. Salmonella can cause diarrhea and vomiting due to foods not cooked enough, but it usually does not require any treatment, except in small children or immunosuppressed individuals when severe infection can occur.
Campylobacter jejuni	Gastrointestinal tract	This is the most common cause of food poisoning in developed countries, causing diarrhea and possibly vomiting within hours to one day after eating contaminated food or water, and is often the cause of "travellers' diarrhea." It is found in animal feces and most commonly associated with undercooked poultry, but can also be transmitted from other livestock, contaminated water, unpasteurized milk, and person-to-person.
Mycobacterium tuberculosis	Respiratory system	Tuberculosis (TB) is transmitted by airborne particles from people with an active TB infection, and it kills approximately 50% of people with active infection, approximately two million people worldwide per year, mostly in developing countries with high malnutrition and poverty. Of those infected, 90% do not have symptoms. Diagnosis is by a TB skin test. Symptoms include persistent blood-tinged cough, fever, and weight loss. TB can be difficult to treat, requiring multiple antibiotics over a long period of time; antibiotic resistance is a problem.
Clostridium tetani	Central nervous system	Tetanus, also known as lock jaw is caused by a toxin produced by the bacteria that blocks inhibitory neurotransmitters and therefore causes continuous contraction of muscles. Infection most commonly occurs through deep puncture wounds and is prevented by tetanus vaccination. Severe infections require treatment in an intensive care facility.

CONCEPT REVIEW 23.1

1. Name the beneficial functions of bacteria.
2. List the functions of the bacteria in our digestive tract.
3. How is an endospore beneficial for the bacterium?
4. What are the three basic bacterial cell shapes?
5. How do antibiotic resistance genes protect bacteria?
6. What are opportunistic infections?
7. Who is at greater risk of acquiring an opportunistic infection?
8. Which bacteria cause food poisoning?
9. Which bacterial infection is multi-resistant to antibiotics?
10. How does *Clostridium tetani* cause disease?
11. Which bacterial infections are opportunistic?
12. Name the two most common sexually transmitted bacterial infections.

23.2 VIRUSES

Viruses are not considered *living* because they possess only a portion of the properties of living organisms. Viruses do not have a cell membrane or cellular organelles; they are composed of proteins and their hereditary material, which can be DNA or RNA. Viruses cannot replicate outside of host cells, and they can only infect specific cell types that contain receptors that the virus can bind to. For example, hepatitis can only infect liver cells, and the tobacco mosaic virus can only infect tobacco plant leaves. Although viruses are not considered living organisms, they may as well be treated as such, since they evolve and change depending on their environment, infect cells, and act as a selection pressure on all living things; millions of viruses have infected bacteria, fungi, plants, and animals for as long as there have been living cells on Earth.

Viruses are extremely small. A typical human cell is approximately 10μm, a bacterium is 10 times smaller at 1μm, and a virus particle is 10 to 100 times smaller than a bacterium (Figure 23.6). Virus particles can take on a variety of structures that contain the genetic material, DNA or RNA, and a surrounding protective protein capsid (Figure 23.7). Viruses that infect bacteria are called **bacteriophages;** these have a complex structure, including a tail and base, where the genetic material is injected into the bacterium and the protein capsule is discarded. Some viruses have a helical structure with a hollow centre containing DNA or RNA, and some of the viruses with this structure can be very long and filamentous, such as Ebola, one of the deadly hemorrhagic viruses. Many viruses have a spherical structure with a type of modified cell membrane, called an envelope, which contains proteins that allow it to bind and infect cells: for example, causing influenza and HIV. See Figure 23.8 for examples of common virus structures.

Unlike any other organism, viruses can have RNA as their genetic material. In fact, most viruses are RNA viruses. When an RNA virus infects a host cell, it must convert the RNA into DNA before it can replicate and produce new virus particles. RNA is converted into DNA by a viral enzyme called reverse transcriptase, which scientists have isolated and used in genetic engineering (Chapter 10). Reverse transcriptase enzymes are notoriously sloppy enzymes and don't copy RNA perfectly into DNA, leaving many point mutations (Chapter 8). The constant small changes in the genetic material of viruses give them the benefit of rapid evolution, which allows them to constantly evade our immune system. For example, influenza and the common cold viruses are RNA viruses that

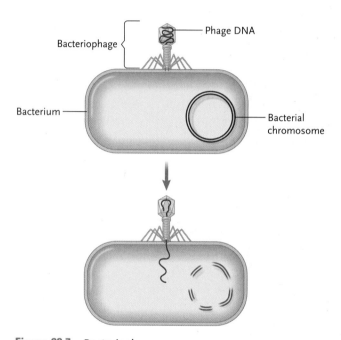

Figure 23.7 **Bacteriophages**

Bacteriophages are viruses that infect bacteria.

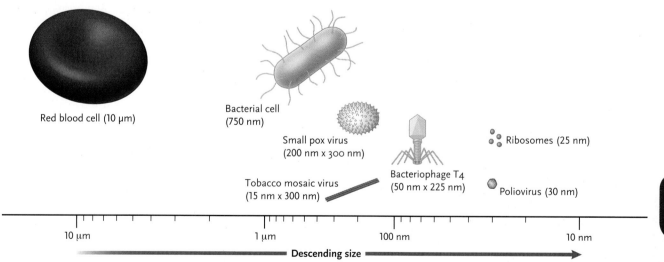

Figure 23.6 Cell Sizes

Size comparison between a human cell, a bacterium, and virus particles.

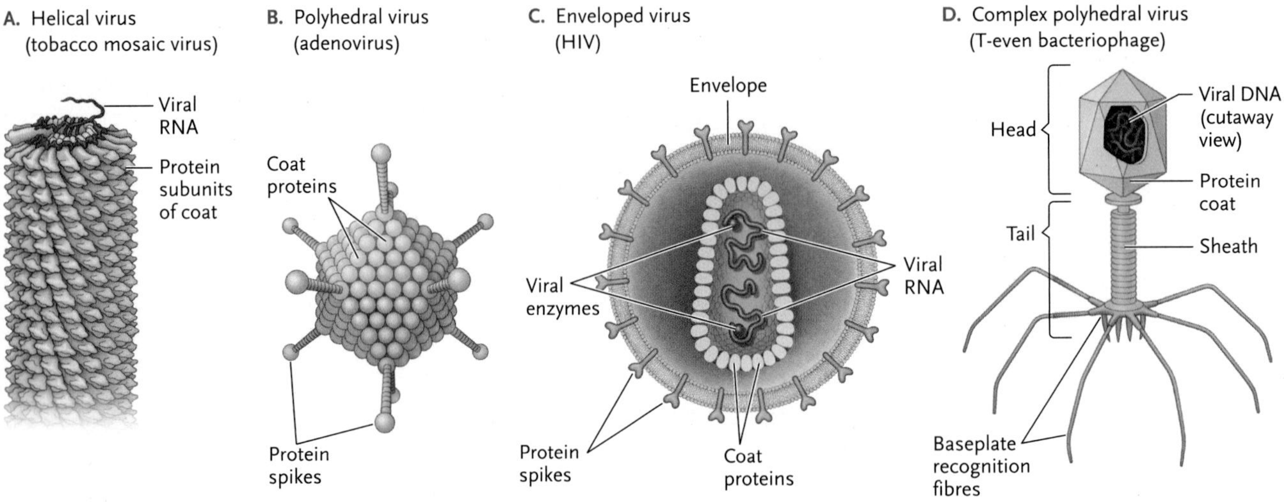

A. Helical virus
(tobacco mosaic virus)

B. Polyhedral virus
(adenovirus)

C. Enveloped virus
(HIV)

D. Complex polyhedral virus
(T-even bacteriophage)

Figure 23.8 Examples of Common Viral Structures

are slightly different every year, causing us to be infected and have symptoms every time we are infected with a new virus. Only the viruses with non-detrimental mutations can continue to infect cells. After our first exposure to a virus that doesn't change rapidly—such as the one that causes chicken pox—we can develop immunity because our memory cells (Chapter 22) recognize the virus and kill it before we get sick again.

How viruses replicate

Viruses can replicate only inside host cells because viruses do not have ribosomes or nucleotides or any of the building blocks and ATP that are required to copy their genetic material and produce new virus particles. Viruses that can infect more than one species are more likely to undergo rapid changes than those that can't. For example, type A influenza can infect humans, birds, and pigs, and it is largely responsible for flu pandemics because there are antigen changes that cause the immune system to mount a stronger response (Chapter 22). Influenza contains two important molecules: **hemagglutinin**, which allows the influenza virus to bind to upper airway cells, and **neuraminidase**, which allows the newly replicated virus particles to be released from the host cell. It is from these two molecules that influenza strains acquire their names, such as H1N1 (Spanish flu in 1918, and swine flu in 2009), or H5N1 (bird flu in 2004).

Viruses have two types of mechanisms for infection and reproduction: a lytic cycle and a lysogenic cycle. The **lytic cycle** occurs when the virus kills the host cell after it reproduces, and the **lysogenic cycle** occurs when the virus integrates its genetic material into the host genome, replicating along with the host cell, but not killing it (Figure 23.9). Looking at the lytic cycle first, the stages of infection involve the following:

1. **Attachment.** The proteins on the virus capsid bind to a membrane protein on the host cell.

2. **Infiltration.** The host cell takes up the virus, usually by endocytosis (Chapter 3).

3. **Replication.** The virus uses host cell organelles, amino acids, nucleotides, and ATP to replicate genetic material and transcribe and translate virus genes into proteins to make new virus particles.

4. **Assembly.** New virus particles form and are released from the host cell, causing lysis and death of the host cell.

When we have a common cold, such as a rhinovirus or adenovirus, the sore throat symptoms come from viral particles going through the lytic cycle and killing our epithelial cells. Our immune system also contributes to the symptoms when it kills infected cells (Chapter 22). Luckily, epithelial cells regrow quickly and replace the dead cells (Table 23.2).

When a virus infects the body's cells and enters the lysogenic cycle, the viral DNA becomes integrated into the chromosomes. The integrated viral DNA then replicates with each cell division, and because new virus particles are being produced, the virus is hidden from the immune system. For example, herpes viruses will alternate between a lytic cycle and a lysogenic cycle. When there are herpes sores—cold sores or genital—the virus particles are in the lytic cycle, killing epithelial cells and causing the sores. When the sores go away, the virus is still there; it retreats into a nearby sensory neuron and enters the lysogenic cycle. Herpes infections tend to stay long term within our cells because the lysogenic cycle allows the virus to go unseen by the immune cells. The same occurs with HIV infection. HIV is an RNA virus that reverse transcribes the RNA into DNA and then integrates into the chromosomes. HIV can also replicate and produce new virus particles that do not kill the cell; this is called **budding** (blue dots in Figure 23.10). HIV binds specifically to immune cells that express the **CD4 membrane receptor** by attachment with the HIV gp120 (glycoprotein with 120 amino acids); therefore, HIV can only infect immune cells (Figure 23.11).

TABLE 23.2

Viral Infections

Virus	Tissue Infected	Characteristics
Human Immunodeficiency Virus (HIV)	Immune cells	HIV is an RNA virus that infects CD4+ immune cells and causes immunodeficiency, with death occurring from opportunistic infections.
Rhinoviruses, adenoviruses, and respiratory syncytial virus	Sinuses and upper airways	Over 100 different viruses cause the common cold and are easily killed by the immune system in healthy people. Various cold viruses infect the epithelial cells of the upper airways. Symptoms include stuffed or runny nose, sneezing, sore throat, and coughing, and usually don't last longer than seven to 10 days.
Influenza	Sinuses and upper airways	There are type A, B, and C influenza viruses, with type A and B being the most common. Symptoms can be similar to a common cold, but there is usually pronounced fatigue, fever, and body aches that accompany the flu. Influenza viruses are RNA viruses and mutate rapidly, which leads to new flu viruses each year. Often influenza infection can lead to a secondary opportunistic infection, such as viral or bacterial pneumonia, which can be fatal in the elderly and very young children.
Herpes	Lips (cold sores) or genitals, and sensory neurons	DNA viruses that cause cold sores or genital herpes are easily transmitted by skin-to-skin contact. Use of condoms does not necessarily prevent infection if sores are in areas not covered by the condom. Herpes infections have a latent period where they enter sensory neurons and then re-enter the skin during the lytic cycle. Outbreaks tend to become less severe over time and can be controlled with antiviral medication.
Human Papilloma Viruses (HPV)	Keratinocytes of the skin	There are approximately 100 different strains of HPV virus that cause warts, with the majority causing no symptoms. Some strains cause warts: 30–40% transmitted through sexual contact, and two strains have been linked with 70% of cervical cancers in women. It is now known that most HPV infections will be resolved by our immune system within one to two years. The HPV viruses that cause cervical cancer do not cause warts or any symptoms, so it is important for women to have a yearly Papanicolaou "pap" test to detect and treat possible cancer cells.
Hepatitis	Liver	Hepatitis is an inflammation of the liver, most commonly caused by five viruses: hepatitis type A, B, C, D, and E. Types A and E are usually acquired through contaminated food or water, mostly in developing countries, and do not cause chronic disease. Types B, C, and D are transmitted through body fluids. Types B and C can cause chronic infections in millions of people around the world and are the most common cause of liver cancer and cirrhosis of the liver. Symptoms range from mild to severe, including jaundice (yellowing of the skin and eyes), extreme fatigue, nausea, vomiting, and abdominal pain. Medical professionals that have high exposure to body fluids can be vaccinated against hepatitis B.
Rotavirus	Gastrointestinal tract	Rotavirus is a common intestinal infection in children, causing vomiting and diarrhea. Almost all children will have at least one infection by age 5, leaving them with antibodies that protect them against future infections. Adults very rarely become infected with rotavirus. Diarrhea can be severe in infants or malnourished children. Rotavirus is usually easily managed with rehydration therapy, but does account for almost 450,000 deaths in developing countries each year.
Epstein Barr Virus (EBV) "Mononucleosis"	B cells and epithelial cells	EBV is a type of herpes virus easily transmitted through saliva. 90% of adults have been infected and acquired immunity to EBV by age 40. It is one of the most common viral infections. Symptoms can range from asymptomatic (in young children) to severe fatigue that can last for several months (more common in adolescents). Other symptoms in the acute phase include fever, sore throat, swollen lymph nodes, and possibly headaches or abdominal pain, which usually disappear within two weeks. It is thought that EBV infection is associated with higher risk of autoimmune diseases such as multiple sclerosis, systemic lupus, and rheumatoid arthritis.

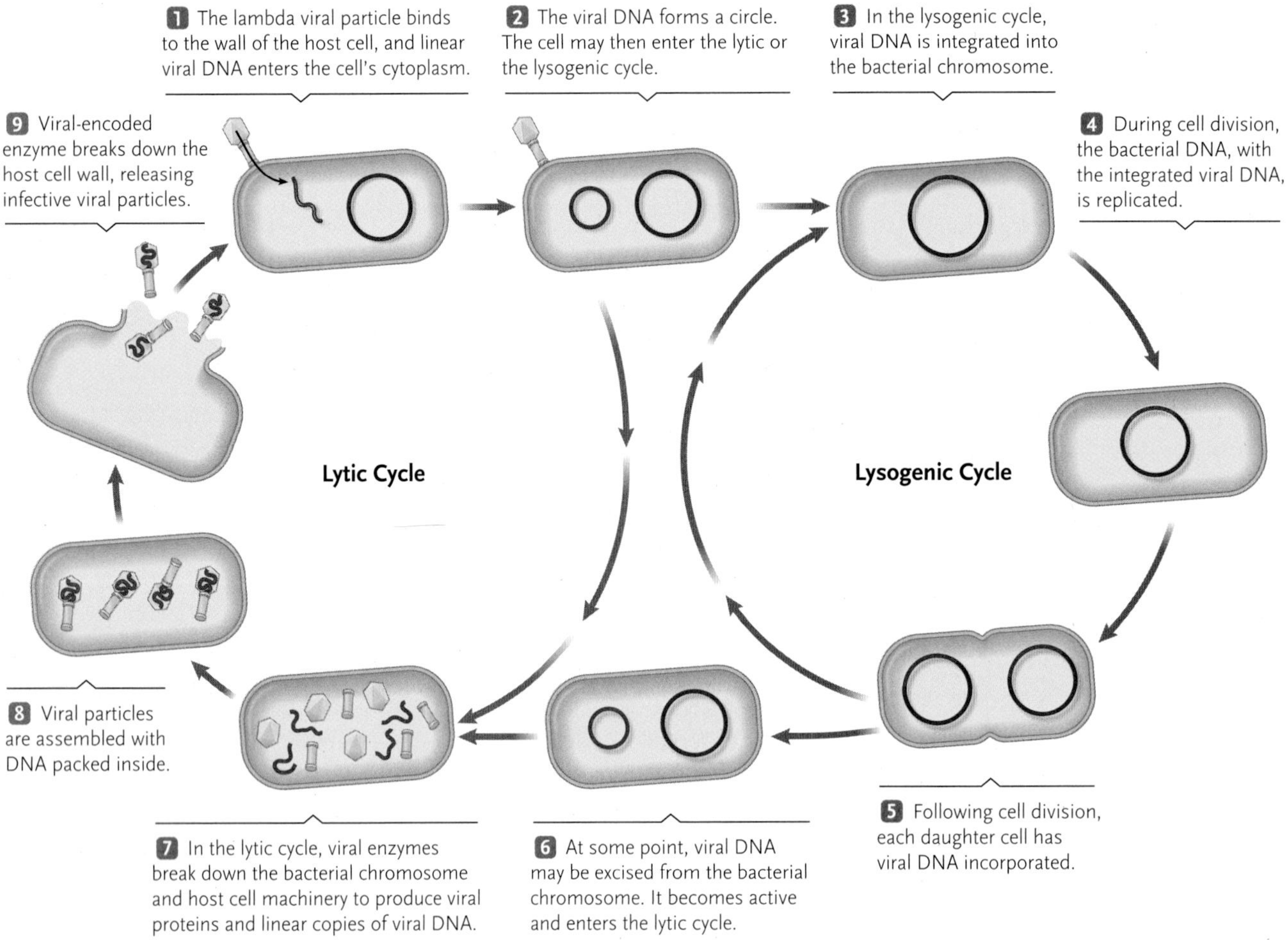

1 The lambda viral particle binds to the wall of the host cell, and linear viral DNA enters the cell's cytoplasm.

2 The viral DNA forms a circle. The cell may then enter the lytic or the lysogenic cycle.

3 In the lysogenic cycle, viral DNA is integrated into the bacterial chromosome.

9 Viral-encoded enzyme breaks down the host cell wall, releasing infective viral particles.

4 During cell division, the bacterial DNA, with the integrated viral DNA, is replicated.

Lytic Cycle

Lysogenic Cycle

8 Viral particles are assembled with DNA packed inside.

5 Following cell division, each daughter cell has viral DNA incorporated.

7 In the lytic cycle, viral enzymes break down the bacterial chromosome and host cell machinery to produce viral proteins and linear copies of viral DNA.

6 At some point, viral DNA may be excised from the bacterial chromosome. It becomes active and enters the lytic cycle.

Figure 23.9   Lytic and Lysogenic Cycles

In the lytic cycle, the virus replicates and kills the host cell. In the lysogenic cycle, the virus integrates its DNA into the host genome.

Once HIV infects and kills T cells, the immune system can no longer function sufficiently to prevent infection, and the person becomes severely immune-compromised and has **AIDS**—acquired immunodeficiency syndrome. With so many new antiviral medications, people can now live for decades with HIV, whereas in the 1980s when the HIV epidemic first began, thousands of people died each year from many opportunistic infections (Section 23.3). HIV is transmitted by sexual intercourse, blood transfusions, and can be transmitted to an infant during childbirth.

DID YOU KNOW?

Stress can activate latent viral infections, such as herpes, because cortisol is produced during stress and inhibits the immune system.

DID YOU KNOW?

The stomach flu is not caused by influenza. Influenza is a respiratory illness that can sometimes cause vomiting in small children that have a high fever. The typical 24-hour stomach illness is most commonly caused by campylobacter bacteria or by noroviruses, such as the Norwalk virus. In children, rotaviruses can last anywhere from one to five days.

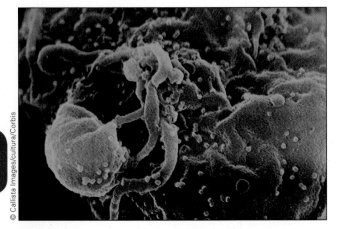

© Callista Images/cultura/Corbis

Figure 23.10   HIV Viruses Budding from a Macrophage

Budding allows HIV to replicate without killing the macrophage.

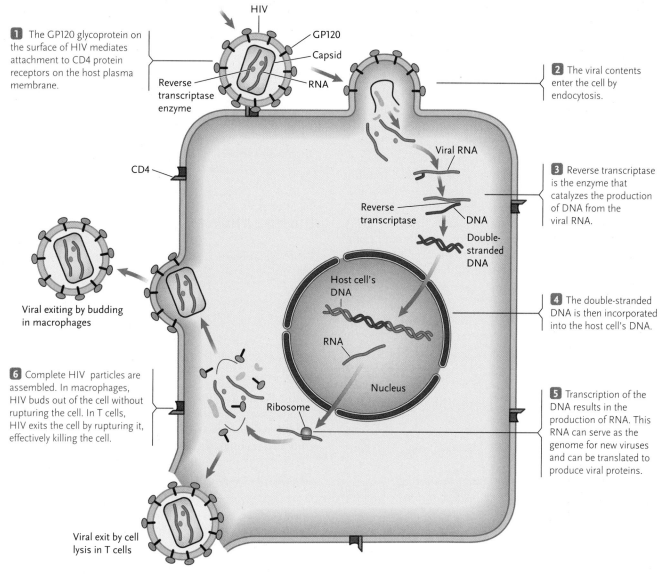

1 The GP120 glycoprotein on the surface of HIV mediates attachment to CD4 protein receptors on the host plasma membrane.

HIV

GP120

Capsid

RNA

Reverse transcriptase enzyme

2 The viral contents enter the cell by endocytosis.

Viral RNA

Reverse transcriptase

DNA

Double-stranded DNA

3 Reverse transcriptase is the enzyme that catalyzes the production of DNA from the viral RNA.

CD4

Host cell's DNA

RNA

Nucleus

4 The double-stranded DNA is then incorporated into the host cell's DNA.

Viral exiting by budding in macrophages

Ribosome

6 Complete HIV particles are assembled. In macrophages, HIV buds out of the cell without rupturing the cell. In T cells, HIV exits the cell by rupturing it, effectively killing the cell.

5 Transcription of the DNA results in the production of RNA. This RNA can serve as the genome for new viruses and can be translated to produce viral proteins.

Viral exit by cell lysis in T cells

Figure 23.11   HIV Infection

HIV infects cells that have a CD1 receptor; it usually enters a lysogenic cycle for a period of years before beginning to lyse helper T cells, causing AIDS.

CONCEPT REVIEW 23.2

1. How are viruses different from living organisms?
2. What is a bacteriophage?
3. How do RNA viruses replicate their genetic material?
4. What is the function of hemagglutinin and neuraminidase in influenza?
5. Explain the difference between the lytic and lysogenic cycles.
6. What are the four stages of a lytic cycle infection?

7. Give an example of a human viral infection that undergoes a lysogenic cycle, making it very difficult for our immune system to get rid of.
8. What cell receptor does HIV bind to? What cell types have this receptor?
9. What viruses can cause liver cancer?
10. Which viral infections can be sexually transmitted?
11. Which virus can cause cervical cancer?
12. Which viral infection is associated with increased risk of autoimmune diseases?

23

23.3 FUNGI

Organisms of the kingdom fungi are eukaryotic and, except for yeast, are multicellular and include organisms such as mushrooms, mould, mildew, and yeasts. Fungi live in soil, on plants, and in the air and water and play an important role in decomposition and nutrient recycling. Many fungi are parasitic and cause extensive damage to food crops. Fungi usually do not cause severe infection in humans unless the immune system is suppressed or there is a breakdown in the bacteria making up the majority of the normal flora, such as after antibiotic treatment. Many fungal infections are opportunistic infections (Table 23.3 and Figure 23.12).

Fungi can reproduce asexually or sexually, and both methods commonly involve the production of spores. Single-celled yeast can reproduce by binary fission, like bacteria. Spores expelled from fungi can travel in the air long distances and infection in humans most often occurs on the skin or in the respiratory tract. In immune-compromised people, fungal infections can be fatal.

Fungi have been used for decades for their antimicrobial properties, for extracting antibiotics that fungi naturally produce to kill bacteria, and for producing vitamins and molecules used to treat cancer and heart disease. Fungal organisms are used to make many foods, such as bread, beer, wine, kombucha, fermented soy such as tempeh, and fermented rice such as miso.

DID YOU KNOW?

Genetic studies have shown that fungi are more similar to animals than to plants.

TABLE 23.3

Fungal Infections

Fungus	Tissue Infected	Characteristics
Candida albicans	Vagina ("yeast infection"), mouth ("thrush"), and also diaper rash	Yeast is a normal flora organism, and infection occurs when it overgrows, such as after antibiotic use that kills protective bacteria, suppressed immune function, and uncontrolled diabetes (increased blood sugar causes yeast to grow rapidly). It is usually easily treated with topical antifungal cream or pills.
Pneumocystis jirovecii (previously but incorrectly called *pneumocystis carnii*, which is a species found only in rats)	Lungs	Pneumocystis pneumonia (PCP) is normally found in the lungs of healthy people and does not cause infection unless the person is severely immune-compromised. Commonly in people with HIV, cancer, or those on immunosuppressant medications. PCP symptoms include unproductive cough and severe shortness of breath, which can be fatal if not treated aggressively.
Aspergillis (over 100 species of mould)	Sinuses and upper respiratory tract	This common mould grows on bread and fruits, and its airborne spores are everywhere in the environment. Aflatoxin grows on raw nuts and is carcinogenic, which is one main reason why nuts are roasted. Normally, Aspergillis organisms do not cause illness, but in people with weakened immune systems, infection can cause fever, cough, sinusitis, and worsening of asthma or other respiratory problems.
Ringworm (Dermatophytosis)	Skin	Ringworm is not caused by a worm; it is a very common fungal infection that feeds on the keratin protein in skin, hair, and nails. It is easily transmitted to children by animals, such as pet cats, or through warm, moist environments, such as public showers. When this fungus grows on the feet, people refer to it as athlete's foot; when it grows in the groin area, it's called jock itch. Symptoms, which appear as raised red rings on the skin or thickened and yellowish nails, are easily treated with topical antifungal ointment.

23

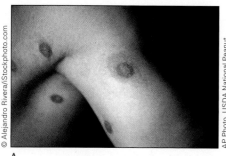

Figure 23.12 **Examples of Fungal Infections**
(A) Ringworm, (B) aflatoxin, and (C) candida

CONCEPT REVIEW 23.3

1. Are fungi prokaryotic or eukaryotic?
2. Name some beneficial characteristics of fungi.
3. How can various fungal species be used in food production?
4. What factors contribute to increased risk of fungal infections?
5. Which fungal infection causes athletes foot?
6. Which fungal organism is responsible for yeast infections or thrush?
7. Which fungal infections can be fatal for people with HIV?

23.4 PARASITES

Parasites are species of the protist and animal kingdoms, and they cause human infection. There are two general categories of parasites: **protozoans** (protists) and **helminths** (worms). *Parasite* is a general term that indicates an organism acquires its nutrients from a host, which means that some bacteria or fungi can also be considered parasitic. **Ectoparasites** include organisms that live on the blood of other organisms, such as lice, fleas, and ticks. This section focuses on the **endoparasites**: protozoans and helminths.

Protozoa

Protozoa are single-celled, eukaryotic protists that have the animal-like ability to move. These characteristics distinguish them from the spore-producing fungus-like protists—such as slime moulds—or the photosynthetic, plant-like protists—such as algae (Chapter 1). Protozoa can move around in a number of ways: by extending **pseudopods**, like amoebas; or by using flagella, such as giardia; or by using cilia, such as paramecium. All protozoans digest food in structures called **vacuoles**, similar to a stomach.

Protozoa that infect animal intestines are usually transmitted by the **fecal-oral route** and are common in regions of poverty where sewage treatment is minimal and water is contaminated by human or animal feces. Protozoa that live in the bloodstream or other tissues are most often transmitted by **insect vectors** (carriers), such as sandflies or mosquitoes; they are common in tropical regions.

Some protozoa can form dormant cysts that allow them to survive in harsh environments, such as extreme heat or lack of nutrients or oxygen, for long periods of time outside of a host. Protozoa can reproduce sexually and asexually and often involve a complex life cycle. For example, **malaria** infects over 200 million people each year and causes over one million deaths, usually in children; it is most prevalent in Africa. Malaria caused by *Plasmodium* species—mostly *P. falciparum* and *P. vivax*—are injected into the bloodstream when a person is bitten by a mosquito carrying the **sporozoite** form of the parasite. From there, they migrate to the liver and reproduce asexually within liver cells. Then thousands of **merozoites** leave the liver cells, killing the liver cells in the process. The merozoites hide from the immune system by wrapping in fragments of liver cell membranes, and then they travel to red blood cells. Once inside a red blood cell, merozoites replicate asexually through cycles of replication, with the result of killing the red blood cell and infecting other red blood cells. Merozoites can develop into gametes that can be taken up by other mosquitoes, where the gametes combine to form new sporozoites. Symptoms of malaria result from damage to the liver and red blood cells, and include severe headache, fever, chills, vomiting, jaundice, and hemoglobin in the urine (Figure 23.13).

23

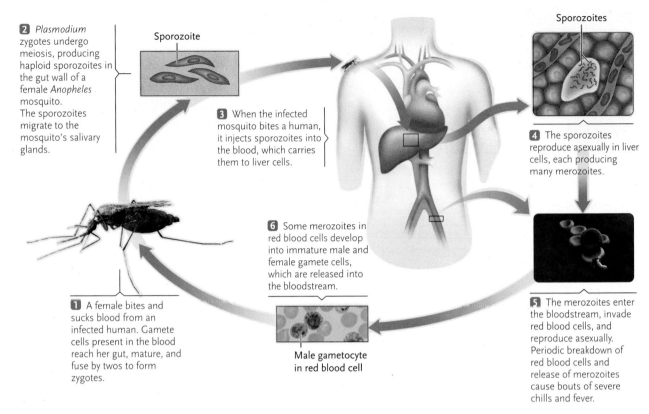

2 *Plasmodium* zygotes undergo meiosis, producing haploid sporozoites in the gut wall of a female *Anopheles* mosquito. The sporozoites migrate to the mosquito's salivary glands.

Sporozoite

3 When the infected mosquito bites a human, it injects sporozoites into the blood, which carries them to liver cells.

Sporozoites

4 The sporozoites reproduce asexually in liver cells, each producing many merozoites.

6 Some merozoites in red blood cells develop into immature male and female gamete cells, which are released into the bloodstream.

1 A female bites and sucks blood from an infected human. Gamete cells present in the blood reach her gut, mature, and fuse by twos to form zygotes.

Male gametocyte in red blood cell

5 The merozoites enter the bloodstream, invade red blood cells, and reproduce asexually. Periodic breakdown of red blood cells and release of merozoites cause bouts of severe chills and fever.

Figure 23.13   Malaria Life Cycle

Malaria has a complicated life cycle that involves infecting the liver and the red blood cells.

Art source: From RUSSELL/HERTZ/STARR/FENTON. Biology, 2E. © 2013 Nelson Education Ltd. Reproduced by permission. www.cengage.com/permissions; Gametocyte image courtesy of Drs. JoAnn Sullivan and William Collins, Division of Parasitic Diseases, Centers for Disease Control and Prevention; Merozoite image, Science Picture Co/Getty Images

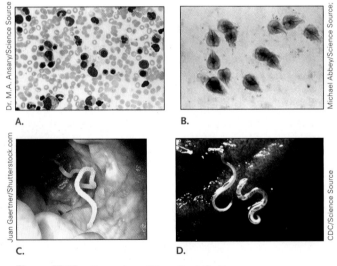

Figure 23.14   Examples of Parasite Infections

(A) Malaria, (B) giardia, (C) tapeworm, and (D) hookworm

Helminths

Helminths are multicellular worms that can be free-living or parasitic. Intestinal worms infect hundreds of millions of people worldwide. There are two categories of worms:

1. **Flatworms** (platyhelminths), which include **tapeworms** (cestodes) and **flukes** (trematodes)

2. **Roundworms** (nematodes)

Helminths infect people through contaminated water, food, or soil, and sometimes insect bites. Undercooked pork, fish, and shellfish can carry flukes. Nematodes such as hookworms can penetrate the skin, usually the feet. Helminths infect people throughout the world, but the risk is much higher in developing countries with poor sanitation or undercooked meat, and among children. Infection with worms stimulates a TH2 immune response (Chapter 22) so that the immune system is less active, which allows the parasites to live in the intestine without being killed by the immune system; it also leaves the infected individual at higher risk of other infections (Figures 23.14 and 23.15).

Early helminth infections are often asymptomatic because of the inhibited immune response and lack of invasion of any other tissues. People with a significant worm load show signs of malnutrition, weakness, intestinal obstruction, abdominal pain, vitamin deficiencies, anemia, cognitive impairment, slowed growth in children, and possibly vomiting or diarrhea that contains worms. Some worms can migrate to other tissues, such as species of *Ascaris* nematodes that migrate to the lungs and can cause hemorrhaging. Helminth infection, as with any other infection, can be particularly detrimental for people with compromised immune systems (Table 23.4).

Public health campaigns and deworming programs began in the early 1900s and have significantly decreased the number of children in North America with helminzhs infections.

DID YOU KNOW?

Parasitic worms have been used to decrease the inflammation caused by autoimmune diseases because helminths trigger the anti-inflammatory immune response. Regions of the world that have a high rate of intestinal parasites also have the lowest rates of autoimmune disease. And developed countries that now have very low intestinal parasites also have the highest incidence of autoimmune diseases. One study suggests that metabolic syndrome (Chapter 17) may be so high in developed countries because of the lack of eosinophil stimulation, which occurs when the immune cells fight parasites (Chapter 22). Eosinophils play a role in preventing insulin resistance.

TABLE 23.4

Parasites

Parasite Species	Tissue Infected	Characteristics
Plasmodium falciparum and *Plasmodium vivax*	Liver and red blood cells	Malaria is a severe protozoan infection (as described above) carried by *Anopheles* mosquitoes. It can be prevented in travellers with prophylactic medication, but it is an endemic infection in most countries of Africa. Usually, infection by malaria parasites can be effectively treated with antimalarial medication, but in a small percent of cases it causes recurring infection. People who are heterozygous for sickle cell anemia (Chapter 6) are much less likely to die from malaria. This is because their damaged red blood cells are broken down by the spleen much more frequently than normal cells; therefore, the parasite population in the body is significantly decreased.
Giardia lamblia	Gastrointestinal tract	Giardia is a flagellated protozoan that infects the digestive tract of many animals, including humans. Giardia infects over 200 million people worldwide through contaminated food or water and can form resistant cysts outside of a host. Symptoms include violent diarrhea, vomiting, abdominal pain, and excessive bloating and gas. Symptoms usually begin within one to two weeks of infection and can last for six weeks or more without treatment; however, the immune system can often kill the infection without treatment. Dehydration can be fatal in small children.
Toxoplasma gondii	Blood, muscle, and nervous system	Toxoplasmosis infects most warm-blooded animals, but particularly cats (from eating infected rodents), which can be transmitted to humans through the fecal-oral route. It can also be transmitted through uncooked pork. As much as 30% of the population has probably had this infection that causes mild flu-like symptoms in healthy people. It can be very dangerous for pregnant women because it can be passed onto an unborn infant and cause encephalitis and neurological diseases. It is important that pregnant women stay away from litter boxes, gardening where cats may defecate, and handling raw pork.
Entamoeba histolytica	Gastrointestsinal tract	Amebiasis is the most common protozoan parasite in the world infecting over 600 million people per year.

(Continued)

TABLE 23.4 (*Continued*)

Parasite Species	Tissue Infected	Characteristics
Tapeworms (many species), *Diphyllobothrium latum* is the longest tapeworm species that infects humans, and can reach 10 metres in length.	Intestines	Tapeworm species can infect all vertebrates. The most common method of infection in humans is ingestion of the larval stage in uncooked fish and beef. Prevalence in the world is highest in populations that eat raw fish. Tapeworms are hermaphrodites, containing both male and female reproductive organs, and can self-fertilize. Infected people are usually asymptomatic, unless a large population of worms develops in the GI tract over a long period of time, which then causes vitamin B12 deficiency and anemia.
Necator americanus and *Ancylostoma duodenale*	Intestines	Hookworm is the most prevalent helminths infection in the world, infecting over 700 million people and causing severe illness in 80 million people, mostly children in developing countries (Figure 23.15). Hookworms are nematodes that enter the body through the skin, usually the feet, and can also be ingested. Hookworm infection can cause iron-deficiency anemia, stunted growth, and impaired cognitive function, but it is rarely fatal. Do not walk barefoot when travelling in regions shown in Figure 23.15.

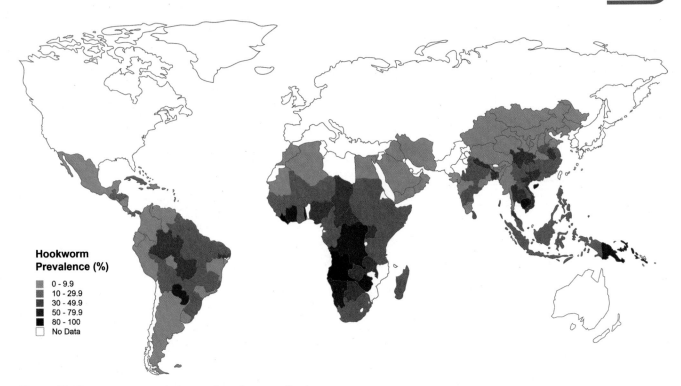

Hookworm Prevalence (%)

- 0 - 9.9
- 10 - 29.9
- 30 - 49.9
- 50 - 79.9
- 80 - 100
- No Data

Figure 23.15   Worldwide Prevalence of Hookworm Infection

Source: Hotez, P.J., Bethony, J., Bottazzi, M.E., Brooker S., and Buss, P. (2005). "Hookworm: The Great Infection of Mankind," PLoS Med 2(3): e67.

23

DID YOU KNOW?

Toxoplasmosis antibodies occur in higher levels in people that have schizophrenia. Minocyline is an antibiotic that happens to be effective in treating toxoplasmosis and has been shown to alleviate schizophrenia symptoms in some people. Other research has shown latent toxoplasmosis infection to be linked with suicidal behaviour, attention deficit disorder, and obsessive-compulsive disorder.

CONCEPT REVIEW 23.4

1. What is the most common method of infection by intestinal protozoans, and by protozoans that live in the bloodstream?

2. What are the two major categories of helminths?

3. How are intestinal worm infections acquired?

4. What populations are more at risk of helminths infection?

5. Which helminths infection can be acquired through the skin?

6. Which populations are at highest risk of tapeworm infection?

Appendix A

CONCEPT REVIEW ANSWERS

Chapter 2 Concept Review Answers

Section 2.1

1. Approximately 55%–60% of the human body is composed of water.

2. The difference between a solute and a solvent is that a solvent is a substance that contains dissolved molecules, which are called the solutes.

3. The difference between hydrophobic and hydrophilic and polar and nonpolar is that hydrophobic means water-fearing molecules and hydrophilic means water-loving molecules. Polar molecules may have a slight charge or a full charge and dissolve easily in water (hydrophilic) and nonpolar molecules do not have a charge and cannot dissolve easily in water (hydrophobic).

4. Surface tension occurs when hydrogen bonds cause water molecules to form droplets and stay close together.

5. Amphipathic molecules have polar and nonpolar regions and hydrophilic and hydrophobic components.

6. Water is biologically important because it is the universal solvent of the body in which important nutrients, ions, and molecules are dissolved inside and outside of the cell. The specific concentration of different molecules within the cell or surrounding the cell is important to the cell's ability to function properly.

Section 2.2

1. Dehydration synthesis is the process by which macromolecules are made from their building blocks by the loss of a water molecule: for example, the combination of amino acids to build proteins. Hydrolysis is the process of breaking down macromolecules into their building blocks by the addition of a water molecule.

Section 2.3

1. There are 20 different amino acids, and nine essential amino acids.

2. The three components of an amino acid are a carboxyl group, an amino group, and a functional group. The carboxyl group and amino group are the common components of every amino acid.

3. Peptide bonds, hydrogen bonds, and disulfide bonds. Peptide bonds hold amino acids together. Hydrogen bonds and disulfide bonds are important for the formation of secondary and tertiary structure.

4. The primary structure is the amino acid sequence. The secondary structure is based on the properties of the functional groups of the amino acids, and the way they fold depends on the interactions between the amino acids. Hydrogen bonds form between polar amino acids, and the hydrophobic amino acids are attracted to other hydrophobic amino acids. The tertiary structure is the three-dimensional shape the protein develops due to the primary and secondary structures. This functional structure of the protein is determined by the amino acid sequence. Polar or charged amino acids are attracted to each other and nonpolar amino acids are also attracted to each other. The quaternary structure is sometimes formed when multiple tertiary structures are combined.

5. Alpha helices are a secondary structure of a protein that forms a helix through hydrogen bonding. Beta sheets are a secondary structure of a protein that forms a twisted sheet through hydrogen bonding.

6. Hemoglobin, DNA polymerase, and microtubules

7. Denaturation, an irreversible reaction, occurs when there is enough energy applied to a protein to break the secondary- and tertiary-structure hydrogen bonds; the covalent peptide bonds remain unbroken, and the protein is no longer able to function.

8. Enzymes are important because they facilitate thousands of chemical reactions in our cells. They can form active sites that catalyze chemical reactions and can increase contact between specific regions of molecules to allow chemical reactions to occur.

Section 2.4

1. A chromosome is a single nucleic acid macromolecule called DNA that contains hundreds to thousands of genes and is found in every cell except mature red blood cells. Humans have 46 chromosomes.

2. The function of DNA is to store information for protein synthesis. DNA is located in the nucleus.

3. There are three components of a nucleotide: a five-carbon sugar deoxyribose or ribose; a phosphate group; and a nitrogen-containing base that

can be double-ring base purines—adenine and guanine—or single-ring base pyrimidines—cytosine, thymine, and uracil.

4. Phosphodiester bonds, glycosidic bonds, and hydrogen bonds

5. TACCGATCAG

6. The function of RNA is to serve as a template to integrate the amino acid sequence of a specific protein.

7. See Table 2.2, DNA compared to RNA

Section 2.5

1. Carbon, hydrogen, and oxygen

2. A monosaccharide has the formula $C_6H_{12}O_6$ and is a single-ring carbohydrate. A disaccharide consists of two monosaccharides paired together by glycosidic bonds. A polysaccharide is a complex carbohydrate that consists of long chains of glucose.

3. Glucose, galactose, and fructose. Two glucose molecules form maltose, glucose and galactose form lactose, glucose and fructose form sucrose.

4. Starch, glycogen, cellulose. Starch is stored energy found in plant cells; our bodies can digest starch for energy. Glycogen is stored energy in animal cells, and it can be stored in the liver and muscle cells if in excess. Cellulose functions in plant structure; it is not used for energy and cannot be digested by the human body.

5. Maltose—grains; sucrose—table sugar; lactose—milk; starch—rice; cellulose—fruits and vegetables.

Section 2.6

1. Lipids are important for the functioning of multiple parts of the body. In the brain, fats are essential for brain function, including learning ability, memory retention, and mood. In the cell, fatty acids and cholesterol aid in cell flexibility and build cell membranes. Most of the heart's energy for contraction comes from burning fats. Fats also compose the material that protects and insulates nerves, isolating electrical impulses and speeding up their transmission. Fat in the form of surfactant allows the lungs to function properly. Fats are also used in digestion by slowing down digestion to give the body more time to absorb nutrients and give a constant level of energy. Fat allows fat-soluble vitamins to be absorbed and keeps the body satiated for long periods of time. Fats cushion and protect internal organs and decrease inflammation, as well as helping the body's metabolism and immune system stay healthy.

2. Adipose tissue consists of groups of adipocyte (fat) cells that contain triglycerides.

3. Saturated triglycerides are solid at room temperature and contain only single-bond carbon atoms; unsaturated triglycerides are liquid at room temperature and contain one or more double-bond carbon atoms.

4. Sources of healthy, saturated fats are meats, avocado, coconut oil, and dairy products. Sources of healthy unsaturated fats are plant oils, flaxseed, nuts, seeds, and vegetables. Trans fats should be minimized in your diet and are found in fast foods and deep fried foods. Trans fats are bad for your health because the enzyme lipase that digests fats is specific only to cis conformation; therefore, trans fats are not broken down, remain circulating in the blood stream, and can cause plaque in blood-vessel walls, which leads to heart disease.

5. Hydrogenation is the process in which hydrogen atoms are added to the fatty acids to break the double bonds in the carbon atoms to make single bonds.

6. Cholesterol is an important constituent of cell membranes; it is used by the liver to make bile, used to produce LDLs to transport fats through the bloodstream, and used to produce steroid hormones.

7. The primary type of fat that makes up cell membranes is phospholipid.

Chapter 3 Concept Review Answers

Section 3.1

1. Cells have microvilli or dendrites or other cell-surface extensions to increase surface area, which increases the amount of molecules that can enter or leave a cell.

2. The cell theory states the following:
   - All living organisms are composed of cells: including bacteria, protists, fungi, plants, and animals, but not including viruses, which are surrounded by protein capsules and are not considered "living."
   - The cell is the basic unit of all living organisms and is the smallest single unit that can survive independently.
   - All cells come from pre-existing cells.
   - All cells contain genetic material—DNA—that is passed on to new cells that arise from cell division and offspring.
   - All cells are either prokaryotic or eukaryotic and are composed of basically the same types of phospholipids, proteins, carbohydrates, and nucleic acid molecules.

Section 3.2

1. Nonpolar molecules are not attracted to water and are also called hydrophobic. Polar molecules are attracted to water and are also hydrophilic.

2.

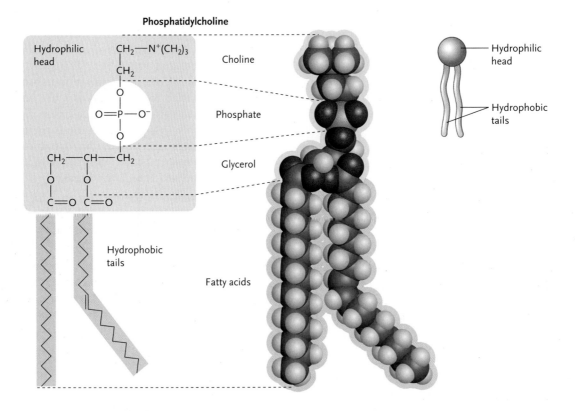

3. The general structure of a cell membrane is termed a fluid mosaic model; it consists of a phospholipid bilayer as well as cholesterol, proteins, glycolipids, and glycoproteins.

4. • Cholesterol—type of fat that increases the strength and decreases the fluidity of the cell membrane.

   • Receptors—proteins that bind regulatory molecules such as neurotransmitters, hormones, cell-signalling molecules, or nutrients

   • Antigens—usually glycoproteins that act as cell markers

   • Enzymes—proteins that act as catalysts for assorted chemical reactions, some of which may occur on cell membranes

   • Ion channels—proteins that span the membrane and allow the transport of specific ions into or out of the cell

   • Carrier proteins—proteins that move substances such as nutrients across cell membranes by changing their conformational structure

   • Adhesion molecules—membrane proteins that play an important role in intercellular interactions

   • Gap junction channels—protein membrane channels that play an important role in intercellular communication

Section 3.3

1. Prokaryotic cells contain some or all of the following (depending on the species): nucleoid region of one circular DNA molecule, cell wall, capsule, cytoplasm, ribosomes, pili, flagella.

2. Circular DNA contains the genetic material; the cell wall maintains cell strength and regulates osmotic pressure; the capsule externally protects the cell from environmental changes; cytoplasm contains dissolved ions and nutrients; ribosomes are essential for protein synthesis; pili help the cell adhere to structures and transfer small amounts of DNA to other cells; flagella are used for locomotion.

3. The cell wall and capsule in prokaryotes are important because they provide extra protection for the cell.

4. Gram positive bacteria stain purple when exposed to crystal violet due to the many layers of peptido-glycan and gram negative bacteria do not stain due

to a smaller layer of peptidoglycan and an extra external layer of lipopolysaccharide.

5. Eukaryotes are approximately 10 times larger than prokaryotes and contain a more complex cell structure and contain membrane-bound organelles.

Section 3.4

1. See Table A.1.

TABLE A.1

Organelle	Function
Centriole	Responsible for organizing spindle fibres during cellular division
Peroxisome	Breaks down lipids for ATP generation in the mitochondria
Lysosome	Breaks down organelles, molecules, and macromolecules as well as bacteria and viruses
Ribosome	Responsible for the production of polypeptides by acting as a site for translation
Mitochondria	Produce ATP through oxidative metabolism
Golgi body	Acts to package and modify proteins
Endoplasmic reticulum	Responsible for the synthesis of proteins on rough ER, as well as carbohydrates and lipids including cholesterol on smooth ER
Nucleus	Houses the genetic information of the cell and is the site of RNA synthesis

2. Mitochondria are thought to have potentially developed as an organelle by means of endosymbiosis due to the fact that they have a double plasma membrane and share many traits in common with bacteria (size, DNA, organelles, and process of replication).

Section 3.5

1. See Table A.2.

TABLE A.2

Cytoskeletal Protein	Function
Microtubule	Maintains cell shape/structure and facilitate intracellular transport
Intermediate filament	Supports cell structure by holding organelles in place
Microfilament	Involved in cell movement

2. See Table A.3.

TABLE A.3

Protein	Function
Integrin	Connects cell membranes to the proteins of the ECF and transmits signals across the membrane
Collagen	Provides strength and resists stretching
Elastin	Allows tissues to stretch and return to original shape
Fibronectin	Connects integrin to ECF proteins

3. Desmosomes hold the cell membranes of adjacent cells together and prevent damage from shearing. Tight junctions hold the cells together, primarily in epithelial cells where only a limited amount of extracellular fluid can pass through. Gap junctions permit the exchange of small particles between cells.

4. Desmosomes are found in stratified epithelial tissue and cardiac muscle. Tight junctions are found in epithelial tissues, such as the blood-brain barrier and gap junctions, are found in cardiac muscle tissue.

Section 3.6

1. Certain molecules are able to pass through while others are not, either due to charge or size of the molecule.

2. Diffusion is the passive movement of particles from areas of high concentration to areas of low concentration. Most small, nonpolar molecules can move easily through the body.

3. Large molecules and polar molecules (except water) typically require proteins to move in and out of the cell.

4. Osmosis is movement of water to an area of low-water concentration from an area of high-water concentration without the input of energy.

5. A solute is a solid substance that is dissolved by the solute. The solute is a liquid substance that when containing a solvent becomes a solution.

6. If the concentration of a solute in a cell is greater than that in the external environment, water will move into the cell by means of osmosis. Conversely, if the concentration of the solute is greater outside the cell, water will move from inside the cell to outside the cell.

7. Hypertonic, hypotonic, and isotonic are relative terms related to different solutions. When a cell is placed in a solution that is hypertonic to the internal fluid of the cell, the water from the cell moves outside the cell, causing the cell to crenate. When the

cell is placed in a solution that is hypotonic to the internal fluid of the cell, water moves from outside the cell to inside, causing the cell to swell and perhaps burst. When the cell is placed in an isotonic solution, the concentrations inside and outside the cell are equal, and there will be no net movement of water.

8. Facilitated diffusion is the movement down a concentration gradient without the use of energy, but aided by a protein channel or carrier. Channels are specific protein openings in the membrane that allow very specific ions to move across; carriers are proteins that bind to a specific molecule and change its shape, which then causes that molecule to move across the membrane.

9. There are no longer any binding sites available for molecules to attach and diffusion will not be aided by the proteins.

10. Active transport is the movement of particles against a concentration gradient by use of proteins. It differs from facilitated diffusion in that the direction of the movement of particles is reversed (against the gradient rather than with the gradient), and this requires energy.

11. The sodium-potassium pump and proton pump are two forms of active transport in the body's cells.

12. Sodium is moved out of cells by means of the sodium-potassium pump. To do this, one ATP is used to alter the shape of the protein that moves three sodium ions out of the cell and two potassium ions into the cell.

13. Sodium-potassium pumps are required for maintaining both concentration and electrical gradients across cell membranes; they are found in all body cells.

14. Proton pumps are found in the mitochondria of cells, and they function to move hydrogen ions into the intermembrane space to create a concentration gradient and electrochemical gradient. The buildup of hydrogen ions causes the protons to move back across the membrane through specific channels, causing ATP synthase to bind a phosphate to ADP creating ATP: a process called chemiosmosis.

15. The natural diffusion of molecules, such as sodium, into the cell creates the energy needed to move larger molecules, such as sugar and amino acids, into the cell.

16. Endocytosis (both phagocytosis and pinocytosis) refers to the movement of large molecules into the cell; exocytosis is the movement of large molecules out of the cell.

17. Phagocytosis is the movement of a solid into a cell. Pinocytosis is the movement of a liquid into a cell.

Receptor-mediated endocytosis is the movement of a specific substance into a cell that binds to a receptor; therefore, the substance can enter only the cells that have the receptor. For example, HIV binds and enters only white blood cells that have the CD4 receptor.

Chapter 4 Concept Review Answers

Section 4.1

1. Adenosine triphosphate (ATP)

2. ATP yields energy when the last phosphate bond is broken and the phosphate then binds to another molecule, converting ATP to ADP. ATP is used by our cells to perform many functions every second.

3. Glucose, fatty acids, amino acids, and nucleotides

4. Oxidation is a reaction in which a molecule loses electrons. Reduction is a reaction in which a molecule gains electrons.

5. Redox reactions are important for cellular respiration because as glucose becomes oxidized and NAD^+ molecules are reduced, forming NADH.

6. A Calorie is a measurement of food energy, also called a kilocalorie. A calorie is the amount of energy required to raise the temperature of 1 ml of water 1°C.

7. The carbon-hydrogen bonds in food molecules are important for the redox reactions because those bonds contain energy, which leads to the production of ATP through cellular respiration.

8. Aerobic respiration is the oxidation of glucose in the presence of oxygen to produce ATP; anaerobic respiration occurs when ATP is produced without oxygen.

Section 4.2

1. Glucose, NAD^+, and two ATP

2. Cytoplasm of all cells

3. Glycolysis is important to produce a small amount of ATP without oxygen.

4. Two pyruvate molecules, two NADH, net two ATP

5. Substrate-level phosphorylation occurs when a specific enzyme in the cytoplasm adds a phosphate to ADP to become ATP during glycolysis.

6. Prokaryotic and eukaryotic organisms can undergo glycolysis.

Section 4.3

1. Pyruvate oxidation and the Krebs cycle occur in the matrix of the mitochondrion.

2. Two acetyl-coenzyme A (acetyl-CoA) molecules

3. Carbon dioxide (CO_2)

4. The production of NADH and $FADH_2$ is the most important result of the Krebs cycle.

5. Two ATP molecules are formed during the Krebs cycle, one for each acetyl-CoA molecule.

6. NADH and $FADH_2$ molecules are important because they provide the hydrogen ions and electrons needed for the electron transport chain.

Section 4.4

1. The purpose of the NADH and $FADH_2$ molecules created during the Krebs cycle is to provide electrons and hydrogen ions required for the electron transport chain to produce ATP.

2. The H^+ ions are transported by a proton pump across the inner mitochondrial membrane (cristae) into the intermembrane space. This forms a high H^+ concentration gradient in the intermembrane space, so the H^+ ions can move down the gradient and back into the matrix, using ATP synthase to produce ATP.

3. NADH and $FADH_2$

4. The electrons combine with hydrogen ions and oxygen molecules (from breathing) to form water molecules at the end of the electron transport chain.

5. ATP synthase is a membrane protein that allows H^+ ions to move back into the matrix of the mitochondrion.

6. Chemiosmosis involves a chemical gradient moving across the membrane and the production of many ATP molecules.

Section 4.5

1. Glucose, fatty acids, glycerol, amino acids, and nucleotides

2. Proteins are broken down into amino acids when digested. Instead of going through glycolysis, amino acids are oxidized by removing a nitrogen group (deamination) to produce pyruvate or acetyl-CoA to enter the Krebs cycle. Nucleic acids are broken down into nucleotides when digested. Like amino acids, nucleotides also contain a nitrogen group and must undergo deamination, not glycolysis, to enter the Krebs cycle. Fats are broken down into fatty acids and glycerol when digested. The carbon-hydrogen bonds in long-chain fatty acids are oxidized into acetyl-CoA, which then enters the Krebs cycle.

3. Deamination is the process of removing the nitrogen group from a molecule in the form of ammonia NH3. Amino acids and nucleotides undergo deamination.

4. CO_2 and water. When amino acids or nucleotides are used to make ATP, ammonia is also a waste product.

Section 4.6

1. Anaerobic respiration in humans: NADH is converted back to NAD^+ through fermentation, and the H^+ is added to pyruvate to become lactate. Anaerobic respiration in yeasts: pyruvate undergoes decarboxylation, producing acetaldehyde. NADH is converted back to NAD^+ through fermentation, and the H^+ is added to acetaldehyde to form ethanol.

2. Anaerobic respiration occurs in human cells when there is a lack of oxygen, such as during high-intensity exercise.

3. Fermentation is the process of moving the H^+ from NADH to pyruvate to create lactic acid or alcohol.

4. Lactate formed in the muscle cells can be converted back to pyruvate once oxygen is available.

Chapter 5 Concept Review Answers

Section 5.1

1. The two main components of bacterial replication are DNA and binary fission, in which the bacterial cell separates into two identical cells.

2. A plasmid is a small extra piece of DNA, in addition to the larger, single, circular DNA molecule found in some bacteria.

3. Step 1. DNA replication begins at the point of origin where enzymes begin to separate the DNA strands.
   Step 2. DNA polymerase adds complementary nucleotides in both directions until two new identical DNA strands are created.

4. Bacteria acquire genetic variation through methods of DNA transfer with other bacteria: conjugation, transduction, and transformation.

5. Conjugation occurs when bacteria transfer DNA through a pilus, as a plasmid or piece of DNA, directly to another bacterium. Transduction occurs when bacteria receive new DNA fragments from viruses. Transformation occurs when bacteria receive DNA fragments or plasmids that are near them (e.g., dead bacteria).

Section 5.2

1. The cell cycle contains two growth phases (G_1 and G_2), DNA replication (S), and cell division (M). Prokaryotes and eukaryotes have a cell cycle.

2. Cells spend the most time in interphase (mostly G_1).

3. In G_1, cells undergo normal cell function and prepare for DNA replication. In S phase, DNA replication occurs. In G_2, cells continue to prepare for mitosis, and DNA is checked for mutations at the G_2 checkpoint. During mitosis, there is no more cell

growth and chromosomes and organelles separate. This results in the formation of two new daughter cells identical to the original cell.

4. Cytokinesis is the process of splitting the cytoplasm.

5. The main purpose of mitosis is to produce two identical daughter cells.

6. Humans have 46 chromosomes in each cell nucleus, 23 chromosomes from each parent.

7. Homologous chromosomes are two similar chromosomes that are not identical. Sister chromatids are the two identical copies of a single chromosome that has been replicated.

8. Karyotype refers to chromosomes and their homologous pairs organized from chromosome 1 to chromosome 23.

Section 5.3

1. Prophase. DNA replicated from interphase is condensed, nuclear membrane and nucleolus breaks down, centrioles migrate to opposite ends of the cell, and spindle fibres form.

   Metaphase. Spindle fibres connect to a kinetochore at the centromere of each set of sister chromatids, which line up at the equatorial plain.

   Anaphase. Spindle fibres pull the chromosomes toward the opposite poles of the cell.

   Telophase. Sister chromatids are at opposite poles of the cell, spindle fibres break down, two nuclear membranes form around each set of chromosomes at opposite ends of the cell, chromosomes begin to uncondense, and nucleolus reforms.

   Cytokinesis. The cytoplasm of the cell divides equally, forming a cleavage furrow in animal cells, and resulting in two somatic, identical, diploid daughter cells.

2. Spindle fibres are microtubules that connect to the sister chromatids at the kinetochore and pull them apart toward opposite ends of the cell.

3. The chromosomes are equally distributed by lining up at the equatorial plane, resulting in the same number of chromosomes moving to opposite ends of the cell when they are pulled apart.

4. Each sister chromatid is identical, so it does not matter which one moves into which new cell.

5. There is a diploid number (2n) of chromosomes in each new cell; this totals 46 in human cells.

6. Examples of somatic cell include liver cells, bone cells, and skin cells.

Section 5.4

1. Apoptosis is programmed cell death.

2. Angiogenesis is the production of new blood vessels.

3. G_1 checkpoint. Cell checks for DNA mutations and, if there are none, proceeds to S phase for DNA replication.

   G_2 checkpoint. Cell checks for mutations that may have occurred during DNA replication, and specific enzymes may correct these mutations.

   M checkpoint. The final checkpoint occurs during metaphase of mitosis to determine whether the cell should complete cytokinesis.

   These three checkpoints are involved in cell regulation to make sure each phase is completed without mistakes before moving on to the next phase. If a mutation is found it will either be fixed by specific enzymes or undergo programmed apoptosis to prevent further mutations.

4. A carcinogen is anything that can cause the development of cancer cells.

5. The factors that can cause cancer include inherited mutations; errors that occur during DNA replication; and environmental factors such as radiation, viruses, and chemicals.

6. Proto-oncogenes regulate the rate of the cell cycle by coding for proteins that increase the cell cycle during normal periods of growth, pregnancy, or repair of damaged tissue. An oncogene, which is a mutated proto-oncogene, abnormally increases the rate of cell division. A tumour-suppressor gene also regulates the rate of the cell cycle by coding for proteins that perform at the three checkpoints to stop the cell cycle if the DNA is damaged. An example of a tumour-suppressor gene is p53.

7. Mutated proto-oncogenes, tumour-suppressor genes, DNA repair enzymes, genes involved in apoptosis, genes that increase angiogenesis, and genes that regulate the immune response can be involved in the development of cancer.

8. The tumour-suppressor gene p53 prevents cancer at development by stopping cell division at the G1 checkpoint if it detects abnormal DNA present before DNA is replicated.

9. Our immune system plays a role in preventing cancer through the presence of natural killer white blood cells that specifically destroy cancer cells.

Section 5.5

1. The purpose of meiosis is to generate haploid sperm or eggs, called gametes, required for an organism to reproduce.

2. Gametes (sperm or eggs) are haploid cells and contain 23 chromosomes. Zygotes and all somatic cells are diploid cells and contain 46 chromosomes.

3. Homologous chromosomes separate during anaphase I, and sister chromatids separate during anaphase II.

4. Gametes are haploid sperm cells formed in male testes and haploid eggs formed in female ovaries.

5. Genetic variation is important because it ensures that every individual is unique and not identical to the parent. A population with high genetic variation is more likely to survive environmental changes.

6. Independent assortment and crossing over.

7. In males there are four sperm cells, each with 23 chromosomes by the end of meiosis II. In females, cell division occurs so that there is more cytoplasm in one cell and the other cell becomes a polar body, which leaves one egg cell. The eggs remain arrested in metaphase II until ovulation.

8. Crossing over is important because it increases genetic variation.

Chapter 6 Concept Review Answers

Section 6.1

1. • Homozygous—alleles the same
   • Heterozygous—alleles different
   • Dominant—allele that is phenotypically prominent (Capital letters)
   • Recessive—allele that is hidden by the dominant allele (Lower case letters)
   • Allele—one copy of a gene in a diploid cell
   • Phenotype—protein that displays a visible trait or function
   • Genotype—genes that code for the phenotype

2. a. purple
   b. one gamete containing a purple allele, and one containing a white allele
   c. PP, Pp, and pp
   d. 25%

Section 6.2

1. See Table A.4.

TABLE A.4

	Y	y
y	Yy	yy
y	Yy	yy

a. 50%
b. 50%
c. 50%
d. 0%

2. A testcross can be used to determine if the purple-flowered plant is PP or Pp by crossing the plant with a homozygous recessive white-flowered plant (pp). If the outcome is 100% purple, the genotype is PP. If the outcome is 50% purple and 50% white, the genotype is Pp.

3. Mom is heterozygous dominant, dad is homozygous recessive, and child is homozygous recessive.

4. Yes, if they are both heterozygous dominant for the trait

5. No.

Section 6.3

1. See Table A.5.

TABLE A.5

	RY	Ry	RY	Ry
Ry	RRYy	Rryy	RRYy	Rryy
Ry	RRYy	Rryy	RRYY	Rryy
ry	RrYy	Rryy	RrYy	Rryy
ry	RrYy	Rryy	RrYy	Rryy

a. RRYy , Rryy
b. Each parent can have four possible gametes
c. RRYy, RRYY, RrYy, Rryy
d. Round and Yellow or Round and green
e. ½ Round and Yellow and ½ Round and green

2. 25% chance of having a child with brown hair and freckles

Section 6.4

1. Polygenic inheritance occurs when a single trait is determined by multiple genes: for example, body height, weight, or skin colour.

2. Pleiotropic refers to one single allele causing multiple phenotypes, example, cystic fibrosis.

3. The heterozygous bird's feathers would be purple. If two heterozygotes mated one-fourth of their offspring would be blue.

4. The dad could be either Larry (ii) or Bill, if Bill is $I^B i$. Mom has blood type $I^A i$ because in order to be type O you must receive i from both parents.

5. The possible genotypes are $I^A i$ or $I^B i$, and the possible phenotype is either type A or type B.

6. Epistasis occurs when the effects of one gene are modified by one or more other genes. A Labrador retriever may have the dominant black fur gene (B) but is yellow because it contains the recessive epistasis allele for pigmentation (e) and conceals the gene for black fur.

7. Examples include temperature of incubating reptile eggs, air temperature that affects fur colour, and soil pH of hydrangeas.

8. They will have a 50% chance of having children that are colour blind.

9. Linkage is when genes are inherited together because they are located closely together on the same chromosome.

Section 6.5

1. The "define" type questions should be removed, that means we have to remove this question in the chapter.

2. Chromosome 21 is affected in Down syndrome.

3. XXX (Trisomy X), XXY (Klinefelter syndrome), XO (Turner syndrome).

4. • Sickle cell anemia—autosomal recessive; abnormal red blood cell formation; and reduced capacity to carry oxygen

   • Tay Sachs—autosomal recessive; brain deterioration leading to death before adolescence

   • Cystic fibrosis—autosomal recessive; excessive mucus production impacting the lungs, liver, pancreas, and sweat glands

   • Huntington's disease—autosomal dominant; brain deterioration beginning in the late 30s

   • Hemophilia—recessive X-linked; lack of blood clotting, resulting in the inability to stop bleeding

   • Duchene muscular dystrophy—recessive X-linked; muscle degeneration that affects children in early childhood

   • Familial hypercholesterolemia—autosomal dominant; possible occurrence of cardiovascular disease by age 30

Chapter 7 Concept Review Answers

Section 7.1

1. Evolution is the change in the characteristics of organisms within a population over time.

2. Survival of the fittest means individuals born with traits that are most beneficial for a particular environment will survive, reproduce, and pass on DNA. For example, predators that are better able to catch prey will have more nutrients, better health, and therefore survive and produce offspring with the same traits.

3. The relationship between evolution and ecology is that environmental changes determine which organisms are the "fittest." Changes in the environment, such as the availability of certain foods, affect which organisms survive and pass on their genes.

4. The major factors involved in natural selection are as follows: genetic variation is beneficial, similarities in DNA sequences are more common in closely related organisms, organisms that survive and reproduce pass on their DNA, all organisms over-reproduce so that the fittest survive, and populations change as the environment changes.

5. Examples of environmental factors that can affect evolution are seasons, weather patterns, ice ages, forest fires, availability of food, the presence of predators, or infectious organisms.

6. Examples of human-inherited traits that would be beneficial for survival include intelligence, being fertile, and having motivation to work and earn money to buy food and shelter. Two hundred thousand years ago, humans needed to be physically fit to hunt for food. People no longer have to be physically fit or be able to hunt to survive because we now live in a technological environment. In different environments, different traits are important for survival.

Section 7.2

1. Bacteria can reproduce every 30 minutes in perfect conditions and therefore can have a very rapid generation time. Bacteria will undergo genetic changes based on changes in the environment, such as lack of a nutrient, change in temperature, or level of moisture, and also recently, exposure to antibiotics. The exposure to antibiotics has created a selection pressure where only those able to withstand exposure to antibiotics are able to survive and therefore pass on the antibiotic resistance gene to future generations of bacteria.

2. As bacterial antibiotic resistance increases, those humans with decreased immune function become more susceptible to infection, and they lack the ability to combat the infections if the antibiotic cannot kill the infection.

3. Microevolution refers to the small genotype and phenotype changes within a species, where macroevolution leads to organisms that are so different that they are a new species.

4. Darwin's theory suggests that giraffes were inherently born with longer necks and, therefore, were better adapted to their environment, passing on the "long-neck gene" to future generations. Lamarck would suggest that as adult giraffes were required to stretch their necks, their neck length increased throughout their lifetime. In response, offspring were born with longer necks and, again, they had to stretch, and thus increased the length of their neck throughout their lifespan.

5. Natural selection is the result of a species' reaction to its environment. Artificial selection is the result of a breeder choosing which plants or animals will reproduce in order to perpetuate specific characteristics.

Section 7.3

1. The finches that Darwin studied are identified as separate species based on the size and shape of their beaks, which are highly adapted to different types of food in South America and the Galapagos. The differences between the birds are enough to make them different species, but there are also a number of similarities between the birds. Darwin hypothesized that microevolution leads to the

macroevolution, creating all of the finch species that exist today as a reaction to the environment and food supply. Different finches will be more "fit" than others to eat certain foods, such as large hard seeds, small seeds in small cracks, berries, worms, or insects and, as a result, pass on those traits to their offspring.

2. Gradualism is the slow development of a new species through a series of small changes that lead to a new species over a great deal of time. Punctuated equilibrium in contrast is the theory that little to no changes take place over a great deal of time, with sudden drastic changes appearing all at once.

3. The three primary methods that provide evidence of evolution are homologous anatomy, the fossil record, and homologous molecules

4. Bats are more closely related to humans than to birds because of the similarity of the bones in the forelimb. Bats and humans both have a humerus, ulna, radius, and phalanges. Birds have a completely different structure in their wings.

5. A selection pressure is something that affects the course of evolution of a species because it creates a condition that permits certain individuals in a population to have an increased likelihood of survival compared to others in the same population.

Section 7.4

1. Factors that maintain Hardy-Weinberg equilibrium include random mating, large populations, the lack of mutations, lack of migration, the lack of ecological changes, and the lack of selection pressures.

2. Answer calculation is as follows:
   - $p = 0.5$
   - $p + q = 1$; therefore, $q = 0.5$
   - $p^2 + 2pq + q^2 = 1$
   - $(0.5)^2 + 2(0.5)(0.5) + (0.5)^2 = 1$
   - p^2 is the number of homozygous dominant mice
   - $2pq$ is the number of heterozygous mice
   - q^2 is the number of homozygous recessive mice
   - Short-tail alleles are recessive and therefore will display the phenotype only if homozygous: therefore, $(0.5)^2$ or 25% of the time.
   - Long-tail alleles are dominant and therefore will display the phenotype if either homozygous or heterozygous: therefore, $p^2 + 2pq$ or $(0.5)^2 + 2(0.5)(0.5)$ or 75% of the time.

3. Any factors that affect the Hardy-Weinberg equilibrium also affect allele frequencies.

4. Natural selection can be affected by means of the following types of selection (see Table A.6):

TABLE A.6

Type of Selection	Function	Example
Directional selection	Benefits one extreme of the population	Antibiotic resistance in bacteria
Stabilizing selection	Benefits the intermediates of the population	Baby birth size
Disruptive selection	Benefits both extremes of the population	Finch beak size

5. Sickle cell anemia acts through stabilizing selection by benefitting the intermediate phenotype of the population.

6. Heterozygote advantage refers to the additive benefits of both the dominant and recessive alleles that result in an increased likelihood of survival for those individuals with both genes.

7. Predation is the primary selection pressure acting on the guppies.

8. The bottleneck effect is random in regard to which individuals it affects—rather than genetics conferring an enhanced survival rate among certain individuals.

9. DNA mutations are the ultimate source of new alleles, though migration may also lead to new alleles in a particular population.

Chapter 8 Concept Review Answers

Section 8.1

1. Adenine, guanine, thymine, cytosine, and uracil

2. • Purines—adenine, guanine
   • Pyrimidines—thymine, cytosine, uracil

3. A purine always binds to a pyrimidine, adenine binds to thymine, and guanine binds to cytosine in DNA.

4. Hydrogen bonds, phosphodiester bonds, glycosidic bonds

5. Chargaff's rule is that the amount of adenine always equals the amount of thymine, and the amount of cytosine always equals the amount of guanine.

6. Franklin, Watson, and Crick discovered the DNA double helix.

7. A nucleoside consists of a purine or pyrimidine base attached to a sugar; a nucleotide is a purine or pyrimidine attached to a sugar and a phosphate group.

8. We acquire nucleosides from our diet when we eat any plant or animal cells (vegetables, fruit, and meat).

9. DNA forms a double helix.

10. DNA has thymine, RNA has uracil. DNA has deoxyribose, RNA has ribose. DNA is double stranded, RNA is single stranded.

11. The carbons of the sugars are named 1, 2, 3, 4, 5. This is important because carbon 3 and 5 bind to the phosphate group to make up the sugar-phosphate backbone. If there are two hydrogen atoms on carbon 2 the sugar is deoxyribose, and if there is a hydroxyl group on carbon 2 it is ribose.

12. Ribose contains a hydroxyl group on carbon 2, while deoxyribose contains only two hydrogen atoms on carbon 2.

13. ATP is the energy molecule of the cell.

14. Telomeres are a repeated sequence at the ends of each chromosome. They are important because they protect the genes on the ends of the chromosomes from deterioration during DNA replication.

15. Stem cells and germ cells produce telomerase. This is important because it replaces lost telomere sequences; therefore, these cells will always be able to divide.

16. Prokaryotes do not contain telomeres because their DNA molecules are circular and can replace all primers with DNA replication, which allows bacteria to divide indefinitely.

Section 8.2

1. Semi-conservative replication means that each of the original strands of DNA is used to produce the two new strands. Every new DNA molecule contains an old strand and a new strand.

2. • Helicase—breaks the hydrogen bond between the bases of the complementary strands of DNA, starting at the origin and separating the two strands, creating a replication fork
   • Single-stranded binding protein—stabilizes the single-stranded region of the DNA molecule
   • Primase—adds a RNA primer, which is a small sequence of nucleotides with a 3′ hydroxyl group
   • Polymerase—adds the complementary nucleotides to each single strand of DNA in the 5′ to 3′ direction only
   • Ligase—forms phosphodiester bonds to seal the gaps between the nucleotides

3. On the leading strand, one RNA primer is needed, and polymerase will follow helicase, adding complementary nucleotides in the 5′ to 3′ direction on the new strand, and this is continuously replicated. On the lagging strand, replication is done in Okazaki fragments by many new RNA primers in order for polymerase to add nucleotides in the 5′ to 3′ direction.

4. For polymerase to add complementary nucleotides, it needs a 3′ hydroxyl group to proceed in the 5′ to 3′ direction. Primase adds an RNA primer that has a free 3′ hydroxyl group, thus allowing polymerase to start adding the complementary nucleotides.

5. See Glossary.

6. Even though DNA replication enzymes are the same in prokaryotes and eukaryotes, replication differs in prokaryotes because the DNA is circular, only one primer is needed on each strand, and polymerase can replicate continuously on both strands. This is different from eukaryotes, where Okazaki fragments are needed on the lagging strand. Also, since DNA is circular in prokaryotes, supercoiling occurs when the DNA unwinds and therefore the enzyme topoisomerase is needed to release the over twisting; this is not required in eukaryotes.

Section 8.3

1. Proofreading occurs when the polymerase enzyme recognizes and removes incorrect base pairs and replaces them with the correct nucleotides.

2. Our cells cope with DNA mutations by proofreading. Polymerase can recognize mistakes made during replication and will stop replicating, reverse in the 3′ to 5′ direction, remove the improper nucleotides by utilizing its exonuclease activity function, and continue replication in the 5′ to 3′ direction. We also have approximately 130 different genes that code for DNA repair enzymes, and they can locate mistakes missed by polymerase during replication. Tumour-suppressor genes also prevent cell division if mutations occur so DNA repair enzymes can fix the mutation. The p53 gene will induce cell apoptosis if the mutation cannot be corrected, thus preventing cell division with the mutation. If mutation still persists through cell division, our immune system produces natural killer cells to recognize and kill cancerous cells.

3. Substitution (mismatch) occurs when polymerase adds the wrong nucleotide.
   Addition (insertion) occurs when polymerase adds an extra nucleotide.
   Deletion occurs when polymerase skips a nucleotide.
   Recombination (transposition) occurs when large sequences of DNA are moved to another location within the genome.

4. A frameshift mutation occurs when a nucleotide is added or removed, causing a shift in all of the following nucleotides, which affects every other codon after that mutation.

5. The consequences of DNA mutations occurring in somatic cells can be harmful and possibly lethal to

the individual person, but they are not passed on to offspring. Mutations in gametes can be passed on to offspring; this results in genetic variation among individuals.

6. Mutations contribute to genetic variation by being passed on to offspring and affecting biochemical processes, development, behaviour, morphological traits, or gene expression.

7. Insertional inactivation is the insertion of a sequence of DNA into a location that disrupts a gene and produces either a non-functional protein or no protein.

8. Transposable elements are sequences of DNA that can move themselves (transpose) to another chromosome within the cell. They do not always cause DNA mutations.

9. We acquire DNA mutations by mistakes during DNA replications, transposition, mutagens or carcinogens, viruses, and inherited mutations.

10. The most common types of carcinogens are radiation, environmental toxins, and viruses.

11. See Table A.7.

TABLE A.7

Virus	Cancer
Human papilloma virus (HPV)	Cervical cancer
Human immunodeficiency virus (HIV)	Kaposi sarcoma
Hepatitis B or C	Liver cancer
Epstein Barr virus (EBV)	Lymphoma

12. See Table A.8.

TABLE A.8

Genetic Disease	DNA Mutation
Cystic fibrosis	Mismatch base in gene that codes for chlorine channels
Huntington's disease	Insertion of multiple CAG repeats in a gene on chromosome 4
Sickle cell anemia	Mismatch in the hemoglobin gene
Cancer	Two or more mutations in genes that code for repair enzymes, cell cycle genes, oncogenes, or tumour-suppressor genes
Phenylketonuria (PKU)	Point mutation in liver enzyme gene that causes brain damage
Nonpolyposis colorectal cancer	Autosomal dominant repeated CA mutation in DNA repair enzyme gene expressed in large intestinal cells

Chapter 9 Concept Review Answers

Section 9.1

1. Adenine and thymine are complementary base pairs and guanine and cytosine are complementary base pairs.

2. The central dogma refers to the flow of information from gene to RNA to protein. DNA is transcribed to mRNA (transcription), which then moves out of the nucleus through nuclear pores to a ribosome for protein synthesis to occur (translation).

3. The sense strand is the strand of DNA that codes for the gene and will have the same sequence as the mRNA. The antisense strand is the opposite strand and is the template for the mRNA.

4. The human genome is made up of 25,000 genes.

Section 9.2

1. The differences in transcription between prokaryotes and eukaryotes are as follows:

   • Prokaryotes do not contain a nucleus; therefore, they can undergo transcription and translation simultaneously.

   • Prokaryotes do not contain introns like eukaryotes and therefore have a simpler process.

   • They both have different mechanisms for regulating gene expression.

2. Transcription takes place inside the nucleus in human cells.

3. RNA polymerase binds to a promoter region of the DNA upstream of the gene that initiates transcription.

4. RNA polymerase "knows" where the beginning of the gene is located because the promoter regions are regulatory regions of the DNA that consist of nucleotide sequences that are called the TATA box or CAAT box.

5. AUG is almost always the first codon in an mRNA molecule.

6. The polymerase is using the template (antisense) strand to form the mRNA molecule and can add nucleotides only in the 5′ to 3′ direction on the mRNA molecule.

7. A terminator sequence prevents RNA polymerase from transcribing nucleotides that are not part of the gene sequence. This causes the mRNA to form a hairpin loop, causing RNA polymerase to dissociate.

8. In RNA processing, eukaryotes contain introns that are sequences of DNA that do not code for amino acids and must be removed. The regions of DNA sequences that do code for amino acids are called exons. To remove the introns, transcription first produces a pre-mRNA molecule to be processed by removing the introns by splicing and adding a 5′ cap

and 3′ tail. Splicing occurs in a spliceosome that consists of pre-mRNA and snRNPs and forms new bonds between the exons. The 5′ cap allows mRNA to begin translation at the ribosome, and the 3′ tail protects it from rapid enzymatic degradation. The new mRNA molecule that consists only of exons can now move out of the nucleus.

9. Only one strand of DNA is a gene sequence (coding strand). RNA polymerases make RNA, and DNA polymerases replicate DNA. The RNA molecule produced from transcription is single stranded, RNA polymerases do not need a primer, and transcription requires processing.

10. The mRNA sequence is 5′ UACGGCUAUG 3′

Section 9.3

1. Ribosomes are composed of a small subunit and a large subunit, which both contain rRNA and proteins.

2. mRNA binds to the small subunit of the ribosome to be translated. rRNA makes it possible for the mRNA to bind to the ribosome. tRNA carries the amino acids to the three binding sites on the large subunit of the ribosome by the anticodon of the tRNA matching the codon on the mRNA.

3. The A site is where the new tRNA enters the ribosome. The P site is where enzymatic activity of the rRNA in the large subunit catalyzes the formation of peptide bonds between amino acids. It is also where the first tRNA molecule binds. The E site is where the empty tRNA molecule moves after the P site and then leaves.

4. Ribosomes are also called ribozymes because the RNA in ribosomes have enzymatic activity.

5. The role of activating enzymes is to recycle the tRNA once it leaves the ribosome in the cytoplasm and attaches the correct amino acid to it.

6. Translation works by synthesizing a protein from an mRNA sequence that binds to a ribosome. tRNA that contains the anticodon then brings the amino acid to the ribosome to bind to the correct codon forming peptide bonds in the P site of the ribosome between the amino acids until the protein is complete.

7. When a stop codon enters the A site, no tRNA molecule can bind, and the ribosome will dissociate, thus completing translation.

8. Protein tertiary structure is complete when translation is complete.

9. If translation occurs on a ribosome of the rough ER, proteins are sent to the Golgi bodies to be processed, such as proteins combined with lipids to produce lipoproteins, or sugars combined with proteins to produce glycoproteins.

10. Leu-Asp-Thr-Glu

11. A silent mutation occurs when a mismatch mutation happens outside a coding region or does not have an effect on the amino acid sequence; therefore, the DNA mutation has no impact on the function of the protein.

Section 9.4

1. Regulation of gene expression is related to homeostasis because molecules involved in regulating body systems—such as blood pressure or utilization of glucose—have to be transcribed and translated in the correct amounts to properly regulate the system and maintain homeostasis.

2. Two types of regulatory proteins that affect transcription in prokaryotes are activators and repressors.

3. The lac operon is found in bacterial species where three genes transcribed together in the presence of lactose.

4. The repressor binds to the operator sequence. The CAP activator binds to the CAP-binding sequence. The RNA polymerase binds to the promoter.

5. The repressor is removed from the DNA when lactose is present and binds to it.

6. Transcription only occurs if lactose is present and glucose is not present.

7. If glucose and lactose are present transcription does not occur. Since glucose is present, the CAP activator does not bind to the CAP binding sequence; therefore, RNA polymerase will not bind to the promoter site, even though lactose is present and the repressor is removed.

8. See Table A.9.

TABLE A.9

Mechanism of Gene Regulation	Description
Methylation	Methyl groups are added to prevent gene transcription, and they are maintained during DNA replication.
RNA processing	Introns are removed to form a mature mRNA molecule, but not necessarily in the same manner every time the gene is transcribed; this is called alternate splicing.
mRNA degradation	Many mRNA transcripts have specific lifespans, and they may be broken down before translation occurs.
Transcription factors	This is the most common mechanism. Many factors can bind to regulatory regions of the DNA and promote transcription of certain genes.

9. Gene silencing occurs when a cell prevents the transcription of a gene.

10. Alternate splicing occurs when only certain exons become part of the mature mRNA molecule after the introns are removed during RNA processing, and other exons at other times. This is important because protein function depends on which exons are used, and this allows over 100,000 different proteins from 25,000 genes.

Chapter 10 Concept Review Answers

Section 10.1

1. Recombinant DNA technology is the process of using DNA from more than one species to create organisms that have useful traits.

2. Recombinant DNA contains DNA sequences from more than one species.

3. Genetically modified organisms are organisms such as plants and animals that have specific DNA sequences inserted into them. For example, the papaya tree is genetically modified to be resistant to the ringspot virus.

4. A gene, restriction endonucleases, a vector, ligase, and a host cell.

Section 10.2

1. Palindromes are short, four-to-six nucleotide, sequences of DNA that are the same on the forward strand as on the reverse strand. Palindromes are sequences of DNA recognized by restriction enzymes.

2. Restriction enzymes come from many species of bacteria.

3. Bacteria protect themselves from cutting their own DNA by using methylase enzymes that block recognition sites on their own chromosome with methyl groups.

4. Restriction enzymes break DNA phosphodiester bonds.

5. Sticky ends are important to allow DNA that was cut from a restriction enzyme to anneal with other DNA that was cut from the same enzyme.

6. Restriction enzymes are useful to genetic researchers because they can be used to cut DNA from two different organisms, such as bacterial DNA and human DNA, to create recombinant DNA.

Section 10.3

1. Vectors have the following important features: able to replicate within the host cell, contain restriction enzyme cut sites, have a selectable marker, be easy to recover from the host cell, and the inserted gene must be expressed as a protein.

2. Plasmids contain an origin of replication; a selectable marker, commonly an antibiotic resistant gene; and multiple restriction enzyme cut sites, which are usually placed into a multiple cloning site sometimes found within another selectable marker.

3. Antibiotic resistance gene and lac Z gene are used as selectable markers because it allows the correct colonies to be located.

4. The lac Z gene is used as a selectable marker because it is responsible for the breakdown of lactose (or the very similar molecule x-galactosidase, x-gal). When a bacterium is grown on x-gal, it turns blue, thus making it a selectable marker.

5. Only white bacterial colonies contain recombinant DNA because if a gene is inserted into the MCS, then the bacteria will be white.

6. • Bacteriophage. Genes involved in lysogenic phase are removed and replaced with foreign DNA, and the recombinant DNA can then infect other bacterial cells: called transduction.
   • Cosmid. This vector is used to carry larger sequences of DNA than plasmids, and is inserted into the bacterial cell by transformation.
   • YAC. These vectors can be grown in yeast cells, which are eukaryotic.

Section 10.4

1. Gene cloning is a method for making millions of copies of a gene using bacteria.

2. Transformation occurs when bacteria take up DNA from their surroundings, such as a recombinant plasmid.

3. A restriction map is useful for determining the appropriate restriction enzymes to cut your DNA with so that you have a whole gene.

4. You can determine which bacterial colonies have the gene of interest by growing the bacteria on plates containing growth media and x-gal. Any bacteria that did not take up the plasmid will not grow on the growth media because the plasmid contains the growth media antibiotic resistant gene. Any bacteria that did not take up the recombinant DNA will turn blue due to the lac Z gene in the presence of x-gal. Therefore, the bacteria with the gene of interest will be white.

5. Hybridization is used to locate a specific colony by creating a probe that is complementary to the gene of interest, and it will need to be made visible by using a radioactive tag such as radioactive phosphate.

6. a. Isolate plasmid DNA and human DNA.
   b. Look at restriction map to determine which restriction enzymes to use that will not ruin your insulin gene.
   c. Cut both plasmid DNA and human DNA with the same restriction enzyme.

d. Mix cut plasmid and human DNA (that contains insulin gene) together with ligase to form recombinant DNA.

e. Mix recombinant DNA with *E. coli* bacterial cells, and heat shock to transform bacteria.

f. Grow transformed bacteria on plate overnight.

g. Use DNA hybridization to screen colonies.

h. Grow specific bacterial colony containing the gene of interest and isolate synthesized protein.

7. • Insulin—type-1 diabetes
   • Erythropoeitin—anemia
   • Growth hormone—prevents dwarfism
   • Anticoagulants—prevents blood clotting
   • Clotting factors—hemophilia
   • Interferons—certain autoimmune diseases

Section 10.5

1. A genomic library is a collection of fragments of DNA from a single organism. The DNA fragments are ligated into plasmids and can be grown and recovered whenever needed.

2. You can find a specific sequence in a library by using hybridization with a specific probe for a specific gene.

3. cDNA is called complementary DNA and is the DNA molecule that contains only the coding region of the gene. It is useful because mRNA cannot be inserted into a plasmid, so it is converted to DNA by reverse transcriptase so it can be inserted into the plasmid without containing the introns.

4. Gel electrophoresis is a method used to separate strands of DNA based on the length of the DNA sequence by using an electrical current. Its main use is to determine if the correct fragment has been inserted into a plasmid to ensure the gene of interest has been cloned.

5. PCR is useful because it generates multiple copies of a specific DNA sequence and amplifies a gene much faster and more reliably than using bacteria.

6. The main ingredients for PCR are template DNA, primers, dNTPs, and taq polymerase. The steps in PCR are denaturation, primer annealing, and primer extension.

7. VNTRs are varying lengths of repeated DNA sequences in the introns and are different in each person. Since VNTRs are unique to each individual person, then we can use them to identify specific people.

8. Gene therapy is the process of inserting a gene into an individual's cells to replace a mutated gene.

9. It is not frequently successful because it is difficult to target specific cell types and have the new gene inserted into the right location in the genome.

Section 10.6

1. Blue roses, blue strawberries, *flavr savr* tomatoes

2. Expression is dependent on the promoter sequence the gene is attached to.

3. They are eukaryotic and can process RNA as well as modify proteins on the Golgi body.

4. a. Isolate the spider silk gene.

   b. Ligate into a plasmid containing a promoter sequence that is involved in the regulation of milk production in goats.

   c. Microinject the recombinant plasmid into stem cells that can be injected into the blastocyst of a goat, produced by in vitro fertilization.

   d. Implant the blastocyst into a pseudopregnant goat. A pseudopregnant goat is a female that has been given hormone injections to cause the uterus to be prepared for implantation.

   e. The goat that is born will have the spider silk gene connected to the promoter involved in lactation. When lactation occurs in that goat, the milk will contain spider silk that can be isolated and used to build many very strong substances. This process is already being used by a company called BioSteel, and they produce bullet-proof vests.

5. Knockout mice are valuable to researchers because genes can be deleted from mouse genomes so that researchers can find out the functions of those genes and determine what may happen to humans with mutations of those genes.

6. To resist herbicides, to produce pesticides, to resist cold temperatures, or to add nutrients

7. bt toxin is produced by a natural soil bacterium; it kills insects that eat it by making pores in their digestive tract, which leads to cell death. Crop plants are genetically engineered to contain bt toxin to kill harmful insects.

8. Potential problems with genetically engineered plants are that insects may become resistant to bt toxin; there may be changes within our ecosystem as the insect population changes, especially pollinating insects; the engineered gene may be transferred into other closely related species; and the implications on human health are not known.

Chapter 11 Concept Review Answers

Section 11.1

1. Diffusion, secretion, filtration, and absorption

2. By cell shape and the number of layers

3. See Tables A.10 through A.17.

TABLE A.10

Simple Squamous Functions	Simple Squamous Location
This type is a single layer of flattened cells that allows for diffusion, osmosis, filtration (glomeruli of kidneys), and secretion (surfactant in the alveoli).	Simple squamous epithelial cells line the heart, blood vessels, lymphatic vessels, alveoli in lungs, and the glomeruli in the kidney. They form part of any serous membrane the covers internal organs such as the heart (pericardium), lungs (pleura), and abdominal organs (peritoneum).

TABLE A.11

Simple Cuboidal Functions	Simple Cuboidal Location
Single layer of cuboidal cells that allows for secretion and absorption.	This type of cells is found primarily in the ducts of exocrine glands, the lumen of the nephron in the kidney, the surface of the ovaries, and parts of the thyroid gland.

TABLE A.12

Simple Columnar Functions	Simple Columnar Location
Ciliated simple columnar epithelial cells function to *move* substances, such as foreign particles trapped by mucus in the upper respiratory tract, or to move the egg along the fallopian tubes. Secretion of mucus.	Ciliated simple columnar epithelial cells are found in the upper respiratory tract, sinuses, fallopian tubes, and uterus.
Non-ciliated simple columnar epithelial cells usually contain specialized extensions called microvilli that allow for absorption of nutrients. This tissue contains goblet cells (considered unicellular glands) that secrete mucus.	Non-ciliated simple columnar epithelial cells are found primarily in digestive tract (stomach to anus), also found in the gallbladder; mucus protects lining of the digestive tract.

TABLE A.13

Pseudostratified Columnar Functions	Pseudostratified Columnar Location
Ciliated pseudostratified columnar epithelial functions to secrete mucus and to trap foreign particles and move substances along the surface of the tissue. Notice how the cells *appear* to be stratified but they are just one layer with some cells not extending to the surface (*pseudo* means "false").	Ciliated pseudostratified epithelial cells are found in similar locations as ciliated simple columnar, primarily the respiratory tract.
Non-ciliated type functions in absorption and secretion.	Non-ciliated pseudostratified epithelial cells can be found in the epididymis and many glands.

TABLE A.14

Stratified Squamous Functions	Stratified Squamous Location
Stratified squamous is the most abundant of the stratified tissue types. The multiple layers function to protect the underlying tissues from abrasion and from microorganisms. Note that the apical cells are squamous and the deeper basal cells are cuboidal.	Keratinized stratified epithelial cells are found in the skin; the outermost layers contain keratin in dead cells which provides a tough barrier (Chapter 12).
	Non-keratinized stratified epithelial cells are located in the mouth, tongue, esophagus, pharynx, and vagina.

TABLE A.15

Stratified Cuboidal Functions	Stratified Cuboidal Location
Stratified cuboidal epithelial cells can be two or more layers, and they function to protect underlying tissues and have some secretion and absorption properties.	Found mainly in esophageal ducts, sweat glands, mammary glands, and salivary glands

TABLE A.16

Stratified Columnar Functions	Stratified Columnar Location
A less common cell type, it functions as protection and secretion.	Found in the ducts of some exocrine glands, such as esophageal glands, and some parts of the urethra

TABLE A.17

Transitional Functions	Transitional Location
Having characteristics of both squamous and cuboidal, transitional epithelial cells allow for a hollow organ to distend and stretch. Cell shape is squamous when stretched and cuboidal when relaxed.	Found in bladder, ureters, and urethra

Section 11.2

1. Connective tissue contains various cell types, protein fibres, and ground substance.

2. Loose—is found in almost every body structure, connects epithelial tissue to muscle layers (skeletal or smooth), gives tissues strength and elasticity, stores fat (loose adipose), acts as a filter in spleen and lymph nodes (loose reticular).

   Dense—forms strong attachments between structures such as ligaments, connecting bones within a joint, and allows organs to stretch and recoil, such as within arteries or the lungs.

   Cartilage—a very strong tissue that acts as a smooth surface for joint movement and gives structures shape, such as the ear.

   Bone—forms our skeleton, stores calcium and other minerals, produces red and white blood cells, stores some fat, and protects internal organs.

   Blood—transports oxygen, nutrients, waste products, and hormones throughout the body.

3. Fibroblasts produce collagen, elastin, reticular fibres and ground substance, and are important in wound healing.

   Adipocytes store fat as triglycerides when extra calories are not utilized within the body.

   Mast cells are immune cells that produce histamine when tissue damage is present or during an allergic reaction.

   Reticular cells produce reticular fibres to filter pathogens and old red blood cells in the lymph, spleen, and liver.

   Chondrocytes make up the cartilage tissue.

   Osteocytes make up bone tissue.

   Blood cells include red blood cells that carry oxygen and carbon dioxide, and white blood cells that are involved in the immune response.

4. Collagen is the most abundant protein in the body, and makes tissue very strong.

   Elastin gives tissues the ability to stretch and recoil to their original shape.

   Reticular fibres are thin collagen fibres surrounded by glycoproteins; they form a framework for many tissues and act as a filter in lymph nodes and spleen.

5. Ground substance is important because it is the fluid part of the connective tissue that supports the functions of the cells and the fibres. It is what oxygen, nutrients, and hormones move through between blood vessels and cells of the body.

6. See Tables A.18 through A.28.

TABLE A.18

Connective Tissue	
Loose Areolar Functions	**Loose Areolar Location**
This is the most widely distributed connective tissue; it functions to strengthen, support, and provide elasticity to every body structure.	This type is found around all epithelial tissues, such as the subcutaneous layer and dermis layer of the skin, mucous membranes, around blood vessels, nerves and all organs.

TABLE A.19

Loose Adipose Functions	Loose Adipose Location
Adipose tissue stores fat as triglycerides that can be used to produce energy (ATP), insulate the body, and protect and cushion organs.	This type is found wherever there is areolar connective tissue: in the subcutaneous layer of the skin and around blood vessels, nerves and organs.

TABLE A.20

Loose Reticular Functions	Loose Reticular Location
This type provides structural support and forms part of the basement membranes that surround organs, blood vessels, and muscles. It acts as a filter to remove old red blood cells in the spleen and pathogens in the spleen and lymph nodes.	This type is found in liver, spleen, lymph nodes, bone marrow, blood vessels, all organs, and smooth muscle.

TABLE A.21

Dense Regular Functions	Dense Regular Location
Dense regular connective tissue provides very strong attachments for connecting muscle to bone (tendons), and connecting bone to bone (ligaments). It contains a high amount of parallel strands of collagen, making it very strong. This tissue does not contain blood vessels, so injuries heal very slowly.	This type is found in tendons, ligaments, and aponeuroses (the sheet-like tendons that connect muscle to muscle) (Chapter 14).

TABLE A.22

Dense Irregular Functions	Dense Irregular Location
This type consists primarily of collagen fibres, like dense regular tissue, except that the fibres are not parallel; they are arranged randomly and form a strong sheet-like connective tissue that provides strength when pulled in many directions.	This type makes up the fascia that surrounds muscles, organs such as the heart (pericardium), digestive tract, bones (periosteum), and forms part of the lower dermis region of the skin.

TABLE A.23

Dense Elastic Functions	Dense Elastic Location
This type allows a great amount of stretch and recoil.	It is found in structures that undergo stretching forces, such as arteries, lungs, trachea, bronchial tubes, and vocal cords.

TABLE A.24

Hyaline Cartilage Functions	Hyaline Cartilage Location
This type provides a strong, smooth surface for joint movement. Cartilage does not contain any blood vessels. Hyaline cartilage is the weakest of the three types and can fracture.	It is found at the ends of long bones, ribs, nose, trachea, larynx, bronchi, and is also the majority of the fetal skeleton.

TABLE A.25

Elastic Cartilage Functions	Elastic Cartilage Location
Provides strength as well as elasticity and shape of certain structures.	It is found in the epiglottis, external ear, and the Eustachian tubes (tubes that connect the inner ear to the throat).

TABLE A.26

Fibrocartilage Functions	Fibrocartilage Location
This is the strongest type of cartilage, and it contains thick collagen fibres that connect certain structures together and act as shock absorbers.	It is found in the intervertebral discs, pubic symphysis, and knee joint.

TABLE A.27

Bone Functions	Bone Location
Bones move when muscles contract and produce movement. Bones are important in mineral homeostasis and provide a reservoir for calcium and phosphorus, and they protect internal organs. Red marrow produces red and white blood cells; yellow marrow stores some fat.	The skeleton includes long bones, skull, ribs, vertebrae, and pelvis. See Chapter 13 for details.

TABLE A.28

Blood Functions	Blood Location
Blood transports oxygen, nutrients, hormones, and waste products throughout the body. Red blood cells transport oxygen and some carbon dioxide (Chapters 17 and 18). White blood cells include a variety of immune cells (Chapter 22). Platelets are important for blood clotting.	Blood is located within blood vessels, although red and white blood cells, as well as platelets are produced in red bone marrow.

Section 11.3

1. Movement, production of heat, posture protection, regulation of movement

2. Skeletal, smooth, and cardiac (see Table A.29)

TABLE A.29

Skeletal Muscle	Connected to bones and allows movement; voluntary contraction or unconscious reflexes
	Multinucleated, striated muscle fibres, fast-twitch or slow-twitch, or intermediate
Smooth Muscle	Found in walls of hollow structures; involuntary contractions
	Single nucleus, non-striated muscle fibres, single unit or multi-unit
Cardiac Muscle	Found only in the walls of the heart and controls blood circulation; involuntary contraction and autorhythmic
	One to three nuclei per cell, branched striated muscle fibres connected by intercalated discs that contain gap junctions for signalling

3. Walls of hollow structures, such as blood vessels, lymphatic vessels, stomach, intestines, airways, fallopian tubes, ureters, bladder, uterus, vas deferens, erector pili muscles in the skin, and iris of the eye

4. Single unit smooth muscle is found in sheets that surround visceral organs where contraction happens simultaneously.

5. Multiunit smooth muscle is found in blood vessels, the eye, erector pili, airways, and is made up of multiple cells that contract individually.

6. The special feature of cardiac muscle is that it contains striated, branched muscle fibres that are connected by intercalated discs that contain gap junctions.

7. Intercalated discs that contain gap junctions are important because they allow signalling molecules to pass quickly between the cells, which is necessary to maintain control and simultaneous contraction of the heart chambers for blood to be circulated throughout the body.

Section 11.4

1. Detect or transmit sensations, interpret environmental stimuli, respond to stimuli

2. Muscle cells, glands, or other neurons

3. Neurons and neuroglia

4. • Multipolar—motor neurons
   • Unipolar—sensory neurons
   • Bipolar—located in the olfactory sensory system and the retina

5. Oligodendrocytes (CNS) and Schwann cells (PNS) produce the myelin covering of neurons.

Astrocytes (CNS) and satellite cells (PNS) maintain the ion concentrations surrounding neurons, and astrocytes form the blood-brain barrier. Microglia are immune cells in the brain and spinal cord. Ependymal cells produce cerebral spinal fluid.

Chapter 12 Concept Review Answers

Section 12.1

1. The components of the integumentary system include the skin, nails, hair, and glands.

2. The integumentary system is responsible for protecting the body from the external environment, regulating body temperature, synthesizing Vitamin D, and enabling the excretion of certain waste products.

3. Vitamin D is used by the body to help with certain processes such as bone mineral regulation, hormone synthesis, and immune system regulation.

4. By constricting blood vessels in the dermal layer of the skin, the body restricts the amount of blood that can transfer heat energy to the external environment. Conversely, the body can increase blood flow by dilating the blood vessels in the dermis to cool down the body; the increased blood flow allows for greater heat loss to the external environment. The skin also produces sweat, which, during the process of evaporation allows the body to dissipate heat.

Section 12.2

1. The layers of the epidermis in order from deepest to most superficial are the stratum basale, the stratum spinosum, the stratum granulosum, the stratum lucidum, and the stratum corneum.

2. The primary types of cells found in the various layers of the epidermis include keratinocytes, Langerhans cells, melanocytes, and Merkel cells.

3. See Table A.30.

TABLE A.30

Cell Type	Function
Merkel cells	Mechanoreceptors that allow for touch sensation
Langerhans cells	A type of immune cell that helps prevent infections by the phagocytosis of various pathogens
Keratinocytes	The most numerous cells in the epidermis; produce keratin—a strong, structural, fibrous protein that gives strength to the tissues and helps protect the body from damage
Melanocytes	In response to ultraviolet light, produces melanin, which acts as a photo-protectant, preventing damage to other skin cells caused by UV light

4. See Table A.31.

TABLE A.31

Product	Function
Keratin	Structural protein the gives strength to the tissue and helps prevent damage
Melanin	Photo-protectant pigmentation that prevents damage caused by a cell's exposure to ultraviolet light
Lamellar granules	Organelles that release lipids and proteins into the extracellular matrix that act as a waterproof barrier and help break down bacterial cell walls

5. In order from deepest to most superficial, the layers of the dermis include the reticular layer and the papillary layer. The dermis includes blood vessels, hair follicles, sebaceous glands, sweat glands, neurons, sensory receptors and the arrector pili muscles.

6. See Table A.32.

TABLE A.32

Receptor Name	Function
Merkel cells	Mechanoreceptors that detect touch
Pacinian corpuscles	Detect pressure
Nerve endings	Detect pain
Thermoreceptors	Detect temperature
Meissner corpuscles	Detect touch and vibration

7. The hypodermis acts primarily to connect the skin to the underlying tissues and to store energy in the form of fat.

8. See Table A.33.

TABLE A.33

Molecule	Effect
Melanin	Black or brown pigment secreted by melanocytes
Hemoglobin	Protein present on red blood cells that becomes bright red when bound to oxygen
Carotene	Orange or yellow pigment in many foods (e.g., carrots and egg yolk)

Section 12.3

1. The two types of sudoriferous glands are eccrine and apocrine. Eccrine are found throughout the body, though particularly on the forehead, the palms of the hands, and soles of the feet. Eccrine glands produce sweat and are particularly important for body temperature regulation. Apocrine glands are found primarily in the axilla, the areola of the nipples, and in the genital regions. The sweat produced in apocrine glands is thicker because it contains fatty acids along with water and ions.

2. Sebaceous glands cover the body surrounding hair follicles. Sebaceous glands release sebum which coats the hair and skin, providing lubrication and helping prevent infection.

3. Acne is caused by increased sebum production, which gets trapped in the gland. When bacteria ingest the sebum, the result is a minor inflammatory response.

4. Ceruminous glands are found in the ear canal and produce ear wax (cerumen). Mammary glands are found in the breast skin tissue and produce milk during lactation.

Section 12.4

1. Burns are divided into three classifications based on severity. Superficial partial thickness (first degree) burns involve only the epidermis and are characterized by redness and slight inflammation. Deep partial thickness (second degree) burns involve damage to the epidermis and the papillary region of the dermis. Full thickness (third and fourth degree) burns result in damage through the epidermal and dermal layers and often include damage to the subcutaneous layer and possibly underlying muscle.

2. See Table A.34.

TABLE A.34

Type of Cancer	Cells Affected
Basal cell carcinoma	Basal cells
Squamous cell carcinoma	Keratinocytes
Melanoma	Melanocytes

3. See Table A.35. Risk factors include the following:

TABLE A.35

Type of Cancer	Risks
Basal cell carcinoma	Rarely spreads, often appears as a raised smooth bump that can sometimes ulcerate
Squamous cell carcinoma	Can spread quickly, may bleed and ulcerate
Melanoma	Most dangerous form of skin cancer, caused by uncontrolled growth of melanocytes

- Exposure to ultraviolet light
- Chronic inflammation
- Continual wound healing
- Use of immunosuppressant medication
- Certain cosmetics, lotions, or sunscreens

Melanoma can be identified using the mnemonic ABCDE:

- A = Asymmetrical
- B = Borders irregular
- C = Colour dark and inconsistent
- D = Diameter greater than 6 mm
- E = Evolving over time

4. Wrinkles, sagging skin, age spots, increased visible blood vessels, skin tags, and increased dryness.

5. Excessive sun exposure, including excessive use of tanning beds; increased exposure to toxins such as cigarette smoke and pollution; as well as lack of sleep, lack of exercise, and poor nutrition.

6. Eat healthy diet, including adequate amount of vitamins A, C, and E; exercise to increase growth hormone production; get enough sleep; don't smoke; avoid toxins and excessive sun exposure.

Chapter 13 Concept Review Answers

Section 13.1

1. Total is 206: 80 bones in the axial skeleton, 126 in the appendicular skeleton.

2. The axial skeleton is made up of the skull, vertebrae sternum, and ribs. The pelvis, shoulders, and limbs make up the appendicular skeleton.

3. The skeletal system is responsible for providing structure; allowing movement; protecting the brain, heart, lungs, and spinal cord; producing blood cells; storing mineral and fat; and hearing.

Section 13.2

1. Bones are made up of connective tissue, bone marrow, blood vessels, nerves, and epithelial tissue. Bone tissue consists of cells, extracellular matrix collagen, and various minerals.

2. Osteoblasts are responsible for bone tissue production. Osteocytes maintain bone tissue. Osteoclasts break down the mineral portion of the extracellular matrix to release calcium, magnesium, and phosphate.

3. Minerals provide the bone with strength, and collagen provides flexibility.

4. Five categories of bone types and examples:
   - sesamoid bones—patella
   - long bones—humerus, femur, etc.
   - short bones—wrists and ankles

   - irregular bones—vertebrae and facial bones
   - flat bones—ribs, sternum, scapulae, etc.

Section 13.3

1. There are four main sutures: (1) coronal, between the frontal bone and the anterior portion of the parietal bones; (2) sagittal, between the medial edges of the parietal bones; (3) lambdoid, between the occipital bone and the posterior edges of the parietal bones; (4) two squamous sutures, between the temporal bone and the lateral parietal bone, as well as the edges of the occipital and sphenoid bones, on each side of the skull.

2. The condylar process of the mandible sits in a grove in the temporal bone, called the mandibular fossa, forming the temporomandibular joint (TMJ).

3. Fontanels are "soft spots" in an infant skull, where the bones have not yet fused together.

4. Scoliosis is an S-shaped lateral curve of the spinal column; kyphosis is an over curvature of the thoracic vertebrae, also called hunchback; and lordosis is an over-curvature of the lumbar region, also called sway back.

Section 13.4

1. The scapula, clavicle, humerus, radium, ulna, carpals (8), metacarpals (5), and phalanges (14) are all found in the upper limbs.

2. The pelvis, femur, tibia, fibula, tarsals (7), metatarsals (5), and phalanges (14) are all found in the lower limbs.

3. The ischia (singular ischium) are the pelvic bones we sit on.

4. The area bound by the sacrum and pubis bones and extending to the pubic symphysis is called the true pelvis.

5. The patella protects the knee joint from insult.

Section 13.5

1. Hormones that increase bone growth include calcitonin, growth hormone, IGF, thyroid hormones, testosterone, estrogen, and insulin. The only hormone that decreases bone growth and causes bone minerals to be released is parathyroid hormone.

2. Osteoblasts are the osteocytes that promote bone growth and production.

3. Osteoclasts are targeted by parathyroid hormone to increase the breakdown of bone.

4. The key minerals in the bone matrix are calcium, magnesium, and phosphorus.

5. Vitamin D improves the absorption of calcium in the digestive tract. Vitamin A is used to regulate

osteoclasts, and Vitamin C is used to help move collagen from the cells to the ECM.

Section 13.6

1. The main classifications of fractures are as follows:

   - transverse fracture—break across the transverse section of a bone
   - spiral fracture—a twisted fraction
   - commuted fracture—crushed bone
   - stress fracture—typically longitudinal cracks, but not full breaks, in a bone
   - greenstick—a fracture where the bone has bent significantly prior to breaking on one side; most commonly found in children

2. During healing from a fracture, phagocytic cells consume the dead cells, chondrocytes form cartilage to temporarily replace the broken section, osteoblasts create new bone, and osteoclasts assist with remodelling the new bone.

3. Diet and exercise are the primary factors involved in preventing osteoporosis.

4. Osteoporosis occurs in menopausal women due to the decreased production of hormones.

Section 13.7

1. Structural classifications include fibrous, cartilaginous, and synovial. Functional classifications refer to syndesmosis or no movement; amphiarthrosis or slight movement; and diarthrosis, meaning highly movable.

2. - Fibrous—for example, skull sutures, teeth and sockets, connections between radius and ulna or tibia and fibula
   - Cartilaginous—for example, joints between ribs and sternum, pubic symphysis, joint between the manubrium and the sternum, joints between the vertebrae and the intervertebral discs
   - Synovial—for example, shoulders, knees, elbows

3. Syndesmosis joints are immovable and include joints such as the suture joints. Amphiarthrosis joints have limited movement and include joints such as the connections between the radius and the ulna. Diarthrosis joints are fully movable and include most joints in the limbs and between phalanges.

4. Synovial fluid provides nutrients, prevents friction between the ends of the connecting bones, and absorbs shock to the joint.

5. Bursa sacs encapsulate synovial fluid, containing it within an area and creating a cushion between bones or tendons and muscle.

6. - Plane joint—includes the carpals of the wrist or tarsals of the ankles
   - Hinge joint—include the knees, ankles, and elbows
   - Pivot joint—includes the radioulnar joints and the atlanto-axial join between C1 and C2 that allow for rotational movement of the head
   - Condyloid—includes the wrists and elbows
   - Saddle joint—joint between the proximal phalange thumb and the metacarpal
   - Ball-and-socket joint—includes the hips and the shoulders

7. Ball-and-socket joints are able to produce angular movements.

8. Pivot joints allow for rotational movement.

9. Pronation and supination are opposing movements that allow a greater degree of dexterity and flexibility.

Chapter 14 Concept Review Answers

Section 14.1

1. The three types of muscle are skeletal, smooth, and cardiac.

2. Muscles function includes the following:
   - Producing body movements
   - Stabilizing body positions and posture
   - Movement of internal organs, such as the constriction of the blood vessels, digestive tract, fallopian tubes, bronchial tubes, and ureters
   - Regulating capillary blood flow and peristalsis
   - Producing heat

Section 14.2

1. Tendons connect muscle to bone.

2. The fascia is dense connective tissue that envelops the muscles and connects them to the subcutaneous layer of the skin.

3. Muscles are held by the endomysium (around each fibre), the perimysium (around each fascile), and the epimysium.

4. Thick and thin filaments—myofibrils—sarcolemma—endomysium—fascicle—perimysium—epimysium

Section 14.3

1. See Table A.36.

2. Lifting the knee up uses the hip flexors, extension of the leg uses the quadriceps, pushing the leg back uses the gluteus maximus, pushing the ground with the toes uses the calves, bringing the leg up behind with the knee bent uses the hamstrings.

TABLE A.36

Muscle	Origin	Insertion	Action
occipitofrontalis frontal belly	epicranial aponeurosis	skin superior to orbit	raises eyebrows
occipitofrontalis occipital belly	occipital and temporal bone	epicranial aponeurosis	draws scalp backward
orbicularis oculi	wall of orbit	eyelid	closes eyes
orbicularis oris	muscles around mouth	skin around mouth	closes and protrudes lips as in kissing
zygomaticus major	zygomatic bone	skin at corners of mouth	draws edges of mouth up as in smiling
platysma	fascia over deltoid	mandible, skin of lower face	depresses mandible as in pouting
masseter	zygomatic arch	mandible	closes the mouth as in chewing
temporalis	temporal bone	mandible	elevates mandible as in chewing
sternocleidomastoid	manubrium and clavicle	Occipital bone and mastoid process	turns head side to side
trapezius	occipital bone and thoracic vertebrae	clavicle and scapula (acromion anteriorly, and spine posteriorly)	raises shoulders as in shrugging

Section 14.4

1. Actin and myosin are responsible for muscle contraction.

2. Troponin and tropomyosin are the two regulatory proteins. Troponin binds with calcium, causing tropomyosin to roll off the myosin-binding sites on the actin filament.

3. Thin filaments and thick filaments slide across each other during contraction. Neither the thick or thin filament shortens, but as the filaments slide across each other, the whole sarcomere shortens. The myosin heads bind to actin, then pivot and pull the actin toward the M line. Each myosin head then pivots toward the M line, pulling the actin of the thin filament toward the M line so that the thick and thin filaments slide across one another.

4. During sarcomere shortening, the H zone and I bands contract while the A bands remain unchanged.

5. Calcium is stored in the sarcoplasmic reticulum.

6. The action potential from the movement of sodium into the muscle causes the release of calcium from the sarcoplasmic reticulum.

7. The following describes the four steps of the cross-bridge cycle:

   a. Myosin is unbound from the actin and contains an ATP molecule in the ATP-binding site on the myosin head. The ATP hydrolyzes, the ADP and phosphate stay bound, and the action of the bond breaking in the ATP energizes the myosin.

   b. The phosphate molecule leaves the myosin, and the energized myosin binds to the myosin-binding site on the actin proteins, as long as calcium is present to bind troponin and move the tropomyosin off the binding sites. The binding of myosin to actin forms the cross-bridge.

   c. The ADP molecule leaves the myosin, and the myosin head moves toward the M line; this is called the power stroke. After the power stroke, the myosin remains bound to the actin.

   d. A new ATP molecule binds to the myosin head, which allows the myosin to be released from the actin. Then the cycle repeats, with ATP hydrolysis in step 1.

8. Muscles use ATP to bind myosin and cause power stroke; they also require ATP for the active transport of calcium back into the sarcoplasmic reticulum.

9. The load required determines the force of contraction and, therefore, how many fibres are activated.

10. Muscle tone prevents muscles from overstretching and helps maintain both balance and posture.

11. Sodium causes action potentials in both nerves and muscles.

12. Calcium causes acetylcholine to be released into the synaptic cleft.

13. Acetylcholine (Ach) causes the sodium channels to open in the muscle membrane, stimulating an action potential.

14. After stimulating an action potential, acetylcholine is broken down and recycled by an enzyme called acetylcholinesterase.

15. The following order of events, beginning with stimulation by the motor neuron, occurs during muscle contraction:

    a. Action potential travels down motor neuron to axon terminal.

    b. Action potential triggers movement of calcium into the axon terminal.

    c. Calcium triggers release of Ach neurotransmitters into synaptic cleft.

    d. Ach binds to specific receptors on the sarcolemma that trigger the opening of Na^+ channels.

    e. Action potential occurs in muscle cell membrane as Na^+ ions cross the membrane from the T-tubules and enter the sarcoplasm.

    f. Action potential in the sarcolemma causes the sarcoplasmic reticulum to release Ca^{2+} ions into the sarcoplasm.

    g. Ca^{2+} ions bind to troponin, which moves tropomyosin off the myosin-binding sites of the actin proteins.

    h. ATP is bound to myosin in the relaxed state. Once an action potential occurs and the myosin-binding sites on the actin are available, ATP hydrolyzes, releasing the phosphate group, and myosin binds to actin—cross-bridge formation.

    i. ADP leaves the myosin, causes the power stroke, and sarcomeres shorten.

    j. A new ATP molecule binds to myosin to release myosin from actin; the cycle will continue as long as the motor neuron stimulates the muscle cell, then the muscle relaxes.

Section 14.5

1. The primary difference between aerobic and anaerobic respiration is the presence of oxygen.

2. Creatine phosphate is stored as creatine kinase and used as a rapid source of energy.

3. Two ATP are used as the net product of anaerobic respiration (glycolysis). In total, four are created in the process, but two are required, resulting in a gain of only two ATP.

4. Approximately 36 ATP result from aerobic respiration starting with one glucose molecule.

5. Aerobic respiration

Section 14.6

1. See Table A.37.

2. The following exercises use the following types of muscle fibre:

    a. squatting 50 lb (fast)

    b. jogging for 1 hour (slow)

    c. running 1500 m (intermediate)

    d. sitting in a chair with good posture (slow)

    e. typing (slow)

    f. eating (slow)

    g. downhill skiing (fast)

    h. walking up a flight of stairs (intermediate)

    i. running long jump (fast)

3. The two primary factors in maintaining a healthy body weight are diet and exercise.

4. Exercise also helps with increasing bone density, preventing osteoporosis, increasing balance and coordination, decreasing the risk of fracture, improving mood (affect), decreasing inflammation in the body, and preventing type-2 diabetes and cardiovascular disease.

TABLE A.37

Characteristic	Slow Oxidative Fibre	Fast Oxidative Fibre	Fast Glycolytic Fibre
Speed of contraction	Slow	Fast	Fast
Resistance to fatigue	High	Intermediate	Low
Oxidative phosphorylation capacity	High	High	Low
Enzymes for anaerobic glycolysis	Low	Intermediate	High
Mitochondria	Many	Many	Few
Capillaries	Many	Many	Few
Colour of fibre	Red	Red	White
Glycogen content	Low	Intermediate	High
Diameter of fibre	Smallest	Intermediate	Largest
Strength of contraction	Weakest	Intermediate	Strongest

Chapter 15 Concept Review Answers

Section 15.1

1. The nervous system is divided into (1) the central nervous system (CNS), composed of the brain and spinal cord; and (2) the peripheral nervous system (PNS), composed of the cranial nerves, spinal nerves, and the neurons outside the CNS.

2. The visceral organs, blood vessels, endocrine glands, and exocrine glands are innervated by the autonomic nervous system.

3. The submucosal plexus stimulates the release of acids, digestive enzymes, and other substances required by the digestive system; the myenteric plexus innervates the smooth muscle needed for peristalsis.

4. The somatic nervous system is primarily involved in voluntary muscle control.

5. See Table A.38.

TABLE A.38

Receptor	Stimulus
Chemoreceptor	Chemical concentrations, such as hormones or neurotransmitters
Osmoreceptor	Osmolarity, concentration of water and ions
Baroreceptor	Blood pressure
Photoreceptor	Light
Mechanoreceptor	Stretch and physical movement
Proprioceptor	Body position
Nociceptor	Pain
Thermoreceptor	Temperature
Tactile receptors in the skin	Touch, pressure, vibration

Section 15.2

1. The two categories of cells that make up nervous tissue are neurons and glial cells.

2. Dendrites extend from the cell body and detect various stimuli by means of membrane receptors. (See Table A.38.) Cell bodies contain the organelles required for typical cell functions such as ATP production and protein synthesis. Axons are the main extensions from the cell body, and serve primarily as a conduit for action potentials ending at the axon terminal. The axon terminal is where neurotransmitters are released and trigger the receptors for muscles, glands, or other neurons.

3. Neurons can be either multipolar, bipolar, or unipolar.

4. See Table A.39.

TABLE A.39

Type of Glial Cell	Function
Astrocyte	• Forms the blood-brain barrier • Regulates ion concentrations • Stores glycogen and releases glucose as needed by the neurons • Removes waste products • Produces scar tissue following insult to the CNS as part of the healing process
Oligodendrocyte	Forms the myelin sheathe around the axons of the neurons
Microglia	Immune cells specifically for the nervous system
Ependymal cell	Secretes cerebrospinal fluid used to provide nutrients to the brain
Satellite cell	Functions similar to astrocytes, but found only in the PNS
Schwann cell	Forms the myelin sheathe around the axons of afferent and efferent neurons in the PNS

Section 15.3

1. In a polarized resting cell, the sodium and chloride are found in higher concentration outside the cell, and potassium is in higher concentration inside the cell.

2. Ion channels that open or close in response to a chemical are considered chemically gated, and those that open or close in response to a change in the charge of a membrane are referred to as voltage-gated.

3. The movement of sodium into the cell causes depolarization as the membrane potential is lost. Repolarization takes place once the sodium channels have closed and potassium leaves the cell.

4. It is important because it results in a refractory period and ensures that the action potential remains unidirectional.

5. The movement of calcium into the cell triggers the exocytosis of neurotransmitters into the synapse.

6. Saltatory—from Latin *saltare*, "to jump"—conduction occurs when the action potential bypasses segments of the axon, resulting in depolarization at the nodes of Ranvier and speeding transmission along the axon.

7. Myelination of the axon and an increased diameter of the axon result in faster transmission along the axon.

8. a. Dopamine plays a large role in the reward pathway of the brain and is largely responsible for addictive behaviours.

b. Serotonin elevates mood/affect.

c. Acetylcholine stimulates muscle contraction.

Section 15.4

1. Pia mater, arachnoid mater, and dura mater.

2. The subarachnoid space contains the cerebrospinal fluid (CSF) and blood vessels; the epidural space contains blood vessels, lymphatic vessels, nerve roots, and fats.

3. White matter is composed of myelinated axons. Grey matter is composed of unmyelinated interneurons, axon terminals, cell bodies, and dendrites.

4. Sensory (afferent) information enters the spinal cord from the dorsal roots and travels up the ascending tracts of the dorsal horn. Motor (efferent) information travels through the ventral horn through the descending tracts and leaves the spinal cord via the ventral roots.

5. The outermost layer of the nerve is the epineurium and is composed of connective tissue. Inside the nerve are bundles of axons referred to as fascicles. Each fascicle is composed of groups of axons wrapped in the endoneurium.

6. There are a total of 31 pairs of spinal nerves. Eight pairs are found in the cervical region, 12 in the thoracic, five in the lumbar, five in the sacral, and one pair of coccygeal nerves.

7. See Table A.40.

TABLE A.40

Nerve	Vertebral Region	Tissue Innervated
Phrenic	Cervical	Diaphragm
Axillary	Brachial	Deltoid, rotator cuff
Musculocutaneous	Brachial	Biceps, brachialis
Median	Brachial	Forearm, thumb, and first three fingers
Radial	Brachial	Thumb and first finger
Ulnar	Brachial	Last two fingers
Obturator	Lumbar	Hip and abductors
Pudendal	Sacral	External genitalia and sphincters of both bladder and rectum
Femoral	Lumbar	Quadriceps and hip flexors
Sciatic	Sacral	Hamstrings
Tibial	Sacral	Calves and foot
Fibular	Sacral	Tibialis anterior
Superior and inferior gluteal	Sacral	Gluteal muscles

Section 15.5

1. A monosynaptic reflex has a single synapse in the spinal cord linking a sensory neuron to a motor neuron. Polysynaptic reflexes have multiple synapses from sensation to response.

2. Reflexes reduce time and cognitive energy required for certain tasks, whether they help prevent damage to the body or facilitate certain tasks.

3. Tapping on the quadriceps muscle causes the muscle to stretch. In response to the stretch, the patellar reflex causes contraction of the quadriceps muscles and the leg extends.

4. Sensory neurons detect a stimulus (heat, pain, etc.) and send the message to interneurons in the spinal cord. A motor neuron then sends a response to the appropriate muscle cell, causing it to contract.

Section 15.6

1. The brain is protected by the blood-brain barrier, skull bones, meninges, and CSF.

2. Ependymal cells produce cerebrospinal fluid in the choroid plexus located in each of the ventricles of the brain.

3. At the arachnoid villi, CSF is reabsorbed into the bloodstream into the superior sagittal sinus.

4. The four regions are the diencephalon, the brainstem, the cerebellum, and the cerebrum. The diencephalon consists of the thalamus, hypothalamus, and pineal gland. The brainstem consists of the midbrain, pons, and medulla oblongata. The cerebellum is located at the base of the brain, posterior to the brainstem. The cerebrum consists of the cerebral cortex (the outermost portion) and the subcortical region located beneath the cortex.

5. The thalamus operates as a relay station through which signals pass to various parts of the brain to be properly interpreted. The pineal gland is responsible for maintaining circadian rhythms and is responsible for the production of the hormone melatonin.

6. The reticular activating system is located in the central core of the brainstem and is responsible for consciousness and the ability to filter extraneous stimuli.

7. Subtantia nigra in the midbrain are responsible for regulating motor movements.

8. The pons is responsible for bridging signals from the motor cortex of the cerebrum to the cerebellum and, in turn, sending signals from the cerebellum to the muscles in a smooth manner. The pons also plays a role in voluntary breathing control.

9. The medulla oblongata is primarily responsible for control of the respiratory and circulatory systems.

10. The cerebellum is responsible for coordinating, learning, and remembering complex motor functions.

11. Basal nuclei regulate skeletal muscle, posture, and complex behaviours. The amygdala is involved in emotions, the memories linked to emotions, and complex social and sexual interactions.

12. The limbic system plays a major role in emotions, learning, memory, and social interactions.

13. Lobe and region:

    a. temporal lobe, auditory cortex

    b. parietal lobe, somatosensory cortex

    c. frontal lobe, Broca's area

    d. frontal lobe, pre-frontal cortex (also called the pre-motor cortex)

    e. parietal lobe, left side

    f. occipital lobe, visual cortex

    g. temporal lobe, hippocampus

    h. frontal lobe

    i. temporal lobe, Wernicke's area

    j. insula

14. See Table A.41.

Section 15.7

1. The somatic nervous system innervates skeletal muscle; the autonomic nervous system innervates glands, smooth muscle, and cardiac muscle.

2. Acetylcholine is released by the preganglionic nerve at the synapse.

3. Norepinephrine and epinephrine are released by the sympathetic postganglionic neuron, and acetylcholine is released by the parasympathetic postganglionic neuron.

4. Five tissues are (1) smooth muscle in bronchi, (2) cardiac muscle, (3) smooth muscle in peripheral blood vessels, (4) sweat glands, and (5) kidneys.

5. The primary role is stimulation of processes used for digestion and relaxation.

6. The sympathetic nervous system triggers the breakdown of glycogen and fat; glycogen production and fat storage are promoted by the parasympathetic system.

7. Preganglionic nerves leave the CNS either through a cranial or spinal nerve.

8. The adrenal gland produces hormones that cause the same effects as the autonomic nervous system, mimicking a sympathetic response but having more long-lasting effects.

Chapter 16 Concept Review Answers

Section 16.1

1. The nervous system and the endocrine system are similar because they communicate with various tissues in the body to maintain homeostasis. They are different because the nervous system uses neurotransmitters as the signalling molecule and the endocrine system uses hormones.

TABLE A.41

Nerve	Function	Origin
I Olfactory	Sense of smell	Cerebrum
II Optic	Vision	Cerebrum
III Oculomotor	Movement of eyelid, eyeball, lens, pupil constriction	Midbrain
IV Trochlear	Movement of eyeball	Midbrain
V Trigeminal (ophthalmic, maxillary, and mandibular branch)	Sensations in head, face, and jaw; motor control of chewing	Pons
VI Abducens	Movement of eyeball	Pons
VII Facial	Taste, facial expression, tears, salivation, sensations	Pons
VIII Vestibulocochlear	Equilibrium and hearing	Lateral to facial nerve
IX Glossopharyngeal	Taste, swallowing, speech, salivation	Medulla oblongata
X Vagus	Taste, sensations from pharynx, swallowing, coughing, voice, GI tract smooth muscle contraction, heart rate reduction, digestive secretion	Medulla oblongata
XI Accessory	Swallowing and movement of head and shoulders	Medulla oblongata
XII Hypoglossal	Speech and swallowing (tongue muscles)	Medulla oblongata

2. Endocrine glands secrete hormones into the body. Exocrine glands are epithelial cells that excrete hormones either to the surface of the body or into a hollow cavity or duct in the body.

3. See Table A.42.

TABLE A.42

Gland	Location
Pituitary gland	Base of the hypothalamus
Pineal gland	Centre of the brain
Thyroid	Anterior to the trachea
Parathyroid	On the surface of the thyroid gland
Thymus	Between the heart and the sternum
Pancreas	Inferior to the stomach
Adrenal glands	Superior to the kidneys
Testes	In the scrotum
Ovaries	In the abdomen

4. General functions of the endocrine system are as follows:

- Homeostatic balance
- Regulation of growth, metabolism, and energy production
- Reproductive functions, such as secondary sex characteristics, lactation, childbirth, development of sperm or eggs
- Stress response
- Regulation of digestion
- Regulation of circadian rhythms such as sleep-wake cycles

Section 16.2

1. Hormones target specific receptors either on the surface of the cell or within the target cell.

2. Water-soluble hormones cannot diffuse through the cell membrane and, therefore, target receptors on the surface of the cell. These membrane receptors then make use of an internal secondary messenger system to alter the cell activity. Fat-soluble hormones can diffuse through the cell and, as a result, can easily bind directly to the target receptor located inside the cell.

3. Secondary messengers relay signals from the membrane receptor to a target molecule inside the cell.

4. Hormones affect transcription and function as transcription factors.

5. Amino acids, small peptides, and proteins make up water-soluble hormones.

6. Cholesterol makes up fat-soluble hormones, called steroids: except does not make up the thyroid hormones.

7. See Table A.43.

TABLE A.43

Water Soluble	Fat Soluble
Catecholamines	Thyroid hormones
Pancreatic hormones	Cortisol
Pituitary hormones	Aldosterone
Hypothalamic hormones	Androgens
	Calcitriol

8. Thyroid hormones are fat-soluble and are made from tyrosine.

Section 16.3

1. Oxytocin and anti-diuretic hormone (ADH) are made in the hypothalamus and released by the posterior pituitary gland.

2. The hypothalamus produces releasing hormones or inhibiting hormones which, in turn, signal the anterior pituitary to produce specific hormones.

3. Oxytocin causes contractions during labour and promotes the release of milk from the mammary glands. ADH acts on the distal tubule and collecting ducts of the kidneys to increase water reabsorption.

4. Growth hormone causes bone growth, protein synthesis (particularly in skeletal muscle), increased metabolism, and stimulation of the immune system. During adulthood, growth hormone maintains healthy cell division to replace old and damaged cells.

5. Excess growth hormone can cause gigantism, and a deficit of growth hormone can result in dwarfism.

6. The thyroid, adrenal glands, and gonads are stimulated by TSH, ACTH, and FSH and LH, respectively.

7. The primary function of prolactin is to stimulate milk production in the mammary glands.

Section 16.4

1. Follicular cells produce T3 and T4 (triiodothyronine and thyronine) and the parafollicular cells produce calcitonin.

2. T3 contains three iodine atoms, while T4 contains four. Most T4 is converted to T3 in the body as T3 is the more active hormone.

3. Tyrosine is used with iodine and other micronutrients to create both T3 and T4.

4. Iodine and selenium are both required for thyroid hormone production. Iodine is found in iodized table salt, and selenium is found in foods such as nuts, fish, wheat bran, pork, shellfish, eggs, and mushrooms.

5. Thyroid hormones are not water soluble and must be transported through the blood attached to plasma proteins.

6. Thyroid hormones have the following functions:
   - Regulating oxygen use by all cells during cellular respiration
   - Increasing basal metabolic rate (BMR) by increasing the production of ATP from carbohydrates and fats
   - Increasing the effects of epinephrine and norepinephrine
   - Stimulating the production of growth hormone
   - Increasing heart rate, breathing rate, and cardiac output

7. Hypothyroidism leads to decreased energy levels and body temperature, and to weight gain, fatigue, and depression. Hyperthyroidism causes insomnia, nervousness, irritability, increased heart rate, weight loss, and protruding eyes.

8. The most common causes of hypothyroidism are iodine or selenium deficiencies (sometimes secondary to anorexia). The most common cause of hyperthyroidism is an autoimmune disease called Graves' disease.

9. Calcitonin inhibits osteoclast activity—prevents breakdown of bone matrix—thereby preventing the increase of calcium levels in the blood.

10. Parathyroid hormone stimulates osteoclasts to release calcium, and kidney cells to increase reabsorption of Ca^{++}, Mg^{++}. It also increases the excretion of HPO_4^{2-} by the kidney and increases the production of active vitamin D (calcitriol) by the kidneys, which then increases the absorption of calcium in the small intestines.

11. Low calcium levels cause increased parathyroid hormone, which stimulates osteoclasts to break down bone to increase blood calcium levels.

12. Dairy products, leafy green vegetables, nuts, beans, and fish all contain calcium.

Section 16.5

1. The adrenal cortex releases cortisol, androgens, and aldosterone. The adrenal medulla releases epinephrine and norepinephrine.

2. The zona glomerulosa secretes mineralcorticoids, mainly aldosterone; the zona fasciculate secretes glucocorticoids, mainly cortisol; and the zona reticularis secretes androgens.

3. The most common mineralcorticoid is aldosterone, the most common glucocorticoid is cortisol; and the most common androgen is dehydroepiandrosterone (DHEA).

4. The ANS stimulates the adrenal medulla to secrete epinephrine and norepinephrine.

5. Aldosterone causes more sodium to be reabsorbed back into the bloodstream, which increases blood pressure and causes nephrons to secrete potassium ions.

6. Cortisol stimulates glycogen and fat breakdown in the liver and fat cells and, therefore, increases blood sugar. It also inhibits the immune response. Cortisol is secreted as part of the stress response.

7. Cushing's syndrome results from too much cortisol and causes the following symptoms: hyperglycemia, muscle weakness, osteoporosis, type-2 diabetes, immune suppression, poor wound healing, thin skin, bruising, depression, moodiness, fat redistribution from limbs to trunk and face, "moon face," "buffalo hump," hair loss, decreased fertility, fatigue, insomnia, menstrual irregularities, water retention, and high blood pressure.

8. Androgens contribute to the development of secondary sex characteristics in males along with sperm production. In females, androgens contribute to libido and are converted into estrogen.

9. Epinephrine and norepinephrine are released as part of the stress response.

10. Epinephrine causes the following symptoms:
    - bronchiole dilation
    - pupil dilation
    - increased blood pressure
    - vasodilation and increased blood flow in liver, skeletal muscles, cardiac muscle, and adipose tissue
    - vasoconstriction in digestive tract
    - increased blood sugar

Section 16.6

1. The exocrine functions include the release of sodium bicarbonate and enzymes into the small intestine.

2. The alpha cells of the pancreas secrete glucagon; the beta cells release insulin.

3. Insulin acts on all cells of the body to increase glucose uptake.

4. The glycemic index refers to the variable amount of insulin triggered by the consumption of different foods.

5. Glucagon is secreted between meals and acts to increase blood sugar by causing the body to break down glycogen and triglycerides in the liver.

6. Glucose is stored as glycogen in both the liver and muscle cells. Fats, glucose, and proteins are stored in both the liver and adipose tissue.

7. Type-1 (or insulin-dependent diabetes mellitus) is an autoimmune disease that leads to a lack of insulin production. Type-2 (or non-insulin-dependent diabetes mellitus) is the result of insulin resistance caused by constant high blood insulin levels.

8. Non-insulin dependent diabetes mellitus (type-2) can be prevented through regulation of diet and exercise.

Section 16.7

1. Testosterone, estrogen, and progesterone are all stimulated by luteinizing hormone.

2. Aromatase

3. Testosterone causes increased muscle mass; increased bone density; bone maturation; development of sex organs and secondary sex characteristics, such as facial and axillary hair growth, and deepening voice. Estrogen causes the LH surge leading to ovulation, growth of uterine lining, increased bone formation, increased blood clotting, and many other functions. Progesterone supports pregnancy and plays many roles in fetal development.

4. FSH promotes spermatogenesis, oogenesis, and egg development.

5. Ovulation is caused by an increase in luteinizing hormone, which is called the LH surge.

Chapter 17 Concept Review Answers

Section 17.1

1. The cardiovascular system functions to transport gases to and from the lungs, distribute nutrients to the tissues, and carry waste to the kidneys. It also delivers defensive cells and signalling molecules, regulates blood pressure, and assists with body temperature regulation.

2. The main components of the cardiovascular system include the heart, blood vessels, and the blood.

3. The pulmonary system carries deoxygenated blood in the arteries to the lungs, whereas the systemic circuit carries oxygenated blood from the heart to the body tissues.

Section 17.2

1. Cardiac muscles tissue is composed of cardiac myocytes, and it is held together by intercalated discs containing desmosomes and gap junctions.

2. Desmosomes provide a physical connection between cells, and gap junctions provide the chemical connection that permits the action potential to move from one cardiac cell to another. This ability to transmit the action potential allows for the coordinated contraction of the heart.

3. Intercalated discs contain the desmosomes and gap junctions.

4. The heart contains specialized cells called the SA node that spontaneously depolarize, acting as a "pacemaker" for the heart.

Section 17.3

1. The external features of the heart include the coronary sinus, which receives blood from the inferior and superior vena cavae. The blood is pumped from the heart through either the pulmonary arteries or into the aortic arch. The heart is surrounded by a membrane structure called the pericardium, which functions to reduce friction during contraction.

2. The coronary arteries are located in the coronary sulcus: a groove between the right and left chambers of the heart.

3. The upper chambers that receive blood from the two circuits are the atria. The right atrium receives blood from the inferior and superior vena cavae of the systemic circuit, and the left atrium receives blood from the pulmonary vein. The lower chambers, called the ventricles, pump blood through the circuits; the right ventricle is connected to the pulmonary arteries, and the left ventricle supplies the aorta.

4. The pericardium attaches to the diaphragm and the great vessels, which secures the heart in the mediastinum.

5. The myocardium and the endocardium are the middle and innermost layers of the heart, respectively.

6. The four valves of the heart are (1) the mitral (bicuspid) valve, also called the left atrioventricular valve; (2) the tricuspid valve, also called the right atrioventricular valve; (3) the aortic semilunar valves; (4) the pulmonary semi-lunar valves.

7. The atrioventricular valves (AVs) are connected to the papillary muscles by chordae tendinae. The chordae tendinae prevent the valves from opening in the wrong direction and thereby prevent backflow during contraction.

8. The first sound—lub—is caused by the sudden closing of the AV valves. This is followed by the second heart sound, dub, which results from the closing of the semilunar valves.

9. The valves are forced open or closed by a pressure gradient created by the force of contraction of the heart and the resisting pressure in the vessels.

Section 17.4

1. Right atrium—tricuspid valve—right ventricle—pulmonary arteries—pulmonary arterioles—pulmonary capillaries—pulmonary venules—pulmonary veins—left atrium—mitral (bicuspid) valve—left

ventricle—aorta—arteries—arterioles—capillaries—venules—veins—vena cava—right atrium

2. The "atrial kick" provided by atrial systole accounts for an increase of 10–20 mL of blood per contraction.

3. Increased pressure in the atria during atrial systole causes the AV valves to open, allowing blood to move into the ventricles.

4. Increased pressure in the ventricles during ventricular systole causes the SL valves to open, allowing blood to move into the aorta and pulmonary arteries.

5. Systole is the contraction of the cardiac muscle. Diastole is the relaxation of the same muscle.

Section 17.5

1. The sinoatrial node, atrioventricular node, bundle of His, bundle branches and purkinje fibres make up the conduction system of the heart.

2. Autorhythmicity is the heart's ability to regulate its own depolarization spontaneously.

3. Sodium, calcium and potassium

4. The pacemaker potential does not have a stable resting membrane potential. Therefore it slowly depolarizes after each action potential, unlike the action potential found in other cells, such as neurons, that depolarize only if stimulated.

5. During the P wave of the ECG, the atria are contracting. During the QRS, the ventricles are contracting and the atria are relaxing; note that the atrial relaxation is hidden by the more prominent electrical impulse in the ventricles. During the T wave, the ventricles are relaxing.

6. The first heart sound is caused by the closing of the AV valves. The second heart sound is caused by the closing of the semilunar valves.

7. Arrhythmias are caused by malfunctions of the conduction system of the heart.

8. During the first isovolumetric phase, the blood volume of the ventricles is not changing as the AV valves are closed and the semilunar valves have not yet opened. During the second isovolumetric phase, the ventricular volume isn't changing because the semilunar valves are closed and the AV valves have not yet opened.

9. Most of the blood enters the ventricles in the time between the relaxation of the ventricles and the contraction of the atria.

10. End diastolic volume is the volume of blood in the ventricles immediately prior to ventricular contraction. The end systolic volume is the volume of blood remaining in the ventricles immediately following ventricular contraction.

Section 17.6

1. The potential increase in cardiac output in response to stimuli is called the cardiac reserve, and it is usually two to three times higher than the resting cardiac output.

2. 130 mL − 50 mL = 80 mL (stroke volume) 80 mL/beat × 75 beats/min = 6000 mL/min OR 6 L/min

3. The sympathetic nervous system increases the rate of cardiac contraction by releasing norepinephrine. Acetylcholine released by the parasympathetic nervous system decreases the heart rate.

4. Epinephrine and thyroid hormones both function to increase heart rate, thereby increasing total cardiac output.

5. Increased vascular resistance leads to increased afterload and a decrease in stroke volume.

6. Preload is the blood volume returning to the atria from the circulation. Afterload is the resistance created by the vasculature.

7. The Frank-Starling law dictates that the stretch of the myocardial fibres is directly proportional to the force of contraction. In essence, increased stretching means a stronger force of contraction.

8. Venous return is affected by the skeletal pump of the body in conjunction with the total fluid volume of the body.

9. The skeletal and respiratory pumps function to move blood toward the heart during muscular contraction.

10. Factors that increase cardiac output include the following:
   • increased sympathetic stimulation (epinephrine and norepinephrine)
   • decreased parasympathetic stimulation
   • exercise
   • increased sodium, potassium, and calcium levels
   • increased body temperature
   • increased preload
   • increased blood volume
   • increased ventricular contractility and distensibility
   • decreased afterload

Section 17.7

1. The three layers of the arteries from innermost to outermost are as follows:
   • tunica intima, composed of simple squamous epithelial tissue (also called the endothelium)
   • tunica media, composed of smooth muscle tissue and elastic connective tissue
   • tunica externa, composed of loose connective tissue

2. Arteries are under greater pressure, have a smaller lumen, and are more elastic. Veins contain valves, have a larger lumen, and are less elastic.

3. Arteries withstand more pressure than veins.

4. Only veins contain valves.

5. Gas exchange occurs in the capillaries.

6. Precapillary sphincters control blood flow and assist with thermo-regulation.

7. The difference is the location. Plasma is located in the vasculature, interstitial fluid is in the tissues, and lymph is found in the lymph ducts and nodes.

8. Edema is the buildup of fluid in the interstitial space.

9. Lymph is moved through the vessels by skeletal pumps, akin to the skeletal pumps in the veins. Larger lymphatic vessels contain smooth muscle that contracts and helps push lymph through the vessels. Lymphatic vessels also have valves, like those in veins, which prevent backflow.

Section 17.8

1. The circle of Willis both regulates the intercranial pressure and provides redundancy should blood vessels in the brain become damaged.

2. The CSF drains through the superior sagittal sinus and the transverse sinus—the two largest veins in the brain—down into the jugular veins.

3. The hepatic portal system allows the blood passing from the digestive system to be filtered by the liver. The liver acts to store and process various nutrients.

Section 17.9

1. Flow = change in pressure/resistance

2. Vessel diameter, constriction and dilation, blood viscosity, and turbulent blood flow all increase resistance.

3. Factors that affect blood flow include the following:
   - metabolic requirements of tissues
   - body temperature
   - total blood volume
   - blood viscosity
   - cardiovascular disease
   - medications

4. When the body temperature decreases, peripheral blood vessels constrict to maintain temperature. When the body temperature increases, blood vessels dilate to dissipate heat.

5. An increased hematocrit increases blood viscosity and therefore increases resistance.

6. Laminar flow is even flow through vessels with a smooth lumen. Turbulent flow results from an uneven surface in the lumen, which causes random blood flow patterns, typically caused by plaque buildup in the vessels: called atherosclerosis. Laminar blood flow can cause blood clots to form.

7. Diuretics decrease blood pressure by decreasing the total blood volume. Vasodilators reduce resistance in the vessels and also cause decreased blood pressure.

Section 17.10

1. A normal, healthy blood pressure is 120/80.

2. A negative feedback loop consists of sensory receptors, sensory neurons, an integrating centre, motor neurons, and effectors.

3. Baroreceptors detect pressure; chemoreceptors detect O_2, CO_2, and H^+; proprioceptors detect body movement.

4. Information from the carotid arteries is sent through the afferent nerves that travel through the glosso-pharyngeal nerve to the brain. Information from the aortic arch is transmitted via the vagus nerve.

5. The cardiovascular centre is located in the medulla oblongata.

6. Neuronal stimulation increases the heart rate and pressure, as well as stimulating the hormonal response that causes epinephrine to be released from adrenal medulla.

7. See Table A.44.

TABLE A.44

Hormone	Where Produced	Effect on Blood Pressure
Antidiuretic hormone (ADH)	Hypothalamus	Increases water reabsorption in the kidneys
Aldosterone	Adrenal cortex	Increases sodium reabsorption in the kidneys, thereby increasing water reabsorption
Epinephrine	Adrenal medulla	Causes peripheral vasoconstriction, meaning increased vascular resistance
Angiotensin II	Converted from angiotensin I by ACE	Causes aldosterone release and also stimulates vasoconstriction
Atrial natriuretic peptide (ANP)	Heart	Lowers blood pressure by causing decreased sodium and therefore water reabsorption in the kidneys

Section 17.11

1. Atherosclerosis is the buildup of plaque, fats, and minerals in the walls of the arteries.

2. Plaque is primarily made up of LDLs ("bad cholesterol"): cholesterol-protein molecules produced by the liver to transport triglycerides from the liver to the tissues. HDLs ("good cholesterol") remove fat from blood vessel walls to be processed in the liver for energy. HDLs can be increased and LDLs decreased with exercise.

3. See Table A.45.

TABLE A.45

Risk Factor	Effect
High blood LDLs and triglycerides	Forms plaque in the vessels, leading to atherosclerosis
Low blood HDL levels	
Obesity	Associated with higher LDL numbers (see above)
Diet	Trans fats increase LDL numbers more than any other type of food (see above)
Lack of physical activity	Contributes to obesity and lower HDL levels
High levels of C-reactive protein	Associated with chronic inflammation
High blood homocysteine levels	High levels linked to CV disease
Smoking	Causes hypertension
Stress	Not clinically proven; may be due to association with other behavioural risk factors
Menopause	Believed due to decreased estrogen levels
Alcohol abuse	Excessive alcohol intake is associated with many diseases, among them CV disease.
Genetic predisposition	Various DNA mutations could affect fat metabolism, transport of fats in the bloodstream, immune response, or any aspect of CV disease.
Ethnicity	Those of African or South Asian descent and also First Nations populations are most likely to suffer from CV disease and diabetes.

4. Atherosclerosis in the coronary vessels can lead to a decreased blow flow to the myocardium. Initially, this can cause ischemia and eventually may lead to myocardial infarction: heart attack. These can also lead to transient ischemic attacks and strokes (also called CVAs (cerebrovascular accident). Chronic hypertension can also lead to kidney failure.

5. Metabolic syndrome is identified as central obesity coupled with high blood triglycerides, low HDLs, hypertension, or hyperglycemia.

6. Angina and a heart attack (MI) are very similar in symptoms, but angina pain subsides when the oxygen demand of the heart decreases.

7. TIAs, also called mini-strokes, have similar symptoms to strokes but tend to resolve in a relatively short period of time.

8. Atherosclerosis can be significantly decreased by eliminating trans fats, daily exercise, controlling blood sugar, not smoking, losing weight, and managing stress.

9. See Table A.46.

TABLE A.46

Medication Type	Effect
Diuretic	Decreases blood pressure by causing increased water excretion
Beta blocker	Inhibits the effects of the sympathetic system, including epinephrine, which decreases the heart rate and force of contraction
Alpha blocker	Decreases the effects of norepinephrine, which decreases vascular tone (leads to vasodilation) and preload
Calcium channel blockers	Decreases force of contraction
Vasodilators	Decreases vascular resistance and preload
ACE inhibitors	Decreases conversion of angiotensin I into angiotensin II (Chapter 20), which causes the production of aldosterone

Section 17.12

1. The blood functions to transport nutrients and oxygen to the tissues, carry waste products to the kidneys, regulate body temperature, and assist with the immune response.

2. Electrolytes are ions used by the cells required to maintain all cell functions.

3. Plasma proteins such as albumins regulate the osmolarity of the blood. Antibodies and complements provide an immune response. Fibrinogen aids in clotting. Hormones are transported in the blood to signal cells throughout the body.

4. Platelets form clots when the vessels become damaged.

5. In a healthy person, approximately 45% of the total blood volume is composed of red blood cells. If the red blood cell count drops below this value, the oxygen-carrying capacity of the blood decreases, leading to decreased oxygen delivery. The decreased oxygen can lead to fatigue.

6. Hematopoiesis is the formation of red blood cells; it is stimulated by the hormone erythropoietin.

7. Red blood cells contain the protein hemoglobin and use the mineral iron to transport oxygen.

8. When red blood cells expire, the majority of the cells parts are recycled to be reused, but the heme group is broken down by the liver, released along with bile, and excreted into the small intestine.

9. See Table A.47.

TABLE A.47

Type of Leukocyte	Function
Neutrophil	Secretes chemicals that kill bacteria in the nearby region
Basophil	Releases histamine during an allergic reaction
Eosinophil	Fights off multicellular parasites and assists with the allergic response
Natural killer cell	Directly kills virus-infected cells and cancer cells
Monocyte	Differentiates into macrophages and dendritic cells, which act as antigen-presenting cells
B Cell	Involved in adaptive immune response and differentiates into plasma cells that release antibodies
T Cell	Involved in adaptive immune response and differentiates into cytotoxic T cells and helper T cells

Chapter 18 Concept Review Answers

Section 18.1

1. Respiratory system functions include the following:
   - taking in oxygen and removing carbon dioxide
   - regulating blood pH
   - warming and moistening air
   - filtering particulate matter from the inhaled air
   - providing a sense of smell
   - producing sound through the vocal cords

Section 18.2

1. The conducting zone transports air, and the respiratory zone is responsible for gas exchange.

2. Nasal conchae help warm the air, possess scent receptors to pick up smells, and cause the air to become turbulent, increasing the amount of filtration that takes place.

3. The pharynx is divided into the nasopharynx, the oropharynx, and the laryngopharynx.

4. Eustachian tubes help equalize pressure within the inner ear.

5. As air passes by the vocal cords it causes them to vibrate, producing sound.

6. Hair and cilia act as physical barriers, filtering various particles from the air prior to the air entering the lungs.

7. The visceral pleura surround the lungs; the parietal pleura line the inside of the thoracic cavity.

8. The fluid in the pleura provides lubrication that reduces friction during the expansion and contraction of the lungs.

9. Surfactant is produced by type II alveolar cells.

10. Surfactant reduces surface tension and prevents alveolar collapse.

Section 18.3

1. Relaxed inspiration is caused by the diaphragm and the internal intercostal muscles. Relaxed exhalation requires no muscular contraction.

2. Accessory muscles may assist the inspiratory effort. These muscles include the sternocleidomastoid muscles, the scalenes, and the pectoralis muscles.

3. The external intercostals, external obliques, rectus abdominis, transverse abdominis, and internal obliques contract during forced exhalation, which increases abdominal and thoracic pressure.

4. Boyle's law states that, in a closed container with a constant number of molecules at a constant temperature, the pressure is inversely proportional to the volume.

5. Volume and pressure are inversely proportional. As the volume increases, the pressure decreases and vice versa.

6. At sea level, the atmospheric pressure is 760 mmHg. If there is no change in volume, the pressure in the lungs—or alveolar pressure—would be 760 mmHg, and the intrapleural pressure would be 756 mmHg.

7. If the intrapleural pressure wasn't lower than the alveolar pressure the lungs would collapse.

8. Compliance is the ability for the lungs to expand and take in air. Someone with decreased compliance has less ability to take in air. Recoil is the elasticity of lungs that facilitates movement of air out of the lungs.

9. The lung volumes are as follows:

    a. tidal volume

    b. inspiratory reserve volume

    c. expiratory reserve volume

    d. residual volume

10. Lung volume is affected by age, sex, altitude, physical fitness, and disease.

11. Anatomical dead space refers to the air that does not contribute to gas exchange.

12. Alveolar ventilation = [tidal volume (mL/breath) − dead space (mL/breath)] × respiration rate (breaths/min)

    a. alveolar ventilation = [600 mL/breath − 180 mL/breath] × 14 breaths/min

    b. alveolar ventilation = [420 mL/breath] × 14 breaths/min

    c. alveolar ventilation = 5880 mL/min OR 5.88 L/min

Section 18.4

1. Dalton states that the sum of the pressures exerted by each gas individually comprises the total pressure. Boyle's law states that the pressure of a gas is inversely proportional to the volume.

2. See Table A.48.

TABLE A.48

Molecule	mmHg
Nitrogen	593.408
Oxygen	158.84
Argon	6.84
Carbon dioxide	0.228
Ozone	0.076
Other trace elements	3.8

3. As the percentage of oxygen from the atmosphere becomes higher than that in the lungs, the oxygen moves toward the alveoli. Once it reaches the alveoli, the pressure in the alveoli is greater than that of the dissolved gas in the capillaries of the lungs, and the oxygen moves into the bloodstream. Once in the bloodstream, the pressure of the dissolved gas is greater than that of the tissues, and the oxygen dissolves into the cells.

4. Atmospheric pressure, ventilation, and metabolism

5. Cells produce carbon dioxide as a waste product, and it moves into the blood to the right side of the heart. Blood circulates to the lungs where the carbon dioxide moves down the concentration gradient to the alveoli and diffuses into the dead space, and is eventually expelled into the atmosphere. The partial pressure of carbon dioxide is higher in venous blood and lower in arterial blood.

6. The following factors cause an increase in the amount of oxygen that dissociates from hemoglobin:

    a. decrease in pH

    b. increase in partial pressure gradient

    c. increase in body temperature

7. Unlike oxygen, carbon dioxide can be transported in three ways (Figure 18.13):

    a. dissolved in plasma, 7–10%

    b. bound to hemoglobin, 25–30%

    c. in the form of bicarbonate ions, 60–65%

8. Chloride shift is important to help stabilize the blood pH.

Section 18.5

1. The dorsal respiratory group controls inspiration, and the ventral respiratory group acts as the pacemaker, managing the breathing rhythm. The ventral respiratory group also signals the muscles used for forceful exhalation.

2. The upper pons smooths the transition from inhalation to exhalation. The lower pons helps regulate the respiratory rate by stimulating the medulla oblongata.

3. Stretch receptors in the lungs signal the medulla oblongata to stop the dorsal respiratory group from further triggering the contraction of inspiratory muscles.

4. The cerebral cortex permits the voluntary control of breathing, and the limbic system controls breathing in response to emotional stimuli. The autonomic nervous system increases the rate and depth of breathing when stimulated by the sympathetic nervous system; conversely, the parasympathetic system decreases the rate and depth of breathing. Increased body temperature can increase respiratory rate. Other factors that can alter breathing include medications, irritants, and a change in the concentrations of oxygen, carbon dioxide, and hydrogen ions in the blood.

5. Peripheral chemoreceptors are located in the aortic arch and carotid arteries. Central chemoreceptors are located in the medulla oblongata.

6. Chemoreceptors detect changes in O_2, CO_2 and H^+ in the blood.

7. • Stimulus—changes in blood oxygen and carbon dioxide levels

    • Receptor—peripheral chemoreceptors detecting changes in blood pH, and central chemoreceptors detecting increased carbon dioxide in the cerebrospinal fluid

- Integrating centre—signals being received by medulla oblongata
- Effectors—motor neurons causing the contraction of inspiratory muscles, alternating with the stimulation of the expiratory muscles

Section 18.6

1. Emphysema and chronic bronchitis are the two primary diseases under the classification of COPD.

2. Emphysema is the breakdown of the alveoli into larger sacs with decreased elasticity. Chronic bronchitis results from chronic inflammation and leads to increased mucus production and a chronic cough.

3. The most common cause of COPD is the inhalation of irritants and pollutants such as cigarette smoke.

4. Respiratory difficulty and the use of additional muscles to aid the respiratory effort (referred to as accessory muscles) are common symptoms of asthma.

5. Bronchodilators and anti-inflammatory medications can be used to treat asthma.

6. Hypoxia can be caused by low hematocrit or hemoglobin, poor circulation, poisoning such as by cyanide, and decreased ventilation caused by disease.

Chapter 19 Concept Review Answers

Section 19.1

1. The organs of the gastrointestinal tract are the esophagus, the stomach, the small intestine, the large intestine, and the anus.

2. The accessory organs are the gallbladder, liver, pancreas, and salivary glands.

3. Ingestion, mechanical digestion, chemical digestion, secretion, absorption, and defecation.

4. Six main processes of the digestive system:
   - physically and chemically breaking down ingested food
   - absorbing digested nutrients into the bloodstream
   - removing undigested waste
   - producing intrinsic factor, which is required for the absorption of vitamin B12
   - housing symbiotic bacteria, the normal flora, mostly in the large intestine

Section 19.2

1. The layers of the GI tract in order are the mucosa, submucosa, muscularis, and serosa.

2. The three components of the mucosa are the epithelial cells that form the mucous membrane, the connective tissue of the lamina propria, and the thin layer of smooth muscle called the muscularis mucosa.

3. The submucosal plexus nerves are responsible for detecting stimuli such as stretching; the myenteric plexus is responsible for coordinating smooth muscle contraction.

4. The circular layer contracts to mix food within the GI tract, and the longitudinal layer acts to move food through the GI tract.

5. The peritoneum is made up of areolar connective tissue covered by simple squamous epithelial tissue.

6. The visceral peritoneum is the outermost layer of the digestive tract, whereas the parietal peritoneum lines the interior of the abdominal cavity.

7. The greater omentum is a collection of fat that provides protection for the viscera (internal organs), and the mesentery holds the organs in place.

Section 19.3

1. The maxilla, the mandible, and the palatine bones make up the mouth.

2. The gingival are commonly known as the gums.

3. Adults have a total of 36 teeth: eight incisors, four canines, eight premolars, 12 molars, and four wisdom teeth.

4. The tongue aids in chewing and swallowing by rolling food into a bolus. The tongue also contains taste receptors that detect food flavours such as sweet, salty, bitter, sour, and savoury (or umami).

5. Peristalsis refers to wavelike contractions that move food through the gastrointestinal tract.

6. The uvula and epiglottis are physical barriers that prevent food from entering the nasopharynx and trachea, respectively.

7. From inner to outer, the muscle layers of the stomach are the oblique, circular, and longitudinal layers.

8. The pyloric sphincter closes to prevent the passage of contents from the stomach to the small intestine.

9. Goblet cells produce mucus, chief cells produce pepsinogen, and parietal cells secrete hydrochloric acid.

10. The liver performs the following functions:
    - synthesizing bile
    - storing nutrients
    - breaking down nutrients
    - breaking down old red blood cells
    - breaking down many organic and inorganic toxins

- excreting excess trace metals
- synthesizing the plasma proteins
- synthesizing fibrinogen
- synthesizing angiotensinogen
- activating vitamin D
- producing insulin-like growth factors (IGFs)
- secreting complement proteins

11. Bile is made of bile salts, water, mucus, and other inorganic salts.

12. The pancreatic endocrine functions include the production and release of insulin and glucagon. The exocrine functions include the production and excretion of pancreatic juice, water, salts, and enzymes.

13. The majority of digestion and absorption of nutrients occurs in the duodenum.

14. Folds, villi, and microvilli line the walls of the small intestine.

15. In order, the large intestine moves from the ileocecal sphincter through the cecum to the ascending colon, transverse colon, descending colon, sigmoid colon, rectum, anal canal, and anus.

16. The large intestine absorbs ions, vitamins, and water, as well as storing waste products until defecation.

17. Normal flora or gut bacteria serve the following functions:
    - fermenting undigested substrates
    - regulating the immune system
    - preventing growth of harmful bacteria
    - producing vitamin K and biotin
    - increasing growth of intestinal cells
    - producing lactic acid
    - converting carbohydrates into short chain fatty acids
    - releasing important minerals from food
    - metabolizing carcinogens

Section 19.4

1. Proteins consist of amino acids, carbohydrates are formed from monosaccharides, triglycerides are broken into fatty acids, and nucleic acids are broken into nucleotides.

2. See Table A.49.

3. Fats take the longest to digest of all macromolecules.

4. Bicarbonate is released to neutralize the acid before it can cause damage to the small intestine.

Section 19.5

1. Water and ions move either into the cells by osmosis, or ion channels, or between epithelial cells.

2. Vitamin D is required for calcium absorption in the small intestine.

TABLE A.49

Enzyme	Location	Food Molecule
Amylase	Saliva	Starch
Lipase	Saliva	Fats
Pepsin	Stomach	Proteins
Pancreatic amylase	Pancreas	Starch
Pancreatic lipase	Pancreas	Fats
Ribonuclease	Pancreas	RNA
Deoxyribonuclease	Pancreas	DNA
Trypsin	Pancreas	Proteins
Chymotrypsin	Pancreas	Proteins
Carboxypeptidase	Pancreas	Peptides
Maltase	Small intestine	Maltose
Sucrase	Small intestine	Sucrose
Lactase	Small intestine	Lactose
Peptidases	Small intestine	Peptides

3. Intrinsic factor is required for the absorption of vitamin B12.

4. Amino acids, glucose, and galactose all enter the cell by means of co-transport with sodium.

5. Sodium-potassium pumps are required to move the sodium back out of the cell after being used for co-transport of glucose and galactose.

6. Fats are broken down into very small droplets called micelles. Once broken down, fats can be absorbed by simple diffusion.

7. Water-soluble vitamins are co-transported across the membrane with sodium, with the exception of vitamin B12, which requires intrinsic factor to be absorbed. Fat-soluble vitamins are absorbed by simple diffusion across the membrane and are packaged into chylomicrons with other fats and enter the lacteal.

8. Injury or inflammation can cause damage to the mucosa and alter the path of absorption, causing triglycerides to be absorbed directly into the bloodstream rather than through the lymph vessels.

9. The thermogenic effect refers to the amount of energy (calories) used by the for digestion.

Section 19.6

1. The cephalic phase is the initiation of the secretion of gastric juices and muscle contractions in anticipation of ingesting food. The gastric phase begins when food enters the stomach and causes gastric secretions and muscle contractions. The intestinal phase triggers the release of pancreatic secretions, bile, intestinal enzyme secretion, and motility.

2. See Table A.50.

TABLE A.50

Hormone	Where Produced	Function
Gastrin	Pyloric region of stomach	Stimulates the production of hydrochloric acid, intrinsic factor, pepsinogen, mucus; also promotes muscular contractions
Secretin	Intestinal mucosa	Stimulates the pancreas to release bicarbonate ions as well as pancreatic enzymes
Cholecystokinin	Intestine	Causes the stomach to slow gastric emptying, causes the gall bladder to release bile, and signals the hypothalamus to create the feeling of fullness

3. Cholecystokinin triggers in the hypothalamus the feeling of being full.

Section 19.7

1. Crohn's disease and ulcerative colitis are chronic inflammatory diseases.

2. A high-fibre diet can help prevent diverticulitis and hemorrhoids.

3. Gastritis and ulcers

Chapter 20 Concept Review Answers

Section 20.1

1. The urinary system functions are as follows:
   - regulating blood, water, and ion composition
   - excreting waste products of metabolism
   - regulating blood pressure
   - regulating blood pH by excreting excess H^+ ions
   - producing calcitriol and erythropoietin

2. The urinary system is made up of the kidneys, the ureters, the bladder, and the urethra.

Section 20.2

1. The cortex is the outer portion of the kidney; the medulla is the inner portion.

2. Urine flows from the collecting ducts through the minor calyx to the major calyx and collects in the renal pelvis. Each renal pelvis drains through the ureters and into the bladder. From there the bladder empties via the urethra.

3. Renal artery—segmental arteries—interlobar arteries—cortical radiate arteries—afferent arteriole—glomerulus—efferent arteriole—peritubular capillaries—cortical radiate veins—arcuate veins—interlobar veins—renal vein

Section 20.3

1. Renal corpuscle—proximal convoluted tubule—descending loop of Henle—ascending loop of Henle—distal tubule—collecting duct

2. The majority of reabsorption takes place in the proximal convoluted tubule.

3. The descending loop of Henle is impermeable to ions, but permits the passage of water. The ascending loop of Henle is impermeable to water, but permits the passage of ions through the epithelial layer.

4. In a healthy individual, 100% of glucose and amino acids are reabsorbed by the kidneys.

5. PTH causes the kidneys to reabsorb calcium. ADH causes the kidneys to reabsorb water. Aldosterone causes the kidneys to reabsorb sodium.

6. Ammonia, urea and uric acid are formed when proteins or nucleic acids are broken down.

7. Secretion occurs in the proximal convoluted tubule, the distal convoluted tubule, and the collecting ducts.

8. Nitrogenous wastes, creatinine, hydrogen ions, potassium ions, and some medications are secreted in the urine.

Section 20.4

1.
   - Filtration. Molecules move from glomerulus into tubule at Bowman's capsule.
   - Reabsorption. Molecules move out of tubule and into capillaries.
   - Secretion. Molecules move from peritubular capillaries into tubule.

2. Water can be reabsorbed by osmosis, through aquaporins, or via paracellular transport.

3. In the apical membrane, glucose and amino acids co-transport with sodium; in the basolateral membrane, glucose and amino acids move by facilitated transport with membrane carrier proteins.

4. The sodium-potassium pump is located on the basolateral membrane of the tubule cells.

Section 20.5

1. See Table A.51 on next page.

2. Renin is used to cleave angiotensinogen, turning it into angiotensin I.

3. Angiotensin-converting enzyme (ACE) converts angiotensin I into angiotensin II.

TABLE A.51

Hormone	Stimulus for Production	Production Site	Target Tissue	Functions
ADH	Low blood pressure	Hypothalamus, released from posterior pituitary gland	Distal tubules and collecting duct	Producing aquaporin membrane proteins to increase water reabsorption
Aldosterone	Increased angiotensin II or high blood potassium	Adrenal cortex (zona glomerulosa)	Distal tubule and collecting duct	Increasing sodium reabsorption, therefore water reabsorption, and increasing potassium secretion
PTH	Low blood calcium or magnesium	Parathyroid gland	Distal tubule	Increasing calcium and magnesium reabsorption
ANP	High blood pressure	Atria	Distal tubule and collecting duct	Decreasing sodium reabsorption; therefore, more water excreted to decrease blood pressure

4. Angiotensin II acts in the following ways:
   - stimulating the adrenal cortex to produce aldosterone
   - causing vasoconstriction to increase blood pressure
   - stimulating the posterior pituitary to release ADH
   - stimulating the sympathetic nervous system to increase blood pressure
   - stimulating the reabsorption of sodium, chloride and, therefore, water in the kidneys
   - stimulating thirst and cravings for salt

Section 20.6

1. Hypertension and diabetes are the most common causes of chronic kidney disease (CKD).

2. Factors that help prevent CKD or mitigate its complications involve maintaining a healthy blood pressure (through diet, exercise, and medications) along with maintaining proper blood sugar levels (through diet, exercise, and medications).

3. Increased salt intake leads to a greater accumulation of ions in the kidneys. The increased ion concentration leads to increased water reabsorption, resulting in higher blood volumes. The increased blood volume will exacerbate the pre-existing hypertensive state.

4. The symptoms of advanced chronic kidney disease include, but are not limited to, the following:
   - nausea
   - weakness
   - fatigue
   - inability to concentrate
   - decreased appetite
   - muscle cramping
   - increased blood pressure
   - itching of the skin
   - edema in the soft tissues, usually including the feet and around the eyes

5. Dialysis and kidney transplant are the only treatments currently available for advanced CKD.

6. Calcium-oxalate and uric acid are the two most common types of kidney stones.

7. Apart from kidney stones, gout can result from a buildup of uric acid.

Chapter 21 Concept Review Answers

Section 21.1

1. The male reproductive system is responsible for the following:
   - spermatogenesis
   - producing steroid hormones, including testosterone
   - storing and transporting sperm
   - producing additional secretions that form semen

   The female reproductive system is responsible for the following:
   - oogenesis
   - producing steroid hormones, primarily estrogen and progesterone
   - storing and transporting ova
   - providing a location for the fertilization of eggs (fallopian tubes)
   - providing a site for the implantation and growth of the fetus
   - producing milk in the mammary glands for feeding the newborn

Section 21.2

1. Seminiferous tubules—rete testis—efferent ductules—epididymis—vas deferens— seminal vesicles—ejaculatory duct—prostate gland—urethra

2. The seminal vesicles secrete an alkaline fluid that helps neutralize the acidic environment of the male urethra. The prostate secretes citric acid, used by the sperm to produce ATP and create prostate-specific antigens that help the sperm enter the cervix. Bulbourethral glands secrete alkaline mucus that neutralizes and lubricates the urethra.

3. Spermatogonia, primary spermatocytes, secondary spermatocytes, spermatids, and finally spermatozoa

4. Spermatozoa, secondary spermatocytes, and spermatids are haploid. Primary spermatocytes and spermatogonia are diploid.

5. Sertoli cells support spermatogenesis in the following ways:
   - forming the blood-testis barrier
   - providing nutrients and secreting fluid that surrounds the developing sperm
   - stimulating and regulating spermatogenesis
   - producing the hormone inhibin, which prevents the release of FSH
   - engulfing improperly developing sperm

6. Leydig cells produce testosterone in response to luteinizing hormone.

7. Lutenizing hormone causes the release of testosterone. Testosterone and FSH work together to stimulate Sertoli cells to release signalling molecules to regulate spermatogenesis.

8. The spermatozoan consists of three sections:
   - Head—contains 23 chromosomes and minimal cytoplasm
   - Mid-piece—contains mitochondria that metabolize fructose for energy
   - Tail—a flagellum composed of microtubules that propel the sperm

Section 21.3

1. Fertilization occurs in the infundibulum of the fallopian tube.

2. From innermost to outermost layer, the endometrium is where the blastocyst implants and is shed each month if no fertilization takes place. The middle layer, or myometrium, is the smooth muscle layer that contracts during childbirth. Finally, the perimetrium is the outermost layer, composed of simple squamous epithelial and areolar connective tissue.

3. The vulva is made up of an adipose layer covering the pubic symphysis, called the mons pubis; the external labia majora; the internal labia minora; and the clitoris.

4. In females, meiosis begins in utero during fetal development.

5. The first meiotic division occurs just before ovulation, and the second takes place only if fertilization occurs.

6. Primary oocytes form in the follicles during fetal development, and undergo their first meiotic division before ovulation. If the oocyte is fertilized, the cell undergoes a secondary division, and the 23 chromosomes combine with the 23 chromosomes from the sperm to form a diploid zygote.

7. A luteinizing hormone surge causes the oocyte to undergo the first meiotic division and start ovulation.

8. Progesterone and estrogen are produced by the corpus luteum.

9. Decreased levels of estrogen and progesterone are caused by a breakdown of the corpus luteum and cause menstruation.

10. LH stimulates the theca cells, and FSH stimulates granulosa cells.

Section 21.4

1. Endometriosis

2. Chlamydia and gonorrhea infections

3. A blood test to look at prostate-specific antigen (PSA) levels.

Chapter 22 Concept Review Answers

Section 22.1

1. Lymphatic vessels, lymph nodes, thymus, spleen, tonsils, the Peyer's patch of small intestine, and bone marrow make up structures of the immune system.

2. The functions of the immune system include the following:
   - returning fluid from the interstitial space to the bloodstream
   - transporting dietary fats and fat-soluble vitamins from the small intestine to bloodstream
   - recognizing and killing infectious organisms
   - recognizing cells and non-harmful foreign molecules, such as food and environmental substances.
   - producing immunological memory cells to prevent infection from the same organism in the future

3. The main infectious organisms are bacteria, viruses, fungi, and parasites.

Section 22.2

1. Lymphatic vessels have closed ends, and fluid is taken up through one-way opening in between the endothelial cells.

2. They are all the same except for the location in which each is found.

3. The right upper body drains into the right internal jugular vein and right subclavian vein to re-enter the circulatory system. Lymph drains from left upper body into the junction of the left internal jugular vein and the left subclavian vein.

4. Edema can be caused by the following:
   - increased capillary pressure or interstitial osmostic pressure
   - improper kidney filtration
   - high venous pressure
   - lack of skeletal muscle use
   - medications that affect blood pressure, vascular resistance, the autonomic nervous system, kidney function, or osmotic balance
   - lack of plasma proteins
   - blocked lymphatic vessels

5. Red and white blood cells are produced in the bone marrow by hematopoietic stem cells.

6. Mesenchymal cells differentiate into adipose cells, osteoblasts, and cartilage.

7. In the thymus, thymocytes (immature T cells) that recognize self-proteins are destroyed by apoptosis.

8. Reticular fibres act as a filter to prevent particles from entering the bloodstream.

9. Phagocytic macrophages, dendritic cells, and lymphocytes are all found in lymph nodes.

10. The adaptive or acquired immune response occurs in the lymph node.

11. The spleen acts to break down red blood cells in the red pulp, and the white pulp of the spleen assists with the immune response. The spleen can also assist with haematopoiesis in embryos.

12. The tonsils, adenoids, appendix, and Peyer's patches (of the small intestine) are all examples of lymphatic nodules.

13. See Table A.52.

TABLE A.52

Cell Type	Function
Neutrophil	Most numerous of all white blood cells, and predominant cell type in pus
	First-responders that produce chemicals involved in the inflammatory response, such as histamine, vasodilators, reactive oxygen species, and cytokines
	Phagocytic cells that engulf many types of microbes, mostly bacteria and opsonized pathogens (have antibodies from B cell response)
	Migrates from blood vessels to infection area within tissues; attracted to damaged cells by process of chemotaxis (migration of immune cells toward an area of infection)
Basophil	Least common type of white blood cell
	Releases histamine and several cytokines involved in the inflammatory response
	Plays a role in allergic reactions and parasitic infections, specifically, ectoparasites such as ticks or lice
Eosinophil	Involved in parasitic infections such as helminths and worms
	Produces many chemicals, including histamine that is involved in allergies, asthma, and inflammation
Mast cell	Primarily involved in allergic reactions and anaphylaxis by releasing histamine if interaction with IgE antibodies occurs
	Plays an important protective role in wound healing
	Prevalent in tissues near openings to the body, such as mouth, nose, skin, lungs, and GI tract
Monocyte	Circulates in the blood and responds to inflammatory signals, migrates into tissues and differentiates into macrophages or dendritic cells
Dendritic cell	Important antigen-presenting cells (APCs) that stimulate lymphocytes (B cells and T cells) in the spleen, lymph nodes, or lymphoid nodules and initiate the adaptive immune response
	Phagocytic cells that can engulf pathogens or opsonized pathogens (when antibodies or complement proteins are bound to the pathogen)
	Present in tissues that are in contact with the environment, such as the skin (Langerhans cells) lungs, and GI tract
Macrophage	Important APCs that stimulate lymphocytes in the lymphatic tissues and initiate the adaptive immune response
	Phagocytic cells that can engulf pathogens or opsonized pathogens (when antibodies or complement proteins are bound to the pathogen)
	Secretes toxic chemicals that directly kill invading organisms
	Secretes cytokines and chemicals involved in the inflammatory response

Natural killer cell	Specifically kills virus-infected cells and cancer cells by lysis with perforin or induction of apoptosis
	Kills cells opsonized with antibodies
	Can produce antigen-specific memory cells important for detecting same infection a second time
B cell	Contains a specific receptor for a specific antigen that is a membrane-bound immunoglobulin (antibody)
	Circulates between blood and lymphatic tissues
	Phagocytic when antigen matches receptor
	Antigen-presenting cell
	Produces cytokines
	Differentiates into plasma cells when activated by helper T cell and then secretes antibodies
	Can produce memory cells
Helper T cell	Do not directly kill any invading microorganisms
	Can only respond to antigen presented by APC (macrophage, dendritic cell, or B cell)
	Produces cytokines that activate B cells, cytotoxic T cells, macrophages, dendritic cells, and NK cells
	Can produce memory cells
Cytotoxic T cell	Directly kills virus-infected cells, cancer cells, and transplant tissue
	Binds to antigen presentation on MHC class I molecules
	Can produce memory cells
Regulatory T cell	Inhibits other immune cells at the end of an infection and prevents auto-reactive immune cells that escaped the selection process in the bone marrow or thymus from damaging tissues

Section 22.3

1. The first line of defense is prevention and includes the chemical and physical barriers that prevent infection:

   - The skin provides a physical barrier that is covered with beneficial normal flora organisms that help fight off infectious organisms.
   - Hairs and mucus in the nose functions to block large particles, and movement of the hairs can stimulate the sneezing reflex.
   - Cilia and mucus in the upper respiratory tract trap particles and propel them up toward the throat to be swallowed, or removed by coughing.
   - Antimicrobial chemicals such as lysozyme are secreted into tears, saliva, and sweat, and break down bacterial cell walls.
   - Sebaceous glands produce acidic oil that prevents the growth of certain bacteria.
   - The stomach produces hydrochloric acid, a very strong acid that kills many organisms taken in with food.
   - Resident beneficial bacteria in the digestive tract play an important role in preventing infection.

2. B Cells, macrophages, dendritic cells and neutrophils are all phagocytic.

3. Cytotoxic T-cells specifically target virus-infected cells.

4. See Table A.53.

5. Redness, pain, swelling, and heat all result from chemicals released by damaged cells.

TABLE A.53

Chemical	Function
Lysozyme	Breaks down bacterial cell walls
Perforin	Causes lysis in cells by puncturing membranes
Interferon	Released when cells are infected by viruses and triggers nearby cells to produce antiviral proteins
Histamine	Causes vasodilation, increased capillary permeability, and attracts phagocytic cells
Complement	Stimulates mast cells to produce histamine, binds to microbes to increase phagocytosis, and forms a membrane attack complex that kills bacteria and fungus
Prostaglandins	Same effects as histamine and also cause pain

   - Redness is caused by increased blood flow.
   - Swelling is caused by increased vascular permeability leading to edema.
   - Pain is caused by the release of prostaglandins and the effect of swelling putting pressure on the nerves.
   - Heat is caused by increased blood flow.

6. Pyrogens are molecules that increase body temperature by increasing the set point determined by the hypothalamus.

Section 22.4

1. An antigen is any molecule on the surface of an infectious organism that can be recognized by any immune cell.

2. Macrophages, B cells, and dendritic cells are all types of antigen-presenting cells.

3. Antigens presented on MHCI allow immune cells to recognize self-proteins and for immune cells to target specific microbial antigens.

4. All nucleated cells express MHCI. MHCII is found only on antigen-presenting cells.

5. Interleukin-I causes T cells to secrete interleukin-2, which causes proliferation of the T cell with that specific T cell receptor.

6 and 7. See Table A.54.

TABLE A.54

Helper T Cell Subtype	Main Role of each Subtype
TH0	Non-reactive, tolerant cells for certain antigens
TH1	Increases activation of cytotoxic T cells
TH2	Increases the activation of B cells
Memory cell	Helps fight future infections of the same type
TH17	Slows down the immune response at the end of the infection

8. CD8+ cytotoxic T cells express CD8. Helper T cells are called CD4+ if they express the CD4 co-receptor.

9. Cytotoxic T cells each have a specific TCR that recognizes only one antigen that, when recognized, binds to an infected body cell, but it will only become fully activated if a helper T cell produces IL-2 and stimulates proliferation.

10. Helper T cells bind to B cells that present the proper matching antigen on MHCII and cause them to proliferate. TH2 helper cells bind to B cells and secrete IL-2

11. B cells may differentiate into plasma cells that secrete antibodies matching the antigen.

12. When an infectious organism is marked by antibodies, it's called opsonized. An opsonized microbe results in the following:
    - increased phagocytosis by macrophages and dendritic cells
    - increased targeting by complement proteins
    - neutralized microbes that cannot produce toxins or bind to body cells
    - immobilized bacteria that cannot spread as easily
    - agglutination—clumping of many antibodies and microbes together so they cannot function

13. Cell-mediated immune responses use T cells; humoral responses use B cells and antibodies.

14. Intracellular infections such as viral infections are fought by cytotoxic T cells, and bacterial or extracellular infections are fought by antibodies and B cells.

15. Memory cells improve the body's ability to rapidly fight off the same, or sometimes similar, infection and may even prevent symptoms from appearing.

16. Viruses, such as those causing the common cold, change surface antigens frequently and this is always seen by the immune system as a new infection.

17. Active immunity occurs when antibodies are produced by the individual. Passive immunity results from antibodies being transferred to the individual either by injection or ingestion.

Section 22.5

1. Edward Jenner recognized that people with cowpox didn't become infected by smallpox.

2. Live attenuated viruses can be produced by producing many generations until a non-virulent strain appears. Examples include measles, mumps, and rubella—known as the MMR vaccine.

3. A killed virus can also be injected directly into a host without causing infection. Examples include the influenza and cholera vaccines.

4. Vaccines for diphtheria, pertussis, and tetanus are created by using inactivated toxins that contain the antigens, but lack the bacteria to cause an infection.

5. DNA technology has been shown to be effective in making vaccines by causing non-pathogenic bacteria, fungi, mammalian cells, or low-virulence viruses to exhibit surface proteins that match those of the actual pathogen.

Section 22.6

1. In any autoimmune disease, the immune cells target a specific type of tissue and attack it as though it were a pathogen, leaving that tissue no longer capable of functioning normally.

2. The main theories of autoimmune disease are as follows:
    - Molecular mimicry. The immune system loses the ability to differentiate between foreign and self-antigens.

- Hygiene theory. Countries with lower rates of infection exhibit a higher rate of autoimmune disease.
- Lack of beneficial bacteria exposure in childhood. Pasteurization and sterilization of foods prevents children from fully developing the normal flora required to fight off infection.
- Genetic factors. Research has shown some variations of certain MHC genes affect antigen presentation.
- Vitamin D. It is known that vitamin D does play an important role in immune regulation, but supplementation alone has not reversed autoimmune disease.

3. Autoimmune diseases are most prevalent in developed countries.

4. Tissue targeted:
   a. beta-islets of the pancreas
   b. myelin on the axons of the ANS
   c. skeletal muscle
   d. thyroid gland

5. Mast cells and basophils are responsible for the release of histamine, while B cells that are activated become plasma cells and then release antibodies.

6. Upon exposure for the first time, no IgE antibodies have been created that are specific to the particular allergen, and as such, no symptoms result.

7. The most severe allergic reactions tend to be caused by venoms and foods.

8. Histamine is the main chemical involved in both allergic reactions and asthma.

Chapter 23 Concept Review Answers

Section 23.1

1. Bacteria provide the benefits of producing oxygen, recycling minerals, fermenting foods, and acting as normal flora

2. The functions of bacteria in the gastrointestinal tract are as follows:
   - fermenting undigested substrates, mostly soluble fibre
   - regulating the immune system to tolerate food molecules
   - preventing growth of harmful bacteria
   - producing vitamin K and biotin
   - increasing growth of intestinal cells and helping maintain a healthy mucosal layer
   - producing lactic acid during fermentation, lowering pH, and preventing growth of pathogenic species
   - converting carbohydrates into short-chain fatty acids
   - releasing important minerals from food, including magnesium, calcium, zinc, and iron
   - metabolizing carcinogens

3. Endospores form a protective physical barrier that protects bacteria from changes to the environment.

4. The three basic bacteria shapes are spherical (cocci), rod (bacilli), and spiral (spirochete).

5. Antibiotic resistance genes work by
   - degrading the antibiotic,
   - modifying the antibiotic into a non-functional form, and
   - forming membrane pumps that remove the antibiotic from the cell.

6. Infections are considered opportunistic when they live in the human body without causing harm or illness until the host's immune system becomes weakened.

7. Children, the elderly, and the immune-suppressed are most at risk for opportunistic infections.

8. *E.coli*, salmonella, and campylobacter are the bacteria most to blame for food poisoning.

9. *Methicillin-resistant staphylococcus aureus* (MRSA) is a multi-resistant strain of a normal flora bacterium.

10. Clostridium tetani releases a toxin that blocks inhibitory neurotransmitters, leading to continuous nervous stimulation of muscle.

11. A number of normal flora bacteria can become pathogenic when the host becomes immune-compromised; therefore, they are considered opportunistic. Staphylococcus is one such example.

12. Chlamydia and gonorrhea are the two most common bacterial sexually transmitted infections (STIs).

Section 23.2

1. Viruses are not true cells as they do not have a cellular membrane or cellular organelles.

2. A bacteriophage is a virus that infects bacteria.

3. RNA viruses make use of an enzyme called reverse transcriptase to convert their RNA into DNA so it can be integrated into their host.

4. Hemagglutinin allows the influenza virus to bind to upper airway cells, and neuraminidase allows the newly replicated virus particles to be released from the host cell.

5. During the lytic cycle the virus kills the host during replication, but in the lysogenic phase the viral

DNA is integrated and replicated along with that of the host cell, without killing it.

6. Attachment—infiltration—replication—assembly

7. The herpes virus undergoes a lysogenic cycle, allowing it to go undetected by the immune system and thereby making it extremely difficult to combat.

8. HIV binds to the CD4 membrane receptors on human immune cells.

9. Hepatitis B and C are the most likely strains to cause liver cancer.

10. Herpes, HPV, hepatitis, and EBV are all examples of sexually transmitted viruses.

11. HPV causes genital warts and two strains have been identified as potential causes of cervical cancer.

12. The Epstein-Barr virus (EBV) has been linked to increased rates of multiple sclerosis.

Section 23.3

1. Fungi are eukaryotic organisms.

2. Fungi have antimicrobial properties and are used in the treatment of heart disease and cancer.

3. Fungi can be consumed whole (mushrooms) or used for the fermentation process of foods and beverages, such as bread, fermented rice, wine, and beer.

4. Immunosuppression, protracted use of antibiotics, or a breakdown of the normal flora can all lead to fungal infections.

5. Dermatophytosis is responsible for conditions such as athlete's foot.

6. Candida albicans

7. Pneumocystis jirovecii

Section 23.4

1. The most likely method of protozoan transmission for intestinal protozoans is the fecal-oral route; protozoans that live in the bloodstream are most often transmitted by insect vectors.

2. The two major categories of helminthes are flatworms and roundworms.

3. While some worms can penetrate the skin, most helminth infections result from consuming undercooked and contaminated soil, water, or food.

4. As with any infection, those with a decreased immune system are most susceptible to helminth infections. Those in developing countries, particularly children, are most at risk for infection from these organisms.

5. Helminths called hookworms are able to penetrate the skin to infect the host.

6. The highest prevalence for tapeworms is in populations that eat raw fish.

Appendix B

PERIODIC TABLE

Key:
Uranium — 92 (Atomic number) — U (Symbol)

Legend:
- MAIN GROUP METALS
- TRANSITION METALS
- METALLOIDS
- NON-METALS
- PROPERTIES UNKNOWN

Group	1	2	3	4	5	6	7	8	9	10	11	12	13	14	15	16	17	18
Period 1	Hydrogen 1 H																	Helium 2 He
Period 2	Lithium 3 Li	Beryllium 4 Be											Boron 5 B	Carbon 6 C	Nitrogen 7 N	Oxygen 8 O	Fluorine 9 F	Neon 10 Ne
Period 3	Sodium 11 Na	Magnesium 12 Mg											Aluminum 13 Al	Silicon 14 Si	Phosphorus 15 P	Sulfur 16 S	Chlorine 17 Cl	Argon 18 Ar
Period 4	Potassium 19 K	Calcium 20 Ca	Scandium 21 Sc	Titanium 22 Ti	Vanadium 23 V	Chromium 24 Cr	Manganese 25 Mn	Iron 26 Fe	Cobalt 27 Co	Nickel 28 Ni	Copper 29 Cu	Zinc 30 Zn	Gallium 31 Ga	Germanium 32 Ge	Arsenic 33 As	Selenium 34 Se	Bromine 35 Br	Krypton 36 Kr
Period 5	Rubidium 37 Rb	Strontium 38 Sr	Yttrium 39 Y	Zirconium 40 Zr	Niobium 41 Nb	Molybdenum 42 Mo	Technetium 43 Tc	Ruthenium 44 Ru	Rhodium 45 Rh	Palladium 46 Pd	Silver 47 Ag	Cadmium 48 Cd	Indium 49 In	Tin 50 Sn	Antimony 51 Sb	Tellurium 52 Te	Iodine 53 I	Xenon 54 Xe
Period 6	Cesium 55 Cs	Barium 56 Ba	Lanthanum 57 La	Hafnium 72 Hf	Tantalum 73 Ta	Tungsten 74 W	Rhenium 75 Re	Osmium 76 Os	Iridium 77 Ir	Platinum 78 Pt	Gold 79 Au	Mercury 80 Hg	Thallium 81 Tl	Lead 82 Pb	Bismuth 83 Bi	Polonium 84 Po	Astatine 85 At	Radon 86 Rn
Period 7	Francium 87 Fr	Radium 88 Ra	Actinium 89 Ac	Rutherfordium 104 Rf	Dubnium 105 Db	Seaborgium 106 Sg	Bohrium 107 Bh	Hassium 108 Hs	Meitnerium 109 Mt	Darmstadtium 110 Ds	Roentgenium 111 Rg	Copernicium 112 Cn	Ununtrium 113 Uut	Flerovium 114 Fl	Ununpentium 115 Uup	Livermorium 116 Lv	Ununseptium 117 Uus	Ununoctium 118 Uuo

Lanthanides:

Cerium 58 Ce	Praseodymium 59 Pr	Neodymium 60 Nd	Promethium 61 Pm	Samarium 62 Sm	Europium 63 Eu	Gadolinium 64 Gd	Terbium 65 Tb	Dysprosium 66 Dy	Holmium 67 Ho	Erbium 68 Er	Thulium 69 Tm	Ytterbium 70 Yb	Lutetium 71 Lu

Actinides:

Thorium 90 Th	Protactinium 91 Pa	Uranium 92 U	Neptunium 93 Np	Plutonium 94 Pu	Americium 95 Am	Curium 96 Cm	Berkelium 97 Bk	Californium 98 Cf	Einsteinium 99 Es	Fermium 100 Fm	Mendelevium 101 Md	Nobelium 102 No	Lawrencium 103 Lr

At the date of publication, elements 113, 115, 117 and 118 had not been named and have been given temporary names.

From MAHAFFY/TACKER/BUCAT/KOTZ/MCMURRY. Chemistry. 2E. © 2015 Nelson Education Ltd. Reproduced by permission. www.cengage.com/permissions.

Appendix C

SYSTÈME INTERNATIONALE/METRIC SYSTEM

Metric Measures and Imperial Equivalents

Unit	Measure	Symbol	Imperial Equivalent
LINEAR MEASURE			
1 kilometre	= 1000 metres	10^3 m km	0.62137 mile
1 metre		10^0 m m	39.37 inches
1 decimetre	= 1/10 metre	10^{-1} m dm	3.937 inches
1 centimetre	= 1/100 metre	10^{-2} m cm	0.3937 inch
1 millimetre	= 1/1000 metre	10^{-3} m mm	Not used
1 micrometre (or micron)	= 1/1 000 000 metre	10^{-6} m μm(or μ)	Not used
1 nanometre	= 1/1 000 000 000 metre	10^{-9} m nm	Not used
MEASURES OF CAPACITY (FOR FLUIDS AND GASES)			
1 litre		L	1.0567 U.S. liquid quarts
1 millilitre	= 1/1000 litre	mL	
	= voume of 1 g of water at stp*		
MEASURES OF VOLUME			
1 cubic metre		m^3	
1 cubic	= 1/1000 cubic metre	dm^3 = L	
decimetre	= 1 litre	L	
1 cubic centimetre	= 1/1 000 000 cubic metre	cm^3 = mL	
	= 1 millilitre (mL)		
1 cubic millimetre	= 1/100 000 000 cubic metre	mm^3	
MEASURES OF MASS			
1 kilogram	= 1000 grams	kg	2.2046 pounds
1 gram		g	15.432 grains
1 milligram	= 1/1000 gram	mg	0.01 grain (about)
1 microgram	= 1/1 000 000 gram	μg (or mcg)	

*stp = standard temperature and pressure

Imperial to Metric Conversions

Length

Imperial		Metric
inch	=	2.54 centimetres
foot	=	0.30 metre
yard	=	0.91 metre
mile (5280 feet)	=	1.61 kilometre

To Convert	Multiply By	To Obtain
inches	2.54	centimetres
feet	30.00	centimetres
centimetres	0.39	inches
millimetres	0.039	inches

Mass/Weight

Imperial		Metric
grain	=	64.80 milligrams
ounce	=	28.35 grams
pound	=	453.60 grams
ton (short) (2000 pounds)	=	0.91 metric ton

To Convert	Multiply By	To Obtain
ounces	28.3	grams
pounds	453.6	grams
pounds	0.45	kilograms
grams	0.035	ounces
kilograms	2.2	grams

Volume and Capacity

Imperial		Metric
cubic inch	=	16.39 cubic centimetres
cubic foot	=	0.03 cubic metre
cubic yard	=	0.765 cubic metres
ounce	=	0.03 litre
pint	=	0.47 litre
quart	=	0.95 litre
gallon	=	3.79 litres

To Convert	Multiply By	To Obtain
fluid ounces	30.00	millilitres
quart	0.95	litres
millilitres	0.03	fluid ounces
litres	1.06	quarts

Appendix D

MOST COMMON AMINO ACIDS

Amino Acid	3-Letter Abbrev.	1-Letter Symbol
Alanine	Ala	A
Arginine	Arg	R
Aspartic acid	Asp	D
Asparagine	Asn	N
Cysteine	Cys	C
Glutamic acid	Glu	E
Glutamine	Gln	Q
Glycine	Gly	G
Histidine	His	H
Isoleucine	Ile	I
Leucine	Leu	L
Lysine	Lys	K
Methionine	Met	M
Phenylalanine	Phe	F
Proline	Pro	P
Serine	Ser	S
Threonine	Thr	T
Tryptophan	Trp	W
Tyrosine	Tyr	Y
Valine	Val	V

Note: The nine essential amino acids are highlighted.

Appendix E

FURTHER READING

Chapter 1

http://www.unep.org/newscentre/default.aspx?DocumentID=2649&ArticleID=8838

http://news.nationalgeographic.com/news/2011/08/110824-earths-species-8-7-million-biology-planet-animals-science/

http://www.pnas.org/content/87/12/4576.long

http://nutritiondata.self.com/facts/vegetables-and-vegetable-products/2482/2

http://www.fda.gov/Food/FoodborneIllnessContaminants/CausesOfIllnessBadBugBook/ucm071092.htm

Chapter 2

http://ajcn.nutrition.org/content/33/1/27.full.pdf

http://www.bio.davidson.edu/genomics/2011/Piper/Background.html

http://www.journals.elsevierhealth.com/periodicals/yjada/article/S0002-8223(03)00294-3/fulltext

http://biology.stackexchange.com/questions/546/whats-the-maximum-and-minimum-temperature-a-human-can-survive

http://www.medbio.info/Horn/Time%201-2/carbohydrate_metabolism.htm

http://www.mayoclinic.com/health/fiber/NU00033

http://culinaryarts.about.com/od/culinaryreference/a/smokepoints.htm

Chapter 3

http://www.niaid.nih.gov/topics/antimicrobialResistance/Examples/gramNegative/Pages/default.aspx

http://faculty.ccbcmd.edu/courses/bio141/lecguide/unit1/prostruct/flag.html

http://www.medicalnewstoday.com/articles/216798.php

http://users.rcn.com/jkimball.ma.ultranet/BiologyPages/J/Junctions.html

Chapter 4

http://www.sciencedirect.com/science/article/pii/S0009279706000998

http://www.mayoclinic.com/health/calories/WT00011

http://www.ncbi.nlm.nih.gov/pubmed/3593498

Chapter 5

http://www.ncbi.nlm.nih.gov/books/NBK10008/

http://www.bscb.org/?url=softcell/centrioles

http://users.rcn.com/jkimball.ma.ultranet/BiologyPages/C/CellCycle.html

http://www.lef.org/magazine/mag2009/nov2009_Horseradish-Protection-Against-Cancer-And-More_01.htm

http://www.sciencedirect.com/science/article/pii/S1044579X98900980

http://www.webmd.com/cancer/features/seven-easy-to-find-foods-that-may-help-fight-cancer

Chapter 6

http://www.rch.org.au/bloodtrans/about_blood_products/Blood_Groups_and_Compatibilities/

http://www.nature.com/scitable/topicpage/rare-genetic-disorders-learning-about-genetic-disease-979

http://www.orangutanssp.org/wp-content/uploads/2013/07/ga_blood_typing.pdf

http://mbe.oxfordjournals.org/content/14/4/399.full.pdf

http://www.redcrossblood.org/learn-about-blood/blood-types

Chapter 7

http://enaliaphysis.org.cy/educational-initiatives/public-lectures/natural-selection/

http://www.sciencedaily.com/releases/2005/10/051026090636.htm

http://mbe.oxfordjournals.org/content/14/4/399.full.pdf

https://www.sciencenews.org/article/human-blood-types-have-deep-evolutionary-roots

Chapter 8

http://www.annualreviews.org/doi/abs/10.1146/annurev.biochem.70.1.369

http://www.uniprot.org/uniprot/P11387

http://www.ucsf.edu/news/2013/09/108886/lifestyle-changes-may-lengthen-telomeres-measure-cell-aging

http://www.sciencemag.org/content/338/6108/758

http://genetics.thetech.org/about-genetics/mutations-and-disease

Chapter 9

http://www.ndsu.edu/pubweb/~mcclean/plsc431/prokaryo/prokaryo2.htm

http://www.nature.com/scitable/topicpage/regulation-of-transcription-and-gene-expression-in-1086

http://www.nature.com/scitable/topicpage/the-role-of-methylation-in-gene-expression-1070

http://www.ncbi.nlm.nih.gov/pubmed/14650573

http://www.ncbi.nlm.nih.gov/pubmed/22155943

Chapter 10

http://www.biotechniques.com/multimedia/archive/00036/BTN_A_04361RR02_O_36116a.pdf

http://www.news-medical.net/health/What-is-Gene-Therapy.aspx

http://www.forbes.com/sites/stevensalzberg/2012/09/24/does-genetically-modified-corn-cause-cancer-a-flawed-study/

http://www.ncbi.nlm.nih.gov/pubmed/12369203

http://www.sciencedaily.com/releases/2011/08/110811201523.htm

http://www.dailymail.co.uk/sciencetech/article-1080042/Meet-Mr-Green-Genes--worlds-glow-dark-cat.html

http://sciencereview.berkeley.edu/bridges-made-of-spider-silk-you-can-thank-the-goats-for-that/

Chapter 11

http://www.mhhe.com/biosci/ap/histology_mh/simpleep.html

Chapter 12

http://www.medindia.net/education/familymedicine/Burns-Classification.htm#

http://www.lpch.org/DiseaseHealthInfo/HealthLibrary/burns/classify.html

http://www.mayoclinic.com/health/melanoma/DS00439/DSECTION=causes

http://www.ncbi.nlm.nih.gov/pubmed/19519827

http://www.nlm.nih.gov/medlineplus/ency/article/004014.htm

http://lpi.oregonstate.edu/infocenter/skin/vitaminE/

http://www.nlm.nih.gov/medlineplus/ency/article/004014.htm

http://www.webmd.com/beauty/aging/myth-vs-reality-on-anti-aging-vitamins

Chapter 13

http://www.diabetes.org/news-research/research/access-diabetes-research/physical-activity-remains-key.html

http://www.joanvernikos.com/pages/sitting-kills-moving-heals.php

http://www.mayoclinic.com/health/tmj-disorders/DS00355

http://www.webmd.com/menopause/osteoporosis-menopause

http://www.niams.nih.gov/Health_Info/Bone/Bone_Health/Nutrition/vitamin_a.asp

http://www.canorth.org/en/patienteducation/Default.aspx?pagename=Long%20Bone%20Fractures

http://www.osteoporosis.ca/

http://www.osteoporosis.ca/osteoporosis-and-you/what-is-osteoporosis/

Chapter 14

http://www.ncbi.nlm.nih.gov/pubmed/7286246

http://lpi.oregonstate.edu/infocenter/othernuts/choline/

http://link.springer.com/article/10.1007%2Fs00109-002-0384-9

http://www.ncbi.nlm.nih.gov/pmc/articles/PMC1133170/pdf/biochemj00125-0011.pdf

http://www.livestrong.com/article/423722-workouts-that-increase-speed-train-fast-twitch-muscles/

Chapter 15

http://pubs.niaaa.nih.gov/publications/aa63/aa63.htm

http://www.health.harvard.edu/press_releases/benefits-of-exercisereduces-stress-anxiety-and-helps-fight-depression

http://pubs.niaaa.nih.gov/publications/arh21-2/144.pdf

http://www.mayoclinic.org/diseases-conditions/glioma/basics/definition/con-20035538

http://www.sciencedirect.com/science/article/pii/000689937490208X

Chapter 16

http://www.ncbi.nlm.nih.gov/pmc/articles/PMC3079864/

http://www.ncbi.nlm.nih.gov/pmc/articles/PMC329619/

http://www.mayoclinic.org/healthy-living/stress-management/in-depth/stress/art-20046037?footprints=mine

http://www.nlm.nih.gov/medlineplus/ency/article/000353.htm

http://onlinelibrary.wiley.com/doi/10.1111/j.1469-445X.2000.tb00014.x/abstract

http://diabetes.diabetesjournals.org/content/43/2/212.short

Chapter 17

http://www.nlm.nih.gov/medlineplus/ency/article/001179.htm

http://www.mayoclinic.com/health/cholesterol-levels/CL00001

http://circ.ahajournals.org/content/111/5/e89.full

http://www.mayoclinic.com/health/trans-fat/CL00032

http://www.heart.org/HEARTORG/Conditions/More/MyHeartandStrokeNews/Menopause-and-Heart-Disease_UCM_448432_Article.jsp

http://www.heartandstroke.com/site/c.ikIQLcMWJtE/b.3484033/k.2811/Heart_disease__Excessive_alcohol_consumption.htm

http://diabetes.niddk.nih.gov/dm/pubs/stroke/#connection

http://www.nhlbi.nih.gov/health/health-topics/topics/hbp/

Chapter 18

http://www.nhlbi.nih.gov/health/health-topics/topics/copd/

http://www.nhlbi.nih.gov/health/health-topics/topics/asthma/causes.html

https://www.lung.ca/diseases-maladies/copd-mpoc/what-quoi/index_e.php

http://www.asthma.ca/

http://www.ncbi.nlm.nih.gov/pubmed/19399004

Chapter 19

http://www.webmd.com/digestive-disorders/understanding-ulcers-basic-information

http://www.ncbi.nlm.nih.gov/pubmed/8822613

http://www.ncbi.nlm.nih.gov/pubmed/23384445

http://www.ncbi.nlm.nih.gov/pubmed/23041079

http://www.ncbi.nlm.nih.gov/books/NBK9847/

http://www.ncbi.nlm.nih.gov/books/NBK54116/#s4.2
http://www.ncbi.nlm.nih.gov/pubmed/12362464
http://www.ccfc.ca/site/c.ajIRK4NLLhJ0E/b.6349429/
http://www.ncbi.nlm.nih.gov/pubmed/19607778

Chapter 20

http://www.kidney.org/kidneydisease/aboutckd.cfm
http://www.mayoclinic.org/diseases-conditions/kidney-stones/
    basics/definition/con-20024829
http://www.med.umich.edu/intmed/nephrology/docs/stones.pdf
http://kidney.niddk.nih.gov/kudiseases/pubs/highblood/

Chapter 21

http://www.ncbi.nlm.nih.gov/pmc/articles/PMC2941592/
http://www.medicinenet.com/endometriosis/article.htm

http://www.cdc.gov/std/pid/stdfact-pid.htm
http://www.prostatecancer.ca/

Chapter 22

http://www.ncbi.nlm.nih.gov/pubmed/15877894
http://www.ncbi.nlm.nih.gov/pubmed/20066002
http://www.ncbi.nlm.nih.gov/pubmed/9553779

Chapter 23

http://www.ncbi.nlm.nih.gov/pubmed/22291721
http://www.cdc.gov/mmwr/preview/mmwrhtml/mm5021a2.htm
http://www.phac-aspc.gc.ca/lab-bio/res/psds-ftss/necator-americanus-
    eng.php
http://sabre.ucsf.edu/docs/Science-2011-Wu-243-7.pdf
http://www.wildfermentation.com/making-sauerkraut-2/

Appendix F

MEDICAL TERMINOLOGY

General Terms

Term	Definition	Example
ab-	away from	abduction
ad-	toward, near	adduction
-algia	pain	neuralgia
ambi-	positioned on both sides	ambidextrous
amphi-	around, on both sides	amphipathic
ana-	upward, backward	anaplasia
anomal/o	irregular	anomaly
ante-	before, forward	anteflexion
anter/o	front	anterior
anthrop/o	man, human being	anthropology
apo-	away, separation	apoptosis
bio-, bi/o	life, living	biology
blast/o, -blast	early embryonic stage, immature	blastocysts
calor/i	heat	calorie
carcin/o	cancer	carcinoma
cata-	down, downward	cataract
caud/o	tail	caudal
-cele	pouching	varicocele
chem(o)	chemistry, drug	chemotherapy
chron/o	time, timing	chronological
circum-	around	circumference
cis	on this side	cis fatty acid
co-	together	coenzyme
contra-	against, opposite	contralateral
corpor/o	body	corporeal
cyt/o, -cyte	cell	erythrocyte
de-	away from or cessation	dehydration
dextr/o	right, on the right side	dextrose
dia-	through, apart	diaphysis
dif-	separation	differentiate
dis-	apart, to separate	dislocate
dors/o	back	dorsal
-dynia	pain	gastrodynia
dys-	difficult, abnormal	dyspnea
ec-, ecto-	outside, out	ectopic
-ectomy	surgical removal	appendectomy
en-, endo-	inside, within	endocrine
epi-	above, over, upon	epigenetics
equi-	equality, equal	equilibrium
eso-	within	esophagus
eti/o	cause	etiology
exo-	outside of, outward	exocytosis
extra-	outside	extracellular
fore-	before, in front of	forehead
-genesis	production, formation	gluconeogenesis
-genic	producing	cardiogenic
hemi-	one half	cerebral hemisphere
hetero-	the other of two, different	heterozygote
hist/o	tissue	histology
homo-	similar	homozygous
hyper-	above, excessive, beyond	hypertonic
hypo-	under, deficient, below	hypothalamus
-ic	pertaining to	hepatic
-icle	small	follicle
infra-	below, beneath	infraspinatus
inter-	between	interstitial
intra-	within	intracellular
ipsi-	same	ipsilateral
juxta-	near	juxtaglomerular
kary/o	nucleus	eukaryote
kel/o	fibrous, growth	keloid
later/o	side	lateral
levo-	left	levodopa
-logist	specialist	neurologist
-logy	study of	cardiology

macr/o	large	macromolecule
media/o	middle	medial
meso-	middle	mesoderm
meta-	after, behind	metacarpal
micro-	small	microorganism
mon/o	single	mononucleosis
morph/o	shape, form	polymorphic
nucle/o	nucleus	nucleoplasm
nutri/o	to nourish	nutrition
para-	alongside, near, beyond, abnormal	parathyroid
per-	through, throughout	periodic table
peri-	around, surrounding	peritubular
physi/o	nature	physiologist
-plasm	formation, growth	neoplasm
-poiesis	formation	hematopoiesis
post-	after, behind	post-translation
poster/o	behind, towards the back	posterior
pre-	before, in front of	premature
pro-	before	prophylactic
proxim/o	near	proximal
retro-	behind, backward	retrolingual
-sect	to cut	section
somat/o	body	somatoscopy
-some	body	chromosome
-stomy	surgical opening	colostomy
sub-	under, beneath	subcutaneous
super-	above, beyond	superior vena cava
supra-	above, beyond	supraorbital
-therapy	treatment	therapeutics
therm/o	heat	thermometer
-tomy	surgical incision	appendectomy
trans-	across	transmembrane
-type, typ/o	class, representative form	phenotype
ventr/o	belly, front of the body	ventrolateral
-verse	turn	transverse
viscer/o	internal organs	visceromegaly

Integumentary System

Term	Definition	Example
acanth/o	thorny, spiny	acanthoma
adip(o)	relating to fat	adipocyte
alb-	white or pale	albino

brom/o	odour	bromoderma
cauter/o	burn, burning	cauterize
chrom/o	colour	chromophore
cili-	eyelid or eyelashes	cilia
cry/o	cold	cryotherapy
cutane/o	skin	subcutaneous
derm/o	skin	dermatologist
erythem/o	flushed, redness	erythema
eschar/o	scab	escharotomy
graph/o	writing	electrocardiograph
hidr/o	sweat	hyperhidrosis
ichthy/o	fish	ichthyosis
iod/o	iodine	iododerm
kerat/o	horny tissue, cornea	keratolysis
lepid/o	flakes, scales	lepidosis
onych/o	nail	onychomycosis
pachy-	thick	pachydermatocele
papul/o	papule, pimple	papulopustular
perspir/o	breathe through	perspiration
phyt/o	plant	phytophotodermatitis
pil/o	hair	pilomotor
pseudo-	false	pseudomonas
psor/o	itching	psoriasis
py/o	pus	pyodermatitis
seb/o	sebum	seborrhea
strata	layer	stratified
trich/o	hair	hypertrichosis
ul/o	scar, scarring	uloid
ungu/o	nail	subungual
verruc/i	wart	verrucosis
xer/o	dry	xeroderma

Skeletal System

Term	Definition	Example
acr/o	extremity, topmost	acromion
arthr/o	joint	arthroscopy
articul/o	joint	articulation
brachi/o	arm	brachiocephalic
burs/o	bursa	bursitis
calcane/o	heel	calcaneodynia
carp/o	wrist	carpoptosis
cephal/o	head	cephaledema
cervic/o	neck, cervix	cervical spine
chondr/o	cartilage	chondrocyte

(Continued)

Skeletal System (continued)

TERM	DEFINITION	EXAMPLE
-clast	break, breaking	osteoclast
cleid/o	clavicle	cleidorrhexis
coccyg/o	coccyx	coccygodynia
condlye	smooth articular surface	occipital condyle
cost/o	rib	costosternal
cox/o	hip	coxarthrosis
crani/o	skull	cranioclast
cubit/o	elbow, forearm	genucubital
-dactyl(o)-	pertaining to a finger or toe	polydactyly
femor/o	femur	ischiofemoral
fibul/o	fibula	fibulocalcaneal
foramen	opening	foramen ovale
fossa	hollow area	fossa ovalis
hamus	hook	hamate
humer/o	humerus	humeroradial
ileo-	ileum	ileocecal
ili/o	ilium	iliolumbar
ischi/o	ischium	ischiodynia
itchy/o	erect, straight	ithylordosis
-itis	inflammation	hepatitis
kyph/o	humpback	kyphoscolilosis
lip/o	fat	liposuction
lord/o	curvature, bending	lordoscoliosis
maxill/o	maxilla	maxillotomy
mega-, megalo-	enlargement	acromegaly
-megaly	enlargement	dactylomegaly
-oid	resembling	osteoid
oste/o	bone	osteochondroma
pan-	all	panarthritis
patell/o	patella	patellofemoral
ped/o	foot, child	pedal
pelv/i	pelvis	pelvimeter
phalang/o	phalanges	phalangitis
-physis	growth, growing	diaphysis
pod/o	foot	podiatrist
pub/o	pubis	pubovesical
sacr/o	sacrum	sacrocoxalgia
scapul/o	scapula	scapulopexy
scoli/o	crooked, twisted	scoliorachitic
skelet/o	skeleton	skeletogenous
spin/o	spinal cord, spine	spinocerebellar
spondyl/o	vertebrae	spondylopyosis

stern/o	sternum	sternocostal
synov/o	synovia, synovial membrane	synovectomy
tal/o	talus	talofibular
tars/o	tarsus, edge of eyelid	tarsoclasis
tempor/o	the temples	temporal bone
tibi/o	tibia	tibiotarsal
xiph/o	sword-shaped, xiphoid	xiphoid process
zyg/o	union, junction	zygomatic bone

Muscular System

TERM	DEFINITION	EXAMPLE
-asthenia, ashten/o	weakness	myasthenia gravis
aux/o	growth, acceleration	auxotrophy
bi	double, twice	biceps
-chalasia	relaxation	achalasia
dynam/o	power, strength	dynamogenesis
erg/o	work	ergometer
fasci/o	fascia	fascia
fibr/o	fibre, fibrous	fibroblast
flex/o	bend	flexor
kinesi/o, kinesia, -kinetic	movement	kinesthesia
lei/o	smooth	leiomyoma
ligament/o	ligament	ligamentopexy
-lysis	dissolution, breakdown	hydrolysis
muscul/o	muscle	musculoskeletal
my/o, myos/o	muscle	myorrhexis
orbis-	circle	orbicularis
pale/o	old	paleokinetic
pyg/o	buttocks	pygalgia
rectus	straight	rectus femoris
rhabd/o	rod	rhabdoid
rhabdomy/o	striated/skeletal muscle	rhabdomyolysis
rot/o, rotat/o	turn, revolve	rotator
-stasis	standing still, standing	homeostasis
sthen/o, -sthenia	strength	sthenometry
-stroma	supporting tissue of an organ	myostroma
syndesm/o	ligament, connective tissue	syndesmectopia
ten/o	tendon	tenorrhaphy
tenont/o	tendon	tenontography

-therapy, therapeut/o	treatment	kinesiotherapy
ton/o	tone, tension	myatonia

Nervous System

TERM	DEFINITION	EXAMPLE
acous/o	hearing	acouesthesia
acoust/o	hearing, sound	acoustics
-algesia, alges/o	pain sensitivity	analgesia
astr/o	star, star shaped	astrocyte
audi/o, audit/o	hearing	audiometer
aur/o, auricul/o	ear	auricular
blephar/o	eyelid	blepharorrhaphy
cerebell/o	cerebellum	cerebellospinal
cerebr/o	cerebrum, brain	cerebrovascular
cochle/o	cochlea	cochletisis
conjunctiv/o	conjunctiva	conjunctinitis
cor/o	pupil	corectasia
corne/o	cornea	corneosclera
-cusis	hearing	presbycusis
cycl/o	ciliary body, circular	cyclodislysis
dacry/o	tear	dacryoadenectomy
dipl/o	double	diploid
drom/o,	running	syndrome
encephal/o	brain	encephalitis
esthesi/o,	sensation, feeling	anesthesia
gangli/o,	ganglion	gangliocytoma
gli/o	glue, neuroglia	gliocyte
hydr/o	water, hydrogen	hydrophobic
hygr/o	moisture	hygroblepharic
hypn/o	sleep	hypnogenic
hypothalam/o	hypothalamus	hypothalamo-hypophysical
ir/o, irid/o	iris	iridemia
kerat/o	cornea, horny tissue	keratomalacia
klept/o	theft, stealing	kleptomania
lacrim/o	tear, lacrimal duct	lacrimotomy
-lemma	confining membrane	epilemmak
-lepsy	seizure	epilepsy
-lexia	speech, word	bradylexia
log/o, log, -logue	word, speech, thought	logorrhea
-lucent	light-admitting	radiolucent
lumin/o	light	luminescence
-mania	madness, obsessive, preoccupation	hypomania

medull/o	medulla, marrow	medulloblast
mening/o	meninges, membranes	meningitis
ment/o	mind	dementia
-mnesia	memory	ecmnesia
myel/o	bone marrow, spinal cord	myeloblast
myring/o	eardrum	myringomycosis
narc/o	numbness	narcolepsy
neur/o	nerve	neurotransmitter
-noia	mind, will	paranoia
ocul/o	eye	oculonasal
ophry/o	eyebrow	ophryitis
ophthalm/o	eye	ophthalmodynia
-opia, -opsia	vision	heteropsia
opt/o	eye, vision	optometer
ot/o	ear	otology
palpebr/o	eyelid	palpebritis
-paresis	partial paralysis	hemiparesis
phac/o	lens	phacocele
phak/o	lens	phakoma
-phobia, phob/o	fear, aversion	hydrophobic
phot/o	light	photophobia
phren/o	mind, diaphragm	tachyphrenia
-piesis	pressure	piesesthesia
platy-	broad, flat	platycoria
-plegia	paralysis	quadriplegia
poli/o	grey (matter)	poliomyelitis
-pore	opening, passageway	neuropore
-praxia	action, activity	parapraxia
psych/o	mind	psychokinesis
psychr/o	cold	psychrophobia
pupill/o	pupil	pupillatonia
radicul/o	nerve root	radiculities
retin/o	retina	retinotoxic
rhiz/o	root	rhizotomy
schiz/o	split, division	schizophrenia
somn/l, -somnia	sleep	insomnia
son/o	sound	sonometer
staped/o	stapes	stapedectomy
syring/o	tube, fistula	syringomyelocele
-taxia, tax/o	arrangement, coordination	dystaxia
-tropia	to turn	anatropia

(Continued)

Nervous System (*continued*)

TERM	DEFINITION	EXAMPLE
tympan/o	eardrum	tympanic membrane
uve/o	uvea	uveoplasty
vag/o	vagus nerve	vagolysis
vitre/o	glassy, vitreous body	vitreocapsulaitis

Endocrine System

TERM	DEFINITION	EXAMPLE
aden/o	gland	adenohypophysis
adren/o	adrenal glands	adrenomegaly
-comat/o	deep sleep	comatose
cortic/o	cortex	corticoadrenal
crin/o	secrete, separate	crinogenic
duct/o	to lead	conduction
ex-	out, away from	exopthalmos
ger/o, geront/o	aged, old age	geriatrics
hirsut/o	hairy	hirsutism
hormon/o	hormone	hormonopoiesis
-ism	condition or disease	hypothyroidism
medull/o	medulla, marrow	medulloadrenal
myx/o	mucus	myxedema
pancreat/o	pancreas	pancreatolihotomy
phe/o	dusky	pheochromoctoma
pineal/o	pineal gland	pinealopathy
pituitar/o	pituitary gland	pituitarism
thym/o	thymus gland	thymotoxin
thyr/o	thyroid gland	thyrocele
tox/o, toxic/o	poison	toxin
-trophy, troph/o	nourishment, growth	hypertrophy

Cardiovascular System

TERM	DEFINITION	EXAMPLE
agglutin/o	clumping	agglutinin
angi/o	vessel	angiogenesis
aort/o	aorta	aorta
arteri/o	artery	arteriosclerosis
arteriol/o	small artery	arteriole
ather/o	fatty substance, plaque	atheroma
bar/o	weight, pressure	baroreceptor
bas/o, basi/o	base, foundation	basophil
brady-	slow	bradycardia
capill-	resembling hair	capillary
cardi/o	heart	cardioptosis

coagul/o	clotting	coagulopathy
coron/o	heart	coronary
-crit	separate	thrombocytocrit
-emia	blood condition	anemia
eosin/o	red	eosinophil
erythr/o	red	erythrocyte
ferr/i, ferr/o	iron	ferrometer
fil/l, fil/o	thread	filopressure
-gram	to record, picture	angiogram
hem/o, hemat/o	blood	hematopoeisis
isch/o	suppress, restrain	ischemia
kal/i	potassium	hyperkalemia
leuk/o	white	leukocytotoxin
-megaly	enlargement	atrimegaly
mi/o	less, smaller	miocardia
-motor	movement, motion	venomotor
myel/o	bone marrow, spinal cord	myelocytosis
natr/o	sodium	hypernatremia
palpit/o, palpitat/o	flutter, throbbing	palpitation
-pheresis	removal	plateletpheresis
-phil, -philia	affinity for, tendency towards	hemophilia
phleb/o	vein	phleborrhexis
-phoresis	bearing, transmission	electrophoresis
-plasty	surgical correction/ repair	angioplasty
rhe/o	flow, current, stream	rheocardiography
schist/o,	split, cleft	schistocyte
scler/o	hard	scleroderma
-sclerosis	hardening	atherosclerosis
ser/o	serum, serous	serosanguineous
sider/o	iron	sideropenia
-spasm	involuntary contraction	vasospasm
spher/o	round, sphere	spherocytosis
sphygm/o	pulse	sphygmascope
-sphyxia	pulse	asphyxia
-stenosis	narrowed, constricted	aortostenosis
strept/o	twisted, curved	streptococcus
tachy-	fast	tachycardia
tel/e	end, distant	telecardiography
-tension	stretched, strained	hypertension

thromb/o	clot, thrombus	thromboelasto-gram
varic/o	varicose veins	varicophlebitis
vas/o	vessel, vas deferens	vasophypotonic
vascul/o	vessel	vasculitis
ven/o	vein	venography
ventricul/o	ventricle of the heart or brain	ventriculogram
venul/o	venule	venular
-volemia	blood volume	normovolemia

Respiratory System

Term	Definition	Example
aspir/o, aspirat/o	inhaling, removal	aspiration
blenn/o	mucus	blennothorax
brachy-	short	brachypnea
bronch/o	bronchus	bronchodilation
bronchiol/o	bronchiole	bronchiolitis
-capnia, capn/o	carbon dioxide	hypercapnia
epiglott/o	epiglottis	epiglottitis
lal/o, -lalia	speech, babble	laliatry
laryng/o	pertaining to the lower throat	larynx
nas/o	nose	nasolabia
osm/o, -osmia	sense of smell, odour, impulse	anosmia
osphresi/o, -osphresia	sense of smell, odour	osphresiometer
ox/o, -oxia	oxygen	hypoxia
pector/o	chest	pectoralgia
phas/o, -phasia	speech	dysphasia
phon/o, -phonia	voice, sound	rhinophonia
phren/o	mind, diaphragm	phrenalgia
-pnea	breathe	hyperpnea
pneum/o	lung, air	pneumopexy
pneumon/o	lung, air	pneumonomy-cosis
-ptosis	prolapse, drooping	laryngoptosis
-ptysis	spitting	hemoptysis
pulmon/o	lung	pulmonologist
respir/o, respirat/o	breathe, breathing	respirator
rhin/o	nose	rhinolithiasis
sept/o	partition	septorhino-plasty
sinus/o	cavity, sinus	sinusotomy
spir/o	breathe	bronchospirom-eter

steth/o	chest	stethoscope
therm/o	heat	thermopolypnea
trache/o	trachea	tracheotomy
traumat/o	trauma, injury, wound	traumatopnea
xen/o	strange, foreign	xenophonia

Digestive System

Term	Definition	Example
-agra	severe pain	cardiagra
amyl/o	starch	amyloid
antr/o	cavity	antrum
-ase	enzyme	lipase
atel/o	incomplete, imperfect	ateloglossia
bucc/o	cheek	buccal
cheil/o, chil/o	lip	cheiloschisis
chol/e	gall, bile	cholecystokinin
choledoch/o	common bile duct	choledochlithi-asis
col/o	colon	colonoscopy
dent/i	tooth	dentist
dips/o	thirst	adipsia
-emesis	vomiting	hematemesis
enter/o	intestines (small intestines)	enteroclysis
gastr/o	stomach	gastric bypass
gingiv/o	gums	gingivoplasty
gloss/o	tongue	glossolalia
gluc/o	glucose, sugar	glucocorticoid
glyc/o	glucose, sugar	glycolysis
hepat/o	liver	hepatospleno-megaly
lapar/o	abdomen, abdominal wall	laparoscope
lingu/o	tongue	retrolingual
odont/o	tooth	anodontia
or/o	mouth	intraoral
-orexia	appetite	anorexia
orth/o	straight, normal, correct	orthodontist
-pepsia	digestion	dyspepsia
phag/o, phagia	eating, ingestion	phagocytosis
-posia	drinking	polypposia
-prandial	meal	postprandial
proct/o	rectum, anus	proctopexy
ptyal/o	saliva	ptyalogenic
pyr/e	portal vein	pylemphraxis
rect/o	rectum	rectocele

(Continued)

Digestive System (continued)

Term	Definition	Example
sial/o	saliva	sialolith
-stalis	contraction	peristalsis
staphyl/o	uvula, grapelike clusters	staphylorrhaphy
stomat/o	mouth	stomatomalacia
-tresia	opening, perforation	proctotresia
typhl/o	cecum, blindness	typhlectasis
uran/o	palate	uranoschisis
zym/o	enzyme, ferment	lysozyme

Urinary System

Term	Definition	Example
a-, an-	without, not	anuria
-astresia	closure, occlusion	urethratresia
atreto-	closed, lacking an opening	astretocystia
calci/o	calcium	hypocalciuria
cali/o	calyx	pyelocaliectasis
cupr/o	copper	cupruresis
cyan/o	blue	urocyanosis
cyst/o	bladder, cyst	cytogram
-ectasis, -ectasia	dilation, expansion	nephrectasia
fusc/o	dark brown	urofuscohematin
glomerul/o	glomerulus	glomerulopathy
hydr/o	water, hydrogen	hydrolysis
iso-	being equal	isotonic
lith/o	stone, calculus	lithotripsy
ly/o	dissolve, loosen	lyophilic
nephr/o	kidney	nephrotoxic
noct/i	night	noctalbuminuria
py/o	pus	pyocalix
pyel/o	renal pelvis	pyelophlebitis
ren/o	kidney	renogastric
-tripsy	to crush, break	lithotripsy
ur/o	urine	urology
-uresis	urination	diuresis
ureter/o	ureter	ureterocolostomy
urethr/o	urethra	urethrorrhagia
-uria	urine condition	ketonuria
uric/o	uric acid	uric acid stones
urin/o	urine	urinalysis
vesic/o	urinary bladder	vesicle

Reproductive System

Term	Definition	Example
amni-	membranous fetal sac	amniocentesis
andr/o	male	androgen
balan/o	glans penis	balanitis
cervic/o	neck, cervix	cervicovaginitis
-cide	killing, agent which kills	spermicide
-classis, -clast	break, breaking	cranioclasis
colp/o	vagina	colporrhaphy
crypt/o	hidden, concealed	cryptorchism
-cyesis	pregnancy	ovariocyesis
embry/o	embryo	embryopathy
epididym/o	epididymis	epididymectomy
episi/o	vulva	episiostenosis
fet/o	fetus	fetography
galact/o	milk	galactose
genit/o	reproduction	genitourinary
gon/o	genitals	Gonorrhea
gonad/o	gonads	gonadogenesis
gravid/o	pregnancy	gravidocardiac
gynec/o	woman/female	gynecology
helc/o	ulcer	helcomenia
hyster/o	uterus	hysterectomy
lact/o	milk	lactorrhea
-lipsis	omit, fail	menolipsis
mamm/o	breast	mammography
mast/o	breast	mastalgia
men/o	menses, menstruation	menorrhagia
nat/o	birth	neonatology
neo-	new	neonatal
nulli-	none	nulliparity
o/o	egg, ovum	oogenesis
obstetr/o	midwife	obstetrician
olig/o	scanty, few, little	oligodendrocyte
oophor/o	ovary	oophorohyster-ectomy
orch/o, orchi/o, orchid/o	testis	orchidopexy
osche/o	scrotum	oscheoplasty
ov/o, ov/i	egg, ovum	ovicide
ovari/o	ovary	ovariocentesis
-partum	childbirth, labour	postpartum
phall/o	penis	phallodynia
prostat/o	prostate gland	prostatocy-totomy
sacchar/o	sugar	monosaccharide
salping/o	fallopian tube	salpingcyesis
semin/i	semen	seminiferous

test/o, testicul/o	testis	testectomy
thel/o	nipple	thelorrhagia
vas/o	vas deferens, vessel	vasovasostomy
venere/o	sexual intercourse	venereologist
vesicul/o	seminal vesicle	vasovesiculitis
viv/i	life, alive	viviparous
vulv/o	vulva	vulvopathy
zo/o	animal	azoospermia

Immune System

Term	Definition	Example
adenoid/o	adenoids	adenoidectomy
anti-	against	antibody
auto-	self	autoimmune
axill/o	armpit	axillary
-coccus	berry-shaped bacterium	staphylococcus
-edema	swelling	lymphedema
-emphraxis	stoppage, obstruction	splenemphraxis
fung/i	fungus, mushroom	fungus
helminth/o	worm	helminthiasis
immun/o	protection, immune	immunogenic
inguin/o	groin	inguinodynia
lien/o	spleen	lienomalacia
lymph/o	lymph	lymphangio-phlebitis

myc/o	fungus, mushroom	mycoplasma
nod/o	knot	nodular
nos/o	disease	nosology
-oma	tumour, mass	carcinoma
onc/o	tumour, mass	oncogenesis
-osis	a condition or disease	osteoporosis
path/o	disease	pathogenic
-penia	deficiency	thrombocyto-penia
peri-	around	perilymphan-gitis
-phylaxis	protection	anaphylaxis
ple/o	more	pleocytosis
pyr/o	fire, fever, heat	pyrogen
sarc/o	flesh	lymphosar-coma
-sepsis	putrefaction	antisepsis
splen/o	spleen	splenectasis
tetan/o	tetanus	tetanophilic
thym/o	thymus	thymectomy
tonsill/o	tonsils	tonsillolith
top/o	particular place or area	splenectopy
verm/i	worm	vermiculite
vir/o	virus	virology

LABELLING PRACTICE

Chapter 1: Introduction to Biology, Anatomy, and Physiology

Label the following diagrams.

1.

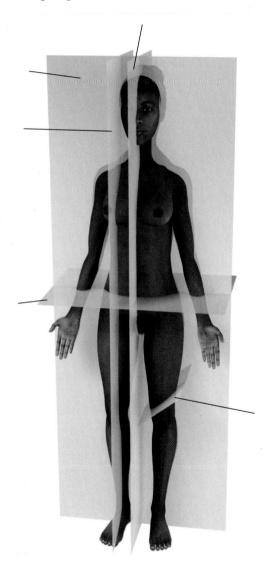

2. a.

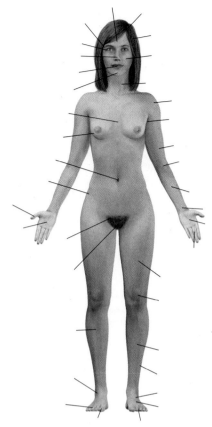

b.

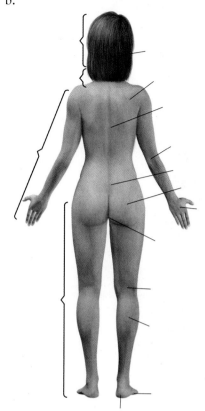

3. a.

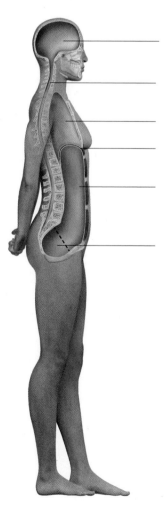

b.

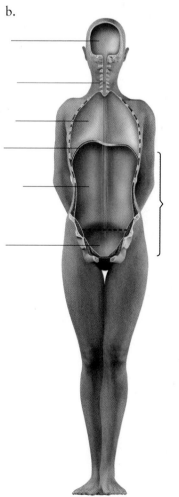

Chapter 2: Biological Molecules

Identify each of the following molecules.

1.

2.

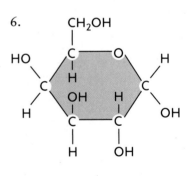

3.

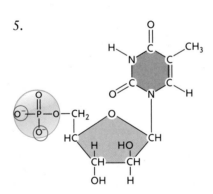

4.

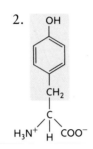

5.

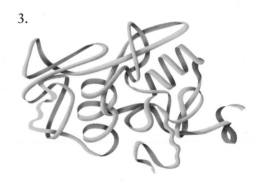

6.

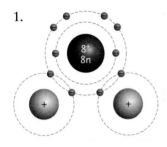

7.

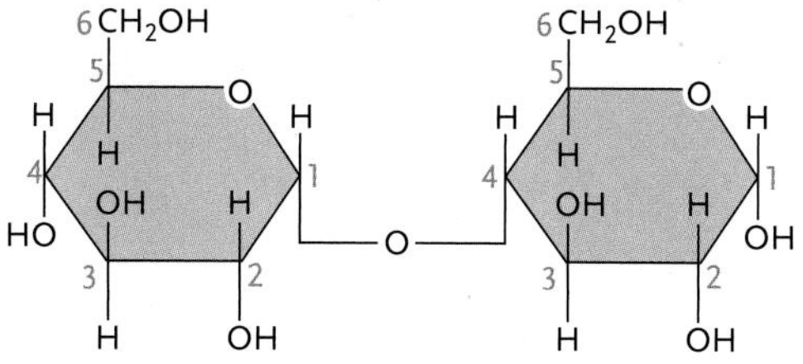

____

8.

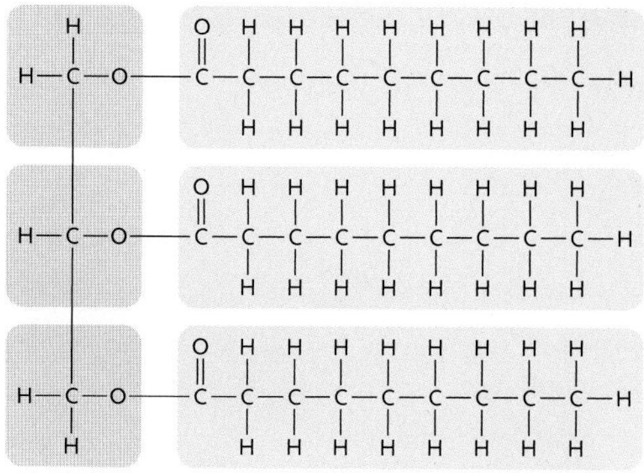

____

9.

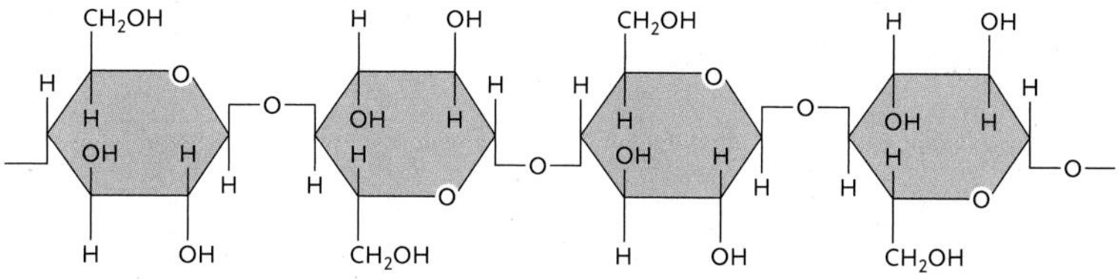

____

10.

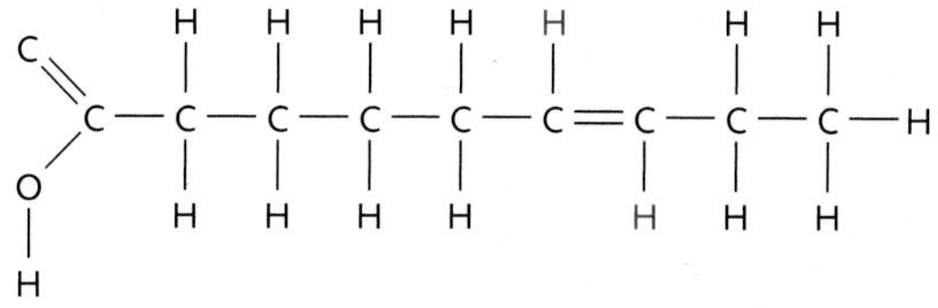

____

11.

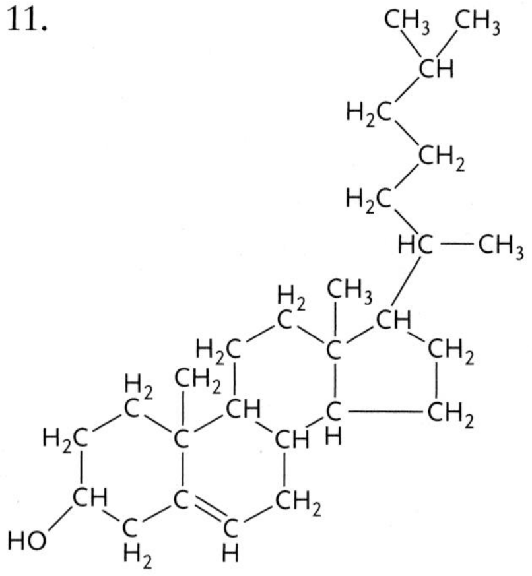

____

12.

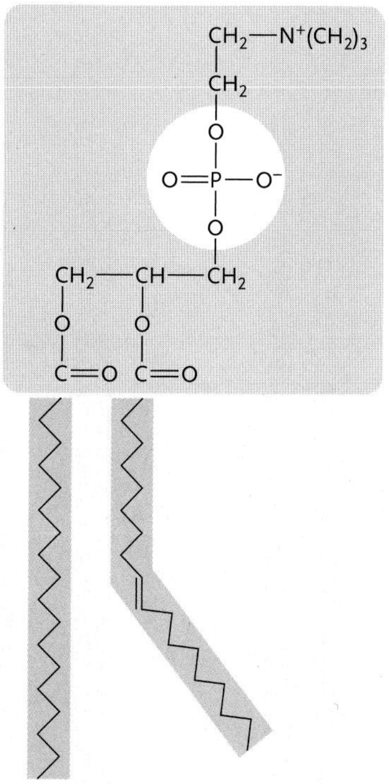

____

Art source: From RUSSELL/HERTZ/STARR/FENTON. Biology, 2E. © 2013 Nelson Education Ltd. Reproduced by permission. www.cengage.com/permissions

Chapter 3: Cell Structure and Function

Label the following diagrams.

1.

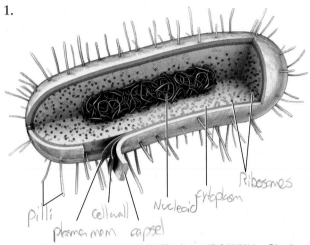

Pilli

Plasma mem   Cell wall   capsel

Nucleoid   Cytoplasm

Ribosomes

2.

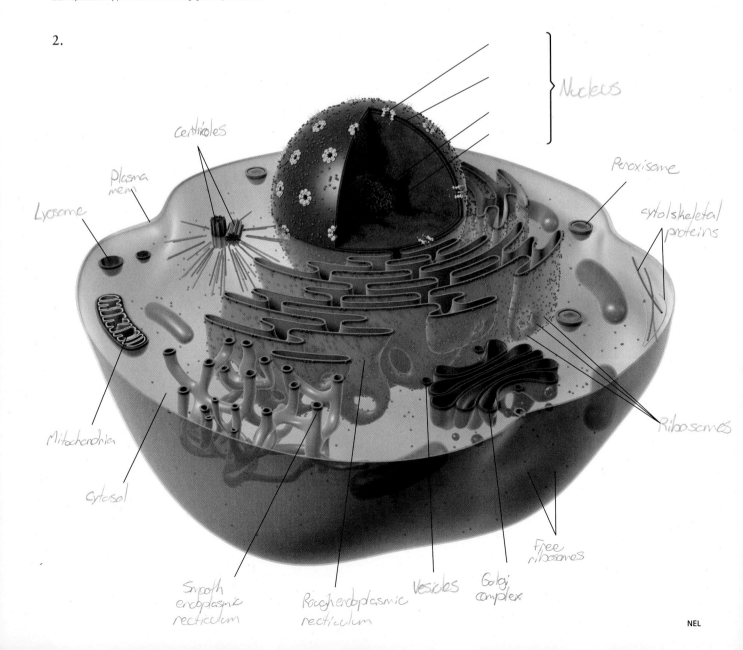

Nucleus

Centrioles

Plasma mem

Lysome

Peroxisome

cytolskeletal proteins

Mitochondria

Cytosol

Ribosomes

Free ribosomes

Smooth endoplasmic recticulum

Rough endoplasmic recticulum

Vesicles

Golgi complex

Chapter 5: Mitosis and Meiosis

Label or identify the following diagrams.

1. Label the first two diagrams and identify the process in each of the remaining diagrams.

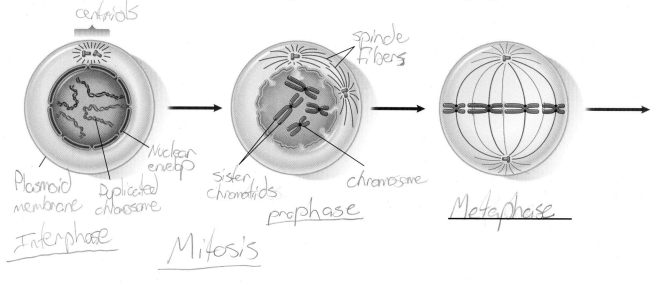

centrioles

spindle fibers

Nuclear envelop

Plasmoid membrane

Duplicated chromosome

sister chromatids

chromosome

Interphase

Mitosis

prophase

Metaphase

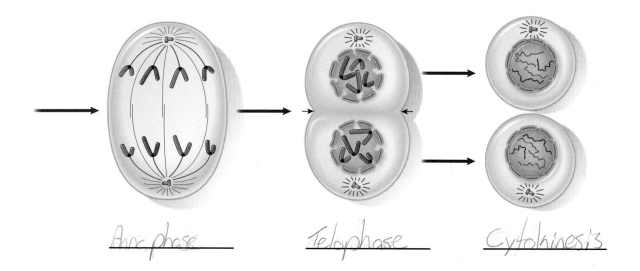

Anaphase

Telophase

Cytokinesis

Art source: From DIGIUSEPPE/FRASER. Biology 11U. © 2011 Nelson Education Ltd. Reproduced by permission. www.cengage.com/permissions

2. Identify the process in each diagram.

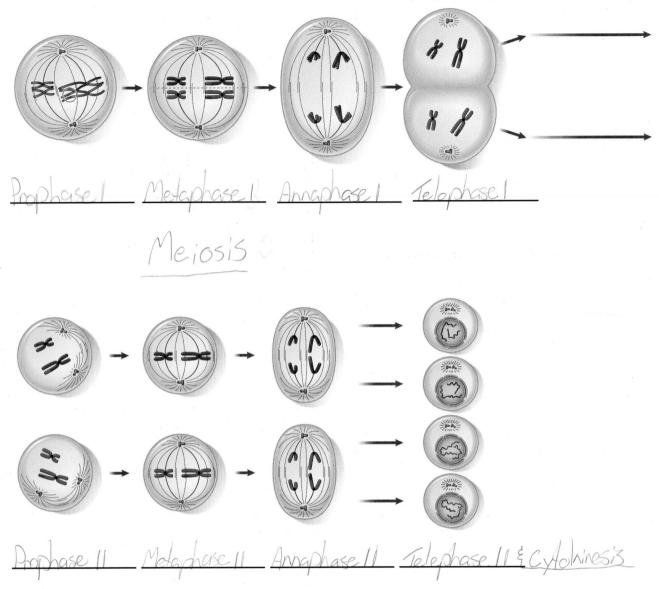

Prophase I   Metaphase I   Anaphase I   Telophase I

Meiosis

Prophase II   Metaphase II   Anaphase II   Telophase II & Cytokinesis

Chapter 9: Gene Expression and Regulation

In each example, state whether glucose and/or lactose is present and whether transcription occurs.

1. a. Label the following diagram and circle Yes or No for each of the following:

Glucose present: Yes (No)
Lactose present: Yes (No)
Transcription occurs: Yes (No)

Cap binding protein

RNA Polymerase

lac repressor

CAP binding site   *Promoter*   *operator*

b. Circle Yes or No:

Glucose present: (Yes) No
Lactose present: (Yes) No
Transcription occurs: Yes (No)

c. Circle Yes or No:

Glucose present: (Yes) No
Lactose present: Yes (No)
Transcription occurs: Yes (No)

d. Circle Yes or No:

Glucose present: Yes (No)
Lactose present: (Yes) No
Transcription occurs: (Yes) No

Chapter 11: Tissues

Label the following diagrams.

1.

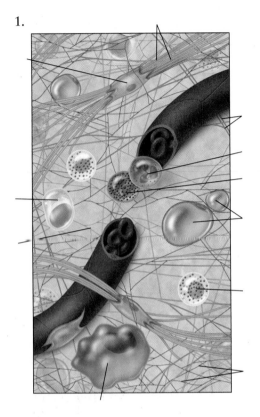

2.

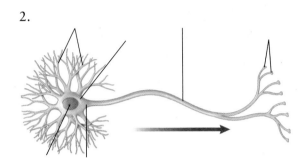

Identify the following neurons.

3. a.   b.   c.

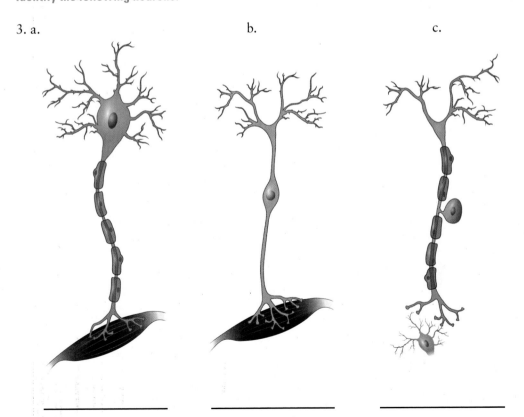

Identify the following tissues.

4.

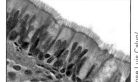

© Dr. Robert Calentine/
Visuals Unlimited/Corbis

_____

5.

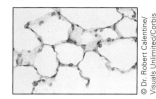

© Carolina Biological/
Visuals Unlimited/Corbis

_____

6.

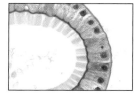

Jose Luis Calvo/
Shutterstock.com

_____

7.

Jose Luis Calvo/
Shutterstock.com

_____

8.

Jose Luis Calvo/
Shutterstock.com

_____

9.

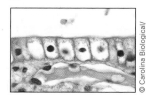

Biophoto Associates/
Science Source

_____

10.

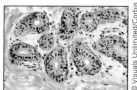

© Visuals Unlimited/Corbis

_____

11.

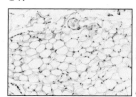

Science Vu, Visuals Unlimited
/Science Photo Library

_____

12.

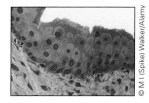

© M I (Spike) Walker/Alamy

_____

13.

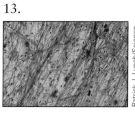

Patrick J. Lynch/Science
Source

_____

14.

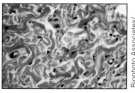

Courtesy of Wendi Roscoe

_____

15.

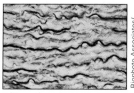

Biophoto Associates/
Science Source

_____

16.

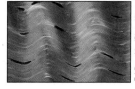

Ed Reschke/Photolibrary/
Getty Images

_____

17.

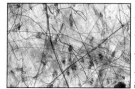

Biophoto Associates/
Science Source

_____

18.

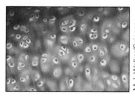

Biophoto Associates/
Science Source

_____

19.

M. I. Walker/Science
Source

_____

20.

Herve Conge, Ism/
Science Photo Library

_____

21.

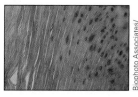

Biophoto Associates/
Science Source

_____

22.

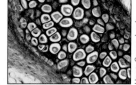

Herve Conge, Ism/
Science Photo Library

_____

23.

Biophoto Associates/
Science Photo Library

_____

24.

Eric V. Grave/Science Source

25.

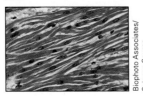

Biophoto Associates/ Science Source

26.

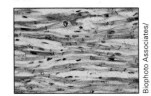

Biophoto Associates/ Science Source

Chapter 12: The Integumentary System

Label the following diagrams.

1.

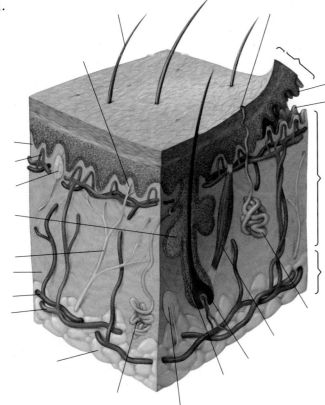

2.

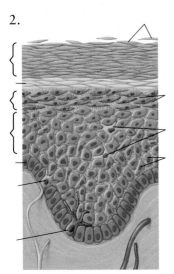

Chapter 13: The Skeletal System

Label the following diagrams.

1. a.

b.

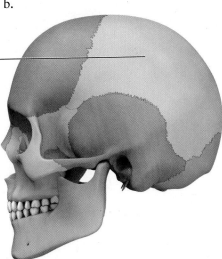

c.

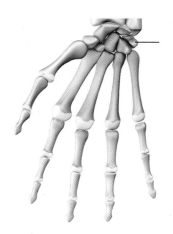

d.

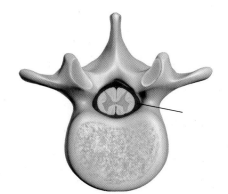

e.

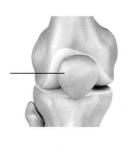

2.

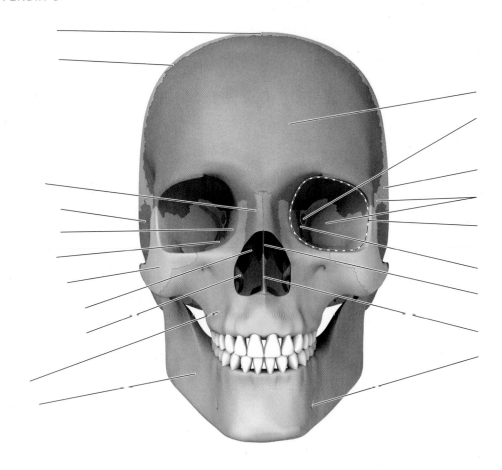

3.

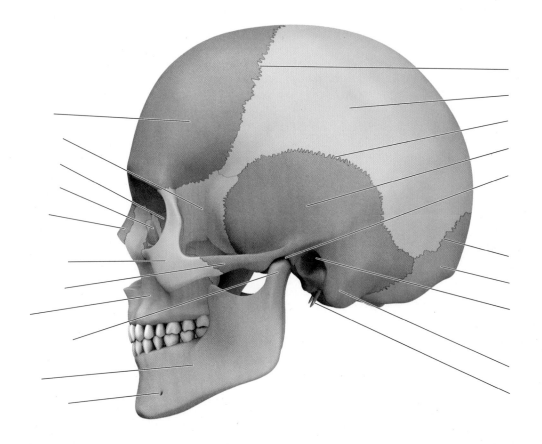

4.

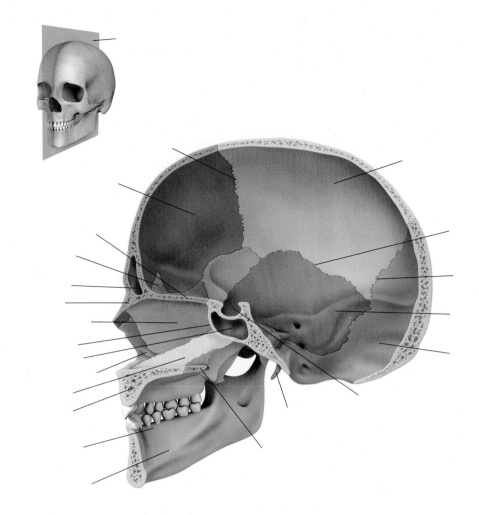

5.

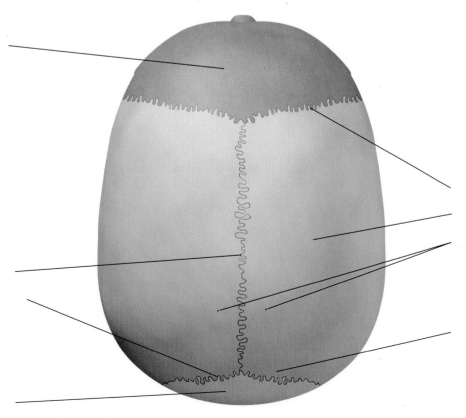

6.

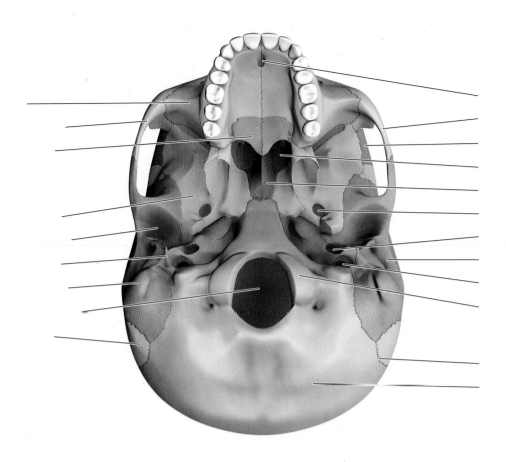

7.

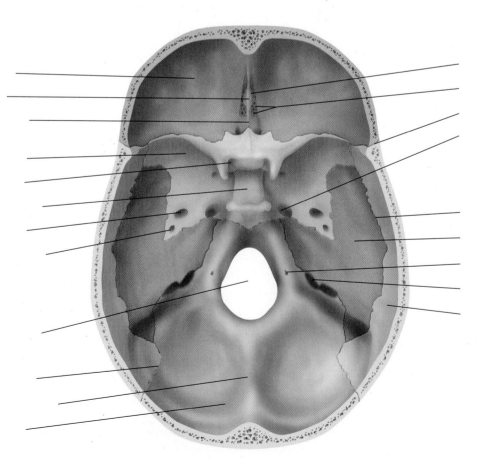

8. a.

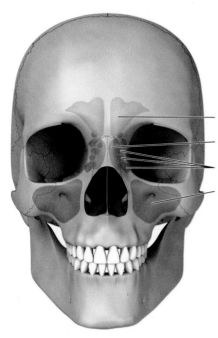

b.

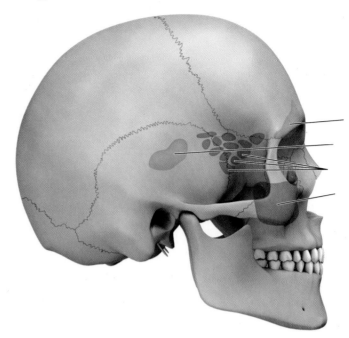

9.

Anterior   **Posterior**

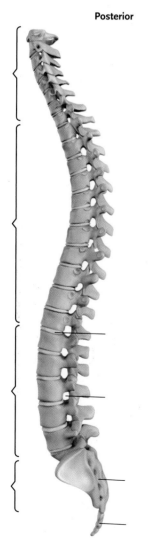

10.

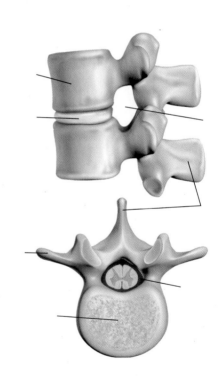

11. a. b.

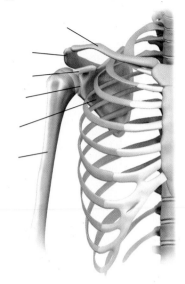
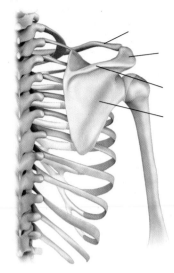

12. a. b. 13.

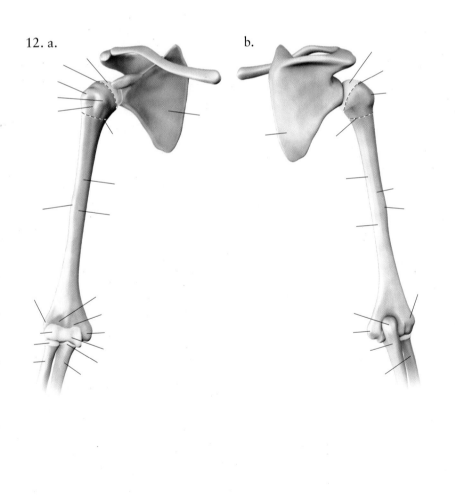
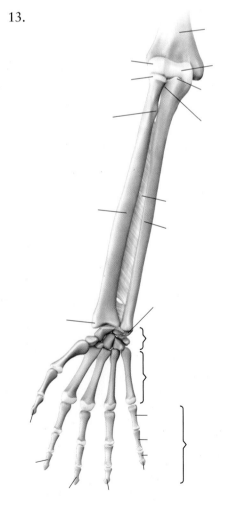

14.

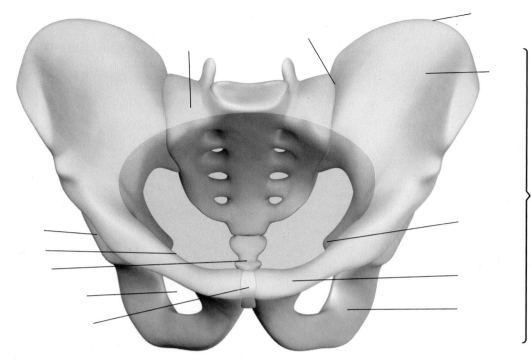

15. a.

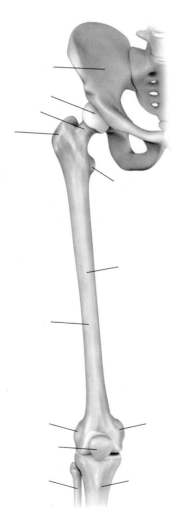

b.

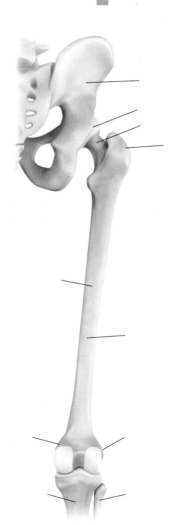

16.

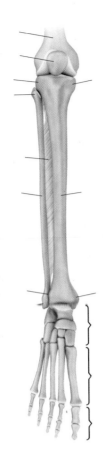

18.

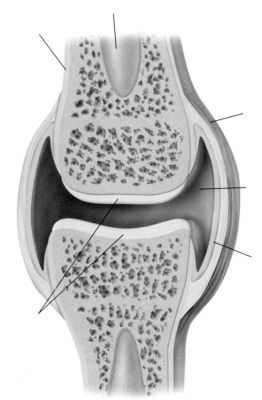

17. a.

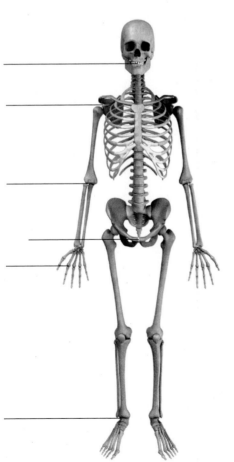

b.

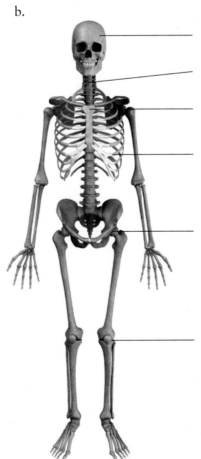

19.

a.

b.

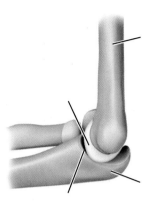

c.

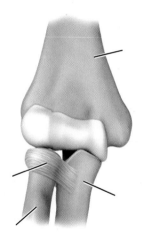

d.

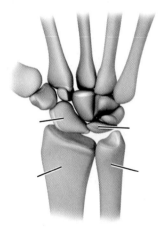

e.

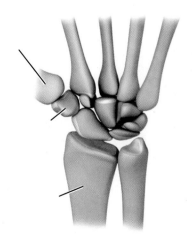

f.

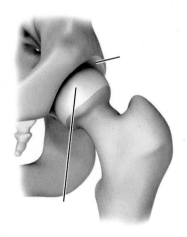

20.

a.

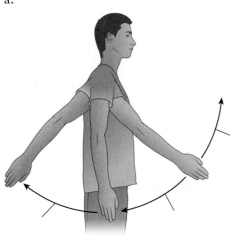

b.

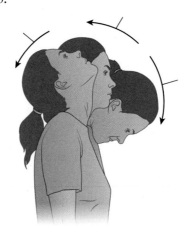

c.

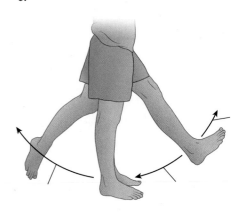

d.

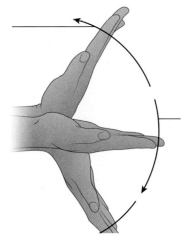

e.

f.

g.

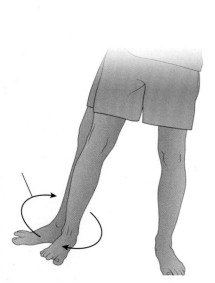

h.

i.

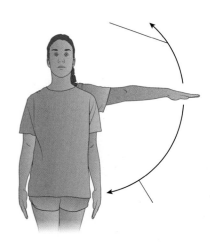

j.

k.

l.

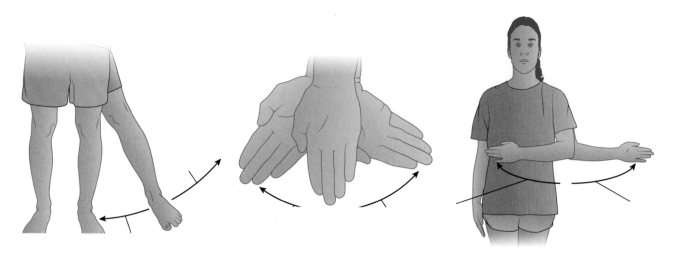

m.

n.

o.

p.

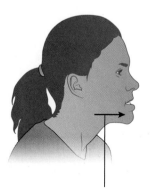

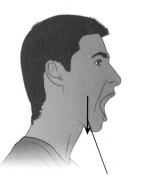

q.

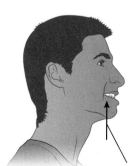

r.

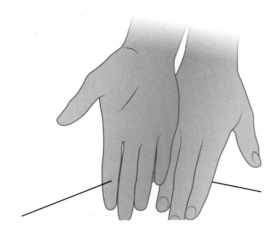

s.

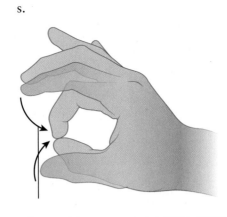

Chapter 14: The Muscular System

Label the following diagrams.

1.

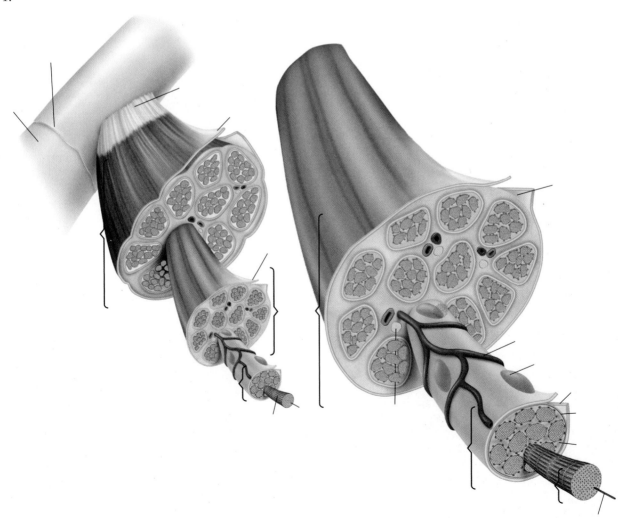

2.

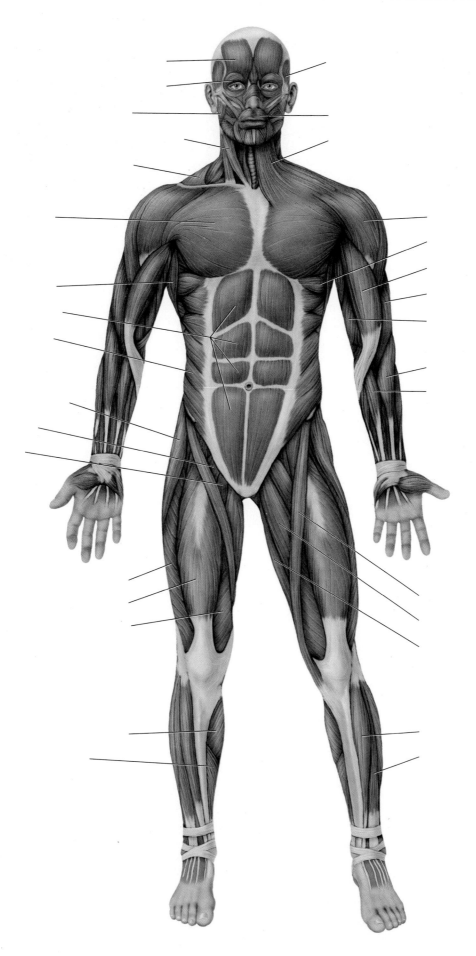

3.

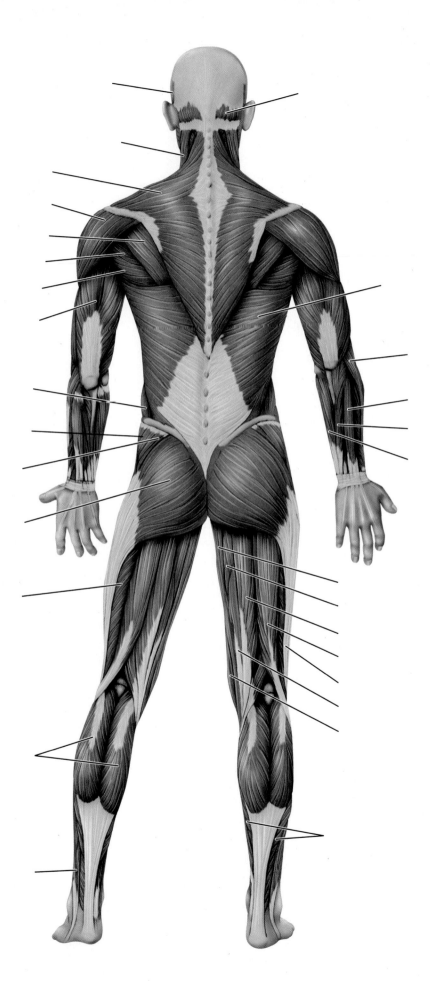

4.

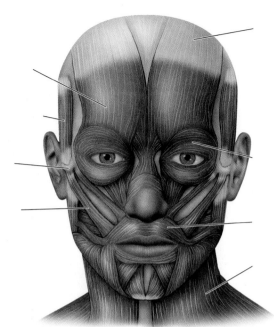

5.

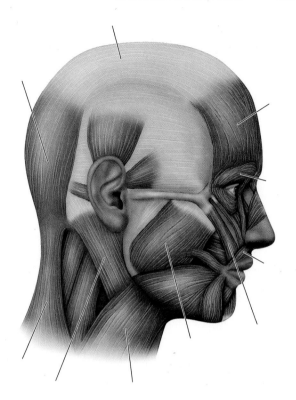

6.

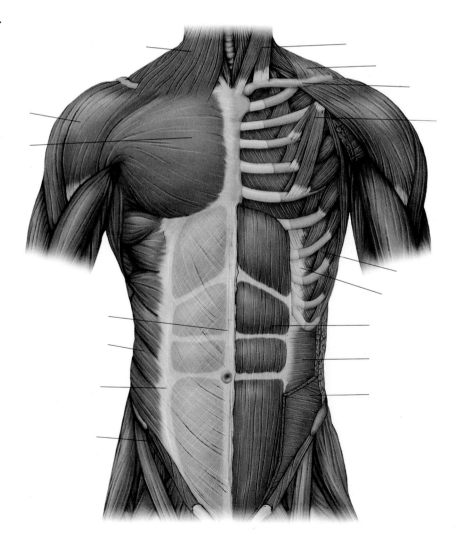

7.

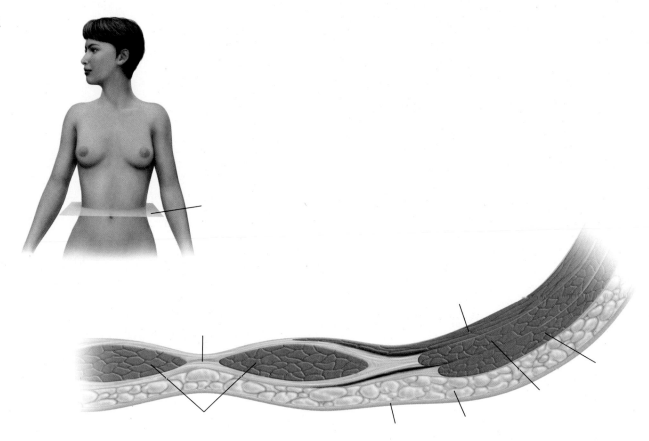

8.

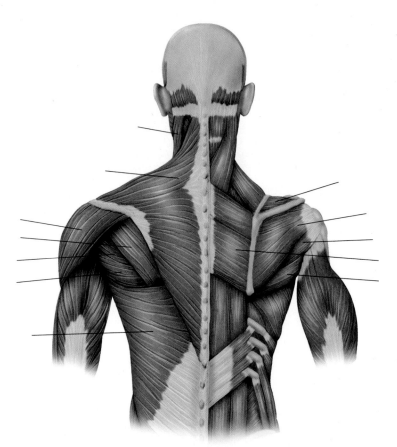

9.

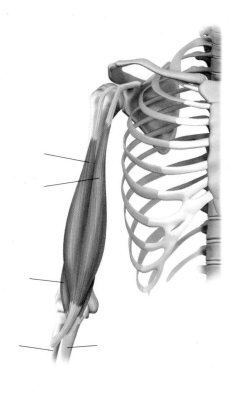

10.

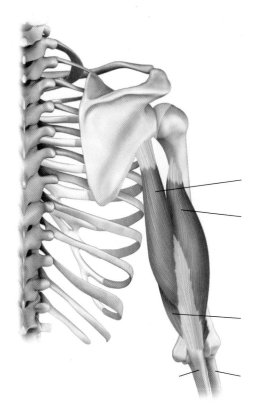

11.

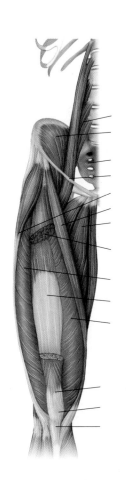

12.

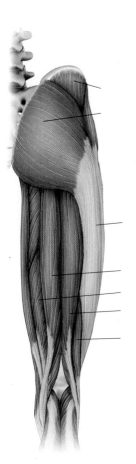

13. a.

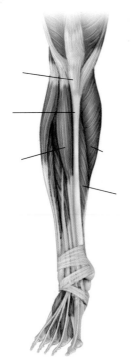

b.

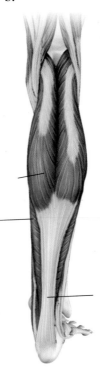

14.

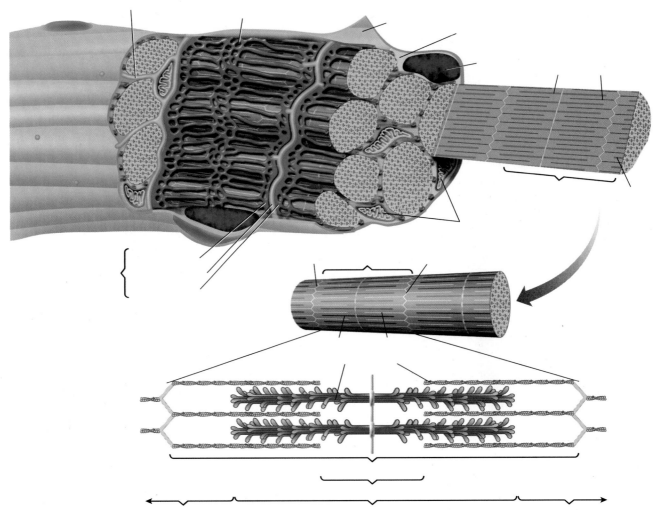

15. a.

b.

c.

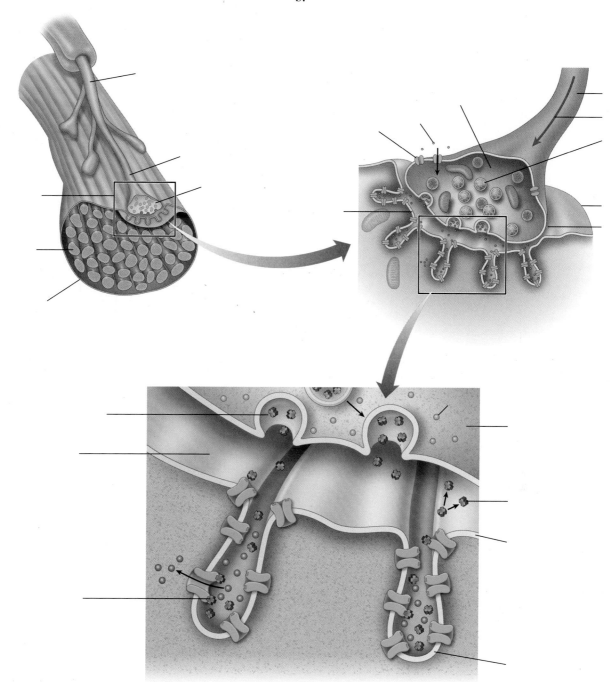

Chapter 15: The Nervous System

Label the following diagrams.

1.

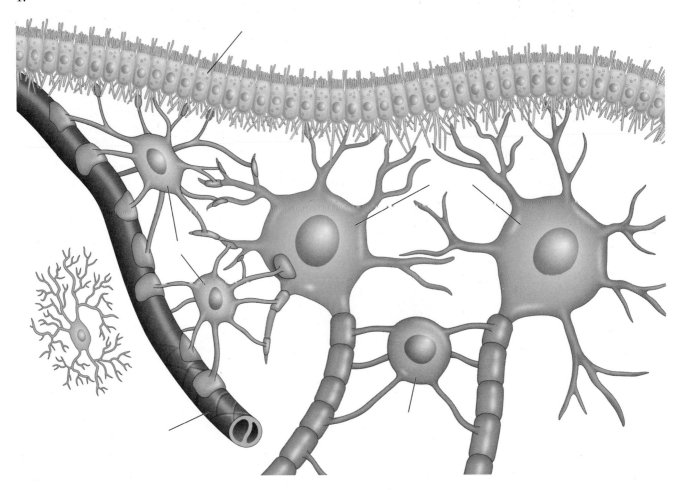

2.

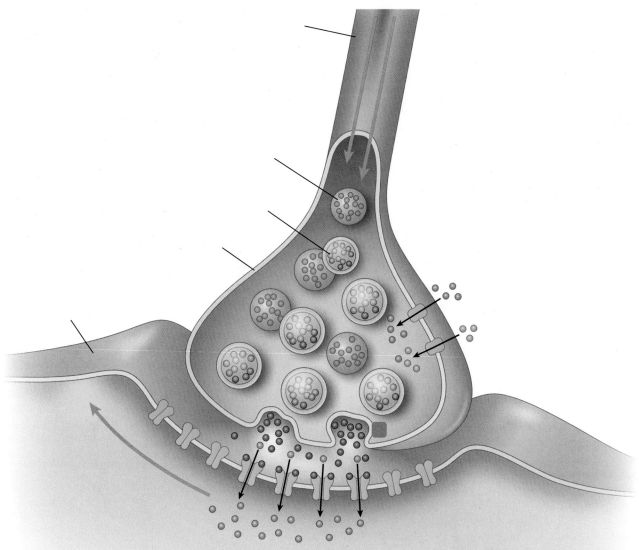

3. a.

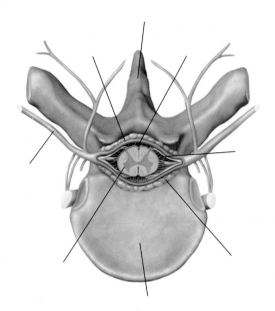

b.

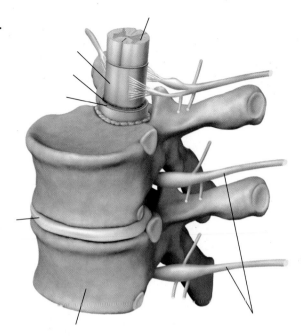

4.

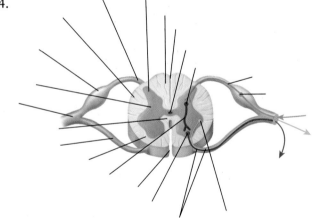

5.

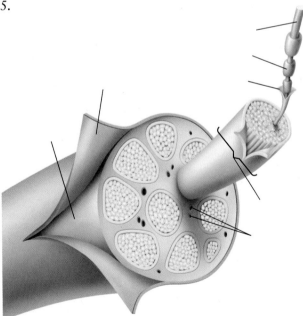

Art source: From DIGIUSEPPE/FRASER. Biology 12 U. © 2012 Nelson Education Ltd.
Reproduced by permission. www.cengage.com/permissions

6.

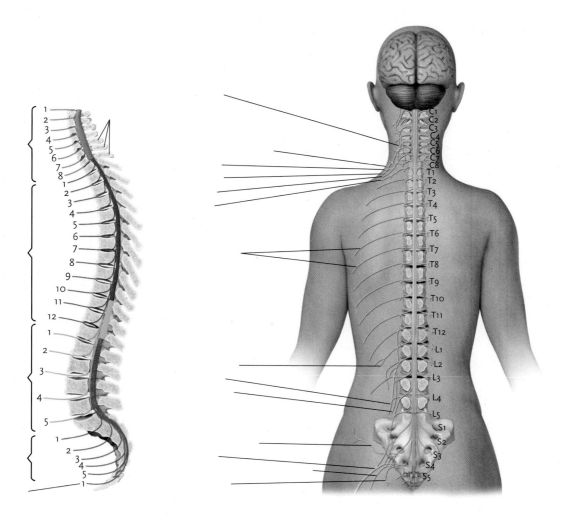

7. a.

b.

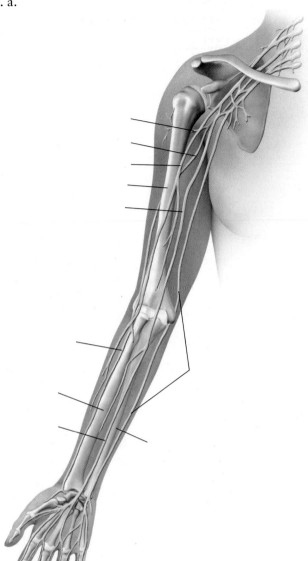

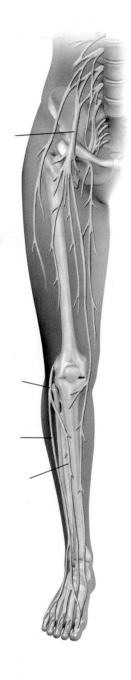

c.

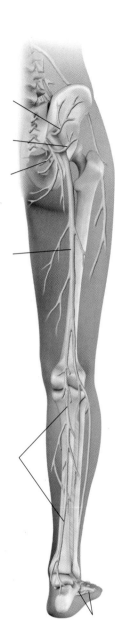

8.

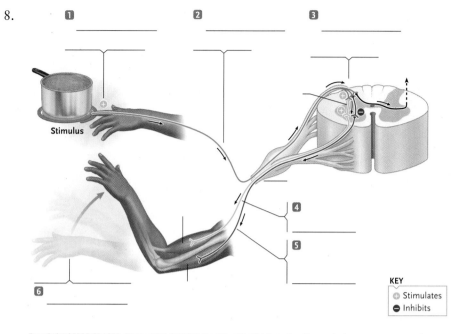

| ❶ _____ | ❷ _____ | ❸ _____ |

Stimulus

❹ _____

❺ _____

❻ _____

KEY
➕ Stimulates
➖ Inhibits

From DIGIUSEPPE/FRASER. Biology 12 U. © 2012 Nelson Education Ltd. Reproduced by permission. www.cengage.com/permissions

9.

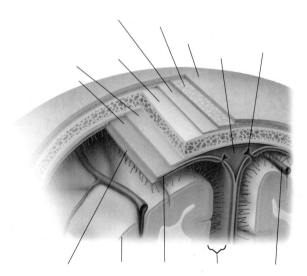

10. a.

b.

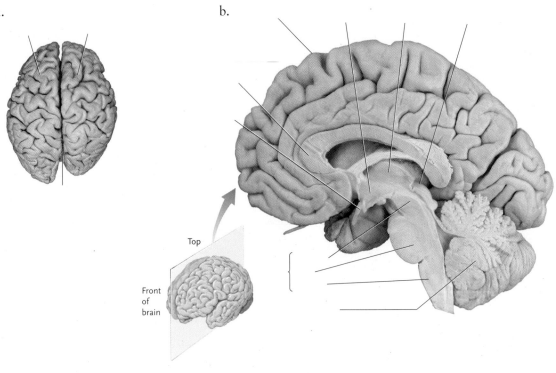

Top

Front
of
brain

11.

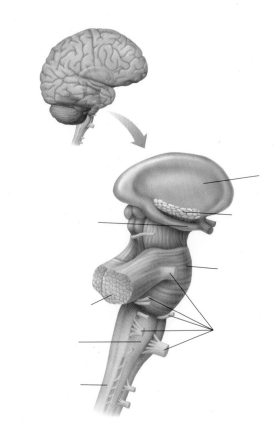

12.

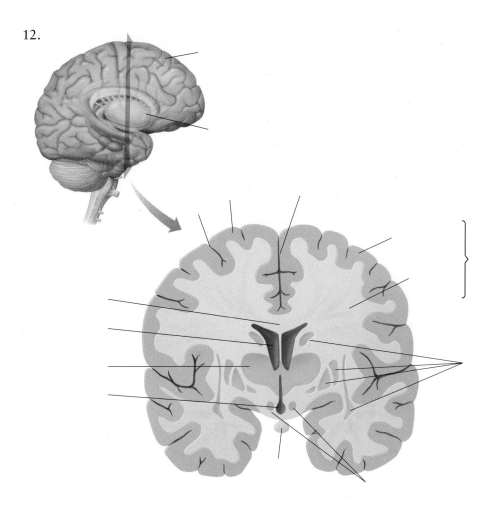

Art source: Courtesy of Mark Nielsen, AnatBooks, LTD

13.

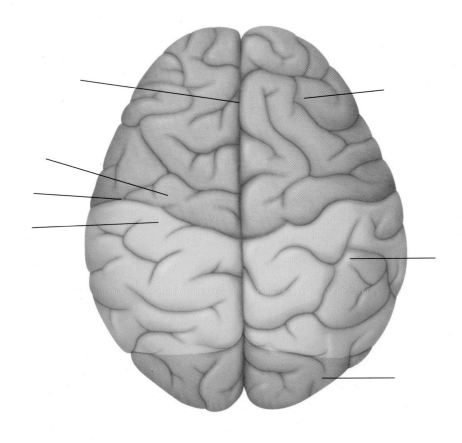

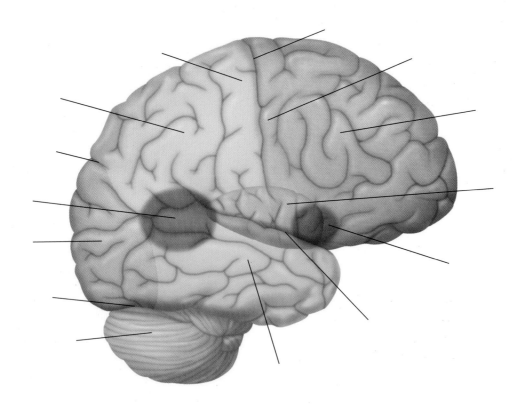

Chapter 16: The Endocrine System

Label the following diagrams.

1. a.   b.   c.

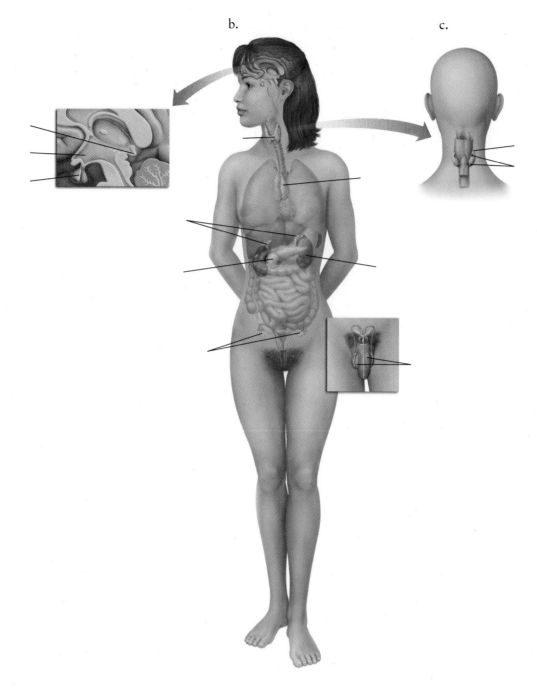

2. a.

b.

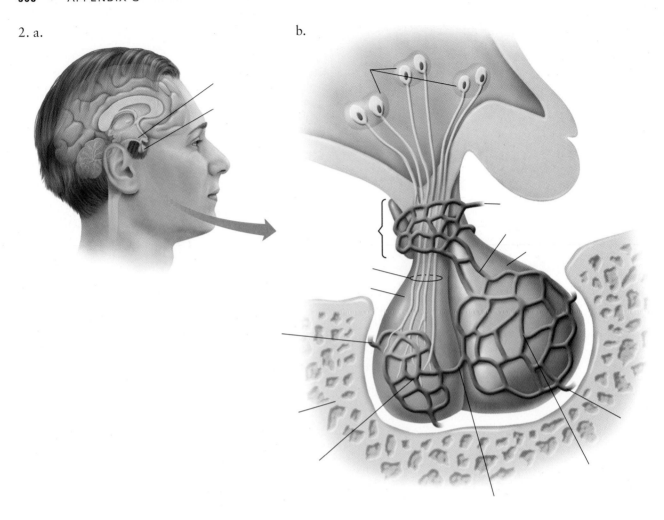

3.

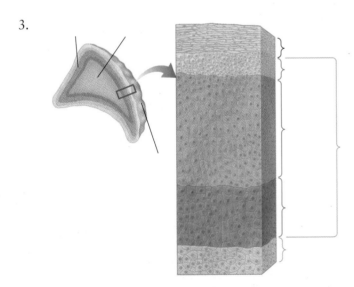

Chapter 17: The Cardiovascular System

Label the following diagrams.

1.

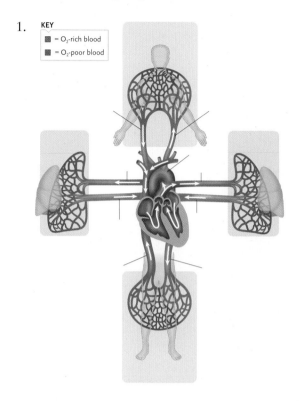

Art source: From RUSSELL/HERTZ/STARR/FENTON. Biology, 2E. © 2013 Nelson Education
Ltd. Reproduced by permission. www.cengage.com/permissions

2.

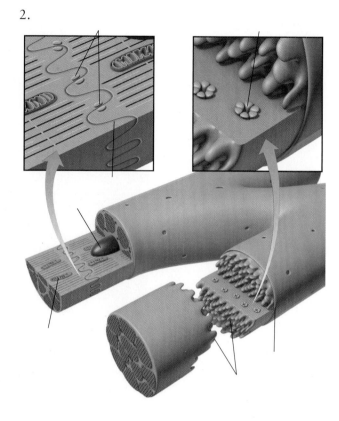

3.

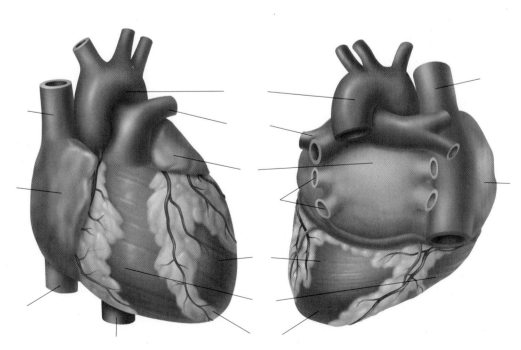

4.

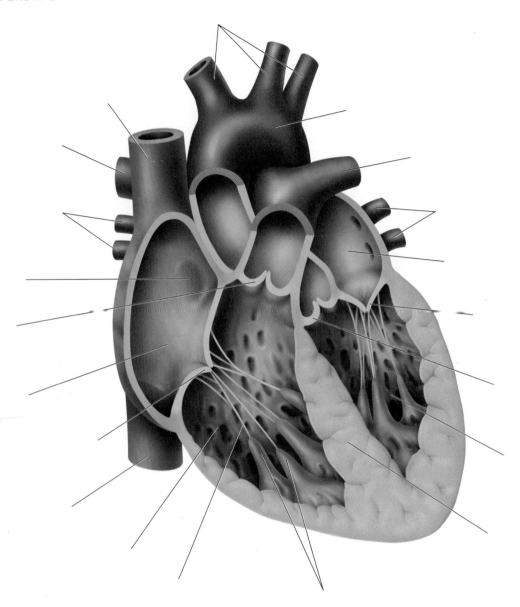

5.

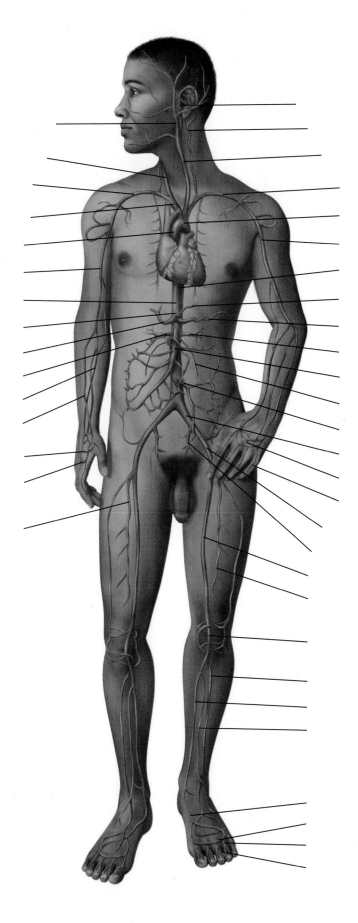

6.

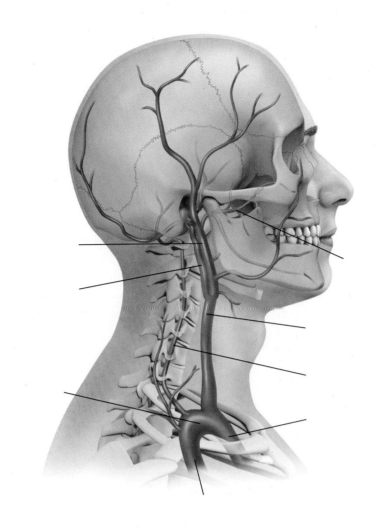

7.

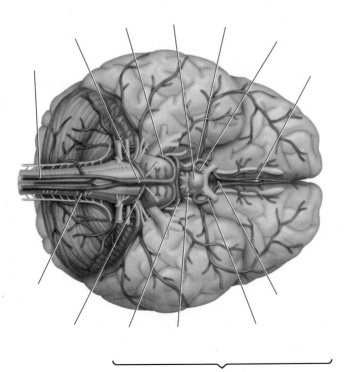

8.

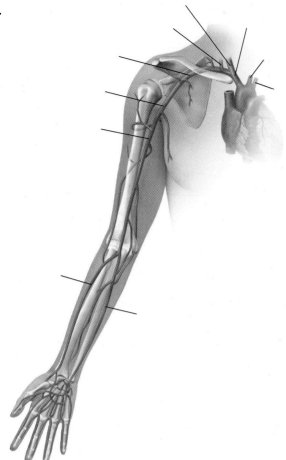

9.

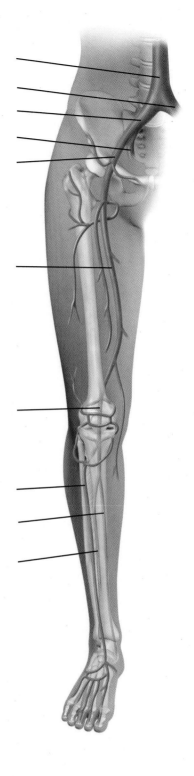

10.

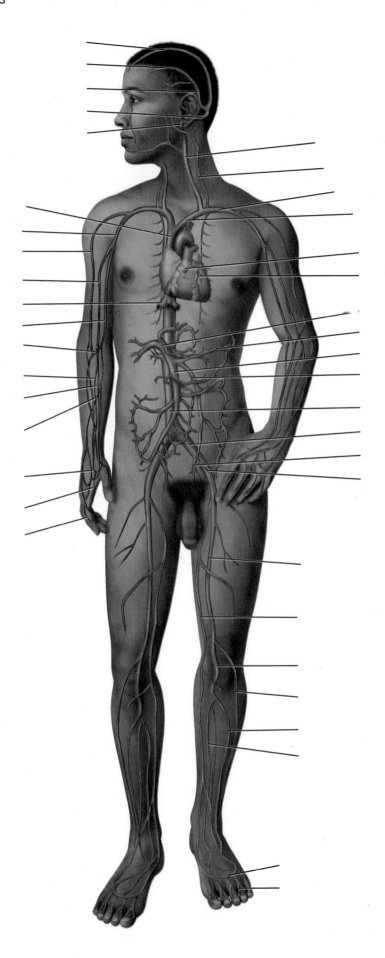

11.

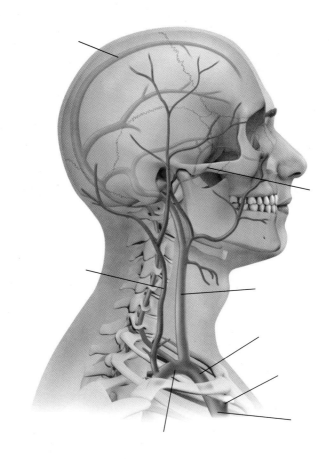

12.

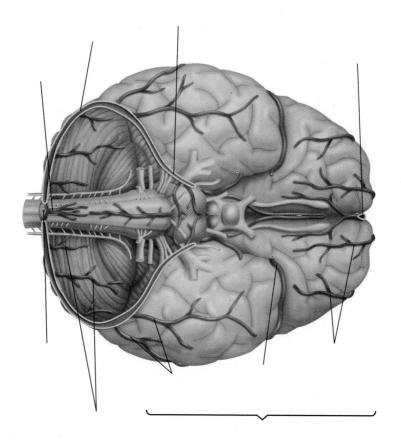

13.

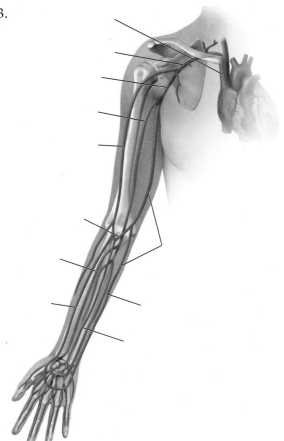

14.

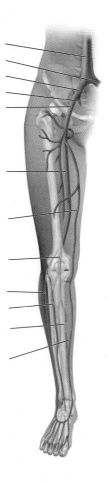

Chapter 18: The Respiratory System

Label the following diagrams.

1.

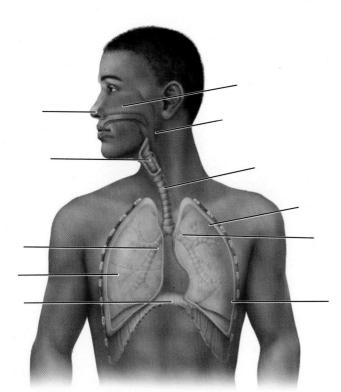

2.

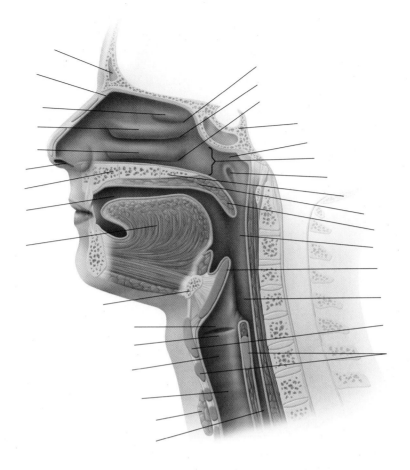

3.

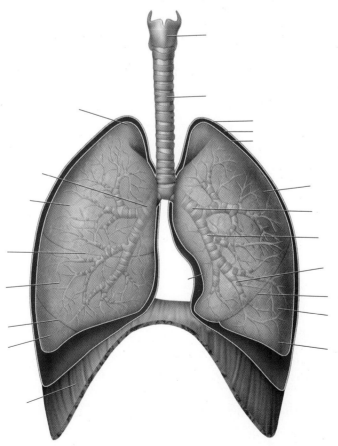

Chapter 19: The Digestive System

Label the following diagrams.

1.

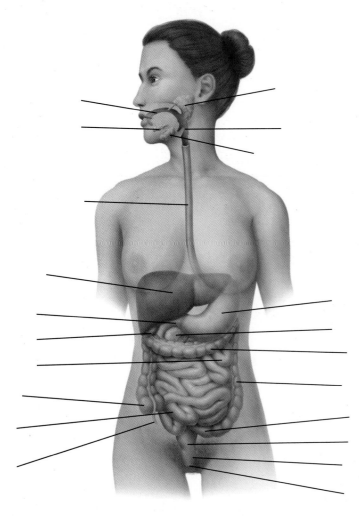

2.

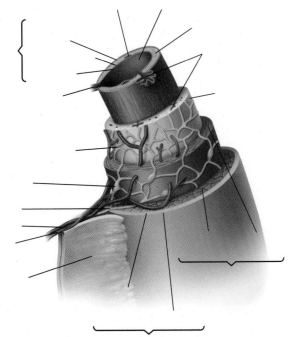

3.

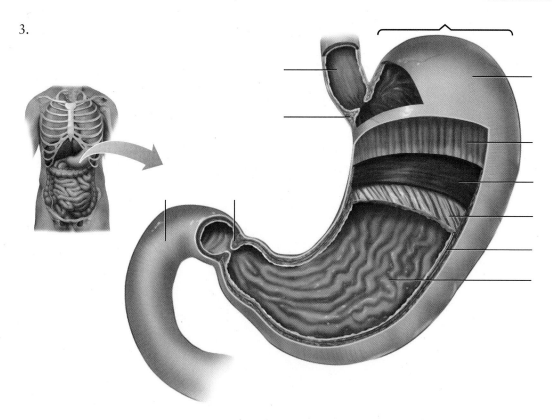

4.

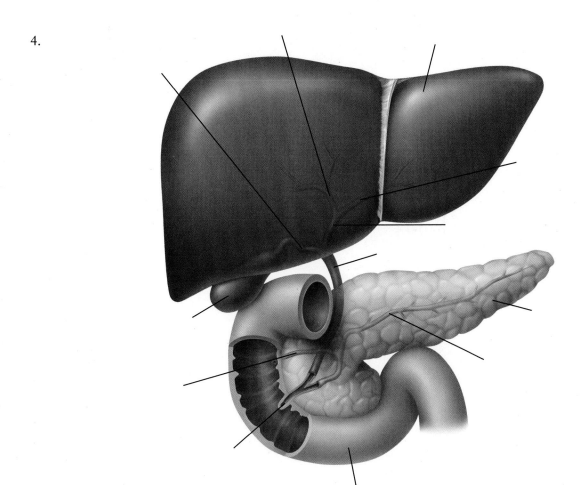

Chapter 20: The Urinary System

Label the following diagrams.

1.

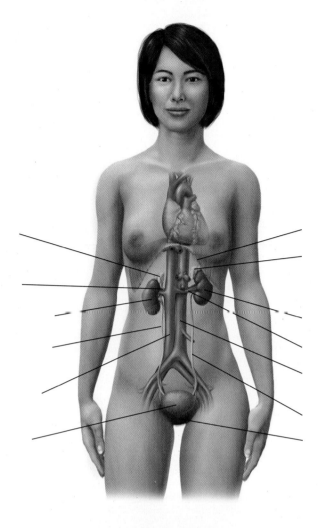

2.

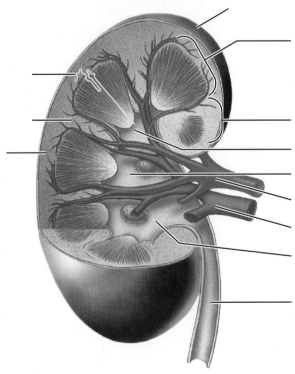

Art source: From DIGIUSEPPE/FRASER. Biology 12 U. © 2012 Nelson Education Ltd.
Reproduced by permission. www.cengage.com/permissions

3.

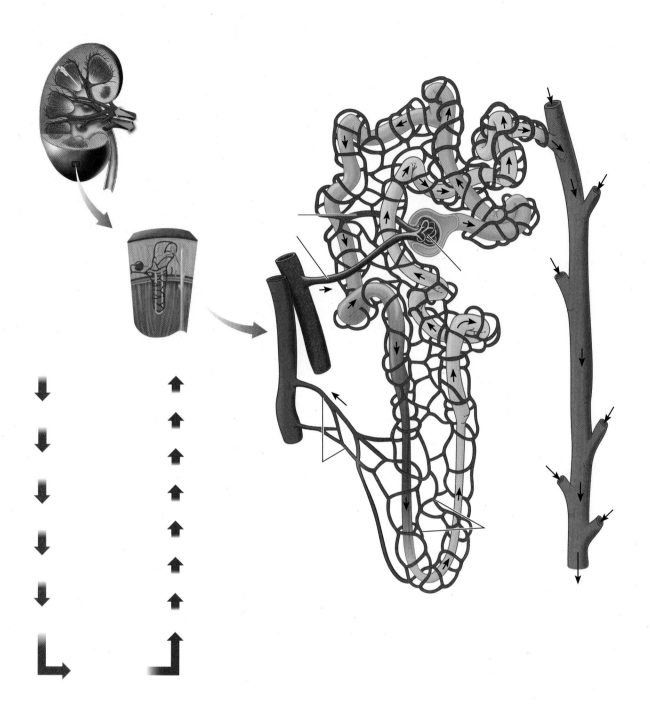

4.

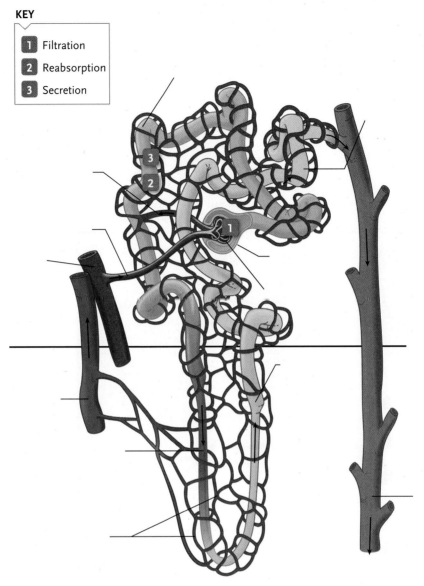

Art source: From DIGIUSEPPE/FRASER. Biology 12 U. © 2012 Nelson Education Ltd. Reproduced by permission. www.cengage.com/permissions

Chapter 21: The Reproductive System

Label the following diagrams.

1. a.

b.

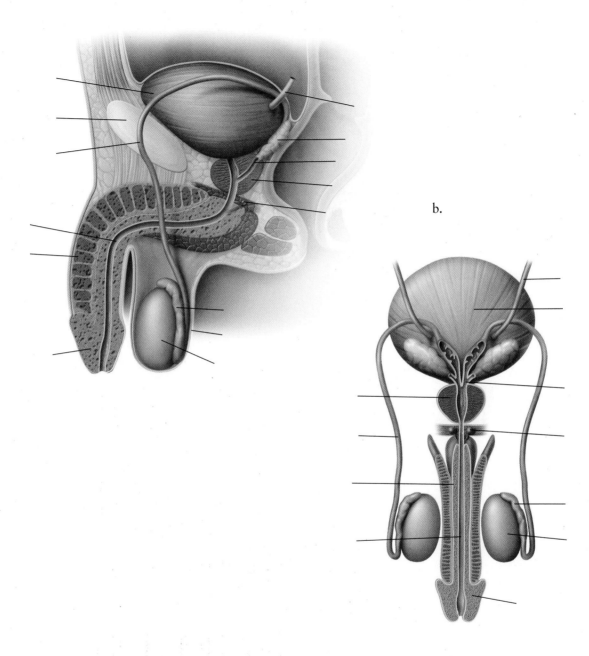

2. a.

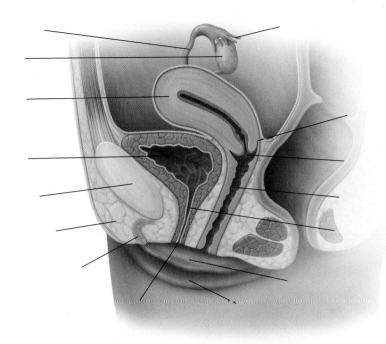

b.

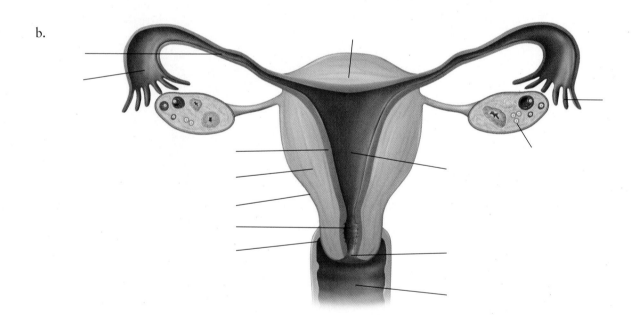

Chapter 22: The Immune System

Label the following diagram.

1.

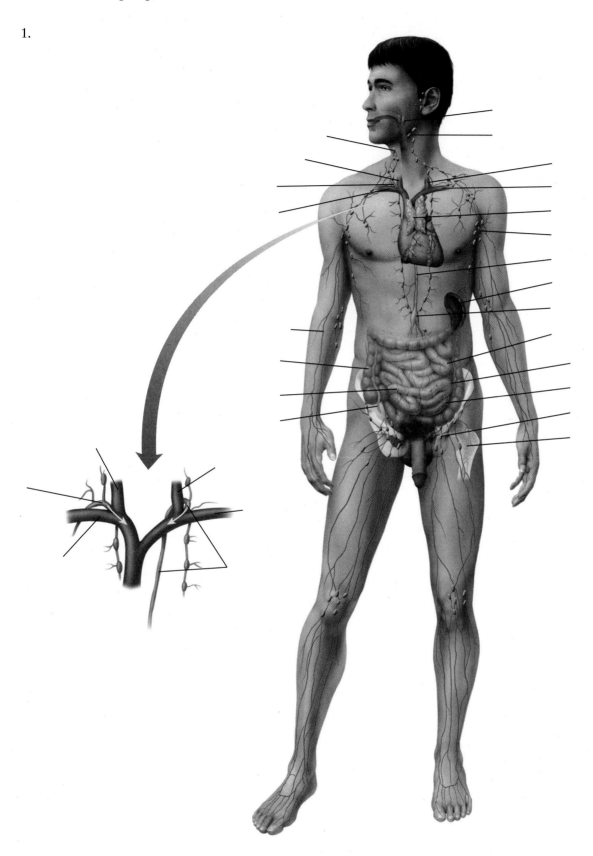

Glossary

3′ tail several hundred adenine nucleotides attached to the mRNA molecule, also known as a poly A tail. p. 147

5′ cap a methylated GTP that enables mRNA to efficiently begin translation at the ribosome. p. 147

A band the overlapping region of the sarcomere. p. 251

ATryn a drug produced in genetically modified goats that is used to treat hereditary antithrombin deficiency. p. 171

abdominal aorta (descending aorta) segment of the aorta located in the abdomen. p. 337

abdominal muscles muscles at the anterior aspect of the abdomen. p. 239

abdominopelvic cavity contains organs of the digestive system, the spleen, bladder, parts of the large intestine, and the internal reproductive organ. p. 18

absorption movement of molecules from the lumen of the digestive tract, or through the skin, into the bloodstream. p. 176

absorptive phase stage in which digestive end products are absorbed from the gut and sent to storage; period after eating when insulin levels are highest. p. 311

accessory glands glands that add to the secretions of the digestive tract or aid in digestion but are not part of the alimentary canal. p. 416

accessory organs organs that secrete subatances required for digestion into the alimentary canal. p. 379

acetaldehyde molecule produced when yeast undergoes fermentation. p. 74

acetyl group a two-carbon molecule that enters the Krebs cycle. p. 69

acetylcholine (ACh) a neurotransmitter. p. 254

acetylcholinesterase enzyme that breaks down acetylcholine. p. 254

acetyl-coenzyme A (acetyl-CoA) final product when pyruvate loses a carbon atom and combines with coenzyme A. p. 69

acini cluster of cells that resembles a berry; found in the pancreas and secretes digestive enzymes. p. 388

acromegaly condition caused by excessive growth hormone after puberty. p. 303

acromion a process at the lateral end of the scapular spine. p. 221

acrosome a caplike structure that covers the head of a sperm cell. p. 419

actin a globular protein that forms microfilaments and is found in all eukaryotic cells. pp. 55, 82, 238

action potential a short-lasting event wherein the voltage of a cell membrane is altered in sequence. p. 254

activating enzyme enzyme that recycles the empty tRNA molecule in the cytoplasm and adds the appropriate amino acid. p. 150

activator a protein that initiates transcription in bacteria. p. 153

active sites grooves on an enzyme where a reaction is catalyzed. p. 32

active transport movement of molecules against a concentration gradient through a membrane protein and with input of energy. p. 71

adaptive immune response also referred to as acquired immune system. p. 431

adductor longus a muscle that helps move the knee toward the midline. p. 246

adenoids a mass of lymphatic tissue located in the pharynx. p. 432

adenosine triphosphate (ATP) molecule used to transfer energy in chemical reactions in every cell. p. 65

adhesion molecules membrane proteins that play an important role in intercellular interactions. p. 50

adipocytes cells that store fat as triglycerides. p. 41

adiponectin a hormone released by adipocytes that increases the effects of insulin. p. 206

adipose tissue composed of adipocytes; also called fatty tissue. p. 41

adjuvant a pharmacological or immunological substance that increases the efficacy of a vaccine or other agent. p. 445

ADP adenosine diphosphate. p. 65

adrenal cortex outer region of the adrenal gland, surrounded by the capsule. p. 303

adrenal glands endocrine glands located superior to the kidneys. p. 307

adrenal medulla innermost part of the kidney that produces epinephrine and norepinephrine. p. 292

adrenocorticotropic hormone (ACTH) a hormone that stimulates the production of glucocorticoids in the adrenal cortex. p. 308

aerobic respiration production of ATP with oxygen. p. 67

afferent arterioles blood vessels that bring blood to the glomerlui. p. 404

afferent neurons neurons that transmit sensory information to the central nervous system. p. 265

afterload pressure of the arteries that the heart must work against to move blood into the vessels. p. 328

agonist the primary muscle contributing to a particular motion. p. 238

agranular leukocytes lymphocytes, monocytes, and plasma cells that don't form membrane granules. p. 353

AIDS (acquired immunodeficiency syndrome) a disease caused by HIV resulting in a suppressed immune response. p. 460

albumin a plasma protein used to help regulate blood pressure. p. 334

aldosterone a hormone produced by the adrenal cortex that causes the kidneys to reabsorb sodium, which causes more water to be absorbed and therefore increases blood pressure. p. 349

allele one of multiple forms of a gene; different alleles of a gene lead to different phenotypes, such as blond hair or brown hair. p. 124

allele frequency proportion of each allele of a gene in a population. p. 124

allergic reactions (hypersensitivity reactions) immune reactions to an otherwise harmless substance. p. 427

alpha blockers medications that block the effects of the sympathetic nervous system and result in decreased blood pressure. p. 352

alpha helices helices formed by hydrogen bonds as part of the secondary structure of a polypeptide. p. 30

alternate splicing some exons forming part of the mature mRNA, other exons doing so at other times, to form alternate amino acid sequences and therefore different proteins. p. 154

alveolar ducts ducts that connect the alveoli to the respiratory bronchioles and allow for gas exchange. p. 360

alveolar pressure pressure held in the alveoli at any point during breathing. p. 365

alveolar processes and sockets contains both the upper and lower teeth. p. 214

alveolar ventilation movement of gas between the alveoli and the external environment during breathing. p. 368

alveoli air sacs that permit gas exchange. p. 360

amino acid organic compounds that contain an amine group, carboxyl group, and a functional group; the building blocks of proteins. p. 27

amino group a nitrogenous functional group consisting of a nitrogen atom bound to two hydrogen atoms. p. 28

ammonia (NH3) molecule that consists of one nitrogen atom and three hydrogen atoms, formed in the body when proteins are broken down; is extrememly alkaline and must be converted into urea in the body to be excreted. p. 73

amniocentesis test that takes a sample of amniotic fluid surrounding an embryo to check for abnormalities in the chromosomes. p. 90

amphiarthrosis slightly movable joint connected by fibrocartilage, such as the articulations between the vertebrae. p. 228

amphipathic molecules that contain hydrophobic and hydrophilic regions. pp. 27, 350

amygdala groups of nuclei in the temporal lobes that control emotion and memory. pp. 285

amylase an enzyme that breaks down starch. pp. 386, 390

anabolic steroids hormones that stimulate the production of molecules in the body as opposed to causing molecules to break down. p. 312

anaerobic respiration production of ATP in the absence of oxygen. p. 67

anal canal terminal part of the large intestine. p. 389

anaphylaxis (anaphylactic shock) a severe allergic reaction characterized by effects to multiple body systems. p. 448

anatomical dead space portion of air in the respiratory tract that in the conducting zones doesn't permit gas exchange. p. 368

anatomical neck narrow groove of the humerus bone where the head meets the body. p. 221

anatomical position human body that is standing up, facing forward, with toes and palms forward. p. 16

anatomy the study of structures within the human body. p. 3

androgen-binding protein (ABP) a protein produced in the seminiferous tubules that binds to testosterone. p. 418

androgens testosterone-like hormones produced in the adrenal glands that stimulate growth. p. 201

anemia a decrease in the number of red blood cells. p. 93

aneuploidy one gamete having more or fewer chromosomes than it is supposed to have. p. 107

angina chest pain brought on by ischemia or hypoxia in the heart. p. 351

angiogenesis growth of new blood vessels. p. 85

angiotensin I a hormone derived from angiotensinogen by cleavage with renin. p. 411

angiotensin II a hormone that helps increase blood pressure by stimulating the production of aldosterone. p. 349

angiotensinogen a precursor hormone formed in the liver. p. 387

angular motion of a synovial joint in a single plane around a fixed point. p. 231

animals eukaryotic, multicellular organisms capable of locomotion. p. 6

anneal formation of complementary base pairs, such as when a primer binds to a single strand of DNA. p. 159

antagonist (muscular) the opposing muscle to a particular motion. p. 239

anterior fontanel the space between the skull bones located at the joining of the frontal bone and both parietal bones; does not close until nearly age two. p. 214

anterior median fissure longitudinal fissure the length of the spinal cord, approximately 3 mm in depth. p. 276

anterior pituitary gland anterior portion of the pituitary gland controlled by the hypothalamus by means of hormones. p. 301

anterior tibial artery artery at the ventral aspect of the tibias. p. 341

anterolateral fontanel the space between the skull bones located at the joining of the temporal and sphenoid bones. p. 215

antibiotic resistance a form of drug resistance where bacteria are not killed by antibiotics. p. 117

antibiotics a drug that kills bacteria. p. 117

antibodies proteins that act as markers for the immune system during a humoral immune response. p. 353

anticipatory response physiological changes initiated by thinking about a change that may occur, such as breathing rate increase prior to exercise that prepares the body for increased oxygen and nutrient demand when exercise occurs. p. 348

anticodon a complementary sequence on the tRNA molecule that binds to the codon on the mRNA. p. 147

anti-diuretic hormone (ADH) a hormone released by the posterior pituitary gland that increases water reabsorption in the nephron. p. 302

antigen substance that triggers an immune response. p. 439

antigen-presenting cell (APC) cell that displays foreign antigens on its surface as markers. p. 439

antigens cell-recognition markers. p. 49

antioxidant prevents free radical molecules from damaging DNA. p. 88

antiparallel two strands of DNA running in opposite directions. p. 132

antipyretic a pharmacological agent that reduces fever. p. 438

antisense strand also known as the template stand. p. 143

anus the ring at the distal end of the gastrointestinal tract. p. 389

aorta primary artery of the body, located where the blood leaves the left ventricle of the heart. p. 320

aortic arch most superior part of the aorta. p. 320

aortic semilunar (SL) valve valve between the aorta and the left ventricle. p. 323

apical side of a cell that is facing the surface or cavity. pp. 176, 408

apneustic centre part of the lower pons that promotes inspiration. p. 373

apocrine glands type of sweat gland found predominantly in the axilla (armpits), areola, and genital regions. p. 198

apoptosis programmed cell death. p. 84

appendicitis inflammation of the appendix. p. 436

appendicular skeleton skeleton that is outside the axial skeleton, and includes the upper and lower limbs, shoulders, and pelvis. p. 220

appendix accessory structure of the large intestine proposed to be a vestigial structure. p. 389

aquaporin a membrane protein that allows the transport of water across a cell membrane. p. 408

arachnoid mater middle layer of the meninges. p. 276

arachnoid villi protrusions of the arachnoid mater through the dura mater. p. 282

archaebacterial prokaryotic unicellular organisms that survive in extreme environments. pp. 4, 451

arcuate veins blood vessels of the renal circulation. p. 404

aromatase an enzyme that converts testosterone into estrogen in females. p. 303

arrhythmia an abnormality in the electrical system of the heart that causes irregular contractions. p. 326

arteries blood vessels that carry blood away from the heart. pp. 320, 340

arterioles small arteries that bridge the flow of blood from arteries to capillaries. p. 323

articular cartilage protective substance that lines the bone at a joint with another bone. p. 208

artificial selection occurs when humans choose the breeding of traits considered the most desirable in plants or animals. p. 118

artificially acquired active immunity immunity granted by exposure to a vaccine. p. 445

artificially acquired passive immunity short-term immunization granted by the transfer of antibodies. p. 444

ascending aorta aorta between the left ventricle and the aortic arch. p. 336

ascending colon the part of the large intestine located in both the right-upper and right-lower quadrants of the abdomen, and leading to the transverse colon. p. 389

ascending loop of Henle the distal portion of the loop of Henle that is impermeable to water, but permeable to ions. p. 407

ascending tracts nerve tracts that carry signals from the body toward the brain. p. 276

asexual reproduction from a single parent to produce an identical offspring. p. 78

asthma disease condition that causes an increase in mucus secretion and bronchoconstriction. p. 376

astrocytes glial cells that maintain the ion concentration surrounding neurons in the CNS and that form the blood-brain barrier. p. 189

atherosclerosis cardiovascular disease that results in the hardening of the blood vessels. p. 350

atlas first cervical vertebra, designated C1. p. 216

atmospheric pressure the pressure created by the gases of the atmosphere. p. 365

ATP adenosine triphosphate; energy. pp. 53, 132

ATP synthase a specific membrane protein that allows protons to move back across the mitochondrial membrane into the matrix of the mitochondrion. p. 61

atrial diastole atrial relaxation. p. 324

atrial natriuretic peptide (ANP) a hormone released by the atria of the heart. p. 349

atrial systole atrial contraction. p. 324

atrioventricular bundle also referred to as the bundle of His. p. 325

atrioventricular node (AV node) a collection of cells that transmit an action potential from the atria to the bundle of His. p. 325

auditory cortex the part of the brain that processes sound. p. 288

AUG first codon of mRNA molecule; codes for the amino acid, methionine. pp. 147, 150

autocrine a substance secreted by a cell that acts on the same cell. p. 433

autoimmune disease a condition in which the immune system targets the body's own cells. p. 311

autonomic involuntary. p. 291

autonomic ganglion a collection of cell bodies in the autonomic nervous system. p. 291

autonomic nervous system (ANS) portion of the nervous system that controls involuntary functions. p. 187

autonomic neurons neurons that provide involuntary control. p. 291

autorhythmic automatic contractions in the heart. p. 187

autorhythmicity ability of the electrically active cells to establish their own action potential. p. 325

autosomal refers to a gene found on chromosomes 1 to 22. p. 108

avascular does not contain blood vessels. p. 176

axial skeleton skeleton composed of the skull, ribs, sternum, and vertebrae. p. 205

axillary area the region beneath the arms. p. 340

axillary artery artery found in the axillary region. p. 340

axillary nerve nerve located in the axillary region. p. 278

axis second cervical vertebra, designated C2. p. 216

axon collaterals branches from the axon that end with multiple axon terminals. p. 266

axon hillock point of connection between the nerve cell body and the axon where action potentials begin. p. 266

axon terminal the end of a nerve cell branch that contains neurotransmitters. p. 254

B cells immune cells. p. 354

bacilli a class of bacteria that is rod-shaped. p. 453

bacteriocidal a substance that kills bacteria. p. 454

bacteriophage a virus that infects bacteria. pp. 158, 457

bacteriostatic a chemical, biological, or physical agent that prevents the replication of bacteria. p. 454

ball-and-socket joint synovial joint that allows the greatest freedom of motion of any joint. p. 230

baroreceptors detect pressure changes. p. 22

basal side of the membrane opposite the apical side. p. 181

basal cell carcinoma cancerous growth composed of cells from the stratum basale. p. 199

basal metabolic rate (BMR) the calorie requirement to maintain basic cellular function. p. 260

basal nuclei collections of nerve cell bodies at the base of the forebrain. p. 284

base pair a coupled purine and pyrimidine. p. 143

basement membrane connective tissue layer underlying epithelial cells that's composed of laminin and integrin proteins and that connects the epithelial cells to the loose connective tissue. p. 176

basilica vein superficial vein of the upper arm. p. 344

basophils a type of white blood cell. p. 353

belly (muscular) middle of the muscle. p. 238

beneficial bacteria *see* normal flora. p. 432

benign tumour a growth of cells that is not cancerous. p. 84

beta blocker medication that blocks the effects of epinephrin and causes a decrease in blood pressure. p. 352

beta sheet a three-dimensional structure formed by the hydrogen bonds between amino acids as part of the secondary structure of a polypeptide. p. 30

bicep muscle that works to flex the forearm at the elbow. p. 246

bifidobacteria a type of normal flora bacteria. p. 432

bilayer a double layer of phsopholipids. p. 48

bile substance produced in the liver that helps break down fats in the small intestine. p. 353

bilirubin a yellow pigment formed by the breakdown of the hemoglobin in red blood cells. p. 353

binary fission method of reproduction in bacteria. p. 78

binding protein a protein that adheres two substances together. p. 135

biology the study of all living things. p. 3

biotechnology the use of living organisms to produce biological products. p. 157

bipolar neurons neurons that produce two extensions branching from the cell body. p. 189

blastocyst an early embryonic structure that contains a fluid-filled space that implants into the uterine wall. p. 420

blood connective tissue that transports oxygen, nutrients, waste products, and hormones throughout the body. p. 323

blood cells include red blood cells that carry oxygen and some carbon dioxide and white blood cells that are involved in the immune response. p. 346

blood glucose amount of glucose in the blood plasma. p. 353

blood pH the measure of alkalinity or acidity in the blood; normal blood pH is 7.35–7.45. p. 353

blood pressure the force of blood pushing against the walls of the blood vessels. p. 318

blood vessels refers to veins, arteries, veules, arterioles, and capillaries. p. 346

blood viscosity thickness of the blood. p. 346

blood volume total amount of blood. p. 329

blood-brain barrier formed by astrocyte foot processes and endothelial cells of the blood vessels; separates the blood from the extracellular fluid of the brain. p. 267

blood-testis barrier a physical barrier between the blood and the seminiferous tubules in animals. p. 417

body cavity a space in the body that contains specific organs or tissues. p. 18

bolus a mass of fluid or semi-fluid. p. 384

bone connective tissue that forms the skeleton, stores calcium and other minerals, produces red and white blood cells, stores some fat, and protects internal organs. p. 213

bottleneck effect random change in allele frequency after a large portion of a population dies, such as in a natural disaster. p. 128

botulism a disease resulting from exposure to the botulinum toxin that causes paralysis. p. 455

Bowman's capsule a part of the nephron surrounding the glomerulus. p. 176

Bowman's space the space between the glomerulus and the Bowman's capsule. p. 406

Boyle's law a physical law that states the volume of a gas in a container is inversely proportional to its pressure. p. 365

brachial arteries arteries located in the upper arms. p. 340

brachial plexus network of nerve fibres running from the spine. p. 278

brachial vein vein located in the upper arm. p. 341

brachialis muscle of the upper arm that works to flex the forearm. p. 246

brachiocephalic trunk artery of the mediastinum that supplies the right arm and head. p. 337

brachiocephalic vein vein in the upper chest formed by the joining of the internal jugular veins and subclavian veins. p. 341

brainstem posterior part of the brain that connects to the spinal cord. p. 283

Broca's area region of the brain's frontal lobe that allows the physical ability to speak. p. 286

bronchi large airway that branches from the trachea. p. 360

bronchiole small airway that branches from the bronchi. p. 360

bronchoconstriction a decrease in the lumen size of the bronchi. p. 376

bronchodilator medication that increases the lumen size of the bronchi. p. 377

brush border the surface of a cuboidal cell covered in microvilli. p. 388

budding production of viral cells that does not kill the infected cell. p. 458

bulbourethral glands glands that produce pre-ejaculate in most mammals. p. 416

bundle branches two branches of the bundle of His. p. 325

bundle of His cells that are part of the fast-conduction system in the septum of the heart and that transmit the action potential to the apex of the heart. p. 325

bursa sacs small sac lined with a synovial membrane and filled with fluid found in synovial joints. p. 229

butyric acid metabolic byproduct of fatty acids released in sweat produced by apocrine glands. p. 198

calcitonin hormone released by the parafollicular glands of the thyroid to decrease blood-calcium levels. p. 304

calcitriol increases calcium levels by increasing the amount of calcium absorbed in the small intestine. pp. 226, 401

calcium channel blockers medications that work to decrease blood pressure by preventing the movement of calcium ions into the heart muscle cells. p. 352

callus a protective mechanism resulting from increased keratin production. p. 196

Calorie kilocalorie measurement of food. p. 66

calorie (kilocalorie) amount of energy required to raise the temperature of 1 ml of water 1°C, which is 1/1000th of a Calorie. p. 66

canaliculi channels that connect lacunae, allowing osteocytes to communicate with one another. p. 207

cancer disease that occurs when mutated cells continue to divide and are not regulated by cellular checkpoints. p. 84

CAP binding protein catabolite activator protein, an activator that binds to the CAP binding sequence and initiates transcription in prokaryotes. p. 153

capillaries small blood vessels that provide oxygen and nutrients to the body's cells and pick up waste. p. 323

capillary pressure also called hydrostatic pressure; pressure exerted on the capillary walls. p. 430

capitulum lateral aspect of the distal end of the humerus that articulates with the radius. p. 221

capsule a protein or carbohydrate external layer that protects bacteria from external environmental changes. p. 50

carbon dioxide molecule composed of one carbon atom and two oxygen atoms, produced during cellular respiration. p. 298

carbon-hydrogen bond a covalent bond that contains energy important for the production of ATP. p. 66

carboxyl group a functional group consisting of a carbon atom with a double bond to oxygen and a single bond to a hydroxyl group. p 28

carboxypeptidase a pancreatic enzyme that breaks peptide bonds. p. 391

carcinogen anything that causes cancer. p. 85

cardia pertaining to the heart. p. 384

cardiac cycle repeated sequence of both systole and diastole for the atria and ventricles of the heart. p. 323

cardiac muscle muscles of the heart. p. 22

cardiac myocytes heart muscle cells. p. 319

cardiac notch a lateral deflection of the anterior wall of the left lung. p. 361

cardiac output the amount of blood pumped by the heart in a given amount of time. p. 327

cardiac reserve residual blood in the ventricles following ventricular systole. p. 327

carotene pigment that the body converts to vitamin A. p. 197

carotenemia condition caused by having a high level of beta-carotene in the blood. p. 197

carotid foramen opening in the styloid process through which the carotid arteries pass. p. 210

carpal tunnel small concave region at the front of the wrist. p. 221

carpals wrist bones. p. 221

carrier proteins membrane proteins that regulate the movement of certain molecules into or out of the cell down a concentration gradient without input of energy. p. 50

cartilage very strong connective tissue that acts as a smooth surface for joint movement and gives structures their shape. p. 180

cartilaginous joints joints composed of, or containing, cartilage at the articulation. p. 228

catalyze increase the speed of chemical reactions. p. 32

cauda equina distal end of the spinal cord. p. 277

caudate nucleus one of the structures of the basal ganglia. p. 285

CD4 membrane receptor a co-receptor that assists in communication with an antigen-presenting cell. p. 458

CD4⁺ T cells mature T cells that express the CD4 membrane receptor. p. 439

CD8⁺ T cells mature T cells that express the CD8 membrane receptor. p. 441

complementary DNA (cDNA) molecule that is made by reverse transcription of mRNA. p. 167

cecum proximal end of the large intestine. p. 389

celiac trunk first major branch of the abdominal aorta. p. 337

cell smallest unit of life. pp. 7, 266

cell membrane membrane that consists of a phospholipid bilayer, contains proteins and cholesterol, and surrounds the cell. p. 43

cell types different varieties of cells. p. 180

cell-mediated immune response immune response that involves cytotoxic T cells. p. 439

cellular metabolism process by which the cell generates energy from nutrients. p. 53

cellular respiration process of oxidizing food molecules and reducing NAD⁺ molecules to produce ATP in various steps. p. 67

cellulose structural plant carbohydrate not used to produce energy. p. 37

central canal (haversian canal) middle of the osteon, which contains blood and lymphatic vessels. p. 207

central chemoreceptors chemical receptors in the medulla oblongata that detect oxygen and carbon dioxide. p. 374

central dogma the flow of information from gene to RNA to protein. p. 145

central nervous system (CNS) brain and spinal cord. p. 263

central sulcus border between the anterior and posterior portions of the brain. p. 285

centrioles found in most eukaryotic cells except plants and fungi, and involved in the organization of spindle fibres. p. 55

cephalic phase first stage of digestion, which involves thinking about eating and causes salivation and the beginning of smooth muscle contractions along the GI tract. p. 395

cephalic vein vein that runs up the lateral side of the arm. p. 341

cerebellar disease damage in the cerebellum resulting in the inability to coordinate muscle movements. p. 285

cerebellum the part of the brain located inferior to the occipital lobe, and that facilitates fine motor control, coordination of multi-muscle movements, and balance. p. 285

cerebral cortex outermost layer of the cerebrum divided into lobes with distinct functions. p. 285

cerebral spinal fluid (CSF) nutrient-rich and protective fluid that nourishes and protects the brain and spinal cord. p. 269

cerebrum largest and most superior part of the brain. p. 285

cerumen waxy substance, found in the auditory canal, that traps foreign particles and inhibits bacterial growth. p. 198

ceruminous glands glands that produce ear wax. p. 198

cervical enlargement region of the spinal cord with increased diameter due to the addition of the nerves for the upper limbs. p. 277

cervical plexus (C1–C5) branched network of nerves in the neck. p. 278

cervical vertebrae the seven vertebrae in the neck. p. 216

cervix most inferior portion of the uterus. p. 420

Charles Darwin credited with the modern theory of natural selection that's based on his book *On the Origin of Species by Means of Natural Selection*. p. 113

checkpoints points at which the cell cycle is controlled to ensure that a previous phase is fully completed before advancing to the next stage. p. 84

chemical digestion chemical breakdown of food and nutrients into smaller components. p. 380

chemical energy energy that results from breaking bonds of ATP in cellular reactions. p. 66

chemical synapse junctions that permit the neurons to pass on a signal to other cells through gap junctions. p. 319

chemically gated channel membrane proteins that open and allow passage in response to a chemical stimulus. p. 270

chemicals individual elements that combine to form molecules. p. 7

chemiosmosis process of using a chemical gradient across a membrane to produce many ATP molecules. p. 61

chemoreceptor receptor that detects chemical concentrations, such as neurotransmitters, drugs, or hormones. p. 22

chemotaxis movement in response to chemical stimuli. p. 433

chief cells secretory cells of the parathyroid gland, which produces parathyroid hormone, and the stomach, which produces pepsinogen. p. 306

chloride shift exchange of bicarbonate and chloride across the membranes of red blood cells. p. 373

chlorophyll pigment found in chloroplasts in plants. p. 6

chloroplast membrane-bound organelle in plant cells that contains the pigment chlorophyll that undergoes photosynthesis. p. 6

cholecystokinin (CCK) hormone that causes the gallbladder to release stored bile into the GI tract. p. 396

cholesterol a type of fat that is a major constituent of cell membranes, used by the liver to make bile, produces LDLs, and is the starting material for steroid hormones. p. 43

chondrocytes cells that make up cartilage connective tissue. p. 180

chondroitin sulfate sulfated chains of polysaccharides found in ground substance of connective tissue, usually bound to proteoglycans. p. 181

choroid plexus location in the ventricles of the brain where cerebrospinal fluid is produced. p. 282

chromatin DNA in loose strands during regular growth phase. p. 52

chromosome a single condensed nucleic acid macromolecule called DNA. p. 32

chronic bronchitis one of the disease conditions that make up chronic obstructive pulmonary disease (COPD). p. 376

chronic inflammatory disease a persistent condition that involves a chronic immune response. pp. 308, 350

chronic kidney disease (CKD) a progressive loss of renal function. p. 412

chronic obstructive pulmonary disease (COPD) a respiratory disease that decreases the body's ability to manage gas exchange. p. 376

chylomicron a type of lipoprotein that delivers fats from the digestive system to the bloodstream. p. 394

chyme partially digested food expelled from the stomach into the small intestine. p. 384

chymotrypsin digestive enzyme that is part of the pancreatic juice that digests proteins. p. 391

cilia cells that contain hairlike extensions. p. 176

circle of Willis a cerebral circular flow of blood vessels that supply the brain. p. 340

circulatory system organ system consisting of the heart, blood vessels, and the blood. p. 317

cis **fatty acids** unsaturated fats in the *cis* conformation. p. 42

cis **Golgi** the side of the Golgi that is closest to the endoplasmic reticulum. p. 53

citrate a six-carbon molecule formed from acetyl-CoA and oxaloacetate. p. 69

citric acid also called citrate. p. 69

citric acid cycle also called the Krebs cycle. p. 70

clavicles bones that comprise what is called the collarbone. p. 221

cleavage furrow occurs when animal cell membranes indent due to contraction of actin microfilaments in order for the cell to divide. p. 82

clitoris a female sex organ and most sensitive erogenous zone. p. 420

cocci bacteria with a spherical shape. p. 453

coccyx vertebrae four fused vertebrae inferior to the pelvis, commonly referred to as the tail bone. p. 215

coding strand DNA strand that codes for a gene. p. 143

codominance both alleles are expressed in a heterozygote. p. 102

codon a three-nucleotide segment that determines the amino acid. p. 145

coenzyme A a coenzyme that combines with a two-carbon molecule during pyruvate oxidation. p. 69

collagen extracellular protein that surrounds connective tissue, cartilage, skin, tendons, and ligaments, and that provides strength to many tissues and resists stretching. p. 56

collecting duct a series of tubules and ducts that connect the nephrons to the ureters. pp. 302, 402

columnar epithelial cells that are rectangular or shaped in columns. p. 176

comminuted fracture fracture that results in two or more separate bone components. p. 227

common bile duct a tubelike structure formed by a combination of the cystic and common hepatic ducts. p. 387

common fibular nerve nerve in the lower leg. p. 279

common hepatic artery artery that provides blood to the liver. p. 337

common iliac arteries large arteries that originate from the aorta. p. 340

compact bone tightly arranged bone composed of osteons. p. 207

complement proteins proteins that exist in the blood and form part of the innate immune system response. p. 387

compliance tendency of an organ to resist returning to its original shape and size. p. 366

concentration gradient the difference in concentration of molecules across a cell membrane. p. 269

concentric contraction (muscular) shortening of the muscle during contraction. p. 239

conducting system path by which the electrical signal of the heart is transmitted. p. 324

conducting zone organs of the respiratory system that filter, warm, and moisturize the air and conduct it toward the lungs. p. 358

condylar joint synovial joint that permits motion in two planes. p. 230

condylar process rounded extension of a bone that articulates with another bone. p. 210

conjugation transfer of DNA directly from one bacterium to another through a pilus. p. 51

connective tissue a biological tissue that is a collection of cells, extracellular matrix, and other components, and that supports, connects, or separates other tissues or organs. p. 175

connexin proteins that make up gap junction channels. p. 57

consciousness awareness of self or surroundings. p. 283

continuous variation occurs when a phenotype shows multiple variations in a population through multiple genes. p. 101

contractile proteins proteins that allow for contraction of a muscle cell (e.g., actin and myosin). p. 250

conus medullaris tapered end of the spinal cord. p. 276

coracoid process protrusion on the scapula. p. 221

core muscles abdominal and back muscles that function to stabilize the body during movement. p. 239

coronal suture suture that connects the frontal bone to the anterior portion of the parietal bones. p. 210

coronary arteries arteries that provide oxygenated blood to the heart. p. 320

coronary sinus collection of veins that returns blood from the heart to the right atrium. p. 320

coronary sulcus (interventricular sulcus) groove separating the atria from the ventricles. p. 320

coronoid fossa indentation in the humerus that articulates with the ulna and increases the range of motion when the arm is bent. p. 221

corpora cavernosa erectile tissue of the penis. p. 416

corpus callosum portion of the brain that connects the two hemispheres of the cerebrum. p. 285

corpus luteum structure made up of granulosa cells after ovulation; secretes progesterone to maintain a potential pregnancy. p. 423

corpus spongiosum *see* corpora cavernosa. p. 416

cortical radiate arteries renal blood vessels. p. 404

cortical radiate veins renal blood vessels. p. 404

cortisol glucocorticoid hormone produced in the adrenal gland during stress. p. 309

cosmid synthetic plasmid containing the *cos* site from the lambda phage. p. 161

costal (hyaline) cartilage cartilage that gives the chest wall elasticity and is found at the anterior end of the ribs. p. 220

coupled transport the use of an electrochemical gradient to move nutrients along with an ion across the plasma membrane into the cell; also called co-transport. p. 60

covalent bond bond in which electrons are shared between two atoms. p. 26

cowpox virus used in humans to build immunity to similar viruses that have more harmful effects. p. 445

cranial cavity area of the head that contains the brain. p. 19

cranial nerves nerves that originate in the brain or brainstem. p. 263

craniofacial bones bones that make up the skull and face. p. 210

c-reactive protein protein produced in the liver during an inflammatory response. p. 350

creatine kinase enzyme used to create creatine phosphate in muscle cells. p. 257

creatine phosphate a molecule used by muscle cells for energy in the first few seconds of high intensity exercise. p. 257

creatinine a breakdown of creatine phosphate. p. 407

cribiform plate bone that forms base of the nasal cavity. p. 213

cristae inner membrane of the mitochondria that is folded and is the location of the electron transport chain. p. 53

cross-bridge cycle that uses the binding of actin and myosin to generate force during muscle contraction. p. 252

crossing-over process where homologous chromosomes pair up and fragments of non-sister chromatids break and are swapped in the same place, resulting in hybrid chromosomes. p. 90

cross-sectional plane *see* transverse plane. p. 19

cuboidal epithelial cells in a shape closely resembling a cube. p. 176

Cushing's syndrome a condition caused by chronically high levels of cortisol. p. 308

cyanobacteria bacteria that undergo photosynthesis and produce oxygen. p. 451

cytokines proteins that function in cell-signalling during an immune response. p. 433

cytokinesis process of splitting the cytoplasm. p. 80

cytoplasm water-filled fluid inside a cell. p. 44

cytoskeleton intracellular proteins that hold organelles in place, give cells shape, and help transport other molecules within the cell. p. 55

cytotoxic T cells immune cell that targets cancer cells, and virus-infected cells. p. 439

Dalton's law principle that states that the total pressure of a mixed gas equals the sum of the partial pressures. p. 368

deamination process in which an amino group is removed from an amino acid. p. 73

decarboxylation process of removing carbon from a molecule as CO_2. p. 69

decomposition breakdown of organic substances into smaller forms. p. 6

defecation elimination of solid or semi-solid waste from the digestive tract. p. 380

dehydration synthesis removal of a water molecule to form a covalent bond. p. 27

deletion a nucleotide skipped during replication. p. 139

deltoid triangular-shaped muscle superior and lateral to the shoulder. p. 239

deltoid tuberosity attachment site for the insertion of the deltoid muscle. p. 221

denaturation irreversible unfolding of a protein in which hydrogen bonds are broken. pp. 31, 168

dendrites branches from the cell body of a neuron that detect stimuli. p. 266

dendritic cells antigen-presenting cells of the immune system. pp. 354, 435

dens protuberance on axis that permits horizontal rotation of the skull. p. 216

dense connective tissue tissue that forms strong attachments between structures, and allows organs to stretch and recoil. p. 199

deoxygenated the removal of oxygen from a molecule. p. 351

deoxyribonuclease an enzyme that breaks down DNA. p. 391

deoxyribose five-carbon sugar in a DNA nucleotide. p. 132

depolarization movement of sodium or calcium into a cell that causes the inside of the cell to become positively charged. p. 270

dermal papillae projections of the dermis that extend into the epidermis to provide increased surface area for diffusion of oxygen and nutrients. p. 196

dermal ridges contiguous papillae that form ridges into the epidermis and form fingerprints. p. 196

dermis middle layer of the skin that contains blood vessels, nerves, and other cell types. p. 195

descending aorta segment of the aorta that's distal to the aortic arch. p. 321

descending colon the part of the large intestine located in the left-upper and left-lower quadrants of the abdomen, leading to the sigmoid colon. p. 389

descending loop of Henle the proximal portion of the loop of Henle that is permeable to water. p. 407

descending tracts collections of axons that transmit efferent signals from the CNS. p. 276

desmosomes proteins that adhere cells together. p. 57

detrimental harmful. p. 293

detrusor muscle smooth muscle in the wall of the bladder. p. 402

dialysis a medical process for removing waste products from the blood when the kidneys no longer function properly. p. 412

diaphragm muscle at the base of the thoracic cavity that facilitates respiration. p. 239

diaphysis middle region, or shaft, of the long bone. p. 209

diarthrosis a joint that is fully movable. p. 228

diastole relaxation of the heart muscles. p. 348

diencephalon located between the brainstem and the cerebrum and includes the thalamus, hypothalamus, and the pineal gland. p. 282

diet food intake. p. 371

diffusion movement of any molecule from an area of high concentration to an area of low concentration until equilibrium is reached and without input of energy. p. 408

dihybrid cross a cross between two individuals that carry two different traits that determines the probability of the traits being passed onto the offspring. p. 99

diplococci paired, spherical bacteria. p. 453

diploid (2n) twice the amount of haploid (n) chromosomes found in somatic and germ-line cells. p. 82

directional selection favouring one extreme trait over another in a population so that the opposite extreme is eliminated. p. 127

directional terms terms used to describe regions of the body in relationship to each other. p. 16

disaccharide two monosaccharides joined by glycosidic bonds. p. 36

disruptive selection favouring each extreme in a population so that the intermediate traits are decreased or eliminated. p. 127

distal epiphysis end of the long bones that's furthest from the core of the body. p. 227

distal tubule section of the nephron located between the ascending loop of Henle and the collecting ducts. p. 302

distensibility ability to be stretched under pressure. p. 328

diuretics medications that increase urination by decreasing water reabsorption in the nephron. p. 352

diversity variety. p. 4

DNA deoxyribonucleic acid. pp. 33, 168

DNA denaturation separation of DNA strands by breaking hydrogen bonds. p. 187

DNA polymerase enzyme that adds complementary nucleotides to DNA in the 5′ to 3′ direction during replication. p. 78

DNA repair enzymes enzymes that correct DNA mutations. p. 141

DNA replication a process in which DNA is copied before cell division. p. 78

DNA transfer method by which genetic variation is acquired among bacteria. p. 78

dNTPs dinucleotide triphosphates, the nucleotides needed to make the new DNA. p. 168

dominant the allele that is phenotypically expressed. p. 95

dopamine a neurotransmitter involved in task-based concentration and the reward pathway. p. 284

dorsal horn dorsal grey matter of the spinal cord. p. 276

dorsal respiratory group (DRG) part of the medulla oblongata responsible for respiratory rhythm. p. 373

dorsal root afferent sensory root of a dorsal nerve that is outside the spinal cord. p. 276

double helix two strands of DNA bound together by nitrogenous bases that forms a helical (spiral) structure. p. 34

Down syndrome delayed development and mental impairment in an individual with trisomy 21. p. 108

duodenum the initial section of the small intestine. p. 388

dura mater outermost layer of the meninges. p. 276

dwarfism condition resulting from decreased levels of growth hormone. p. 303

dystrophin muscle protein that holds myofibrils together to prevent muscle degeneration. p. 109

early antral follicle follicle formed by granulosa cells that surrounds the fluid-filled cavity referred to as the antrum. p. 423

eccentric contraction lengthening of the muscle during contraction. p. 257

eccrine glands type of sweat gland, found predominantly on the forehead, palms, and soles of the feet. p. 198

ecology the study of the relationship between organisms and their environment. p. 115

ectoparasites parasites that live on the surface of the host. p. 463

edema an accumulation of fluid in the interstitial space. p. 334

effectors any part of the body that responds to stimulation from the integrating centre that will bring an altered factor back to homeostasis. p. 22

efferent ductules connect the rete testis to the epididymis. p. 416

efferent neurons neurons that transmit a signal away from the central nervous system. p. 265

eicosanoids signalling molecules involved in inflammation. p. 437

ejaculatory duct a duct formed from the union of the vas deferens and the seminal vesicle. p. 416

elastin extracellular matrix protein that allows tissues to be flexible and stretchy, and to reform to their original shape after being stretched. p. 56

electrical gradient difference in electric potential across a cell membrane. p. 60

electrocardiogram (ECG) a recording of the electrical activity of the heart. p. 325

electrolytes ions in the fluids of the body that conduct an electrical charge. p. 352

electron transport chain electrons transferred through membrane proteins, producing many ATP molecules. p. 71

emphysema one of the disease conditions that make up chronic obstructive pulmonary disease (COPD). p. 376

encapsulated refers to benign tumours surrounded by a layer of cells. p. 84

end diastolic volume (EDV) maximum volume of blood in the ventricles after diastole. p. 326

end stage renal disease progressive loss of kidney function to the point where the kidneys no longer function; caused primarily by chronic hypertension. p. 351

end systolic volume (ESV) volume of blood in the ventricles after systole. p. 327

endocardium innermost layer of the heart. p. 321

endocrine gland an organ that secretes hormones. p. 22

endocytosis movement of large molecules or fluid into the cell. p. 61

endometriosis a gynecological condition where endometrial cells grow outside of the uterus. p. 425

endometrium innermost layer of the uterus. p. 277

endomysium connective tissue that surrounds individual muscle cells. p. 237

endoneurium connective tissue that surrounds individual neurons. p. 278

endoparasites parasites that live inside their host. p. 463

endoplasmic reticulum membrane network within the cytoplasm of every cell type that can be smooth or rough. p. 52

endospore a highly resistant state that some bacteria enter where they are dormant and can survive harsh environments. p. 453

endosteum dense, irregular, connective tissue that lines the surface of bones and contains bone precursor cells. p. 209

endosymbiosis a process whereby primitive bacteria were engulfed by another cell, giving rise to eukaryotic organisms with organelles. p. 53

endothelium layer of cells that lines the interior surface of vessels. p. 321

enteric nervous system nervous system that controls the digestive system. p. 265

enzymes globular proteins that facilitate thousands of chemical reactions within cells. pp. 32, 49

eosinophils white blood cells that fight parasitic infections. p. 353

ependymal cells glial cells that produce cerebral spinal fluid. p. 189

epicardium outermost layer of the heart, also known as the visceral pericardium. p. 321

epidermal ridges extensions of dermal ridges into the epidermis, which form fingerprints and footprints. p. 196

epidermis most superficial layer of the skin. p. 195

epididymis a part of the male reproductive system connecting the efferent ducts to the vas deferens. p. 416

epidural space anatomical space outside the dura mater. p. 276

epigastric region central-upper region of the abdomen. p. 389

epiglottis a small cartilaginous structure that prevents the passage of solids or fluids into the trachea during swallowing. p. 359

epinephrine a stimulatory hormone produced in the adrenal medulla in response to sympathetic nervous system stimulation. p. 260

epineurium outermost layer of connective tissue surrounding neurons. p. 277

epiphyseal line line of bone that replaces the epiphyseal plate once bone growth is complete. p. 209

epiphyseal plate layer of hyaline cartilage where chondrocytes produce extracellular matrix until mineralization occurs during bone growth. p. 228

epistasis effects of one gene modified by effects of one or more other genes. p. 104

epithelial tissue type of tissue that lines and covers organs and body cavities, and also makes up outermost layer of the skin. p. 175

equatorial plane place where chromosomes line up in the centre of a cell during cell division. p. 82

erector spinae deep muscles that extend the length of the back. p. 245

erythrocytes red blood cells. p. 353

erythropoietin hormone that stimulates the production of red blood cells. pp. 353, 401

esophageal arteries a group of arteries that supply the esophagus. p. 337

esophagus segment of the alimentary canal between the pharynx and the stomach. p. 384

essential amino acids nine amino acids that cannot be synthesized by the body and must be acquired in the diet. p. 28

essential fatty acids fats that cannot be synthesized by the body and must be acquired in the diet. p. 43

estrogen a sex hormone produced in the ovaries. pp. 201, 312

ethanol two-carbon alcohol molecule. p. 74

ethmoid bone bone that forms the anterior portion of the base of the skull and forms the nasal septum. p. 194

eubacteria prokaryotic unicellular organisms that survive in many different environments. pp. 4, 451

eukaryotic complex cellular structure that contains membrane-bound organelles. p. 4

Eustachian tube tube that connects the inner ear to the pharynx to balance pressure in the ear. p. 359

evolution the change in the inherited traits in a population of organisms over time. p. 20

excitatory post-synaptic potentials (EPSPs) graded membrane potentials that increase the likelihood of an action potential in the post-synaptic neuron. p. 272

excretion process of eliminating waste from an organism. p. 405

exhalation (or expiration) movement of air from the lungs into the external environment. p. 358

exocrine substance such as sweat or enzymes that is secreted through a duct into a hollow organ or to the outside of the body. p. 388

exocrine glands glands that secrete substances either outside the body or into a hollow organ. p. 298

exocytosis the movement of large molecules out of vesicles and across the cell membrane. p. 61

exon shuffling occurs when certain exons are translated into a protein, also known as alternate splicing. p. 154

exons sequences of DNA that translate into amino acid sequences for protein synthesis. p. 147

exonuclease activity an enzyme function that allows polymerase to reverse direction and remove incorrect nucleotides. p. 121

exophthalmos refers to the bulging appearance of the eyes as seen in conditions such as Graves' disease; literally, "from the eye". p. 305

expiratory reserve volume amount of air in the lungs following normal exhalation. p. 367

expression vector a piece of DNA that has a promoter region that allows transcription and translation of a foreign gene in another organism, such as bacteria or yeast. p. 161

extensors muscles in the posterior of the forearm that allow extension of the hand at the wrist. p. 246

external auditory meatus opening in the skull where sound waves enter the inner ear. p. 210

external environment everything outside the body that is not regulated, such as temperature, nutrients in food, content of air inhaled; also includes the inside of the GI tract, urethra, vagina, ear canal, nasal and oral cavities, and respiratory tract. p. 21

external iliac a pair of arteries that originate at the aorta and branch in the pelvic region. p. 340

external intercostal muscles the superficial muscles between the ribs that contract to increase thoracic volume during inspiration. p. 364

external jugular veins large veins that receives blood from the head. p. 341

external obliques largest and outermost muscle of the lateral and anterior abdomen; allows the body to twist and move forward during contraction. p. 365

external urethral sphincter a ring of muscle that can be voluntarily controlled to contract and prevent urination. p. 402

extracellular outside the cell. p. 441

extracellular fluid fluid outside the cell, includes interstitial fluid and blood plasma. p. 61

extracellular matrix the proteins and connective tissue located outside of a cell. p. 180

extracellular proteins proteins that are produced by each cell, secreted or integrated into the cell membrane, and involved with intercellular interactions such as adhesion, signalling, or cellular communication. p. 56

facilitated diffusion movement of molecules across a membrane, through a membrane protein, down a concentration gradient, and without input of energy. p. 59

fallopian tube tubes that lead eggs cells from the ovaries to the uterus. p. 420

fascia layer of fibrous tissue that surrounds muscles and groups of muscles. p. 237

fascicle a bundle or cluster of neurons or muscle cells. p. 237

fasting phase stage of digestion between meals. p. 311

fats nonpolar lipids that have many important functions in the body, such as production of cell membranes and steroid hormones; can be broken down to provide energy. p. 39

fat-soluble hormones nonpolar hormones that can move into the target cell and bind to a receptor that affects gene expression. p. 299

fatty acid chains of carbon and hydrogen; three fatty acid chains plus a glycerol molecule form a triglyceride. p. 41

fecal-oral route infections acquired through consuming contaminated food or water. p. 463

femoral nerve nerve located along the femur. p. 278

femur longest bone in the human body; located in the thigh. p. 224

fermentation process of anaerobic respiration in animals, bacteria, and yeast that allows small amounts of ATP to be produced in the absence of oxygen. p. 74

fever an elevation in body temperature, often in response to infection. p. 437

fibre *see* cellulose. pp. 37, 237

fibrinogen a protein that helps form clots. p. 353

fibroblasts most abundant connective tissue cell type, which produces collagen, elastin, reticular fibres, and ground substance. p. 180

fibrocartilage type of cartilage composed of fibrous connective tissue. p. 227

fibronectin protein that connects integrins to other extracellular matrix proteins. p. 56

fibrous joints structural joints that are connected by fibrous connective tissue and are immovable, such as the joint between the radiu and the ulna. p. 228

fibula the lateral bone in the lower leg. p. 224

filtrate blood plasma that has passed through the glomerulus into the Bowman's capsule. p. 406

filtration movement of substances from the glomerular capillaries into the Bowman's capsule in the kidney. p. 404

fimbriae fringed tissue surrounding the opening of the fallopian tube. p. 420

fixators muscles that stabilize the primary muscle of a motion. p. 239

flagella external structures composed of microtubules; used for locomotion and feeding (singular, flagellum). p. 50

flat bones bones such as the sternum, ribs, and scapulae. p. 208

flatworms (platyhelminths) a type of animal parasite, such as tapeworms. p. 464

flexors muscles in the anterior of the forearm that allow flexion of the hand at the wrist. p. 246

flow movement of a fluid. p. 345

fluid mosaic model term used to describe a typical cell membrane composed of unsaturated phospholipids that are liquid rather than solid, cholesterol, and proteins. p. 49

flukes (trematodes) a type of animal parasite. p. 464

FOIL acronym (first, outer, inner, last) used to determine the possible gametes in a dihybrid cross. p. 101

follicle stimulating hormone (FSH) a sex hormone that stimulates the production of testosterone and estrogen in males and females. p. 312

follicles an egg surrounded by granulosa cells. p. 422

follicular cells cells of the thyroid gland that produce triiodothyronine and thyroxine—T3 and T4, respectively. p. 304

follicular phase phase in the menstrual cycle in which the follicles of the ovaries mature. p. 423

food energy energy from digested food, such as glucose that can be used to make ATP. p. 66

foramen any opening in a bone that allows blood vessels and nerves to pass through. p. 216

foramen magnum opening through which the spinal cord exits the skull. p. 213

foramen ovale (1) opening in the base of the skull that allows the mandibular nerve to travel through the skull; (2) a fetal cardiac opening that allows blood to enter the left atrium from the right atrium. pp. 213, 321

force (muscular) a power that's generated by a muscle to overcome the load. p. 253

fossa ovalis a depression in the right atrium of the heart, the remnant of a thin fibrous sheet that covered the foramen ovale during fetal development. p. 321

frameshift mutation a change in multiple codons that results from an insertion or deletion mutation. p. 139

Frank-Starling law the principle that increased stretch in the muscle cells will increase the force of contraction. p. 328

frontal bones bones that form the forehead. p. 210

frontal lobe the anterior lobe of the brain. p. 286

frontal plane an anatomical division of the body into anterior and posterior sections. p. 17

fructose monosaccharide found in fruit. p. 36

full thickness burn (third and fourth degree burn) burn that extends either into or through the subcutaneous layer of the skin to the underlying muscle. p. 199

fundus part of a hollow organ opposite from its opening, such as the left portion of the body of the stomach. p. 384

fungi eukaryotic multi-cellular organisms (except yeast) that are not photosynthetic and absorb nutrients from other organisms. p. 435

G1 phase first phase of interphase; the main gap or growth phase when the cell undergoes normal cell functions and prepares for DNA replication. p. 79

G2 checkpoint point at which DNA is checked for DNA mutations after replication. p. 80

G2 phase third phase of interphase; the second gap or growth phase where the cell continues to prepare for mitosis. p. 80

galactose monosaccharide found in milk. p. 36

gallbladder digestive organ that stores bile. p. 386

gametes sperm or egg cells that contain a haploid number of chromosomes. p. 82

gametogenesis formation of the gametes. p. 415

ganglia a collection of neuron cell bodies. p. 266

gap junction channels protein membrane channels that play an important role in intercellular communication by allowing the movement of molecules directly between adjacent cells. p. 50

gap phase also known as growth phase of the cell. p. 79

gas exchange movement of gases in and out of the blood in relation to the. p. 333

gastric phase a phase of digestion that involves digestion in the stomach. p. 396

gastrin a hormone released by the stomach that stimulates the production of hydrochloric acid. p. 396

gastrocnemius calf muscle in the lower leg that allows for flexion of the foot at the ankle. p. 248

gastrointestinal tract organs that form the alimentary canal. p. 379

gel electrophoresis method that uses an electrical current to separate strands of DNA based on the length of the DNA sequence. p. 167

gene DNA sequence of hundreds of nucleotides that code for a specific protein. p. 115

gene cloning replication of a gene sequence in bacteria. p. 161

gene expression transcription and translation of a gene that results in the synthesis of a protein. p. 145

gene flow occurs when phenotypes in a population become similar. p. 128

gene pool all the genes and allele variations in a population. p. 124

gene silencing addition of methyl groups to prevent gene transcription. p. 154

gene splicing the removal of introns from mRNA after transcription. p. 157

genetic code universal chart used to determine the amino acid based on the mRNA codon. p. 150

genetic engineering *see* recombinant DNA technology. p. 157

genetic modification any kind of alteration in the genome, including deleterious mutations caused by radiation or chemicals. p. 158

genetic variation a means for individuals to acquire new alleles through sexual reproduction. p. 114

genetically modified organisms (GMOs) organisms such as plants or animals that contain recombinant DNA. p. 158

genome all genes in an organism. p. 158

genomic DNA library an archive of the possible sequences in the human genome contained within plasmids within bacteria. p. 166

genotype genes that code for a phenotype. p. 124

germ cells cells found in the testes or ovaries that contain a diploid number of chromosomes and produce gametes during meiosis. p. 82

gigantism condition resulting from increased levels of growth hormone in childhood. p. 303

gingiva mucosal tissue that lines the inside of the mouth. p. 384

gingivitis inflammation of the gingiva. p. 384

glenoid cavity socket in which the humerus sits. p. 221

glial cells cells that support the neuron functions. p. 266

gliding movement of a joint in a single plane running parallel to the articulating structure. p. 231

glomerular filtration rate (GFR) the amount of fluid moved from the glomerulus to the Bowman's capsule in a set amount of time. p. 406

glomerulus a network of capillaries where fluid leaves and enters the Bowman's capsule during filtration. p. 404

glossopharyngeal nerve the ninth cranial nerve that carries motor and sensory information to the mouth and tongue. p. 349

glucagon a hormone produced in the pancreas that stimulates the liver to break down glycogen and increase blood sugar. pp. 260, 388

glucocorticoid a group hormones produced in the adrenal cortex in response to stress, including cortisol, the primary glucocorticoid, that stimulates the breakdown of fat to increase blood sugar and inhibits the immune response. p. 307

gluconeogenesis formation of glucose from the breakdown of fat or amino acids. p. 387

glucosamine sulfate a sugar-amino acid found in ground substance in connective tissues. p. 181

glucose monosaccharide that is the building block of starch, also found in fruit and table sugar. p. 36

GLUT membrane protein receptor that facilitates the transport of glucose across a membrane. p. 392

gluteal tuberosity attachment site for the gluteal muscles. p. 224

gluteus maximus largest and strongest muscle in the body; located on the posterior pelvis. p. 248

glycemic index a measure of the rate at which various foods prompt the release of insulin. p. 311

glycerol a sugar alcohol that connects three fatty acids in a triglyceride. p. 41

glycogen polysaccharide that is stored energy in liver and muscle cells. p. 37

glycogenesis the production of glycogen from excess glucose in the liver and muscle cells. p. 387

glycogenolysis the breakdown of glycogen into glucose. pp. 257, 387

glycolysis 10-step process of chemical reactions that occurs in the cytoplasm; includes the oxidation of glucose and reduction of NAD$^+$, resulting in two pyruvate molecules and two ATP. p. 607

glycoprotein a molecule composed of protein and sugar. pp. 49, 150

glycosidic bond bond between a sugar and another molecule, such as the base in a nucleotide. p. 34

glycosylation enzymes enzymes in the Golgi that cause the addition of oligosaccharides to some proteins. p. 53

goblet cells cells in the respiratory and digestive tract that produce mucus. p. 360

Golgi bodies a separate membrane network found in the cell near the endoplasmic reticulum, where newly synthesized proteins are further processed and packaged into vesicles. p. 150

gomphosis joint joint that houses the teeth. p. 228

gonad reproductive organ where gametes are produced through meiosis; the ovaries in females and the testes in males. p. 415

gonadal arteries arteries that provide oxygenated blood to the gonads. p. 340

gonadotropin-releasing hormone (GnRH) a hypothalamic hormone that stimulates the anterior pituitary gland to produce FSH and LH in males and females. p. 418

graded potential changes in membrane potential that vary in magnitude depending on the stimulus. p. 271

gradualism rate of evolution that is a slow and gradual process where very small changes occur in a population over millions of years. p. 119

gram negative bacteria bacteria that have a small layer of peptidoglycan and an external layer of lipopolysaccharide that does not stain when exposed to crystal violet. p. 50

gram positive bacteria bacteria that have many layers of peptidoglycan but do not have a lipopolysaccharide layer, and stain purple when exposed to crystal violet. p. 50

granular leukocytes a type of white blood cells that contain granules and are vital to the immune system. p. 353

granulosa cells cells that surround the egg, forming the follicle, and produce sex hormones and growth factors. p. 423

Graves' disease an autoimmune disease that is caused by antibodies binding to the TSH receptor, causing increased production of thyroid hormones. p. 305

great saphenous vein a large vein in the leg. p. 344

greater omentum a large, protective fold of the visceral peritoneum. p. 383

greater sciatic notch indentation in the ilium. p. 224

greater trochanter lateral projection of the femur that acts as an attachment site for muscles. p. 224

greenstick fractures bone fractures where the bone bends, and partially breaks; more common in children than adults. p. 227

grey matter tissue of the brain that appears grey and contains cell bodies, dendrites, and axon terminals of neurons. p. 276

ground substance fluid part of the connective tissue that supports the functions of cells and fibres. p. 180

growth hormone (GH) hormone produced by the anterior pituitary gland that stimulates growth of bone and muscle and the regeneration of cells throughout the body. p. 201

growth phase see G1 phase. p. 79

gut-associated lymphoid tissue (GALT) lymphatic nodules and lymph nodes located along the digestive; important for preventing and fighting infections that enter via the GI tract. p. 432

gyrus a ridge of the cerebral cortex. p. 285

H zone central region of the sarcomere, containing the central portion of the thick filaments but no thin filaments. p. 251

hamstrings muscles that flex the leg at the knee. p. 248

haploid (n) the number of chromosomes found in gametes, which is half of the diploid number. p. 82

Hardy-Weinberg equilibrium condition when a population is stable and not evolving and allele frequencies stay the same over successive generations. p. 125

heart murmur an extra heart sound during the cardiac cycle made by blood flowing backward through a valve. p. 323

helicase an enzyme that breaks hydrogen bonds between the bases on complementary strands of DNA, beginning at the origin of replication. p. 135

helminths (worms) a type of animal parasite. pp. 463, 464

helper T cells part of the adaptive immune response that stimulate other immune cells to fight bacterial and viral infections, but do not themselves directly kill infections. p. 439

hemagglutinin a substance that causes red blood cells to agglutinate, and is also the membrane protein that influenza viruses bind to during infection. p. 458

hematocrit level percentage of the blood that is red blood cells; normal level, 45%. p. 353

hematoma an accumulation of blood, commonly called a bruise. p. 227

hematopoiesis process of creating new blood cells. p. 206

hematopoietic stem cells stem cells that differentiate into blood cells and are located in the bone marrow. p. 430

heme chemical group that contains Fe^{2+} that binds oxygen. p. 353

hemoglobin blood protein that contains heme. p. 108

hemorrhagic stroke a stroke caused by a ruptured blood vessel in the brain. p. 351

hepatic portal vein blood vessel that takes blood from the GI tract to the liver. p. 340

hepatic sinusoids low-pressure channels that take blood from the hepatic artery and portal vein and deliver it to the central vein and then to the hepatic vein; includes Kupffer cells inside the sinusoids that take up and destroy foreign material. p. 386

hepatopancreatic duct duct that carries secretions from both the common bile duct and the pancreatic duct to the small intestine during digestion. p. 387

heredity the passing of genes from parents to offspring. p. 93

heterozygote advantage the benefits of having a heterozygote genotype compared to having homozygous dominant or recessive genes; for example, the benefit of being heterozygous for sickle cell anemia when living in an environment with high rate of malaria infection. p. 128

heterozygous alleles that are different. p. 124

high-density lipoprotein (HDL) smallest major group of lipoproteins, and are considered the "good" blood cholesterol. p. 350

hinge joint joint where the articular surfaces allow for motion only in a single plane, such as the elbow. p. 230

hip flexors (iliopsoas) a group of muscles that work to flex the thigh at the hip, raising the knee up toward the body. p. 246

hippocampus a part of the temporal lobe of the brain associated with memory. p. 288

histamine substance that causes vasodilation in an inflammatory response or allergic reaction. p. 353

histology the study of tissues. p. 175

HIV (human immunodeficiency virus) a virus that targets the cells of the immune system that express the CD4 receptor. p. 439

homeostasis the regulation of multiple factors within the internal environment of the body, such as temperature, blood sugar, ions, water, and pH. p. 10

homocysteine a non-protein amino acid that is associated with an increased risk for cardiovascular disease. p. 353

homologous chromosomes two similar, but not identical, chromosomes where one comes from the mother and one comes from the father; also called homologues. p. 80

homologous structures term used by evolutionists to compare anatomy among organisms to show that similarities indicate a common ancestor. p. 122

homologues *see* homologous chromosomes. p. 80

homozygous alleles that are the same. p. 124

horizontal fissure separates the superior and middle lobe of the lung. p. 361

horizontal plane *see* transverse plane. p. 18

hormones chemical messengers secreted by endocrine glands. p. 297

host cell a cell that's used to take up the recombinant DNA. p. 158

humerus bone of the upper arm. p. 221

humoral immune response a response involving the secretion of antibodies by plasma cells. p. 439

hyaluronic acid component of extracellular matrix of connective, epithelial, and nerve tissue that plays a key role in signalling cells to proliferate; often a component in anti-aging creams to stimulate the growth of new skin cells. p. 181

hyaluronidase enzyme that breaks down hyaluronic acid. p. 181

hybridization a method for binding two complementary strands of DNA, used to locate specific DNA sequences in bacterial colonies. p. 163

hydrochloric acid (HCl) an acid formed by the elements hydrogen and chlorine dissolved in water, produced by parietal cells in the stomach. p. 386

hydrogen bond a slightly positive charge on the hydrogen of one molecule attracting a slightly negative charge on another molecule, most often oxygen. p. 26

hydrogen peroxide H_2O_2, used to breakdown substances such as alcohol and formaldehyde and can be effective in destroying infectious organisms. p. 55

hydrogenation process in which hydrogen is added to fatty acid molecules that contains double bonds and converts them to single bonds, such as the production of margarine from vegetable oil. p. 42

hydrolysis the process of breaking covalent bonds with the addition of a water molecule. p. 37

hydrolytic enzyme enzyme that aids in hydrolysis. p. 27

hydrophilic water loving. p. 26

hydrophobic water fearing. p. 26

hydrostatic pressure mechanical pressure generated by a fluid on the wall of its container. pp. 333, 406

hyoid bone attachment site for the tongue and neck muscles; does not articulate with any other bones. p. 214

hypercapnia increased amount of carbon dioxide in the blood, which can result from hypoventilation or lung disease. p. 370

hyperpolarization increased charge of a cell beyond the normal resting membrane potential. p. 271

hypertension condition of increased blood pressure. p. 350

hyperthyroidism over-production of thyroid hormones. p. 305

hypertonic concentration of the solute higher in the interstitial fluid than inside the cell. p. 59

hypodermis innermost layer of the dermis. p. 194

hypocapnia lower than normal concentration of carbon dioxide in the blood, resulting from hyperventilation. p. 370

hypothalamus a portion of the brain located superior to the brainstem, part of the diencephalon; is responsible for regulating the autonomic nervous system and controls the endocrine functions of the pituitary gland. p. 22

hypothyroidism under-production of thyroid hormones. p. 305

hypotonic concentration of the solute lower in the interstitial fluid than inside the cell. p. 59

hypoxia decreased oxygen level. pp. 351, 353

I band region between the A bands of adjacent sarcomeres. p. 251

ileocecal sphincter a ring of muscle that controls the opening to the large intestine from the small intestine. p. 388

ileum the final section of the small intestine. p. 388

iliac crest top of the hip bones. p. 221

iliacus one of the hip flexor muscles. p. 246

iliocostalis muscles muscles of the back that attach the ribs. p. 245

ilium largest and most superior bones in the pelvic girdle. p. 221

immortalize cells that do not die, which occurs in cancer cells. p. 85

immune response when the immune system fights infections. p. 439

immune suppression medications medications that decrease the ability of the immune system to function and are used to treat chronic inflammatory conditions such as rheumatoid arthritis. p. 447

immunoglobulins proteins made by the immune system that attach to foreign substances in the bloodstream and neutralizes them. p. 442

inactivated toxin a toxin that has been altered either through heat or chemical means to no longer be pathogenic. p. 445

independent assortment random separation of homologues into gametes during meiosis. p. 90

inferior vena cava inferior portion of the largest vein in the body; brings blood to the right atrium. p. 320

inflammatory response part of an initial reaction to damaged tissues that fights infection and involves vasodilation, swelling, redness, and pain. p. 436

infundibulum funnel-shaped cavity or organ, such as the stalk that connects the hypothalamus with the posterior pituitary gland. p. 420

ingestion consumption of food. p. 380

inhalation (or inspiration) movement of air from the external environment into the lungs. p. 358

inhibin a hormone that inhibits FSH secretion in males and females. p. 418

inhibitory neurotransmitters neurotransmitters that reduce the likelihood of an action potential occurring in the postsynaptic neuron by causing IPSPs. p. 272

inhibitory post-synaptic potentials (IPSPs) graded membrane potentials that decrease the likelihood of an action potential in the post-synaptic neuron. p. 272

initial segment the part of the axon hillock that is the starting point for the action potential in a neuron. p. 266

innate immune response a non-specific immune response that acts as the second line of defense against infection and involves cells such as neutrophils, macrophages, dendritic cells, and natural killer cells. p. 433

insect vectors (carriers) insects that carry or transmit a pathogen, such as mosquitoes that transmit malaria. p. 463

insertion (DNA replication) also known as addition, an extra nucleotide added during replication. p. 139

insertion (muscular) connection point of a muscle to a moving bone. p. 238

insertional inactivation occurs when a transposable element inserts into a DNA sequence in the middle of a gene, producing a nonfunctional protein. p. 140

inspiratory capacity total amount of air that can be drawn into the lungs; includes the tidal volume. p. 366

inspiratory reserve volume amount of air that can be drawn into the lungs after normal inspiration, not including the tidal volume. p. 366

insula the part of the brain believed to be involved in consciousness. p. 288

insulin hormone that is secreted by the pancreas and lowers blood sugar after a meal. p. 60

insulin-like growth factors (IGFs) a hormone, stimulated by GH and secreted by the liver, that stimulates growth of many cell types, such as bone and muscle. p. 226

insulin resistance when cells no longer respond to insulin to take up blood sugar. p. 311

integrating centre part of the body that interprets signals coming from the receptors and determines if there is any deviation from the set point. p. 22

integrins membrane-spanning proteins involved in attaching the cell membrane to proteins in the extracellular matrix and in transferring signals from outside to inside the cell, and vice versa. p. 56

integumentary system the skin, which is the organ system that prevents damage to the body and provides a physical barrier to outside the body. p. 193

interatrial septum the barrier between the right and left atria of the heart. p. 321

intercalated discs feature of cardiac muscle that connects the branched, striated muscle fibres and contains desmosomes and gap junctions. p. 187

intercondylar fossa indentation at the distal end of the posterior femur. p. 224

intercostal arteries arteries that provide oxygen to the area between the ribs. p. 337

interferon (INF) a cytokine released by virus-infected cells that cause nearby uninfected cells to block viral replication. p. 435

interleukin-1 (IL-1) a cytokine produced by antigen-resenting cells that signal helper T cells to proliferate. p. 440

interleukin-2 (IL-2) a cytokine-signalling molecule that signals other immune cells, such as cytotoxic T cells and B cells to proliferate. p. 440

interlobar arteries blood vessels that carry oxygenated blood in the kidneys. p. 404

interlobar veins blood vessels that carry deoxygenated blood in the kidneys. p. 404

intermediate filaments intracellular cytoskeletal proteins composed of intertwined proteins that support cell structure and help organelles remain in place. p. 55

intermediate trait a trait expressed by a heterozygote that is a variation of the homozygous dominant and homozygous recessive traits. p. 101

intermembrane space space between the inner membrane and the outer membrane of the mitochondria. p. 71

internal environment everything that is in close contact with the body's cells and regulated by homeostasis, such as interstitial fluid and blood. p. 21

internal iliac artery main artery of the pelvis. p. 340

internal intercostal muscles deep muscles between the ribs. p. 239

internal jugular veins veins that collect blood from the brain, superficial face, and neck. p. 341

internal obliques innermost antero-lateral abdominal oblique muscle, which acts as an antagonist to the diaphragm in forced exhalation. p. 365

internal urethral sphincter a ring of muscle that controls the flow of urine. p. 404

interneurons short neurons used to transmit signals within the central nervous system. p. 267

interosseous membrane membrane between bones composed of connective tissue, such as between the radius and ulna, and the tibia and fibula. p. 221

interphase growth phase of cell, consisting of the G1, S, and G2 phases that occur before mitosis. p. 79

interstitial fluid fluid that is between cells. p. 44

intertubercular groove (bicipital groove) groove in the humerus that houses the biceps tendon. p. 221

interventricular septum barrier between the right and left ventricles of the heart. p. 321

intervertebral disc cylindroid structure made of cartilage that separates and cushions the vertebrae. p. 216

intestinal phase a phase of digestion that occurs in the intestines. p. 396

intracellular inside the cell. p. 441

intrapleural pressure pressure in the pleural cavity (usually negative compared to the alveolar and environmental pressures) to prevent lungs from collapsing. p. 365

intrinsic factor a protein secreted by the stomach that is required for the absorption of vitamin B12 in the ileum. p. 386

introns sequences of DNA in genes of eukaryotes that do not translate into part of an amino acid sequence. p. 147

involuntary movement movement that is unconsciously controlled, such as the knee jerk reflex. p. 187

iodine an element on the periodic table that is required for the production of thyroid hormones. p. 304

ion channels proteins that span the membrane and allow the transport of specific ions into or out of the cell. p. 50

irregular bones bones that do not belong to the classification of long, short, flat, or sesamoid; include the vertebrae and the bones of the face and skull. p. 208

ischemia decreased blood supply. p. 351

ischemic stroke a stroke caused by an occlusion of a blood vessel in the brain. p. 351

ischial spines protrusions that form the narrowest portion of the pelvic opening. p. 222

ischial tuberosities protrusions at the inferior and posterior aspects of the pelvis; the bones that we sit on. p. 221

islet cells endocrine cells of the pancreas that produce either glucagon (alpha islet cells) or insulin (beta islet cells). p. 388

isokinetic contraction change in tension of a contraction without changing the velocity. p. 239

isometric contraction muscle contraction that doesn't result in a change of muscle length. p. 239

isotonic contraction contraction that causes muscle to shorten with even tension throughout the motion. p. 239

isotonic concentration of the solute the same both in the interstitial fluid and inside the cell. p. 239

jaundice yellow pigmentation that builds up in the skin—caused by an increased level of bilirubin (produced from the breakdown of red blood cells)—that occurs if the liver is not functioning properly. p. 353

Jean Baptiste de Lamarck French scientist who believed traits were acquired by individuals during their lifetime, depending on their behaviours. p. 116

jejunum the middle component of the small intestine. p. 388

joint an articulation between bones. p. 228

juxtaglomerular cells cells in the kidney that synthesize, store, and secrete renin when blood pressure decreases. p. 406

karyotype organization of chromosomes and their homologous pairs from chromosome 1 to 23, from longest to shortest. p. 80

keratin intermediate-filament protein found in hair and nails. p. 55

keratinized cells that contain the protein keratin, found in the outermost layer of the skin, hair, and nails. p. 178

keratinocytes skin cells identified by their ability to produce the protein keratin. p. 195

ketone an organic compound produced when fat is broken down to produce ATP. p. 387

kidney stones solid, crystalline accumulations that form in the kidneys and may prevent the passage of urine. p. 412

kinetochore protein located at the centromere region that holds sister chromatids together. p. 82

kingdom categorization of all living organisms; six kingdoms—archaeabacteria, eubacteria, protists, fungus, plants, and animals. p. 4

knockout mice mice in which a gene has been blocked or deleted so a specific functional protein is not produced. p. 171

Kupffer cells specialized macrophages in the liver that engulf foreign material and cell debris. p. 387

kyphosis convex curvature of the thoracic vertebrae commonly referred to as hunchback. p. 218

labia majora longitudinal folds of skin that extend between the mons pubis and the perineum. p. 420

labia minora extensions of skin on either side of the vaginal opening that are situated between the labia majora. p. 420

lac operon three genes involved in lactose metabolism in bacteria. p. 153

lacZ gene gene found in the *E. coli* operon that's used as a selectable marker in genetic engineering. p. 160

lacrimal bones smallest facial bone, which forms the medial portion of the eye sockets. p. 210

lactase enzyme that breaks down lactose. p. 36

lactate three-carbon molecule produced from pyruvate during anaerobic respiration. p. 74

lactation process of producing milk in mammary glands. p. 303

lacteal lymphatic vessels that absorb fats from the small intestine. p. 388

lactic acid a by-product of anaerobic metabolism in humans derived from pyruvate. p. 74

lactobacilli bacteria a type of gram-positive, rod-shaped bacteria that is part of our normal flora. p. 432

lactose a dissaccharide formed from glucose and galactose. p. 36

lacunae small spaces inside the bone that house the osteocytes and are connected by canaliculi. p. 207

lambdoid sutures suture that connects the occipital bone to the posterior edges of the parietal bones. p. 210

lamella ring of bone matrix. p. 207

lamellar granules organelles that release compounds into the extracellular fluid to waterproof the skin and prevent infection. p. 196

lamina propria a layer of dense connective tissue under the epithelial layer of mucus membranes. p. 382

laminar flow a property of fluids such that individual particles move parallel to one another, such as blood flowing through a healthy blood vessel. p. 347

Langerhans cells immune cells found in the epidermis. p. 196

lanugo fur that humans have in certain stages of embryonic development. p. 122

large intestine segment of the intestine consisting of the cecum, and the ascending, transverse, descending, and sigmoid colon. p. 379

laryngopharynx cavity at the posterior aspect of the upper respiratory tract. p. 359

larynx commonly referred to as the voice box. p. 359

lateral condyles lateral attachment site that articulates with the fibula. p. 224

lateral horns one of three grey matter columns within the spinal cord involved with the sympathetic response. p. 276

latissimus dorsi largest muscle in the back. p. 245

law of independent assortment Mendel's law that homologous pairs separate into gametes completely randomly, so any possible combination can occur. p. 99

law of segregation Mendel's law that the two alleles of a trait separate during the formation of gametes, so that half the gametes carry one copy and half carry the other copy. p. 99

LDL (low-density lipoprotein) a type of cholesterol molecule made by the liver to transport fats to the adipose tissue, also known as "bad" blood cholesterol. p. 350

LDLR (low-density lipoprotein receptor) the receptor on fat cells that recognizes and takes in LDLs. p. 110

leading strand DNA strand in which only one RNA primer is required for polymerase to follow helicase in the 5′ to 3′ direction. p. 135

left carotid artery artery in the left side of the neck that feeds the brain. p. 337

left gastric artery artery that supplies the superior portion of the stomach. p. 340

left subclavian artery artery beneath the left clavicle. p. 337

left subclavian vein vein beneath the left clavicle. p. 335

leukocytes white blood cells. p. 353

Leydig cells cells that produce testosterone in the presence of luteinizing hormone. p. 312

LH surge an increase in luteinizing hormone that triggers ovulation in females. p. 423

ligament connective tissue that connects bone to bone. p. 229

ligand-gated *see* chemically gated. p. 254

ligase an enzyme that forms phosphodiester bonds after the RNA primers have been removed during DNA replication. p. 136

limbic system a set of brain structures located under the cerebrum. p. 285

linked genes located close together on the same chromosome. p. 106

lipase an enzyme that breaks down lipids. pp. 386, 390

lipogenesis formation of lipids from Acetyl-CoA. p. 387

lipopolysaccharide (LPS) a polysaccharide that contains a fat, such as the external cell wall layer of gram-negative bacteria. p. 452

lipoprotein lipase (LPL) a group of enzymes that break down fat into components that can be used for energy. p. 260

lipoproteins proteins that have been processed and bound to lipids. pp. 150, 353

live attenuated vaccine a virus that has had the virulence decreased, but still prompts an immune response. p. 445

liver organ in the upper right quadrant of the abdomen. p. 386

load (muscular) the resistance to a muscle contraction. p. 253

lobe anatomical or functional division of an organ. p. 360

lobule a division of a tissue that is only visible histologically, such as the lobules of the liver. p. 361

locomotion movement from one location to another. p. 6

long bones type of bone found in the arms, legs, fingers, and toes. p. 208

longissimus muscles that connect the transverse processes of various vertebrae to the sacrum. p. 245

longitudinal fissure a deep groove that separates the two hemispheres of the brain. p. 285

longitudinal layer smooth muscle running lengthwise along the intestine. p. 384

loose connective tissue tissue that connects to muscle layers, gives strength and elasticity, stores fat, or acts as a filter in the spleen and lymph nodes. p. 180

lordosis concave curvature of the spine also known as sway back. p. 218

low-density lipoprotein (LDL) made from cholesterol and used to transport fats. p. 350

lumbar enlargement enlargement of the spinal cord at the separation to the lower limbs. p. 277

lumbar plexus branched connections of nerves in the lower back. p. 278

lumbar vertebrae five vertebrae inferior to the thoracic region that separate the thoracic vertebrae from the sacrum. p. 216

luteal phase the phase of the menstrual cycle that occurs after ovulation, days 14–28. p. 423

luteinizing hormone (LH) a sex hormone that stimulates the production of sex hormones in the gonads of males and females, also causes ovulation in women. pp. 312, 418

lymph fluid within the lymphatic system. p. 333

lymph node an oval-shaped organ of the lymphatic system that contains lymphocytes, the T cells and B cells. p. 430

lymphocyte the B cells and T cells. p. 439

lysogenic cycle the viral cycle in which the host is not killed, but the viral DNA is intergrated into the host DNA. p. 458

lysosome small membrane-bound organelle that contains digestive enzymes that break down molecules, macromolecules, and old organelles inside the cell. p. 53

lysozyme enzyme that damages bacterial cell walls; found in human tears, sweat, and saliva. p. 433

lytic cycle the viral cycle in which the host is destroyed and new viral particles are produced. p. 458

M phase mitosis occurs, with no more protein synthesis and cell growth. p. 80

macroevolution large-scale changes that may give rise to a new species and that occurs over a very long period of time. p. 113

macromolecule a combination of small molecules; the four macromolecules that make up cells are proteins, carbohydrates, fats, and nucleic acids. pp. 7, 27

macrophages white blood cells that engulf bacteria, viruses, and any dead cells by phagocytosis to break down and destroy invading microorganisms. pp. 61, 354

macula densa an area of cells that line the distal convoluted tubule of the nephron in the kidney; detects sodium levels and regulates blood vessel diameter to regulate blood pressure. p. 406

major calyx the point at which multiple minor calyces converge. p. 402

major histocompatibility (MHCII) a class of molecules found on all cells of the body that are the major reason why the immune cells can recognize our own cells versus foreign cells, such as pathogens or transplant tissue. p. 439

malaria a disease resulting from a parasite commonly found in Africa that infects the liver and red blood cells and can be fatal. p. 463

malignant tumour growth of cells that are cancerous and can spread. p. 87

maltase an enzyme that breaks down maltose. p. 391

maltose a disaccharide formed from two glucose molecules and found in grains. p. 36

mammary glands glands that produce milk in females. p. 198

mandible forms the lower jaw bone, the largest and strongest of the facial bones. p. 210

mandibular fossa groove in the temporal bone that houses the condylar process of the mandible. p. 210

manubrium the broad superior portion of the sternum. p. 220

mast cells immune cells that produce histamine when tissue injury or allergic reaction occurs. p. 180

maxiallae bones that form the upper jaw bone and house the upper teeth. p. 210

mechanical digestion the physical breakdown of food into smaller components, such as chewing. p. 380

mechanoreceptors detect stretching. p. 22

medial and lateral epicondyles medial and lateral processes of the elbow. p. 221

median antebrachial vein vein that drains the venous plexus of the hand. p. 341

median cubital vein superficial vein of the upper limb. p. 344

median nerve one of the main veins originating at the brachial plexus. p. 278

mediastinum anatomical central region of the thoracic cavity. p. 18

medulla oblongata the part of the brain responsible for maintaining basic life functions such as breathing and heart rate. pp. 22, 276

medullary cavity hollow portion in the middle of long bones. p. 190

meiosis the process of cell division that produces haploid gametes from diploid germ cells. p. 114

Meissner corpuscles sensory receptors that detect touch and vibration in the skin. p. 197

melanin skin pigment produced by melanocytes in response to sunlight; protects the skin cells from DNA damage caused by ultraviolet light. p. 196

melanocytes skin cells identified by their ability to produce melanin. p. 196

melanoma a tumour originating from mutated melanocytes. p. 199

melatonin a hormone produced by the pineal gland in low light; stimulates sleep. p. 283

membrane attack complex (MAC) formed from complement proteins circulating in the blood that can directly kill bacteria and fungi. p. 436

memory cells B cells and T cells that recognize antigens from previous infections. p. 427

meninges protective covering of the brain and spinal cord. p. 276

Merkel cells touch receptors in the skin. p. 196

merozoites one of the life-cycle stages of malaria. p. 463

mesenchymal cells cells that can differentiate into a variety of cell types. p. 430

mesentery a double layer of the peritoneum. p. 383

messenger RNA (mRNA) the product of transcription of a gene that will be the template for protein synthesis on a ribosome. p. 145

metabolic syndrome a number of conditions that combine to increase the risk of heart disease, obesity, and type-2 diabetes. p. 351

metabolism process of producing energy through the breakdown of nutrients. p. 304

metaphysis region of the bone that grows during development. p. 208

metastasis the spread of cancerous cells to other regions of the body. p. 56

methionine (Met) first amino acid in the protein molecule that is coded by AUG mRNA codon. p. 150

methylase enzyme enzyme that adds methyl groups to DNA that affect the regulation of transcription. p. 158

micelles a collection of lipids, often spherical, that have outer hydrophilic heads aggregating around a hydrophobic core. p. 392

microevolution the change in inherited traits in a population of organisms within a single species over time. p. 113

microfilaments the smallest intracellular cytoskeletal proteins, composed of actin proteins; have a role in cellular movement, such as the "crawling" movement of immune cells through the tissues. p. 55

microglia glial cells that function as immune cells in the CNS. p. 269

microtubules the largest intracellular cytoskeletal proteins, composed of large, hollow tubes; used to transport molecules throughout the cell. p. 55

microvilli microscopic extensions of the cellular membrane that act to increase surface area. p. 176

micturition urination. p. 404

midbrain upper portion of the brainstem primarily associated with motor control, but also involved in consciousness (contains the reticular formation), vision, hearing, and temperature regulation. p. 284

midsagittal plane an anatomical division of the body into equal left and right sections. p. 18

minor calyx the part of the kidney that collects urine from the collecting ducts. p. 402

mitochondria membrane-bound organelle that produces the cells energy in the form of ATP in the presence of oxygen. p. 53

mitosis process of cell division in somatic cells to produce two new identical diploid daughter cells. p. 79

mitral valve valve between the left atrium and the left ventricle. p. 323

M line centre of the H zone of the sarcomere. p. 251

modifier genes genes that modify the effects of another gene. p. 104

molecular mimicry theory that an immune cell may target a self-cell if the body antigens are similar enough to the antigens of a pathogen. p. 447

monococci a form of cocci that exist as individual cells rather than clusters or chains. p. 453

monocytes immune cells that differentiate into macrophages during an infection. p. 354

monohybrid cross use of a Punnett square to determine the possible phenotype ratios of a single trait in offspring. p. 97

monomers single molecules. p. 27

monosaccharide simple carbohydrate monomer occurring as three types: glucose, fructose, and galactose. p. 36

monosynaptic reflex reflex consisting of one sensory neuron and one synapse to the motor neuron. p. 280

monounsaturated fat fats that contain one double-bonded carbon in the fatty acid chain. p. 42

mons pubis a mass of fatty tissue over the pubic symphysis, more prominent in females. p. 420

motor cortex area of the brain that initiates muscle control. p. 254

motor neurons neurons that transmit information from the central nervous system to muscles or glands, and are multipolar. p. 267

motor neurons *see* efferent neurons. p. 265

mouth the opening through which most animals take in food. p. 384

MRSA (methicillin-resistant *staphylococcus aureus*) an antibiotic-resistant form of the bacterium, *staphylococcus aureus*. p. 453

mucosa mucus membrane covered in epithelial cells that are often involved in secretion or absorption. p. 382

mucus a sticky, fat-containing substance that is secreted in the digestive system and respiratory system to protect epithelial cells and trap foreign particles. p. 360

multinucleated cells having several nuclei, such as skeletal muscle. p. 187

multiple cloning site (MCS) usual location in a plasmid where the restriction enzyme cut sites are placed, usually within another selectable marker. p. 160

multipolar neurons neurons that produce multiple dendrites branching from the cell body. p. 189

multiunit smooth muscle smooth muscle that consists of many cells contracting individually, which allows for precise control. p. 187

multipolar neurons a type of neuron that has multiple dendrites attached to the cell body and receives stimuli. p. 189

muscle fibre single muscle cell. p. 187

muscle tone residual tension in muscles during relaxation. p. 254

muscular system organ system composed of smooth, cardiac, and skeletal muscle. p. 9

muscularis refers to the muscular area of a tissue or organ. p. 382

muscularis mucosa the muscular layer of a mucus membrane. p. 382

musculocutaneous nerve nerve that originates in the brachial plexus and innervates the lateral arm. p. 278

myelin sheath protective covering of the axon made of lipids that increases the rate of transmission of action potentials. p. 267

myenteric plexus part of the enteric nervous system that controls smooth muscle contractions during digestion. p. 266

myocardial infarction a condition of ischemia or hypoxia in a coronary blood vessel leading to heart-muscle cell death; also called heart attack,. p. 350

myocardium middle layer of the heart consisting of the cardiac muscle tissue. p. 321

myofacial pain muscular facial pain that is a common symptom of the improper alignment of the condylar process in the mandibular fossa. p. 194

myofibrils bundles of contractile and regulatory proteins in muscle cells. p. 238

myoglobin a muscle protein that binds to oxygen, similar to hemoglobin. p. 257

myometrium muscular layer of the uterus. p. 420

myosin a protein contained in the thick filaments of the myofibrils that binds to actin during contraction. p. 238

Na+/K+ pump an active transport membrane protein, found in every cell of the body, that maintains the electrical and ion concentration gradient across the cell membrane by moving three sodium ions out of the cell for every two potassium ions that move in. p. 269

nasal cavity a large, air-filled space located above the palate and behind the nose. p. 213

nasal conchae long narrow, curved bones in the nasal passageway that provide a pathway for inhaled air; warms and moistens air before it enters the lungs. p. 359

nasal septum bone that separates the right and left sides of the nasal cavity. p. 213

nasopharynx most superior portion of the pharynx. p. 359

natural killer cells white blood cells that recognize cancer cells and virus-infected cells and destroy them. pp. 87, 141

natural selection means by which a population evolves based on environmental influences; makes organisms most "fit" in a particular environment more likely to survive and reproduce. p. 114

naturally acquired active immunity immunological memory formed after exposure to an infectious pathogen. p. 445

naturally acquired passive immunity immunity passed on from a mother to a fetus or baby as antibodies are transferred through the placenta or milk. p. 445

negative feedback a process by which the body reacts to a deviation from a normal set point to return the body to within normal limits. p. 21

nephron the functional unit of the kidney. p. 402

nerve a collection of neurons surrounded by connective tissue. pp. 263, 277, 278, 279

neuraminidase a type of enzyme expressed by the influenza virus that allows the virus to reproduce inside respiratory cells. p. 458

neurofilament intermediate filament found in the axons of neurons. p. 55

neuroglia non-neuronal cells located within the central and peripheral nervous systems; include oligodendrocytes, microglia, astrocytes, ependymal cells, Shwann cells, and satellite cells. p. 188

neurons cells of the nervous system that conduct action potentials. pp. 188, 265

neurotransmitter a chemical-signalling molecule that is released from axon terminals and allows communication between neurons and other cells. pp. 254, 272

neutrophils white blood cells that are the most abundant type of white blood cell and are usually the first responders during an infection, killing pathogens by secreting neutralizing chemicals. p. 353

niacin vitamin B3, produces NAD+. p. 69

nitrogen group amino group that contains nitrogen and hydrogen. p. 73

nitrogenous wastes nitrogen-containing substances that are excreted by the body once they are no longer usable, including ammonia, urea, and uric acid. p. 407

nociceptors detect pain; also known as nerve endings. p. 22

non-ciliated epithelial cells that do not contain cilia. pp. 158, 177

nondisjunction failure of chromosomes to separate correctly during meiosis I or meiosis II. p. 107

nonpolar molecules that do not have a charge and cannot dissolve easily in water. p. 26

non-striated muscle fibres that are not striated, such as smooth muscle. p. 187

norepinephrine (NE) a neurotransmitter released from ANS neurons that is a hormone produced in the adrenal medulla in response to sympathetic nervous stimulation; also called noradrenaline. p. 292

normal flora collection of beneficial bacteria found on the skin and in the gastrointestinal tract. p. 451

nuclear envelope the double-layer membrane that surrounds the nucleus and contains pores that allow large molecules, such as mRNA, to move out of the nucleus. p. 52

nuclear membrane *see* nuclear envelope. p. 52

nuclei plural of nucleus. p. 266

nucleolus a small non-membrane-bound organelle found in the nucleus; contains DNA that transcribes ribosomal RNA that is used to make ribosomes. p. 52

nucleoplasm fluid within the nucleus. p. 52

nucleoside a molecule that contains a sugar and a base. p. 132

nucleoside triphosphate nucleoside with three phosphates added by DNA polymerase to the 3′ end of a nucleotide on the new strand; includes the nucleoside triphosphate ATP. p. 135

nucleotide an organic monomer of DNA or RNA, made up of a five-carbon ring sugar, a phosphate group, and a nitrogen-containing base. pp. 34, 131

nucleus membrane-bound organelle that contains genetic material, DNA. p. 52

nucleus accumbens part of the basal ganglia; plays an important role in laughter and reward. p. 285

oblique fissure a depression or groove that separates the inferior lobe of the lungs from the rest of the lungs. p. 361

oblique plane an anatomical division of the body on an angle. p. 18

obturator foramen openings in the ischium that allow the passage of nerves and blood vessels. p. 222

obturator nerve nerve that receives sensory information from the medial thigh and is responsible for motor innervation of the lower extremity. p. 278

occipital bone posterior portion of the skull that protects the occipital lobe of the brain. p. 213

occipital condyles two processes at base of the skull that sit on the first cervical vertebra (atlas) and allow movement of the head. p. 213

occipital lobe most posterior lobe of the brain that contains the primary visual cortex. p. 287

occludens proteins that make up tight junctions. p. 57

olecranon (elbow) a process at the proximal end of the ulna that articulates with the humerus. p. 221

olecranon fossa a large indentation in the posterior of the humerus that allows the arm to straighten. p. 221

olfactory foramen opening for the olfactory nerve. p. 213

oligodendrocytes glial cell that produce the protective myelin that covers neurons in the CNS. p. 170

oligosaccharide small chain of monosaccharides. p. 53

omega 3 fatty acid essential fatty acid, also called alpha-linoleic acid. p. 43

omega 6 fatty acid essential fatty acid, also called linoleic acid. p. 43

omega 9 fatty acid essential fatty acid, also called oleic acid. p. 43

oncogenes mutated proto-oncogene that increases the cell cycle rate and causes cancer. p. 85

oncology the study of cancer. p. 85

oocyte a female gamete. p. 420

oogenesis maturation of egg cells in the ovary. p. 312

opportunistic infections infections that remain dormant unless the affected individual is immunosupressed. p. 435

opsonize the process by which a pathogen is marked for destruction, such as when an antibody binds to an infectious organisms and increases the ability of other cells to target and kill the infection. p. 442

optic foramen opening for the optic nerve. p. 213

organ combination of tissues that function as a unit, such as the stomach or the heart. p. 7

organ system organs that function together with other organs and tissues, such as the digestive system or the cardiovascular system. p. 7

organic molecules molecules of a living organism, such as a plant or animal. p. 27

organism a living entity composed of one or more cells. p. 7

origin the connection point of a muscle to a non-moving bone. p. 238

origin of replication a specific location in a genome where DNA replication begins. p. 135

oropharynx part of the pharynx bordered by the uvula and the hyoid bone. p. 359

osmoreceptor receptor that detects changes in osmolarity. p. 22

osmosis movement of water across a semipermeable membrane to an area of low water concentration (higher ion concentration) without input of energy. p. 59

osmotic composition components of the blood that determine its osmolarity. p. 401

osmotic concentration the concentration of a solute in the interstitial fluid surrounding a cell relative to the concentration inside the cell. p. 59

osmotic gradient difference in concentration between two solutions across a membrane. p. 402

osmotic pressure the pressure created by the movement of water due to an osmotic gradient. pp. 334, 406

osteoblasts immature osteocytes that build bone. p. 206

osteocalcin secreted by osteoblasts to stimulate production of bone matrix. p. 206

osteoclast mature osteocytes that function to break down bone. pp. 206, 306

osteocytes mature bone cells. p. 180

osteon smallest unit of bone tissue, composed of concentric rings of bone matrix that surround a central canal and contain osteocytes. p. 207

osteoporosis a disease characterized by the loss of bone density that can lead to increased risk of fractures. p. 306

ova (eggs) egg cells. p. 415

ovarian ligament ligament that connects the ovaries to the uterus. p. 420

ovary female reproductive organ where meiosis occurs. p. 415

ovulation part of the menstrual cycle where the egg passes from the ovary into the fallopian tubes. p. 313

oxaloacetate a four-carbon molecule in Krebs cycle. p. 69

oxidation reactions in which electrons are donated from one molecule to another. p. 66

oxidative phosphorylation phosphorylation of ADP using oxygen, which occurs in the mitochondria. p. 67

oxidized molecule that loses electrons. p. 67

oxygenation diffusion of oxygen into the blood and tissues. p. 334

oxyhemoglobin compound formed by the binding of oxygen to the heme groups of hemoglobin. p. 370

oxytocin a hormone released by the posterior pituitary gland that causes uterine contractions and is involved in bonding between mom and baby. p. 302

P wave a positive inflection on an ECG that indicates atrial depolarization. p. 325

p53 tumour-suppressor gene that prevents cell division when DNA mutations are recognized, recruits DNA repair enzymes, and can initiate apoptosis during the G2 phase. p. 80

pacemaker cells specialized cells of the heart that control the rate of contraction. p. 187

pacemaker potential a slow change in voltage across the cell's membrane caused by sodium and calcium ions. p. 325

pacinian corpuscles a type of mechanoreceptor that detects pressure in the skin. p. 197

palatine bones bones that form the posterior portion of the hard palate, which is the roof of the mouth. p. 214

palindrome a specific symmetrical sequence of nucleotides that are the same in opposite directions. p. 158

pancreas an endocrine gland that secretes insulin and glucagon, and is also an exocrine gland that secretes digestive enzymes into the small intestine. p. 22

pancreatic duct a vessel through which pancreatic enzymes are secreted into the gastrointestinal tract. p. 310

pancreatic juice a liquid secreted by the pancreas containing a combination of enzymes and bicarbonate. p. 388

pantothenic acid vitamin B5, produces coenzyme A. p. 69

papillary layer superficial layer of the dermis that connects the dermis to the epidermis. p. 195

paracellular transport the movement of molecules *between* cells as opposed to *through* cells, which is transcellular transport. p. 410

paracrine a form of cellular communication where cells produce a signalling molecule that affects nearby cells without circulating through the bloodstream. p. 433

parafollicular cells cells of the thyroid gland that produce calcitonin. p. 304

paranasal sinuses openings in the skull that are lined with mucosal membranes that warm and moisturize the incoming air before it enters the lungs. p. 213

parasagittal plane an anatomical division of the body into unequal left and right sections. p. 17

parasite an organism that benefits at the expense of its host. p. 463

parasympathetic nervous system a division of the autonomic nervous system that is involved in resting and digesting. p. 291

parathyroid hormone (PTH) hormone that acts as an antagonist to calcitonin by raising blood calcium levels. p. 226

parietal bones bones that form the superior and lateral portions of the skull. p. 210

parietal lobe the superior-lateral lobe of the brain. p. 286

parietal pericardium outer layer of the pericardium. p. 321

parietal peritoneum segment of the peritoneum that's bound to the wall of the abdominal cavity. p. 383

parietal pleura segment of the pleura that is bound to the chest wall and surrounds the lungs. p. 361

Parkinson's disease a degenerative disorder of the central nervous system that involves the loss of dopaminergic neurons that regulate muscle movements. p. 284

parotid salivary gland largest and most posterior of the salivary glands. p. 386

partial pressure pressure created by one particular gas as a part of the total air composition. p. 368

partial thickness burn (second degree burn) deep burn that extends into the dermis, characterized by the formation of blisters. p. 199

patella sesamoid bone located between and anterior to the distal femur and the proximal tibia, also called the knee cap. p. 224

patellar ligament a continuation of the quadriceps tendon that inserts on the tibia. p. 224

patellar reflex a monosynaptic reflex arc, also called the knee jerk relex. p. 280

pathogen-associated molecule patterns (PAMPs) molecules associated with pathogens that are recognized by the immune system. p. 439

pathogenic bacteria bacteria that cause harm or suffering to the host. p. 454

pathogens things that cause harm (e.g., harmful bacteria, viruses, fungi or parasites). p. 427

pattern recognition receptors (PRRs) proteins expressed by immune cells that identify pathogen-associated molecular patterns. p. 439

pectoralis the primary chest muscles. p. 239

pelvic brim opening formed by the pubis bones and the sacrum. p. 222

pelvic inflammatory disease (PID) refers to the condition of inflammation of the uterus, fallopian tubes, and ovaries and can cause infertility. p. 425

pepsin an enzyme that breaks proteins down into component peptides. p. 386

pepsinogen a compound activated by hydrochloric acid to become pepsin. p. 384

peptidases enzymes that break down polypeptides. p. 391

peptide bond covalent bond that links amino acids together. p. 28

peptidoglycan eubacteria cell wall composed of sugars and amino acids. pp. 50, 452

perforating canal opening in the bone matrix that allows blood and lymph vessels to connect with vessels outside the bone. p. 207

perforin a protein that opens a pore in the membrane of a target cell, which allows water to rush into the cell and this kills it. p. 435

pericardial cavity space between the pericardial membranes that surround the heart. p. 18

pericardium a double-walled connective tissue membrane that surrounds the heart. p. 321

perimetrium outermost layer of the uterus. p. 420

perimysium sheath of connective tissue that surrounds a bundle of muscle fibres. p. 237

perineum area between the anus and the scrotum (in males) or the anus and the vagina (in females). p. 420

perineurium protective sheath around a bundle of nerves. p. 277

periosteum dense, irregular connective tissue that covers the bones. p. 208

peripheral chemoreceptors chemical receptors located in the aortic arch and carotid bodies that detect changes in the chemical composition of the blood. p. 374

peripheral nervous system (PNS) all nerves outside the central nervous system. p. 263

peristalsis rhythmic contraction of smooth muscle that causes movement through a tube. p. 384

peritubular capillaries small blood vessels that surround the nephron. p. 404

peroxisomes organelles that break down oxygen radicals. p. 54

Peyer's patch a bundle of lymphatic tissue surrounding the digestive system. p. 432

phagocytosis endocytosis that involves the movement of particles into the cell. p. 61

pharynx an organ that is part of the digestive system and the respiratory system, also known as the throat. p. 359

pharyngeal pouches grooves in the lateral neck region in all vertebrate embryos that disappear in mammals later in embryonic development but differentiate into gills in fish. p. 122

phenotype an organism's visible traits. p. 124

pheromones molecules that act to attract a mate in most species. p. 198

phosphodiester bond bond between a sugar and a phosphate in nucleic acids. p. 34

phospholipid primary fats found in cell membranes and containing two nonpolar fatty acid tails and a polar head. p. 43

phosphorylation the addition of a phosphate group to any organic molecule, such as ATP. p. 53

phosphorylation enzymes enzymes in the cell that add phosphate groups to some molecules. p. 53

photoreceptors detect light and are found only in the retina. p. 22

photosynthesis process in which plants combine carbon dioxide from the air with hydrogen from water and energy from the sun to produce glucose. p. 6

phrenic nerve nerve that innervates the diaphragm. p. 278

physiology the study of how anatomical parts function. p. 3

pia mater innermost layer of the meninges. p. 276

pili hairlike structures that helps bacteria adhere to structures and transfer small pieces of DNA(singular, pilus). p. 50

piloerection contraction of the arrector pili muscles that cause the phenomenon commonly called goose bumps. p. 197

pineal gland endocrine gland found in the brain that produces melatonin. p. 283

pinocytosis endocytosis that involves the movement of fluids into a cell. p. 61

pituitary gland endocrine gland located at the base of the brain that controls many important functions such as the stress response, growth, metabolism, and reproduction. p. 283

pivot joint joint that permits only rotation. p. 230

planar joint synovial joint that only permits a gliding motion. p. 229

plants eukaryotic, multicellular organisms that are photosynthetic. p. 6

plasma the fluid component of the blood. p. 333

plasma membrane *see* cell membrane. p. 48

plasmid extra piece of DNA in some bacteria. p. 454

platelets cell fragments that assist the blood clotting process. p. 353

pleiotropic a single allele causing multiple phenotypes in a population. p. 101

pleura fluid-containing membranes that surround the lungs. p. 361

pleural cavity space between the pleural membranes that surround the lungs. p. 18

pneumotaxic centre area of the pons that regulates the respiratory rate. p. 373

podocytes cells in the Bowman's capsule that wrap around the capillaries in the glomerulus. p. 406

point mutation one or a few nucleotides that are incorrectly placed in the DNA sequence during replication. p. 139

polar molecules that contain a charge and dissolve in water. p. 26

polar body a small cell produced during meiosis in females and that undergoes apoptosis. p. 422

poly A tail see 3' tail. p. 167

poly T primer a sequence of thymines that can be used to copy RNA in PCR reactions. p. 167

polyadenylation addition of many adenine nucleotides to mRNA to prevent mRNA from degradation. p. 147

polygenic multiple genes contributing to one trait, such as height or weight. p. 101

polymer combination of monomers, such as amino acids combining to produce a protein. p. 27

polymerase chain reaction (PCR) technique used to generate multiple copies of a specific DNA sequence—faster and more reliable than using bacteria for this process. p. 168

polypeptide protein with many amino acids covalently bonded together by peptide bonds. p. 28

polysaccharide complex carbohydrates composed of long chains of glucose. p. 36

polysynaptic reflex a reflex that involves more than two neurons. p. 280

polyunsaturated fat fats that contain two or more double-bonded carbons. p. 42

pons the part of the brain that connects the thalamus to the medulla oblongata and that regulates breathing. p. 284

popliteal artery an artery located in the knee. p. 340

population a group of the same species of organisms that live in the same geographical area. p. 114

positive feedback a process that continues to increase the response caused by an initial stimulation, such as during labour. p. 21

post-central gyrus the portion of the parietal lobe that is the sensory cortex. p. 285

posterior fontanel space between the skull bones located at the joining of the occipital bone and both parietal bones. p. 215

posterior median sulcus a narrow, longitudinal groove that runs along the posterior aspect of the spinal cord. p. 276

posterior pituitary posterior portion of the pituitary gland controlled by nervous stimulation. p. 302

posterior tibial artery artery located at the dorsal aspect of the lower leg. p. 341

posterolateral fontanel space between the skull bones located where the temporal and occipital bones meet. p. 215

postganglionic nerve fibres that carry signals from the ganglion to the effector in the autonomic nervous system. p. 292

post-synaptic neuron neuron that receives neurotransmitters at the synapse. p. 271

post-translational modification any change made to a protein after translation, such as methylation or addition of sugars to form a glycoprotein. p. 150

power stroke movement of the myosin head toward the M line in a sarcomere; causes muscle contraction. p. 253

pre-central gyrus the portion of the frontal lobe that is the motor cortex. p. 285

pre-diabetic the condition when a person cannot properly regulate blood sugar because cells are becoming insulin resistant. p. 311

pre-frontal cortex anterior part of the frontal lobe of the brain involved in planning muscle movements. p. 286

preganglionic nerve fibres that carry signals from the central nervous system to the ganglion in the autonomic nervous system. p. 291

preload pressure generated to return blood to the heart. p. 328

pressure force generated against an area. p. 345

pre-synaptic neuron neuron that forms the axon teminal at the synapse and releases the neurotransmitter(s). p. 272

primary oocytes an egg cell that has not yet undergone the second meiotic division and has only a single layer of granulosa cells. p. 420

primary spermatocytes sperm cells that have not yet undergone the second meiotic division. p. 417

primase an enzyme that adds a short sequence of nucleotides that initiates DNA replication. p. 135

primer a short sequence of RNA that is required for the initiation of DNA replication. p. 168

primer annealing the binding of primers to single-stranded DNA. p. 168

primer extension occurs when DNA polymerase replicates the DNA using the primers as a starting point. p. 168

primordial follicle first stage of folliculogenesis in females. p. 422

probability the likelihood that some event will occur. p. 96

probe approximately 20 nucleotides, complementary to the gene of interest, that contain a selectable marker, such as radioactive phosphate. p. 163

progesterone a sex hormone produced in the ovaries that is involved in regulation of the menstrual cycle, pregnancy, and development of the embryo. p. 312

prokaryotes simple, unicellular organisms that do not contain membrane-bound organelles. p. 451

prolactin a hormone that stimulates milk production in mammals. p. 303

promoter region a region of DNA upstream from the gene that is not transcribed and that RNA polymerase binds to. p. 147

proofreading a function in which polymerase is able to recognize some mistakes that occur during replication. p. 140

prophylactic a preventative measure or means. p. 454

proprioceptors receptors that help determine body position. p. 22

prostaglandins a subclass of eicosanoids involved in inflammation. p. 437

prostate cancer cancer of the prostate gland. p. 425

prostate gland a ring-shaped exocrine gland found in males, which produces an alkaline fluid that makes up 50–75% of seminal fluid. p. 416

protein a sequence of amino acids that has a specific function, examples include enzymes, contractile and transport proteins. pp. 33, 250, 387

proteoglycan a glycosylated protein that makes up a large part of the extracellular matrix of most tissues. pp. 53, 181

proteome all the proteins produced in an organism. p. 158

protists eukaryotic unicellular organisms. pp. 5, 463

proton pump an active transport membrane protein that moves protons against its concentration gradient with input of energy. p. 60

proto-oncogenes genes that normally increase the rate of cell division. p. 85

protozoan single-celled, eukaryotic protists that have locomotion. p. 463

proximal convoluted tubule a part of the nephron that collects filtrate from the Bowman's capsule. p. 405

proximal epiphysis end of the long bones that's closest to the core of the body. p. 208

pseudopods extensions of the cytoplasm that resemble fake feet. p. 463

pseudostratified cells that are a single layer but appear to be stratified, and the nuclei may appear layered. p. 176

psoas major one of the hip flexor muscles. p. 246

pubic symphysis connection between the right and left pubis bones that form the anterior portion of the pelvic girdle. p. 221

pulmonary arteries arteries that take blood from the right ventricle to the lungs. p. 320

pulmonary circulation the circuit by which blood is pumped to the lungs for gas exchange before returning to the systemic circulation. p. 318

pulmonary semilunar valve valve between the pulmonary artery and the right ventricle. p. 323

pulmonary trunk blood vessel that divides to form the left and right pulmonary arteries. p. 336

pulmonary veins veins that return blood from the lungs to the heart. p. 320

punctuated equilibrium an evolutionary process where periods of no change are followed by periods of rapid change. p. 120

Punnett square method used to predict the phenotype ratios for any combination of genotypes. p. 96

purines nitogenous bases—adenine and guanine—that have a double-ring structure. p. 34

Purkinje fibres extensions of highly excitable cells that pass through the ventricles from the bundle of His. p. 325

pyloric sphincter sphincter between the distal end of the stomach and the beginning of the small intestine. p. 384

pylorus region of the stomach that connects to the duodenum. p. 384

pyrimidines nitrogenous bases—cytosine, thymine, and uracil—that have a single-ring structure. p. 34

pyrogens molecules that promote a fever. p. 438

pyruvate three-carbon molecule formed through glycolysis. p. 69

pyruvate dehydrogenase an enzyme that removes the carbon atom from pyruvate. p. 69

QRS section of the ECG that indicates the depolarization of the ventricles and contains the (hidden) repolarization of the atria. p. 325

radial artery artery that runs along the medial forearm. p. 340

radial fossa indentation in the humerus that articulates with the radius and increases the range of motion when the arm is bent. p. 221

radial nerve nerve that runs along the radius in the forearm. p. 278

radial notch point at which the head of the radius articulates with the ulna. p. 221

radial vein vein located in the lateral aspect of the forearm. p. 341

radioactive phosphate a common radioactive tag used to make DNA probes visible since nucleotides contain phosphate. p. 163

radioulnar joint joint between the head of the radius and the ulna. p. 221

radius bone of the forearm that's located laterally in anatomical position. p. 221

reabsorption the movement of molecules back into the bloodstream. p. 405

receptors extensions of a sensory neuron or a specialized cell that detects stimuli, for example, thermoreceptors that detect changes in temperature. pp. 22, 49

recessive the allele that is hidden by the dominant trait. p. 95

reciprocal innervation occurs when one muscle contracts and the antagonistic muscle on the opposite side of the body relaxes. p. 281

recoil the return of a tissue to its original state after being stretched. p. 366

recombinant bovine somatotropin (rBST) bovine growth hormone, grown in bacteria and given to cows to increase their milk production or growth rate. p. 164

recombinant DNA also known as a hybrid molecule, contains nucleotide sequences from two different species. p. 157

recombinant DNA technology process of moving specific sequences of DNA from one organism to another, also known as genetic engineering or gene splicing. p. 157

recombination mutation large sequences of DNA moved to another location within the genome. p. 139

rectum most distal part of the large intestine. p. 389

rectus abdominis paired muscles running vertically at the anterior aspect of the abdomen. p. 365

rectus femoris the largest of the quadriceps muscles in the thigh. p. 246

red blood cells biconcave-shaped cells that contain the protein hemoglobin and carry oxygen through the bloodstream. p. 353

red bone marrow bone marrow responsible for the production of blood cells. p. 207

red pulp a segment of the spleen that breaks down old red blood cells. p. 431

redox reaction oxidation-reduction reaction. p. 67

reduction reactions in which a molecule accepts electrons from another molecule. p. 66

reflex an involuntary and nearly instant reaction to a stimulus. p. 280

reflex arc pathway taken by a nerve impulse from sensation to effect. p. 280

refractory period time following an action potential during which the cell membrane cannot undergo another depolarization. p. 272

regulatory proteins proteins that regulate the function of other molecules (e.g., regulation of muscle contraction by troponin and tropomyosin). p. 237

regulatory region region of DNA not transcribed, but that helps regulate transcription, such as a promoter region. p. 147

regulatory T cells (TH17) a population of T cells that decrease the immune response after an infection. p. 441

releasing hormones hormones secreted by the hypothalamus that stimulate the anterior pituitary. p. 301

renal arteries arteries that supply blood to the kidneys. p. 402

renal capsule outermost connective tissue layer of the kidney. p. 402

renal columns a medullary extension of the renal cortex in between the renal pyramids. p. 402

renal corpuscle the part of the nephron containing the glomerulus and the Bowman's capsule p. 405

renal failure a disease state in which the kidneys fail to filter waste from the blood. p. 412

renal hilum a recessed central fissure in the kidney. p. 402

renal lobe a segment of the kidney. p. 402

renal pelvis a funnel-like, dilated, proximal part of the ureter. p. 402

renal pyramid cone-shaped tissue of the kidneys. p. 402

renal vein vein that drains the kidney. p. 404

renin enzyme secreted by juxtaglomerular cells of the kidney that cleaves angiotensin to produce angiotensin I and is involved in the production of aldosterone, which increases blood pressure. p. 411

renin-angiotensin system process by which the body retains fluid to increase blood pressure. p. 307

replicate to produce identical copies. p. 159

replication fork open portion of the DNA molecule during replication. p. 135

repolarization movement of potassium ions out of the cell to bring the cell membrane back to its resting membrane potential. p. 271

repressor a protein found in bacteria that prevents transcription of a gene. p. 153

residual capacity amount of air in the lungs following normal exhalation. p. 367

residual volume amount of air in the lungs following forced exhalation. p. 367

resistance force generated in opposition to pressure. p. 117

respiration the transportation of oxygen and carbon dioxide between cells and the external environment. p. 358

respiratory acidosis a decrease in blood pH due to an increase in CO_2 and hydrogen ions. p. 370

respiratory alkalosis an increase in blood pH due to a decrease in CO_2 and hydrogen ions. p. 370

respiratory bronchioles bronchioles that are involved in gas exchange. p. 360

respiratory zone area of the lungs where gas is exchanged. p. 358

resting membrane potential electrical gradient of a cell, usually –70mV. p. 60

restriction endonuclease enzymes that cut DNA at specific palindrome sequences. p. 158

restriction enzyme cut sites *see* restriction endonuclease. p. 159

restriction map a map of all restriction enzyme sequences in a DNA sequence. p. 163

rete testis a network of tubules in the hilum of the testes. p. 416

reticular cells produce reticular fibres in loose reticular connective tissue and filter pathogens and old red blood cells in the spleen and lymph nodes. p. 181

reticular fibres a thin collagen fibre produced by reticular cells, surrounded by glycoproteins that form a framework for many tissues, and acting as a filter in the spleen and lymph nodes. p. 431

reticular formation region in the brainstem that is involved in consciousness. p. 283

reticular layer deep layer of the dermis, which contains many elastin and fibrin layers that give the skin strength and elasticity. p. 195

reverse transcriptase enzyme found in retroviruses that converts RNA into DNA. pp. 167, 457

riboflavin vitamin B2, produces FAD^+. p. 69

ribonuclease an enzyme that breaks down RNA. p. 391

ribose five-carbon sugar contained in RNA, differentiated from deoxyribose by a hydroxyl group bound to carbon 2. p. 132

ribosomal RNA (rRNA) makes up part of the ribosome and ensures that mRNA can bind to the ribosome. p. 147

ribosome an organelle composed of a large and a small subunit of protein and ribosomal RNA and is the site of translation. p. 50

ribozymes alternative name for ribosomes because of the enzymatic activity of rRNA. p. 150

rickets a bone-deforming disease caused by a vitamin D deficiency in childhood. p. 226

right carotid artery artery located in the right side of the neck. p. 337

right lymphatic duct lymphatic duct that drains fluid from the superior and right portions of the body. p. 335

right subclavian artery artery running along the inferior portion of the right clavicle. p. 337

right subclavian vein vein running along the inferior portion of the right clavicle. p. 335

RNA ribonucleic acid. p. 35

RNA polymerase an enzyme that transcribes DNA into mRNA p. 147

rotation movement around a central axis. p. 231

rotator cuff muscles that allow for the full range of motion of the shoulder joint. p. 239

rough ER endoplasmic reticulum that contains ribosomes where protein synthesis occurs. p. 52

roundworms (nematodes) an animal parasite. p. 464

rugae a series of ridges in the wall of an organ, such as in the stomach. p. 384

S phase second phase of interphase when DNA replication occurs. p. 79

sacral plexus collection of branched nerves near the sacrum. p. 278

sacrum vertebrae the five fused vertebrae that form the posterior aspect of the pelvic girdle. p. 215

saddle joint synovial joint that permits motion in two planes and results from the articular surfaces being reciprocally concave and convex. p. 230

sagittal plane an anatomical division of the body into left and right sections. p. 17

sagittal suture suture that connects the medial edges of the two parietal bones. p. 210

salivation process of secreting saliva. p. 386

saltatory conduction transmission of an action potential along a myelinated axon. p. 269

sarcolemma cell membrane of a muscle cell. pp. 237, 250

sarcomere smallest contractile unit of the myocyte. p. 238

sarcoplasm fluid, similar in nature to the cytoplasm of non-muscle cells. p. 237

sarcoplasmic reticulum smooth endoplasmic reticulum of muscle cells that contains calcium. p. 237

sartorius muscle extends the knee and rotates the lower leg. p. 246

satellite cells glial cells that maintain the ion concentration surrounding neurons in the PNS. p. 189

saturated fat fats that do not contain double-bonded carbons and are solid at room temperature. p. 41

saturation when membrane proteins are full and cannot transport more molecules into or out of the cell. p. 60

scalene muscle of the neck that contracts during forced inhalation. p. 365

scapula bones located posterior to the ribs; also called the shoulder blades. p. 220

Schwann cells glials cell that produce the protective myelin that covers neurons in the PNS. p. 189

sciatic nerve longest nerve in the human body, extending from the spine along the posterior thigh. p. 224

scoliosis lateral curvature of the spine. p. 218

scrotum a dual-chambered mammalian body part that contains the testes. p. 416

sebaceous glands glands that produce the oil sebum. p. 198

sebum an oil produced in the sebaceous glands. p. 194

secondary immune response an immune response to a previously encountered infectious agent that is much more rapid than the primary immune response. p. 443

secondary oocyte a female gamete immediately following the second meiotic division. p. 420

secondary spermatocyte male gamete immediately following the second meiotic division. p. 417

secretin a hormone that regulates secretions in the stomach and pancreas. p. 396

secretion movement of substances out of the blood or the body. p. 176

segmental arteries branches of the renal arteries. p. 404

selectable marker method used to identify the host cells that contain the recombinant DNA. p. 159

selection pressure an environmental factor that kills a portion of a population, such as predators. p. 117

selective breeding *see* artificial selection. p. 158

selenium an element on the periodic table that is required for the conversion of T4 to T3. p. 304

sella turcica bone surrounding the pituitary gland. p. 213

semen a fluid that contains sperm and secretions from the prostate and bulbourethral glands. p. 416

semi-conservative replication method of DNA replication where the original strands of DNA separate and act as a template for two new strands seminal vesicles tubes that secrete the fluid that becomes semen. p. 135

seminiferous tubules tubules in the testes where meiosis occurs. p. 416

semi-permeable membrane a membrane that allows only certain molecules to easily cross the hydrophobic tails of the inner membrane. p. 58

senescent a cell unable to divide. p. 133

sense strand also known as the coding strand of DNA. p. 143

sensory neurons neurons that transmit sensory information to the brain. p. 267

serosa smooth membrane composed of a thin layer of cells that secrete serous fluid. p. 383

serous fluid a lubricating fluid contained within a membrane. p. 321

Sertoli cells cells that act to nourish developing sperm cells. p. 417

serum the fluid portion, plasma, of the blood that does not contain fibrinogen. p. 353

sesamoid bones bones that develop within a tendon, such as the patella. p. 208

sex chromosomes X and Y chromosomes that determine gender. p. 108

sex hormones hormones produced by the gonads. p. 226

short bones bones of the ankles and wrists. p. 208

sigmoid colon distal, S-shaped portion of the large intestine. p. 389

silent mutation mutation that has no effect on phenotype. p. 152

simple refers to single-layered epithelial cells. p. 227

single unit smooth muscle smooth muscle, found in sheets, that surrounds visceral organs, where many cells contract simultaneously. p. 187

single-strand binding protein protein that stabilizes the single-stranded region of the DNA molecule during replication. p. 135

sinoatrial node (SA node) region of highly excitable cells found in the right atrium of the heart; regulates heart rate. p. 319

sister chromatid identical replicated chromosomes. p. 80

skeletal muscle striated muscle connected to bone and usually under voluntary control. p. 22

sliding filament theory a theory to describe the movement of actin sliding across myosin during muscle contraction. p. 251

small intestine organ consisting of the ileum, jejunum, and duodenum; where absorption of nutrients occurs. p. 379

smallpox a viral infection caused by a strain of Variola virus that infects the blood vessels and skin. p. 445

small saphenous vein superficial vein of the upper leg. p. 344

smooth ER endoplasmic reticulum that contains no ribosomes and is where lipids and carbohydrates are produced. p. 52

smooth muscle involuntary, non-striated muscle. p. 22

snRNPs small nuclear ribonucleoprotein particles that consist of small nuclear RNA molecules associated with proteins involved in splicing introns. p. 147

sodium-potassium pump a membrane protein that uses one molecule of ATP to move three sodium ions out of the cell and two potassium ions into the cell. p. 60

soleus muscle in the lower leg that allows for plantar flexion of the foot, such as standing on your toes. p. 248

solute molecules that are dissolved in a solvent. p. 25

solvent fluid that contains dissolved molecules. p. 25

somatic nerves under voluntary control. p. 263

somatic cells cells of the body not destined to become sperm or eggs. p. 82

somatic nervous system (SNS) the collection of nerves responsible for voluntary control. p. 265

somatosensory cortex the main sensory receptive area of the parietal lobe for touch sensation. p. 286

species a group of organisms that can breed o nly with each other and produce viable offspring. p. 116

spermatid gamete formed by the division of the secondary spermatocytes. p. 417

spermatogenesis formation of sperm cells. p. 312

spermatogonia an undifferentiated male germ cell. p. 417

spermatozoa (sperm) male gametes. p. 415

sphenoid bone bone that forms the middle-base of the skull and is behind the eye sockets. p. 213

sphincter specialized muscle tissue that regulates the movement of substances within the body. p. 187

sphincter of Oddi a muscular ring that controls the flow of pancreatic juices and bile. p. 387

spinalis muscles that originate and insert into the spinous processes of the vertebrae, connecting them. p. 245

spindle fibres microtubules that transport chromosomes during cell division. p. 82

spinous processes posterior extensions of the vertebrae. p. 216

spiral fracture fracture that's caused by torsional forces along the length of the bone. p. 227

spirochete helically shaped bacteria. p. 454

spleen an organ that functions to filter the blood and break down old red blood cells; contains immune cells that fight infections. p. 340

spliceosome six different snRNPs and their associated proteins that remove introns and form new bonds between exons. p. 147

spongy bone (trabecular bone) made up of trabeculae, but not arranged into osteons in the same way as compact bone. p. 207

sporozoite a stage in the life cycle of the *plasmodium* species. p. 463

squalene a 30-carbon oil that is used in vaccines to stimulate the immune system. p. 198

squamous flattened epithelial cells. pp. 176, 210

squamous cell carcinoma cancerous growth originating in the keratinocytes of the stratum spinosum. p. 199

squamous sutures sutures that connect the temporal bones to the lateral portion of the parietal bones. p. 210

stabilize (muscular) additional support provided by muscles that allows the primary muscles to work more effectively. p. 135

stabilizing selection favouring the average of the common traits in a population. p. 126

staphylococci round clusters of bacteria. p. 453

sternal notch visible depression at the base of the neck. p. 220

sternocleidomastoid muscles that allow for flexion of the neck. p. 365

sternum a flat, long bone in the centre of the chest that's joined with the ribs to form the anterior portion of the rib cage, also called the breastbone. p. 220

steroid a hormone created from cholesterol, such as testosterone, estrogen, aldosterone, progesterone, and cortisol. p. 300

sticky ends the overhanging ends of non-paired nucleotides. p. 158

stratified refers to multi-layered epithelial cells. p. 176

stratum basale deepest layer of the epidermis, composed of a single layer of rapidly dividing keratinocytes. p. 195

stratum corneum most superficial layer of the epidermis that's constantly being replaced; the cytoplasm is replaced by keratin. p. 196

stratum granulosum middle layer of the epidermis. p. 196

stratum lucidum extra layer of the epidermis, found in areas of the body that endure more stress, and composed of dead keratinocytes and large amounts of keratin. p. 196

stratum spinosum greatest proportion of the epidermis, primarily composed of keratinocytes. p. 196

streptobacilli rod-shaped bacteria that grow in chains. p. 453

streptococci small circular bacteria that live in chains and can cause throat infections. p. 453

stress fractures small cracks in the bone resulting from overuse. p. 227

stretch receptors receptors that detect mechanical pull on the cells. p. 373

stretch reflex muscular contraction in response to stretching. p. 280

striated sequence of muscle fibres that have a light and dark banding pattern. p. 187

striations dark and light stripes in skeletal and cardiac muscle cells, as seen with a microscope. p. 251

stromal cells connective tissue cells in an organ. p. 430

styloid process extension of the temporal bone that contains a small opening to permit the passage of the carotid artery. p. 210

subarachnoid space anatomical space between the arachnoid mater and the pia mater. p. 276

subclavian vein a vein located inferior to the clavicle. p. 341

subcortical region structures located inferior to the cortex. p. 285

subcutaneous fat adipose tissue found in the subcutaneous layer under the dermis of the skin. p. 41

subcutaneous layer deepest layer of the skin that contains fat. p. 195

sublingual salivary gland salivary gland found under the tongue that produces saliva. p. 386

submandibular salivary gland major salivary gland in the base of the mouth. p. 386

submucosa connective tissue that supports the mucosa. p. 382

submucosal plexus part of the enteric nervous system. p. 265

substantia nigra structure in the brain that plays an important role in regulating body movements and is involved in Parkinson's disease. p. 284

substitution also known as a mismatch, an incorrect nucleotide added during replication. p. 139

substrate-level phosphorylation process in which a kinase enzyme in the cytoplasm adds a phosphate to ADP, which occurs during glycolysis. p. 69

sucrase an enzyme that breaks down sucrose. p. 391

sucrose formed by dehydration synthesis reaction between glucose and fructose. p. 36

sudoriferous glands *see* sweat glands. p. 198

sulcus a fissure in the surface of the brain. p. 285

superficial burn (first degree burn) burn that affects only the epidermal layer of the skin p. 199

superior and inferior mesenteric arteries arteries that supply the small and large intestines as well as the pancreas. p. 337

superior and inferior articular facets articular processes on vertebrae where they connect with adjacent vertebrae and have hyaline cartilage to prevent friction. p. 216

superior and inferior gluteal nerves nerves that signal the gluteal muscles. p. 279

superior phrenic arteries arteries that supply the superior portion of the diaphragm. p. 337

superior sagittal sinus region that collects blood and CSF from the brain. p. 282

superior vena cava the superior portion of the largest vein in the body that brings blood to the right atrium. p. 320

surface tension the adhering force created by hydrogen bonds formed between water molecules. p. 26

surfactant an amphipathic molecule the reduces surface tension in the lungs. p. 361

surgical neck part of the humerus most commonly fractured. p. 221

survival of the fittest individuals born with traits that are most beneficial for them to survive, reproduce, and pass on DNA within a particular environment. p. 114

sutures immovable joints between bones, categorized as either fibrous or synarthrosis. p. 210

sweat glands help maintain body temperature by releasing sweat, which evaporates on the skin and increases heat loss. p. 22

sympathetic nervous system a division of the autonomic nervous system that controls the fight-or-flight response. p. 291

synapse the connection between the axon terminal and the target cell. p. 254

synaptic cleft gap between the axon terminal and the target cell. p. 271

synaptic end bulb part of the axon terminal that stores the vesicles containing neurotransmitters. p. 254

synarthrosis a type of joint that produces little to no movement. p. 210

syndesmosis an immovable articulation where bones are connected by ligaments. p. 228

synergistic (muscular) muscles working together to produce a specific motion. p. 239

synergistic effect increased efficacy of one hormone when paired with another—greater than the sum of the two hormones individually. p. 305

synovial cavity space between bones containing synovial fluid. p. 228

synovial fluid thick, viscous fluid that lubricates joints and helps absorb shock. p. 228

synovial joints highly movable joints, also called diarthrosis, that contain synovial fluid; include the shoulder, knee, and elbow. p. 228

systemic circulation the circuit of blood from the heart to the tissues and back to the heart. p. 318

systole refers to the contraction portion of the cardiac cycle. p. 348

T cells Lymphocytes—immune cells—that are involved in the acquired immune response and react to specific antigens. p. 354

T wave portion of an ECG that represents ventricular repolarization. p. 325

tactile receptors receptors that detect touch, pressure, and vibration in the skin. p. 22

tapeworms (cestodes) an animal parasite. p. 464

taq polymerase polymerase enzyme obtained from a heat-tolerant strain of bacteria. p. 168

taste receptors receptors that facilitate the sensation of taste. p. 384

telomerase an enzyme found in stem cells and germ cells that prevents the loss of telomeres during replication. p. 134

telomere repeated DNA sequence at the end of each eukaryotic chromosome that protects genes from deterioration. p. 133

template strand the strand of DNA that is copied or transcribed into mRNA. p. 136

temporal bones bones that form the lateral portions of the skull inferior to the parietal bones and protect the temporal lobes of the brain. p. 210

temporal lobe inferior-lateral lobe of the brain that contains the auditory cortex and the hippocampus, which is involved with long-term memory. p. 287

temporomandibular joint (TMJ) the articulation between the mandible and the temporal bones. p. 210

tendonitis inflammation of the tendons. p. 436

tendon connective tissue between muscles and bone. p. 229

terminal bronchioles section of the bronchiole directly connected to the alveoli. p. 360

terminator sequence a sequence of DNA at the end of the gene that causes the mRNA molecule to form a hairpin loop, causing the polymerase to dissociate from DNA. p. 147

testcross method devised by Mendel to determine the genotype of individuals that express the dominant phenotype. p. 98

testes male gonads in animals. p. 415

testosterone a sex hormone produced in the testes that is involved in development of male secondary sex characteristics and spermatogenesis. p. 312

TH0 cells tolerant helper T cells. p. 441

TH1 cells a type of T helper cell that stimulates proliferation and activation of cytotoxic T cells. p. 441

TH2 cells a type of T helper cell that stimulates the proliferation and activation of B cells. p. 441

thalamus part of the brain responsible for sense perception and motor function; considered the "sensory relay station." p. 282

theca layer layer of ovarian tertiary follicle cells that indirectly plays a role in the production of estrogen. p. 423

thermogenic effect term that describes the energy required for digestion and absorption of nutrients in the digestive tract. p. 394

thermoreceptors receptors that detect changes in temperature. p. 22

thick filaments part of a myofibril that contains myosin proteins. p. 238

thin filaments part of a myofibril containing actin, troponin, and tropomyosin. p. 238

thoracic pertaining to the chest. p. 430

thoracic cavity space in the thoracic region that contains the lungs, heart, trachea, upper esophagus, pleural cavity, pericardial cavity, and mediastinum. p. 320

thoracic duct largest vessel of the lymphatic system. p. 216

thoracic vertebrae the 12 vertebrae that are connected to ribs and form the posterior aspect of the rib cage. p. 216

thymocytes cells that differentiate into mature T lymphcytes. p. 430

thymus a specialized organ of the immune system where T cells mature; located posterior to the sternum. p. 427

thyroid follicles subunits of the thyroid gland where thyroid hormones are produced. p. 304

thyroid gland an endocrine gland located in the neck. p. 304

thyroid hormones (T3 and T4) hormones produced by the thyroid that regulate metabolic rate. p. 226

thyroid stimulating hormone (TSH) a hormone released by the anterior pituitary that promotes the release of T3 and T4 from the thyroid. p. 303

thyroxine (T4) a thyroid hormone containing four atoms of iodine. p. 304

tibia the large anterior bone in the lower leg. p. 224

tibial nerve nerve that runs along the tibia in the lower leg. pp. 279, 341

tibial tuberosity attachment site on the tibia for the patellar ligament. p. 224

tibialis anterior muscle in the lower leg that causes dorsiflexion of the foot, such as lifting toes up when walking or running. p. 249

tibiofibular joint point of articulation between the tibia and the fibula at the distal end of the leg. p. 224

tidal volume the amount of air moved during normal inhalation and exhalation. p. 366

tight junctions connections between endothelial cells that prevent the movement of molecules between endothelial cells, such as in the digestive tract or the blood-brain barrier. p. 57

tissue a group of cells that is a combination of cells and proteins that function together in a specialized role. p. 7

tolerance term that describes the condition when the immune system does not react to a foreign molecule it is exposed to in high amounts, such as food molecules in the digestive tract. p. 427

tongue a muscular organ in the mouth used for talking and moving food around during chewing. p. 384

tonsils collection of lymphoid tissue in the pharynx. p. 432

topoisomerase an enzyme in prokaryotes that relieves the supercoils of the unwinding DNA during replication. p. 136

total lung capacity maximum volume of the lungs. p. 367

toxin a poisonous substance. p. 455

trabeculae osteocytes inside lacunae and surrounded by matrix; found in spongy bone. p. 207

trachea largest part of the lower respiratory tract and part of the conducting zone. p. 360

trans fats incompletely hydrogenated fats in the *trans* conformation. p. 42

***trans* Golgi** the side of the Golgi that is farthest from the endoplasmic reticulum. p. 53

transcription process in which DNA produces a single-strand nucleic acid called RNA. p. 145

transcription factors factors that promote or inhibit the transcription of RNA in a eukaryotic cell. pp. 155, 299

transduction process by which bacteria can acquire new DNA from viruses. p. 161

transfer RNA (tRNA) molecules that carry or transfer amino acids to the ribosome during protein synthesis. p. 147

transformation process by which bacteria can acquire DNA that is near them. p. 160

transient ischemic attack (TIA) a temporary condition that shows signs and symptoms similar to a stroke; also called "mini stroke". p. 351

transitional epithelial cells epithelial cells that can change shape, such as the cells of the bladder that change as the bladder fills with urine. pp. 176, 402

translation protein synthesis. p. 54

transposable elements sequences of DNA that can move themselves to another chromosome within a cell; also called transposons or jumping genes. p. 139

transverse abdominis a muscle of the abdomen that runs horizontally. p. 365

transverse colon the part of the large intestine that runs parallel to the transverse plane from the right to left and leads to the descending colon. p. 389

transverse fracture fracture that's perpendicular to the length of the bone. p. 227

transverse plane an anatomical division of the body into superior and inferior sections; also called a cross-sectional or horizontal plane. p. 17

transverse process lateral protuberance of the vertebrae. p. 216

transverse sinus areas beneath the brain that allow blood to drain from the back of the head. p. 341

transverse tubules (T-tubules) tunnel-like extensions of the cell membrane that extend into muscle cells and allow an action potential to penetrate into the cell. p. 237

tricep muscle that works to extend the forearm at the elbow. p. 246

tricuspid the valve that separates the right atrium from the right ventricle. p. 321

triglyceride a fat composed of glyerol and three fatty acid chains. p. 66

triiodothyronine (T3) thyroid hormone containing three atoms of iodine. p. 304

trisomy 21 a nondisjunction mutation where the cells have an extra chromosome 21, also called Down syndrome. p. 108

trochlea medial aspect of the distal end of the humerus that articulates with the ulna. p. 221

tropomyosin a regulatory protein that binds to the myosin-binding sites on the actin and prevents contraction unless calcium is present. p. 238

troponin a regulatory protein of the thin filaments that binds calcium and moves tropomyosin. p. 238

true pelvis bones that protect the internal, lower abdominal organs. p. 222

trypsin a pancreatic enzyme that breaks down proteins. p. 391

tubulin protein that forms microtubules. p. 55

tumour a growth of cells that is either benign or malignant. p. 84

tunica externa outermost layer of the blood vessels. p. 331

tunica intima innermost layer of the blood vessels. p. 331

tunica media middle layer of the blood vessels. p. 331

turbinates osseus processes that cause changes in patterns of air movement. p. 359

turbulent flow movement of a fluid that undergoes irregular mixing, unlike laminar flow. p. 328

type I alveolar cells simple epithelial cells that form the alveolar wall. p. 361

type II alveolar cells specialized epithelial cells that secrete surfactant. p. 361

type-1 diabetes an autoimmune disease affecting the beta cells of the pancreas so that insulin can't be formed and injections of insulin are required for life. p. 311

type-2 diabetes an acquired condition where the body develops an insulin resistance and can no longer regulate blood sugar. p. 311

tyrosine an amino acid that is used as the precursor for the production of certain hormones, such as dopamine, epinephrine, and the thyroid hormones. p. 300

ulna bone of the forearm that's located medially in anatomical position. p. 221

ulnar artery artery running along the ulna in the forearm. p. 340

ulnar vein vein running along the ulna in the forearm. p. 341

unipolar neurons *see* sensory neurons. p. 188

unsaturated fat fats containing one or more double-bonded carbons and are liquid at room temperature. p. 41

urea waste molecule produced from ammonia and carbon dioxide, and excreted by the kidneys. pp. 73, 407

ureters vessels that carry urine from the kidneys to the urinary bladder. p. 402

urethra vessel that carries urine from the urinary bladder to the external environment. p. 404

urine a nitrogenous, liquid-waste product in animals. p. 401

uterus a female sex organ. p. 420

vaccine an agent that stimulates immunity to a particular infectious disease. p. 445

vacuole a membrane-bound organelle. p. 463

valve a physical barrier that controls or directs the movement of a fluid. p. 429

variable amplitude differences in membrane potential. p. 272

variable number tandem repeats (VNTRs) varying lengths of repeated DNA sequences in the introns; unique to every individual and used in DNA fingerprinting. p. 168

varicose veins veins that have become large or distended because of damaged valves in the vein. p. 331

vas deferens ducts that connect the right and left epididymis to the ejaculatory ducts. p. 416

vascular permeability ability of a blood vessel wall to permit the flow of small molecules in and out of the vessel. p. 437

vasoconstriction a decrease in the lumen size in the blood vessels. p. 332

vasodilation an increase in the lumen size in the blood vessels. p. 332

vasodilators medications that work to decrease blood pressure by dilating the blood vessels. p. 352

vastus intermdius one of the quadriceps muscles; located underneath the rectus femoris. p. 246

vastus lateralis one of the quadriceps muscles located in the lateral thigh. p. 246

vastus medialis one of the quadriceps muscles in the medial thigh near the knee. p. 246

vector also known as a vehicle, is what carries the inserted gene, such as a plasmid or a virus. p. 158

veins blood vessels that carry blood toward the heart. pp. 320, 341

venous return the return of blood to the heart through the venous system. p. 328

ventilation movement of air between the lungs and the external environment. p. 358

ventral horn columns of grey matter at the anterior portion of the spinal cord. p. 276

ventral respiratory group (VRG). a column of neurons in the medulla oblongata that control breathing. p. 373

ventral root the anterior of the two nerve roots branching off the spinal cord. p. 276

ventricular diastole relaxation of the ventricles. p. 324

ventricular systole contraction of the ventricles. p. 324

venules small veins that bridge the flow of blood from the capillaries to the veins. p. 323

vernix an oil that protects a fetus while surrounded by amniotic fluid. p. 198

vertebrae 33 bones that make up the spine, sacrum, and coccyx. p. 215

vertebral cavity body region that contains the spinal cord. p. 18

vertebral foramen opening in the vertebrae through which the spinal cord is able to extend. p. 216

vesicles sacs within a cell that are surrounded by a membrane and contain cellular products such as newly formed proteins, hormones, or neurotransmitters. pp. 53, 150

vestibule an entrance or passageway. p. 420

villi extensions from the lining of the intestinal wall. p. 388

vimentin most common intermediate filament found in most cells. p. 55

viruses small infectious agents that reproduce inside living cells. p. 427

visceral pericardium segment of the pericardium attached to the heart. p. 321

visceral peritoneal membrane (peritoneum) layer of the peritoneum that covers the internal organs. p. 383

visceral pleura segment of the pleura connected to the outer surface of the lungs. p. 361

vital capacity maximum volume of air that can be expelled after a full inhalation. p. 367

vitamin A derivative of beta-carotene, used by the body to maintain healthy skin and mucus membranes and improve vision; also called retinol. p. 197

vitamin D a compound that can be synthesized by the body when exposed to sunlight and be absorbed from the diet; required for calcium absorption in the small intestine and important for immune regulation. p. 226

vocal cords membranes that vibrate to generate sound. p. 359

voltage-gated membrane proteins that open and allow passage in response to an electrical stimulus. p. 254

voluntary muscles controlled consciously. p. 187

vomer a small part at the base of the skull that is a component of the nasal septum. p. 210

vulva external female genital organs. p. 420

water-soluble hormones polar hormones that act by binding to a receptor on the cell membrane, rather than inside the cell as fat-soluble hormones do. p. 299

Wernicke's area portion of the brain responsible for word recognition. p. 288

white matter segments of the nervous system composed of myelinated axons. p. 276

white pulp segments of adenoid tissue in the spleen that contain white blood cells. p. 431

withdrawal reflex a spinal reflex designed to prevent damage to the body. p. 280

xiphoid process inferior portion of the sternum. p. 220

X-linked genes on the X chromosome unrelated to sex characteristics. p. 106

yeast artificial chromosome (YAC) chromosome that's easily grown in eukaryotic yeast cells and is used as a vector in genetic engineering because it can carry large sequences of DNA and yeast cells can process RNA. p. 161

yellow bone marrow bone marrow composed primarily of fat cells. p. 430

Z discs ends of a sarcomere. p. 251

zona fasciculata middle region of the adrenal cortex that produces glucocorticoids, such as cortisol. p. 303

zona glomerulosa the most superficial region of the adrenal cortex; produces aldosterone. p. 307

zona pellucida membrane that surrounds the plasma membrane of an oocyte. p. 422

zona reticularis the innermost region of the adrenal cortex responsible for producing androgens, such as DHEA. p. 307

zygomatic arch part of the temporal bone that connects to the zygomatic bones to form the cheek bones. p. 210

zygomatic bones bones that articulate with the temporal bones to form the cheek bones. p. 210

zygote the initial cell formed when two gametes combine. p. 88

Index

Note: Page numbers with "f" indicate figures; those with "t" indicate tables.